EXS 64

DNA Methylation

Molecular Biology and Biological Significance

Edited by J. P. Jost
H. P. Saluz

Birkhäuser Verlag
Basel · Boston · Berlin

Editors' addresses:

Dr. Jean-Pierre Jost
Friedrich Miescher-Institut
P.O. Box 2543
CH-4002 Basel
Switzerland

Dr. Hans-Peter Saluz
IRBM
Via Pontina Km 30,600
I-00040 Pomezia (Roma)
Italy

Library of Congress Cataloging-in-Publication Data

DNA methylation: molecular biology and biological significance / edited by J.P. Jost, H.P. Saluz.
-- (EXS; 64)
Includes bibliographical references and index.
ISBN 3-7643-2778-2 (acid-free paper) —ISBN 0-8176-2778-2 (acid-free paper)
1. DNA—Methylation. I. Jost, J. P. (Jean-Pierre), 1937–
II. Saluz, H. P., 1952– . III. Series.
[DNLM: 1. DNA—metabolism. 2. Methylation. W1 E65 v. 64 / QU 58 D629]
QP624.D63 1993
574.87′3282—dc20

Deutsche Bibliothek Cataloging-in-Publication Data

DNA methylation: molecular biology and biological significance / ed. by J. P. Jost; H. P. Saluz. – Basel; Boston; Berlin: Birkhäuser, 1993
(EXS; 64)
ISBN 3-7643-2778-2 (Basel . . .)
ISBN 0-8176-2744-8 (Boston)
NE: Jost, Jean-Pierre [Hrsg.]; GT

Printed in Germany on acid-free paper

ISBN 3-7643-2778-2
ISBN 0-8176-2778-2

"A beautiful model or theory may not be right, but an ugly one must be wrong".

Jacques Monod

in: *Symmetry and function of biological systems at the macromolecular level. Nobel Symposium 11, 1969.*

..."Juger des problèmes devenus mûrs pour l'analyse, décider quand il est temps d'explorer à nouveau un vieux territoire, reprendre des questions naguère considérées comme résolues ou insolubles, tout cela constitue l'une des qualités majeures d'un scientifique"...

François Jacob

in: *Le Jeu des Possibles, Essai sur la diversité du vivant.*

Contents

Preface

The occurrence of 5-methylcytosine in DNA was first described in 1948 by Hotchkiss (see first chapter). Recognition of its possible physiological role in eucaryotes was first suggested in 1964 by Srinivasan and Borek (see first chapter). Since then work in a great many laboratories has established both the ubiquity of 5-methylcytosine and the catholicity of its possible regulatory function. The explosive increase in the number of publications dealing with DNA methylation attests to its importance and makes it impossible to write a comprehensive coverage of the literature within the scope of a general review. Since the publication of the 3 most recent books dealing with the subject (DNA methylation by Razin A., Cedar H. and Riggs A. D., 1984 Springer Verlag; Molecular Biology of DNA methylation by Adams R. L. P. and Burdon R. H., 1985 Springer Verlag; Nucleic Acids Methylation, UCLA Symposium suppl. 128, 1989) considerable progress both in the techniques and results has been made in the field of DNA methylation. Thus we asked several authors to write chapters dealing with aspects of DNA methylation in which they are experts. This book should be most useful for students, teachers as well as researchers in the field of differentiation and gene regulation.

We are most grateful to all our colleagues who were willing to spend much time and effort on the publication of this book. We also want to express our gratitude to Yan Chim Jost for her help in preparing this book.

Jean-Pierre Jost
Hans-Peter Saluz

DNA Methylation: Molecular Biology and Biological Significance
ed. by J. P. Jost & H. P. Saluz

A Chronicle of DNA methylation (1948–1975)

Arthur Weissbach

Roche Institute of Molecular Biology, Nutley, N.J. 07110, U.S.A.

> "What do you know about this business?" the King said to Alice.
> "Nothing," said Alice.
> "Nothing whatever?" persisted the King.
> "Nothing whatever," said Alice.
> "That's very important," the King said . . . .
>
> Lewis Carroll in *Alice in Wonderland*

1 Introduction

It is obvious that any attempt by a non-professional historian to survey three to four decades in any area, however limited the area may be, is of questionable value. Recollecting past events and influences in science is distorted by ignorance, imperfect memories, personal bias, and perceptions altered by the trends and concepts which dominate our present thinking. With these caveats in place, one can then attempt to identify some of the events and conceptual mileposts which influenced research in DNA methylation between 1948 and 1975.

As one of the offspring of molecular biology, the study of the methylation of nucleic acids, both DNA and RNA, has depended on the vast outpouring of ideas, data, and techniques from the mother lode. In turn, DNA methylation research has offered elegant solutions to several key areas, including bacterial restriction/modification systems and the importance of post-replicative modification in cell processes. Paradoxically, in eukaryotic systems, the roles of DNA methylation remain elusive, despite an imposing number of studies linking it with the control of gene expression, genomic imprinting, replication and development. In part this is caused by the widespread use of restriction endonucleases which has offered a facile, valuable glimpse of DNA methylation profiles. However, the correlation of the methylation states of the gene with genetic expression, based on restriction endonuclease analysis, has been too gross, in general, to offer clear insights. Furthermore, because of the vagueness arising from the use of this technique, there are almost as many negative correlative studies as positive ones. Until genomic sequencing (or another tool which can pinpoint changes in specific, targeted methylation sites in DNA) becomes the standard tool, the true picture of the flux of DNA methyl groups in cells will remain obscure. The well-known and dramatic example of how important genomic sequencing can be is illustrated by the Saluz-Jost studies (Saluz and Jost, 1986) of the activation of the chicken vitellogenin gene. The subtleties and insights revealed in their work require new mechanisms and thinking about DNA methylation and demethylation during gene activation.

2 Early states

The discovery of the first modified base, 5-methylcytosine (5-meC) in calf thymus DNA by Rollin Hotchkiss in 1948 (Hotchkiss, 1948) occurred in a period when the structure and function of nucleic acids were mysteries, their relationship to genes and the "genetic" material tantalizing but unproven, and the techniques for handling macromolecules and their constituents were just being developed. Hotchkiss used paper chromatography in his early work with the intent of detecting amino acids in his DNA preparations. He hoped to answer persistent criticisms that the Rockefeller laboratory DNA samples were not pure since he was aware that the smart money was betting at that time that nucleoproteins rather than pure DNA were the true genetic material. He found no amino acids on his chromatograms which, though not silencing doubters, provided workers with a rapid technique for separating purine and pyrimidines and also uncovered 5-meC. At about this time he became aware that Vischer and Chargaff at Columbia were also developing qualitative and quantitative paper chromatography for the analysis of DNA constituents (Vischer and Chargaff, 1949). Hotchkiss prudently "retired from that field to do more biological things" (Srinivason et al., 1979), feeling that he would "have needed a steamroller and hardhat" to compete with Chargaff's laboratory (ibid, pg. 358). It is worth noting that both Hotchkiss and Chargaff were relying on the 1944 studies of Consden et al. (Consden et al., 1973) which used paper chromatography to separate amino acids and to the appearance of the Beckmann quartz spectrophotometer which appeared on the market shortly after the second World War and quickly became the *sine qua non* for all nucleic acid chemists. In any event, Wyatt et al. in 1951 (Wyatt et al., 1951) confirmed the Hotchkiss observation and also found 5-meC in wheat germ. The discovery of another, seemingly minor, methylated base, N6-methyladenine (N6-MeAd) in bacterial DNA by Dunn and Smith in 1955 (Dunn and Smith, 1955) was later to have an important role in phage research.

One can realize how obscure and incomprehensible the data about DNA, much less DNA methylation, was during the period from 1930 to 1952 by noting that the Levene tetranucleotide hypothesis for DNA (Levene and Bass, 1931) which implied a uniform structure for DNA was still being considered during this period. The near proof that DNA was the "genetic material" did not occur until the Hershey-Chase "Waring blender" experiment in 1952 (Hershley and Chase, 1952).

Table 1. Parallel histories

DNA as the genetic material	
1944, 1946	– Avery, McLeod, and McCarty (Avery et al., 1944; McCarty and Avery, 1946) – Pneumococcal transformation by DNA
1952	– Hershey and Chase experiment (Hershey and Chase, 1952) – transfer of Bacteriophage T2 DNA without phage protein to host cells
1952–1953	– Luria and Human (Luria and Human, 1952) and Bertani and Weigele (Bertani and Weigle, 1953) – Host controlled modification and restriction in bacteriophage infection
1953	– Watson, Crick (Watson and Crick, 1953) – The DNA double helix
1958	– Meselson, Stahl (Meselson and Stahl, 1958) – DNA replication is semiconservative
1958	– Lehman et al. (Lehman et al., 1958) – The isolation of *E. coli* DNA polymerase I
1961	– Nirenberg and Matthei (1961) – Solving the genetic code
1961	– Crick et al. (1961) – General nature of the genetic code
1961	– Brenner et al. (1961) – Identification of messenger RNA
1972	– Mertz and Davis (1972) – *E. coli* restriction endonuclease generates sticky ends
1972	– Jackson et al. (1972) – First recombinant DNA (SV-40)
1973	– Cohen et al. (1973) – Construction of biologically active recombinant DNA plasmids
1975	– Asilomar Conference
DNA methylation	
1948	– Hotchkiss (1948) – Identification of 5-MeC in calf thymus DNA
1951	– Wyatt (1951) – Finds 5-MeC in animal and plant cells
1955	– Dunn and Smith (1955) – N6-MeAD in prokaryotic DNA
1959	– Kornberg et al. (1959) – Suggest methyl groups are added to DNA by a postreplicative enzymic reaction
1962	–Arber and Dussoix (1962) – DNA methylation is involved in host-induced modification/restriction
1964	– Gold and Hurwitz (1964) – Identify a DNA methyltransferase in *E. coli*
1964	– Srinivasan and Borek (1964) – DNA methylation has a role in gene regulation
1967	– Burdon et al. (1967) – DNA methyltransferase associated with mouse ascites chromatin
1968	– Meselson and Yuan (1968) – Isolation of *E. coli* restriction endonuclease
1968	– Sheid et al. (1968), Kalousek and Morris (1968) – Solubilize DNA methyltransferase from rat liver nuclei
1968	– Lark (1968) and Billen (1968) – Demonstrate DNA methylation is post-replicative
1969	– Kuehnlein et al. (1969) – Purification of *E. coli* B DNA methyl transferase
1970	– Kelly and Smith (1970) – Base sequence of the restriction site of *H. influenzae*
1971	– Scarano (1971) – Deamination of 5-methylcytosine to thymidine in DNA leads to a mutagenic change
1974	– Arber (1974) – Maintenance of methylation patterns during replication of cellular DNA
1975	– Riggs (1975) and Holliday and Pugh (1975) – Emphasize the concept of maintenance methylation; discuss DNA methylation and X chromosome inactivation

Soon after, in 1953, the double helix structure of DNA was published (Watson and Crick, 1953). Though the pneumococcal transformation experiments of Avery, MacLeod and McCarty (Avery et al., 1944; McCarty and Avery, 1946) already well known by 1946, were in everybody's mind, the possibility of protein and other contaminants in their DNA preparations remained a serious issue and prompted the 1948 experiments of Hotchkiss who had been working in their laboratory. Just how troubling these uncertainties were is illustrated by the doubts of S. E. Luria who, as late as 1952, was of the mind that both protein and DNA were involved in the transfer of genetic information (Luria and Human, 1952). Fortunately, these widely held incertitudes did not detour his Ph.D. student, J. D. Watson, who, though sharing Luria's ambivalence, by that time was in Cambridge doing post-doctoral work with Francis Crick.

The Watson-Crick double helix structure of DNA led to the central dogma outlining the flow of genetic information (Watson and Crick, 1953) (now much modified) and to the dramatic elucidation of the enzymology of DNA synthesis by Kornberg (Kornberg, 1960). With the spectacular Nirenberg-Matthei experiment of 1961 (Nirenberg and Matthei, 1961), which deciphered the language of the genetic code, molecular biology came of age and research in DNA methylation was poised to start. A temporal comparison of signal events in our understanding of DNA, its role in genetics and parallel developments in the field of DNA methylation is shown in Table 1.

3 Restriction modification systems

Any consideration of the beginnings of DNA methylation research must start with the elucidation of the bacterial restriction/modification (restr/mod) systems. The discovery of host restriction mechanisms in bacteriophage infections (Luria and Human, 1952; Bertani and Weigde, 1953) led to a series of genetic and biochemical investigations which were among the earliest breakthroughs in Molecular Biology/Genetics. The success of understanding the restriction/modification systems is a graphic demonstration of scientific accomplishment and led to Nobel prizes for several of the main players. Among other things, these studies also led to the discovery of restriction endonucleases and the first defined role of site-specific DNA methylation.

The studies of both the modification and restriction enzymic systems still provide a conceptual and experimental framework for scientists interested in both prokaryotic and eukaryotic DNA methylation. From the discovery of host induced modification in 1942 by Luria (Luria and Human, 1952) until the complete unfolding of the biochemistry and molecular biology of the processes took about 30 years, whereas a comparable understanding of even minor eukaryotic systems which involve DNA methylation remains unsolved. This despite the fact that the enzymes involved in prokaryotic restriction/modification systems were identified at about the same time that the eukaryotic DNA methyltransferases were recognized. Gold and Hurwitz, in 1964 (Gold and Hurwitz, 1964), were the first to posit a DNA methyltransferase in *Escherichia coli*. Later Kuehnlein et al., in 1969, isolated an enzyme from *E. coli* which could methylate unmodified DNA at specific sites and confer resistance to the homologous restriction endonuclease. Over the next few years this enzyme was purified some two hundred fold (Smith et al., 1972; Kuehnlein and Arber, 1972). Similarly, during 1967 and 1968, DNA methyltransferases had already been found in murine chromatin (Burdon et al. 1967), and solubilized from rat liver (Sheid et al., 1968; Kalousek and Morris, 1968; Kerr and Borek, 1973). By 1975, a human DNA methyltransferase had been purified several hundred fold and characterized (Roy and Weissbach, 1975).

The significance of the isolation of restriction endonucleases, particularly the type II enzymes (for reviews see Yuan and Hamilton, 1984; Smith and Kelly, 1984; Yuan and Smith, 1984), is seen in every molecular biology laboratory today. From a historical point of view, the discovery of Mertz and Davis (Mertz and Davis, 1972) that endonuclease Eco Rl generated sticky ends when it cut DNA, thus making it easy to insert other DNA fragments, and the dramatic paper of Jackson et al. (Jackson et al., 1972) in which the first recombinant DNA (involving SV40 DNA) was made, represent the start of recombinant DNA technology. Herbert Boyer and Stanley Cohen seized upon these enzymes to make the first recombinant plasmid DNA molecules and vectors (Cohen et al., 1973). Boyer recognized from the very beginning (Boyer et al., 1972) that the replicon theory of Francois Jacob (Jacob et al., 1963) predicted that, if it contained the proper origin of replication, any piece of DNA could be copied by the cell's replication machinery. Plasmid and viral vectors immediately became, and have remained, the preferred vehicles for the new technology. The speed with which scientists adopted the technology to make recombinant DNAs was extraordinary.

By 1975 the exploding pace of recombinant DNA constructions forced a call for guidelines in the construction of recombinant DNA molecules because of both public and scientific fears. Scientists responded with the Asilomar conference which examined many of the parameters and potentials of the technology in genetic engineering.

4 Prokaryotes vis-à-vis eukaryotes

To the relatively narrow circle of scientists interested in eukaryotic DNA methylation the prokaryotic modification/restriction systems may provide, in a broad sense, some analogies to consider in vertebrate systems.

A) DNA methylation in bacteria is not an essential system for viability since null mutants are viable (Marinus, 1984). A survey through the invertebrate and vertebrate kingdoms would suggest the same since most invertebrates, fungi and viruses seem to lack these enzymes (Low et al., 1969; Bird, 1984; Urieli-Shoval et al., 1982; Diala and Hoffmann, 1982). DNA methyltransferases thus may be added-on enzymes introduced into certain bacteria by phage or plasmids during evolution. This exoinsertion prospect has not been considered in eukaryotic systems but must remain a possibility because of the diverse eukaryotes which do not show DNA methylation. However, the role of DNA methylation in "Protecting" genetic information in bacteria from being contaminated by other invading DNAs might be, with considerable imagination, likened to a putative role in eukaryotes wherein DNA methylation "protects" DNA and prevents it from being transcribed and expressed in higher organisms. Bestor (Bestor, 1990) has expressed a similar view, i.e., that transcription regulation by DNA methylation is a compensation mechanism in expanded eukaryotic genomes. Plant systems, with their extraordinarily high level of methylated cytosine (Gruenbaum et al., 1981) may offer the best experimental models for viewing the domains wherein DNA methylation functions in gene expression.

B) The recognition of the molecular sites where restriction and modification occur and the isolation and characterization of the systems involved in *E. coli* and *Hemophilus influenzae* (Kuehnlein et al., 1969; Kelly and Smith, 1970; Smith et al., 1972) represented a breakthrough in the fledging field of molecular biology. Unfortunately, as mentioned before, the same cannot be said about our knowledge of the controls operative in eukaryotic DNA methylation; so that some 30 to 35 years after the great harvest in prokaryotic DNA methylation, very little is known about enzyme mechanisms, specific sites of methylation and demethylation, and the various factors which control or regulate the methylation processes. For instance, though the by now obvious possibility that methylation affects the binding of regulatory proteins (and RNA molecules) to DNA (Inambar et al., 1991), there is little solid experimental evidence and much uncertainty to back this attractive hypothesis and no data to show that such putative interactions are involved in gene regulation. Furthermore, with the exception of a few studies (Saluz and Jost, 1986; Saluz et al., 1986; Ward et al., 1990; Pfeifer et al., 1990; Kochanek et al., 1991; Woelfl et al., 1991), the dynamics of specific CpG methylation and demethylation reactions in gene expression are unclear and, instead, we have, as alluded before, a large body of data showing "global" changes in the 5-MeC content of the genome or changes in the restriction endonuclease cleavage patterns in development and/or gene expression (Razin and Riggs, 1980; Doerfler, 1983; Taylor, 1984). It is not surprising that these early approaches reveal many negative and positive correlations when a range of expressed genes are surveyed.

C) The genetic analysis of the *E. coli* host induced restriction/modification systems revealed the presence of the two (Type II or III systems) or three (Type I systems) genes involved in the process (for review see Yuan and Smith, 1984). Clearly, an analogous breakthrough in eukaryotic cells is unlikely in the near future though the cloning of a murine DNA methyltransferase by Bestor et al. (Bestor et al., 1988) is an important advance, which follows their extensive purification of multiple species of murine DNA methyltransferases (Bestor and Ingram, 1985; Bestor and Ingram, 1983). Furthermore, the intense interest in X chromosome inactivation (Riggs, 1975) and genomic imprinting (Sapienza et al., 1987; Swain et al., 1987; Bartolomei et al., 1991; Ferguson-Smith et al., 1991) should reveal some clue concerning the genetic control of sex-linked DNA methylation processes.

D) The elucidation of DNA methylation reactions in prokaryotes led to the concept of the maintenance of genomic methylation patterns as a post replicative event. Arber, in fact, neatly outlined the steps in the process (Arber, 1974) which was later referred to a maintenance methylation by Riggs (Riggs, 1975) and Holliday and Pugh (Holliday and Pugh, 1975). As useful and attractive as the concept of maintenance methylation is in visualizing how the genome maintains methylation markers during the cell cycle, it does little to help us understand the flux of methyl groups in and out of DNA during differentiation and development, and, perhaps, in the replication process itself. In this respect, there has been considerable speculation that there are separate "maintenance" and "*de novo*" DNA methyltransferases even though every eukaryotic enzyme studied catalyzes both processes (Bolden et al., 1986). It is safe to say that the factors and processes which control DNA methylation will be much more complicated and interactive than simple explanations based on the presence of two, or more, different DNA methyltransferases. The elucidation of the enzymic mechanisms, the participating factors and other governing processes now seem to be a necessary prerequisite for our understanding of DNA methylation.

Perhaps some of our present difficulties in understanding DNA methylation in higher organisms reflect the pervasive human desire to seek, above all, the "why" of something. The continuing cataloging of gross methylation changes at specific CpG sites in genes and genomes is driven, in part, by our hope that this data will tell us the real "why" of DNA methylation. Yet the strength of the scientific system is that it teaches us to ask "how" rather than "why" and as Stephan Jay Gould (reflecting on a comment of Francis Crick) put it in "Male Nipples and Clitoral Ripples", "we must first establish 'how' in order to know whether or not we should be asking 'why' at all." In a sense, this is the lesson in the bacterial restriction/modification story wherein understanding the "how" of the systems, i.e. the enzymic reactions and the associated genetics, removed the need to know why.

Arber, W. (1974) Progr. Nucl. Acids Res. Mol. Biol., ed. W. E. Cohn. *14*, 1–34.

Arber, W., and Dussoix, D. (1962) Host specificity of DNA produced by *Escherichia coli*. J. Mol. Biol. *5*, 18–36.

Avery, O. T., MacLeod, C. M., and McCarty, M. (1944) J. Exp. Med. *79*, 137–158.

Bartolomei, M. S., Zemel, S., and Tilgham, S. M. (1991) Nature *351*, 153–155.

Bertani, G., and Weigle, J. J. (1953) Host controlled variation in bacterial viruses. J. Bacteriol. *65*, 113–121.

Bestor, T. H., and Ingram, V. M. (1985) Growth-dependent expression of multiple species of DNA methyltransferase in murine erythroleukemia cells. Proc. Natl. Acad. Sci. USA *82*, 2674–2678.

Bestor, T. H., and Ingram, V. M. (1983) Two methyltransferases from murine erythroleukemia cells: purification, sequence specificity, and mode of interaction with DNA. Proc. Natl. Acad. Sci. USA *80*, 5559–5563.

Bestor, T. H. (1990) DNA methylation: evolution of a bacterial immune function into a regulator of gene expression and genome structure in higher eucaryotes. Phil. Trans. R. Soc. Lond. B *326*, 179–187.

Bestor, T. H., Laudano, A., Mattaliano, R., and Ingram, V. M. (1988) Cloning and sequencing of a cDNA encoding DNA methyltransferase of mouse cells. The carboxyl-terminal domain of the mammalian enzymes is related to bacterial restriction methyltransferases. J. Mol. Biol. *203*, 971–983.

Billen, D. (1968) J. Mol. Biol. *31*, 477–486.

Bird, A. P. (1984) in: Methylation of DNA, pp. 130–141. Ed. T. A. Trautner, Springer Verlag, Heidelberg & Berlin.

Bolden, A. H., Ward, C. A., Nalin, C. M., and Weissbach, A. (1986) The primary DNA sequence determines *in vitro* methylation by mammalian DNA methyltransferases. Prog. Nucl. Acids Res. Mol. Biol. *33*, 231–250.

Boyer, H. W., Betlach, M., Bolivar, F., Rodriguez, R. L., Heynecker, H. L., Shine, J., and Goodman, H. M. (1977) in: Recombinant Molecules: Impact on Science and Society, pp. 9–20. Eds R. F. Beers, Jr. and E. G. Basset. Raven Press, New York.

Brenner, S., Jacob, F., and Meselson, M. (1961) An unstable intermediate carrying information from genes to ribosomes for protein synthesis. Nature *190*, 576–581.

Burdon, R. H., Martin, B. T., and Lal, B. M. (1967) Synthesis of low molecular weight ribonucleic acid in tumor cells. J. Mol. Biol. *28*, 357–371.

Cohen, S. N., Chang, A., Boyer, H., and Helling, R. (1973) Construction of biologically functional bacterial plasmids *in vitro*. Proc. Natl. Acad. Sci. USA *70*, 3240–3244.

Consden, R., Gordon, A. H., and Martin, A. J. P. (1944) Qualitative analysis of proteins: a partition chromatographic method using paper. Biochem. J. *38*, 224–232.

Craig, I. (1991) Nature *349*, 742–743.

Crick, F. H. C., Barnett, L., Brenner, S., and Watts-Tobin, R. J. (1961) General nature of the genetic code for proteins. Nature *192*, 1227–1232.

Diala, E. S., and Hoffmann, R. M. (1982) Hypomethylation of HeLa cell DNA and the absence of 5-methylcytosine in SV40 and adenovirus (type 2) DNA: analysis by HPLC. Biochem. Biophys. Res. Commun. *107*, 19–26.

Doerfler, W. (1983) DNA methylation and gene activity. Ann. Rev. Biochem. *52*, 93–124.

Dunn, D. B., and Smith, J. D. (1955) The occurrence of 6-methylcytosine in deoxyribonucleic acids. Biochem. J. *68*, 627–636.

Ferguson-Smith, A. C., Cattanach, B. M., Barton, S. C., Beechey, C. V., and Surani (1991) Embryological and molecular investigations of parental imprinting on mouse chromosome 7. Nature *351* 667–670.

Gold, M., and Hurwitz, J. (1964) The enzymatic methylation of ribonucleic acid and deoxyribonucleic acid. J. Biol. Chem. *239*, 3858–3865.

Gruenbaum, Y., Naveh-Many, T., Cedar, H., and Razin, A. (1981) Sequence specificity of methylation in higher-plant DNA. Nature *292*, 860–862.

Hershey, A. D., and Chase, M. (1952) Independent functions of viral protein and nucleic acid in growth of bacteriophage. J. Gen. Physiol. *36*, 39–56.
Holliday, R., and Pugh, J. E. (1975) DNA modification mechanisms and gene activity during development. Science *187*, 226–232.
Hotchkiss, R. D. (1948) The quantitative separation of purines, pyrimidines, and nucleosides by paper chromatography. J. Biol. Chem. *168*, 315–332.
Inambar, N. M., Ehrlich, K. C., and Ehrlich, M. (1991) CpG methylation inhibits binding of several sequence-specific DNA-binding proteins from pea, wheat, soybean and cauliflower. Plant. Mol. Biol. *17*, 11–123.
Jackson, D., Symons, R., and Berg, P. (1972) Biochemical method for inserting genetic information into DNA of Simian virus 40; circular SV40 DNA molecules contain lambda phage genes and the galactose operon of *E. coli*. Proc. Natl. Acad. Sci. USA *69*, 2904–2909.
Jacob, F., Brenner, S., and Cuzin, F. (1963) Cold Spring Harbor Symp. Quant. Biol. *28*, 329–348.
Kalousek, F., and Morris, N. R. (1968) Deoxyribonucleic acid methylase activity in rat spleen. J. Biol. Chem. *243*, 2440–2443.
Kelly, T. J. Jr., and Smith, H. O. (1970) A restriction enzyme from *Hemophilus influenzae*. J. Mol. Biol. *51*, 393–401.
Kerr, S. J., and Borek, E. (1973) in: The enzymes, vol. IX, pp. 167–195. Ed. P. Boyer. Academic Press, New York.
Kochanek, S., Radbruch, A., Tesch, H., Renz, D., and Doerfler, W. (1991) DNA methylation profiles in the human genes for tumor necrosis factors alpha and beta in subpopulations of leukocytes and in leukemias. Proc. Natl. Acad. Sci. USA *88*, 5759–5763.
Kornberg, A. (1960) Science *131*, 1503–1508.
Kornberg, A., Zimmerman, S. B., Kornberg, S. R., and Josse, J. (1959) Enzymatic synthesis of deoxyribonucleic acid. VI. Influence of bacteriophage T2 on the synthetic pathway in host cells. Proc. Natl. Acad. Sci. USA *45*, 772–776.
Kuehnlein, U., and Arber, W. (1972) Host specificity of DNA produced by *Escherichia coli*. XV. The role of nucleotide methylation in in vitro B-specific modification. J. Mol. Biol. *63*, 9–19.
Kuehnlein, U., Linn, S., and Arber, W. (1969) Host specificity of DNA produced by *Escherichia coli*. Proc. Natl. Acad. Sci. USA *63*, 556–562.
Lark, K. (1968) Studies on the *in vivo* methylation of DNA in *Escherichia coli* 15T. J. Mol. Biol. *31*, 389–399.
Lehman, I. R., Bessman, M. H., Simms, E. C. and Kornberg, A. (1958) Enzymatic synthesis of deoxyribonuclei acid. J. Biol. Chem. *233*, 163–170.
Levene, P. A., and Bass, L. W. (1931) The Nucleic Acids. The Chemical Catalog Co., New York.
Low, M., Hay, J., and Keir, H. M. (1969) DNA of herpes simplex virus is not a substrate for methylation in vivo. J. Mol. Biol. *46*, 205–207.
Luria, S. E., and Human, M. L. (1952) A nonhereditary, host-induced variation of bacterial viruses. J. Bacteriol. *64*, 557–569.
Marinus, M. G. (1984) in: DNA methylation, pp. 81–110. Eds A. Razin, H. Cedar and A. D. Riggs. Springer Verlag, Heidelberg & Berlin.
McCarty, M., and Avery, O. T. (1946) Studies on the chemical nature of the substance inducing transformation of pneumococcal types. J. Exp. Med. *83*, 89–96.
Mertz, J., and Davis, R. W. (1972) Cleavage of DNA by R1 endonuclease generates cohesive ends. Proc. Natl. Acad. Sci. USA *69*, 337–3374.
Meselson, M., and Stahl, F. (1958) The replication of DNA in *Escherichia coli*. Proc. Natl. Acad. Sci. USA *44*, 671–682.
Meselson, M., and Yuan, R. (1968) DNA restriction enzyme from *E. coli*. Nature *217*, 1110–1114.
Nirenberg, M. W., and Matthei, J. H. (1961) The dependence of cell-free protein synthesis in *E. Coli* upon naturally occurring or synthetic polyribonucleotides. Proc. Natl. Acad. Sci. USA *47*, 1588–1602.
Oberle, I., Russeau, F., Heitz, D., Kretz, C., Devys, D., Hanauer, A., Boue, J., Bertheas, M. F., and Mandel, J. L. (1991) Science *252*, 1097–1102.
Pfeifer, G., Steigerwald, S. D., Hansen, R. S., Gartler, S. M., and Riggs, A. (1990) Polymerase chain reaction-aided genomic sequencing of an X chromosome-linked CpG island: methyla-

tion patterns suggest clonal inheritance, CpG site autonomy, and an explanation of activity state stability. Proc. Natl. Acad. Sci. USA *87*, 8252–8256.

Razin, A., and Riggs, A. D. (1980) DNA methylation and gene function. Science *210*, 604–610.

Riggs, A. D. (1975) X-inactivation, differentiation, and DNA methylation. Cytogenet. Cell Genet. *14*, 9–11.

Roy, P. H., and Weissbach, A. (1975) DNA methylase from HeLa cell nuclei. Nucl. Acids Res. *2*, 1669–1684.

Saluz, H. P. and Jost, J. P. (1986) Optimized genomic sequencing as a tool for the study of cytosine methylation in the regulatory region of the chicken vitellogenin II gene. Gene *42*, 151–157.

Saluz, H. P., Jiricny, J., and Jost, J. P. (1986) Genomic sequencing reveals a positive correlation between the kinetics of strand-specific DNA demethylation of the overlapping estradiol/glucocorticoid-receptor binding sites and the rate of avian vitellogenin mRNA synthesis. Proc. Natl. Acad. Sci. USA *83*, 7167–7171.

Sapienza, C. A., Peterson, A. C., Rossaint, J., and Balling, R. (1987) Degree of methylation of transgenes is dependent on gamete of origin. Nature *328*, 251–254.

Scarano, E. (1971) The control of gene function in cell differentiation and embryogenesis. Adv. Cytopharmacol. *1*, 13–23.

Sheid, B. S., Srinivasan, P. R., and Borek, E. (1968) Deoxyribonucleic acid methylase of mammalian tissues. Biochemistry *7*, 280–285.

Smith, H. O., and Kelly, S. V. (1984) in: DNA Methylation, pp. 39–71. Eds A. Razin, H. Cedar, and A. D. Riggs. Springer Verlag, Heidelberg & Berlin.

Smith, J., Arber, W., and Kuehnlein, U. (1972) DNA modification mechanisms and gene activity during development. J. Mol. Biol. *63*, 1–8.

Srinivasan, P. R., and Borek, E. (1964) Enzymatic alteration of nucleic acid structure. Science *145*, 548–553.

Srinivasan, P. R., Fruton, J. S., and Edsall, J. T. (1979) The origins of modern biochemistry. The New York Acad. Sci. (1979) pp. 321–342.

Swain, J. L., Stewart, T. A., and Leder, P. (1987) Parental legacy determines methylation and expression of an autosomal transgene: a modulator mechanism for parental imprinting. Cell *50*, 719–727.

Taylor, J. H. (1984) DNA methylation and cellular differentiation. Springer Verlag, Heidelberg & Berlin.

The genetic code. Cold Spring Harbor Symp. Quant. Biol. (1967) Cold Spring Harbor Laboratory, Cold Spring Harbor, N.Y.

Urieli-Shoval, S., Gruenbaum, Y., Sedar, G., and Razin, A. (1982) The absence of detectable methylated bases in *Drosophila melanogaster* DNA. FEBS Lett. *146* 148–152.

Vischer, E., Zamenhof, S., and Chargaff, E. (1949) Microbial nucleic acids: the desoxypentose nucleic acids of avian tubercle bacilli and yeast. J. Biol. Chem. *177*, 429–438.

Vischer, E., and Chargaff, E. (1947) The separation and characterization of purines in minute amounts of nucleic acid hydrolysates. J. Biol. Chem. *168*, 781–782.

Ward, C., Bolden, A., and Weissbach, A. (1990) Genomic sequencing of the 5′-flanking region of the mouse beta-globin major gene in expressing and nonexpressing mouse cells. J. Biol. Chem. *265* 32030–3033.

Watson, J. D., and Crick, F. H. C. (1953) Genetical implications of the structure of deoxyribonucleic acid. Nature *171*, 964–967.

Watson, J. D., and Crick, F. H. C. (1953) Molecular structure of nucleic acids. Nature *171*, 737–738.

Woelfl, S., Schrader, M., and Wittig, B. (1991) Lack of correlation between DNA methylation and transcriptional inactivation: the chicken lysozyme gene. Proc. Natl. Acad. Sci. USA *88*, 271–275.

Wyatt, G. R. (1951) Recognition and estimation and 5-methylcytosine in nucleic acids. Biochem. J. *48*, 581–584.

Yuan, R., and Hamilton, D. L. (1984) in: DNA Methylation, pp. 12–37. Eds A. Razin, H. Cedar and A. D. Riggs. Springer Verlag, Heidelberg & Berlin.

Yuan, R., and Smith, H. O. (1984) in: DNA Methylation, pp. 73–80. Eds A. Razin, H. Cedar and A. D. Riggs. Springer Verlag, Heidelberg & Berlin.

DNA Methylation: Molecular Biology and Biological Significance
ed. by J. P. Jost & H. P. Saluz

Major techniques to study DNA methylation

Hanspeter Saluz and Jean-Pierre Jost*

*IRBM, Via Pontina km 30.600, 00040-Pomezia (Roma), Italy, and *Friedrich Miescher-Institute, P.O. Box 2543, 4002 Basle, Switzerland*

1 Introduction

The major aim of this chapter is to give a brief overview of the most commonly used procedures to study minor DNA bases, especially 5-methylcytosine, and to refer the interested reader to the relevant methodological literature.

The techniques to analyze DNA methylation can be divided into two categories, one of which leads to sequence-unspecific results (sequence-unspecific methods) and the other allows a precise location of some or all methylated sites within a given DNA target sequence (sequence-specific procedures).

The first category can be used to analyze the different types of modified bases and to quantify them in an organism, organ, tissue, etc. but it does not provide any information about the precise location of the modified site within a given nucleic acid sequence. Into this category fall immunological (Sano et al., 1980; Achwal et al., 1984), chromatographic (Vischer and Chargaff, 1948; Marshak and Vogel, 1951; Randerath, 1961; Stahl, 1967; Kuo et al., 1980; Wagner and Capesius, 1981), electrophoretic (Clotten and Clotten, 1962; Zweig and Whitaker, 1967) or spectroscopic (Razin and Cedar, 1977) procedures upon a complete chemical (Marshak and Vogel, 1951; Kochetkov and Budowskii, 1974; Ford et al., 1980) or enzymatic hydrolysis (Ford et al., 1980) of the target DNA. The chromatographic techniques applied most often are high-pressure liquid (Eick et al., 1983) and thin-layer-chromatography (Stahl, 1967 and references therein). The latter one can be combined with, for example, high-voltage electrophoresis (Zweig and Whitaker, 1967), the former one with, for example, mass-spectrometry (Razin and Cedar, 1977).

Another sequence-unspecific approach involves the use of methylation-sensitive restriction endonucleases where different digested genomic DNAs are compared upon separation by gel electrophoresis (Adams and Burdon, 1985). A comparison of the length and intensities of the resulting digestion patterns on the gel, leads to some information about the amount of methylated sites.

The second category enables the researcher to analyze the precise location of such modified bases within a known DNA sequence. These studies are still often performed by means of pairs of isoschizomeric restriction enzymes, one of which is methylation sensitive, the other not. In combination with Southern-blot analysis or polymerase chain reaction (PCR) the methylated sites can be analysed in a sequence-specific manner, but only if they are located within the recognition sequence of such an enzyme and usually also only if they are fully methylated (i.e. in both strands) or fully unmethylated. The analysis of hemimethylated sites is usually not possible by this method (Gruenbaum et al., 1981).

A way to investigate *in vitro* methylated sites or very small genomes, for example from certain viruses, is to chemically sequence (Maxam and Gilbert, 1980) the end-labeled DNA. This procedure, however, has a severe drawback in that it can only be applied to short DNA stretches.

In 1984, Church and Gilbert combined the above sequencing procedure with the method of indirect end-labeling (Church and Gilbert, 1984) which directly allowed the study of uncloned DNA in its native state. This method was called genomic sequencing. To date there are several different strategies available for sequencing native, uncloned DNA (Church and Gilbert, 1984; Saluz and Jost, 1986; Mueller et al., 1989; Pfeifer et al., 1989; Saluz and Jost, 1989; Frommer et al., 1991). Such techniques became the approach of choice in many laboratories since, in contrast to the above procedures, all cytosines within a given uncloned DNA stretch could be analysed in a sequence- and strand-specific manner.

2 Sequence-unspecific methods

2.1 Immunological approaches

Some investigators applied immunological methods to study DNA methylation (Sano, 1980; Achwal, 1984). Hereby, the DNA was first immobilized on filter membranes, and the methylation analysis was performed by means of rabbit anti 5-methylcytosine antibodies. These were then visualized with, for example, ^{125}I-goat anti-rabbit IgG (Sano, 1980) followed by autoradiography.

2.2 Analysis of methylation after complete DNA hydrolysis

For most sequence-unspecific methods to detect DNA methylation, first a complete hydrolysis of the target DNA has to be performed. For the hydrolysis of DNA, usually chemical (Kochetkov and Budowskii, 1974) or enzymatic (Ford et al., 1980) ways are chosen. Hydrolytic cleavage of nucleic acids results in purine- and pyrimidine-bases, carbohydrate phosphates and free carbohydrates. In addition, a hydrolysis always produces secondary products, for example, by deamination reactions of aminopurines and aminopyrimidines or of their nucleosides and nucleotides. Deoxyribonucleic acids are barely affected by a treatment with basic solutions. An acid treatment of DNA, however, usually results in the cleavage of the N-glycosidic purine to carbohydrate bonds, while other purine-nucleoside and nucleotide bonds are relatively resistant towards acids. The phosphoric acid residue of the purine nucleotides is more easily cleaved when compared to pyrimidine residues. Good results for a hydrolysis of DNA are obtained according to the procedures of Vischer and Chargaff (1948), Wyatt (1951) or Marshak and Vogel (1951). In the former procedure the heat-dried DNA is treated with 98% formic acid for 30 min at 175°C and subsequently dried in a dessicator in the presence of potassium hydroxide. The residue is then dissolved in HCl before analyzing the bases. In the second procedure 72% perchloric acid is used to hydrolyze the heat-dried nucleic acids (100°C, 1 h).

For an enzymatic DNA hydrolysis several different enzyme types, such as DNase I of pancreas, snake venom phosphodiesterase, spleen phosphodiesterase or micrococcal nucleases can be applied (Cantoni and Davis, 1966; Linn and Roberts, 1982). Hydrolysis with the first two enzymes results in mononucleotides with a 5′ phosphate whereas the latter two produce 3′-phosphorylated mononucleotides.

The hydrolysis products are usually analyzed by one of the following procedures:

2.2.1 Paper- (PC) and thin layer-chromatography (TLC). Paper- and thin layer chromatography are often used to separate nitrogenous bases, nucleosides, mononucleotides and also oligonucleotides. For example, in the most simple case, a spot or streak of a solution containing the substances to be separated is applied near the bottom of a strip of filter paper or thin layer plate. The bottom of this strip is then dipped into a nonpolar solvent, which slowly rises through the paper or TLC matrix by capillary action. Chromatographic separation takes place according to the affinity of the solutes. The greater the affinity of the compound to the solvent the closer it will be to the moving edge of the solvent. Paper chromatography was already applied towards the end of the forties for the analysis of nucleic acid compounds (Vischer and Chargaff, 1948; Marshak and Vogel, 1951) and Randerath was the first scientist to apply TLC for such an analysis (Randerath, 1961). Matrices for TLC analysis (Stahl, 1967) were usually Kieselgel, Cellulose, Dextran-gel or ionic exchange matrices, such as DEAE or ECTEOLA-Cellulose (Coffey and Newburgh, 1963; Jacobson, 1964).

Cedar et al. (1979), for example, labeled the 5′ ends of DNA stretches which arised from a cleavage with the restriction endonuclease MspI (recognition sequence: CCGG; enzyme not affected by 5-mC at the second C position) with radioactive phosphorus. Upon hydrolysis of the labeled restricted DNA, they performed 2-dimensional TLC to separate dCMP for 5-mdCMP. By this means they were able to measure the proportion of CCGG- and $C^{m}CGG$-sequences, respectively.

2.2.2 High-performance liquid chromatography. The development of high-performance liquid chromatography (HPLC) allows the use of resins composed of very small beads for base analysis; for example Partisil SCXK 218 [Reeve Angel], Durrum-DC-1A, Aminex A6 or A10 [BioRad] giving high resolution in a very short time. This technique is mainly intended for analytical purposes and therefore for amounts larger than a few milligrams of sample, the size of column required necessitates reduction in pressure (moderate-pressure liquid chromatography involving very fine beads but slower flow rates).

Eick et al. (1983), for example, developed a procedure to separate major and minor bases using reverse-phase high-performance liquid chromatography. They found the distribution of 5-mC in genomic hanster cell DNA to be nonrandom in that different 5′-CpG-3′-containing restriction sites were methylated to different extents.

2.2.3 Electrophoretic analysis on solid supports. Electrophoresis is the name given to the movement of charged macromolecules in an electric field. The electrical force causes the macromolecule to migrate towards the electrode of opposite sign. Often a streak of a solution, containing a mixture of materials to be separated, in our case a hydrolysate of DNA, is made on a piece of filter paper saturated with buffer. The ends of the paper are dipped in buffer solutions connected to electrodes. When an electric current is passed through the system, the components of the sample separate into zones which migrate distances depending in part on the electrophoretic mobilities (directly proportional to the charge of the molecule and indirectly proportional to its friction) of the components which are detected by staining or by other methods (Clotten and Clotten, 1962 and references therein).

Another possibility is to separate the DNA hydrolysis products by two-dimensional separation. The first dimension uses, for example, high-voltage electrophoresis (>50 V/cm) on cellulose acetate strips. This is followed by a transfer of the electrophoresed materials to DEAE thin-layer plates and chromatography with organic solvents for separation in the second dimension (Zweig and Whitaker, 1967 and references therein).

The above methods allow the detection and estimation of the ratio of modified/unmodified DNA bases.

2.2.4 Spectroscopic procedures. Spectroscopic techniques are used for the quantitative determination or the qualitative identification of 5-mC residues.

Mass spectrometry, for example, was applied to measure the content of 5-methylcytosine in DNA (Razin and Cedar, 1977). The distribution of methylated cytosines was studied by this means in the chromatin of various tissues of chickens and the results led to the conclusion that 5-methylcytosine was nonrandomly distributed with respect to the nucleoproteins.

Spectroscopic procedures were also applied to study the influence of 5-mC on DNA structure. A well-suited technique for such investigations proved to be nuclear magnetic resonance spectroscopy (NMR; Wuethrich, 1986; Van De Ven and Hilbers, 1988). Two major developments in NMR technology, the employment of multiple-pulse two-dimensional (2D) Fourier-transform (FT) methods and the use of high-field superconducting magnets, led to the breakthrough of high resolution NMR needed for structural studies of biopolymers in solution. The species $d(C^{m}GC^{m}GC^{m}G)_2$, for example, has been studied extensively in the context of methylation and its influence on Z-DNA formation (Feigon et al., 1984; Orbons and Altona, 1986; Genset et al., 1987). It was found that Z-DNA formation was induced by increasing the concentration of organic solvent (CH30D) and decreasing the temperature.

3 Analysis with methylation-sensitive restriction endonucleases and agarose gel electrophoresis

The close relationship between DNA methylation and DNA restriction makes the restriction endonucleases a frequently used tool for the analysis of methylation patterns (Bird and Southern, 1978; Doerfler, 1984). There are several isoschizomeric restriction endonucleases commercially available (Table 1) recognizing either only the unmethylated,

Table 1. A few examples of methylation-sensitive restriction endonucleases

Enzyme	Sensitivity	Recognition sequence	Modified refractory sequence
BsuRI	5-mC	GG/CC	$GG^{m}CC$
DpnI	N^6mA	$G^{m}A/TC$	GATC
DpnII	N^6mA	GA/TC	$G^{m}ATC$
EcoRI	N^6mA	G/AATTC	$GA^{m}ATTC$
HhaI	5-mC	GCG/C	$G^{m}CGC$
HpaII	5-mC	C/CGG	$C^{m}CGG$
MboI	N^6mA	/GATC	$G^{m}ATC$
MspI	5-mC	C/CGG	$^{m}CCGG$
TaqI	N^6mA	T/CGA	$TCG^{m}A$
ThaI	5-mC	CG/CG	$^{m}CG^{m}CG$

Isochizomers are HpaII/MspI and DpnI/DpnII/MboI.

or both the unmethylated and the methylated recognition sequence. Methylation-sensitive restriction endonucleases can be used for an estimation of the distribution of, for example, CG dinucleotides and also to receive a rough idea about the proportion of methylated to unmethylated CGs (Fig. 1). Here the genomic DNA of different tissues or organisms is digested with methylation-sensitive restriction endonucleases, and upon separation on agarose gels their respective patterns of cleavage are compared.

Bird et al. (1979), for example, showed, on the basis of sensitivity to the methylated mCpG sensitive restriction endonucleases HpaII, HhaI and AvaI, that the genomic DNA of sea urchin could be divided into two compartments. One of these compartments was highly methylated and therefore resistant to the enzymes (indicated by the high molecular weight of the digested genomic DNA). The DNA of the other compartment, due to the few methylated sites, was not resistant to these enzymes and therefore cleaved to a low molecular weight. In addition it could be shown that transcription occurred mainly in the unmethylated compartment (Bird et al., 1979).

4 Sequence-specific methods

4.1 Digestion with methylation-sensitive restriction endonucleases and Southern-blot analysis

Methylation-sensitive restriction endonucleases are often used for a comparison of the resulting DNA restriction fragments by Southern blot analysis (Southern, 1975; Bird and Southern, 1978; Waalwijk and Flavell, 1978) which allows the analysis of about 10% of all potentially methylated sites of a given genome (Winnacker, 1984).

This procedure has, however, the severe drawback that hemimethylation often remains undected (Gruenebaum et al., 1981).

Another strategy involving restriction enzymes and Southern-blot analysis is to compare the patterns of cleavage of native and cloned DNA (Doerfler, 1981), digested with individual methylation-sensitive restriction endonucleases.

An interesting third approach was introduced by Singer-Sam et al. (1990), called the “quantitative PCR assay for methylation studies”. They cleaved the total DNA to completion with methylation-sensitive restriction endonucleases, for example Hpall, and upon heat inactivation they performed a polymerase chain reaction (PCR) using specific primers, flanking the restriction site, which allowed the exponential amplification of the intact DNA only. The amplification products were then separated on appropriate agarose gels and analyzed by densitometry.

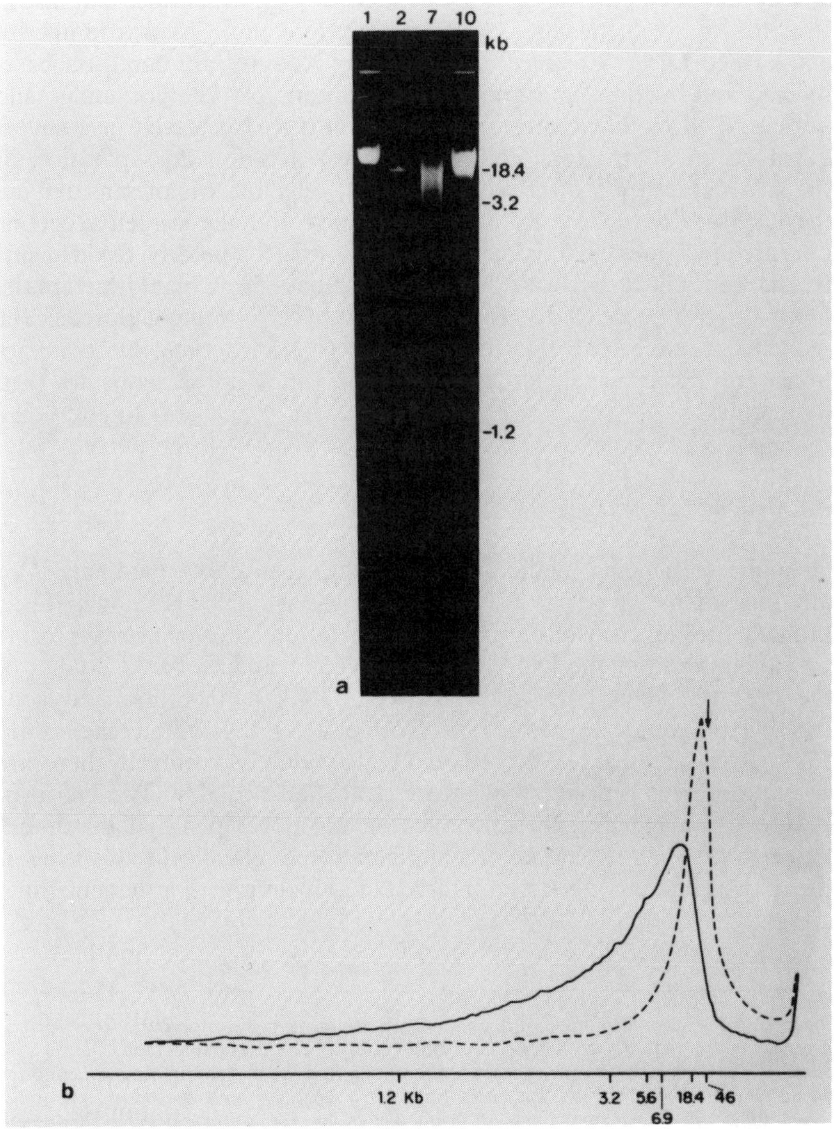

Figure 1. Methylation-sensitive restriction endonucleases can be used for an estimation of the distribution of CG dinucleotides and to gain information about the proportion of methylated to unmethylated CGs. In this example HpaII- or MspI-digested DNA from mouse liver was separated by electrophoresis. a) Channel 1: Undigested control DNA; channel 7: DNA cleaved with MspI; channel 10: DNA cleaved with HpaII; channel 2: marker DNA. b) Densitometer tracing of channel 7 (solid line) and channel 10 (dotted line). The arrow indicates the position of uncleaved DNA. (From Singer et al., 1979). Copyright by the AAAS.

4.2 Chemical sequencing

A way to investigate *in vitro* methylated sites or small genomic DNA stretches is to chemically sequence (Maxam and Gilbert, 1980) the end-labeled DNA. Chemical sequencing procedures are based on base-specific, but random and limited modifications of the pyrimidines and purines. The modified sites are then quantitatively cleaved in a subsequent chemical reaction. Following these reactions and separation of the DNA fragments on a sequencing gel, only the fragments carrying the label are detectable by autoradiography and the sequence can be interpretated directly. Hydrazine which is used to modify the cytosine residues does not react with the methylated base 5-mC and therefore the DNA can not be cleaved at those positions. The resulting gap within the sequencing ladder can thus be identified (if appropriate control reactions are performed, i.e. reactions with unmodified cytosines and thymidines).

4.3 Genomic sequencing

Genomic sequencing methods were developed to allow the analysis of any kind of native genomic DNA where the sequence is known. Their primary use lies in determining the methylation state of every cytosine on either strand of the DNA, which proved invaluable in the studies of the observed tight correlation between DNA methylation and gene inactivity (Saluz et al., 1986, 1988; Toth et al., 1989; 1990; Pfeifer et al., 1990; Rideout et al., 1990; Ward et al., 1990). At present there are several different genomic sequencing strategies available. The first procedure, as indicated in the Introduction, was developed by Church and Gilbert (1984) and is based on a combination of chemical sequencing of the native genomic DNA and indirect end-labeling with a sequence-spe-

---

Figure 2. Schematic diagram of the original genomic sequencing procedure. The DNA is digested to completion with a restriction endonuclease. The restriction fragments are subjected to the chemical sequencing reactions, resulting in a set of subfragments covering the whole sequence base by base. The subfragments are separated on a sequencing gel, transferred to a nylon membrane and covalently linked by UV-irradiation. The filter-bound subfragments containing the target sequence are hybridized with a specific single-stranded radioactively labeled DNA probe which is complementary to one end of the target fragments. Upon processing of the membrane and exposure to an X-ray film the target sequence can be visualized. The position of 5-methylcytosine is revealed as a gap (indicated by an arrowhead to the right of the X-ray film box) in the C-ladder of the genomic sequence (C_{gen}), compared to that of the cloned DNA (C). The thick lines represent the target sequence, while its probe region is shown as an open box at one end of the target sequence. The thin lines are the remaining unrecognized DNA fragments. (From Saluz et al., 1991). Copyright by Elsevier.

cific radioactively labeled single-stranded probe (Fig. 2). The purified genomic DNA is digested to completion with a restriction endonuclease which provides fragments of a defined length containing the target sequence. The DNA is then subjected to chemical sequencing reactions (Maxam and Gilbert, 1980) to produce specific cleavage products covering the whole sequence base by base, which can be separated by size on

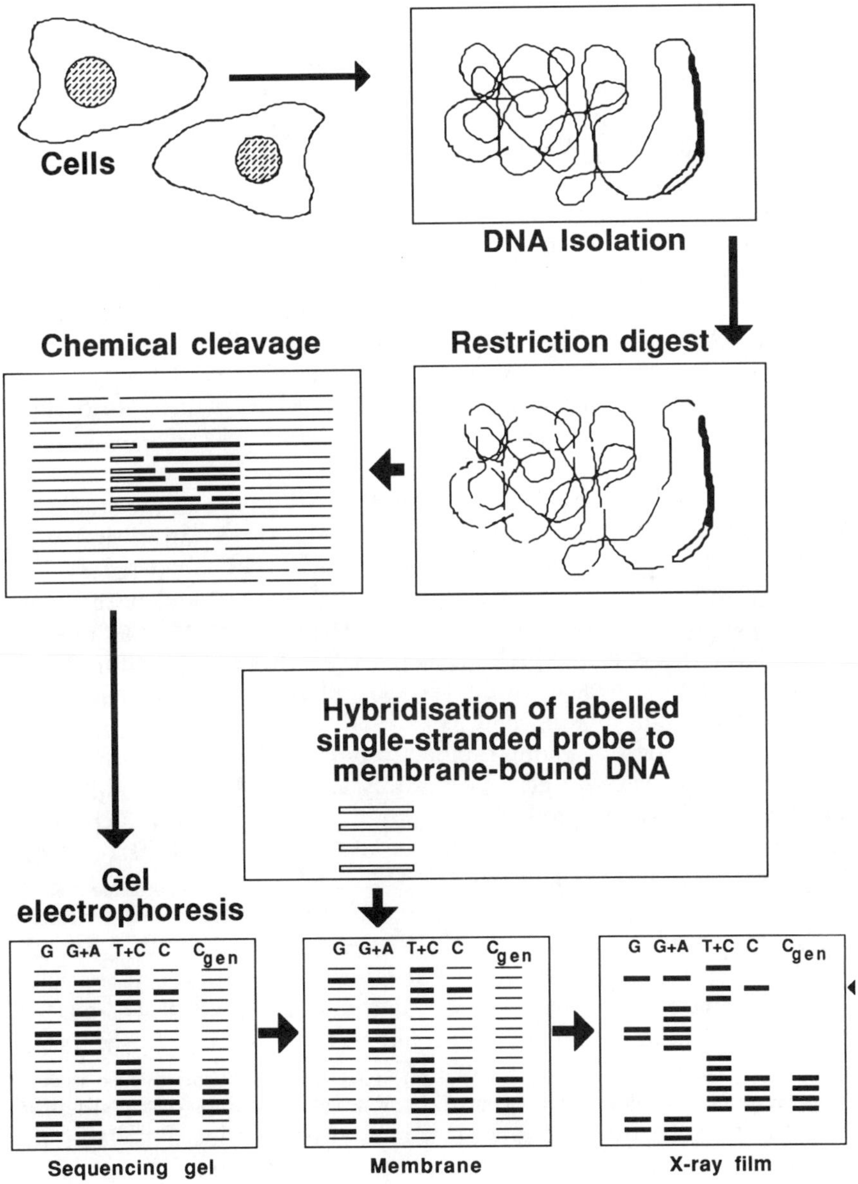

a sequencing gel. Subsequently they were electrophoretically transferred to a nylon membrane and covalently fixed by means of UV-irradition. The filterbound subfragments containing the target sequence are specifically hybridized with a single-stranded radioactively labeled DNA- or RNA probe complementary to one end of the target fragments. After processing of the membrane and exposure to an X-ray film the sequence of interest can be visualized. Five-methylcytosine, which is not modified by the C-specific reagent, therefore appears as a gap within the sequencing ladder when compared with that of cloned DNA. This and an optimized procedure of it (Saluz and Jost, 1986) was successfully used in several methylation studies of genomes of higher eucaryotes (Saluz et al., 1986, 1988; Toth et al., 1989, Nick et al., 1986; Lois et al., 1989; Ephrussi et al., 1985; Ward et al., 1990).

Since these procedures were relatively complex, in that they involved several difficult steps, such as electrotransfer, hybridization to a very small amount of target molecules or the tedious cloning work for the probe production, other procedures were developed which retained the benefits of the original technique but were simpler to implement. One of the new strategies involved the use of linear amplification with Taq polymerase and a radioactive synthetic primer (Fig. 3; Saluz and Jost, 1989). The purified genomic DNA was digested with a restriction endonuclease to reduce its viscosity and then subjected to chemical cleavage reactions. The target DNA sequence was used as a template for the linear amplification. The amplification products were separated directly on a sequencing gel, the gel was fixed, dried and exposed to an X-ray film to visualize the sequence of interest. The advantage of this strategy was that only a very few steps were involved and most of the time-consuming parts were circumvented. However, to produce a signal of a single-copy gene of a higher eucaryote which was sufficiently strong to visualize the sequence within 12–48 h, about 25–50 μg of genomic DNA per reaction mixture had to be used.

Primed extension of a long radioactively labeled probe, hybridized to the target sequence, was also successfully applied (Becker et al., 1989). Similar to the linear amplification procedure, fewer critical steps were involved than in the original technique of Church and Gilbert (1984).

Another technique where less DNA had to be used was also developed in 1989 (Fig. 4; Mueller et al., 1989). Here ligation-mediated polymerase chain reactions (LMPCR) were used where the genomic DNA was chemically cleaved, heat-denatured and with a first synthetic primer extended to produce blunt-ended double-stranded target DNA fragments. Commercially available linkers were ligated to the target fragments thus producing a template for PCR involving two additional complementary genomic primers. After several rounds of exponential amplification another end-labeled primer was introduced to label the amplification products by primer extension. The products were sepa-

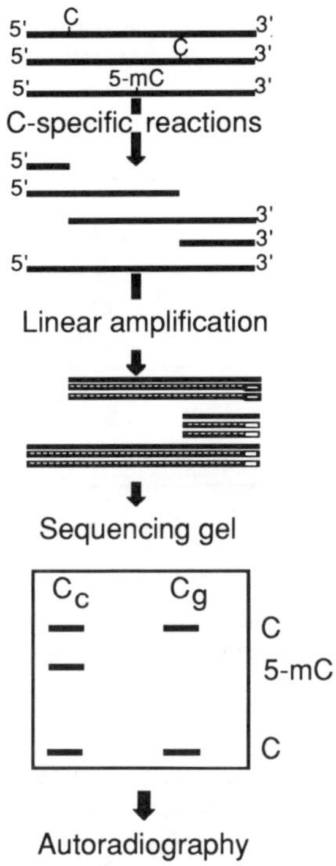

Figure 3. Reaction scheme of genomic sequencing using linear amplification with a radioactively labeled primer. In the example, the genomic DNA (Cg) is subjected to a C-specific reaction, where only unmethylated cytosines are chemically modified and subsequently cleaved. Upon repeated cycles of extension with a radiolabeled primer and separation on a sequencing gel, no sequencing fragment corresponding to the position of the methylated cytosine is observed, resulting in a characteristic gap (5-meC beside Cg track) in the sequencing ladder when compared to that of the unmethylated control DNA (Cc).

rated on a sequencing gel and visualized by autoradiography. A slight modification of this approach was published in the same year (Pfeifer et al., 1989) using a combination of this PCR-based procedure with indirect end-labeling detection as previously described for the "classical" procedure (Church and Gilbert, 1984; Saluz and Jost, 1986). One of the big advantages of these latter techniques was their sensitivity due to the use of PCR. However, because of the many steps involved and the potential selectivity of ligation reactions, these techniques, if improperly used, may cause difficulty in certain cases.

One major disadvantage of all the above genomic sequencing procedures is the fact that they are based on chemical cleavage reactions and therefore many target molecules are required to obtain a complete sequence profile (at least 10^3 to 10^4 target molecules are required to produce a sequence ladder of 300 readable nucleotides). Therefore, these techniques cannot be used to study one single gene of one single cell in

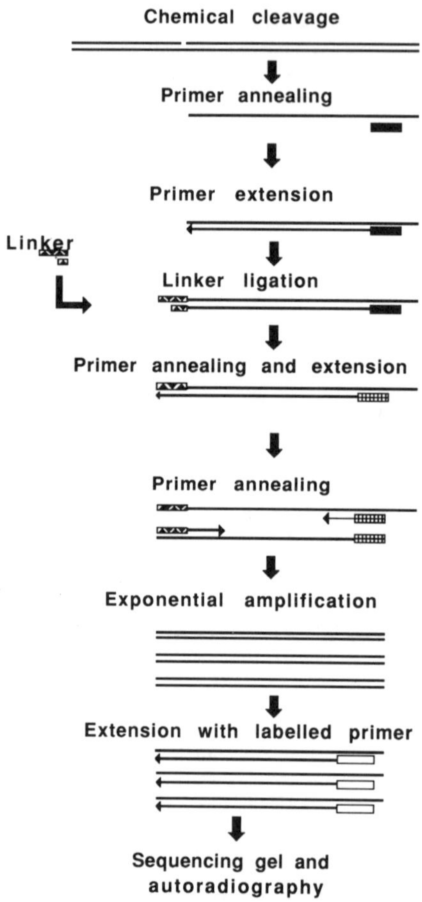

Figure 4. Scheme of ligation-mediated PCR. In this example we show the principle for only one chemically cleaved target fragment. In practice, all fragments would be similarly amplified. Boxes represent the different primers. Solid black box is the first gene-specific primer (i.e. annealed to the genomic DNA). Zig-zag patterned boxes are the long and short linker primers. Hatched boxes indicate the second gene-specific primer, positioned with its extending end 3′ to that of the first primer so as to increase specificity. White boxes represent a radioactively end-labeled primer to visualize the sequence upon electrophoresis and autoradiography.

its native state. Now, seven years after the first publications (Church and Gilbert, 1984; Becker and Wang, 1984) on genomic sequencing further new developments which circumvent chemical cleavage reactions are expected, allowing the researchers to study any gene independent of the number of target-molecules per cell. One procedure, circumventing chemical cleavage, was recently developed by Frommer et al. (1992). It is based on bisulfite-induced modification of the genomic DNA under conditions where cytosine is converted to uracil and 5-methyl-

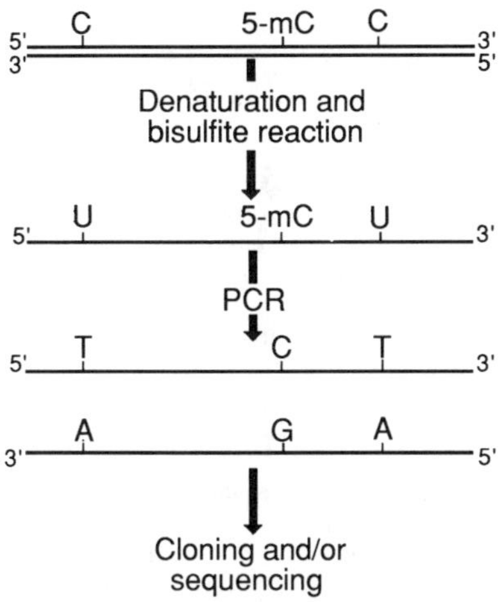

Figure 5. Strategy for genomic sequencing involving bisulfite: three cytosines, one of which is methylated, are indicated on the upper strand of a double-stranded DNA stretch. Upon heat denaturation, the now single-stranded DNA is reacted with bisulfite. This causes a deamination of the unmethylated cytosines converting them to uracil. The 5-meC residue is resistant to this treatment and remains unmodified. An exponential amplification (PCR) of the treated DNA results in copy-fragments in which all U and T residues have been amplified as A/T base-pairs and only 5-meC residues as G/C pairs.

cytosine remains unchanged (Fig. 5). The target sequence is then exponentially amplified using polymerase chain reactions (PCR) and strand-specific primers. Upon sequencing of the amplified DNA, all uracil and thymine residues become detectable as thymine and only 5-methylcytosines as cytosines. By a comparison of the modified sequence with the unmodified one it becomes possible to analyze the methylated cytosines. The amount of genomic DNA necessary for one experiment is still relatively high (approx. 10 μg). This might be due to unspecific strand-breakage which should soon be possible to overcome. This procedure will most probably become the method of choice for the analysis of 5-methylcytosines in any given genomic target sequence.

Acknowledgments. We would like to thank Drs Andrew Wallace for critical reading of the manuscript and Marianne Frommer for providing us with the manuscript of genomic sequencing involving bisulfite-induced modification of the DNA.

Achwal, C. W., Iyer, C. A., and Chandra, H. S. (1984) Estimation of the amount of 5-methylcytosine in *Drosphilia melanogaster* DNA by amplified ELISA and photoacoustic spectroscopy. EMBO J. *3*, 263–266.

Adams, R. L. P., and Burdon, R. H. (1986) Molecular Biology of DNA Methylation. Springer Verlag, New York, Berlin, Heidelberg, Tokyo.

Becker, M. M., and Wang, J. C. (1984) Use of light for footprinting DNA in vivo. Nature *309*, 682–687.

Becker, M. M., Wang, Z., Grossmann, G., and Becherer, K. A. (1989) Genomic footprinting in mammalian cells with ultraviolet light. Proc. Natl. Acad. Sci. USA *86*, 5313–5319.

Bird, A. P., and Southern, E. M. (1978) Use of restriction enzymes to study eucaryotic DNA methylation. J. Mol. Biol. *118*, 27–47.

Bird, A. P., Taggart, M. H., and Smith, B. A. (1979) Methylated and unmethylated DNA compartments in the sea urchin genome. Cell *17*, 889–901.

Cantoni, G. L., and Davis, D. R. (1966) Procedures in Nucleic Acid Research. Harper and Row, New York–London.

Cedar, H. Solage, A., Glaser, G., and Razin, A. (1979) Direct detection of methylated cytosine in DNA by use of the restriction enzyme MspI. Nucl. Acids Res. *6*, 2125–2132.

Church, G. M., and Gilbert, W. (1984) Genomic sequencing. Proc. Natl. Acad. Sci. USA *81*, 1991–1995.

Clotten, R., and Clotten, A. (1962) Hochspannungselektrophorese, Georg Thieme Verlag, Stuttgart.

Coffey, R. G., and Newburgh, R. W. (1963) The effect of calcium sulfate as the binder in DEAE-cellulose thin layer chromatography for separating nucleic acid degradation products. J. Chromatog. *11*, 376–382.

Doerfler, W. (1981) DNA methylation – a regulatory signal in eucaryotic gene expression. J. Gen. Virol. *57*, 1–20.

Doerfler, W. (1984) DNA Methylierung. Angew. Chem. *12*, 917–1004.

Eick, D., Fritz, H. J., and Doerfler, W. (1983). Quantitative determination of 5-methylcytosine in DNA by reverse-phase high-performance liquid chromatography. Anal Biochem. *135*, 165–171.

Ephrussi, A., Church, G. M., Tonegawa, S., and Gilbert, W. (1985) B lineage-specific interactions of an immunoglobulin enhancer with cellular factors in vivo. Science *227*, 134–140.

Feigon, J., Wang, A. H.-J., Van der Marel, G. A., Van Boom, J. H., and Rich, A. (1984) A one- and two-dimensional NMR study of the B to Z transition of $(m^5dC\text{-}dG)_3$ in methanolic solution. Nuc. Acids Res. *12*, 1243–1263.

Ford, J. P., Coca-Prados, M., and Hsu, M.-T. (1980) Enzymatic analysis of 5-methylcytosine content in eucaryotic DNA. J. Biol. Chem. *255*, 7544–7547.

Frommer, M., McDonald, L. E. Millar, D. S., Collis, C. M., Watt, F., Grigg, G. W., Molloy, P. L., and Paul, C. L. (1992) A genomic sequencing protocol which yields a positive display of 5-methylcytosine residues in individual DNA strands. Proc. Natl. Acad. Sci. USA *89* 1827–1831.

Genest, D. Mazeau, K., and Ptak, M. (1987) Two-dimensional 1H NMR study of d(br5C-G)3 in the Z-form. Self association and flexibility of the left-handed double helix. J. Biomol. Struct. Dyn. *5*, 67–78.

Gruenbaum, Y., Cedar, H., and Razin, A. (1981) Restriction enzyme digestion of hemimethylated DNA. Nucl. Acids Res. *9*, 2509–2515.

Jacobson, K. (1964) Chromatographic separation of nucleotides and nucleosides. J. Chromatog. *14*, 542–543.

Kochetkov, N. K., and Budowskii, E. I. (1974) in: Organic Chemistry of Nucleic Acids. Part B, pp. 477–527. Plenum Publishing Company Ltd, London, New York.

Kuo, K. C., McCune, R. A., and Gehrke, C. W. (1980) Quantitative reversed-phase high performance liquid chromatographic determination of major and modified deoxyribonucleosides in DNA. Nucl. Acids Res. *8*, 4763–4776.

Linn, S. M., and Roberts, R. J. (1982) Nucleases. Cold Spring Harbor Laboratory, Cold Spring Harbor, New York.

Lois, R., Dietrich, A., Hahlbrock, K., and Schultz, W. (1989) A phenylalanine ammonia-lyase gene from parsley: structure, regulation and identification of elicitor and light responsive cis-acting elements. EMBO J. *8*, 1642–1648.

Maxam, A. M., and Gilbert, W. (1980) Sequencing end-labeled DNA with base-specific chemical cleavages. Meth. Enzymol. *65*, 499–560.

Marshak, A., and Vogel, H. J. (1951) Microdetermination of purines and pyrimidines in biological materials. J. Biol. Chem. *189*, 597.

Meijlink, F. C. P. W., Philipsen, J. N. J., Gruber, M., and Geert, A. B. (1983) Methylation of the chicken vitellogenin gene: influence of estradiol administration. Nucl. Acids Res. *11*, 1361–1373.

Mueller, P. R., and Wold, B. (1989) In vivo footprinting at a developmentally regulated enhancer. Science *246*, 780–786.

Nick, H., Bowen, B., Terl, R. J., and Gilbert, W. (1986) Detection of cytosine methylation in the maize alcohol dehydrogenase gene by genomic sequencing. Nature *319*, 243–246.

Orbons, L. P. M., and Altona, C. (1986) The B and Z forms of the d(m^5C-G)$_3$ and d(br^5C-G)$_3$ hexamers in solution. Eur. J. Biochem. *160*, 131–139.

Pfeifer, G. P., Steigerwald, S. D., Mueller, P. R. Wold, B., and Riggs, A. D. (1989) Genomic sequencing and methylation analysis by ligation mediated PCR. Science *246*, 810–813.

Pfeifer, G. P., Tanguay, R. L., Steigerwald, S. D., and Riggs, A. D. (1990) PCR-aided DNasel footprinting of single copy gene sequences in permeabilized cells. Genes Dev. *4*, 1277–1287.

Randerath, K. (1961) Duennschichtchromatographie an lonentauscher-Schichten. Angew. Chem. *73*, 674–676.

Razin, A., and Cedar, H. (1977) Distribution of 5-methylscytosine in chromatin. Proc. Natl. Acad. Sci. USA *74*, 2725–2728.

Rideout, W. M., Coetzee, G. A., Olumi, A. F., and Jones, P. A. (1990) 5-methylcytosine as an endogenous mutagen in the human LDL receptor and p53 genes. Science *249*, 1288–1290.

Saluz, H. P., and Jost, J. P. (1986) Optimized genomic sequencing as a tool for the study of cytosine methylation in the regulatory region of the chicken vitellogenin II gene. Gene *42*, 151–157.

Saluz, H. P., Jiricny, J., and Jost, J. P. (1986) Genomic sequencing reveals a positive correlation between the kinetics of strand-specific DNA methylation of the overlapping estradiol/glucocorticoid-receptor binding sites and the rate of avian vitellogenin mRNA synthesis. Proc. Natl. Acad. Sci. USA *85*, 6697–6700.

Saluz, H. P., Feavers, I. M., Jiricny, J., and Jost, J. P. (1988) Genomic sequencing and in vivo genomic footprinting of an expression-specific DNase I hypersensitive site of avian vitellogenin II promoter reveal a demethylation of a mCpG and a change in specific interactions of proteins with DNA. Proc. Natl. Acad. Sci. USA *85*, 6697–6700.

Saluz, H. P., and Jost, J. P. (1989) Genomic footprinting with Taq polymerase. Proc. Natl. Acad. Sci. USA *86*, 2602–2606.

Saluz, H. P., and Jost, J. P. (1990) A Laboratory Guide for In Vivo Studies of DNA Methylation and Protein/DNA Interaction, pp. 129–214; BioMethods series, Vol. 3. Birkhäuser, Basel–Boston.

Saluz, H. P., Wiebauer, K., and Wallace, A. (1991) Studying DNA modifications and DNA-protein interactions in vivo. TIG *7*, 207–211.

Sano, H., Royer, H.-D., and Sager, R. (1980) Identification of 5-methylcytosine in DNA fragments immobilized on nitrocellulose paper. Proc. Natl. Acad. Sci. USA *77*, 3581–3585.

Singer, J., Roberts-Ems, J., and Riggs, A. D. (1979) Methylation of mouse liver DNA studied by means of the restriction enzymes MspI and HpaII Science *203*, 1019–1021.

Singer-Sam, T. P., Yang, N. Mori, R. L., Tanguay, J. M., Le Bon, J. C., and Riggs, A. D. (1990) DNA methylation in the 5′ region of the mouse PGK-1 gene and a quantitative PCR assay for methylation, in: Nucleic Acid Methylation. New Book Series, UCLA Symposia on Molecular and Cellular Biology, vol. 1, p. 28. Eds G. Clawson, D. Willis, A. Weissback and P. Jones. Alan R. Liss Inc., New York.

Southern, E. M. (1975) Detection of specific sequences among DNA fragments separated by gel electrophoresis. J. Mol. Biol. *98*, 503–517.

Stahl, E. (1967) in: Duennschichtchromatographie, pp. 749–769. Springer-Verlag, Berlin, Heidelberg, New York.

Toth, M., Lichtenberg, U., and Doerfler, W. (1989) Genomic sequencing reveals a 5-methylcytosine-free domain in active promoters and the spreading of preimposed methylation patterns. Proc. Natl. Acad. Sci. USA *86*, 3728–3732.

Toth, M., Mueller, U., and Doerfler, W. (1990) Establishment of de novo DNA methylation patterns. Transcription factor binding and deoxycytidine methylation at CpG and non-CpG sequences in an integrated adenovirus promoter. J. Mol. Biol. *214*, 673–683.

Van De Ven, F. J. M., and Hilbers, C. W. (1980) Nucleic acids and nuclear magnetic resonance. Eur. J. Biochem. *178*, 1–38.

Vischer, E., and Chargaff, E. (1948) The composition of the pentose nucleic acids of yeast and pancreas. J. Biol. Chem. *176*, 715–734.

Waalwijk, C., and Flavell, R. A. (1978) MspI, an isoschizomer of HpaII which cleaves both unmethylated and methylated HpaII sites. Nucl. Acids Res. *5*, 3231–3236.

Wagner, I., and Capesius, I. (1981) Determination of 5-methylcytosine from plant DNA by high-performance liquid chromatography. Biochim. Biophys. Acta *654*, 52–56.

Ward, C., Bolden, A., and Weissbach, A. (1990) Genomic sequencing of the 5′-flanking region of the mouse beta-globin major gene in expressing and nonexpressing mouse cells. J. Biol. Chem. *265*, 3030–3033.

Winnacker, E. (1984) in: Gene und Klone. Verlag Chemie, Weinheim, Florida and Basel.

Wuethrich, K. (1986) in: NMR of Proteins and Nucleic Acids. J. Wiley, New York.

Wyatt, G. R. (1951) The purine and pyrimidine composition of deoxypentose nucleic acids. Biochem. J. *48*, 584–590.

Zweig, G., and Whitaker, J. R. (1967) Paper Chromatography and Electrophoresis, vol. 1. Academic Press, New York and London.

DNA Methylation: Molecular Biology and Biological Significance
ed. by J. P. Jost & H. P. Saluz

Methylation of cytosine influences the DNA structure

Wolfgang Zacharias

Department of Biochemistry, Schools of Medicine & Dentistry, University of Alabama at Birmingham, UAB Station, Birmingham, AL 35294-0005, USA

Introduction

In the past decade it has become increasingly clear that DNA has the potential to adopt a variety of unusual secondary structures which deviate from the classical right-handed B-DNA form. Intrinsic properties of the DNA molecule (primary base sequence and base modifications, degree of supercoiling) determine whether such an unusual structure is possible. Environmental factors (like ionic strength or pH of the medium, temperature, solvent polarity, interactions with proteins or drugs) have a strong influence on whether the structure is actually formed or not.

Several of these structures have been extensively characterized *in vitro* with a variety of physicochemical techniques, enzymatic or chemical probing, topological and immunological assays and theoretical considerations (reviewed in Rich et al., 1984; McLean et al., 1988; Wells, 1988; Zimmerman, 1982). The main focus here shall be on the non-B-DNA structures formed by inverted repeat sequences (cruciforms), alternating purine-pyrimidine sequences (left-handed Z-DNA), homopurine-homopyrimidine sequences (triple-helical DNA), and oligo-dA tracts (curved DNA regions).

It has become increasingly important to identify and analyze potential biological functions of these unusual structures. Such a functional analysis requires the concomitant structural analysis of the DNA inside the cell, which is a much more complex task than an *in vitro* characterization. However, very recently some progress has been made toward the identification of unusual DNA structures *in vivo* (Wells, 1988; Zacharias, 1992).

Enzymatic methylation of cytosines in DNA is believed to have fundamental roles in basic gene regulatory or signaling functions (reviewed in other chapters of this book). In addition, it has been known for some time that methylation of bases can affect the physical properties as well as the structural features of the DNA double helix. In this chapter, the influences of CpG as well as cytosine methylation on the formation and properties *in vitro* of curved DNA regions, DNA triple helices, cruciforms, and left-handed Z-DNA shall be discussed.

2 Cytosine methylation and DNA structures

2.1 Cruciform structures

Cruciform structures are extruded in supercoiled circular DNA domains at inverted repeat sequences when intrastrand pairing of adjacent base sequences is possible, leaving a short unpaired loop region at the center of the structure (Fig. 1). Such cruciform structures have been found in

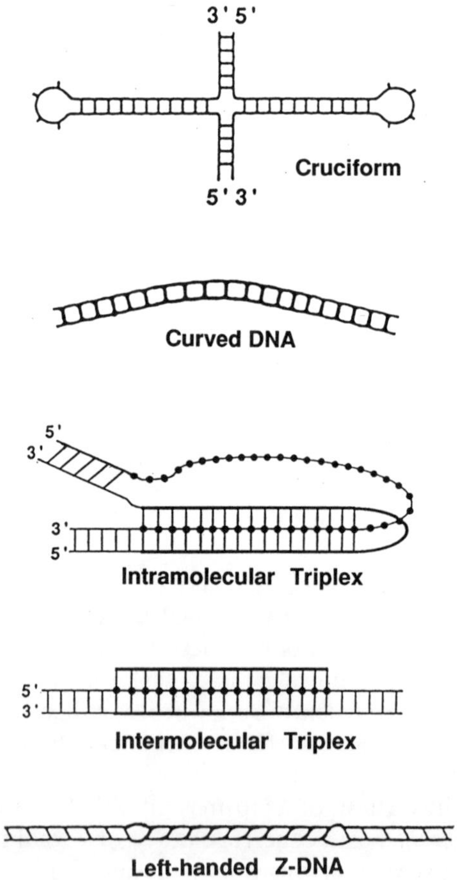

Figure 1. Structural alternatives to the right-handed B-form of DNA.

a number of upstream regulatory regions of genes. They may play a role in transcriptional regulation and RNA polymerase interactions with sequence elements on the DNA template. On the other hand, cruciform structures are stereochemically equivalent to intermediate structures in recombination events, thus allowing conclusions about mechanistic aspects of this process (reviewed in Lilley et al., 1988).

The extrusion of cruciform structures usually is initiated by the rate-limiting base pair opening at the central region, followed by unpairing in both directions and new intra-strand pairing of bases, until the cruciform is fully extruded. Thus, the kinetics of cruciform formation strongly depend on the helix stability of the central potential loop region (Murchie and Lilley, 1987).

Recently, the effects of base methylation at the centers of inverted repeats on cruciform extrusion kinetics were studied (Murchie and Lilley, 1989). Methyl groups were introduced at the N-6 position of adenine and/or the C-5 position of cytosine in the potential loop region by bacterial DNA methyltransferases. The rate of cruciform extrusion in supercoiled plasmids containing these modified sequences was measured by S1 nuclease cleavage at the unpaired central loop region.

It was found that methylation at C-5 of cytosine residues in the inverted repeat center caused an approximately two- to three-fold decrease in extrusion rates, whereas methylation at N-6 of adenine residues at this location resulted in a four-fold rate enhancement. In a sequence that contained both C-5 and N-6 methyl groups simultaneously, the effects were additive and could be treated independently. These results show that the effects of base methylations on cruciform extrusion rates directly parallel their influences on the helix stability of the parent double-stranded DNA segments, thus further confirming the proposed mechanism for cruciform formation.

2.2 Curved DNA regions

In DNA sequences with short (a few base pairs) runs of oligo-dA·oligo-dT, a significant curvature of the molecule is found when such oligo A units are repeated at approximately 10 base pair intervals, i.e., in phase with the helix periodicity of the DNA (Fig. 1). This property can be observed and quantitated by polyacrylamide gel electrophoresis, where curved DNA fragments migrate with a mobility slower than the one expected based on the size of the molecule (Koo and Crothers, 1987). This phenomenon of intrinsic curvature of some DNA sequences, however, has to be distinguished from protein-induced or ligand-induced bends in DNA regions which are otherwise not bent (Travers, 1990).

Currently, two models are under discussion to rationalize the behavior of curved DNA molecules (reviewed by Lilley, 1986). The "wedge model" by Trifonov et al. (1986) suggests that the ApA dinucleotide units have a wedge-like structural deviation from the average B-helix conformation, causing an overall curvature when in phase with the helix periodicity. The "junction model" by Crothers et al. (1986) proposed that oligo-dA tracts possess a different, non-B-DNA structure, and that the helix axis changes direction at the interphase of this non-B structure with the B-form helix.

A few reports have analyzed the effects of methylation on the curvature of such molecules. In one study, short DNA segments of 10 base pairs, containing the recognition sites for bacterial DNA methyltransferases, were ligated to generate oligomeric ladders (Diekmann, 1987). The relative gel mobilities of these fragments, indicating the degrees of curving in the molecules, were measured by gel electrophoresis. Methylation at N-6 of adenine by M*Eco*RI in the sequence CGGAATTCCG further increased the curving in the molecule that was already curved without methylation. In the non-curved fragment CGGGATCCCG, N-6 methylation of adenine or C-5 methylation of cytosine in the most central position both induce slight curvature, whereas the sequence CCATCGATGG is unaffected by methylation at either base.

With a similar approach, position effects of methyl groups on the curvature of self-ligated oligonucleotide families were analyzed (Hagerman, 1990). The sequences oligo-dA·oligo-dT or oligo-dI·oligo-dC were mutated by replacing dT with dU or dC with 5-methyl-dC, respectively, at various positions in the oligopyrimidine tracts. It was found that removal of methyl groups at the 5′ end of the tracts resulted in reduction of curvature, whereas removal at the 3′ end of the tracks caused a significant increase in curvature of the molecule.

Taken together, these experiments show that base methylation, especially methylation at the 5-position of pyrimidines, can cause either a decrease, an enhancement, or even an induction of curvature, depending on the location of the base modification within the oligopyrimidine tract.

2.3 DNA triple helices

It has been known for several decades that DNA polymers of appropriate sequences can form triple-stranded helix structures by binding of a single homopyrimidine strand into the major groove of a homopurine·homopyrimidine duplex (Fig. 1) (reviewed in Wells et al., 1988). This occurs by forming pyrimidine-purine-pyrimidine base triads with Hoogsteen pairing between the third strand and the parent double helix. In recent years this triple-helical DNA structure has drawn renewed attention, since it was suggested that triplex segments, formed as intermolecular complex with a short single-stranded oligopyrimidine strand (Fig. 1), can inhibit gene expression (Wells et al., 1988; Maher III et al.,

1989). In addition, homopurine·homopyrimidine segments with mirror symmetry and sufficient length can form intramolecular triplex structures in supercoiled plasmids at pH 4 to pH 5 (Fig. 1) (reviewed in Wells et al., 1988).

Since a moderately acidic medium (pH below 6) seemed to be a general requirement for triplex formation, it was important to identify other factors that could stabilize triple-helical DNA segments. Thermal melting experiments as a function of pH with the synthetic DNAs poly(dT-dC)·poly(dG-dA) and its cytosine-methylated analog revealed that the methylated polymer can form a triple-helical structure at a pH slightly below 8.0, whereas the unmethylated polymer required pH values below 6.0 for triplex formation (Lee et al., 1984).

These results were confirmed by circular dichroism and calorimetric measurements (Plum et al., 1990) as well as by pH titrations of UV melting curves and gel electrophoresis (Xodo et al., 1991) on triplex structures formed between short oligonucleotide duplexes and single-stranded homopyrimidines as third strand. Replacement of cytosine by 5-methylcytosine in the homopyrimidine component allowed intermolecular triplex formation at pH 6.8, compared to pH 5.8 for the unmethylated third strand. It also increased the thermal stability of the triple helix formed by approximately 10°C at any given pH value.

The stabilizing effect of cytosine methylation on triple-stranded DNA segments has also been applied to functional analyses of triplex structures under conditions where proteins or enzymes remain fully active. It could be shown that triplex formation at the target sites for restriction enzymes, bacterial DNA methyltransferases, or the SP1 transcription factor prevented interactions of these proteins with their recognition sites (Maher III et al., 1989). The concept of triplex-mediated inhibition of a site-specific DNA methyltransferase was also used to cleave yeast chromosomal DNA at a single site (Strobel et al., 1991). In these examples, the use of 5-methylcytosine-containing single-stranded components strongly facilitated triplex formation at neutral pH and at approximately 50-fold lower oligomer concentrations than that required for an unmethylated strand.

It seems clear now that replacement of cytosine by 5-methylcytosine increases the thermal stability of intermolecular triple helices. It also enables triplex formation at neutral pH, which is crucial for *in vivo* applications of targeted triplex formation on gene inhibition studies. No data are available yet on the effects of cytosine methylation on intramolecular triplex formation in supercoiled plasmids.

2.4 Left-handed helix forms

2.4.1 DNA oligomers and polymers. The discovery of left-handed DNA helix structures in double-stranded DNA oligonucleotides and polymers resulted in numerous investigations, performed with a broad variety of physical, optical, enzymatic, and theoretical approaches, on the behavior and properties of this unusual DNA structure (reviewed in Zimmerman, 1982; Rich et al., 1984). Left-handed helices in linear DNA molecules are stabilized by various ionic and solvent conditions in sequences comprised mainly of alternating dG-dC and/or dA-dC dinucleotide repeats (Rich et al., 1984; Wells et al., 1983) (Fig. 1). Although some significant progress has been made in recent years concerning the *in vivo* existence of Z-DNA (McLean et al., 1988; Wells, 1988; Zacharias et al., 1990; Zacharias, 1992), the potential biological functions are still under investigation.

An important finding in this direction was the observation by UV and CD spectroscopy that cytosine methylation stabilizes a Z-helix in poly(dG-5-mC)·poly(dG-5-mC) at drastically lower ionic strengths than in the unmethylated polymer (0.7 M versus 2.5 M NaCl or 0.6 mM versus 700 mM $MgCl_2$, respectively) (Behe and Felsenfeld, 1981). Further investigations on left-handed dG-5-mC sequences by X-ray crystallography (Fujii et al., 1982), proton NMR spectroscopy (Feigon et al., 1984; Orbons and Altona, 1986), ^{31}P and ^{13}C NMR spectroscopy (Chen et al., 1983), transient electric dichroism (Chen et al., 1982), CD and UV spectroscopy (Hacques and Marion, 1986) provided a detailed characterization of the Z-helix structures adopted by this sequence.

As for the unmethylated dG-dC polymer, the left-handed structure of poly(dG-5-mC)·poly(dG-5-mC) has a dinucleotide repeat unit, alternating phosphodiester backbone structure, *anti* and *syn* base orientation at the cytidine-sugar and guanosine-sugar moieties, respectively, with concomitant C2′-endo and C3′-endo sugar conformations. Other helix parameters like twist angle and rise per residue differ slightly between methylated and unmethylated Z-helices (Feigon et al., 1984; Fujii et al., 1982; Chen et al., 1983).

The changes in helical repeat associated with the B to Z transition in poly(dG-dC)·poly(dG-dC) and poly(dG-5-mC)·poly(dG-5-mC) were also measured with an enzymatic approach (Behe et al., 1981). The polymers were absorbed on solid calcium phosphate or oxalate crystals, and the cutting patterns after limited digestion with different nucleases were determined. The B-form of each polymer gave a helix repeat of 10.5 base pairs per turn, whereas the Z-forms had a helix repeat of approximately 13.6 base pairs per turn. However, it remains somewhat surprising in this experiment that the nucleases can digest both the right-handed and left-handed forms of DNA polymers.

The stabilizing effects of the methyl group at the C-5 position of cytosine were attributed mainly to a relative destabilization of the right-handed B-form, and an increased hydrophobic interaction with the Z-helix. These interpretations were supported by a number of theoretical approaches. Quantum chemical calculations revealed a substantially higher anti/syn base rotation barrier for methylcytosine compared to cytosine, as well as possible hydrophobic interactions between the methyl group and the CH_2-component of an adjacent deoxyribose unit (Van Lier et al., 1983). Calculations of the electrostatic stabilities of the B- and Z-forms of methylated and unmethylated dG-dC polymers suggested that the methyl group induces a minor structural change in the Z-helix which allows better accessibility of charges by counterions in the solvent (Pack et al., 1986). A free energy perturbation approach with statistical mechanics treatment determined that the methyl group contributes approximately −0.4 kcal/(mol-base pair) to the stability of the Z-helix mainly because of methyl-solvent interactions (Pearlman and Kollman, 1990).

A number of calorimetric measurements, UV melting curve determinations, and temperature jump kinetics experiments were performed in order to analyze the B-Z equilibrium properties of poly(dG-5-mC)·poly(dG-5-mC). The van't Hoff enthalpy changes for this equilibrium were in the order of 70-200 kcal/mol (Klump and Loeffler, 1985; Chaires and Sturtevant, 1986), and were strongly dependent on the chain lengths of the polymer preparations (Walker and Aboul-ela, 1988). Under conditions of low ionic strength and external pressure, the melting temperature of left-handed poly(dG-5-mC)·poly(dG-5-mC) was at 120–130°C (Klump and Loeffler, 1985; Chaires and Sturtevant, 1986). The cooperativity length for the B to Z transition in this polymer were also determined, but they differed significantly between different reports, varying from approximately 100–150 base pairs (Klump and Loeffler, 1985; Chaires and Sturtevant, 1986) to approximately 1,000 base pairs (Walker and Aboul-ela, 1988). It has to be pointed out that the substantial variations in the thermodynamic parameters obtained was probably caused by different ionic conditions, by the strong influence of polymer lengths and the associated end effects, and possibly also by various trace amounts of metal ions present in the DNA preparations (see below).

Although the unmethylated poly(dG-dC)·poly(dG-dC) is clearly in a right-handed B-type conformation under low-salt conditions, there has been some controversy about the helix sense of poly(dG-5-mC)·poly(dG-5-mC) at very low ionic strength. Based on vacuum CD spectra, ^{31}P-NMR experiments and sedimentation measurements, it was concluded that the methylated polymer adopts a left-handed Z-like structure in a few millimolar NaCl (Krueger and Prairie, 1985; Feuerstein et al., 1985; Behe, 1986). However, more stringent analyses, including atomic absorption spectroscopy, revealed that the Z-form under those conditions was caused by trace amounts of divalent metal ions which very efficiently promoted left-handed helix formation (Woisard et al., 1986; Devarajan and Shafer, 1986; Chen, 1986). Surprisingly, it was also found that poly(dG-5-mC)·poly(dG-5-mC) adopts an acid-induced Z-like helix structure at approximately pH 4 under very low ionic strength conditions which favor the B-form at neutral pH (Chen, 1986).

In summary, it is clear that CpG methylation in alternating dC-dG oligonucleotides and polymers strongly facilitates left-handed Z-DNA formation and allows B to Z transitions under near physiological ionic strength conditions.

2.4.2 Recombinant plasmids. In circular plasmids with sufficient negative superhelical strain, regions of predominantly alternating dG-dC or dC-dA sequences can invert into a left-handed helix structure. This supercoil-induced B to Z transition results in the relaxation of some of the negative supercoils, the establishment of B-Z junction regions and, of course, the drastic change in the helix structure at the region undergoing the transition (Zacharias et al., 1988, and references therein). The B to Z transition in plasmids can be monitored by B-Z junction-specific chemical or enzymatic probes, Z-DNA-specific antibodies, and two-dimensional gel electrophoresis (McLean and Wells, 1988; Rich et al., 1984). Since the B to Z switch can also occur inside bacterial cells (Zacharias et al., 1990), it was important to identify biological factors that could enhance Z-DNA formation inside living cells.

Based on the stabilizing effects of methylation on Z-DNA formation in DNA polymers, the influences of cytosine methylation on the supercoil-induced B to Z transition in cloned alternating dC-dG inserts were analyzed by one- and two-dimensional agarose gel electrophoresis (Klysik et al., 1983; Zacharias et al., 1988). Methylation of the dC-dG inserts was achieved by modification with M*Hha*I DNA methyltransferase reaction; the two-dimensional plasmid topoisomer distributions were analyzed with a statistical mechanics treatment.

Upon methylation at C-5 of cytosine, the helical twist parameters of the alternating dC-dG helix in the right-handed form changed from 10.5 to 10.7 base pairs per turn, and also the left-handed helices changed from 11.5 to 12.8 base pairs per turn. The free energy difference per base pair between right-handed and left-handed helix structure decreased three-fold from 0.32 to 0.11 kcal/(mol-base pair). On the other hand, the energy costs to establish B-Z junction regions were largely unaffected. In other words, for a given length of dC-dG region, less supercoiling is required after methylation for the B to Z transition to occur.

These results show that, as in the case of synthetic DNA polymers, cytosine methylation also facilitates the supercoil-induced B to Z transition in circular DNA domains. Thus, this base modifications can potentially serve as a modulator for the B-Z switch *in vivo* as well as for DNA supercoiling and topology in general.

2.5 *Other helix structures*

A variety of other DNA sequences and helix forms have been described in which the structures and properties are influenced by cytosine methylation. Several of these polymers have been investigated mainly for basic structural or mechanistic analyses and do not represent natural DNA substrates.

A systematic study described the methylcytosine-induced changes in the physical properties (UV absorbance, fluorescence, melting, buoyant density) of a series of homopolymeric or alternating copolymeric DNA sequences of the types poly(purine)·poly(pyrimidine) or poly(purine-pyrimidine)·poly(purine-pyrimidine) (Gill et al., 1974). Other base and sequence variations to be mentioned here are the alternating dG-5EthyldC polymer which, quite similar to the methylcytosine-containing analog, also forms a low salt Z-helix (Vorlickova and Sagi, 1989), poly(dI-5-mC)·poly(dI-5-mC) which can adopt a Z-form as well as an unidentified X-DNA form under certain ionic conditions (Vorlickova and Sagi, 1991), and poly(dA-5-mC)·poly(dA-5-mC), in which the methyl group enables Z-DNA formation despite the fact that the unmethylated polymer forms a Z-helix only under very drastic conditions (McIntosh et al., 1983). Unexpected transitions between Z-form and A-form structures have been observed in a hybrid poly(rG-5-mC)·poly(rG-5-mC) (Wu and Behe, 1985) and in poly(dG-5-mC)·poly(dG-5-mC) treated with Terbium(III) ions (Chatterji, 1988). On the other hand, methylation of cytosines in the homopolymeric polydG·polydC causes an A-form to B-form transition (Sarma et al., 1986).

3 Summary: The interplay between methylation and DNA structure

DNA-protein interactions constitute a crucial level in the regulation of gene expression and other key cellular processes. Unorthodox DNA structures, like cruciforms, curved regions, triple-helical segments, or left-handed Z-helices, can provide new and unique interaction sites with structure-specific proteins which are not present in right-handed B-form helices. It has become clear that cytosine methylation can influence these unusual structures by either enhancing or inhibiting their formation. Thus, this base modification could indirectly affect gene regulatory events *in vivo* by altering the equilibrium between such unusual DNA structures and regular B-helix regions.

Several of the non-B-form DNA structures (cruciforms, triplexes, Z-DNA) are, on one hand, induced by DNA supercoiling and, on the other hand, also relax some of the supercoils as a consequence of their formation. Since cytosine methylation can influence the formation of these unusual structures, the methylation event could in this way essentially alter the degrees of superhelicity in circular plasmid or chromosomal loop domains, thus affecting supercoil-dependent DNA-protein interactions even at large distances. This may constitute an additional mechanism for the modulation of regulatory events by base methylation.

With the current progress in the field we are learning that DNA structure, DNA methylation, and cellular regulation appear to be linked by a complex interplay at different levels, and we are only at the beginning of understanding this complexity.

Behe, M. J. (1986) Vacuum UV CD of the low-salt Z-forms of poly(rG-dC)·poly(rG-dC), and poly(dG-m^{5}dC)·poly(dG-m^{5}dC). Biopolymers *25*, 519–523.

Behe, M., and Felsenfeld, G. (1981) Effects of methylation on a synthetic polynucleotide: the B-Z transition in poly(dG-m^{5}dC)·poly(dG-m^{5}dC). Proc. Natl. Acad. Sci. USA *78*, 1619–1623.

Behe, M., Zimmerman, S., and Felsenfeld, G. (1981) Changes in the helical repeat of poly(dG-m^{5}dC)·poly(dG-m^{5}dC) and poly(dG-dC)·poly(dG-dC) associated with the B-Z transition. Nature *293*, 233–235.

Chaires, J. B., and Sturtevant, J. M. (1986) Thermodynamics of the B to Z transition in poly(m^{5}dG-dC). Proc. Natl. Aca. Sci. USA *83*, 5479–5483.

Chatterji, D. (1988) Terbium(III) induced Z to A transition in poly(dG-m^{5}dC). Bipolymers *27*, 1183–1186.

Chen, C.-W., Cohen, J. S., and Behe, M. (1983) B to Z transition of double-stranded poly[deoxyguanylyl(3′-5′)-5-methyldeoxycytidine] in solution by phosphorus-31 and carbon-13 nuclear magnetic resonance spectroscopy. Biochemistry *22*, 2136–2142.

Chen, F.-M. (1986) Conformational lability of poly(dG-m^{5}dC)·poly(dG-m^{5}dC). Nucl. Acids Res. *14*, 5081–5097.

Chen, H. H., Charney, E., and Rau, D. C. (1982) Length changes in solution accompanying the B-Z transition of poly(dG-m^{5}dC) induced by $Co(NH_3)6^{3+}$. Nucl Acids Res. *10*, 3561–3571.

Devarajan, S., and Shafer, R. H. (1986) Role of divalent cations on DNA polymorphism under low ionic strength conditions. Nucl. Acids Res. *14*, 5099–5109.

Diekmann, S. (1987) DNA methylation can enhance or induce DNA curvature. EMBO J. *6*, 4213–4217.

Feigon, J., Wang, A. H.-J., van der Marel, G. A., Van Boom, J. H., and Rich A. (1984) A one- and two-dimensional NMR study of the B to Z transition of (m^{5}dC-dG)$_3$ in methanolic solution. Nucl. Acids Res. *12*, 1243–1263.

Feuerstein, B. G., Marton, L. J., Keniry, M. A., Wade, D. L., and Shafer, R. H. (1985) New DNA polymorphism: evidence for a low salt, left-handed form of poly(dG-m^{5}dC). Nucl. Acids Res. *13*, 4133–4141.

Fujii, S., Wang, A. H.-J., van der Marel, G., van Boom, J. H., and Rich, A. (1982) Molecular structure of (m^{5}dC-dG)$_3$: the role of the methyl group on 5-methyl cytosine in stabilizing Z-DNA. Nucl. Acids Res. *10*, 7879–7892.

Gill, J. E., Mazrimas, J. A., and Bishop, C. C. Jr. (1974) Physical studies on synthetic DNAs containing 5-methyl-cytosine. Biochim. Biophys. Acta *335*, 330–348.

Hacques, M.-F., and Marion, C. (1986) DNA polymorphism: spectroscopic and electro-optic characterizations of Z-DNA and other types of left-handed helical structures induced by Ni^{2+}. Biopolymers *25*, 2281–2293.

Hagerman, P. J. (1990) Pyrimidine 5-methyl groups influence the magnitude of DNA curvature. Biochemistry *29*, 1980–1983.

Klump, H. H., and Löffler, R. (1985) Reversible helix-coil transitions of left-handed Z-DNA structures. Biol. Chem. Hoppe-Seyler *366*, 345–353.

Klysik, J., Stirdivant, S. M., Singleton, C. K., Zacharias, W., and Wells, R. D. (1983) Effects of 5 cytosine methylation on the B-Z transition in DNA restriction fragments and recombinant plasmids. J. Mol. Biol. *168*, 51–71.

Koo, H. S., Wu, H. M. and Crothers, D. M. (1986) DNA bending at adenine-thymine tracts. Nature *320*, 501–506.

Koo, H.-S., and Crothers, D. M. (1987) Chemical determinants of DNA bending at adenine-thymine tracts. Biochemistry *26*, 3745–3458.

Krueger, W. C., and Prairie, M. D. (1985) A low-salt form of poly(dG-5M-dC)·poly(dG-5M-dC). Biopolymers *24*, 905–910.

Lee, J. S., Woodsworth, M. L., Latimer, L. J. P., and Morgan, A. R. (1984) Poly(pyrimidine)·poly(purine) synthetic DNAs containing 5-methylcytosine form stable triplexes at neutral pH. Nucl. Acids Res. *12*, 6603–6614.

Lilley, D. M. J. (1986) Bent molecules – how and why? Nature *320*, 487–488.

Lilley, D. M. J., Sullivan, K. M., Murchie, A. I. H., and Furlong, J. C. (1988) Cruciform extrusion in supercoiled DNA – mechanisms and contextual influence, in: Unusual DNA Structures, pp. 55–72. Eds R. D. Wells and S. C. Harvey. Springer Verlag, New York.

Maher III, L. J., Wold, B., and Dervan, P. B. (1989) Inhibition of DNA binding proteins by oligonucleotide-directed triple helix formation. Science *245*, 725–730.

McIntosh, L. P., Grieger, I., Eckstein, F., Zarling, D. A., van de Sande, J. H., and Jovin, T. M. (1983) Left-handed helical conformation of poly[d(A-m^{5}C)·d(G-T)]. Nature *304*, 83–86.

McLean, M. J., and Wells, R. D. (1988) The role of sequence in the stabilization of left-handed DNA helices *in vitro* and *in vivo*. Biochim. Biophys. Acta *950*, 243–254.

Murchie, A. I. H., and Lilley, D. M. J. (1987) The mechanism of cruciform formation in supercoiled DNA: initial opening of central base pairs in salt-dependent extrusion. Nucl. Acids Res. *15*, 9641–9654.

Murchie, A. I. H., and Lilley, D. M. J. (1989) Base methylation and local DNA helix stability: effect on the kinetics of cruciform extrusion. J. Mol. Biol. *205*, 593–602.

Orbons, L. P. M., and Altona, C. (1986) Conformational analysis of the B and Z forms of the d(m^{5}C-G)$_3$ and d(Br^{5}C-G)$_3$ hexamers in solution: a 300-MHz and 500-MHz NMR study. Eur. J. Biochem. *160*, 141–148.

Pack, G. R., Prasad, C. V., Salafsky, J. S., and Wong, L. (1986) Calculations on the effect of methylation on the electrostatic stability of the B- and Z-conformers of DNA. Biopolymers *25*, 1697–1715.

Pearlman, D. A., and Kollman, P. A. (1990) The calculated free energy effects of 5-methyl cytosine on the B to Z transition in DNA. Biopolymers *29*, 1193–1209.

Plum, G. E., Park, Y.-W., Singleton, S. F., Dervan, P. B., and Breslauer, K. J. (1990) Thermodynamic characterization of the stability and the melting behaviour of a DNA triplex: a spectroscopic and calorimetric study. Proc. Natl. Acad. Sci. USA *87*, 9436–9440.

Rich, A., Nordheim, A., and Wang, A. H.-J. (1984) The chemistry and biology of left-handed Z-DNA. Ann. Rev. Biochem. *53*, 791–846.

Sarma, M. H., Gupta, G., and Sarma, R. H. (1986) 500–MHz ^{1}H NMR study of poly(dG)·poly(dC) in solution using one-dimensional nuclear Overhauser effect. Biochemistry *25*, 3659–3665.

Strobel, S. A., and Dervan, P. B. (1991) Single-site enzymatic cleavage of yeast genomic DNA mediated by triple helix formation. Nature *350*, 172–174.

Travers, A. A. (1990) Why bend DNA? Cell *60*, 177–180.

Trifanov, E. N., and Sussman, J. L. (1980) The pitch of chromatin DNA is reflected in its nucleotide sequence. Proc. Natl. Acad. Sci. USA *77*, 3816–3820.

Van Lier, J. J. C., Smits, M. T., and Buck, H. M. (1983) B-Z transition in methylated DNA: a quantum-chemical study. Eur. J. Biochem. *132*, 55–62.

Vorlicková, M., and Sági, J. (1991) Transitions of poly(dI-dC), poly(dI-methyl^{5}dC) and poly(dI-bromo^{5}dC) among and within the B-, Z-, A- and X-DNA families of conformation. Nucl. Acids Res. *19*, 2343–2347.

Vorlicková, M., and Sági, J. (1989) Divalent cations are not required for the stability of the low-salt Z-DNA conformation in poly(dG-ethyl^{5}dC). J. Biomol. Struct. Dyn. *7*, 329–334.

Walker, G. T., and Aboul-ela, F. (1988) B-Z cooperativity and kinetics of poly(dG-m^{5}dC) are controlled by an unfavorable B-Z interface energy. J. Biomol. Struct. Dyn. *5*, 1209–1219.

Wells, R. D., Brennan, R., Chapman, K. A., Goodman, T. C., Hart, P. A., Hillen, W., Kellogg, D. R., Kilpatrick, M. W., Klein, R. D., Klysik, J., Lambert, P. F., Larson, J. E., Miglietta, J. J., Neuendorf, S. K., O'Connor, T. R., Singleton, C. K., Stirdivant, S. M., Veneziale, C. M., Wartell, R. M., and Zacharias, W. (1983) Left-handed DNA helices, supercoiling, and the B-Z junction. Cold Spring Harbor Symp. Quant. Biol., Vol. XLVII, pp. 77–84.

Wells, R. D. (1988) Unusual DNA structures. J. Biol. Chem. *263*, 1095–1098.
Wells, R. D., Collier, D. A., Hanvey, J. C., Shimizu, M., and Wohlrab, F. (1988) The chemistry and biology of unusual DNA structures adopted by oligopurine·oligopyrimidine sequences. FASEB J. *2*, 2939–2949.
Woisard, A., Guschlbauer, W., and Fazakerley, G. V. (1986) The low ionic strength form of the sodium salt of poly(dm^{5}C-dG) is a B DNA. Nucl. Acids Res. *14*, 3515–3519.
Wu, H.-Y., and Behe, M. J. (1985) Salt induced transitions between multiple conformations of poly(rG-m5dC)·poly(rG-m5dC). Nucl. Acids Res. *13*, 3931–3940.
Xodo, L. E., Manzini, G., Quadrifoglio, F., van der Marel, G. A., and van Boom, J. H. (1991) Effect of 5-methylcytosine on the stability of triple-stranded DNA – a thermodynamic study. Nucl. Acids Res. *19*, 5625–5631.
Zacharias, W. (1992) DNA methylation *in vivo*. Methods Enz. *212*, 336–346.
Zacharias, W., Jaworski, A., and Wells, R. D. (1990) Cytosine methylation enhances Z-DNA formation *in vivo*. J. Bacteriol. *172*, 3278–3283.
Zacharias, W., O'Connor, T. R., and Larson, J. E. (1988) Methylation of cytosine in the 5-position alters the structural and energetic properties of the supercoil-induced Z-helix and of B-Z junctions. Biochemistry *27*, 2970–2978.
Zimmerman, S. B. (1982) The three-dimensional structure of DNA. Ann. Rev. Biochem. *51*, 395–427.

DNA Methylation: Molecular Biology and Biological Significance
ed. by J. P. Jost & H. P. Saluz

Methylation of DNA in Prokaryotes

Mario Noyer-Weidner and Thomas A. Trautner

Max-Planck-Institut für molekulare Genetik, Ihnestr 73, DW-1000 Berlin 33, Germany

1 Introduction

A much wider variety of biological functions of postreplicative DNA methylation is observed in prokaryotes than in eukaryotes. In eukaryotes DNA methylation is primarily a means of the control of gene expression. Many chapters of this book are devoted to various aspects of this function. In prokaryotes, DNA methylation affects such diverse phenomena as determination of accessibility of DNA to digestion by endonucleases, control of initiation of DNA replication, and the definition of origins of packaging in the maturation of phage DNA, which will be dealt with in this article. We shall also be concerned with the enzymes, which facilitate methylation, the DNA methyltransferases. In the eukaryotes, as far as we know at this time, the various DNA methyltranferases encountered represent a rather homogeneous group, whereas in prokaryotes, we find a very diverse set of DNA methyltransferases. Beyond their biological significance, DNA methyltransferases represent a remarkable class of enzymes in their own right. Not only are they paradigms for sequence specific DNA binding proteins, but they also show specificity in their catalytic interaction with defined DNA sequences. Furthermore, their universal distribution, the multitude of enzymes with different or identical specificities observed among prokaryotes and the obligatory coexistence of isospecific restriction and methylating enzymes in restriction/modification systems make DNA methyltransferases choice candidates for evolutionary studies.

Abbreviations: A: adenine, bp: base pair(s); C: cytosine; ENase: restriction endonuclease; G: guanine; 5-mC: 5-methylcytosine; MTase: DNA methyltransferase; N: non defined base or nucleotide; N6-mA: N6-methyladenine; N4-mC: N4-methylcytosine; R: purine; SAM: S-adenosyl-methionine; SAH: S-adenosyl-homocysteine; T: thymine; TRD: target recognizing domain; U: uracil; wt: wild type; Y: pyrimidine

The nomenclature of restriction and modification enzymes is that of Smith and Nathans, 1973. DNA sequences are written in 5′ to 3′ direction from left to right, if not indicated otherwise. Methylated bases are underlined.

Presenting the large amount of information available about prokaryotic methylation in this review required focusing on certain areas at the expense of others. The weight which was given to the description of individual topics of prokaryotic DNA methylation is not free from personal biases, but was primarily determined by the availability of other reviews. A review concerned with the biological role of methylation in *E. coli* has been provided by Marinus (1987). Many reviews address "classical" restriction/modification systems, in which DNA methylation by "modification" methyltransferases prevents the attack of cognate "restriction" endonucleases (Roberts, 1990; Nelson and McClelland, 1991; Roberts and Macelis, 1991; Szybalski et al., 1991; Wilson, 1991; Wilson and Murray, 1991; Trautner and Noyer-Weidner, 1992). Examples of the reverse situation, methylation of DNA serving as a signal to elicit restriction, have been treated by Raleigh (1992). The functions of *E. coli* Dam methylation have been described in minireviews by Messer and Noyer-Weidner (1988) and Barras and Marinus (1989).

2 Inventory

2.1 Methylation affecting accessibility of DNA to endonucleolytic cleavage

2.1.1 Methylation interfering with cleavage

2.1.1.1 Restriction/modification (R/M) systems. Among the prokaryotes by far the most widely distributed and also most abundant kind of methylation is that associated with R/M systems of bacteria. R/M systems were discovered in the early days of molecular biology and at first defined operationally (Luria and Human, 1952; Bertani and Weigle, 1953). Their presence in bacteria causes a "restriction" of growth of phages on such strains in comparison to strains without R/M systems. Phage progeny recovered from growth on bacteria with R/M systems is "modified" in that such phages will now be insensitive to restriction. Modification is an epigenetic phenomenon: phages maintain their modification when replicating on R/M proficient strains, but rapidly lose this property upon growth on bacteria without this R/M system. Later work provided the biochemical basis for R/M (reviewed by Arber and Linn, 1969). It was shown that all R/M systems include two sequence-specific enzymatic activities: (1) a restriction endonuclease (ENase) activity, which cleaves DNA endonucleolytically, provided it contains an accessible target site, characteristic for each enzyme; (2) a modification methyltransferase (MTase) activity, which recognizes the same target site as the "cognate" restriction activity within which it methylates a specific base. This methylation-based modification interferes with restriction. It serves to protect the chromosomal DNA of the bacterium against "self" restriction and is also responsible for the transient modification of those phages which escape restriction.

There is an important asymmetry in this dual configuration of enzymatic activities. Whereas the ENase activity of an R/M system cannot usually exist in a cell in the absence of its cognate MTase activity, since this would cause suicidal degradation of the cell's DNA, the reciprocal situation – solitary existence of a MTase – is often encountered. Several cases of solitary MTases, which fulfill functions different from protection against restriction are discussed later in this review. The asymmetric relationship between the components of R/M systems is of practical significance: the cloning of genes responsible for restriction and their overexpression requires in many cases the presence of the modification activity, whereas no precautions are needed for the cloning of MTase genes, provided the host strains used in cloning do not have restriction activities directed against methylated DNA (see section 2.1.2).

Table 1 and Figure 1 summarize the most important properties of various R/M systems. These have been classified as types I, II, IIs, and III, primarily on the basis of their structure, cofactor requirements, nature of the recognition site, and the spatial relationship between recognition- and catalytic sites. The significant differences between the types of R/M systems point to separate evolutionary pathways, which have led to their generation.

In type II and type IIs-systems the MTases and ENases are separate proteins, which can act independently of each other. In this situation the parameters of methylation can readily be studied in the absence of restricting activities, in distinction from type I and III R/M systems, where methylation and restriction are performed by the same enzyme complex.

Type II MTases with the capacities to produce 5-mC, N4-mC and N6-mA (Fig. 2) have been identified. The type II MTases are active as monomeric enzymes. They methylate their targets, which usually have to be in double-stranded configuration, in two steps, with a hemimethylated target site as an intermediate reaction product (Rubin and Modrich, 1977; Herman and Modrich, 1982). The activities of prokaryotic type II MTases are not markedly different with non-methylated or hemimethylated targets, in contrast to some prokaryotic type I MTases (Suri et al., 1984) or to eukaryotic type II "maintenance" MTases, which strongly favor hemimethylated DNA (Gruenbaum et al., 1982; Bestor and Ingram, 1983).

An unusual case of a type II MTase has been identified in the *Dpn* II R/M system, which comprises two A-N6-MTases, *Dpn*M (also referred to as M.*Dpn*II) and *Dpn*A, both specific for the target sequence GATC (Lacks et al., 1986; de la Campa et al., 1987; Cerritelli et al., 1989). The "conventional" MTase *Dpn*M methylates double-stranded GATC

Table 1. Some properties of restriction/modification systems

Type of R/M System	Structure	Co-factor requirement(s) R	M	Types of methylation observed	Location of target recognizing domain(s)	Site of cleavage with respect to recognition sequence	Comments
II	ENase and MTase re-present separate enzymes	Mg^{++}	SAM	5-mC, N4-mC N6-mA	Within individual enzymes	Within the target	
II_s		Mg^{++}	SAM	5-mC, N6-mA		At defined location within 20 bp on one side of the target	
III	Enzyme complex consisting of ENase and MTase subunits	ATP SAM Mg^{++}	SAM (ATP, Mg^{++})	N6-mA	Within MTase subunit	At a defined distance of 25–27 bp from the 3′ site of the target	ENase requires two targets in opposite orientation for scission
I	Enzyme complex consisting of ENase, MTase and specificity determining subunits	ATP SAM Mg^{++}	SAM (ATP Mg^{++})	N6-mA	Within specificity determining subunit	Remote from target	

References to entries into the table are found in the text. Cofactors shown in parentheses indicate that their presence stimulates the methylation (M) reaction.

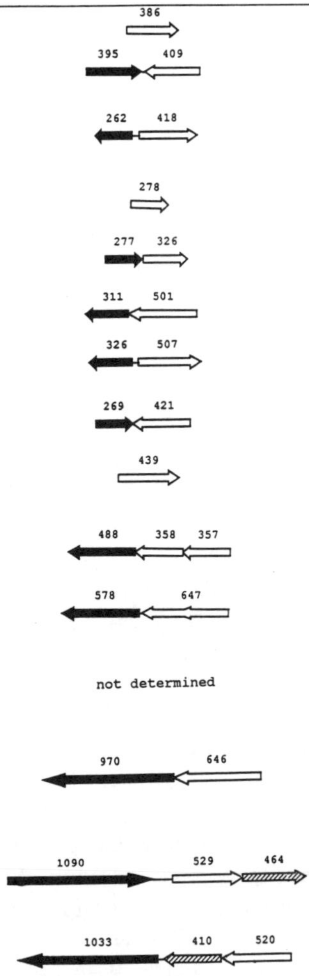

Type	System	Specificity	Gene Arrangement	References
II	*Sss*I	CG	386	Renbaum et al., 1990
	*Bsu*FI	CCGG	395 409	Kapfer et al., 1990 Walter et al., 1990
	*Msp*I	CCGG	262 418	Walder et al., 1983 Nwankwo and Wilson, 1988 Linn et al., 1989
	Dam	GATC	278	Herman and Modrich, 1982 Brooks et al., 1983
	*Eco*RI	GAATTC	277 326	Greene et al., 1981 Newman et al., 1981
	*Bsu*BI	CTGCAG	311 501	Xu et al., unpublished
	*Pst*I	CTGCAG	326 507	Walder et al., 1981 Walder et al., 1984
	*Sfi*I	GGCCN$_5$GGCC	269 421	Van Cott et al., quoted in Wilson, 1991
	SPR	CC(A/T)GG, CCGG, GGCC	439	Buhk et al., 1984
II$_s$	*Hga*I	GACGC CTGCG	488 358 357	Sugisaki et al., 1991
	*Fok*I	GGATG CCTAC	578 647	Kita et al., 1989 Looney et al., 1989 Stankevicius et al., quoted in Wilson, 1991
	*Eco*31I	GGTCTC CCAGAG	not determined	Mitake et al., quoted in Wilson, 1991
III	*Eco*P1	AGACC TCTGG	970 646	Hümbelin et al., 1988
I	*Eco*K12	AACN$_6$GTGC TTGN$_6$CACG	1090 529 464	Borck et al., 1976 Sain and Murray, 1980 Gough and Murray,1983 Loenen et al., 1987
	*Eco*124/3	GAAN$_7$RTCG GTTN$_7$RAGC	1033 410 520	Price et al., 1989

Figure 1. Prototype target sequences and gene structures of solitary DNA methyltransferases or DNA methyltransferases organized in R/M systems. The specificity of the R/M systems shown is indicated, with the methylated bases underlined. The description of gene arrangements follows the convention of Wilson (1991). The arrows give the direction of gene transcription. Open, solid, and shaded arrows describe genes for MTases, ENases and the specificity determining subunit of type I R/M systems. The numbers of amino acids of the corresponding proteins is given above the arrows.

targets. The *Dpn*A MTase, however, methylates the target both in the double-stranded and single-stranded configuration. This is the only case of a MTase with the capacity for site-specific methylation of single-stranded DNA.

Methylation by type II MTases requires SAM. The function of SAM is in most of the cases limited to serving as a methyl group donor. In the

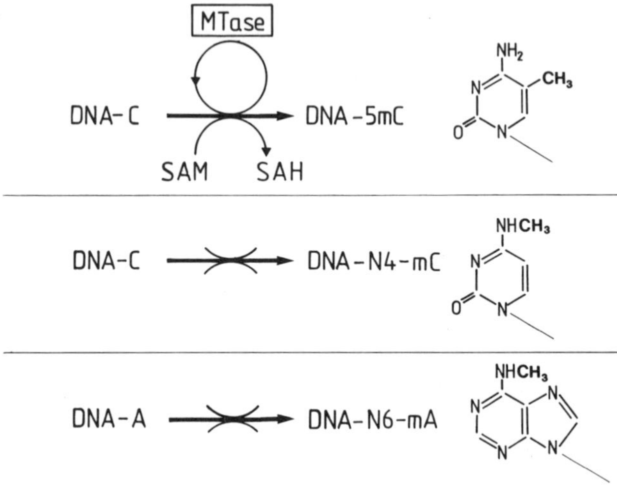

Figure 2. Schematic presentation of the types of methylation catalyzed by C5-, C-N4- and A-N6-DNA methyltransferases.

case of the *E. coli* Dam MTase, however, it also affects binding of the enzyme to the target site (Bergerat and Guschlbauer, 1990). In the absence of SAM M.Dam binds loosely to its target site, GATC. The binding is increased 5-fold in the presence of SAM, which is assumed to cause an allosteric change of the enzyme resulting in a higher target affinity. The functions of SAM as activator and as methyl group donor in the catalytic process can be distinguished by their differential sensitivity to SAM analogues: only the catalytic process is sensitive to sinefungin or SAH. NMR studies suggest that the dual function of SAM in the case of M.Dam implicates two enzyme-binding sites of SAM, one allosteric and one catalytic site (Bergerat et al., 1991). It has to be elucidated whether such a dual role of SAM is also of significance in other type II MTases or correlates with the specific biological function of M.Dam (see section 4).

Methylation and endonucleolytic scission by type II MTases and ENases occur within the same DNA target. DNA targets are symmetric (palindromic) sequences consisting of up to eight bp (Fig. 1). The palindromic nature of the targets and their size confinement define a finite number of possible recognition sequences. Comparing this number with the frequencies of specificities of different R/M systems actually observed, one finds a heavy bias towards those systems which recognize G/C rich targets. There is a pronounced discrimination against targets with a dominant A/T content. On top of this compositional bias one finds predominantly targets of R/M systems which contain G's within their 5′ halves. These biases are important in considering the evolution of R/M systems (Klump, 1987).

The type IIs R/M systems (reviewed in Szybalski et al., 1991) also have separate ENases and MTases. Their target sites, however, are asymmetric, non-palindromic sequences (see Fig. 1). The endonucleolytic scission by type IIs ENases occurs at enzyme-specific locations within 20 bp outside the target sequence. Modification of the two complementary, but non-identical, sequences representing the DNA target requires methylation following *different* recognition events usually by two different MTases, one for each strand of the target. Although MTases providing two different specificities are required to modify an asymmetric target site, only one restriction enzyme is involved in DNA scission dependent on such DNA targets. This situation reflects the basically different catalytic activities of ENases and MTases, following recognition of the same DNA target sequence.

From the compilation of R/M systems in Wilson (1991) type II systems are roughly ten times as frequent as type IIs systems. This is at first surprising, considering that with a given size of a nucleotide target sequence, palindromic sequences represent a minority. With the apparent requirement for methylation, however, of *both* strands within a target sequence, R/M systems acting on non-palindromic targets must have two, rather than one MTase. With this, the observed frequency distribution of the two types of R/M systems is possibly a consequence of an evolutionary balance between the requirement for enzymatic complexity and the difference in availability of palindromic vs. non-palindromic sequences.

The enzymes representing type I and type III R/M systems are complex multi-subunit structures with the capacity to mediate both modification and restriction (Yuan and Hamilton, 1984; Bickle, 1987). SAM serves in these systems not only as donor of the methyl group but is also required for cleavage by the complex (Yuan and Hamilton, 1984; Bickle, 1987). A-N6 is the only methylated base originating from type I or III modification.

In type III enzymes – four of which have been identified (Kauc and Piekarowicz, 1978; Wilson, 1991) – one MTase subunit (M) is associated with one subunit (R) responsible for restriction. The M subunit can be active on its own as a MTase, i.e. it comprises functional elements responsible for target recognition and modification (Bächi et al., 1979; Hadi et al., 1983). Restriction, on the other hand, requires the association of the R with the M subunit. Target sequences of type III R/M enzymes are asymmetric and contain A only in one strand (Piekarowicz et al., 1981; Hadi et al., 1983; Meisel et al., 1991). Therefore only one strand of any target sequence can be methylated. This leads during semiconservative DNA replication to the transient generation of totally unmodified sites. Such a situation is biologically tolerable since type III ENases require the presence of *two* unmethylated target sites in opposite directions in order to cleave DNA (Hadi et al.,

1979; Meisel et al., 1992). Evidence for the requirement of a single target for modification and a duplicate target (in the configuration described) for restriction came from work with the type III *Eco*P15 R/M system (Meisel et al., 1991; Meisel et al., 1992). Type III restriction ENases cut DNA about 25 nucleotides away from the double target site. The activities of the MTase and ENase are apparently not coordinated, such that due to the obligatory presence of SAM, which is required as a cofactor also in restriction, no complete DNA digests are obtained.

Type I enzyme complexes have in addition to subunits responsible for restriction (R) and modification (M) one subunit (S) which determines the specificity for both restriction and modification (Yuan and Hamilton, 1984). Based on interchangeability of subunits, immunological cross-reaction and sequencing of some of their genes, type I R/M systems have been classified into three families: K, A, and R124 (Boyer and Roulland-Dussoix, 1969; Glover and Colson, 1969; Bullas and Colson, 1975; Murray et al., 1982; Wilson and Murray, 1991). Comparing the sequences of subunit genes (compiled in Wilson, 1991; e.g. *Eco*A and *Eco*E: Fuller-Pace and Murray, 1986; Cowan et al., 1989; *Eco*K: Loenen et al., 1987; *Eco*R124/3: Price et al., 1989) indicated that the M subunits within one family have at least 90% identical amino acids, whereas similarity between the S subunits, whose structure will be discussed in 3.3.1, amounts to only some 40% amino acid identity. Type I enzymes will bind to their targets irrespective of their state of methylation (Yuan et al., 1975). If methylated, the complex will dissociate, provided ATP is present (Burckhardt et al., 1981). The unmodified strand within a hemimethylated target becomes rapidly methylated, whereas the preferential interaction of the enzyme with unmethylated DNA is its cleavage (Bickle et al., 1978). Such cleavage occurs at a random location some 1000 bp distant from the target (Studier and Bandyopadhyay, 1988). It is not understood what determines in the interaction of type I enzymes with unmodified DNA the balance between (*de novo*) methylation and DNA scission.

2.1.1.2 Solitary phage MTases counteracting restriction. The most obvious function of R/M systems is to discriminate against foreign DNA. R/M systems represent barriers for the uptake of double-stranded, unmodified DNA as encountered in bacteriophage infection or transduction. They do not generally interfere, however, with those DNA uptake processes, in which double-stranded DNA becomes converted to a single-stranded form, which rapidly synapses with the recipient chromosome, as e.g. in transformation (summarized in Trautner and Noyer-Weidner, 1992). The concept that the primary role of restriction is to prevent phage infection is underlined by the finding that many bacteriophages have gained the potential to infect cells with R/M systems by sophisticated mechanisms overcoming host restriction (reviewed by Krüger and Bickle, 1983). This represents an interesting case of molecularly interpretable host/parasite coevolution.

"Self methylation" by phage-encoded MTases is one of the mechanisms realized. A number of virulent and temperate bacteriophages of a variety of phylogenetically different hosts encode solitary type II MTases which recognize the same target as that of the R/M system of the host which they infect (e.g. *B. amyloliquefaciens* phage H2, Connaughton et al., 1990). A new class of solitary type II MTase was found to be encoded by some temperate *B. subtilis* phages. These are "multi-specific" MTases, which have the capacity to methylate *several different* DNA target sequences. One of these multiple specificities may be identical to the R/M system present in the bacterial host lysogenized by the phages (Kiss and Baldauf, 1983; Noyer-Weidner et al., 1983; Buhk et al., 1984; Tran-Betcke et al., 1986; Behrens et al., 1987; Lange et al., 1991b; summarized in Trautner and Noyer-Weidner, 1992). The multi-specific MTases were of particular relevance in the functional characterization of C5-MTases (3.2.3).

2.1.2 Methylation eliciting restriction

In addition to the capacity to protect DNA against cleavage by restriction ENases, methylation can also provide a signal eliciting restriction. Several prokaryotic restriction ENases have been identified which cleave DNA invading a cell provided it is methylated at specific DNA sequences. It is self evident that such sequences must not be methylated by cells encoding this type of restriction activity. Therefore, ENases acting on methylated DNA are solitary enzymes.

2.1.2.1 Restriction of methylated DNA in Streptococcus pneumoniae. The restriction ENase *Dpn*I, which resembles typical type II ENases (Lacks, 1980), is unusual as it cleaves DNA only when it is methylated at the A residue of GATC sites (Lacks and Greenberg, 1975; Lacks and Greenberg, 1977). The *Dpn*I restriction activity is found only in some strains of *S. pneumoniae* (Muckerman et al., 1982). Others encode the *Dpn*II restriction activity directed against DNA carrying unmethylated GATC sites (Lacks and Greenberg, 1977). Restriction of DNA by the *Dpn*II ENase is prevented by A methylation at GATC sites (Lacks and Greenberg, 1977). As evident, the *Dpn*I and *Dpn*II systems are complementary and mutually exclusive. Consequently, infection of phages produced in cells of the opposite phenotype is restricted to $<10^{-5}$ by the *Dpn*I or *Dpn*II ENases (Muckerman et al., 1982). While the *Dpn*I and *Dpn*II restriction activities provide protection of individual cells against phage infection, their complementary phenotypes might specifically contribute to the survival of the species rather than the individual cell in the case of infection. Thus, in a mixed population of *S. pneumoniae* strains, consisting of *Dpn*I and *Dpn*II encoding cells, phages escaping restriction in one cell type are sensitive to restriction by the other cell type in a subsequent infection cycle. The nonrelated genes encoding the *Dpn*I and *Dpn*II systems occupy the same position in the chromosome and, hence, are flanked at either side by extended homologous regions (Lacks et al., 1986). As proposed by Lacks et al. (1986) this might be of importance in the context discussed. Cells killed by phage infection would release DNA containing the restriction gene "cassette" corresponding to the susceptible phenotype. Subsequent transformation of cells of the complementary phenotype by such free DNA accompanied by the occasional exchange of restriction gene cassettes might contribute to maintaining both restriction phenotypes in an infected cell population.

2.1.2.2 Restriction of methylated DNA in Escherichia coli and other species. The observation that MTase encoding genes could not be cloned in some *E. coli* strains led to the discovery of three *E. coli* K-12 restriction systems directed against methylated DNA (Noyer-Weidner et al., 1986; Raleigh and Wilson, 1986; Heitman and Model, 1987). Due to their significance for the cloning of heterologous DNA in *E. coli*, these activities have found widespread interest.

Restriction of methylated DNA by *E. coli* K-12 depends on the products of the *mcrA* gene, the *mcrBC* operon and the *mrr* gene (Noyer-Weidner et al., 1986; Raleigh and Wilson, 1986; Heitman and Model, 1987; Raleigh et al., 1989; Ross et al., 1989a; Ross et al., 1989b; Dila et al., 1990; Hiom and Sedgwick, 1991; Kelleher and Raleigh, 1991; Kretz et al., 1991; Waite-Rees et al., 1991; Krüger et al., 1992). The restriction phenotypes associated with *mcrA*, *mcrBC* and *mrr* gene expression are complex (Table 2). Depending on the modification type eliciting restriction, they are generally referred to as Mcr, if restriction depends on the presence of modified cytosine in DNA, or Mrr, if restriction depends on adenine modification (for details of the nomenclature see Raleigh et al., 1991). Further subspecification of phenotypes by addition of capital letters to the general phenotypic designations (e.g., McrA, McrBC or McrF; Table 2) reflect differences in the specificities of the *mcrA*, *mcrBC* and *mrr* encoded restriction activities for DNA substrates carrying an identical type of modified base (e.g. 5-mC).

Considering the genotypes and phenotypes related to restriction of methylated DNA in *E. coli* K-12 (Table 2) a few points should be stressed:

(1) Most restriction phenotypes associated with the *mcrBC* operon depend on the products of both genes of the operon.
(2) So far only methylation of C residues in DNA has been shown to elicit restriction by the *mcrA* or *mcrBC* encoded activities. In contrast methylation of C *and* A residues has potential signal

Table 2. Genes and phenotypes related to restriction of methylated DNA in *E. coli* K-12

Genes		*mcrA*	*mcrBC*	*mrr*
Associated restriction phenotypes		McrA	McrBC[4]	McrF/Mrr
Methylated base with potential signal character		5-mC	5-mC/N4-mC	5-mC/N6-mA
Restriction of DNA(+) modified by MTases with the specificities indicated[1,2]	CG	+	+	+
	CCGG	+	−	−
	CCGG	−	+	−
	GCGC	−	+	+
	GGCC	−	+	−
	AGCT	−	+	−
	GAGCTC[3]	−	+	−
	CATG	−	−	+
	TCGA	−	−	+
	TGCA	−	−	+
	GANTC	−	−	+
	GTTAAC	−	−	+

[1]Data are taken from Noyer-Weidner et al., 1986; Raleigh and Wilson, 1986; Heitman and Model, 1987; Kelleher and Raleigh, 1991; Kretz et al., 1991; Waite-Rees et al., 1991). The list is incomplete. For more extensive compilations see e.g. Wilson and Raleigh (1986) and Waite-Rees et al. (1991).

[2]Note that the specificity of an MTase whose activity elicits restriction is not identical with that of the restricting activity. Detailed information on the methylated sequence specifically recognized has only been provided for McrBC (Sutherland et al., 1992; see text).

[3]The target GAGCTC is recognized by the C-N4-MTase *Pvu*II (Tao et al., 1989). M.*Pvu*II methylation was shown to elicit McrBC restriction (Blumenthal, 1985; Wilson and Raleigh, 1986).

[4]In addition to phenotypes depending on both the McrB and McrC proteins, McrB permits restriction of DNA methylated at CCGG or CG sites in the absence of McrC when expressed from a plasmid in an *mcrBC* deletion mutant background (Dila et al., 1990; Kelleher and Raleigh, 1991). *In vitro* restriction of such modified DNA, however, requires McrB *and* McrC. This suggests that *in vivo* some other protein might substitute for McrC in the specific situation described.

character for the *mrr* gene product. Two phenotypes, termed McrF and Mrr, are, hence, associated with expression of the *mrr* gene.

(3) The specificities of the McrA, McrBC and McrF restriction activities for 5-mC containing DNA targets are partially overlapping. Hence, depending on its modification pattern, “foreign” DNA invading *E. coli* K-12 may be subject to restriction by only one or several of these systems. Methylation at CG sites, as e.g. catalyzed by eukaryotic MTases, can affect DNA targets recognized by all of these activities.

(4) The sequence specificities of most *E. coli* MTases, including M.Dam, M.Dcm and MTases associated with type I R/M systems, are such that they will not generate signals for restriction by the methylation dependent activities. However, methylation related to two type II R/M systems of *E. coli* and two-type II R/M systems of

the related species *Citrobacter* has been found to elicit McrBC restriction (Povilionis et al., 1989).

(5) The modification-dependent *E. coli* K-12 restriction activities are non-essential. *E. coli* K-12 mutant strains lacking the capacity to restrict methylated DNA due to the deletion of all relevant genes (*mcrA*, *mcrBC*, *mrr*) are viable (Woodcock et al., 1989; Kretz et al., 1991; Kelleher and Raleigh, 1991).

The identification of the *mcrA* and *mcrBC* genes as specifying restriction of methylated DNA was actually a rediscovery of these genes. The *mcrA* and *mcrBC* loci are identical with the *rglA* and *rglB* loci (Noyer-Weidner et al., 1986; Raleigh and Wilson, 1986; Raleigh et al., 1989), previously described in the context of T-even phage restriction (reviewed by Revel, 1983). T-even phages are potentially sensitive to restriction by the *mcrA*- and *mcrBC*-encoded activities as their DNA contains 5-hydroxymethylcytosine (5-hmC), a signal eliciting restriction, instead of C residues. Despite the presence of 5-hmC T-even phage DNA is normally resistant to modification-dependent restriction. This is due to the glucosylation of 5-hmC residues in the phage DNA, mediated by phage encoded glucosyl-transferases. The fact that a nonglucosylated state of T-even phage DNA is required to render it sensitive to modification-dependent restriction is reflected in the original designation of the relevant genes as *rgl* (*r*estriction of *gl*ucoseless phages). While the term *mcr* has meanwhile been adopted instead of *rgl* in gene designations, the term Rgl is still used when specifically discussing phenotypes related to T-even phage restriction. Different Rgl phenotypes are associated with the products of the *mcrA* gene and *mcrBC* operon. RglA restriction, dependent on the *mcrA* gene product, affects non-glucosylated T2, T4 and T6 DNA, while RglB restriction, dependent on the products of both genes of the *mcrBC* operon, is limited to T2 and T4 DNA. The difference between the RglA and RglB restriction phenotypes does not necessarily reflect a fundamental difference in the specificities of the RglA (McrA) and RglB (McrBC) activities for 5-hmC containing DNA substrates. This follows from the efficient cleavage of nonglucosylated T6 DNA by the McrBC proteins *in vitro* (Sutherland et al., 1992). The apparent phenotypic difference might be due to a T6 encoded anti-restriction function specifically directed against the RglB (McrBC) activity. A weak anti-RglB (McrBC)-restriction activity, termed Arn, protecting superinfecting phages from RglB restriction has been reported for phage T4 (Dharmalingam and Goldberg, 1976; Dharmalingam et al., 1982).

The nucleotide sequences of the *mcrA* (Hiom and Sedgwick, 1991), *mcrBC* (Ross et al., 1989a; Dila et al., 1990) and *mrr* genes (Waite-Rees et al., 1991) have been established. The gene sequences and the derived amino acid sequences of the corresponding proteins did not reveal pronounced similarities among each other, though a very distant relationship between McrC and Mrr has been suggested (Waite-Rees et al., 1991). Except for amino acid motifs identified in McrB and McrC, which resemble nucleotide binding sites (Ross et al., 1989a; Dila et al., 1990), no significant similarities to other proteins of known sequence have been detected.

So far only McrBC restriction has been characterized biochemically (Sutherland et al., 1992). DNA cleavage by McrBC depends on the presence of the general sequence $5'R^{m}C(N_{40\text{-}80})R^{m}C3'$ with (N40-80) indicating the allowed range of spacing between methylated cytosines by nondefined bases. In response to two appropriately spaced methylated cytosines, even if present in only one strand of DNA, the McrBC activity catalyzes double-strand breaks at multiple positions of the intervening region. A unique feature of McrBC restriction is its absolute dependence on GTP which is apparently hydrolyzed in the reaction. ATP, the cofacor of most nucleotide-dependent nucleases, including type I and type III restriction ENases (see Table 1), inhibits McrBC restriction (Sutherland et al., 1992). Nevertheless, considering its composition of different subunits, the complex prerequisites for DNA recognition and cleavage and the requirement for a nucleotide triphosphate as a cofactor in the cleavage reaction, the McrBC ENase appears to be more similar to type I or type III restriction enzymes than to type II restriction enzymes.

The *mcrA*, *mcrBC* and *mrr* genes are chromosomally located. Their positions in the *E. coli* K-12 chromosome (Fig. 3) have been precisely defined (for references see legend to Fig. 3). *mcrA* is carried by the excisable prophage like element e14, which is integrated into the chromosome at about 25 min on the *E. coli* K-12 standard map. The *mcrBC* and *mrr* genes flank the genetic determinants (*hsdRMS*) of the *E. coli* K-12 type I R/M-system *Eco*K, located at about 98.5 min (Figure 3). Due to the clustering of genes (*mrr*, *hsdRMS*, *mcrBC*) specifying restriction functions at 98.5 min, this region of the *E. coli* K-12 chromosome has occasionally been referred to as "immigration control region" (ICR) (Raleigh et al., 1989).

E. coli strains show pronounced variability with respect to the presence of the loci discussed here. Absence of the *mcrA* gene is frequently due to an SOS-induced loss of the e14 element from the chromosome (Greener and Hill, 1980; Hiom and Sedgwick, 1991). For potential reasons underlying variability in the ICR, specifically as observed among *E. coli* K, B, and C strains, see Raleigh (1992).

What might be (or might have been) the biological function of the obviously nonessential *E. coli* restriction activities directed against methylated DNA? A plausible role would be protection of *E. coli* strains against infection by phages which escape "conventional" host restriction by incorporation of modified bases into their DNA (reviewed by

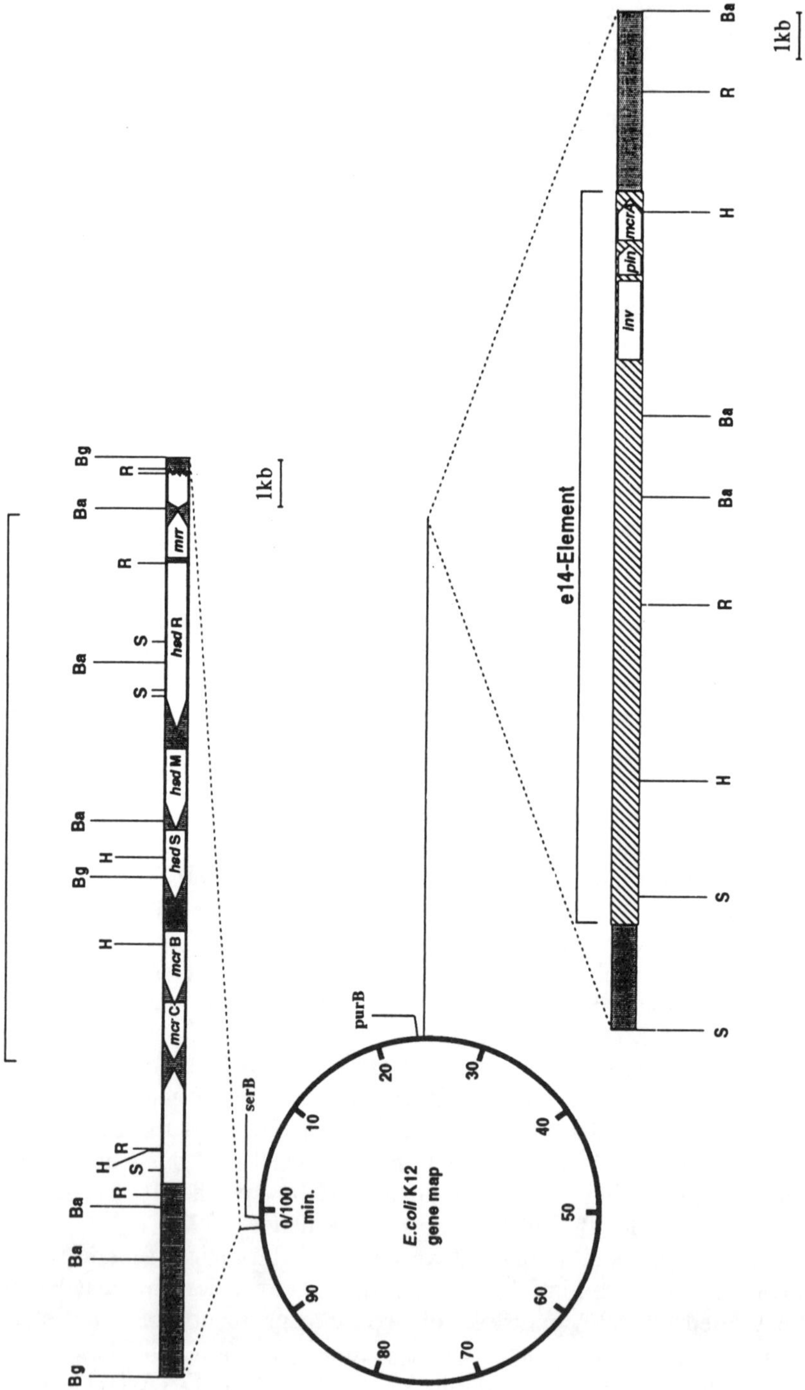

I C R
Bg
Ba
R
S
hsd R
hsd M
H
hsd S
mcr B
mcr C
mrr
serB
purB
0/100 min.
E.coli K12 gene map
10
20
30
40
50
60
70
80
90
1kb
e14-Element
inv
pin
mcrA

Krüger and Bickle, 1983). The development of an anti-RglB (McrBC) restriction activity by phage T4 (and presumably T6), which protect their DNA against most host-encoded restriction systems by incorporation of 5-hmC (Krüger and Bickle, 1983), might be a reflection of this functional potential. As the T-even phages, probably at a later stage in evolution, protected their DNA efficiently against RglA and RglB restriction by glucosylation of 5-hmC residues, the RglA and RglB activities appear to be evolutionary relics, at least in the particular relationship between T-even phages and their *E. coli* hosts (Krüger and Bickle, 1983).

The capacity of the McrA, McrBC and McrF activities to restrict 5-mC containing DNA raised speculations that they might serve in removing 5-mC residues from DNA (Wilson and Murray, 1991). In this way these activities might fulfill a repair function by counteracting the mutagenic potential deriving from spontaneous deamination of 5-mC to thymine (see section 5). At least in case of the McrBC activity, however, such a role is hard to reconcile with the complex prerequisites for DNA recognition and cleavage by this enzyme and the fact that is catalyzes DNA double-strand breaks in response to recognized 5-mC residues (Sutherland et al., 1992).

While, at present, one can only speculate about the function of the *E. coli* activities directed against methylated DNA, the recent description of similar activities in a variety of other species including *Bacillus thuringensis* (Macaluso and Mettus, 1991), multiple *Streptomyces* strains (MacNeil, 1988) and the mycoplasma *Acholeoplasma laidlawii* (Sladek et al., 1986; Sladek and Maniloff, 1987), suggests that such activities are of general biological significance, perhaps in contexts which have yet to be elucidated.

Figure 3. Chromosomal location of genes determining restriction of modified DNA in *E. coli* K-12. The circle represents the linkage map of *E. coli* K-12 (Bachmann, 1987) with some relevant markers indicated. Chromosomal regions comprising the "immigration control region", ICR (Raleigh et al., 1989), and the e14 element (Greener and Hill, 1980) are enlarged (the S = *Sal*I sites of the region containing e14 are proximal to *pur*B). The position, extension and orientation of relevant genes in both chromosomal regions, as indicated by open arrows, derives from mapping experiments (Revel, 1967; Ravi et al., 1985; Heitman and Model, 1987; Raleigh et al., 1989), the analysis of cloned DNA segments (Sain and Murray, 1980; van de Putte et al., 1984; Brody and Hill, 1985; Ross et al., 1987; Noyer-Weidner and Reiners-Schramm, 1988; Raleigh et al., 1989; Hiom and Sedgwick 1991; Kelleher and Raleigh, 1991; Waite-Rees et al., 1991) and nucleotide sequence information (Gough and Murray, 1983; Loenen et al., 1987; Ross et al., 1989a; Dila et al., 1990; Hiom and Sedgwick, 1991; Waite-Rees et al., 1991). The ICR is flanked by partially (Waite-Rees et al., 1991) or completely (Krüger and Noyer-Weidner, unpublished) sequenced ORFs of unknown function, whose position and orientation is indicated by non-designated open arrows. Within the e14 element the positions of *pin*, a gene encoding a site-specific inversion recombinase, and a segment, termed *inv*, which is invertible by this activity, are shown in addition to that of *mcrA*. Positions of the following restriction sites are indicated: Ba = *Bam*HI, Bg = *Bgl*II, H = *Hin*dIII, R = *Eco*RI, S = *Sal*I. *Bgl*II sites are only shown for the ICR comprising chromosome region, as their determination in the e14 region is incomplete.

Irrespective of their biological role, the modification dependent *E. coli* restriction activities are of considerable practical significance, since they constitute a severe barrier to cloning of heterologous DNA in *E. coli* K-12 (see e.g. Orbach et al., 1988; Whittaker et al., 1988; Kretz et al., 1989; Woodcock et al., 1989; Graham et al., 1990; Grant et al., 1990). This holds in particular for DNA of eukaryotes encoding CpG-specific MTase activities (see Table 2). This problem could be overcome by the use of mutant strains deficient in the restricting activities discussed here (for compilations of mutant strains see e.g. Raleigh et al., 1988; Woodcock et al., 1989). For most cloning experiments employing modified heterologous DNA the use of strains in which all of the relevant genes are deleted is recommended (Woodcock et al., 1989; Kelleher and Raleigh, 1991; Kretz et al., 1991), as some other mutant alleles still specify residual restriction activity (Krüger et al., 1992). Mutant *E. coli* K-12 strains *differentially* affected in the methylation-dependent restriction activities may, however, provide tools to identify differences in methylation patterns of DNA isolated from eukaryotic cells, a consequence of the different target specificity of the relevant restriction functions. As shown by Grant et al. (1990), plasmids comprising a hybrid insulin transgene were rescued with different efficiencies in different *E. coli* K-12 mutant backgrounds when isolated from different transgenic mouse lineages. The differences correlated with differential expression of the transgene in the various lineages. Efficient rescue of plasmids from lineages expressing the transgene early in development required only McrA and McrBC deficiency of the recipient strain, while additional McrF deficiency was required in efficient retrieving of plasmids from lineages expressing the transgene late in development.

The modification-dependent restriction activities have also been used to facilitate gene cloning. The McrBC activity served to establish a *E. coli* K-12 vector system allowing the positive selection of recombinant plasmids (Noyer-Weidner and Reiners-Schramm, 1988). The isolation of strains encoding temperature-sensitive McrA and McrBC activities (Piekarowicz et al. 1991a) has further provided the basis for the construction of strains facilitating the identification of recombinant DNA encoding non *E. coli* specific DNA-methyltransferases (Piekarowicz et al., 1991b).

2.2 Methylation involved in regulatory or unknown functions

The *E. coli* Dam MTase, a solitary type II A-N6-MTase, is involved in the control of many cellular processes in *E. coli*, which will be discussed in section 4. Many *E. coli* phages have been identified which encode a MTase methylating the M.Dam target sequence GATC (see section 4). Except for the involvement of the phage P1 MTase in phage DNA packaging (see section 4.7), the function of phage encoded Dam MTases is not clear.

No function has up to now been attributed to the solitary type II C5-MTase of *E. coli*, M. Dcm (Marinus and Morris, 1973; Hanck et al., 1989; Sohail et al., 1990). The activity of this enzyme has, however, been of relevance in the analysis of the role of 5-mC as a potential endogeneous mutagen (section 5).

Another enzyme of unidentified function is M.*Sss*I. This MTase was detected in *Spiroplasma sp.* (Nur et al., 1985) and cloned in *E. coli* (Renbaum et al., 1990). It is unusual among prokaryotes in that it methylates the diad sequence CG. It is thus an isomethylomer of eukaryotic MTases. Due to this property it may provide a valuable tool in the analysis of the functional role of methylation in eukaryotes.

3 Structural and evolutionary aspects of DNA methyltransferases

The large spectrum of different MTases identified in the prokaryotes can be attributed to three factors related to the specificity of these enzymes:

(1) There are three categories of MTases: those which generate 5-mC, N4-mC and N6-mA (see Fig. 2).
(2) MTases of the three categories methylate the characteristic base only when it is part of a DNA "target" sequence specifically recognized by each enzyme. A few hundred different target sequences have been identified.
(3) There are many isomethylomeric MTases, i.e. enzymes from different sources with the same specificity.

This scenario identifies MTases as operationally related proteins which share the property for highly specific interaction with DNA. The primary amino acid sequences of about one hundred different MTases have been determined (summarized in Wilson, 1991). This has facilitated comparisons of MTases with the aim to distinguish between conserved domains, which might be related to enzymatic regions mediating the universal steps in the methylation reaction, and variable sequence elements, possibly involved in the definition of the enzymes' specificities. Such comparisons can also contribute to answer evolutionary questions: How have MTases with different specificities evolved? Are MTases, particularly those with the same specificity, but stemming from different organisms, related? How do the isospecific ENases and MTases of a type II R/M system compare? What is the genomic organization of the corresponding genes? We shall address these and other questions on structure/function relationships and the evolution of MTases in the following chapters.

3.1 Categories of DNA methyltransferases and their genes

Comparing primary sequences of type II MTases revealed a remarkable similarity among the C5-MTases. These enzymes were readily distinguishable from the two other categories of MTases. These – the C-N4- and A-N6-MTases – also shared similarities among each other which were, however, not as pronounced as amongst the C5-MTases. It is the difference in the type of methylation to which the structural diversities between these MTases have been attributed. In C5-methylation, the process of DNA methylation is an addition of the methyl group to the ring carbon 5, whereas in C-N4- and A-N6-methylation the methyl group becomes attached to exocyclic amino groups.

3.1.1 Type II and type IIs C5-methyltransferases

A comparison of some 30 monospecific and multispecific C5-MTases indicated that amino acid sequences, which were highly conserved amongst all enzymes, alternated with less conserved regions. This pattern of enzyme structure was recognized independently in several laboratories (Som et al., 1987; Lauster et al., 1989; Posfai et al., 1989). In the absence of uniform criteria in the definition of conserved and variable sequence motifs, different consensus patterns have been proposed by these laboratories in the description of conserved and variable elements. In Figure 4A we present an alignment of five C5-MTases and give the subdivisions made by the different research groups. Differences between these schemes are partly due to the recognition of two sequence motifs by Posfai et al. (1989), termed II and III in their alignment pattern, which had not been considered in the other two laboratories as conserved motifs. They are further due to a subdivision of a large region taken as one conserved element (CE II) by Lauster et al. (1989) into several shorter conserved regions in the other alignment schemes. As is evident from Figure 4A, there is no substantial disagreement between the different groups on the building plan of C5-MTases, particularly not for the assignment of the largest variable region ("V" in Fig. 4A) containing the target recognizing domain (TRD) of C5-Mtases (see 3.2.4). In this review we shall follow the assignment of conserved and variable regions proposed by Lauster et al. (1989), where four different "conserved" sequence elements (CEs) alternate with five regions of variable length and amino acid composition. This arrangement is schematically shown in Figure 4B for the gene of M.SPRI, a phage-encoded multispecific MTase. As will be discussed in chapter 5 of this book and as shown in Figure 4A, the mammalian CG MTases also carry this arrangement of sequence elements within their carboxy terminal part (Bestor et al., 1988; Bestor, personal communication). The consensus sequence of C5-MTases established with type II MTases also obtains for type IIs C5-MTases, which recognize asymmetric target sequences. This has been verified for the two composite C5-MTases of the *Hga*I R/M system (Sugisaki et al., 1991), which methylate the target 5′GACGC/3′CTGCG.

The order of the four CEs was found to be the same in all enzymes studied and is apparently crucial for the functioning of C5-MTases.

It is also maintained in the remarkable case of M.*Aqu*I, a C5-MTase consisting of a heterodimer of two subunits (Fig. 1) (Karreman and de Waard, 1990). One subunit of 248 amino acids includes the amino terminal elements, the other of 139 amino acids the carboxy terminal elements of the canonical C5-MTases. Only the combination of the subunits provides an active MTase. The genes encoding these subunits are represented by two unidirectionally transcribed reading frames which partially overlap within the equivalent of the large variable region.

Similar to this natural situation, complementation was observed when separately expressed amino and carboxy terminal segments of the closely related monospecific isomethylomeric MTases M.*Bsp*RI and M.*Bsu*RI were combined (Posfai et al., 1991). Whereas the individual amino and carboxy terminal segments were inactive, some homologous and heterologous combinations yielded MTase activity. These findings demonstrate the need for the presence of the entire domain ensemble of C5-MTases for enzymatic activity. They also indicate that a division of these protein elements into two peptides may be tolerable for enzymatic activity. These experiments and particularly the finding of natural complementation in the case of M.*Aqu*I demonstrate that the enzyme segments on either side of the variable region represent independent functional entities, which must not necessarily be covalently joined to provide MTase activity.

The distinction within the primary amino acid sequence of conserved and variable regions suggested a modular structure of C5-MTases, in which a core of conserved regions determining universal steps in the DNA methylation reaction had become associated with different variable regions. These might either represent regions whose variability is tolerable for enzymatic activity or they could be responsible for variable parameters, most importantly target recognition.

In support of this concept two universal biochemical functions of MTases, the capacities for methyl group transfer and for association with the methyl donor SAM could be assigned to CE II and CE I of C5-MTases, respectively. To understand the catalysis of methyl group transfer by C5-MTases, Wu and Santi (1987) analyzed the reaction mechanism of methylation by M.*Hha*I. This led to the proposal that methylation proceeded through the formation of a transient, covalent MTase/DNA complex between the cytosine to become methylated and the cysteine contained in the highly conserved motif PCXXXS within

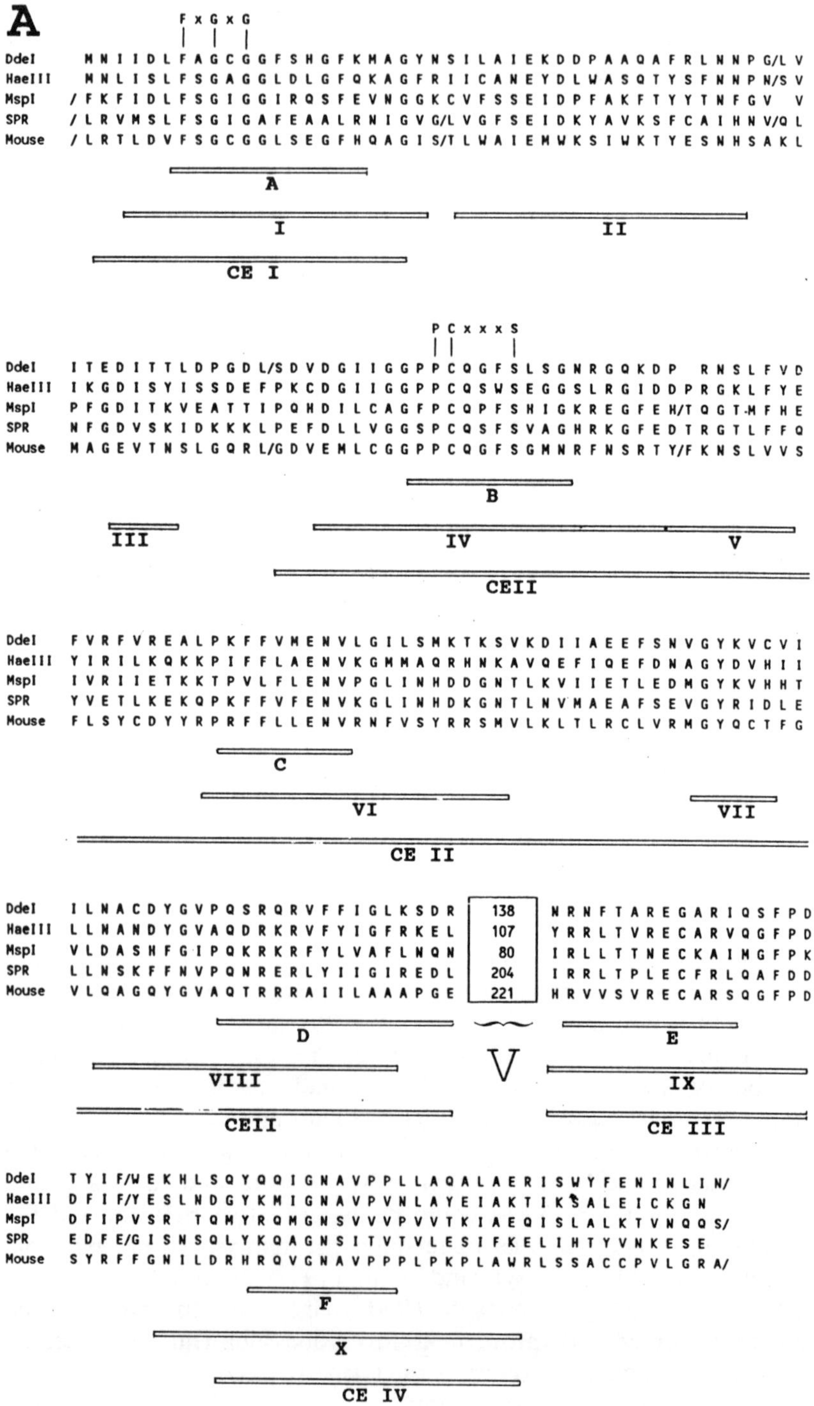
A
F x G x G
DdeI M N I I D L F A G C G G F S H G F K M A G Y N S I L A I E K D D P A A Q A F R L N N P G/L V
HaeIII M N L I S L F S G A G G L D L G F Q K A G F R I I C A N E Y D L W A S Q T Y S F N N P N/S V
MspI / F K F I D L F S G I G G I R Q S F E V N G G K C V F S S E I D P F A K F T Y Y T N F G V V
SPR / L R V M S L F S G I G A F E A A L R N I G V G/L V G F S E I D K Y A V K S F C A I H N V/Q L
Mouse / L R T L D V F S G C G G L S E G F H Q A G I S/T L W A I E M W K S I W K T Y E S N H S A K L
A
I
II
CE I
P C x x x S
DdeI I T E D I T T L D P G D L/S D V D G I I G G P P C Q G F S L S G N R G Q K D P R N S L F V D
HaeIII I K G D I S Y I S S D E F P K C D G I I G G P P C Q S W S E G G S L R G I D D P R G K L F Y E
MspI P F G D I T K V E A T T I P Q H D I L C A G F P C Q P F S H I G K R E G F E H/T Q G T M F H E
SPR N F G D V S K I D K K K L P E F D L L V G G S P C Q S F S V A G H R K G F E D T R G T L F F Q
Mouse M A G E V T N S L G Q R L/G D V E M L C G G P P C Q G F S G M N R F N S R T Y/F K N S L V V S
B
III
IV
V
CEII
DdeI F V R F V R E A L P K F F V M E N V L G I L S M K T K S V K D I I A E E F S N V G Y K V C V I
HaeIII Y I R I L K Q K K P I F F L A E N V K G M M A Q R H N K A V Q E F I Q E F D N A G Y D V H I I
MspI I V R I I E T K K T P V L F L E N V P G L I N H D D G N T L K V I I E T L E D M G Y K V H H T
SPR Y V E T L K E K Q P K F F V F E N V K G L I N H D K G N T L N V M A E A F S E V G Y R I D L E
Mouse F L S Y C D Y Y R P R F F L L E N V R N F V S Y R R S M V L K L T L R C L V R M G Y Q C T F G
C
VI
VII
CE II
DdeI I L N A C D Y G V P Q S R Q R V F F I G L K S D R 138 N R N F T A R E G A R I Q S F P D
HaeIII L L N A N D Y G V A Q D R K R V F Y I G F R K E L 107 Y R R L T V R E C A R V Q G F P D
MspI V L D A S H F G I P Q K R K R F Y L V A F L N Q N 80 I R L L T T N E C K A I M G F P K
SPR L L N S K F F N V P Q N R E R L Y I I G I R E D L 204 I R R L T P L E C F R L Q A F D D
Mouse V L Q A G Q Y G V A Q T R R R A I I L A A A P G E 221 H R V V S V R E C A R S Q G F P D
D
E
V
VIII
IX
CEII
CE III
DdeI T Y I F/W E K H L S Q Y Q Q I G N A V P P L L A Q A L A E R I S W Y F E N I N L I N/
HaeIII D F I F/Y E S L N D G Y K M I G N A V P V N L A Y E I A K T I K S A L E I C K G N
MspI D F I P V S R T Q M Y R Q M G N S V V V P V V T K I A E Q I S L A L K T V N Q Q S/
SPR E D F E/G I S N S Q L Y K Q A G N S I T V T V L E S I F K E L I H T Y V N K E S E
Mouse S Y R F F G N I L D R H R Q V G N A V P P P L P K P L A W R L S S A C C P V L G R A/
F
X
CE IV

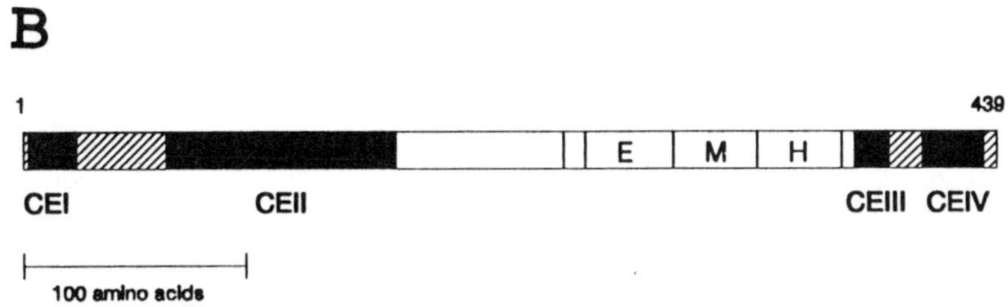

Figure 4. Organization of the primary structure of C5-DNA methyltransferases. (A) Alignment of C5-MTases. Slashes (/) within the amino acid sequences mark positions where parts of the individual sequences were omitted to achieve maximal alignment. The box interrupting the aligned sequences represents the "variable" region "V" in the MTases compared, which includes the target-recognizing domains (TRDs) (see text). The numbers of amino acids of the variable region are indicated. The bars below the sequences describe the extension of conserved sequence elements as defined by Som et al. (1987) top, Posfai et al. (1989) center, and Lauster et al. (1989) bottom. The sequences shown were determined by Sznyter et al. (1987) (M.*Dde*I); Slatko et al., (1988) (M.*Hae*III); Lin et al., (1989) (*Msp*I); Buhk et al. (1984) (M.SPR); Bestor et al. (1988) and Bestor, T., personal communication (mouse). (B) Schematic representation of the building plan of C5-MTases, as exemplified by the multispecific MTase M.SPR. This enzyme has 439 amino acids. The variable region "V" responsible for target recognition is presented as an open bar. Sections designated as E, M, and H within this region are TRDs recognizing the targets $CC^{A}/_{T}GG$, CCGG, GGCC, respectively. Ten and six amino acids bracketing the ensemble of TRDs are dispensable for MTase activity (Trautner et al., 1992). Solid blocks mark conserved elements, hatched bars variable regions other than that carrying the TRDs.

CE II of C5-MTases (Fig. 4A). In this intermediate complex, where the enzyme and cytosine are joined through a thioether bond between carbon 6 of C and the cysteine of the enzyme, the pyrimidine C5 would be converted to a carbanion, which would through a nucleophilic attack on enzyme bound SAM induce the transfer of its methyl group to C5. β-elimination would initiate the resolution of the ternary complex into methylated DNA, free enzyme and SAH. Support for this model came from the observation that mutations in MTases, leading to an exchange of the conserved cysteine by other amino acids, destroyed MTase activity (Wilke et al., 1988; Wyszynski et al., 1992). It was also shown that the ternary complex could not be resolved with DNA in which cytosine was substituted by 5-fluorocytosine or 5-azacytosine, which prevents β-elimination (Santi et al., 1983; Friedman, 1985; Friedman, 1986). Direct evidence for the involvement in the methylation process of the cysteine of the PCXXXS motif of CE II was recently provided by Chen et al. (1991), who characterized the ternary complex formed in a reaction between M.*Hae*III, SAM and a fluorinated target C contained in an oligonucleotide.

The SAM binding site was assigned to the highly conserved sequence motif (F,G)XGXG contained in CE I. This assignment is indirect and follows from the identification of this consensus sequence in numerous DNA-, RNA-, and protein methyltransferases which all share the use of

SAM as methyl group donor (Ingrosso et al., 1989). The motif has not been observed in the thymidylate synthases, which attach a methyl group to the C5-position of dUMP in a reaction, which is similar to that catalyzed by C5-MTases but employs tetrahydrofolate instead of SAM as the methyl group donor (Hardy et al., 1987). Direct attempts to determine the peptide involved in SAM binding used photoaffinity labeling of MTases either by radioactive SAM or azido-SAM (Som and Friedman, 1990). None of these experiments gave indications that the suspected consensus sequence was involved in SAM-binding. In experiments with the *Eco*RII MTase and [methyl^{3}H] SAM as methyl donor, peptides containing the cysteine of the PCXXXS motif as ^{3}H-methyl cysteine could be identified. This was taken as an indication that the catalytic site of the enzyme is in close vicinity to the SAM-binding site (Som and Friedman, 1991).

No functional assignments could be made for the CEs of C5-MTases located on the carboxy terminal side of the variable region. In particular, none of them showed similarity to established DNA binding motifs (Harrison, 1991), which could be visualized to be important in the first – possibly nonspecific – encounter of MTases with DNA. It is conceivable that at least some of the conserved amino acid sequences of these CEs are essential in the generation of the three-dimensional structure of active enzymes. They could also be involved in other general and so far unidentified reactions of the MTases, and/or might contribute to the reactions described before.

3.1.2 Type II and type IIs C-N4- and A-N6-methyltransferases

C-N4- and A-N6-MTases are found only in prokaryotes. Experiments aimed at understanding the biochemistry of methylation of the extracyclic NH_2-group of C and A by these enzymes have shown that methylation leading to N6-mA (and probably N4-mC) occurs by direct addition of the methyl group to the NH_2-group, rather than through an alternative pathway implicating the intermediate methylation at C1 (Pogolotti et al., 1988). No indication has been provided, however, of the generation of a covalent MTase/DNA complex as formed during C5-methylation.

In an overall comparison, sequence similarities are less pronounced among the C-N4- and A-N6-MTases, and no general building plan of these enzymes, as for the C5-MTases, could be established. However, significant sequence similarities, defining so far five "families" of MTases (Table 3), were apparent from several pairwise comparisons of A-N6-MTases (Lauster et al., 1987; Lauster, 1989; and references quoted in Table 3). Often similarity of MTases of one family is correlated with their capacity to recognize the same or related targets, which may be of interest in understanding the mechanism of target recognition

Table 3. Families of A-N6-DNA methyltransferases

Methyltransferase	Target sequence	Reference
M.*Dam*	GATC	Brooks et al. (1983)
M.*Dam*T2	GATC	Miner and Hattman (1988)
M.*Dam*T4	GATC	Macdonald and Mosig (1984)
M.*Dpn*II	GATC	Manarelli et al. (1985)
M.*Eco*RV	GATATC	Bougueleret et al. (1984)
M.*Hha*II	GANTC	Schoner et al. (1983)
M.*Hin*fI	GANTC	Chandrasegaran et al. (1988)
M.*Dpn*A	GATC	Lacks et al. (1986)
M.*Cvi*RI	TGCA	Stefan et al. (1991)
M.*Cvi*BIII	TCGA	Narva et al. (1987)
M.*Taq*I	TCGA	Slatko et al. (1987)
M.*Tth*HB8I	TCGA	Barany et al. (1992)
M.*Nla*III	CATG	Labbé et al. (1990)
M.*Fok*I	GGATG	Looney et al. (1989)
M.*Fok*I	CATCC	Kita et al. (1989)
M.LlaI	not determined	Hill et al. (1991)
M.*Mbo*II	GAAGA	Bocklage et al. (1991)
M.*Rsr*I	GAATTC	Kaszubska et al. (1989)
M.*Pst*I	CTGCAG	Walder et al. (1984)
M.*Bsu*BI	CTGCAG	Xu et al. (1992)

The description of families is based on computer-aided and visual comparisons of amino acid sequences of A-N6-MTases.

by these enzymes. MTases which recognize the same sequence may, however, also fall into different families like the MTases of the *Dpn*II system, M.*Dpn*II and M.*Dpn*A, whose genes are adjacent to each other. On the other hand the amino acid sequences of the fused amino and carboxy terminal parts of the M.*Fok*I MTase (see below) have 41% amino acid identity in spite of recognizing different targets. This points to gene duplication and diversification of the two components in the evolution of M.*Fok*I. Furthermore, the similarity between M.*Mbo*II (type IIs) and M.*Rsr*I (type II) demonstrates that type IIs and type II R/M systems employ basically identical MTases. It is remarkable that members of families of MTases may derive from phylogenetically distant organisms as illustrated e.g. for the CTGCAG MTases, which are derived from *P. stuartii* (*Pst*I) (Walder et al., 1984) and *B. subtilis* (*Bsu*BI) (Xu et al., 1992), or the TCGA family which includes MTases from *Chlorella* viruses (M.*Cvi*B III, Narva et al., 1987; M.*Cvi*RI, Stefan et al., 1991) and a bacterium *Thermus aquaticus* (M.*Taq*I; Slatko et al., 1987).

Only two short sequence motifs shared by NH_2-group methylating enzymes could be identified in overall comparisons (Chandrasegaran and Smith, 1988; Klimasauskas et al., 1989; Smith et al., 1990). One is the sequence (D,N,S)PP(Y,F). The other one, which may be highly degenerate in some of these enzymes, is the putative "SAM binding" motif (F,G)XGXG, discussed before. These motifs may occur at diverse relative positions to the molecular ends and may have a different order among various MTases, although they occupy the same relative positions within the MTases of the same family. Initially it was assumed that all C-N4-MTases carried the sequence SPP(Y,F) and all A-N6-MTases the sequence (D,N)PP(Y,F) (Klimasauskas et al., 1989). This correlation did, however, not obtain as the primary sequences of more MTases became known (Brooks et al., 1991).

The importance of the presence of the (D,N,S)PP(Y,F) motif is illustrated by the unusual type IIs MTase M.*Fok*I (Fig. 1), which methylates both complementary strands of the asymmetric target 5′GG<u>A</u>TG/3′CCT<u>A</u>C (Kita et al., 1989; Looney et al., 1989). In this MTase, which has approximately twice the molecular weight of other type II or type IIs MTases, the two conserved motifs occur twice and are distributed such that the enzyme can be considered as a fusion product of originally two separate A-N6 MTases. Site-directed mutageneses of the *Fok*I MTase gene causing a change in either of the (D,N,S)PP(Y,F) motifs destroyed the capacity of the enzyme to methylate either one of the asymmetric target sites (Sugisaki et al., 1989). In addition to showing that the (D,N,S)PP(Y,F) containing motif is crucial for enzyme activity, these data allowed the assignment of the amino and carboxy terminal halves of the enzyme to 5′GG<u>A</u>TG and 3′CCT<u>A</u>C methylation, respectively (Sugisaki et al., 1989). The same conclusion was reached by Looney et al. (1989) who separately expressed the two moieties of the *Fok*I MTase.

The role of the (D,N,S)PP(Y,F) motif in methylation has not been determined. From similarities recognized in the alignment of sequences surrounding this motif in the C-N4- and A-N6-MTases with those around the PCXXXS motif of C5-MTases (Klimasauskas et al., 1989), it was inferred that the (D,N,S)PP(Y,F) sequence corresponded to the PC motif and was involved in the methylation of the exocyclic amino group. Further biochemical work is, however, required to fully understand the biochemical role played by this part of the C-N4- and A-N6-MTases.

3.1.3 Type III and type I A-N6-methyltransferases

The amino acid sequences of the type III A-N6-MTases M.*Eco*P1 (target AGACC) and M.*Eco*P15 (target CAGCAG) are very similar (Hümbelin et al., 1988). They consist of two contiguous highly conserved segments of about equal size at their amino- and carboxy terminals. These are separated by a variable region which represents about one half of the size of the whole enzyme. The putative SAM-binding motif, represented here as FAGSG, is located within this variable region whereas the unique motif DPPY is found within the conserved amino terminal region.

In type I MTases, the methylating enzyme subunit (M) must be associated with a specificity determining subunit S to be active as a MTase. Analyzing amino acid sequences of the M and S subunits, convincing sequence homology to the FXGXG motif is not observed (Klimasauskas et al., 1989; Smith et al., 1990). On the other hand, the M subunit of *Eco*K has the sequence NPPF, characteristic for all other A-N6-MTases (Loenen et al., 1987). The fact that this motif is located in type I R/M systems in the M-subunit, rather than in the DNA target recognizing S-subunit of the complex, indicates that this sequence plays a general role in the transfer of methyl groups to A residues rather than in the recognition of targets containing the A to be methylated. This understanding of the function of the M subunit as a modular component providing methylation as directed by the S subunit is consistent with the observation that the M protein is interchangeable between type I R/M enzymes with different specificities, provided they are of the same family.

3.2 Specificity of DNA methyltransferases as determined by variable domains

Highly specific interaction of MTases with their DNA targets has two fascinating aspects:

(1) What are the enzymatic structures which recognize the DNA target?
(2) How do these structures interact with the various functionally different domains of MTases in the methylation process?

At this time, particularly in the absence of information on the three-dimensional structure of MTases, definite answers to these questions cannot be given. There is, however, a wealth of information on the enzyme regions involved in target recognition. This derives from comparisons of the primary structures of various MTases, from construction of chimeric MTases, and from mutagenesis experiments. Such data will be discussed in the following.

3.2.1 Target recognition in type I R/M systems

Only in the multisubunit type I R/M systems can one protein be singled out, which is exclusively responsible for target recognition. This assignment has motivated extensive studies on the structure and comparison of the target-recognizing S proteins of the various families. The targets of type I enzymes (see Fig. 1) are asymmetric and bipartite. They consist of three and four defined bp separated, depending on the enzyme, by six to eight random bp. The distribution of A/T base pairs between the defined sequences of the target is such that A methylation, as mediated by type I MTases, affects the three bp sequence in one and the four bp sequence in the complementary strand. How are these split targets recognized?

The S polypeptides from the three type I R/M enzymes all share the same basic building plan, in which two regions of variable amino acids alternate with regions whose sequence is conserved between the S subunit of one family (A family: Kannan et al., 1989; K family: Gough and Murray, 1983; *Eco*R124 family: Price et al., 1989). The variable regions of the S proteins of individual enzymes of one family show minor differences in the number of their amino acids. The variable regions at the amino termini have about 150 amino acids, the carboxy variable regions about 180 amino acids (Cowan et al., 1989; Kannan et al., 1989).,

Two lines of evidence have been provided, showing that the amino and carboxy terminal variable regions of the S proteins recognize the 5′ three bp and the 3′ four bp sequences of the canonical type I targets, respectively:

(1) Pairwise comparisons of the sequences of S proteins within one family which recognize the same 5′ sequence, e.g. *Eco*A ($\underline{\text{GAG}}\text{N}_7\text{GTCA}$) and *Eco*E ($\underline{\text{GAG}}\text{N}_7\text{ATCG}$) (Cowan et al., 1989) or *Eco*K ($\underline{\text{AAC}}\text{N}_6\text{GTGC}$) and *Sty*SP ($\underline{\text{AAC}}\text{N}_6\text{GTRC}$) (Fuller-Pace and Murray, 1986), have shown extensive similarities only between their amino terminal variable sequences, which are not observed between enzymes recognizing different three bp target components. The same rule was found in a comparison between two enzymes from different families, *Eco*A ($\underline{\text{GAG}}\text{N}_7\text{GTCA}$) and *Sty*SB($\underline{\text{GAG}}\text{N}_6\text{RTAYG}$) (Cowan et al., 1989).

(2) Reciprocal recombination between the S genes of *Sty*SB and *Sty*SP, leading to exchanges of DNA corresponding to variable regions, generated two novel S subunits in which the composite elements of target recognition had become reassembled (Fuller-Pace et al., 1984; Nagaraja et al., 1985; Gann et al., 1987; Cowan et al., 1989).

Changes of the specificities of the S proteins have in addition to recombination within the S gene been attributed to alterations in the size of the conserved region separating the component TRDs. The S protein of the R/M system *Eco*R124 has within the conserved region two repeats of the four amino acid sequence TAEL. It is specific for the target GAAN_6RTCG. An allelic type I R/M system, *Eco*R124/3, recognizes the same defined nucleotides of the target sequence with one additional spacer nucleotide (GAAN_7RTCG). The corresponding S protein is identical to that of *Eco*R124 except for the presence of *three* rather than two TAEL elements (Price et al., 1989). This observation of natural generation of a new specificity – probably by slippage recombination within the DNA corresponding to the TAEL sequence – invited systematic studies on the effects of changes in the composition and size of the TAEL repeat region (Gubler and Bickle, 1991). Among several mutant derivatives analyzed, only those which occur naturally were efficient R/M systems. No additional specificities arose as a consequence of such mutagenesis, indicating rather stringent requirements, fulfilled in the natural forms of the S subunits, for the structure of functional subunits.

In spite of the unambiguous assignment of some TRDs to the S proteins of type I R/M systems it is not clear how they combine with DNA, nor is it known how TRD and M subunit combine to achieve specific methylation. The M subunit, beyond serving in S subunit directed methylation, also contributes to the differential interaction of the enzyme with hemi- or unmethylated targets: Mutations obliterating the rather stringent substrate requirement of *Eco*K for hemimethylated DNA have all been localized within the M gene of this system (Kelleher et al., 1991).

3.2.2 Target recognition in type III R/M systems

Little is known about target recognition in these systems, where the MTase serves also as a specificity determining protein for the ENase. Targets of type III MTases are contiguous asymmetric sequences of five or six base pairs, which have a total strand bias for A and T (Fig. 1), such that methylation occurs in only one strand. One would argue from the similarity between different enzymes of their amino and carboxy terminal regions (Hümbelin et al., 1988) that the TRD is located within the central region. This variable region, however, also mediates general functions. Two amino acid exchanges in this region of M.*Eco*P15 led to methylation-deficient, but still restriction-proficient, enzymes (Hümbelin et al., 1988), suggesting that the mutations affected neither target recognition nor the association of MTase and ENase. As is discussed in 2.1.1, type III R/M systems are distinguished by the feature that methylation can occur of a single target, whereas restriction requires the presence of two targets in opposite orientation (Meisel et al., 1992). This remarkable property cannot be rationalized from the primary sequences and must await further analysis of target/enzyme interaction.

3.2.3 Target recognition by type II and type IIs MTases

Within the autonomously acting type II and type IIs MTases, the TRDs must be represented by amino acids of the same polypeptide, which is also responsible for all other steps of methylation. Within the C5-MTases with their highly conserved building plan, the TRDs, or at least a significant part of them, could be assigned to the largest variable enzyme region, located in all C5-MTases between CE II and CE III (see Fig. 4).

The first indication for this assignment came from comparative analyses of the primary amino acid sequences of multispecific C5-MTases of some bacteriophages of *Bacillus subtilis* (Buhk et al., 1984; Tran-Betcke et al., 1986; Behrens et al., 1987; Lange et al., 1991b). These enzymes, whose building plan followed that of monospecific C5-Mtases (Fig. 4B), showed a particularly high degree of sequence similarity, which included also sequences considered as non conserved in overall comparisons. Sequence differences between multispecific MTases with different specificities were primarily observed within the large variable region separating CEs II and III, suggesting its role in target recognition.

Direct evidence for the location of specificity determinants within this region was provided by the construction of chimeric, multispecific MTases in which composite parts derived from MTases with different specificities (Balganesh et al., 1987; Trautner et al., 1988). Relating the methylation specificity of the chimeric MTases with the location of the junction of the two composite enzyme parts indeed assigned all TRDs to the largest variable region.

Two other approaches provided a much more refined view on the organization of the variable region of multispecific MTases.

(1) Because of their multispecificity it was possible to distinguish two classes of mutations which led to an altered phenotype within the MTase genes (Wilke et al., 1988). Class I mutations destroyed the overall methylation capacity of a MTase. Class II mutations had a phenotype in which either one of the multiple methylation capacities of multispecific MTases had been lost. Class I mutations therefore marked domains responsible for catalyzing general steps of the methylation reaction, while class II mutations identified those regions contributing to target recognition. Analyzing at first random mutations in the gene of the multispecific MTase M.SPRI, class I mutations exclusively affected parts representing the conserved enzyme regions, whereas class II mutations were confined to that part of the gene encoding the carboxy terminal half of the large variable region (Fig. 4B) (Wilke et al., 1988). This distribution was verified by site-directed mutagenesis (Wilke et al., 1988; Trautner et al., 1992). Analysis of class II mutations affecting either one of the three methylation capacities of M.SPRI permitted a precise description of the confinement of the TRDs. They are nonoverlapping contiguous sequence motifs of 38 (TRD E), 37 (TRD M), and 40 (TRD H) amino acids (Fig. 5) separated by one spacer amino acid.

TRD	Sequence recognized	
E	CC(A/T)GG	V E Q V G N I N P S G N G M N G N V Y N S S G L S P T I T T N K G E G L K I
F	GCNGC	V R E P L M V G H V D L K G H D A I K R V Y S P E G L S P T L T T M G G G H R E P K I
B	G(G/A/T)GC(C/T/A)C	V G G I N V G K R W L N N G K T Y S R N F K Q G N R V Y D S N G I A T T L H S Q S V G G L G G Q T S L Y K
M	CCGG	V E Y S R K S G L G R E L A V S H T L S A S D W R G L N R N Q K Q N A V V
H	GGCC	V R P V L T P E R G E K R Q N G R R F K D D G E P A F T V N T I D R H G V A V G

Figure 5. Comparison of the TRDs of multispecific DNA methyltransferases. TRDs E, M, H of M.SPR, F of M.Ø3TI and B of M.H2I were aligned at the conserved T(L,I,V) dipeptide. Identical or similar amino acids at equivalent locations are boxed. The confines of TRDs E, M, H were determined by mutagenesis experiments (Trautner et al., 1992), those of the others are putative.

In addition to the *active* TRDs, characterized by these analyses, inactive TRDs – termed pseudodomains – could be identified in some multispecific MTases. These may be considered as natural class II mutant enzymes. Following appropriate mutagenesis, they could be activated to confer an additional methylation activity on the enzyme, within which they occur (Lange et al., 1991a). The amino terminal half of the variable region not accounted for by the TRDs (see Fig. 4B), is very similar only among the different multispecific MTases. Apparently it has a general function in methylation, since two mutations in this region were of a class I phenotype (Wilke et al., 1988).

(2) The information obtained from mutagenesis experiments on the organization of the variable region made TRD swapping experiments feasible. Using recombinant DNA technology, it was possible to excise, add, or permute TRDs within one or between the various multispecific MTases, generating enzymes with novel combinations of specificities (Walter et al., 1992). These experiments in addition to providing an independent definition of a TRD demonstrate the functional independence of TRDs within the enzymatic context analyzed. They also demonstrate the insensitivity of the multispecific MTases to alterations in the size and composition of the region encompassing the TRDs. In fact, it was possible to substitute protein segments unrelated to TRDs for one TRD without affecting the methylation activity of the chimeric MTase as determined by the residual TRDs.

Using these various approaches, eleven active TRDs, providing five different specificities, and three pseudodomains representing inactive TRDs, have been characterized (Fig. 5) (Trautner et al., 1992). Comparing their amino acid sequences, we have invariably observed in all TRDs, including putative TRDs of monospecific MTases, a dipeptide T(L,I,V) and in several pairwise comparisons additional "consensus" amino acids (Lauster et al., 1989; Lange et al., 1991a). The different degrees of conservation of amino acids betwen TRDs of different specificities (Fig. 5) suggest that the highly variable amino acids define the specificity of the TRD, whereas the consensus amino acids determine general parameters in target recognition or catalysis. Among more than one hundred point mutations tested, not a single mutation was observed which caused a defined change in the specificity of a TRD (Wilke et al., 1988; Trautner et al., 1992). This makes it highly probable that the specificity of a TRD is provided by several amino acids. Sequence comparisons and secondary structure predictions of TRDs of multispecific MTases have not indicated the presence of classical DNA binding motifs. If these TRDs would directly bind to their DNA targets and not have the effect of modulating other parts of the enzyme to bind to DNA, they would represent a new kind of DNA binding structure.

The TRDs as defined here are significantly smaller than the variable regions of the S proteins of type I R/M enzymes. There is no sequence similarity between these structures, suggesting that target recognition by type I and type II or type IIs MTases proceeds by different mechanisms.

To obtain information on TRDs of monospecific MTases, Klimasauskas et al. (1991) also used the approach to construct chimeric MTases involving the monospecific MTases M.*Hpa*II (recognition sequence CCGG) and M.*Hha*I (recognition sequence GCGC). A total of eleven different chimeric enzymes with joining points at different locations were constructed. In those constructs, which had enzymatic activity, only one methylation specificity was expressed. This was consistently that of the MTase from which the variable region derived.

With the data presented so far, the evolution of various type II and IIs C5-MTases with different specificities can be interpreted as a process in which the same enzyme "core" had become associated with interchangeable "modules" representing individual specificity determining TRDs. Several experiments, however, contradict that model in this simple form and suggest that at least in some cases there must be a "fine tuning" between TRDs and the conserved parts of the MTases, which would set limits to the free assortment of TRDs and enzyme cores. This was obvious from the experiments of Klimasauskas et al. (1991), referred to earlier, in which only five of eleven chimeric constructs of M.*Hha*I and M.*Hpa*II had MTase activity. Also the substitution by genetic engineering of one of the three TRDs of M.SPR by the entire variable region of the monospecific MTase M.*Bsu*FI generated a MTase which expressed the residual activities of M.SPRI but not of the insert (Walter et al., 1992).

Certainly modular shuttling of TRDs is not the only element in the evolution of C5-MTases. Lauster (1989) and Kapfer et al. (1991) by performing intramolecular sequence comparisons of a number of MTases presented evidence that gene duplication is another element which must have been operative in the evolution of MTases.

Further information on the structure/function relationship of TRDs was derived from comparisons of the amino acid sequences of MTases from different sources which would recognize the same target (isomethylomeric enzymes, see Fig. 1). An informative case is that of four enzymes which recognize the target CCGG (Walter et al., 1990). Three of the MTases compared methylate the outer C of the target: M.*Bsu*FI (*B. subtilis*) (Walter et al., 1990), M.*Msp*I (from a nonspecified gram-negative bacterium, originally thought to be *Moraxella*) (Lin et al., 1989) and M.SPRI (a multispecific MTase from *B. subtilis* phage SPR) (Buhk et al., 1984; Pósfai et al., 1984). M.*Hpa*II (Card et al., 1990) methylates the *inner* C of the target. Within all enzymes the general building plan of C5-MTases (see Fig. 4B) was conserved. In primary sequence comparisons, the highest percentage of identical amino acids was observed between M.*Bsu*FI and M.*Msp*I. In these two MTases similarity included the otherwise variable region, separating CEs II and III, showing that target recognition by these monospecific bacterial MTases was performed by virtually identical sequence(s). In contrast, aligning the sequence of the CCGG specific TRD of M.SPRI at the T(L,I,V) motif with those of the monospecific bacterial MTases revealed little similarity between the phage and bacterial MTase sequences. Apparently there are distinct requirements for TRDs with identical specificities to be functional, depending on whether they are part of a monospecific or multispecific MTase. The variable region of M.*Hpa*II, which methylates the inner C of the same target, has only spurious similarity to that of the other monospecific enzymes, although an exceptionally high similarity is seen between the other regions of the enzymes. This finding implies that the TRD containing region combines two inseparable functions: *recognition* of a target, but also the *directing* of a methyl group to only one defined base within the TRD, when there is the option of methylating several bases.

In comparisons of many pairs of other isomethylomeric C5-MTases, the sequence similarities, between the variable regions were not significantly greater than between pairs of MTases recognizing different targets (Walter et al., 1990). This suggests that the same target can be recognized by different TRDs.

The recognition of isospecific MTases in bacteriophages and their bacterial hosts raises the question about the evolutionary relationship between these enzymes. In the case of the *B. subtilis* phages discussed above, the low similarity between the host and virus MTases suggests independent evolution. Hill et al. (1991) have, however, reported on a case, where the significant similarity between a phage borne MTase domain and that contained in a plasmid encoded MTase strongly supported a genetic exchange between phage and host.

No assignments of TRDs could be made within the N6-mA or N4-mC MTases. Apparently these enzymes follow a construction principle which is basically different from that of the C5-MTases. It is conceivable that here target recognition is not facilitated by a contiguous sequence of amino acids but by the generation of a TRD composed of amino acids dispersed within the primary amino acid sequence.

3.2.4 Relationship between isospecific restriction and modification enzymes

Invariably the genes encoding the enzymatic components of R/M systems are closely linked. They may abut each other or even overlap (Fig.

1). The selective forces leading to this gene arrangement are not obvious. It is all the more surprising as cloning experiments have shown that linkage of two genes constituting an R/M system is not obligatory for their functioning. Proficient R/M systems have been constructed by separating ENase and MTase genes within one cell on two different genetic carriers (Kiss et al., 1985; Brooks et al., 1989; Düsterhöft et al., 1991; Kapfer et al., 1991).

In type I systems genes specifying the S and M subunits are always joined, whereas that encoding the R subunit can be detached by a few hundred bp from them. It is recalled that these genes are located within a segment of the chromosome, the ICR-region, which also harbors genes for methylation directed endonucleases (see 2.1.2 and Fig. 3). All subunit genes are transcribed in the same direction. In the case of the type I *Eco*K system, the R subunit gene is transcribed by one promoter, p_{res}, the two other subunit genes by another promoter (p_{mod}) (Loenen et al., 1987). It is not known how the activity of these promoters is regulated to achieve a coordinated expression of these genes.

The MTase and ENase genes of the type III *Eco*P1 R/M system are organized as one operon with only two base pairs separating their reading frames (Hümbelin et al., 1988).

A wide spectrum of combinations of the MTases with their ENases is realized in the type II systems. Also here the question of a coordinated regulation of expression of the two component genes is not clear in most cases. In some systems, however, *C*-("controller")-genes have been found (Brooks et al., 1991; Tao et al., 1991) next to the R genes, which control expression of only the R or the R and M genes.

Comparing type II and type IIs isospecific ENases and MTases it was plausible to ask whether target recognition in both enzymes involved a related sequence motif. In no case have such comparisons revealed any significant similarity between such enzyme pairs. This observation suggests that in the type II R/M systems target recognition is related to the catalytic function of each component enzyme. Apparently the DNA target must be brought into totally different configurations whether it is to be cleaved or methylated. From structural studies of R.*Eco*RI (Rosenberg, 1991) and R.*Eco*RV ENases (Winkler, 1992) in interaction with their targets one can rationalize that the complexes between ENase and DNA cleavage site would facilitate hydrolysis of the characteristic phosphodiester bond, but not methylation.

The absence of sequence similarity between type II ENases and MTases suggests that these enzymes have different evolutionary origins. In cases where both the ENases and MTases of isospecific R/M systems have extensive similarities, evolution from a common ancestor and/or horizontal gene transfer might have occurred. However, also in these cases evolutionary independence in the assembly of ENase and MTase genes has been observed. This situation e.g. is documented in the isospecific R/M systems *Bsu*BI and *Pst*I. Here extensive sequence similarity has been observed both between the two MTases, but also between the ENases. The relative orientation of these genes is, however, different in both systems. Whereas the genes of the *Pst*I system show a tail to tail arrangement with divergent transcription, the two genes of the *Bsu*BI systems overlap by five bp and have a tandem head to tail arrangement with unidirectional transcription (Fig. 1) (Xu et al., 1992).

4 Dam methylation in *E. coli*

4.1 Introduction

Dam methylation in *E. coli*, mediated by a solitary type II MTase, affects the N6-position of A in GATC sites (Lacks and Greenberg, 1977; Hattman et al., 1978; Geier and Modrich, 1979). Dam methylation is the only documented case of prokaryotic methylation involved in the regulation of cellular processes.

The mechanistic basis for Dam methylation-mediated regulation is in virtually all cases basically different from that underlying the function of eukaryotic DNA methylation. The regulatory capacity of eukaryotic DNA methylation rests on the formation of methylation patterns within DNA. In contrast, M.Dam methylates close to all GATC sites ($>99\%$) in the *E. coli* genome (Geier and Modrich, 1979). Its regulatory potential derives in general from a transient undermethylation of newly replicated DNA.

4.2 Instruction of postreplicative mismatch repair

Postreplicative mismatch repair, the subject of several recent reviews (Claverys and Lacks, 1986; Radman and Wagner, 1986; Modrich, 1987; Modrich, 1989), has evolved to counteract replication errors escaping DNA-polymerase proofreading. To be efficient, postreplicative mismatch-repair depends on instructive information directing the repair process towards the newly synthesized (error-prone) DNA strand. The finding that *E. coli dam*$^-$ mutants (Marinus and Morris, 1973) have a mutator phenotype generating a 10- to 100-fold increase in spontaneous mutations compared to isogenic *dam*$^+$ strains (Marinus and Morris, 1974; Marinus and Morris, 1975; Bale et al., 1979; Glickman, 1979) suggested that, at least in *E. coli*, DNA methylation might be involved in instructing postreplicative mismatch repair. As first proposed by Wagner and Meselson (1976), transient undermethylation of the newly synthesized DNA strand provides a means for its discrimination from the parental (methylated) template strand (see Fig. 6A).

Direct evidence for such a role was provided in transfection experiments employing artificially constructed heteroduplex molecules, whose constituent DNA strands were in defined states of methylation at GATC sites (Pukkila et al., 1983; Lu et al., 1983; Kramer et al., 1984; Wagner et al., 1984). With hemimethylated heteroduplexes mismatch repair was strongly biased toward the unmethylated strand, both *in vivo* (Pukkila et al., 1983; Kramer et al., 1984; Wagner et al., 1984) and *in vitro* (Lu et al., 1983). Unmethylated heteroduplex molecules were repaired without strand preference. Completely methylated heteroduplexes were refractory to repair, with exceptions relying on the activity of other *E. coli* repair pathways.

Experiments of this type (Nevers and Spatz, 1975; Rydberg, 1978; Bauer et al., 1981; Pukkila et al., 1983; Schaaper, 1988) and other genetic evidence (Glickman and Radman, 1980; McGraw and Marinus, 1980) revealed the implication of the products of several genes other than *dam*, mutations in which generate a mutator phenotype, *mutH*, *mutL*, *mutS*, and *mutU* (*mutU* is also referred to as *uvrD*, *uvrE*, or *recL*; Kushner et al., 1978), in methyl-directed mismatch repair. MutH, MutL, and MutS apparently function in the initiation of the methyl-directed mismatch repair process (Fig. 6B; reviewed in detail by Modrich, 1989). The product of *mutU*, DNA helicase II (Hickson et al., 1983; Kushner et al., 1983; Kumura and Sekiguchi, 1983), and other proteins involved in the mismatch repair process (Modrich, 1989; Lahue et al., 1989) act at later stages, such as excision-resynthesis of the strand to be repaired, which may extend over several kilobases (Wagner and Meselson, 1976; Lu et al., 1984), and religation.

Sensitivity of the overall repair process to Dam methylation is mediated by the activity of MutH in initiation of repair (Fig. 6B). MutH has the capacity to generate single-strand breaks immediately 5′ to unmethylated GATC sites (Welsh et al., 1987), a prerequisite for methyl-directed repair (Lu et al., 1984). While the GATC specific nicking activity of MutH is extremely weak in near homogeneous MutH preparations, it is stimulated in a mismatch-dependent manner in a reaction requiring MutS, MutL and ATP (Modrich, 1989; Fig. 6B). This way information on the presence of a mismatch and the state of Dam methylation within a mismatch containing DNA segment is integrated via the specific properties of MutH to direct the repair process to the non-methylated strand. The role of MutH explains why covalently closed heteroduplexes are refractory to repair when they lack GATC sites or contain only fully methylated GATC sites (Lahue et al., 1987; Laengle-Rouault et al., 1986, Laengle-Rouault et al., 1987; Lahue et al., 1989).

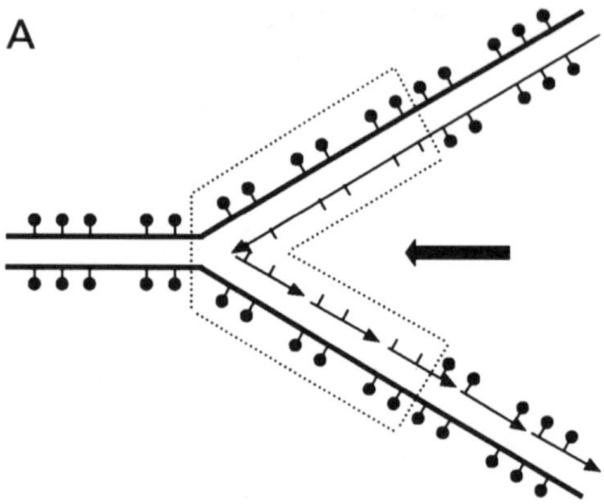
A

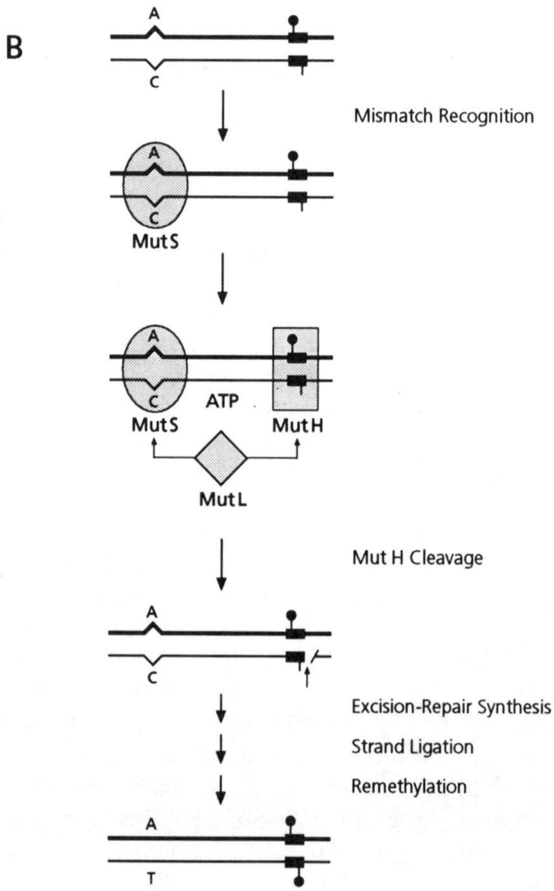
B
A
C
Mismatch Recognition
A
C
Mut S
A
C
ATP
Mut S
Mut H
Mut L
Mut H Cleavage
A
C
Excision-Repair Synthesis
Strand Ligation
Remethylation
A
T

The complementary action of M.Dam and MutH in instructing postreplicative mismatch repair provides an explanation for various phenotypes which accompany Dam deficiency in *E. coli*. The increased spontaneous mutation rate can be attributed to the loss of strand bias in repair. Other phenotypes, such as hyperrecombination (Marinus and Konrad, 1976), a slight induction of the SOS system (Peterson et al., 1985), decreased viability in the presence of some mutagens (Glickman et al., 1978; Jones and Wagner, 1981) and the incompatibility of *dam* and *recA*, *recB*, *recC*, *polA* or *lexA* mutations can be accounted for by elevated levels of MutH-mediated single-strand breaks at GATC sites and by the MutH-dependent generation of double-strand breaks which may occur when mismatch provoked repair initiates in both strands of DNA (for a detailed discussion see Claverys and Lacks, 1986; Radman and Wagner, 1986; Modrich 1987).

Dam overproduction also causes an increased rate of spontaneous mutations (Herman and Modrich, 1981, Marinus, 1984). Enhanced methylation, resulting from overproduction, shortens the persistence of hemimethylation at GATC sites after passage of the replication fork (Campbell and Kleckner, 1990). This interferes with MutH scissions at unmethylated GATC sites, required to allow excision repair synthesis in response to a base mispair.

The antagonistic action of M.Dam (DNA modification at GATC sites) and MutH (DNA strand scission at GATC sites) raised speculations that the Dam/MutH system might derive from an ancestral type II R/M system (Manarelli et al., 1985). The separate positions of the

---

Figure 6. The "time window" generated by hemimethylation of Dam sites following replication and its significance in the instruction of postreplicative mismatch repair in *E. coli*. (A) Schematic representation of the methylation status at Dam sites in a DNA region undergoing replication. The situation at a replication fork, migrating in the direction indicated by the solid arrow, is shown. Parental DNA strands are drawn as heavy lines, daughter strands as thin lines. Continuity or discontinuity of daughter strands symbolizes leading or lagging strand synthesis, with the direction of synthesis indicated by the arrow heads. Methyl groups at Dam sites are represented by filled circles. Due to a lag of Dam methylation with respect to replication Dam sites located in newly replicated DNA will be transiently hemimethylated. This generates a migrating window (symbolized by the area surrounded by the dotted line) within which DNA is hemimethylated and thus distinct from the rest of the genome. (B) Schematic representation of the methyl directed DNA mismatch repair pathway. The model shown follows a proposal of Modrich (1989) on the initiation of the repair process. The top shows a segment of newly replicated DNA (the parental strand is represented by a heavy line, the daughter strand by a thin line) containing a base mismatch and a hemimethylated Dam site. In a primary step of the repair process MutS interacts with the mispair. This triggers binding of MutH and MutL to the heteroduplex. MutH becomes bound at the hemimethylated Dam site. MutL serves in signal transduction between MutS and MutH which stimulates the latter to generate a single-strand break in the unmethylated strand at the hemimethylated GATC site. Such nicks are a prerequisite for the subsequent steps of the repair process indicated. For different models concerning the details of these steps see Modrich (1989).

dam and *mutH* genes in the *E. coli* chromosome (*mutH* maps at 61 min, *dam* at 74 min of the *E. coli* K-12 standard gene map; Marinus, 1973; Nestman, 1978; Bachmann, 1987), however, argues against the descendence of the Dam/MutH system from a type II R/M system (see 3.2.4.).

It is a remarkable fact that the DNA of many *E. coli* phages, e.g. T7, ∅X174 and M13, shows a pronounced bias against GATC sites (McClelland, 1985; Deschavanne and Radman, 1991). Several lytic *E. coli* phages, such as T1 (Schneider-Scherzer et al., 1990), T2 (Miner and Hattman, 1988) and T4 (Schlagman and Hattman 1983; Macdonald and Mosig, 1984) encode M.Dam-like modification activities. As no *bona fide* GATC-specific restriction ENase has been identified in *E. coli*, the reasons for counterselection against GATC sites or for their modification by the T1, T2, and T4 encoded MTases remain unclear.

4.3 Control of E. coli chromosome replication and chromosome segregation

During the past few years evidence has accumulated that Dam methylation plays an important role in *E. coli* in the regulation of the initiation of chromosomal replication and may also contribute to the controlled segregation of replicated chromosomes to daughter cells.

In *E. coli* initiation of DNA replication is a tightly controlled process. Replication of the *E. coli* chromosome is initiated at a unique site of the chromosome, the origin of replication, *oriC*, precisely once per cell cycle (reviewed by von Meyenburg and Hansen, 1987). Furthermore, if multiple origins are present in a cell, replication is initiated simultaneously at all origins and only once per origin (Skarstad et al., 1986; Helmstetter and Leonard, 1987; Koppes and von Meyenburg, 1987). Dam methylation is required for this coordination of the replication process. In Dam-deficient or Dam-overproducing *E. coli* strains replication initiation occurs at random times and, if there are multiple origins, asynchronously (Boye et al., 1988; Bakker and Smith, 1989; Boye and Løbner-Olesen, 1990). As detailed in the following, the regulatory effect of Dam methylation on replication is mainly due to the coordinated oscillation of the *oriC* and *dnaA* gene regions between fully methylated and hemi-methylated states. DnaA is required for the first and most crucial step in initiation (Bramhill and Kornberg, 1988). In the fully methylated state both regions are active, i.e. the *dnaA* gene is efficiently transcribed under these conditions and replication will initiate at *oriC* provided sufficient amounts of DnaA are available. *oriC* and *dnaA* will attain a hemimethylated state following replication. The transition to the hemimethylated state occurs roughly simultaneously for both regions due to their close vicinity in the chromosome (they are located within a distance of 50 kb). Hemimethylation triggers inactivation of both regions, following their specific sequestering by membrane proteins. The sequestering of *oriC* prevents initiation of new rounds of replication even when the levels of DnaA protein are still high. The simultaneous transcriptional inactivity of the *dnaA* gene results in a decrease of the DnaA protein concentration, preventing immediate reinitation of replication after reconversion of the sequestered regions into the fully methylated state by the action of M. Dam. An oscillation of the *oriC* and *dnaA* regions between fully methylated (non-sequestered, functional) and hemimethylated (sequestered, non-functional) states cannot take place in a *dam*$^-$ strain. Dam overproduction interferes with the periodicity of the two states as characteristic for a wt cell. This explains the uncoordinated initiation of replication under both conditions.

A function of Dam methylation in the regulation of *oriC* activity was suggested by the abundance of GATC sites in this region. Within the 245-bp minimal *oriC* (Oka et al., 1980) eleven GATC sites are found instead of only one as expected on a statistical basis (Fig. 7A). Eight of these sites are conserved at equivalent positions in five other phylogenetically related species (Zyskind and Smith, 1986; Fig. 7A).

The first experimental evidence that Dam methylation might regulate *oriC* function in Dam proficient strains was provided by transformation experiments with minichromosomes, i.e. plasmids with *oriC* as the only origin of replication. When they were modified at Dam sites, they transformed *dam*$^-$ strains very inefficiently compared to *dam*$^+$ strains (Smith et al., 1985; Messer et al., 1985; Russel and Zinder, 1987), due to the fact that *hemi*methylated *oriC*s, generated in replication, are inefficient in initiation (Russel and Zinder, 1987).

Observations reported by Ogden et al., (1988) and Campbell and Kleckner (1990) are crucial in explaining the inert state of hemimethylated *oriC*. Ogden et al. (1988) showed that a fraction of proteins of the outer membrane bound specifically to *oriC* when it was hemimethylated. Neither unmethylated nor fully methylated *oriC* became bound. The interaction of the membrane proteins with *oriC* sequestered it from remethylation as evident from the prolongated persistence of hemi-methylation at GATC sites in *oriC* compared to GATC sites in flanking regions. This was subsequently confirmed by Campbell and Kleckner (1990) in studies on the remethylation kinetics at individual GATC sites located in *oriC*, in the promoter region of the *dnaA* gene and in several other regions of the chromosome. Whereas most of the sites investigated were rapidly remethylated after passage of the replication fork, GATC sites located within *oriC* were sequestered from remethylation, remaining in a hemimethylated state for about 30 to 40% of the cell cycle. At the end of this period remethylation of these GATC sites occurred rapidly. Importantly, the promoter

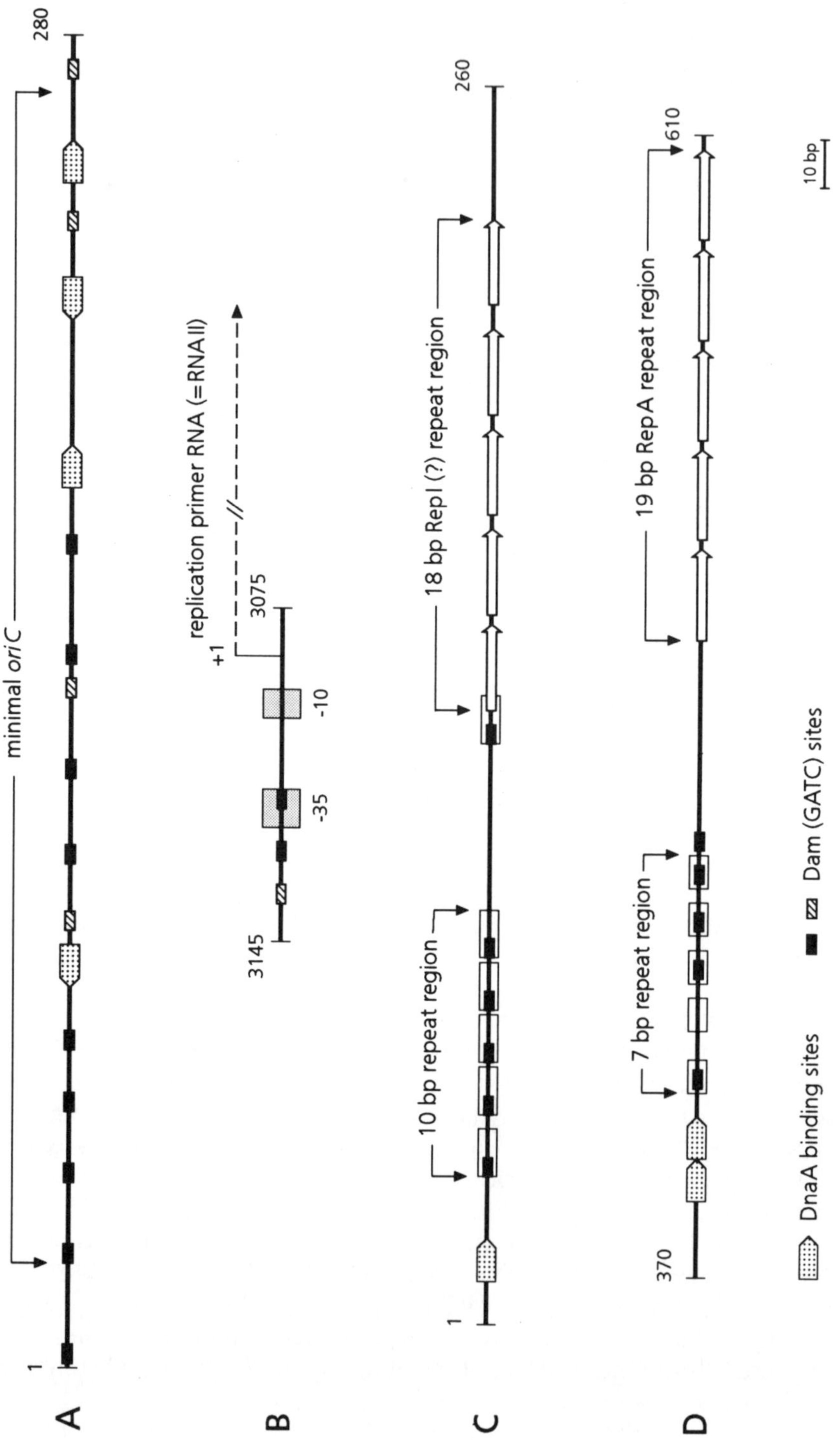
A
1
280
minimal oriC
B
3145
-35
-10
+1
3075
replication primer RNA (=RNAII)
C
1
260
10 bp repeat region
18 bp RepI (?) repeat region
D
370
610
7 bp repeat region
19 bp RepA repeat region
10 bp
DnaA binding sites
Dam (GATC) sites

region of the *dnaA* gene, whose product is required for replication initiation (Bramhill and Kornberg, 1988), was also sequestered from remethylation roughly simultaneously with *oriC*. During this period transcription of the *dnaA* gene was reduced to about 10% or less of the steady state level.

These findings allowed to propose the model initially discussed. One of the basic elements of this model is sequestering of *oriC* in a period of the cell cycle when the DnaA concentration is high. This prevents premature reinitiation of replication. As pointed out by Campbell and Kleckner (1990), this mechanism would allow that the levels of initiation proteins must not be precisely controlled but may raise beyond a critical threshold permitting efficient and simultaneous initiation at all origins in the cell (as observed in *dam*$^+$ strains) without the risk of secondary initiations. The coordinate sequestering of the *dnaA* region, accompanied by a drastic reduction in *dnaA* gene expression, on the other hand, would lead to a reduction in the initiation potential permitting release of the *oriC* region after about 30 to 40% of the cell cycle without immediate initiation of replication at the released origins. Reinitiation of replication would occur only after the synthesis of sufficient amounts of DnaA after release of the *dnaA* gene.

Strong support for this model has been provided by *in vitro* replication experiments. While in initial *in vitro* experiments no specific inefficiency in replication of *hemimethylated* minichromosomes, provided or generated in the experiment, could be detected (Hughes et al., 1984; Messer et al., 1985; Smith et al., 1985; Landoulsi et al., 1989), a specific inhibition, dependent on the preincubation of hemimethylated minichromosomal DNA with outer membrane fractions, was demonstratred later (Landoulsi et al., 1990). Interestingly, when hemimethylated DNA was preincubated with DnaA subsequent additions of even

---

Figure 7. Schematic representation of the distribution of Dam, DnaA, and other protein binding sites in regions serving as origins of DNA replication or controlling origin activity. (A) Origin of replication of the *E. coli* K-12 chromosome, *oriC*. (B) RNA II promoter region of pBR322. (C) REPI replication origin of ColV plasmids, *ori*. (D) Origin of replication of the P1 plasmid replicon, *ori*R. The extension of the different regions is indicated by solid lines. Numbers at the beginnings and ends of the regions represented refer to the corresponding nucleotide positions in the established DNA sequences (A: Meijer et al., 1979; B: Sutcliffe, 1979; C: Perez-Casal et al., 1989; D: Abeles et al., 1984). Dam and DnaA binding sites are indicated as shown. Hatching of some Dam sites in *ori*C (A) or the pBR322 RNA II promoter region (B) indicates that these sites were found to be non-conserved in comparative analysis with closely related replicons (Zyskind and Smith, 1986; Patnaik et al., 1989). Dam sites are surrounded by larger open boxes, if they are part of repeated DNA sequences as found in the REPI and P1 plasmid origins. Open arrows in the REPI and P1 origins indicate repeated DNA sequence motifs presumed or known to become bound by replicon-encoded proteins required for replication. The −10 and −35 regions of the pR322 RNA II promoter are represented by large shaded boxes.

high levels of outer membrane fractions caused no inhibition of DNA synthesis. This provides direct evidence that sequestering of hemimethylated *oriC* correlates with the capacity to prevent access of DnaA to *oriC* and, consequently, further initiations of replication at newly replicated *oriC*.

What triggers the release of the sequestered regions? Several findings (Messer et al., 1985; Campbell and Kleckner, 1990) suggested that M. Dam directly contributes to the release by the slow reconversion of hemimethylated GATC sites in these regions to the fully methylated state. The most direct evidence for such a role has been provided by Landoulsi et al. (1990), who showed that addition of M.Dam to preformed hemimethylated minichromosome – outer membrane complexes, which are inactive in *in vitro* replication in the absence of M. Dam, led to an almost complete reactivation. This demonstrates that hemimethylated GATC sites in membrane *oriC* complexes are accessible to Dam MTase and that their remethylation is sufficient to restore optimal DNA replication.

In addition to the major effects of Dam methylation on regulation of replication, it has been assumed to facilitate strand separation at *oriC* in the initiation of replication. As shown by Yamaki et al. (1988), the melting temperature at *oriC* is lowered by Dam methylation. This is presumably a consequence of the reduced strength of A/T base pairing following A-N6-methylation (Collins and Myers, 1987; Engel and von Hippel, 1978; Murchie and Lilley, 1989).

As speculated by Ogden et al. (1988), the discussed binding of membrane proteins to *oriC* could also serve to ensure proper segregation of daughter chromosomes, as envisioned in the replicon model proposed by Jacob et al. (1963). Although the attachment of DNA to the membrane via *oriC* clearly does not persist through the entire cell cycle, the association of the DNA with the membrane for a period of about 30 to 40% of the cell cycle (Campbell and Kleckner, 1990) would provide enough time to establish two membrane domains separating the daughter chromosomes. A particular significance of membrane binding to hemimethylated *oriC* for controlled chromosome partitioning, however, has been questioned by the recent finding that *dam*$^-$ strains do not produce anucleated cells at significantly elevated frequencies compared to *dam*$^+$ strains (Vinella et al., 1992).

4.4 Control of plasmid replication

Except for ColE1-type plasmids, autonomous plasmid minireplicons include DNA region(s) encoding at least one protein, termed Rep, which is essential for plasmid replication (reviewed by Scott, 1984). As shown during the past years, replication of ColE1-type plasmids and some members of Rep encoding plasmids is affected by Dam methylation, although in apparently different ways. Dam methylation is, however, not of general significance for plasmid replication in *E. coli*, as e.g. replication of pSC101 (Russel and Zinder, 1987) or of F miniplasmids (Abeles and Austin, 1987) is unaffected by Dam proficiency or deficiency of the host strain.

The plasmid ColE1 contains three GATC sites located upstream and within the promoter of the replication primer transcript, the so-called RNAII. Two of these sites are positionally conserved in the primer-promoter region of all plasmids related to ColE1 (Patnaik et al., 1990), e.g. pBR 322 (Fig. 7B) which contains the origin region of the ColE1-related plasmid pMB1 (Bolivar et al., 1977). Various lines of evidence indicated that specifically pBR322 DNA hemimethylated at GATC sites is ineffcient in replication (Messer et al., 1985; Russel and Zinder, 1987). The inhibitory effect of hemimethylation on pBR322 replication is entirely mediated by the three GATC sites located in the primer promoter region of the origin containing DNA fragment. Disruption of all three GATC sites is necessary and sufficient to completely alleviate the hemimethylation imposed inhibition of pBR322 replication *in vivo* (Patnaik et al., 1990).

The mechanism underlying these effects is similar to that described for hemimethylation-dependent inhibition of chromosomal replication. As shown by Malki et al. (1992), the outer membrane fraction interacting with hemimethylated *oriC* also binds, albeit with lower affinity, to the hemimethylated RNAII promoter region of pBR322. It also specifically inhibits *in vitro* replication of hemimethylated pBR322 DNA. These findings suggest that *oriC* and the RNAII promoter region of ColE1-type plasmids possess common structural elements which serve as binding sites for outer membrane proteins. The significance of sequestering of the RNAII promoter region of ColE1-type plasmids, in which replication is initiated randomly (Bazaral and Helinski, 1970), remains to be elucidated.

Replication of plasmids containing the REPI replicon, such as the bacterial virulence plasmid pColV-K30, also depends on Dam proficiency of the host (Gammie and Crosa, 1991). The REPI replicon consists of a gene encoding RepI, a protein essential for replication (Perez-Casal et al., 1989) and an origin of replication, *ori*, the site of RepI action. The origin contains a potential DnaA binding site, five 18-bp tandem repeats presumed to be the binding sites of RepI and five AT-rich 10-bp repeats, each containing a GATC site (Fig. 7C). Methylation at these GATC sites is essential for plasmid replication, as plasmids consisting of only the REPI replicon and a selectable marker can only be maintained in *dam*$^+$ but not in *dam*$^-$ strains. In contrast to the observations made with minichromosomes or ColE1-type plasmids, methylated DNA transformed into *dam*$^-$ strains does not accumulate in hemimethylated but in a completely unmethylated form. This indicates that both fully methylated and hemimethylated REPI plasmids can be replicated in *dam*$^{-}$ strains, while unmethylated DNA cannot undergo further rounds of replication. The five 10-bp repeats containing the GATC sites in the REPI *ori* are spaced such that they are located on one surface of the DNA helix. Manipulation interfering with the specific wild type conformation of the 10-mer containing region abolishes *ori* function (Gammie and Crosa, 1991). A model on the role of Dam methylation in the initiation of REPI was based on this observation, the description of amino acid similarities between REPI and various proteins interacting with GATC or related DNA targets and the documented contribution of Dam methylation to helix instability in the REPI origin (Gammie and Crosa 1991). The model invokes two binding sites of RepI, one for the 18-mers and one for the properly methylated 10-mers contained in the *ori* region. Binding to both sets of origin repeats would allow specific strand opening at *ori*, with strand melting being facilitated by Dam methylation. According to this hypothesis methylation at the Dam sites in the origin would play a dual role in the initiation of replication by contributing to both required protein-DNA interactions and helix destabilization.

Effects of Dam methylation on replication of phage P1 in its prophage state (Abeles and Austin, 1987) are similar to those reported for the REPI replicon. In its prophage state P1 is maintained in *E. coli* as a plasmid whose copy number is about 1 per chromosome (Prentki et al., 1977). The minimal P1 replicon contains a gene encoding RepA, which, together with the host protein DnaA, is required for replication, a copy control locus, *incA*, and the origin of replication, *oriR* (Abeles et al., 1984). *oriR* has a similar organization as the RepI origin (Fig. 7D): It contains two DnaA boxes, five 19-bp repeats, the binding sites for RepA (Abeles, 1986) and five 7-bp repeats located in between the DnaA and RepA binding regions. Four of the 7-bp repeats include the sequence GATC. A fifth GATC site found in *oriR* is located in the immediate neighborhood (Fig. 7D). Importance of methylation of at least some of these GATC sites was indicated by the observation that P1 miniplasmids cannot be established in *dam*$^{-}$ strains (Abeles and Austin, 1987). Further, within an *in vitro* system which specifically replicated DNA containing the fully methylated *oriR*, nonmethylated *oriR* was completely inert. A detailed mutational analysis of the *oriR* region (Brendler et al., 1991) revealed that each of the five 7-bp repeats, four of which include GATC sites, are important for *oriR* function. Mutations in these repeats, abolishing *oriR* function, included, but were not limited to the GATC tetranucleotides. This suggests that the 7-bp repeats have a functional potential beyond that mediated by methylation of the GATC sites they contain. In a model on the initiation of P1 replication (Brendler et al., 1991) Dam methylation within the 7-bp repeats is proposed to facilitate required interactions between proteins and the 7-bp repeats and to contribute to helix destablization (Yamaki et al., 1988), this way favoring strand opening.

4.5 Transposition

Transposition of several transposable elements, including IS*10* and the IS*10* associated transposon Tn*10*, is regulated by Dam methylation. Tn*10* (Fig. 8; for a review see Kleckner, 1989) has two invertedly oriented copies of IS*10*, IS*10*-Left and IS*10*-Right, at its ends. Only IS*10*-Right, whose single gene specifies an active transposase, is fully functional (and, hence, termed IS*10* without any further designation when considered as an individual element). IS*10*-Left specifies an inactive transposase. Consequently it cannot transpose independently, although the ends of the element, which are contacted by transposase in transposition, are functional.

Two GATC sites are found in IS*10* (Fig. 8). One of these sites, located near the outer end of IS*10*, overlaps the -10 region of the transposase promoter, pIN. The other site is located 7–10 bp from the inner terminus of IS*10* (Fig. 8) in a region contacted by the transposase.

An involvement of Dam methylation in the control of IS*10*/Tn*10* transposition was suggested by the observation that several Tn*10* promoted events occur with increased frequency in *dam*$^-$ strains (Roberts et al., 1985). All effects caused by the absence of Dam methylation are mediated by the two GATC sites of IS*10*. As revealed by analyses including IS*10* mutants altered in either or both of these sites, the pIN promoter and the inner terminus are each activated in the unmethylated state (Roberts et al., 1985). While the increased level of transposase expression is relevant for all IS*10*/Tn*10* associated transposition events, inner terminus activation is significant for only some of these processes (Kleckner, 1989). Therefore, Tn*10* transposition mediated via two methylation-insensitive outer ends is increased only 10-fold in a *dam*$^-$ background, while IS*10* transposition which involves one methylation-sensitive inner end and one outer end is increased about 100-fold. Tn*10* promoted deletions and DNA rearrangements, which employ two inner ends, are increased to an even higher extent (at least 500-fold or more) (Roberts et al., 1985).

In a Dam proficient *E. coli* cell, IS*10* will never become fully unmethylated. Transiently hemimethylated IS*10*, however, will be generated during replication or in conjugal DNA transfer between *E. coli* cells, in which DNA invades the recipient in single-stranded form. Roberts et al. (1985), therefore, specifically addressed the question of whether IS*10* would also become activated in the hemimethylated form. This is indeed the case. Of particular importance with respect to the

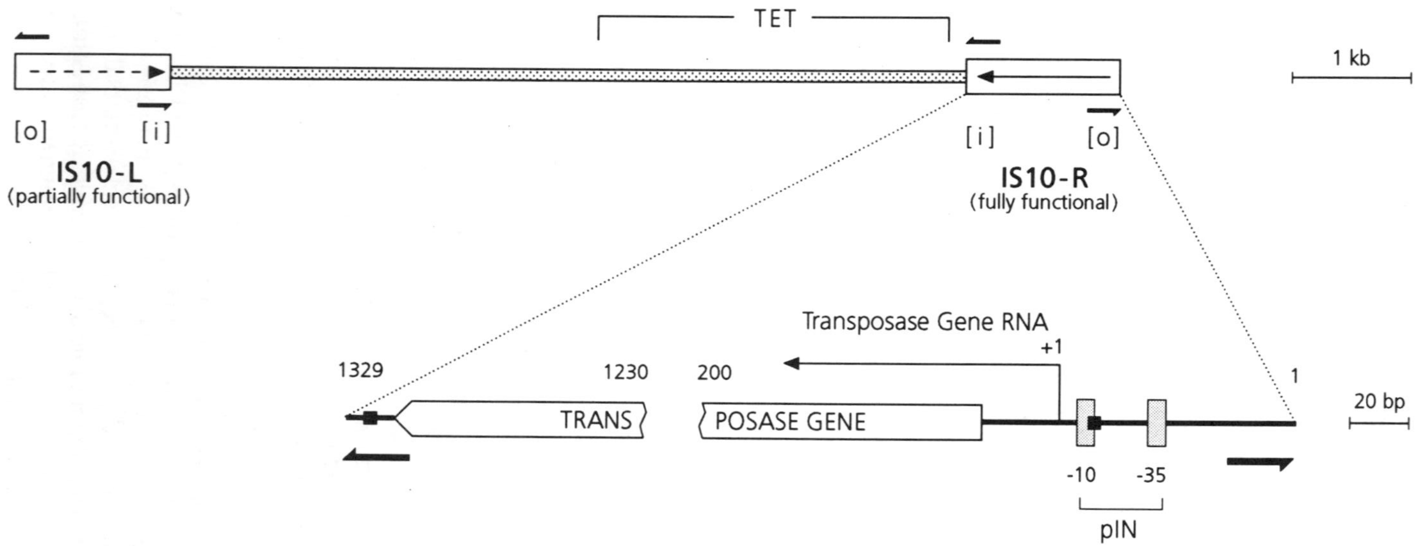

Figure 8. Structure of Tn*10* and IS*10*-Right. The upper part of the figure shows the general organization of Tn*10*. Tn*10* comprises two inverted copies of IS*10*, IS*10*-Left (L) and IS*10*-Right (R). The intervening sequence specifies tetracycline resistance, with the relevant region marked TET. The broken or solid arrows within the boxes representing IS*10*-L and IS*10*-R, indicate the transcription of defective or fully functional transposase genes in the two IS*10* copies, respectively. Short arrows above and below the termini of the IS*10* elements, not drawn to scale, symbolize nearly perfect inverted repeats. The ends of IS*10*-L and IS*10*-R are referred to as outside (o) or inside (i) according to their positions in intact Tn*10*. The lower part of the figure shows details of the organization of IS*10*-R, consisting of 1329 bp. Numbers at the ends of the element and at the borders of the interruption of the transposase gene refer to corresponding positions in the nucleotide sequence of Halling et al. (1982). The -10 and -35 regions of the transposase gene promoter, pIN, are indicated as shaded boxes. The only two Dam sites of IS*10*-R, located within pIN and the inner terminal repeat, are indicated as small black boxes.

biological significance of this observation was the finding that the IS*10* species methylated at the transposase coding strand (species I) transposes much more frequently than the species methylated at the non-template strand (species II). Several factors contribute to the differential activity of the two hemimethylated species (Roberts et al., 1985). Firstly, the transposase gene promoter, pIN, is activated to a higher degree in species I than in species II. Secondly, inner end activation is detectable only in species I but not in species II. A third factor, increasing the individual contributions of hemimethylation at pIN and the inner terminus to species I activity, relies on the temporal coupling between transposase expression and action, achieved by the simultaneous activiation of pIN and the inner terminus. As a consequence of these effects species I transposes about 2400 to 60,000 times more frequently than fully methylated IS*10*, while species II shows only a moderately (about 12 times) elevated transposition frequency (Roberts et al., 1985; Kleckner 1989).

The biological significance of the functional asymmetry of the two hemimethylated IS*10* species has been interpreted (Roberts et al. 1985; Kleckner, 1989) in considering the mode of IS*10*/Tn*10* transposition which occurs by nonreplicative excision from the donor molecule leaving behind a double-stranded gap. After passage of the replication fork only the highly activated IS*10* species is likely to transpose. In the case of transposition the cell will retain at least one copy of the critical chromosomal region in its original form. This would permit cell survival, if the molecule donating the transposing element is not repaired. It might further facilitate restoration of the donor molecule by recombinational double strand gap repair.

This view is supported by analyses performed with IS*50* and the IS*50* associated transposon Tn*5*, whose mode of transposition (reviewed by Berg, 1989) resembles that of IS*10*/Tn*10*. Although not explicitly studied with respect to the situation in the hemimethylated state, IS*50*/Tn*5* transposition seems to be affected by Dam methylation in a way similar to that documented for IS*10*/Tn*10* transposition (Yin et al., 1988; Makris et al., 1988). Among the various GATC sites present in Tn*5*, two tandemly arranged sites are found in the −10 region of the promoter of the transposase gene carried by IS*50* (among the Tn*5* component IS*50* elements only IS*50*-R specifies a functional transposase). An additional two sites are located within a 19-bp region at the inner end of IS*50* contacted by the transposase. Both the promoter and the inner terminus are activated in a *dam*$^{-}$ background. Different transposition frequencies observed in a *dam*$^{+}$ background with engineered transposons, which carry the transposase gene in different orientations relative to the outer and inner ends of IS*50*, further suggested a functional nonequivalence of the two hemimethylated IS*50* forms, similar as described for IS*10* (Dodson and Berg, 1989).

Transposition of other elements containing GATC sites at critical regions, such as Tn*903*/IS*903* or IS*3*, is also elevated in a *dam*$^-$ background (Roberts et al., 1985; Spielmann-Ryser et al., 1991), while other elements, such as IS*1* and Tn*3*, lacking GATC sites at critical positions, transpose at equal frequencies in *dam*$^+$ and *dam*$^-$ hosts (Roberts et al., 1985).

4.6 Gene expression

As shown in the preceding sections transient hemimethylation following passage of the replication fork is used to regulate the activity of the *dnaA* gene and the IS*10* transposase gene. The methylation-dependent regulation of the *dnaA* gene, resulting in a decreased transcription rate, is primarily due to sequestering of the *dnaA* gene region by membrane proteins. Binding of these proteins presumably prevents accessibility of the *dnaA* gene promoters by RNA polymerase. In the case of the IS*10* transposase gene, whose activity is increased in the hemimethylated state, the methylation status of a GATC site within the gene promoter, pIN, appears to *directly* affect the affinity of the promoter for RNA polymerase. Both indirect and direct mechanisms relying on transient hemimethylation of GATC sites could also contribute to the regulation of other chromosomal genes, whose activity would thus be linked to the cell cycle.

Convincing evidence for an altered level of expression in the hemimethylated state has, however, not been provided for any other gene. For some chromosomal genes whose regulatory sequences contain GATC sites (for a compilation see Plumbridge, 1987; Marinus, 1987; Barras and Marinus, 1989) moderately different levels of expression in *dam*$^+$ and *dam*$^-$ backgrounds have been reported.

Assuming that the level of gene expression in the hemimethylated state is similar to the level of gene expression in the unmethylated state, differential expression of *E. coli* genes in *dam*$^+$ and *dam*$^-$ strains has frequently been taken as an indication that these genes are subject to a cell cycle linked regulation. As the proposed benefits of cell cycle linked regulation of these genes (reviewed by Plumbridge, 1987; Marinus, 1987; Barras and Marinus, 1989) are not always readily evident, it remains questionable whether the underlying assumption is justified.

In the following we will address two special cases of gene regulation affected by Dam methylation, the expression of the *mom* gene of bacteriophage Mu and of the *papA* gene of *E. coli*.

The *mom* gene of bacteriophage Mu encodes an unusual DNA modification activity which, by converting adenine residues in specific DNA sequence contexts to acetoamido-adenine (Swinton et al., 1983; Kahmann, 1984), renders the phage DNA resistant to many host restriction activities (Toussaint, 1976; Kahmann, 1984). Transcription of the *mom* operon, consisting of the *com* and *mom* genes (Fig. 9A) (*com* encodes a posttranscriptional activator of *mom* gene expression; Kahmann et al., 1985) depends on both a Mu-specific trans-activator, the product of gene *C* (Hattman et al., 1985; Heisig and Kahmann, 1986) and Dam proficiency of the host strain. Critical for the dependence of *mom* gene expression on Dam methylation is a region upstream of the *mom* operon promoter, termed region I, which contains three GATC sites whose methylation is required for transcription to initiate (Fig. 9A; Hattman, 1982). The requirement for Dam methylation in *mom* gene expression can be alleviated by various types of mutations affecting the integrity of region I (Seiler et al., 1986), consistent with the proposal that in its unmethylated or hemimethylated state region I becomes bound by a cellular repressor preventing *mom* expression (Hattman and Ives, 1984). By isolating and characterizing mutants permitting *mom* gene expression in a dam^- background, the postulated repressor, MomR, could be identified and shown to be identical with OxyR (Bölker and Kahmann, 1989), a regulatory protein which in response to oxidative stress induces various genes involved in stress tolerance (Christman et al., 1989). The 43 bp binding site of MomR/OxyR in the regulatory region of the *mom* operon (Bölker and Kahmann, 1989) coincides with region I, with binding being prevented by methylation at the GATC sites. At present it is unclear why binding of OxyR, which normally acts as a positive regulator of gene activity, to region I prevents *mom* gene transcription. As the OxyR-binding site overlaps 4 base pairs with the binding site of the Mu-encoded transactivator protein C (Fig. 9A), which immediately precedes the *mom* promoter (Bölker et al., 1989), two explanations appear likely (Bölker and Kahmann, 1989): OxyR binding could block access of protein C to its binding site, thus preventing *mom* gene expression. Alternatively both proteins might bind simultaneously with protein C, however, due to the interaction with OxyR, having lost the capacity to activate transcription.

It was previously assumed that the methylation dependent regulation of *mom* expression might delay Mom synthesis to a late stage in phage development, in which toxic effects of Mom-specific modification can no longer interfere with phage development (Kahmann et al., 1985). This view has been questioned by the finding that phage Mu burst sizes are not significantly altered in MomR/OxyR deficient mutants (Bölker and Kahmann, 1989). The biological significance of the MomR/OxyR mediated effect of Dam methylation on *mom* expression, hence, remains unclear.

Another mechanism underlying Dam methylation mediated gene control has been described in the context of Pap pili phase variation (Blyn et al., 1990, Braaten et al., 1991). In this case differential regulation is

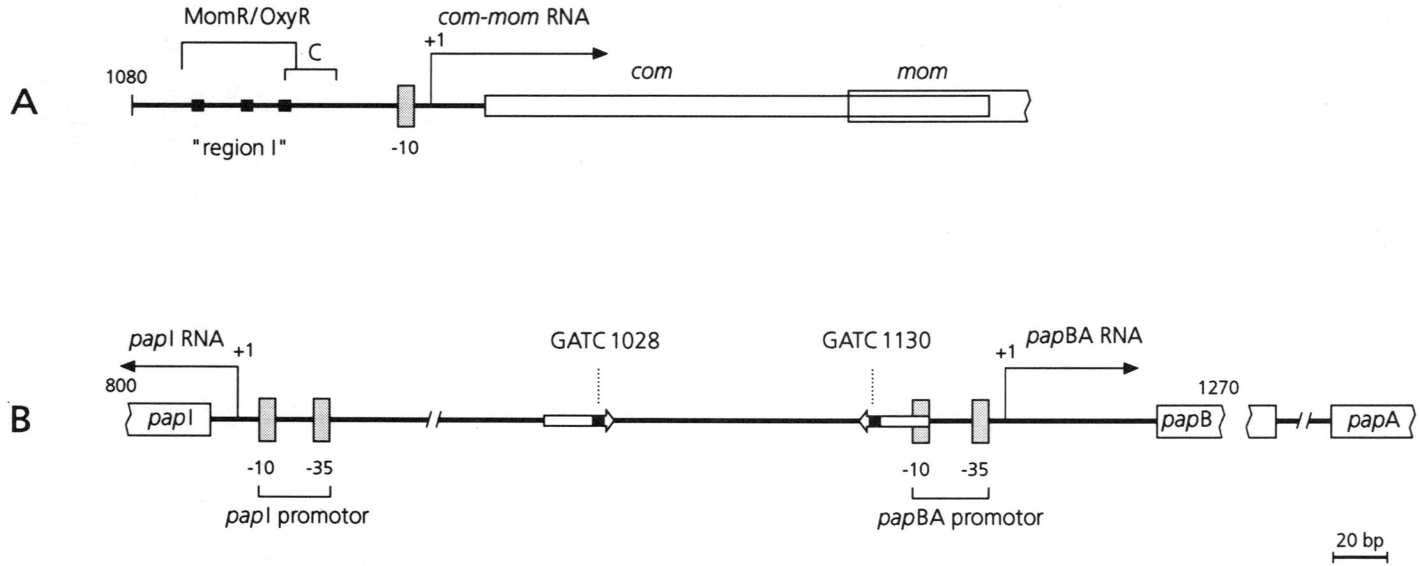

Figure 9. Schematic representation of the organization of the *mom* operon of phage Mu and the *pap*BA operon of uropathogenic *E. coli.* (A) Control region of the phage Mu *mom* operon. The control region preceding the overlapping *com* and *mom* genes, represented by open bars, is drawn as a solid line. Three Dam sites, located within the so-called region I, whose methylation is required for *mom* operon expression, are represented by black boxes. The site which becomes bound by the MomR/OxyR repressor of the *mom* operon when it is unmethylated and the binding site of a phage Mu encoded transactivator, C, required for expression, are indicated. The position of a sequence corresponding to a −10 promoter region is shown by a shaded box. A region with similarity to the consensus of −35 regions of *E. coli* promoters was not identified in the *mom* operon control region. The number at the left side of the region represented refers to the nucleotide sequence of this region (Kahmann, 1983). (B) Control region of the *pap*I gene and the *pap*BA operon of *E. coli.* Genes are represented by accordingly designated open bars. The control region of the *pap* I genes and the *pap* BA operon is represented by a solid line. −10 and −35 regions of the *pap*I gene and *pap*BA operon promoters are indicated by shaded boxes. Two nearly perfect inverted repeats of 27 bp, containing Dam sites (indicated as black boxes) relevant for *pap*BA expression, are represented by open arrows. Numbers at the beginning and the end of the truncated *pap*I and *pap*B genes, respectively, refer to the numbering of the corresponding nucleotide sequence position as used by Blyn et al. (1990).

achieved by changing a "pattern" of methylated Dam sites, a mechanism resembling in some aspects gene expression control attributed to eukaryotic DNA methylation.

Uropathogenic *E. coli* strains encoding pyelonephritis associated pili (Pap) alternate between expression (phase on) and non-expression (phase-off) states (Low et al., 1987). Oscillation of Pap pili expression between phase on and phase off states occurs at the level of transcription with the phase state being inheritable (Blyn et al., 1989). Alternations in phase states are not accompanied by DNA rearrangements (e.g. inversions as found in phase variation of *Salmonella typhimurium* flagella synthesis or expression of *E. coli* type 1 pili (Silverman et al., 1979; Freitag et al., 1985; Abraham et al., 1987) or base pair alterations in the control region of the *papBA* operon, which includes the Pap pilin encoding gene *papA* (Blyn et al., 1990). Further examination of the control region relevant for phase variation (Fig. 9B) revealed that it contained two GATC sites, termed $GATC_{1028}$ and $GATC_{1130}$, whose methylation status correlated with the phase state. In DNA of phase off cells $GATC_{1028}$ is fully methylated, while $GATC_{1130}$ is completely unmethylated. In DNA of phase on cells the pattern is converse (Blyn et al., 1990). A role of Dam methylation in the control of phase variation, suggested by these observations, was confirmed by the finding that the intracellular level of Dam methylation activity is critical for phase variation (Blyn et al., 1990). Strains deficient in M. Dam activity displayed a uniform phase off phenotype, overproduction of M. Dam prevented phase off to phase on transitions.

Two proteins affect accessibility of $GATC_{1028}$ or $GATC_{1130}$ to Dam methylation (Braaten et al., 1991). The product of the *papI* gene (Fig. 9B), a positive regulator of *pap* pilin transcription (Baga et al., 1985), contributes to inhibition of methylation at $GATC_{1028}$. The product of a chromosomal locus termed *mbf* (methylation blocking factor gene), mapping at 19.6 min in the *E. coli* standard map, is required for methylation protection of both $GATC_{1028}$ and $GATC_{1130}$ (Braaten et al., 1991). Mbf is furthermore essential for Pap pilus expression by positively regulating *pap* pilin transcription (Braaten et al., 1991). Mbf may be identical with Lrp, a global regulator of gene expression in *E. coli* (Newman et al., 1992). In the case of *pap* pilin expression, the positive action of Mbf appears to specificially correlate with preventing methylation at $GATC_{1028}$. Details on the mechanism implicating Mbf in the change of the methylation pattern at $GATC_{1028}$ and $GATC_{1130}$ are not yet known. Protein binding to hemi- or unmethylated $GATC_{1028}$ or $GATC_{1130}$ has been invoked in a model proposing a replication associated switch in the methylation status at these two GATC sites (Blyn et al., 1990).

It should be noted that different levels of gene expression observed in comparing *dam*$^+$ and *dam*$^-$ strains do not necessarily reflect a direct effect of Dam methylation on promoter or other gene control regions. Several genes are expressed at elevated levels in *dam*$^-$ mutant strains as a consequence of a slight induction of the SOS response due to increased DNA damage in such strains (Peterson et al., 1985).

4.7 Control of the initiation of phage P1 DNA packaging

Packaging of P1 DNA into phage capsids is initiated with the cleavage of phage DNA concatemers, consisting of repeated units of P1 DNA within a specific region termed *pac* (Bächi and Arber, 1977; Sternberg and Coulby, 1987a, 1987b). Starting at the site of *pac* cleavage, a headful of DNA is then packaged unidirectionally into the procapsid (Sternberg and Coulby, 1987a). Thereafter consecutive packaging events follow which initiate at the cleavage site terminating the preceding packaging event. As a headful of phage DNA represents DNA molecules with a size of 10 to 15% in excess of the complete viral genome (Yun and Vapnek, 1977; Sternberg, 1990), encapsidated P1 DNA molecules are terminally redundant, i.e. contain the same sequence at both ends. This redundancy is essential for vegetative phage growth, as it permits cyclization of the phage DNA injected into the host by host or phage encoded recombination functions (Segev et al., 1980; Segev and Cohen, 1981; Sternberg et al., 1986). Evidently, cleavage at every *pac* region in a phage DNA concatemer, which would produce unit length phage genomes lacking terminal redundancy, must be prohibited. Therefore the question arises as to what distinguishes the *pac* region at which packaging commences from all others.

The finding that the P1 *pac* region, consisting of about 160 bp, contains seven GATC sites (Sternberg and Coulby, 1987a, 1987b) and the fact that P1 encodes its own Dam MTase (Coulby and Sternberg, 1988), suggested that Dam methylation might be involved in the control of *pac* cleavage. This involvement was documented by several findings (Sternberg and Coulby, 1990). Firstly, the *pac* region could not be cleaved *in vivo* in the absence of P1- or host encoded Dam methylation activity. Secondly, the unmethylated *pac* region could not be cleaved *in vitro* by a "pacase extract", while fully methylated *pac* was readily cleavable. Thirdly, uncut *pac* regions contained in P1 virions, are, in contrast to the cut region, considerably undermethylated and, hence, essentially refractory to cleavage by the pacase extract. Based on these findings, Sternberg and Coulby (1990) proposed a model for Dam methylation dependent regulation of *pac* cleavage. Replication of P1 DNA generates hemi- or unmethylated *pac* DNA, resistant to cleavage. Methylation of one *pac* region to its fully methylated state defines the site at which DNA packaging initiates. Methylation of the majority of *pac* sequences is prevented by a host or P1 encoded protein binding to undermethylated *pac*. Pacase itself is suggested to be a candidate protein as it binds to, but does not cleave, hemimethylated *pac* sites (Sternberg, unpublished work, cited in Sternberg and Coulby, 1990). Cleavage of the single packaging initiating sequence within a P1 DNA concatemer might be accounted for by the competition between methylation and *pac* site binding to produce one functional *pac* site for every three to five nonfunctional sites.

4.8 The significance of the intracellular level of Dam methyltransferase

In general, Dam-mediated regulation depends on the transient undermethylation of replicated DNA. Such transient undermethylation critically depends on the intracellular level of Dam MTase (about 130 molecules are contained in a rapidly growing *E. coli* wt cell, Boye et al., 1992). This is evident from the regulatory perturbations discussed in the previous sections caused by M.Dam overproduction. It is therefore conceivable that expression of the *dam* gene is subject to specific control mechanisms ensuring its maintenance at the level characteristic for a wt cell. Although details on the regulation of *dam* gene expression are not yet known, a recent report (Løbner-Olesen et al., 1992) suggests a complex situation.

The *dam* gene is part of an operon including the *aroK* and *aroB* genes and another open reading frame, *urf74.3*, whose function is unknown (Løbner-Olesen et al., 1992). The gene order, starting at the beginning of the operon, is *aroK*, *aroB*, *urf74.3* and *dam*. According to Løbner-Olesen et al. (1992) five promoters, P1 to P5, located in front or within the operon contribute to *dam* gene expression. With the exception of P5, the -10 and -35 regions of all promoters closely resemble the consensus of -10 and -35 regions in promoters recognized by the "conventional" σ^{70}-RNA polymerase. The finding that the different promoters are not related beyond the -10 to -35 regions has been taken as an indication that they might be subject to different types of control. It will be interesting to see what type of intracellular information is integrated via this regulatory scenario to govern *dam* gene expression.

4.9 Evolution of Dam methylation

Screening a large number of bacteria, whose genealogy had independently been determined, the capacity for Dam methylation was found to be confined to two non-related clusters of gram-negative eubacteria (Barbeyron et al., 1984). One cluster is represented by cyanobacteria, the other by the related families *Enterobacteriaceae* (which includes *E.*

coli), *Parvobacteriaceae* and *Vibrionaceae*. Dam methylation apparently evolved independently in the two clusters, as it is mediated by different genes (Brooks et al., 1983), and originated late in the *E. coli* lineage (Barbeyron et al., 1984).

The confinement of Dam methylation to only a few families of gram-negative bacteria raises the question of how other species regulate processes whose control in *E. coli* depends on Dam methylation. An obvious possibility would be that MTases of different specificity might play similar roles in other species as M. Dam in *E. coli*. However, no function comparable to that of M. Dam has as yet been documented for any other prokaryotic MTase. Further, processes as those discussed above also have to be regulated in organisms that lack any detectable methylation.

An alternative to the Dam/MutH mediated instruction of postreplicative mismatch repair in *E. coli* has been described in the analysis of mismatch repair in *S. pneumoniae*. Mismatch repair in this organism has been found to be directed by nonspecific single strand nicks, leading to the suggestion that free termini in newly synthesized DNA strands might generally provide a means to direct mismatch repair (Lacks et al., 1982; Manarelli et al., 1985; Claverys and Lacks, 1986). The recent finding that mismatch repair by nuclear extracts of *Drosophila melanogaster* or human cells, which resembles by various criteria postreplicative strand specific repair in bacteria, can be efficiently directed by DNA nicks (Holmes et al., 1990) suggests that in species other than *E. coli* free termini in newly synthesized DNA may indeed provide the required signals for strand discrimination in mismatch repair.

Initiation of DNA replication, another elementary process regulated by Dam methylation in *E. coli*, may be controlled in other species by means similar to those documented for the methylation-independent stringent control of replication of low-copy-number plasmids, e.g. the *E. coli* F factor (Scott, 1984). Alternatives to regulate other less elementary processes, such as phase variation in gene expression or initiation of phage DNA packaging, have been described in *E. coli* and other species.

Considering these alternatives one may ask why Dam methylation gained such a prominent role in *E. coli*. A possible evolutionary scenario might be that Dam methylation had at some stage in evolution become involved in the regulation of one elementary process in *E. coli*, e.g. postreplicative mismatch repair or control of replication. This role would have provided a strong selective pressure for maintaining Dam methylation (in fact it is the most conserved methylation activity among *E. coli* strains and related species). As the activity of M. Dam, even if initially relevant for control of only one cellular process, is not confined to a specific genome region or genetic entity, it might have been implicated in subsequent steps of evolution in the control of multiple cellular activities. The great variety of processes affected by Dam methylation may thus represent an example par excellence of the functional potential of epigenetic information imposed by DNA methylation.

5 Deamination of 5-methylcytosine to thymine: Very short patch (VSP) repair in *E. coli*

DNA methylation not only provides a means to superimpose secondary information on the fixed primary information of the DNA sequence, it can also be the cause of alterations in its primary informational content. This is due to the inherent instability of cytosine and its methylated derivative 5-methylcytosine in DNA, which undergo deamination to uracil and thymine, respectively, at a considerable rate under physiological conditions (Lindahl and Nyberg, 1974, Ehrlich et al., 1986). U/G mismatches resulting from C deamination can be corrected by a repair process initiating with the removal of uracil by the action of uracil-glycosylases (Lindahl, 1974). Deamination of 5-mC, however, generates thymine and thus T/G base mispairs consisting of two natural DNA bases. Therefore, instruction of repair pathways counteracting the mutational potential of 5-mC deamination cannot derive from the DNA atypic quality of one of the bases constituting the mismatch. Neither can it derive from general features discriminating the mutationally altered from the non-altered strand (as in postreplicative mismatch repair), as T/G mismatches resulting from 5-mC deamination largely occur in nonreplicating DNA and would actually give rise to stable mutations upon replication.

Information may, however, derive from the fact that 5-mC and also T/G mispairs resulting from its deamination, are usually present in specific DNA sequence contexts. These are determined by the target specificity of the MTase which initially catalyzed the formation of 5-mC in DNA. An example of how the specific sequence context can be used to direct repair of the type of T/G mismatches discussed has been provided by analyses of *E. coli* K-12.

In *E. coli* K-12 cytosine methylation is due the activity of the Dcm methyltransferase, which modifies the inner cytosine of the DNA target sequence CC(A/T)GG (Marinus and Morris, 1973; Schlagman et al., 1976). The function of this MTase is not known.

A report by Coulondre et al. (1978) showed that the methylated cytosine in Dcm target sites is subject to spontaneous deamination. The analysis of single base mutations generating nonsense codons in the *E. coli lacI* gene revealed mutation hotspots of C/G to T/A transitions, which coincided with the second cytosine in Dcm sites of the *lacI* gene. No such hotspots were observed in *E. coli* B which lacks Dcm methylation. Analyses of the anomalously frequent recombination of certain amber mutations in the phage λ repressor gene, *cI*, with nearby mutations pointed to the existence of a specific repair pathway correcting T/G mismatches in the Dcm target sequence CC(A/T)GG and in the related sequences CAGG and CCAG (Lieb, 1983; Lieb, 1985; Lieb et al., 1986). This repair pathway is characterized by very short patches of DNA repair synthesis (with repair tracts comprising about 10 nucleotides) and was, hence, termed VSP (Lieb, 1983; Lieb, 1985; Lieb et al., 1986). Subsequent experiments aimed at identifying functions involved in VSP repair revealed that the process depended on DNA polymerase I (Dzidic and Radman, 1989) and was strongly stimulated by MutS and MutL (Jones et al., 1987; Lieb, 1987; Zell and Fritz, 1987), two proteins also acting in postreplicative mismatch repair (see 4.2). Other components of the postreplicative mismatch repair pathway, e.g. MutH and MutU (DNA helicase II) were not required, consistent with the fact that VSP repair occurred preferentially in nonreplicating DNA and extended only over very short nucleotide tracts. Of specific relevance for VSP repair is the product of a recently identified gene termed *vsr* (Sohail et al., 1990). *vsr* is located immediately 3′ to the *dcm* gene with both genes, organized as an operon, actually overlapping for a short stretch. Integrity of the *vsr* gene was absolutely required to restore VSP repair in repair deficient mutants (Sohail, et al., 1990).

The *vsr* gene product has recently been purified and shown to be a strand- and sequence-specific DNA mismatch endonuclease (Hennecke et al., 1991). It generates single-strand breaks immediately 5′ to the underlined T in CT(AT)G or T(A/T)GG sequence provided the T is mispaired with a G. The DNA target specificity of Vsr is entirely consistent with previous genetic data indicating that the Dcm target site CC(A/T)GG and related sites lacking either of the two peripheral base pairs are subject to VSP repair if 5-mC at the position underlined has been deaminated to T (Lieb et al., 1986). The cleavage specificity of the Vsr endonuclease ensures that VSP repair occurs in favor of the G-containing strand. Thus an essential element in directing repair of T/G mismatches arising from deamination of 5-mC, which may also be operative in other repair pathways counteracting the mutagenic potential of 5-mC in DNA, appears to be the specificity of an endonuclease initiating repair for a DNA sequence which is the target of a C5-MTase and, hence, prone to contain the type of mismatch discussed.

Acknowledgements. A first draft of this manuscript was written during the authors' stay at Stift Geras, Austria. We thank the abbot of the monastery Prälat Professor D. Dr. J. Angerer for his generous and understanding hospitality. We thank J. Brooks, A. Kiss, D. Krüger, M. Marinus and N. Murray for critical comments on the manuscript and G. Wilson for the communication of unpublished results. We acknowledge critical comments on the manuscript from members of our department. We thank Regina Bachman for editorial help. Research in the authors' laboratories was in part funded by the Deutsche Forschungsgemeinschaft (Tr 10/1-3 and SFB 344) and the EEC (Contract No. ST2J-0034-2-D).

Note added in proof

After submission of this manuscript, G. Wilson (personal communication) has informed us that he discovered open reading frames with similarity to the *vsr* gene localized in the vicinity of the *dcm* gene (see chapter 5) next to the C5-MTase genes of R/M systems *Alu*I (AGCT), *Bsu*RI (GGCC), and *Hpa*II (CCGG). These observations suggest that also in cases of C5-MTases other than Dcm mismatch repair enzymes have co-evolved, which specifically counteract the mutagenic effect resulting from deamination of 5-mC.

Abeles, A. (1986) P1 plasmid replication: purification and DNA-binding activity of the replication protein RepA J. Biol. Chem. *261*, 3548–3555.

Abeles, A. L., and Austin, S. J. (1987) P1 plasmid replication requires methylated DNA. EMBO J. *6*, 3185–3189.

Abeles, A., Snyder, K., and Chattoraj, D. (1984) P1 plasmid replication: replication structure. J. Mol. Biol. *173*, 307–324.

Abraham, J. M., Freitag, C. S., Clements, J. R., and Eisenstein, B. I. (1985) An invertible element of DNA controls phase variation of type 1 fimbriae of *Escherichia coli*. Proc. Natl. Acad. Sci. USA *82*, 5724–5727.

Arber, W., and Linn, S. (1969) DNA modification and restriction. Ann. Rev. Biochem, *38*, 467–500.

Bächi, B., and Arber, W. (1977) Physical mapping of *Bgl*II, *Bam*HI, *Eco*RI, *Hin*dIII and *Pst*I restriction fragments of bacteriophage P1 DNA. Mol. Gen. Genet. *153*, 311–324.

Bächi, B., Reiser, J., and Pirrotta, V. (1979) Methylation and cleavage of sequences of the *Eco*P1 restriction-modification enzyme. J. Mol. Biol. *128*, 143–163.

Bachmann, B. J. (1987) Linkage map of *Escherichia coli* K-12, in: *Escherichia coli* and *Salmonella typhimurium*. Cellular and Molecular Biology, pp. 807–876. Eds F. C. Neidhardt, L. Ingraham, K. Brooks Low, B. Magasanik, M. Schaechter and H. E. Umbarger. American Society for Microbiology, Washington, D. C.

Baga, M., Goransson, M., Normark, S., and Uhlin, B. E. (1985) Transcriptional activiation of a *pap* pilus virulence operon from uropathogenic *Escherichia coli*. EMBO J. *4*, 3887–3893.

Bakker, A., and Smith D. W. (1989) Methylation of GATC sites is required for precise timing between rounds of DNA replication in *Escherichia coli*. J. Bacteriol. *171*, 5738–5742.

Bale, A., d'Alarcao, M., and Marinus, M. G. (1979) Characterization of DNA adenine methylation mutants of *Escherichia coli* K12. Mutation Res. *59*, 157–165.

Balganesh, T. S., Reiners, L. Lauster, R., Noyer-Weidner, M., Wilke, K. and Trautner, T. A. (1987) Construction and use of chimeric SPR/Ø3T DNA methyltransferases in the definition of sequence recognizing enzyme regions. EMbo J. *6*, 3543–3549.

Barany, F., Danzits, M., Zebala, J., and Mayer, A. (1992) Cloning and sequencing of genes encoding the *Tth*HB81 DNA restriction and modification enzymes: comparison with the isoschizomeric *Taq*I enzymes. Gene *112*, 3–12.

Barbeyron, T., Kean, K., and Forterre, P. (1984) DNA adenine methylation of GATC sequences appeared recently in the *Escherichia coli* lineage. J. Bacteriol. *160*, 586–590.

Barras, F., and Marinus, M. G. (1989) The great GATC:DNA methylation in *E. coli*. Trends Genet. *5*, 139–143.

Bauer, J., Krämmer, G., and Knippers, R. (1981) Asymmetric repair of bacteriophage T7 heteroduplex DNA. Mol. Gen. Genet. *181*, 541–547.

Bazaral, M., and Helinski, D. R. (1970) Replication of a bacterial plasmid and an episome in *Escherichia coli*. Biochemistry *9*, 399–406.

Behrens, B., Noyer-Weidner, M., Pawlek, B. Lauster R., Balganesh, T. S., and Trautner, T. A. (1987) Organization of multispecific DNA methyltransferases encoded by temperate *Bacillus subtilis* phages. EMBO J. *6*, 1137–1142.

Berg, D. E. (1989) Transposon Tn5, in: Mobile DNA, pp. 186–210. Eds D. E. Berg and M. M. Howe. American Society for Microbiology, Washington, D. C.

Bergerat, A., and Guschlbauer, W. (1990) The double role of methyl donor and allosteric effector of S-adenosyl-methionine for Dam methylase of *E. coli*. Nucl. Acids Res. *18*, 4369–4375.

Bergerat, A., Guschlbauer, W., and Fazakerley, V. (1991) Allosteric and catalytic binding of S-adenosylmethionine to *Escherichia coli* DNA adenine methyltransferase monitored by ^{3}H NMR. Proc. Natl. Acad. Sci USA *88*, 6394–6397.

Bertani, G., and Weigle, J. J. (1953) Host controlled variations in bacterial viruses. J. Bacteriol. *65*, 113–121.

Bestor, T. H., and Ingram, V. M. (1983) Two DNA methyltransferases from murine erythroleukemia cells: purification, sequence specificity, and mode of interaction with DNA. Proc. Natl. Acad. Sci. USA *80*, 5559–5563.

Bestor, T., Laudano, A., Mattaliano, R., and Ingram, V. (1988) Cloning and sequencing of a cDNA encoding DNA methyltransferase of mouse cells: the carboxy-terminal domain of the mammalian enzyme is related to bacterial restriction methyltransferases. J. Mol. Biol. *203*, 971–983.

Bickle, T. A. (1987) DNA restriction and modification systems. In: *Escherichia coli* and *Salmonella typhimurium*. Cellular and Molecular Biology, pp. 692–696. Eds F. C. Neidhardt, L. Ingraham, K. Brooks Low, B. Magasanik, M. Schaechter and H. E. Umbarger. American Society for Microbiology, Washington, D. C.

Bickle, T. A., Brack, C., and Yuan, R. (1978) ATP-induced conformational changes in the restriction endonuclease from *Escherichia coli* K-12. Proc. Natl. Acad. Sci. USA *75*, 3099–3103.

Blumenthal, R. M., Gregory, S. A., and Cooperider, J. S. (1985) Cloning of a restriction-modification system from *Proteus vulgaris* and its use in analyzing a methylase-sensitive phenotype in *Escherichia coli*. J. Bacteriol. *164*, 501–509.

Blyn, L. B., Braaten, B. A., and Low, D. A. (1990) Regulation of *pap* pilin phase variation by a mechanism involving differential Dam methylation states. EMBO J. *9*, 4045–4054.

Blyn, L. B., Braaten, B. A., White-Ziegler, C. A., Rolfson, D. A., and Low, D. A. (1989) Phase-variation of pyelonephritis-associated pili in *Escherichia coli*: evidence for transcriptional regulation. EMBO J. *8*, 613–620.

Bocklage, H., Heeger, K., and Müller-Hill, B. (1991) Cloning and characterization of the *Mbo*II restriction-modification system. Nucl. Acids Res. *19*, 1007–1013.

Bölker, M., and Kahmann, R. (1989) The *Escherichia coli* regulatory protein OxyR discriminates between methylated and unmethylated states of the phage Mu *mom* promoter. EMBO J. *8*, 2403–2410.

Bölker, M., Wulczyn, F. G., and Kahmann, R. (1989) Role of bacteriophage Mu C protein in activiation of the *mom* gene promoter. J. Bacteriol. *171*, 2019–2027.

Bolivar, F., Rodriguez, R. L., Greene, P. J., Betlach, M. C., Heynecker, H. L., Boyer, H. W., Crosa, J. H., and Falkow, S. (1977) Construction and characterization of new multipurpose cloning vehicles. II. A multipurpose cloning system. Gene *2*, 95–113.

Borck, K. Beggs, J. D., Brammar, W. J., Hopkins, A. S., and Murray, N. E. (1976) The construction *in vitro* of transducing derivatives of phage lambda. Mol. Gen. Genet. *146*, 199–207.

Bougueleret, L., Schweizstein, M., Tsugita, A., and Zabeau, M. (1984) Characterization of the genes coding for the *Eco*RV restriction and modification system of *Escherichia coli*. Nucl. Acids Res. *12*, 3659–3676.

Boye, E., Løbner-Olesen, A., and Skarstad, K. (1988) Timing of chromosomal replication in *Escherichia coli*. Biochim. Biophys. Acta **951,** 359–364.

Boye, E., and Løbner-Olesen, A. (1990) The role of *dam* methyltransferase in the control of DNA replication in *E. coli*. Cell *62*, 981–989.

Boye, E., Marinus, M. G., and Løbner-Olesen, A. (1992) Quantitation of *dam* methyltransferase in *Escherichia coli*. J. Bacteriol. *174*, 1682–1685.

Boyer, H. W., and Roulland-Dussoix, D. (1969) A complementation analysis of the restriction and modification of DNA in *Escherichia coli*. J. Mol. Biol. *41*, 459–472.

Braaten, B. A., Blyn, L. B., Skinner, B. S., and Low, D. A. (1991) Evidence for a methylation-blocking factor (*mbf*) locus involved in *pap* pilus expression and phase variation in *Escherichia coli*. J. Bacteriol. *173*, 1789–1800.

Bramhill, D., and Kornberg, A. (1988) A model for initiation at origins of DNA replication. Cell *54*, 915–918.

Brendler, T., Abeles, A., and Austin, S. (1991) Critical sequences in the core of the P1 plasmid replication origin. J. Bacteriol. *173*, 3935–3942.

Brody, H., and Hill, C. W. (1985) Attachment site of the genetic element *e14*. J. Bacteriol. *170*, 2040–2044.

Brooks, J. E., Blumenthal, R. M., and Gingeras, T. R. (1983) The isolation and characterization of the *Escherichia coli* DNA adenin methylase (*dam*) gene. Nucl. Acids Res. *11*, 837–851.

Brooks, J. E., Benner, J. S., Heiter, D. F., Silber, K. S., Sznyter, L. A., Jager-Quinton T., Moran, L. S., Slatko, B. E., Wilson, G. G., and Nwankwo, D. O. (1989) Cloning the *Bam*HI restriction modification system. Nucl. Acids Res. *17*, 979–997.

Brooks, J. E., Nathan, P. D., Landry, D., Szynter, L. A., Waite-Rees, P., Ives, C. L., Moran, L. S., Slatko, B. E., and Benner, J. S. (1991) Characterization of the cloned *Bam*HI restriction modification system: its nucleotide sequence, properties of the methylase, and expression in heterologous hosts. Nucl. Acids Res. *19*, 841–850.

Buhk, H.-J., Behrens, B., Tailor, R., Wilke, K., Prada, J. J., Günthert, U., Noyer-Weidner, M., Jentsch, S., and Trautner, T. A. (1984) Restriction and modification in *Bacillus subtilis*: nucleotide sequence, functional organization and product of the DNA methyltransferase gene of bacteriophage SPR. Gene *29*, 51–61.

Bullas, L. R., and Colson, C. (1975) DNA restriction and modification systems in *Salmonella*. III. SP, a *Salmonella potsdam* system allelic to the SB system in *Salmonella typhimurium*. Mol. Gen. Genet. *139*, 177–188.

Burckhardt, J., Weisemann, J., Hamilton, D. L., and Yuan, R. (1981) Complexes formed between the restriction endonuclease *Eco*K and heteroduplex DNA. J. Mol. Biol. *153*, 425–440.

de la Campa, A. G., Kale, P., Springhorn, S. S. and Lacks, S. A. (1987) Proteins encoded by the *Dpn*II restriction gene cassette. Two methylases and an endonuclease. J. Mol. Biol. *196*, 457–469.

Campbell, J. L., and Kleckner, N. (1990) *E. coli oriC* and the *dna*A gene promoter are sequestered from *dam* methyltransferase following the passage of the chromosomal replication fork. Cell *62*, 967–979.

Card, C. O., Wilson, G. G., Weule, K., Hasapes, J., Kiss, A., and Roberts, R. J. (1990) Cloning and characterization of the *Hpa*II methylase gene. Nucl. Acids Res. *18*, 1377–1383.

Cerritelli, S., Springhorn, S. S., and Lacks, S. A. (1989) *Dpn*A, a methylase for single-stranded DNA in the *Dpn*II restriction system and its biological function. Proc. Natl. Acad. Sci. USA *86*, 9223–9227.

Chandrasegaran, S., and Smith, H. O. (1988) Amino acid sequence homologies among twenty-five restriction endonucleases and methylases, in: Structure and Expression, vol. 1, pp. 149–156. Eds R. H. Sarma and M. H. Sarma. Adenine, Guilderland, N.Y.

Chandrasegaran, S., Lunnen, K. D., Smith, H. O., and Wilson, G. G. (1988) Cloning and sequencing of the *Hin*fI restriction and modification genes. Gene *70*, 387–392.

Chen, L., MacMillan, A. M., Chang, W., Ezaz-Nikpay, K., Lane, W. S., and Verdine, G. L. (1991) Direct identification of the active-site nucleophile in a DNA (Cytosine-5)-methyltransferase. Biochemistry *30*, 11018–11025.

Christman, M. F., Storz, G., and Ames, B. N. (1989) OxyR, a positive regulator of hydrogen peroxide-inducible genes in *Escherichia coli* and *Salmonella typhimurium*, is homologous to a family of bacterial regulatory proteins. Proc. Natl. Acad. Sci. USA *86*, 3484–3488.

Claverys, J.-P., and Lacks, S. A. (1986) Heteroduplex deoxyribonucleic acid base mismatch repair in bacteria. Micobiol. Reviews *50*, 133–165.

Collins, M., and Myers, R. (1987) Alterations in DNA helix stability due to base modifications can be evaluated using denaturing gradient electrophoresis. J. Mol. Biol. *198*, 737–744.

Connaughton, J. F., Kaloss, W. D., Vanek, P. G., Nardone, G. A., and Chirikjian, J. G. (1990) The complete sequence of the *Bacillus amyloliquefacians* proviral H2, *Bam*HI methylase gene. Nucl. Acids Res. *18*, 4002.

Coulby, J. N., and Sternberg, N. L. (1988) Characterization of the phage P1 *dam* gene. Gene *74*, 191.

Coulondre, C., Miller, J. H., Farabaugh, P. J., and Gilbert, W. (1978) Molecular basis of substitution hotspots in *Escherichia coli*. Nature *274*, 775–780.

Cowan, G. M., Gann, A. A. F., and Murray, N. E. (1989) Conservation of complex DNA recognition domains between families of restriction enzymes. Cell *56*, 103–109.

Deschavanne, P., and Radman, M. (1991) Counterselection of GATC sequences in enterobacteriophages by the components of the methyl-directed mismatch repair system. J. Mol. Evol. *33*, 125–132.

Dharmalingam, K., and Goldberg, E. B. (1976) Phage coded protein prevents restriction of unmodified progeny T4 DNA. Nature *260*, 454–456.

Dharmalingam, K., Revel, H. R., and Goldberg, E. B. (1982) Physical mapping and cloning of the bacteriophage T4 anti-restriction endonuclease gene. J. Bacteriol. *149*, 694–699.

Dila, D., Sutherland, E., Moran, L., Slatko, B., and Raleigh, E. A. (1990) Genetic and sequence organization of the *mcrBC* locus of *Escherichia coli* K-12. J. Bacteriol. *172*, 4888–4900.

Dodson, K. W., and Berg, D. E. (1989) Factors affecting transposition activity of IS*50* and Tn*5* ends. Gene *76*, 207–213.

Düsterhöft, A., Erdmann, D., and Kröger, M. (1991) Stepwise cloning and molecular characterization of the *Hgi*DI restriction-modification system from *Herpetosiphon giganteus Hpa2*. Nucl. Acids Res. *10*, 1049–1056.

Dzidiz, S., and Radman, M. (1989) Genetic requirements for hyper-recombination by very short patch mismatch repair: involvement of *Escherichia coli* DNA polymerase I. Mol. Gen. Genet. *217*, 254–256.

Ehrlich, M., Norris, K. F., Wang, R. Y.-H., Kuo, K. C., and Gehrke, C. W. (1986) DNA cytosine methylation and heat-induced deamination. Biosci. Rep. *6*, 387–393.

Engel, J. D., and von Hippel, P. H. (1978) Effects of methylation on the stability of nucleic acid conformations: studies at the polymer level. J. Biol. Chem. *253*, 927–934.

Freitag, C. S., Abraham, J. M., Clements, J. R., and Eisenstein, B. I. (1985) Genetic analysis of the phase variation control of expression of type 1 fimbriae in *Escherichia coli*. J. Bacteriol. *162*, 668–675.

Friedman, S. (1985) The irreversible binding of azacytosine-containing DNA fragments to bacterial DNA (cytosine-5)methyltransferases. J. Biol. Chem. *260*, 5698–5705.

Friedman, S. (1986) Binding of the *Eco*RII methylase to azacytosine-containing DNA. Nucl. Acids. Res. *14*, 4543–4556.

Fuller-Pace, F. V., Bullas, L. R., Delius, H., and Murray, N. E. (1984) Genetic recombination can generate altered restriction specificity. Proc. Natl. Acad. Sci. USA *81*, 6095–6099.

Fuller-Pace, F. V., and Murray, N. E. (1986) Two DNA recognition domains of the specificity polypeptides of a family of type I restriction enzymes. Proc. Natl. Acad. Sci. USA *83*, 9368–9372.

Gann, A. A. F., Campbell, A. J. B., Collins, J. F., Coulsond, A. F. W., and Murray, N. E. (1987) Reassortment of DNA recognition domains and the evolution of new specificities. Mol. Micobiol. *1*, 13–22.

Gammie, A. E., and Crosa, J. H. (1991) Roles of DNA adenine methylation in controlling replication of the REPI replicon of plasmid pColV-K30. Mol. Microbiol. *5*, 495–503.

Geier, G. E., and Modrich, P. (1979) Recognition sequence of the *dam* methylase of *Escherichia coli* K12 and mode of cleavage of the *Dpn*I endonuclease. J. Biol. Chem. *254*, 1408–1413.

Glickman, B. W. (1979) Spontaneous mutagenesis in *Escherichia coli* strains lacking 6-methyladenine residues in their DNA. An altered mutational spectrum in *dam* mutants. Mutat. Res. *61*, 153–162.

Glickman, B. W., and Radman, M. (1980) *Escherichia coli* mutator mutants deficient in methylation-instructed DNA mismatch correction. Proc. Natl. Acad. Sci. USA *77*, 1063–1067.

Glickman, B., Van den Elsen, P., and Radman, M. (1978) Induced mutagenesis in *dam*$^-$ mutants of *Escherichia coli*: a role for 6-methyladenine residues in mutation avoidance. Mol. Gen. Genet. *163*, 307–312.

Glover, S. W., and Colson, C. (1969) Genetics of host-controlled restriction and modification in *Escherichia coli*. Genet. Res. *13*, 227–240.

Gough, J. A., and Murray, N. E. (1983) Sequence diversity among related genes for recognition of specific targets in DNA molecules. J. Mol. Biol. *166*, 1–19.

Graham, M. W., Doherty, J. P., and Woodcock, D. M. (1990) Efficient construction of plant genomic libraries requires the use of *mcr*$^-$ host strains and packaging mixes. Plant Mol. Biol. Rep. *8*, 33–42.

Grant, S. G. N., Jessee, J., Bloom, F. R., and Hanahan, D. (1990) Differential plasmid rescue from transgenic mouse DNAs into *Escherichia coli* methylation-restriction mutants. Proc. Nat. Acad. Sci. USA *87*, 4645–4649.

Greene, P. J., Gupta, M., Boyer, H. W., Brown, W. E., and Rosenberg, J. M. (1981) Sequence analysis of the DNA encoding the *Eco*RI endonuclease and methylase. J. Biol. Chem. *256*, 2143–2153.

Greener, A., and Hill, C. W. (1980) Identification of a novel genetic element in *Escherichia coli* K-12. J. Bacteriol. *144*, 312–321.

Gruenbaum, Y., Cedar, H., Razin, A. (1982) Substrate and sequence specificity of a eukaryotic DNA methylase. Nature *295*, 620–622.

Gubler, M., and Bickle, T. A. (1991) Increased protein flexibility leads to promiscuous protein-DNA interactions in type IC restriction-modification systems. EMBO J. *10*, 951–957.

Hadi, S. M., Bächi, B., Iida, S., and Bickle, T. A. (1983) DNA restriction-modification enzymes of phage P1 and plasmid p15B. Subunit functions and structural homologies. J. Mol. Biol. *165*, 19–34.

Hadi, S. M., Bächi, B., Shepherd, J. C. W., Yuan, R., Ineichen, K., and Bickle, T. A. (1979) DNA recognition and cleavage by the *Eco*P15 restriction endonuclease. J. Mol. Biol. *134*, 655–666.

Halling, S. M., Simons, R. W., Way, J. C., Walsh, R. B., and Kleckner, N. (1982) DNA sequence organization of Tn*10*'s IS*10*-Right and comparison with IS*10*-Left. Proc. Natl. Acad. Sci. USA *79*, 2608–2612.

Hanck, T., Gerwin, N., and Fritz, H. (1989) Nucleotide sequence of the *dcm* locus of *Escherichia coli* K12. Nucl. Acids Res. *17*, 5844.

Hardy, L. W., Finer-Moore, J. S., Montfort, W. R., Jones, M. O., Santi, D. V., and Stroud, R. M. (1987) Atomic structure of thymidylate synthase: target for rational drug design. Science, *235*, 448–455.

Harrison, S. C. (1991) A structural taxonomy of DNA-binding domains. Nature *353*, 715–719.

Hattman, S. (1978) Sequence specificity of the wild-type (dam^{+}) and mutant (dam^{h}) forms of bacteriophage T2 DNA adenine methylase. J. Mol. Biol. *119*, 361–376.

Hattman, S. (1982) DNA methyltransferase-dependent transcription of the phage Mu *mom* gene. Proc. Natl. Acad. Sci. USA *79*, 5518–5521.

Hattman, S., and Ives, J. (1984) S1 nuclease mapping of the phage Mu *mom* gene promoter: a model for the regulation of *mom* expression. Gene *29*, 185–198.

Hattman, S., Brooks, J. E., and Masurekar, M. (1978) Sequence specificity of the P1-modification methylase (M.*Eco*P1) and the DNA methylase (M.*Eco*dam) controlled by the *E. coli dam* gene. J. Mol. Biol. *126*, 367–380.

Hattman, S., Ives, J., Margolin, W., and Howe, M. M. (1985) Regulation and expression of the bacteriophage Mu *mom* gene; mapping of the transactivation (Dad) function to the *C* region. Gene *39*, 71–76.

Heisig, P., and Kahmann, R. (1986) The sequence and *mom*-transactivation function of the *C* gene of bacteriophage Mu. Gene *43*, 59–67.

Heitman, J., and Model, P. (1987) Site-specific methylases induce the SOS DNA repair response in *Escherichia coli*. J. Bacteriol. *169*, 3243–3250.

Helmstetter, C. E., and Leonard, A. C. (1987) Coordinate initiation of chromosome and minichromosome replication in *Escherichia coli*. J. Bacteriol. *169*, 3489–3494.

Hennecke, F., Kolmar, H., Bründl, K., and Fritz, H.-J. (1991) The *vsr* gene product of *E. coli* K-12 is a strand- and sequence-specific DNA mismatch endonuclease. Nature *353*, 776–778.

Herman, G. E., and Modrich, P. (1981) *Escherichia coli* K-12 clones that overproduce *dam* methylase are hypermutable. J. Bacteriol. *145*, 644–646.

Herman, G. E., and Modrich, P. (1982) *Escherichia coli dam* methylase. Physical and catalytic properties of the homogeneous enzyme. J. Biol. Chem. *257*, 2605–2612.

Hickson, I. D., Arthur, H. M., Bramhill, D., and Emmerson, P. T. (1983) The *E. coli uvr*D gene product is DNA helicase II. Mol. Gen. Genet. *190*, 265–270.

Hill, C., Miller, L. A., and Klaenhammer, T. R. (1991). *In vivo* genetic exchange of a functional domain from a type IIs methylase between lactococcal plasmid PTR 2030 and a virulent bacteriophage. J. Bacteriol. *173*, 4363–4370.

Hiom, K., and Sedgwick, S. G. (1991) Cloning and structural characterization of the *mcrA* locus of *Escherichia coli*. J. Bacteriol. *173*, 7368–7373.

Holmes, Jr., J., Clark, S., and Modrich, P. (1990) Strand-specific mismatch correction in nuclear extracts of human and *Drosophila melanogaster* cell lines. Proc. Natl. Acad. Sci. USA *87*, 5837–5841.

Hümbelin, M., Suri, B., Rao, D. N., Hornby, D. P., Eberle, H., Pripfl, T., Kenel, S., and Bickle, T. A. (1988) Type II DNA restriction and modification systems *Eco*P1 and *Eco*P15: Nucleotide sequence of the *Eco*P1 operon, the *Eco*P15 *mod* gene and some *Eco*P1 *mod* mutants. J. Mol. Biol. *200*, 23–29.

Hughes, P., Squali-Houssaini, F.-Z., Forterre, P., and Kohiyama, M. (1984) *In Vitro* replication of a *dam* methylated and non-methylated *ori*-C plasmid. J. Mol. Biol. *176*, 155–159.

Ingrosso, D., Fowler, A. V., Bleibaum, J., and Clarke, S. (1989) Sequence of the D-Aspartyl/L-Isoaspartyl protein methyltransferase from human erythrocytes. J. Biol. Chem. *264*, 20131–20139.

Jacob, F., Brenner, S., and Cuzin, F. (1963) On the regulation of DNA replication in bacteria. Cold Spring Harbor Symp. Quant. Biol. *28*, 329–348.

Jones, M., and Wagner, R. (1981) N-methyl-N′-nitro-N-nitrosoguanidine sensitivity of *E. coli* mutants deficient in DNA methylation and mismatch repair. Mol. Gen. Genet. *184*, 562–563.

Jones, M., Wagner, R., and Radman, M. (1987) Mismatch repair of deaminated 5-methyl-cytosine. J. Mol. Biol. *194*, 155–159.

Kahmann, R. (1983) Methylation regulates the expression of a DNA-modification function encoded by bacteriophage Mu. Cold Spring Harbor Symp. Quant. Biol. *47*, 639–646.

Kahmann, R. (1984) The *mom* gene of bacteriophage Mu. Curr. top. Microbiol. Immunol. *108*, 29–47.

Kahmann, R., Seiler, K., Wulczyn, F. G., and Pfaff, E. (1985) The *mom* gene of bacteriophage Mu: a unique regulatory scheme to control a lethal function. Gene *39*, 61–70.

Kannan, P., Cowan, G. M., Daniel A. S., Gann, A. A. F., and Murray, N. E. (1989) Conservation of organization in the specificity polypeptides of two families of type I restriction enzymes. J. Mol. Biol. *209*, 335–344.

Kapfer, W., Walter, J., and Trautner, T. A. (1991) Cloning, characterization and evolution of the *Bsu*FI resriction endonuclease gene of *Bacillus subtilis* and purification of the enzyme. Nucl. Acids Res. *19*, 6457–6463.

Karreman, C., and de Waard, A. (1990) *Agmenellum quadruplicatum* M·*Aqu*I, a novel modification methylase. J. Bacteriol *172*, 266–272.

Kaszubska, W., Aiken, C., O'Connor, C. D., and Gumport, R. I. (1989) Purification, cloning and sequence analysis of *Rsr*I DNA methyltransferase: lack of homology between two enzymes, *Rsr*I and *Eco*RI, that methylate the same nucleotide in identical recognition sequences. Nucl. Acids Res. *17*, 10403–10425.

Kauc, L., and Piekarowicz, A. (1978) Purification and properties of a new restriction endonuclease from *Haemophilus influenzae* Rf. Eur. J. Biochem. *92*, 417–426.

Kelleher, J. E., and Raleigh, E. A. (1991) A novel activity in *Escherichia coli* K-12 that directs restriction of DNA modified at CG dinucleotides. J. Bacteriol. *173*, 5220–5223.

Kelleher, J. E., Daniel, A. S., and Murray, N. E. (1991) Mutations that confer *de novo* methyltransferase activity. J. Mol. Biol. *221*, 431–440.

Kiss, A., and Baldauf, F. (1983) Molecular cloning and expression in *Escherichia coli* of two modification methylase genes of *Bacillus subtilis*. Gene *21*, 111–119.

Kiss, A., Posfai, G., Keller, C. C., Venetianer, P., and Roberts, R. J. (1985) Nucleotide sequence of the *Bsu*RI restriction-modification system. Nucl. Acids Res. *13*, 6403–6421.

Kita, K., Kotani, H., Sugisaki, H., and Takanami, M. (1989) The FokI restriction-modification system. I: Organization and nucleotide sequences of the restriction and modification genes. J. Biol. Chem. *264*, 5751–5756.

Kleckner, N. (1989) Transposon Tn*10*, in: Mobile DNA, pp. 229–268. Eds D. E. Berg and M. M. Howe. American Society for Microbiology, Washington, D. C.

Klimasauskas, S., Nelson, J. L., and Roberts, R. J. (1991) The sequence specificity domain of cytosine-C5 methylases. Nucl. Acids Res. *19*, 6183–6190.

Klimasauskas, S., Timinskas, A., Menkevicius, S., Butkienė, D., Butkus, V., and Janulaitis, A. (1989) Sequence motifs characteristic of DNA [cytosine-N4] methyltransferases: similarity to adenine and cytosine-C5 DNA-methylases. Nucl. Acids Res. *17*, 9823–9832.

Klump, H. H. (1987) Ordnung in der Vielfalt. Organisation und Evolution eines Restriktions-Codes. Futura *4*, 10–16.

Koppes, L. J. H., and von Meyenburg, K. (1987) Nonrandom minichromosome replication in *Escherichia coli* K-12. J. Bacteriol. *169*, 430–433.

Kramer, B., Kramer, W., and Fritz, H.-J. (1984) Different base/base mismatches are corrected with different efficiencies by the methyl-directed DNA mismatch-repair system of *E. coli*. Cell *38*, 879–887.

Kretz, P. L., Kohler, S. W., and Short, J. M. (1991) Identification and characterization of a gene responsible for inhibiting propagation of methylated DNA sequences in *mcr*A *mcr*B1 *Escherichia coli* strains. J. Bacteriol. *173*, 4707–4716.

Kretz, P. L., Reid, C. H., Greener, A., and Short, J. M. (1989) Effect of lambda packaging extract *mcr* restriction activity on DNA cloning. Nucl. Acids Res. *17*, 5409.

Krüger, D. H., and Bickle, T. A. (1983) Bacteriophage survival: Multiple mechanisms for avoiding the deoxyribonucleic acid restriction systems of their hosts. Microbiol. Rev. *47*, 345–360.

Krüger, T., Grund, C., Wild, C., and Noyer-Weidner, M. (1992) Characterization of the *mcr*BC region of *Escherichia coli* K-12 wild-type and mutant strains. Gene *114*, 1–12.

Kumura, K., and Sekiguchi, M. (1983) Identification of the *uvr*D product of *Escherichia coli* as DNA helicase II and its induction by DNA-damaging agents. J. Biol. Chem. *259*, 1560–1565.

Kushner, S. R., Shepherd, J., Edwards, B., and Maples, V. F. (1978) *uvrD*, *uvrE* and *recL* represent a single gene in: DNA Repair Mechanisms, pp. 251–254. Eds P. C. Hanawalt, E. Friedberg and C. F. Fox. Academic Press, New York.

Kushner, S. R., Maples, V. F., Easton, A., Farrance, I., and Peramachi, P. (1983) Physical biochemical, and genetic characterization of the *uvrD* gene product. ICN-UCLA Symp. Mol. Cell. Biol. *11*, 153–159.

Labbé, D., Höltke, H. J., and Lau, P. C. K. (1990) Cloning and characterization of two tandemly arranged DNA methyltransferase genes of *Neisseria lactamica*: An adenine-specific M.*Nla*III and a cytosine-type methylase. Mol. Gen. Genet. *224*, 101–110.

Lacks, S. A., (1980) Purification and properties of the complementary endonucleases *Dpn*I and *Dpn*II. Meth. Enzymol. *65*, 138–146.

Lacks, S., and Greenberg, B. (1975) A deoxyribonuclease of *Diplococcus pneumoniae* specific for methylated DNA. J. Biol. Chem. *250*, 4060–4066.

Lacks, S., and Greenberg, B. (1977) Complementary specificity of restriction endonucleases of *Diplococcus pneumoniae* with respect to DNA methylation. J. Mol. Biol. *114*, 153–168.

Lacks, S. A., Dunn, J. J., and Greenberg, B. (1982) Identification of base mismatches recognized by the heteroduplex-DNA-repair system of *Streptococcus pneumoniae*. Cell *31*, 327–336.

Lacks, S. A., Mannarelli, B. M., Springhorn, S. S., and Greenberg, B. (1986) Genetic basis of the complementary *Dpn*I and *Dpn*II restriction systems of *S. pneumoniae*: An intercellular cassette mechanism. Cell. *46*, 993–1000.

Laengle-Rouault, F., Maenhaut-Michel, G., and Radman, M. (1986) GATC sequence and mismatch repair in *Escherichia coli*. EMBO J. *5*, 2009–2013.

Laengle-Rouault, F., Maenhaut-Michel, G., and Radman, M. (1987) GATC sequences, DNA nicks and the MutH function in *Escherichia coli* mismatch repair. EMBO J. *6*, 1121–1127.

Lahue, R. S., Su. S.-S., and Modrich, P. (1987) Requirement for d(GATC) sequences in *Escherichia coli mutHLS* mismatch correction. Proc. Natl. Acad. Sci. USA *84*, 1482–1486.

Lahue, R. S., Au, K. G., and Modrich, P. (1989) DNA mismatch correction in a defined system. Science *245*, 160–164.

Landoulsi, A., Hughes, P., Kern, R., and Kohiyama, M. (1989) *dam* methylation and the initiation of DNA replication on *oriC* plasmids. Mol. Gen. Genet. *216*, 217–223.

Landoulsi, A., Malki, A., Kern, R., Kohiyama, M., and Hughes, P. (1990) The *E. coli* cell surface specifically prevents the initiation of DNA replication at *oriC* on hemimethylated DNA templates. Cell. *63*, 1053–1060.

Lange, C., Jugel, A., Walter, J., Noyer-Weidner, M., and Trautner, T. A. (1991a) 'Pseudo' domains in phage-encoded DNA methyltransferases. Nature *352*, 645–648.

Lange, C., Noyer-Weidner, M., Trautner, T. A., Weiner, M., and Zahler, S. A. (1991b) M.H2I, a multispecific 5C-DNA methyltransferase encoded by *Bacillus amyloliquefaciens* phage H2. Gene *100*, 213–218.

Lauster, R. (1989) Close relationship between the *Hin*fI and DpnA DNA-methyltransferase. Nucl. Acids Res. *17*, 4402.

Lauster, R. (1989) Evolution of type II DNA methyltransferases. A gene duplication model. J. Mol. Biol. *206*, 313–321.

Lauster, R., Kriebardis, A., and Guschlbauer, W. (1987) The GATATC-modification enzyme *Eco*RV is closely related to the GATC-recognizing methyltransferases *Dpn*II and *dam* from *E. coli* and phage T4. FEBS Lett. *220*, 167–176.

Lauster, R., Trautner, T. A., and Noyer-Weidner, M. (1989) Cytosine-specific type II DNA methyltransferases. A conserved enzyme core with variable target-recognizing domains. J. Mol. Biol. *206*, 305–312.

Lieb, M. (1983) Specific mismatch correction in bacteriophage lambda crosses by very short patch repair. Mol. Gen. Genet. *191*, 118–125.

Lieb, M. (1985) Recombination in the λ repressor gene: evidence that very short patch (VSP) mismatch correction restores a specific sequence. Mol. Gen. Genet. *199*, 465–470.

Lieb, M. (1987) Bacterial genes *mutL*, *mutS*, and *dcm* participate in repair of mismatches at 5-methylcytosine sites. J. Bacteriol. *169*, 5241–5246.

Lieb, M., Allen, E., and Read, D. (1986) Very short patch mismatch repair in phage lambda: repair sites and length of repair tracts. Genetics *114*, 1041–1060.

Lin, P. M., Lee, C. H., and Roberts, R. J. (1989) Cloning and characterization of the genes encoding the *Msp*I restriction modification system. Nucl. Acids Res. *17*, 3001–3011.

Lindahl, T. (1974) An N-glycosidase from *Escherichia coli* that releases free uracil from DNA containing deaminated cytosine residues. Proc. Natl. Acad. Sci. USA *71*, 3649–3653.

Lindahl, T., and Nyberg, B. (1974) Heat-induced deamination of cytosine residues in deoxyribonucleic acid. Biochemistry *13*, 3405–3410.

Linder, P., Doelz, R., Gubler, M., and Bickle, T. A. (1990) An anticodon nuclease gene inserted into a *hsd* region encoding a type I DNA restriction system. Nucl. Acids. Res. *18*, 7170.

Løbner-Olesen, A., Boye, E., and Marinus, M. G. (1992) Multiple promoters control expression of the *Escherichia coli dam* gene. Mol. Microbiol. *6*, 1841–1851.

Loenen, W. A. M., Daniel, A. S., Braymer, H., D., and Murray, N. E. (1987) Organization and sequence of *hsd* genes of *Escherichia coli* K-12. J. Mol. Biol. *198*, 159–170.

Looney, M. C., Moran, L. S., Jack, W. E., Feehery, G. R., Benner, J. S., Slatko, B. E., and Wilson, G. G. (1989) Nucleotide sequence of the FokI restriction-modification system: separate strand specificity domains in the methyltransferase. Gene *80, 193–208.*

Low, D., Robinson, E. N., McGee, Z. A., and Falkow, S. (1987) The frequency of expression of pyelonephritis-associated pili is under regulatory control. Mol. Microbiol. *1*, 335–346.

Lu, A.-L., Clark, S., and Modrich, P. (1983) Methyl-directed repair of DNA base-pair mismatches *in vitro*. Proc. Natl. Acad. Sci. USA *80*, 4639–4643.

Lu, A. L., Welsh, K., Clark, S., Su, S. S., and Modrich, P. (1984) Repair of DNA base-pair mismatches in extracts of *Escherichia coli*. Cold Spring Harbor Symp. Quant. Biol. *49*, 589–596.

Luria, S. E., and Human, M. G. (1952) A nonhereditary, host-induced variation of bacterial viruses. J. Bacteriol. *64*, 557–569.

Macaluso, A., and Mettus, A. M. (1991) Efficient transformation of *Bacillus thuringiensis* require nonmethylated plasmid DNA. J. Bacteriol. *173*, 1353–1356.

Macdonald, P. M., and Mosig, G. (1984) Regulation of a new bacteriophage T4 gene, *69*, that spans an origin of DNA replication. EMBO J. *3*, 2863–2871.

MacNeil, D. J. (1988) Characterization of a unique methyl-specific restriction system in *Streptomyces avermitilis*. J. Bacteriol. *170*, 5607–5612.

Makris, J. C., Nordmann, P. L., and Reznikoff, W. S. (1988) Mutational analysis of insertion sequence *50*, (IS*50*) and transposon 5 (Tn5) ends. Proc. Natl. Acad. Sci. USA *85*, 2224–2228.

Malki, A., Kern, R., Kohiyama, M., and Hughes, P. (1992) Inhibition of DNA synthesis at the hemimethylated pBR322 origin of replication by a cell membrane fraction. Nucl. Acids Res. *20*, 105–109.

Manarelli, B. M., Balganesh, T. S., Greenberg, B., Springhorn, S., and Lacks, S. A. (1985) Nucleotide sequence of the *Dpn*II methylase gene of *Stereptococcus pneumoniae* and its relationship to the *dam* gene of *Escherichia coli*. Proc. Natl. Acad. Sci. USA *82*, 4468–4472.

Marinus, M. G. (1973) Location of DNA methylation genes on the *Escherichia coli* K-12 genetic map. J. Bacteriol. *127*, 47–55.

Marinus, M. G. (1984) Methylation of prokaryotic DNA in: DNA Methylation, pp. 81–109. Eds A. Razin, H. Cedar and A. D. Riggs. Springer-Verlag, New York.

Marinus, M. G. (1987) DNA methylation in *Escherichia coli*. Ann. Rev. Genet. *21*, 113–131.

Marinus, M. G., and Konrad, E. B. (1976) Hyper-recombination in *dam* mutants of *Escherichia coli* K-12. Mol. Gen. Genet. *149*, 213–277.

Marinus, M. G., and Morris, N. R. (1973) Isolation of deoxyribonucleic acid methylase mutants of *Escherichia coli* K-12. J. Bacteriol. *114*, 1143–1150.

Marinus, M. G., and Morris, N. R. (1974) Biological function for 6-methyladenine residues in the DNA of *Escherichia coli* K12. J. Mol. Biol. *85*, 309–322.

Marinus, M. G., and Morris, N. R. (1975) Pleiotropic effects of a DNA adenine methylation mutation *dam*-3 in *Escherichia coli* K-12. Mutat. Res. *28*, 15–26.

McClelland, M. (1985) Selection against *dam* methylation sites in the genomes of DNA of enterobacteriophages. J. Mol. Evol. *21*, 317–322.

McGraw, B. R., and Marinus, M. G. (1980) Isolation and characterization of *dam*$^+$ revertants and suppressor mutations that modify secondary phenotypes of *dam*-3 strains of *Escherichia coli* K12. Mol. Gen. Genet. *178*, 309–315.

Meijer, M., Beck, E., Hansen, F. G., Bergmans, H. E., Messer, W., von Meyenburg, K., and Schaller, H. (1979) Nucleotide sequence of the origin of replication of the *E. coli* K-12 chromosome. Proc. Natl. Acad. Sci. USA *76*, 580–584.

Meisel, A., Krüger, D. H., and Bickle, T. A. (1991) M.*Eco*P15 methylates the second adenine in its target recognition sequence. Nucl. Acids. Res. *19*, 3997.

Meisel, A., Bickle, T. A., Krüger, D. H., and Schröder, C. (1992) Type III restriction enzymes need two inversely oriented recognition sites for DNA cleavage. Nature *355*, 467–469.

Messer, W., and Noyer-Weidner, M. (1988) Timing and targeting: The biological functions of Dam methylation in *E. coli*. Cell *54*, 735–737.

Messer, W., Bellekes, U., and Lother, H. (1985) Effect of *dam* methylation on the activity of the *E. coli* replication origin, *ori*C. EMBO J. *4*, 1327–1332.

von Meyenburg, K., and Hansen, F. G. (1987) Regulation of chromosome replication in: *Escherichia coli* and *Salmonella typhimurium*: Cellular and Molecular Biology, pp. 1555–1577. Eds F. C. Neidhart, J. L. Ingraham, K. Brooks Low, B. Magasanik, M. Schaechter and H. E. Umbarger. American Society of Microbiology, Washington, D.C.

Miner, Z., and Hattman, S. (1988) Molecular cloning, sequencing and mapping of the bacteriophage T2 *dam* gene. J. Bacteriol. *170*, 5177–5184.

Modrich, P. (1987) DNA mismatch correction, Ann. Rev. Biochem. *56*, 435–466.

Modrich, P. (1989) Methyl-directed DNA mismatch correction. J. Biol. Chem. *264*, 6597–6600.

Muckerman, C. C., Springhorn, S. S., Greenberg, B., and Lacks, S. A. (1982) Transformation of restriction endonuclease phenotype in *Streptococcus pneumoniae*. J. Bacteriol. *152*, 183–190.

Murchie, A. I. H., and Lilley, M. J. (1989) Base methylation and local DNA helix stability: effect on the kinetics of cruciform extrusion. J. Mol. Biol. *205*, 593–602.

Murray, N. E., Gough, J. A., Suri, B., and Bickle, T. A. (1982) Structural homologies among type I restriction-modification systems. EMBO J. *1*, 535–539.

Nagaraja, V., Shepherd, J. C. W., and Bickle, T. A. (1985) A hybrid recognition sequence in a recombinant restriction enzyme and the evolution of DNA sequence specificity. Nature *316*, 371–372.

Narva, K. E., Wendell, D. L., Skrdla, M. P., and Van Etten, J. L. (1987) Molecular cloning and characterization of the gene encoding the DNA methyltransferase, M.*Cvi*BIII, from *Chlorella* virus NC-1A. Nucl. Acids Res. *15*, 9807–9823.

Nelson, M., and McClelland, M. (1991) Site-specific methylation: effect on DNA modification methyltransferases and restriction endonucleases. Nucl. Acids Res. *19*, 2045–2071.

Nestman, E. R. (1978) Mapping by transduction of mutator gene *mutH* in *Escherichia coli*. Mutat. Res. *49*, 421–423.

Nevers, P., and Spatz, H. C. (1975) *Escherichia coli* mutants *uvrD* and *uvrE* deficient in gene conversion of λ-heteroduplexes. Mol. Gen. Genet. *139*, 233–243.

Newmann, A. K., Rubin, R. A., Kim, S.-H., and Modrich, P. (1981) DNA sequences of structural genes for *Eco*RI DNA restriction and modification enzymes. J. Biol. Chem. *256*, 2131–2139.

Newman, E. B., D'Ari, R., and Lin, R. T. (1992) The leucine-Lrp regulon in *E. coli*: A global response in search of a raison d'être. Cell *68*, 617–629.

Noyer-Weidner, M., and Reiners-Schramm, L. (1988) Highly efficient positive selection of recombinant plasmids using a novel *rglB*-based *Escherichia coli* K-12 vector system. Gene *66*, 269–278.

Noyer-Weidner, M., Diaz, R., and Reiners, L. (1986) Cytosine-specific DNA modification interferes with plasmid establishment in *Escherichia coli* K12: Involvement of *rglB*. Mol. Gen. Genet. *205*, 469–475.

Noyer-Weidner, M., Jentsch, S., Pawlek, B., Günthert, U., and Trautner, T. A. (1983) Restriction and modification in *Bacillus subtilis*: DNA methylation potential of the related bacteriophages Z, SPR, SPß, Ø3T and ρ11. J. Virol. *46*, 446–453.

Nur, I., Szyf, M., Razin, A., Glaser, G., Rottem, S., and Razin, S. (1985) Procaryotic and eucaryotic traits of DNA methylation in Spiroplasmas (Mycoplasmas). J. Bacteriol. *164*, 19–24.

Nwankwo, D. O., and Wilson, G. G. (1988) Cloning and expression of the *Msp*I restriction and modification genes. Gene *64*, 1–8.

Ogden, G. B., Pratt, M. J., and Schaechter, M. (1988) The replicative origin of the *E. coli* chromosome binds to cell membranes only when hemimethylated. Cell *54*, 127–135.

Oka, A., Sugimoto, K., Takanami, M., and Hirota, Y. (1980) Replication origin of the *Escherichia coli* K-12 chromosome: the size and structure of the minimum DNA segment carrying the information for autonomous replication. Mol. Gen. Genet. *178*, 9–20.

Orbach, M. J., Schneider, W. P., and Yanofsky, C. (1988) Cloning of methylated transforming DNA from *Neurospora crassa* in *Escherichia coli*. Molec. Cell. Biol. *8*, 2211–2213.

Patnaik, P. K., Merlin, S., and Polisky, B. (1990) Effect of altering GATC sequences in the plasmid ColE1 primer promoter. J. Bacteriol. *172*, 1762–1768.

Perez-Casal, J. F., Gammie, A. E., and Crosa, J. H. (1989) Nucleotide sequence analysis and expression of the minimum REPI replication region and incompatibility determinants of pColV-K30. J. Bacteriol. *171*, 2195–2201.

Peterson, K. R., Wertman, K. F., Mount, D. W., and Marinus, M. G. (1985) Viability of *Escherichia coli* K-12 DNA adenine methylase (*dam*) mutants requires increased expression of specific genes in the SOS regulon. Mol. Gen. Genet. *201*, 14–19.

Piekarowicz, A., Yuan, R., and Stein, D. C. (1991a) Isolation of temperature-sensitive McrA and McrB mutations and complementation analysis of the McrBC region of *Escherichia coli* K-12. J. Bacteriol. *173*, 150–155.

Piekarowicz, A., Yuan, R., and Stein, D. C. (1991b) A new method for the rapid identification of genes encoding restriction and modification enzymes. Nucl. Acids Res. *19*, 1831–1835.

Piekarowicz, A., Bickle, T. A. Shepherd, J. C. W., and Ineichen, K. (1981) The DNA sequence recognised by the *Hinf*III restriction endonuclease. J. Mol. Biol. *146*, 167–172.

Plumbridge, J. (1987) The role of *dam* methylation in controlling gene expression. Biochimie *69*, 439–443.

Pogolotti, A. L., Ono, A., Subramaniam, R., and Santi, D. V. (1988) On the mechanism of DNA-adenine methylase. J. Biol. Chem. *263*, 7461–7464.

Pósfai, J., Bhagwat, A. S., Pósfai, G., and Roberts, R. J. (1989) Predictive motifs derived from cytosine methyltransferases. Nucl. Acids Res. *17*, 2421–2435.

Pósfai, G., Kim, S. C., Szilák, L., Kovács, A., and Venetianer, P. (1991) Complementation by detached parts of GGCC-specific DNA methyltransferases. Nucl. Acids Res. *19*, 4843–4847.

Pósfai, G., Baldauf, F., Erdei, S., Pósfai, J., Venetianer, P., and Kiss, A. (1984) Structure of the gene coding for the sequence specific DNA-methyltransferase of *B. subtilis* phage SPR. Nucl. Acids Res. *12*, 9039–9049.

Povilionis, P. L., Lubys, A. A., Vaisvila, R. I., Kulakauskas, S. T., and Janulaitis, A. A. (1989) Investigation of methyl-cytosine specific restriction in *Escherichia coli* K-12. Genetika *25*, 753–755.

Prentki, P., Chandler, M., and Caro, L. (1977) Replication of the prophage P1 during the cell cycle of *Escherichia coli*. Mol. Gen. Genet. *152*, 71–76.

Price, C., Linger, J., Bickle, T. A., Firman, K., and Glover S. W. (1989) Basis for changes in DNA recognition by *Eco*R124 and *Eco*R124/3 type I DNA restriction and modification enzymes. J. Mol. Biol. *205*, 115–125.

Pukkila, P. J., Peterson, J., Herman, G., Modrich, P., and Meselson, M. (1983) Effects of high levels of DNA adenine methylation on methyl-directed mismatch repair in *Escherichia coli*. Genetics *104*, 571–582.

Radman, M., and Wagner, R. (1986) Mismatch repair in *Escherichia coli*. Ann. Rev. Genet. *20*, 523–538.

Raleigh, E. A. (1992) Organization and function of the *mcrBC* genes of *E. coli* K-12. Mol. Microbiol. *6*, 1079–1086.

Raleigh, E. A., and Wilson, G. (1986) *Escherichia coli* K-12 restricts DNA containing 5-methylcytosine. Proc. Natl. Acad. Sci. USA *83*, 9070–9074.

Raleigh, E. A., Trimarchi, R., and Revel, H. (1989) Genetic and physical mapping of the *mcrA* (*rglA*) and *mcrB* (*rglB*) loci of *Escherichia coli* K 12. Genetics *122*, 279–296.

Raleigh, E. A., Murray, N. E., Revel, H. Blumenthal, R. M., Westaway, D., Reith, A. D., Rigby, P. W. J., Elhai, J., and Hanahan, D. (1988) McrA and McrB restriction phenotypes of some *E. coli* strains and implications for gene cloning. Nucl. Acids Res. *16*, 1563–1575.

Raleigh, E. A., Benner, J., Bloom, F., Braymer, H. D., DeCruz, E., Dharmalingam, K., Heitman, J., Noyer-Weidner, M., Piekarowicz, A., Kretz, P. L., Short, J. M., and Woodcock, D. (1991) Nomenclature relating to restriction of modified DNA in *Escherichia coli*. J. Bacteriol. *173*, 2707–2709.

Ravi, R. S., Sozhamannan, S., and Dharmalingam, K. (1985) Transposon mutagenesis and genetic mapping of the *rglA* and *rglB* loci of *Escherichia coli*. Mol. Gen. Genet. *198*, 390–392.

Renbaum, P., Abrahamove, D., Fainsod, A., Wilson, G. G., Rottem, S., and Razin, A. (1990) Cloning, characterization, and expression in *Escherichia coli* of the gene coding for the CpG DNA methylase from *Spiroplasma* sp. strain MQ1 (M·*Sss*I). Nucl. Acids Res. *18*, 1145–1152.

Revel, H., (1967) Restriction of nonglucosylated T-even bacteriophage: properties of permissive mutants of *Escherichia coli* B and K-12. Virology *31*, 688–701.

Revel, H., (1983) DNA modification: glycosylation. In: Bacteriophage T4. pp. 156–165. Eds C. K. Mathews, E. M. Kutter, G. Mosig and P. Berget. American Society for Microbiology. Washington, D. C.

Roberts, D., Hoopes, B. C., McClure, W. R., and Kleckner, N. (1985) IS*10* transposition is regulated by DNA adenine methylation. Cell *43*, 117–130.

Roberts, R. J. (1990) Restriction enzymes and their isoschizomers. Nucl. Acids Res. *18*, 2331–2361.

Roberts, J. R., and Macelis, D. (1991) Restriction enzymes and their isoschizomers. Nucl. Acids. Res. *19*, 2077–2109.

Rosenberg, J. M. (1991) Structure and function of restriction endonucleases. Curr. Opin. Struct. Biol. *1*, 104–113.

Ross, T. K., Achberger, E. C., and Braymer, H. D. (1987) Characterization of the *Escherichia coli* modified cytosine restriction (*mcrB*) gene. Gene *61*, 277–289.

Ross, T. K., Achberger, E. C., and Braymer, H. D. (1989a) Nucleotide sequence of the *mcrB* region of *Escherichia coli* K-12 and evidence for two independent translational initiation sites at the *mcrB* locus. J. Bacteriol. *171*, 1974–1981.

Ross, T. K., Achberger, E. C., and Braymer, H. D. (1989b) Identification of a second polypeptide required for *mcrB* restriction of 5-methylcytosine-containing DNA in *Escherichia coli* K-12. Mol. Gen. Genet. *216*, 402–407.

Rubin, R. A., and Modrich, P. (1977) *Eco*RI methylase. Physical and catalytic properties of the homogeneous enzyme. J. Biol. Chem. *252*, 7265–7272.

Russell, D. W., and Zinder, N. D. (1987) Hemimethylation prevents DNA replication in *E. coli*. Cell. *50*, 1071–1079.

Rydberg, B. (1978) Bromouracil mutagenesis and mismatch repair in mutator strains of *Escherichia coli*. Mutation Res. *52*, 11–24.

Sain, B., and Murray, N. E., (1980) The *hsd* (host specificity) genes of *E. coli* K12. Mol. Gen. Genet. *180*, 35–46.

Santi, D. V., Garrett, C. E., and Barr, P. J. (1983) On the mechanism of inhibition of DNA-cytosine methyltransferases by cytosine analogs. Cell *33*, 9–10.

Schaaper, R. M. (1988) Mechanisms of mutagenesis in the *Escherichia coli* mutator *mutD5*: Role of DNA mismatch repair. Proc. Natl. Acad. Sci. USA *85*, 8126–8130.

Schlagman, S., and Hattman, S. (1983) Molecular cloning of a functional *dam*$^+$ gene coding for phage T4 DNA adenine methylase. Gene *22*, 139–156.

Schlagman, S., Hattman, S., May, M. S., and Berger, L. (1976) *In vivo* methylation by *Escherichia coli* K-12 *mec*$^+$ deoxyribonucleic acid-cytosine methylase protects against *in vitro* cleavage by the RII restriction endonuclease (R·*Eco*RII) J. Bacteriol. *126*, 990–996.

Schneider-Scherzer, E., Auer, B., deGroot, E. J., and Schweiger, M. (1990) Primary structure of a DNA (N6-adenine)-methyltansferase from *Escherichia coli* virus T1. J. Biol. Chem. *265*, 6086–6091.

Schoner, B., Kelly, S., and Smith, H. O. (1983) The nucleotide sequence of the *Hha*II restriction and modification genes from *Haemophilus haemolyticus*. Gene *24*, 227–236.

Scott (1984) Regulation of plasmid replication. Microbiol. Rev. *48*, 1–23.

Segev, N., and Cohen, G. (1981) Control of circularization of bacteriophage P1 DNA in *Escherichia coli*. Virology *114*, 333–342.

Segev, N., Laub. A., and Cohen, G. (1980) A circular form of bacteriophage P1 DNA made in lytically infected cells of *Escherichia coli*. Virology *101*, 261–271.

Seiler, A., Blöcker, H., Frank, R., and Kahmann, R. (1986) The *mom* gene of bacteriophage Mu: the mechanism of methylation-dependent expression. EMBO J. *5*, 2719–2278.

Silverman, M., Zieg, J., Hilmen, M., and Simon, M. (1979) Phase variation in *Salmonella*: Genetic analysis of a recombinational switch. Proc. Natl. Acad. Sci. USA *76*, 391–395.

Skarstad, K., Boye, E., and Steen, H. B. (1986) Timing of initiation of chromosome replication in individual *Escherichia coli* cells. EMBO J. *5*, 1711–1717.

Sladek, T. L., and Maniloff, J. (1987) Endonuclease from *Acholeplasma laidlawii* strain JA1 associated with *in vivo* restriction of DNA containing 5-methylcytosine. Isr. J. Med. Sci. *23*, 423–426.

Sladek, T. L., Nowak, J. A., Maniloff, J. (1986) Mycoplasma restriction: identification of a new type of restriction specificity for DNA containing 5-methylcytosine. J. Bacteriol. *165*, 219–225.

Slatko, B. E., Benner, J. S., Jager-Quinton, T., Moran, L. S., Simcox, T. G., Van Cott, E. M., and Wilson, G. G. (1987) Cloning, sequencing and expression of the *Taq*I restriction-modification system. Nucl. Acids Res. *15*, 9781–9796.

Slatko, B. E., Croft, R., Moran, L. S., and Wilson, G. G. (1988) Cloning and analysis of the *Hae*III and *Hae*II methyltransferase genes. Gene *74*, 45–50.

Smith, D. W., Garland, A. M., Herman, G., Enns, R. E., Baker, T. A., and Zyskind, J. W. (1985) Importance of state of methylation of *oriC* GATC sites in initiation of DNA replication in *Escherichia coli*. EMBO J. *4*, 1319–1326.

Smith, H. O., and Nathans, D. (1973) A suggested nomenclature for bacterial host modification and restriction systems and their enzymes. J. Mol. Biol. *81*, 419–423.

Smith, H. O., Annau, T. M., and Chandrasegaran, S. (1990) Finding sequence motifs in groups of functionally related proteins. Proc. Natl. Acad. Sci. U.S.A. *87*, 826–830.

Sohail, A., Lieb, M., Dar, M., and Bhagwat, A. S. (1990) A gene required for very short patch repair in *Escherichia coli* adjacent to the DNA cytosine methylase gene. J. Bacteriol. *172*, 4214–4221.

Som, S., and Friedman, S. (1990) Direct photolabeling of the *Eco*RII methyltransferase with S-adenosyl-L-methionine. J. Biol. Chem. *265*, 4278–4283.

Som, S., and Friedman, S. (1991) Identification of a highly conserved domain in the *Eco*RII methyltransferase which can be photolabeled with S-adenosyl-L-[methyl-^{3}H]methionine. J. Biol. Chem. *266*, 2937–2945.

Som, S., Bhagwat, A. S., and Friedman, S. (1987) Nucleotide sequence and expression of the gene encoding the *Eco*RII modification enzyme. Nucl. Acids Res. *15*, 313–323.

Spielmann-Ryser, J., Moser, M., Kast, P., and Weber, H. (1991) Factors determining the frequency of plasmid cointegrate formation mediated by insertion sequence IS3 from *Escherichia coli*. Mol. Gen. Genet. *236*, 441–448.

Stefan, C., Xia, Y., and Van Etten, J. L. (1991) Molecular cloning and characterization of the gene encoding the adenine methyltransferase M.*Cvi*RI from *Chlorella* virus XZ-6E. Nucl. Acids Res. *19*, 307–311.

Sternberg, N. (1990) Bacteriophage P1 cloning system for the isolation, amplification, and recovery of DNA fragments as large as 100 kilobase pairs. Proc. Natl. Acad. Sci. USA *87*, 103–107.

Sternberg, N., and Coulby, J. (1987a) Recognition and cleavage of the bacteriophage P1 packaging site (*pac*). I. Differential processing of the cleaved ends *in vivo*. J. Mol. Biol. *194*, 453–468.

Sternberg, N., and Coulby, J. (1987b) Recognition and cleavage of the bacteriophage P1 packaging site (*pac*). II. Functional limit of *pac* and location of *pac* cleavage termini, J. Mol. Biol. *194*, 469–479.

Sternberg, N., and Coulby, J. (1990) Cleavage of the bacteriophage P1 packaging site (*pac*) is regulated by adenine methylation. Proc. Natl. Acad. Sci. USA *87*, 8070–8074.

Sternberg, N., Sauer, B., Hoess, R., and Abremski, K. (1986) Bacteriophage P1 *cre* gene and its regulatory region. Evidence for multiple promoters and for regulation by DNA methylation. J. Mol. Biol. *187*, 197–212.

Studier, F. W., and Bandyopadhyay, P. K. (1988) Model how type I restriction enzymes select cleavage sites in DNA. Proc. Natl. Acad. Sci. USA *85*, 4677–4681.

Sugisaki, H., Kita, K., and Takanami, M. (1989) The *Fok*I restriction-modification system. II Presence of two domains in *Fok*I methylase responsible for modification of different DNA strands. J. Biol. Chem. *264*, 5757–5761.

Sugisaki, H., Yamamoto, K., and Takanami, M. (1991) The *Hga*I restriction-modification system contains two cytosine methylase genes responsible for modification of different DNA strands. J. Biol. Chem. *266*, 13,952–13,957.

Suri, B., Nagaraja, V., and Bickle, T. A. (1984) Bacterial DNA modification. Curr. Top. Microbiol. Immunol. *108*, 1–9.

Sutcliffe, J. G. (1979) Complete nucleotide sequence of the *Escherichia coli* plasmid pBR322. Cold Spring Harbor Symp. Quant. Biol. *43*, 77–90.

Sutherland, E., Coe. L., and Raleigh, E. A. (1992) McrBC: a multisubunit GTP-dependent restriction endonuclease. J. Mol. Biol. *225*, 327–348.

Swinton, D., Hattman, S., Crain, P. F., Cheng, C.-S., Smith, D. L., and McCloskey, J. A. (1983) Purification and characterization of the unusual deoxynucleoside, a-*N*-(9-β-D-2′-deoxyribofuranosylpurin-6-yl)glycinamide, specified by the phage Mu modification function. Proc. Natl. Acad. Sci. USA *80*, 7400–7404.

Sznyter, L. A., Slatko, B., Moran, L., O'Donnell, K. H., and Brooks, J. E. (1987) Nucleotide sequence of the *Dde*I restriction-modification system and characterization of the methylase protein. Nucl. Acids Res. *15*, 8249–8266.

Szybalski, W., Kim, S. C., Hasan, N., and Podhajska, A. J. (1991) Class-IIS restriction enzymes – a review. Gene *100*, 13–26.

Tao, T., Bourne, J. C., and Blumenthal, R. M. (1991) A family of regulatory genes associated with type II restriction-modification systems. J. Bacteriol. *173*, 1367–1375.

Tao, T., Walter, J., Brennan, K. F., Cotterman, M. M., and Blumenthal, R. M. (1989) Sequence, internal homology and high-level expression of the gene for a DNA-(cytosine *N4*)-methyltransferase, M.*Pvu*II. Nucl. Acids Res. *17*, 4161–4175.

Toussaint, A. (1976) The DNA modification function of temperate phage Mu-1. Virol. *70*, 17–27.

Tran-Betcke, A., Behrens, B., Noyer-Weidner, M., and Trautner, T. A. (1986) DNA methyltransferase genes of *Bacillus subtilis* phages: comparison of their nucleotide sequences. Gene *42*, 89–96.

Trautner, T. A., and Noyer-Weidner, M. (1992) Restriction/modification- and methylation systems in *Bacillus subtilis*, related species and their phages. In: *Bacillus subtilis*, the Model Gram-Positive Bacterium: Physiology, Biochemistry and Molecular Genetics. Eds J. A. Hoch, R. Losick and A. L. Sonenshein. American Society for Microbiology, Washington, D.C. (in press).

Trautner, T. A., Balganesh, T. S., and Pawlek, B. (1988) Chimeric multispecific DNA methyltransferases with novel combinations of target recognition. Nucl. Acids Res. *16*, 6649–6657.

Trautner, T. A., Pawlek, B., and Behrens, B. (1992) The size and arrangement of individual target recognizing domains of the multispecific DNA-C5-methyltransferase M.SPRI. To be submitted.

Vinella, D., Jaffé, A., D'Ari, R., Kohiyama, M., and Hughes, P. (1992) Chromosome partitioning in *Escherichia coli* in the absence of *dam*-directed methylation. J. Bacteriol. *174*, 2388–2390.

van de Putte, P., Plasterk, P., and Kujpers, A. (1984) A Mu *gin* complementing function and an invertible DNA region in *Escherichia coli* K-12 are situated on the genetic element *e14*. J. Bacteriol. *158*, 417–422.

Wagner, R., and Meselson, M. (1976) Repair tracts in mismatched DNA heteroduplexes. Proc. Natl. Acad. Sci. USA *73*, 4135–4139.

Wagner, R., Dohet, C., Jones, M., Doutriaux, M. P., Hutchinson, F., and Radman, M. (1984) Involvement of *Escherichia coli* mismatch repair in DNA replication and recombination. Cold Spring Harbor Symp. Quant. Biol. *49*, 611–615.

Waite-Rees, P. A., Keating, C. J., Moran, L. S., Slatko, B. E., Hornstra, L. J., and Benner, J. S. (1991) Characterization and expression of the *Escherichia coli* Mrr restriction system. J. Bacteriol. *173*, 5207–5219.

Walder, R. Y., Hartley, J. L., Donelson, J. E., and Walder, J. A. (1981) Cloning and expression of the *Pst*I restriction-modification system in *Escherichia coli*. Proc. Natl. Acad. Sci. USA *78*, 1503–1507.

Walder, R. Y., Langtimm C. J., Chatterjee, R., and Walder, J. A. (1983) Cloning of the *Msp*I modification enzyme. The site of modification and its effects on cleavage by *Msp*I and *Hpa*II. J. Biol. Chem. *258*, 1235–1241.

Walder, R. Y., Walder, J. A., and Donelson, J. E. (1984) The organization and complete nucleotide sequence of the *Pst*I restriction-modification system. J. Biol. Chem. *259*, 8015–8026.

Walter, J., Noyer-Weidner, M., and Trautner, T. A. (1990) The amino acid sequence of the CCGG recognizing DNA methyltransferase M. *Bsu*FI: implications for the analysis of sequence recognition by cytosine DNA methyltransferases. EMBO J. *9*, 1007–1013.

Walter, J., Trautner, T. A., and Noyer-Weidner, M. (1992) High plasticity of multispecific

DNA methyltransferases in the region carrying DNA target recognizing domains. EMBO J., in press.
Welsh, K. M., Lu, A.-L., Clark, S., and Modrich, P. (1987) Isolation and characterization of the *Escherichia coli mutH* gene product. J. Biol. Chem. *262*, 15,625–15,629.
Whittaker, P. A., Campbell, A. J. B., Southern, E. M., and Murray, N. E. (1988) Enhanced recovery and restriction mapping of DNA fragments cloned in a new λ vector. Nucl. Acids Res. *16*, 6725–6736.
Wilke, K., Rauhut, E., Noyer-Weidner, M., Lauster, R., Pawlek, B., Behrens, B., and Trautner, T. A. (1988) Sequential order of target-recognizing domains in multispecific DNA-methyltansferases. EMBO J. *7*, 2601–2609.
Wilson, G. G. (1991) Organization of restriction-modification systems. Nucl. Acids Res. *19*, 2539–2566.
Wilson, G. G. and Murray, N. E. (1991) Restriction and modification systems. Annu. Rev. Genet. *25*, 585–627.
Winkler, F. K., (1992) Structure and function of restriction endonucleases. Curr. Opin. Struct. Biol. *2*, 93–99.
Woodcock, D. M., Crowther, P. J., Doherty, J., Jefferson, S., DeCruz, E., Noyer-Weidner, M., Smith, S. S., Michael, M. Z, and Graham, M. W. (1989) Quantitative evaluation of *Escherichia coli* host strains for tolerance to cytosine methylation in plasmid and phage recombinants. Nucl. Acids Res. *17*, 3469–3478.
Wu, J. C., and Santi, D. V. (1987) Kinetic and catalytic mechanism of *HhaI* methyltransferase. J. Biol. Chem. *262*, 4776–4786.
Wyszynski, M. W., Gabbara, S., and Bhagwat, A. S. (1992) Substitutions of a cysteine conserved among DNA cytosine methylases result in a variety of phenotypes. Nucl. Acids Res. *20*, 319–326.
Xu, G., Kapfer, W., Walter, J., and Trautner, T. A. (1992) *Bsu*BI: an isospecific restriction and modification system of *PstI* – characterization of the *Bsu*BI genes and enzymes. Submitted to Nucl. Acids Res.
Yamaki, H., Ohtsubo, E., Nagai, K., and Maeda, Y. (1988) The *ori*C unwinding by *dam* methylation in *Escherichia coli*. Nucl. Acids Res. *16*, 5067–5073.
Yin, J. C. P., Krebs, M. P., and Reznikoff, W. S. (1988) Effect of *dam* methylation on Tn*5* transposition. J. Mol. Biol. *199*, 35–45.
Yuan, R. (1981) Structure and mechanism of multifunctional restriction endonucleases. Ann. Rev. Biochem. *50*, 285–315.
Yuan, R., and Hamilton, D. L. (1984) Type I and type III restriction-modification enzymes in: DNA methylation, pp. 11–37. Eds A. Razin, H. Cedar and A. D. Riggs. Springer-Verlag, New York.
Yuan, R., Bickle, T. A., Ebbers, W., and Brack, C. (1975) Multiple steps in DNA recognition by restriction endonuclease from *E. coli* K. Nature *256*, 556–560.
Yun, T., and Vapnek, D. (1977) Electron microscopic analysis of bacteriophages P1, P1Cm, and P7. Determination of genome sizes, sequence homology, and location of antibiotic resistance determinants. Virology *77*, 376–385.
Zell, R., and Fritz, H.-J. (1987) DNA mismatch-repair in *Escherichia coli* counteracting the hydrolytic deamination of 5-methyl-cytosine residues. EMBO J. *6*, 1809–1815.
Zyskind, J. W., and Smith, D. W. (1986) The bacterial origin of replication, *ori*C. Cell *46*, 489–490.

DNA Methylation: Molecular Biology and Biological Significance
ed. by J. P. Jost & H. P. Saluz

Structure, function and regulation of mammalian DNA methyltransferase

Heinrich Leonhardt and Timothy H. Bestor

Department of Anatomy and Cellular Biology, Laboratory of Human Reproduction and Reproductive Biology, Harvard Medical School, 45 Shattuck St, Boston, Massachusetts 02115, USA

1 Introduction

The haploid mammalian genome contains $\sim 5 \times 10^7$ CpG dinucleotides (Schwartz et al., 1962), about 60% of which are methylated at the 5 position of the cytosine residue (Bestor et al., 1984). The unmethylated fraction of the genome is exposed to diffusible factors in nuclei (Antequera et al., 1989), perhaps due to the action of proteins which bind to methylated sequences and induce their condensation (Meehan et al., 1989). Methylation may therefore control the availability of regulatory sequences for interaction with the transcriptional apparatus. Activation of tissue-specific genes is often accompanied by the disappearance of methyl groups from promoter regions, and differentiated cell types display characteristic unique methylation patterns. It has been argued that the selective advantage of such a regulatory mechanism would be expected to be most pronounced for those organisms with large genomes, and in fact 5-mC is absent from the DNA of most organisms with genomes smaller than 5×10^8 base pairs but essentially universal among organisms having genomes above this size (Bestor, 1990).

Methylation patterns undergo sweeping reorganization during gametogenesis and early development (Monk, 1990). Measurements of bulk 5-mC levels (Monk et al., 1987) and studies of an imprinted transgene (Chaillet et al., 1991) have shown that the DNA of primordial germ cells has a very low 5-mC content, and that sperm DNA is relatively more methylated than oocyte DNA. Methylation levels actually decline significantly in the preimplantation embryo (with the paternal DNA being more affected) to reach a minimum at the blastocyst stage. Methylation levels increase in the postimplantation embryo and adult levels of 5-mC are attained only after completion of gastrulation (Monk, 1990). It must be pointed out that these findings are averaged over a very large number of CpG sites, and that the behavior of many individual DNA sequences may be quite different than the genome-size average. It should also be noted that the number of methylated CpG sites exceeds the number of genes by a factor of about 50, and it is likely that the methylation status of a significant proportion of CpG dinucleotides is not subject to close regulation; this is consistent with the finding that many CpG sites show partial methylation in clonal cell populations. Strain-specific modifiers affect the methylation status of imprinted transgenes in mice and are likely to influence the methylation status of endogenous DNA sequences as well (Engler et al., 1991), and patterns of methylated CpG sites around certain genes undergo changes in aging mice (Uehara et al., 1989). The above findings make it clear that vertebrate methylation patterns are dynamic and subject to genetic and developmental control.

Several contributors to this volume discuss the role of methylation patterns in a variety of biological processes. Here we will be concerned with the mechanisms which establish and maintain patterns of methylated cytosine residues in the vertebrate genome. Because the only characterized component of the undoubtedly complex DNA methylating system is DNA methyltransferase itself, this enzyme will be the focus of attention.

2 Purification of mammalian DNA MTase

There is a long history of attempts to purify and characterize DNA (cytosine-5)-methyltransferase (DNA MTase) and numerous and often contradictory sizes and biochemical properties have been reported over the years (for a list of reported sizes, see Adams et al. (1990)). Recent purification and antibody studies have most frequently given an apparent M_r on SDS-polyacrylamide gel electrophoresis of around 190,000 for DNA MTase extracted from a number of proliferating human and murine cell types and tissues (Bestor and Ingram, 1985; Pfeifer and Drahovsky, 1986), including preimplantation mouse embryos (Howlett and Reik, 1991). DNA MTase is very sensitive to proteolysis, especially within the ~N-terminal 350 amino acids (Bestor, 1992), and smaller but enzymatically active cleavage products accumulate during purification. Proteolysis is presumably responsible for the smaller forms of DNA MTase that have been observed *in vivo* in non-dividing Friend murine erythroleukemia (MEL) cells, where a DNA MTase species of M_r 150,000 is found (Bestor and Ingram, 1985), and in full-term human placenta, where forms of DNA MTase of various smaller sizes have been identified (Pfeifer et al., 1985; Zucker et al., 1985). However, MEL cells have amplified the DNA MTase gene and express high levels of DNA MTase (Bestor et al., 1988), and full-term human placenta is an unusual non-proliferating tissue. At the present time it has not been proven that forms of DNA MTase smaller than M_r 190,000 are not the result of proteolysis either *in vivo* or during purification. It is most likely that the sole or predominant form of DNA MTase in normal somatic tissues and proliferating cell types has an apparent M_r of 190,000. The open reading frame in the cloned DNA MTase cDNA yields a calculated mass for the primary translation product of about 170,000, and expression of the cloned cDNA in COS cells yields a protein of about this apparent size (Czank et al., 1991). This observation suggests that DNA MTase normally undergoes a post-translational modification in mouse cells which retards its rate of migration on SDS-polyacrylamide gels. The nature of the modification is not yet known.

3 Sequence and structure of DNA MTase

The cDNA for DNA MTase from murine erythroleukemia cells was cloned by means of a degenerate synthetic oligonucleotide probe whose sequence was based on the amino acid sequence of a fragment of the purified enzyme (Bestor et al., 1988). The cDNA sequence revealed that DNA MTase consists of a 1,000 amino acid N-terminal domain linked to a C-terminal domain of about 500 amino acids that is closely related to bacterial type II DNA C5 methyltransferases. About 30 of the bacterial enzymes have been sequenced, and all contain 10 conserved motifs in invariant order (Lauster et al., 1989; Posfai et al., 1989). For reasons that are not clear none of the known DNA C5 methyltransferases have recognition sequences of 6 bp. All also contain a variable region between conserved motifs VIII and IX (Fig. 1) which has been shown by mutagenesis experiments to confer sequence specificity to the transmethylation reaction (Klimasauskas et al., 1991; Lange et al., 1991). Figure 1 shows the organization of conserved motifs in the C-terminal domain of DNA MTase compared to M. *DdeI* (the most closely related bacterial enzyme; Szynter et al., 1987; Bestor et al., 1988) and M. *SssI*, a *Spiroplasma* methyltransferase whose recognition sequence is the dinucleotide CpG (Renbaum et al., 1990). Despite the fact that DNA MTase and M. *SssI* recognize the same DNA sequence, the variable region of DNA MTase is dissimilar in amino acid sequence and more than twice as large as that of M. *SssI*, and in fact is the longest of the monospecific C5 DNA methyltransferases. As discussed elsewhere it is likely that mammalian DNA MTase is the result of fusion between genes for a prokaryotic-like restriction methyltransferase and an unrelated DNA binding protein (Bestor, 1990).

The C-terminal methyltransferase domain of DNA MTase is joined to the N-terminal domain by a run of 13 alternating glycyl and lysyl residues. In the center of the N-terminal domain is a cluster of 8 cysteinyl residues which has been shown to bind Zinc ions (Bestor,

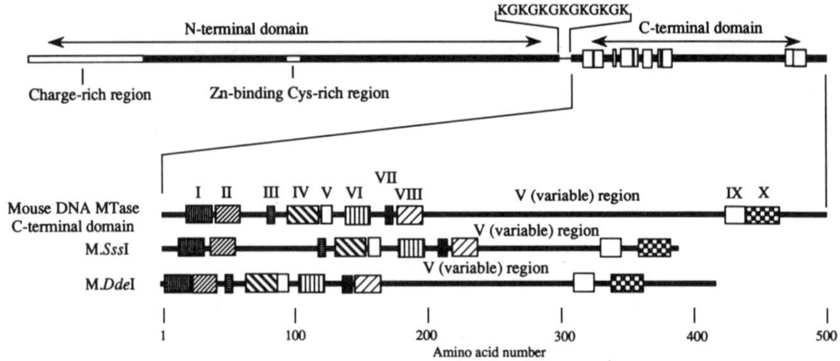

Figure 1. Sequence features and conserved motifs in mammalian DNA MTase. At top is a diagram of sequence features in DNA MTase; below is a depiction of elements in the C-terminal domain conserved between bacterial and mammalian DNA C5 methyltransferases. Boxes with common fill patterns indicate conserved motifs and are numbered I through X. Motif I is the putative S-adenosyl L-methionine binding site (Ingrosso et al., 1989), IV is the prolylcysteinyl active center (Wu and Santi, 1987; Chen et al., 1991), and the variable region is involved in sequence recognition (Lange et al., 1991; Klimasauskas et al., 1991). Note that the order of the conserved motifs is invariant and the variable region of DNA MTase is much longer than that of M. *SssI*, which recognizes the same sequence. M. *DdeI* methylates the cytosine residue in the sequence CTNAG.

1992). As described below, the N-terminal domain is involved in the discrimination of unmethylated and hemimethylated DNA, and the Zinc binding site is likely to be involved in this function. The first 200 amino acids of the N-terminal domain are very rich in charged and polar amino acids, and the first ~350 amino acids are very sensitive to proteolysis. Deletion of these sequences does not affect *in vitro* enzymatic activity or preference for hemimethylated sites (Bestor and Ingram, 1985).

4 *De novo* and maintenance methylation

Riggs (1975) and Holliday and Pugh (1975) predicted that vertebrate methylation patterns could be transmitted by clonal inheritance through the action of a DNA methyltransferase that was strongly stimulated by or dependent on hemimethylated DNA, which is the product of semiconservative DNA replication. This led to the expectation of two types of DNA methyltransferases: *de novo* enzymes, which would establish tissue-specific methylation patterns during gametogenesis and early development (in concert with a system that erased methylation patterns in the germline), and maintenance enzymes, which would ensure the clonal transmission of lineage-specific methylation patterns in somatic tissues. Razin and collaborators (Gruenbaum et al., 1982) showed that a DNA

MTase activity in extracts of somatic nuclei preferred hemimethylated substrates by a large factor, although *de novo* methylation was also observed. It was later shown that the *de novo* and maintenance activities reside in the same protein and that the preference for hemimethylated sites was 30–40 fold higher (Bestor and Ingram, 1983; Pfeifer et al., 1983; Bolden et al., 1984). Somatic cells do have the capacity to perform *de novo* methylation; methylation patterns are slowly restored after treatment with the demethylating drug 5-azacytidine (Flatau et al., 1984), and *de novo* methylation of the promoter regions of tissue-specific genes is observed in cells in long-term culture (Antequera et al., 1990). These findings confirm that *de novo* methylation is not confined to cells of the germline or early embryo, although *de novo* methylation of foreign DNA does appear to be much more efficient in embryonic cells (Jahner and Jaenisch, 1985). While the prediction of a distinct class of *de novo* DNA methyltransferases has not been confirmed, the existence of such enzymes cannot yet be excluded. It should soon be possible to answer the question definitively through use of a sensitive, versatile, and highly specific probe for DNA C5 methyltransferases recently introduced by Gregory Verdine's laboratory (Chen et al., 1991). Oligonucleotides containing the modified nucleoside 5-fluorodeoxycytidine (FdC) have been shown to trap a covalent transition state intermediate between DNA and DNA methyltransferases in a form that is stable to strong denaturing conditions, as predicted by Santi et al. (1983). If the FdC-containing oligonucleotide is radioactive, the covalent complexes with DNA methyltransferases can be visualized by autoradiography after electrophoresis on SDS-polyacrylamide gels. This mechanism-based probe and inhibitor should provide sub-femtomol sensitivity, and it will be possible to test lysates of cell populations in which *de novo* methylation are occurring (especially germ cells and cells of the preimplantation embryo) for species of DNA methyltransferase distinct from the known M_r 190,000 form. Immobilization of the FdC-containing oligonucleotides on a solid support should allow rapid purification of any new species, and amino acid sequencing of proteins purified in this way will allow cloning.

5 Discrimination of hemimethylated and unmethylated CpG sites

Bacterial and mammalian DNA methyltransferases differ most markedly in that the type II bacterial enzymes do not discriminate between hemimethylated and unmethylated recognition sequences. Adams and colleagues (Adams et al., 1983) observed an increased rate of *de novo* methylation after treatment of a crude DNA MTase preparation with trypsin and concluded that the enzyme must contain a protease-sensitive domain that makes contacts with the C5 methyl group of hemimethylated sites. In double-stranded B form DNA the C5 positions of cytosine residues in CpG sites are separated by only a few Ångstroms in the major groove, and analysis of bacterial restriction methyltransferases have suggested that at least 3 regions of the protein must be very close to the target cytosine (Fig. 1); these are the S-adenosyl L-methionine binding site near the N-terminus (Ingrosso et al., 1989), the prolylcysteinyl dipeptide at the catalytic center (Wu and Santi, 1987), and a region near the C-terminus that mediates sequence-specific DNA binding (Lange et al., 1991; Klimasaukas et al., 1991). All these regions are within the C-terminal domain of mammalian DNA MTase. While contacts between the methyl group and any of these motifs might be expected to mediate discrimination of unmethylated and hemimethylated sites, the results of recent proteolysis experiments indicate that the discrimination is carried out by distant sequences in the N-terminal domain of DNA MTase. Protease V8 cleaves DNA MTase between the N- and C-terminal domains, as shown by microsequencing of fragments. Cleavage caused a large stimulation in the rate of *de novo* methylation without significant change in the rate of methylation of hemimethylated DNA; this demonstrates that the N-terminal domain inhibits the *de novo* activity of the C-terminal methyltransferase domain (Bestor, 1992). The finding was unexpected, as the close proximity of the methyl group in a hemimethylated CpG site to the C5 position of the target cytosine imposes severe steric constraints and it seems unlikely that an additional protein structural element could be accommodated near the target cytosine in the major groove. This and other lines of evidence (Bestor, 1992) lead to the conclusion that it is methylation-dependent structural alterations in DNA, rather than direct contact of the protein with major groove methyl groups, that is responsible for discrimination of unmethylated and hemimethylated CpG sites. This conclusion is not without precedent; DNase I preferentially cleaves methylated CpG sites (Fox, 1986), and yet this enzyme makes contacts only in the minor groove of B form DNA (Lahm and Suck, 1991). DNase I must therefore sense cytosine methylation indirectly through alterations of DNA structure rather than via direct major groove contacts. However, the physical separation between the catalytic and regulatory regions of DNA MTase suggests that the mechanism used by DNA MTase in the discrimination of unmethylated and hemimethylated CpG sites is fundamentally different than any known type of DNA:protein interaction.

Cleavage between the N- and C-terminal domains stimulates *de novo* methylation, and because most purification schemes measure *de novo* activity in assays, the purification method which gives the best apparent yield will be that which most favors proteolysis. The sensitivity of DNA MTase to proteolysis and the fact that most biochemical characterization of the enzyme has involved partially purified enzyme preparations with unknown extents of proteolysis is likely to be part of the cause for the wide range of enzymatic properties ascribed to DNA MTase.

6 *De novo* sequence specificity

Little is known of how sequence-specific methylation patterns are established in the mammalian genome. The sequence specificity of purified DNA MTase does not extend much past the CpG dinucleotide (Gruenbaum et al., 1981; Simon et al., 1983; Hubrich et al., 1989; Bestor and Ingram, 1983), and cell types with different methylation patterns contain species of DNA MTase that are identical by all criteria, including *de novo* sequence specificity (Bestor et al., 1988).

There are several candidate mechanisms for sequence-specific methylation. First, as mentioned earlier it is possible that tissue- and sequence-specific *de novo* methyltransferases are expressed at specific stages of development and that the altered methylation patterns are maintained in somatic tissues through the maintenance activity of the known form of DNA MTase. While there is no evidence for a family of DNA methyltransferases, their existence remains a possibility. The sensitive and versatile FdC-oligonucleotide probes described earlier should provide an answer to the question of multiple species of DNA methyltransferases in mammals. Second, *de novo* methylation may be relatively indiscriminate during certain stages of development, and critical CpG sites might be protected from methylation by sequence-specific masking proteins. At such times the *de novo* activity of DNA MTase might be stimulated by proteolytic cleavage between the N- and C-terminal domains or interaction with a factor which counteracts the inhibitory effects of the N-terminal domain. The masking model cannot be looked on with much favor, as it is precisely the unmethylated CpG sites which are accessible to diffusible factors in nuclei (Antequera et al., 1989), and genomic sequencing has not shown a bias in the sequences flanking methylated and unmethylated CpG sites (Jost et al., 1990). Sequence-specific masking proteins would be expected to leave some evidence of a consensus sequence around unmethylated CpG sites. Third, a family of specificity factors, analogous to the specificity subunits of bacterial type I restriction-modification systems, might interact with DNA MTase to confer sequence specificity while enhancing *de novo* methylation activity. This possibility suffers the same problem as the masking proteins: there is no evidence of a consensus sequence around methylated or unmethylated CpG sites. Furthermore, proteins that interact strongly with DNA MTase have not been identified. Fourth, it is possible that *de novo* methylation is indiscriminate and that tissue-specific methylation patterns are established by sequence-specific demethylation. Sequence-specific demethylation, presumably through a mechanism related to excision repair, has been documented in the case of the chicken vitellogenin gene (Jost et al., 1990) and could be widespread. It is sobering to recognize that at the present time it is not known whether tissue-specific methylation patterns are established by sequence-specific *de novo* methylation, by indiscriminate *de novo* methylation and sequence-specific demethylation, or by some combination of the two.

7 Targeted disruption of the DNA MTase gene in mice and in mouse cells

The regulatory role of DNA methylation remains controversial, in large part because reversible, tissue-specific methylation patterns are restricted to large-genome organisms such as vertebrates and vascular plants in which genetic approaches are limited. It has recently become possible to introduce predetermined mutations in any mouse gene for which cloned probes are available by gene targeting in embryonic stem (ES) cells (Mansour et al., 1988). This approach has been used to disrupt both alleles of the DNA MTase gene in ES cells with a construct which introduces a short deletion-replacement at the translational start site (Li et al., 1992). The mutation is a partial loss of function allele which produces trace amounts of a slightly smaller protein, as established by gel electrophoresis and immunoblotting. Net enzyme activity *in vitro* assays is about 5% of wildtype. This is limiting, and the homozygous mutant ES cells and embryos have about one-third of the wildtype level of 5-mC in their DNA. The homozygous mutant ES cells show no discernible phenotype even after prolonged passage *in vitro*. The mutation has also been established in the germline of mice. Homozygous mutant embryos complete gastrulation and the early stages of organogenesis but are stunted, delayed in developmental stage, and fail to develop past the 20 somite stage. Histological analysis shows that many cells in the mutant embryos contain fragmented, pycnotic nuclei which are typical of apoptosis rather than necrosis. It was interesting to find that reduced 5-mC levels are lethal at the stage where normal embryos attain adult levels of 5-mC in their DNA (Monk, 1990). In addition to the partial loss of function mutation, a second independent mutation was constructed by means of a targeted insertion mutation in sequences downstream of the region targeted by the first construct. This presumptive severe loss of function mutation causes homozygous embryos to die at earlier stages and to have less 5-mC in their DNA than does the partial loss of function mutation, and embryos with one copy each of the partial and severe loss of function mutation die at intermediate stages and have intermediate levels of 5-mC in their DNA.

Embryos homozygous for the partial loss of function mutation retain $\sim 1 \times 10^7$ methylated CpG sites per genome, one-third of the wildtype level. This finding shows that even fairly modest reductions in 5-mC content which have no apparent effect on the phenotype of cultured ES cells completely prevent normal development past midgestation. The cause of the developmental block is not known, but an attractive and testable hypothesis is inappropriate gene expression as a result of the activation of genes that are normally repressed by methylation.

Embryos homozygous for the partial loss of function mutation complete gastrulation and the early stages of organogenesis. They will therefore serve as a robust test system for hypotheses regarding the importance of DNA modification in developmental gene control, X inactivation, genomic imprinting, virus latency, and other biological phenomena in which DNA methylation has been proposed to play a role.

Acknowledgments. Supported by the National Institutes of Health and the March of Dimes Birth Defects Foundation.

Adams, R. L., Burdon, R. H., McKinnon, K., and Rinaldi, A. (1983) Stimulation of *de novo* methylation following limited proteolysis of mouse ascites DNA methylase. FEBS Lett. *163*, 194–198.

Adams, R. L. P., Bryans, M., Rinaldi, A., Smart, A., and Yesufu, H. M. I. (1990) Eukaryotic DNA methylases and their use for *in vitro* methylation. Phil. Trans. R. Soc. Lond. B *326*, 11–21.

Antequera, F., Boyes, J., and Bird, A. (1990) High levels of *de novo* methylation and altered chromatin structure at CpG islands in cell lines. Cell *62*, 503–14.

Antequera, F., Macleod, D., and Bird, A. P. (1989) Specific protection of methylated CpGs in mammalian nuclei. Cell *58*, 509–517.

Bestor, T. (1990) DNA methylation: evolution of a bacterial immune function into a regulator of gene expression and genome structure in higher eukaryotes. Phil. Trans. R. Soc. Lond. B *326*, 179–187.

Bestor, T. (1992) Activation of mammalian DNA methyltransferase by cleavage of a Zinc-binding regulatory domain. EMBO J., *11*, 2611–2617.

Bestor, T., Hellewell, S. B., and Ingram, V. M. (1984) Differentiation of two mouse cell lines is accompanied by demethylation of their genomes. Mol. Cell. Biol. *4*, 1800–1806.

Bestor, T., Laudano, A., Mattaliano, R., and Ingram, V. (1988) Cloning and sequencing of a cDNA encoding DNA methyltransferase of mouse cells. The carboxyl-terminal domain of the mammalian enzymes is related to bacterial restriction methyltransferases. J. Mol. Biol. *203*, 971–83.

Bestor, T. H., and Ingram, V. M. (1983) Two DNA methyltransferases from murine erythroleukemia cells: purification, sequence specificity, and mode of interaction with DNA. Proc. Natl. Acad. Sci. USA *80*, 5559–63.

Bestor, T. H., and Ingram, V. M. (1985) Growth-dependent expression of multiple species of DNA methyltransferase in murine erythroleukemia cells. Proc. Natl. Acad. Sci. USA *82*, 2674–8.

Bolden, A., Ward, C., Siedlecki, J. A., and Weissbach, A. (1984) DNA methylation. Inhibition of *de novo* and maintenance methylation *in vitro* by RNA and synthetic polynucleotides. J. Biol. Chem. *259*, 12437–43.

Chaillet, J. R., Vogt, T. F., Beier, D. R., and Leder, P. (1991) Parental-specific methylation of an imprinted transgene is established during gametogenesis and progressively changes during embryogenesis. Cell *66*, 77–83.

Chen, L., MacMillan, A. M., Chang, W., Ezaz, N. K., Lane, W. S., and Verdine, G. L. (1991) Direct identification of the active-site nucleophile in a DNA (cytosine-5)-methyltransferase. Biochemistry *30*, 11018–25.

Czank, A., Häuselman, R., Page, A. W., Leonhardt, H., Bestor, T. H, Schaffner, W., and Hergersberg, M. (1991) Expression in mammalian cells of a cloned gene encoding murine DNA methyltransferase. Gene *109*, 259–263.

Engler, P., Haasch, D., Pinkert, C. A., Doglio, L., Glymour, M., Brinster, R., and Storb, U. (1991) A strain-specific modifier on mouse chromosome 4 controls the methylation of independent transgene loci. Cell *65*, 939–947.

Flatau, E., Gonzales, F. A., Michalowsky, L. A., and Jones, P. A. (1984) DNA methylation in 5-aza-2′-deoxycytidine-resistant variants of C3H 10T1/2 C18 cells. Mol. Cell. Biol. *4*, 2098–102.

Fox, K. R. (1986) The effect of HhaI methylation on DNA local structure. Biochem. J. *234*, 213–216.

Gruenbaum, Y., Stein, R., Cedar, H., and Razin, A. (1981) Methylation of CpG sequences in eukaryotic DNA. FEBS Lett. *124*, 67–71.

Gruenbaum, Y., Cedar, H., and Razin, A. (1982) Substrate and sequence specificity of a eukaryotic DNA methylase. Nature *295*, 620–622.

Holliday, R., and Pugh, J. (1975) DNA modification mechanisms and gene activity during development. Science *187*, 226–232.

Howlett, S. K., and Reik, W. (1991) Methylation levels of maternal and paternal genomes during preimplantation development. Development *113*, 119–127.

Hubrich, K. K., Buhk, H. J., Wagner, H., Kroger, H., and Simon, D. (1989) Non-C-G recognition sequences of DNA cytosine-5-methyltransferase from rat liver. Biochem. Biophys. Res. Commun. *160*, 1175–1182.

Ingrosso, D., Fowler, A. V., Bleibaum, J., and Clarke, S. (1989) Sequence of the D-aspartyl/L-isoaspartyl protein methyltransferase from human erythrocytes. Common sequence motifs for protein, DNA, RNA, and small molecule S-adenosylmethionine-dependent methyltransferases. J. Biol. Chem. *264*, 20131–20139.

Klimasauskas, S., Nelson, J. L., and Roberts, R. J. (1991) The methylase specificity domain of cytosine C5 methylases. Nucl. Acids Res. *19*, 6183–6190.

Jahner, D., and Jaenisch, R. (1985) Chromosomal position and specific demethylation in enhancer sequences of germ line-transmitted retroviral genomes during mouse development. Mol. Cell. Biol. *5*, 2212–2220.

Jost, J.-P., Saluz, H.-P., McEwan, I., Feavers, I. M., Hughes, M., Reiber, S., Liang, H. M., and Vaccaro, M. (1990) Tissue specific expression of avian vitellogenin gene is correlated with DNA hypomethylation and *in vivo* specific protein-DNA interactions. Phil. Trans. Roy. Soc. Lond. B *326*, 53–63.

Lahm, A., and Suck, D. (1991) DNase I-induced DNA conformation. J. Mol. Biol. *221*, 645–667.

Lange, C., Jugel, A., Walter, J., Noyer, W. M., and Trautner, T. A. (1991) 'Pseudo' domains in phage-encoded DNA methyltransferases. Nature *352*, 645–648.

Lauster, R., Trautner, T. A., and Noyer-Weidner, M. (1989) Cytosine-specific type II DNA methyltransferases. A conserved enzyme core with variable target-recognizing domains. J. Mol. Biol. *206*, 305–312.

Li, E., Bestor, T. H., and Jaenisch, R. (1992) Targeted mutation of the DNA methyltransferase gene results in embryonic lethality. Cell *69*, 915–926.

Mansour, S. L., Thomas, K. R., and Capecchi, M. R. (1988) Disruption of the protooncogene int-2 in mouse embryo-derived stem cells: a general strategy for targeting mutations to non-selectable genes. Nature *336*, 348–352.

Meehan, R. R., Lewis, J. D., McKay, S., Kleiner, E. L., and Bird, A. P. (1989) Identification of a mammalian protein that binds specifically to DNA containing methylated CpGs. Cell *58*, 499–507.

Monk, M. (1990). Changes in DNA methylation during mouse embryonic development in relation to X-chromosome activity and imprinting. Phil. Trans. Roy. Soc. Lond. B. *326*, 179–187.

Monk, M., Boubelik, M., and Lehnert, S. (1987) Temporal and regional changes in DNA methylation in the embryonic, extraembryonic, and germ cell lineages during mouse embryo development. Development *99*, 371–382.

Pfeifer, G. P., and Drahovsky, D. (1986) DNA methyltransferase polypeptides in mouse and human cells. Biochim. Biophys. Acta *868*, 238–242.

Pfeifer, G. P., Grunwald, S., Palitti, F., Kaul, S., Boehm, T. L., Hirth, H. P., and Drahovsky, D. (1985) Purification and characterization of mammalian DNA methyltransferases by use of monoclonal antibodies. J. Biol. Chem. *260*, 13787–13793.

Pfeifer, G. P., Grunwald, S., Boehm, T. L., and Drahovsky, D. (1983) Isolation and characterization of DNA cytosine 5-methyltransferase from human placenta. Biochim. Biophys. Acta *740*, 323–330.

Posfai, J., Bhagwat, A. S., Posfai, G., and Roberts, R. J. (1989) Predictive motifs derived from cytosine methyltransferases. Nucl. Acids Res. *17*, 2421–2435.

Renbaum, P., Abrahamove, D., Fainsod, A., Wilson, G. G., Rottem, S., and Razin, A. (1990) Cloning, characterization, and expression in Escherichia coli of the gene coding for the CpG DNA methylase from Spiroplasma sp. strain MQl (M. *SssI*). Nucl. Acids Res. *18*, 1145–1152.

Riggs, A. D. (1975) X inactivation, differentiation, and DNA methylation. Cytogenet. Cell Genet. *14*, 9–25.

Santi, D. V., Garrett, C. E., and Barr, P. J. (1983) On the mechanism of inhibition of DNA-cytosine methyltransferases by cytosine analogs. Cell *33*, 9–10.

Simon, D., Stuhlmann, H., Jahner, D., Wagner, H., Werner, E., and Jaenisch, R. (1983) Retrovirus genomes methylated by mammalian but not bacterial methylase are non-infectious. Nature *304*, 275–277.

Schwartz, M. N., Trautner, T. A., and Kornberg, A. (1962) Enzymatic synthesis of DNA. J. Biol. Chem. *237*, 1961–1967.

Sznyter, L. A., Slatko, B., Moran, L., O'Donnell, K. H., and Brooks, J. E. (1987) Nucleotide sequence of the DdeI restriction-modification system and characterization of the methylase protein. Nucl. Acids Res. *15*, 1846–1866.

Uehara, Y., Ono, T., Kurishita, A., Kokuryu, H., and Okada, S. (1989) Age-dependent and tissue-specific changes of DNA methylation within and around the c-*fos* gene in mice. Oncogene *4*, 1023–1028.

Wu, J. C., and Santi, D. V. (1987) Kinetic and catalytic mechanism of HhaI methyltransferase. J. Biol. Chem. *262*, 4778–4786.

Zucker, K. E., Riggs, A. D., and Smith, S. S. (1985) Purification of human DNA (cytosine-5)-methyltransferase. J. Cell. Biochem. *29*, 337–349.

DNA Methylation: Molecular Biology and Biological Significance
ed. by J. P. Jost & H. P. Saluz

Regulation of *de novo* methylation

R. L. P. Adams, H. Lindsay, A. Reale, C. Seivwright, S. Kass, M. Cummings and C. Houlston

Department of Biochemistry, University of Glasgow, Glasgow G12 8QQ, Scotland

Abbreviations. M is used to represent methylcytosine when in nucleotide sequences. AdoMet and AdoHcy refer to S-adenosyl methionine and S-adenosyl homocysteine respectively.

1 Introduction to the *in vivo* situation

It has become dogma that vertebrate DNA, and indeed the DNA of most animals, contains methylcytosine that is restricted to the dinucleotide MG (where M is 5-methylcytosine). Plants have a higher proportion of DNA cytosines methylated largely because they have additional methylcytosine in the trinucleotide sequence MNG (Belanger and Hepburn, 1990). Not all CG (nor CNG) sequences are methylated and yet those that are, are maintained in the methylated form from cell generation to generation in a tissue specific manner. This maintenance of methylation pattern is believed to be an intrinsic property of the enzyme catalysing the transfer of methyl groups from S-adenosyl methionine (AdoMet) to the DNA; an enzyme known as a DNA-(5-cytosine) methyltransferase or DNA methylase. In other words, the methylase acts on newly synthesized DNA to add a methyl group to cytosines on the daughter strand which are paired with guanines in MG dinucleotide (or MNG trinucleotide) sequences present on the parental strand.

This dogma fails to explain how patterns of methylation can be changed. In the next sections we will review evidence showing that in many instances methyl groups are 'lost' from DNA whilst in other instances they are added to DNA, at sites in which neither strand was previously methylated, by a process known as *de novo* methylation. Are these processes of loss and *de novo* methylation brought about by a different enzymic machinery; by a failure of the DNA methylase to function reliably; or by a regulated alteration in intracellular conditions which leads to a modulation of the maintenance activity? Although all three suggestions are pertinent in different situations, this paper will concentrate on the ways in which the DNA methylase can carry out *de novo* methylation and the controls that might be brought to bear on this activity.

1.1 Maintenance of methylation patterns in established cells and tissues

It is over 10 years since Wigler et al. (1981) introduced, into mouse cells, a cloned chicken thymidine kinase gene methylated *in vitro* with *Hpa*II methylase. The clones obtained maintained the introduced pattern of methylation for over 25 generations with just less than 100% fidelity. In this case the expression of the gene was under positive selection but, in a similar study, Stein, Razin and Cedar (1982) showed that the gene for adenine phosphoribosyltransferase maintained its introduced pattern of methylation even in the absence of selection or expression. These studies showed that, in an established cell culture line, a stable pattern of methylation could be maintained, i.e. in a constant environment maintenance methylation is effective and *de novo* methylation insignificant.

More recent experiments carried out using more natural systems have confirmed these findings. Thus Doerfler's group (Behn-Krappa et al., 1991) have shown that the pattern of methylation of particular genes in lymphocytes and other blood cells has been maintained throughout human evolution and, furthermore, this pattern does not change when the lymphocytes are stimulated to divide. The pattern is different, however, from that found in cultured human cell lines. Also, Pfeifer et al. (1990a and b) have shown that 118 out of 120 CG dinucleotides in the promoter proximal region of the human *PGK-1* gene on the inactive X-chromosome are maintained in a fully, or almost fully, methylated state whereas one pair is maintained unmethylated. All these sites remain unmethylated on the active X-chromosome.

It is thus clear that, once established, maintenance of methylation patterns can be efficient. The question that these studies do not address is whether or not this maintenance is a property of the DNA methylase acting in isolation or whether it is a consequence of the action of other factors which direct the enzyme to specific sites. In a constant environment such factors would be present in a stable milieu and could regulate the action of a *de novo* DNA methylase such that it might seem to be acting as a maintenance enzyme. Such is the situation with the prokaryotic Type II restriction methylases and the related Dam and Dcm methylases which, *de facto*, maintain the pattern of methylation following replication, though it is known in these cases that the enzyme is not affected by the methylation status of the non-substrate strand (Rubin and Modrich, 1977). Furthermore, the action of the Dam methylase is blocked when substrate sites are membrane bound resulting in these sites remaining hemimethylated for over 10 min following their replication (Campbell and Kleckner, 1990).

1.2 Situations in which methylation patterns are changed

In contrast to the above examples where methylation patterns are stable, there are many examples known in which the patterns of DNA methylation change; either by gain or loss of specific methyl groups. Such changes are seen most clearly during gametogenesis and early development (Cedar and Razin, 1990) but are also apparent during differentiation and changes in expression of genes following imposition or removal of selection or on transformation of cells (Antequera et al., 1990; Achten et al., 1991; Jones et al., 1990).

When selection pressures are changed, there are often, but not always, associated changes in DNA methylation patterns. For example, the HSV *tk* gene present in transformed cells growing under HAT selection is not methylated and does not normally become methylated when selection is removed even though expression ceases (Pellicer et al., 1980; Davies et al., 1982). In these cases reexpression is a frequent event when selection is reimposed. In a few instances, however, inactivation is associated with *de novo* methylation and in these instances reexpression is uncommon but its frequency can be increased by treatment of cells with the demethylating agent, 5-azacytidine (Christy and Scanglos, 1982).

Kruczek and Doerfler (1982) have shown that the viral DNA in adenovirus transformed hamster cells can become methylated at all regions except the promoter proximal sequence of the early, transforming genes. These genes, whose expression is selected for, remain unmethylated in these 5′ regions and this clearly demonstrates an interference with *de novo* methylation at these sites. Although in the initial tumour the level of methylation may be low, there is a marked increase when tumour cells are explanted into culture. This increase in methylation applies only to the integrated viral DNA and not to other cellular genes tested. How the *de novo* methylation is initiated and how it spreads along the integrated viral DNA has recently been investigated (Toth et al., 1990; Orend et al., 1991). There seems to be a favoured site for the initiation of the *de novo* methylation which then spreads in both directions. The nature of the factors or sequences that predispose a site for initiation of *de novo* methylation are unknown as is the mechanism of spreading. The situation, however, is reminiscent of the initiation and spreading of X-chromosome inactivation (and *de novo* methylation) seen in the early embryogenesis of female mammals (Migeon, 1990; Lyon, 1991). A similar spreading mechanism was proposed in 1975 by Holliday and Pugh to explain how a biological clock might work. Their mechanism involved an enzyme capable of maintenance methylation and of *de novo* methylation at an adjacent site and while such activity is possible (see below) we have found no evidence that the mouse DNA methylase walks along the DNA efficiently adding methyl groups.

Indeed, as we show below, this enzyme appears to work in a distributive manner and it is likely that spreading of methylation patterns involves factors, other than the methylase, present in the chromatin. A similar conclusion has been reached by Doerfler's group (Doerfler et al., 1990; Toth et al., 1990) who showed that the spreading of methylation was interrupted by proteins bound to the DNA.

Introduction of genes into the cells derived from early mouse embryos or into teratocarcinoma cells can also be associated with *de novo* methylation (Jähner et al., 1982) and this may reflect the fate of many endogenous genes at this time in development (Monk et al., 1987; Chaillet et al., 1991). This *de novo* methylation is facilitated by the high levels of DNA methylase present in the egg and early embryo (Monk et al., 1991; Howlett and Reik, 1991). However, it does not seem to apply to all genes and housekeeping genes are amongst those excluded from this *de novo* methylation (Cedar and Razin, 1990) and some even undergo demethylation at this time (Frank et al., 1991). As most clearly exemplified by the allele-specific methylation of one X-chromosome in female mammals, exclusion from *de novo* methylation cannot be the consequence of the activity of the DNA methylase but must rely on other factors and, moreover, factors that do not readily exchange between identical alleles on the two chromosomes within a nucleus.

It is thus clear that there must be chromatin associated factors that act to limit *de novo* methylation in a tissue and allele specific manner. It is surprising, however, that as yet no definite evidence has been produced to show that *de novo* methylation is regulated by transcription factors bound to DNA. Yet imprinting is believed to be one consequence of the differential action of these factors during spermatogenesis and oogenesis and unmethylated DNA is more likely to be associated with transcription factors *in vivo* (e.g. Pfeifer et al., 1990a). By contrast, several examples are known (see Adams, 1990a) in which DNA methylation interferes with transcription factor binding, and at least one example is known (Vaccaro et al., 1990) where methylation is important for transcription factor binding.

It should also be borne in mind that, although there may be an overall inverse correlation between methylation and gene expression, there are many examples where demethylation of a gene is not accompanied by increased transcription pointing to the regulation of expression as well as of methylation by extrinsic chromosomal proteins.

2 Results of *in vitro* experiments

The results presented in this section are largely those that we have obtained with *in vitro* studies with the DNA methylase obtained from mouse ascites tumour cells and, to a lesser extent, the enzyme from pea shoot tips. There will be a sustained attempt, however, to relate our results with those of others in the field.

2.1 Evidence that one enzyme molecule can carry out both maintenance and de novo methylation

Bestor et al. (1988) have cloned the mouse DNA methylase cDNA and report that the gene is a single copy gene and only one size of mRNA is seen (Fig. 1 and Szyf et al., 1991). The size of the mRNA that we detect in mouse erythroleukaemia cells is rather larger (about 7000 nucleotides) than that reported by Bestor et al. (1988). Both enzyme activity and levels of mRNA are low in resting (serum arrested) cells

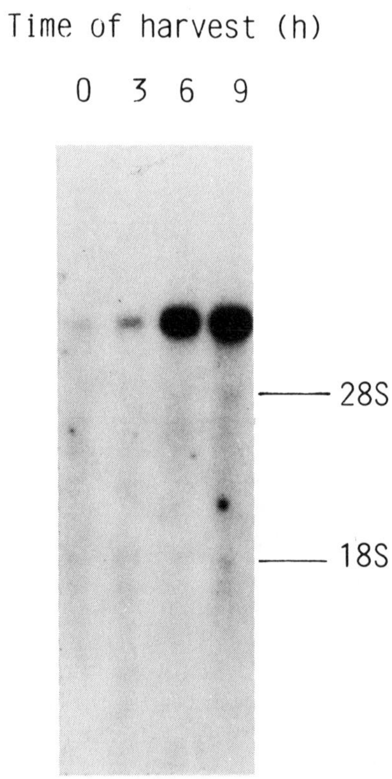

Figure 1. Northern blot analysis of DNA methylase mRNA levels in cells at different phases of the cell cycle. Mouse erythroleukaemia cells were fractionated by centrifugal elutriation (Adams, 1990b) and G1-phase cells selected and replaced in culture at zero time. Samples were harvested after 0, 3, 6 and 9 h and total RNA isolated. 10 μg was applied to each lane of a 1% agarose gel and the Northern blot probed with a random primed insert from the mouse DNA methylase plasmid (Bestor et al., 1988). The arrows indicate the positions of ribosomal 28S and 18S RNA.

(Szyf et al., 1991); and the same is true for cells in G1-phase (Adams, 1990 and Fig. 1). The levels of enzyme and mRNA rise dramatically as cells enter S-phase showing that there is control of transcription and also a rapid turnover of the enzyme as cells divide. This confirms the idea that control of expression is both at transcriptional and post-transcriptional levels. However, even though there is only one gene and one size of mRNA there are several sizes of protein on SDS gels that react to antisera directed against the enzyme (see below).

The intact enzyme as isolated from rapidly growing mouse cells such as ascites, mastocytoma or erythroleukaemia cells (Adams et al., 1986; Pfeifer et al., 1985b; Bestor and Ingram, 1985) or eggs or early embryos (Howlett and Reik, 1991) is a monomer of about 195 kDa but the enzyme isolated from rat or mouse liver or human placenta is smaller probably as a result of partial proteolysis *in vivo* (Zucker et al., 1985; Simon et al., 1978; Adams et al., 1989; Pfeifer et al., 1985b). The question is not yet resolved as to whether different processed fragments carry out different reactions in the cell though we have previously shown that limited proteolysis does increase the *in vitro de novo* methylase activity of the enzyme (Adams et al., 1983).

There is a similar controversy about the size of the plant enzyme. We (Yesufu et al., 1991) have isolated the enzyme from pea shoot tips and find a protein of about 160 kDa but smaller enzymes have been isolated from wheat and rice (Theiss et al., 1987; Giordano et al., 1991). Whether this represents a phylogenetic difference or the partial breakdown of the enzyme from monocotyledons is not known for certain but the pea enzyme is very susceptible to proteolysis on storage.

2.2 Reactions catalysed by DNA methylase

It is important to clarify the sort of reaction that can be catalysed by the enzyme. Table 1 shows that relative activities of a partially purified preparation of mouse ascites DNA methylase with several different model DNA substrates present at 14 μg per ml. An extremely good substrate is the double-stranded polynucleotide containing alternating I and C bases. This is a much better substrate than a similar duplex containing G and C bases and this is believed to be the result of increased stability of the latter though other contributory factors are the possible formation of Z-DNA or triple or four-stranded structures. Clearly methylation of (poly dI-dC)·(poly dI-dC) is, at least initially, *de novo* methylation. However, the hemimethylated analogue (prepared by extensive replication of priming fragments of (poly dI-dC)·(poly dI-dC) in the presence of dCTP and dMTP) is not much better when used under these conditions (see below). This is surprising as a hemimethylated oligonucleotide duplex containing only one CG (SVpro-ds

Table 1. Effect of different substrates on methyltransfer. The indicated substrates were incubated with the partially purified mouse DNA methylase (18 μg) and tritiated AdoMet for 1 h at 37° in a volume of 70 μl and the incorporation of tritium into DNA measured.

DNA substrate (1 μg per assay)	Incorporation (cpm)	cpm/pmol CG
none	41	—
(poly dI-dC)·(poly dI-dC)	49,700	34.8
(poly dI-dC)·(poly dI-dM)[a]	44,400	62.1
(poly dI-dC)·(poly dI-dM)[b]	56,800	107.5
$(CAG)_7(CTG)_7$	148	0.11[c]
$(CAG)_7(MTG)_7$	392	0.73[c]
SVpro-ss[e]	37	0
SVpro-ds (CG/CG)	1,122	8.6[d]
SVpro-ds (CG/MG)	17.240	265[d]
SVpro-ds (CNG/MNG)	5,340	41.1[d]
SVpro-ds mut	140	—

1000 cpm is approximately equal to the transfer of 1 pmol methyl group.
(a) This hemimethylated substrate was prepared by annealing equal amounts of (poly dI-dC)·(poly dI-dC) with $(dI\text{-}dM)_3$.
(b) This hemimethylated substrate was prepared using Klenow DNA polymerase and a 3/1 ratio of dCTP/dMTP.
(c) As there are no CGS in these molecules the calculation is based on Cs in CNG sites.
(d) These figures are based on all the methylation being in CG sites.
(e) SVpro is a 22-base oligonucleotide from the promoter region of SV40 DNA. It contains one CG.

(CG/MG)) accepts 30 times more methyl groups per substrate cytosine than does its unmethylated analogue.

A single strand of SVpro is an extremely poor acceptor of methyl groups. In contrast, it has been shown previously that single-stranded *M. luteus* DNA is a better substrate than duplex *M. luteus* DNA. The distinction probably lies in the ability of the bacterial DNA to form duplex regions, however transiently, whereas the oligonucleotide does not contain any significant, self-complementary regions. These findings are similar to those reported by Carotti et al. (1986) and Smith et al. (1991) and indicate that the enzyme must interact with both strands of a duplex molecule at the site of methyl transfer.

A further finding shown in Table 1 is the fact that the enzyme is able to methylate oligonucleotides lacking CG dinucleotides. The fact that the enzyme is not specific for cytosines in CG dinucleotides is shown by the activity with SVpro-mut a duplex in which the

CG has been replaced with TG
GC AC

This substrate accepts methyl groups at 5–9% the rate shown by the control duplex, SVpro-ds(CG/CG). Furthermore, cytosines in a 21mer duplex having $(CAG)_7$ on one strand and $(CTG)_7$ on the other are methylated at 0.3% the rate of (poly dI-dC)·(poly dI-dC) and this rate is increased 6.6-fold when $(CTG)_7$ is replaced by $(MTG)_7$.

These results show that, *in vitro*, the mouse DNA methylase is able to methylate sequences other than CG. Similar conclusions have been reached previously by Simon et al. (1980) who showed methylation of CA and CT dinucleotides and by Pfeifer et al. (1985a) who concluded that the 5′ cytosine in CCGG sequences could be methylated by the enzyme from mouse mastocytoma cells. Such findings are compatible with the evidence of Woodcock (1987), that at least half the methylcytosines in eukaryotic DNA are not in MG dinucleotides, and of Toth et al. (1990) who, by genomic sequencing, have identified methylcytosine in MA and MT dinucleotides. This is also consistent with the analysis of Smith et al. (1991) who showed that the base following the substrate cytosine is of little consequence although it would normally be paired with cytosine or methylcytosine.

Other studies (Hepburn and Tisdale, 1991) have shown that the methyltransferase also interacts strongly with the O^6 position of guanine residues (whether in CG dinucleotides in DNA or in polyG) which implies a partial opening of the double helix as this oxygen is normally involved in base pairing. Although interaction with the O^6 position of guanine residues still leaves two hydrogen bonds between the CG base pair, a similar interaction with (poly dI-dC)·(poly dI-dC) would reduce the CI base pair to a single hydrogen bond. Santi et al. (1983) have shown that DNA-5-cytosine methyltransferases saturate the 5:6 double bond in substrate cytosines by forming a covalent link between carbon-6 and a cysteine residue at the active site of the enzyme. The implications of these studies are that both maintenance and *de novo* methylation involve the interaction of the enzyme with both the C and the G of a CG base pair and that the template strand should have a cytosine (or preferably a methylcytosine) closely 5′ to the G:

i.e. 5′ CN or possibly 5′ CNN
GC 5′ GNC 5′

Introduction into SVpro-ds of two methylcytosines in one strand to generate two MNG sequences enhances incorporation fivefold even though no hemimethylated CGs are present. One of the methylcytosines is present in the duplex sequence

MCG
GGC

As we have shown above that the hemimethylated $(CAG)_7 \cdot (MTG)_7$ duplex sequence is a better acceptor than the unmethylated $(CAG)_7 \cdot (CTG)_7$ duplex sequence, it is likely that the enzyme would first methylate this trinucleotide duplex region to give

MCG
GGM

which contains a hemimethylated CG that would be further methylated to give

MMG
GGM

The implications of this are twofold. Firstly, it confirms that the enzyme is not entirely specific for CG and, secondly, aberrant methylations particularly at the 5′ cytosine in CCG sequences can be very effective at enhancing the chance of *de novo* CG methylation.

These results with the mouse enzyme should be contrasted with the results of parallel experiments carried out with the enzyme from pea seedlings (Houlston et al., 1992). In this case the hemimethylated $(CAG)_7 \cdot (MTG)_7$ sequence is by far the best substrate and we have evidence that the enzyme interacts with this DNA with very high affinity. A very stable complex is quickly formed which withstands dissociation with all but the strongest denaturating agents. This complex is formed much more readily than are complexes with $(CAG)_7 \cdot (CTG)_7$ or with (poly dI-dC)·(poly dI-dC). Strangely, partially hemimethylated (poly dI-dC)·(poly dI-dC) is a very poor substrate for the pea enzyme despite the fact that the hemimethylated duplex SVpro-ds(CG/MG) is very much better than the unmethylated SVpro-ds(CG/CG).

2.3 Covalent and non-covalent linkage of the DNA substrate to the enzyme

Figure 2 shows the effect of DNA concentration on the relative rates of incorporation of methyl groups into unmethylated and hemimethylated (poly dI-dC)·(poly dI-dC) catalysed by the mouse DNA methyltransferase. At the lower concentrations where the amount of (poly dI-dC)·(poly dI-dC) is limiting and when initial rates are considered (Fig. 2b), the rate of *de novo* methylation is about 14-fold greater than the rate of maintenance methylation. Estimates of affinity of the enzyme for the DNA show no significant difference, both giving a Km of about 4.3 μg per ml.

It is clear that, as previously reported by ourselves and others, there is inhibition at higher concentrations of DNA. (Palitti et al., 1987; Adams et al., 1989). The inhibition is seen even at short incubation times and with either unmethylated or hemimethylated DNA. Palitti et al. (1987) did not see significant inhibition with high concentrations of (poly dI-dC)·(poly dI-dC); but concluded that inhibition was the result of formation of a tight complex between DNA and enzyme. We have now shown that this inhibition is a result of covalent binding of the substrate to the enzyme for which neither AdoMet, nor CG dinucleo-

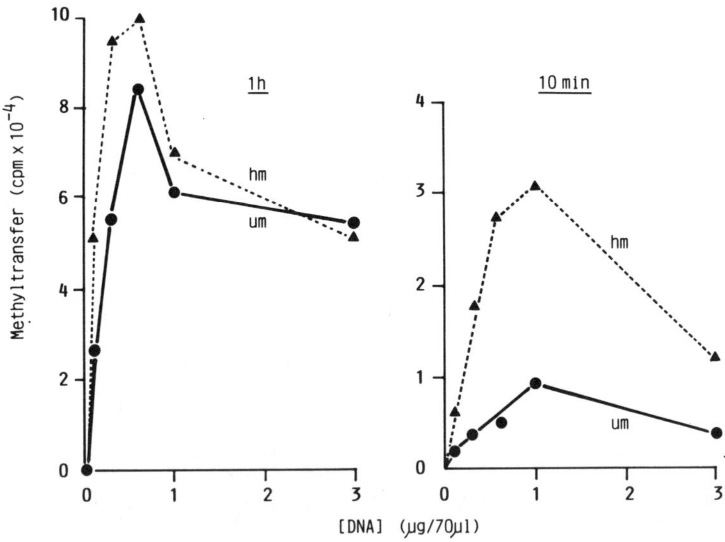

Figure 2. DNA concentration curves for mouse DNA methylase assays carried out with a 1-h (left) or a 10-min incubation. 18 µg partially purified enzyme was present in each incubation.

tides are essential (Fig. 3a) but the complex formation is inhibited by N-ethylmaleimide showing that it involves the reaction of an enzyme cysteine with the DNA. Binding occurs to both the unmethylated strand and the methylated strand in a hemimethylated duplex as well as to unmethylated and fully methylated duplex DNA. Binding occurs with the non-CG containing oligonucleotide (SVpro-ds mut) and, although this interaction is compatible with the finding that such a substrate can act as a methyl group acceptor (Table 1), the relative strengths of the two activities are unrelated. Binding takes place slowly over 24 h as an increasing proportion of the substrate becomes immobilised (Fig. 3b). As some degradation of the enzyme is occurring during the incubation, an increasing proportion of DNA-bound enzyme fragments appear on the gel with increasing time of incubation. The pattern of DNA-bound fragments seen in Figure 3 is similar to that seen with NEM and AdoMet binding (Fig. 6) and with some Western blots (see below) and probably represents binding to active-site-containing fragments of differing size. Even though undetectable amounts of the 195 kDa enzyme remain in the preparation of trypsinised enzyme used, it is clear that this polypeptide is involved prominently in the binding reaction showing that this is not just an artefact of the degraded enzyme (although binding to a partially denatured enzyme cannot be ruled out). We envisage that the enzyme is reacting with the DNA substrate in a substantially normal manner but at sites where neither methyl transfer nor complex dissociation can occur efficiently. These sites may be

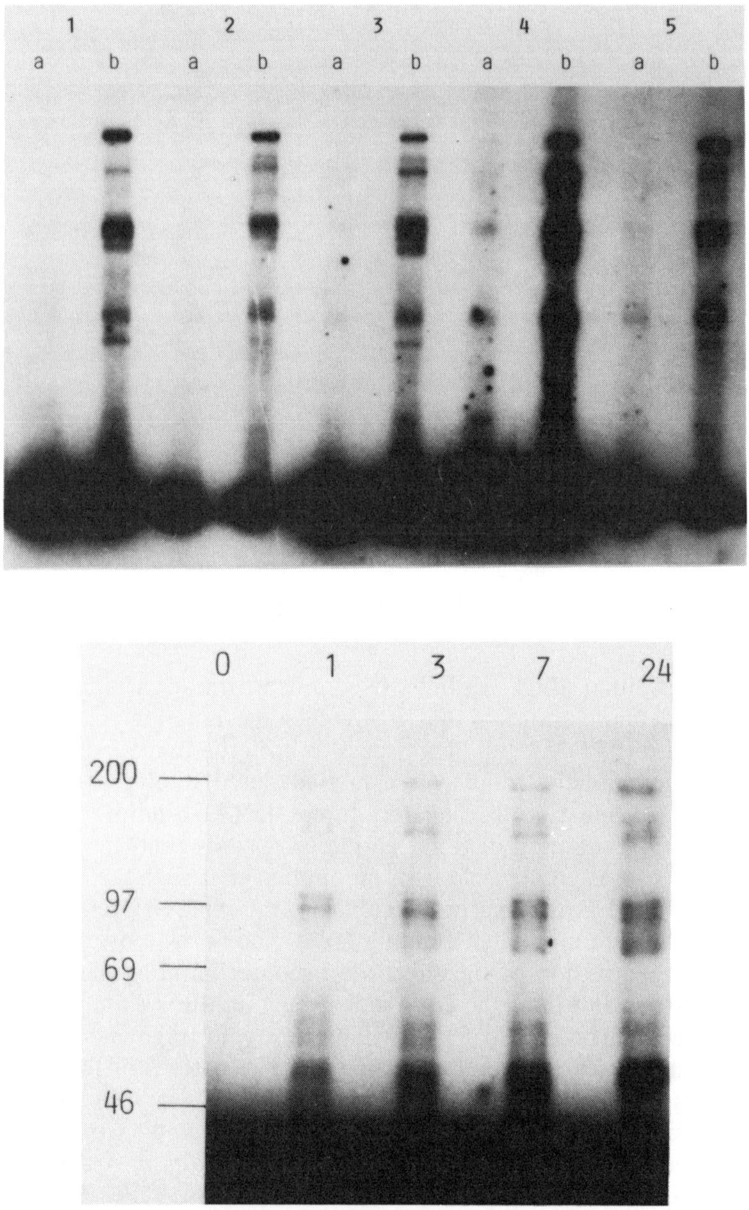

Figure 3. Covalent binding of DNA methylase to DNA. A shows the results of SDS polyacrylamide gel electrophoresis of incubation mixtures containing 18 μg enzyme protein and 20 ng ^{32}P-labelled oligonucleotide. In part A the incubation time at 37° was for zero (a) or 3 h (b). The oligonucleotides used were: 1, SV40pro-ds(CG/MG) in which the unmethylated strand was labelled; 2, SV40pro-ds(CG/CG); 3, SV40pro-ds(MG/MG); 4, SV40pro-ds mut; and 5, SV40pro-ds(CG/MG) in which the methylated strand was labelled. In part B the incubation was with SV40pro-ds(CG/MG) for the indicated number of hours. The size (in kDa) of marker proteins is indicated down the left hand side of the diagram.

cytosines in sequences other than CG or possibly at thymines (Palitti et al., 1987) or even guanines (Hepburn and Tisdale, 1991) and the complexes formed may be similar to those formed with azacytosine containing DNA (Santi et al., 1984). Formation of such complexes probably arises through interaction of the enzyme with nucleotides that are unable to accept a methyl group and illustrates the lack of an absolute specificity in the binding of DNA methylase to DNA.

Although like covalent binding occurs at short incubation times (Fig. 3b), these experiments were done at very low DNA concentrations (1 μg per ml) with a thousand-fold excess of enzyme protein. Were the ratio to be reversed, most of the enzyme could quickly become DNA-bound and this could be a major factor causing the inhibition of methyltransfer by high DNA concentrations that is seen even at 10 min incubation (Fig. 2b).

In addition to this covalent binding, gel retardation studies have shown that the partially proteolized enzyme can react non-covalently with oligonucleotide duplexes containing a CG dinucleotide (Saluz, H., Adams, R. L. P., and Jost, J.-P., unpublished). When methyltransfer has been effected, this enzyme:DNA complex dissociates such that the free DNA in the incubation is fully methylated (as indicated by Maxam Gilbert sequencing) whereas the unmethylated DNA is still protein bound. It may be the failure of the release mechanism that leads to a small proportion of the enzyme molecules becoming covalently bound to the DNA.

Several reports have shown that the initial interaction of the enzyme with duplex DNA is inhibited by concentrations of NaCl in the 100–200 mM range, but this is manifest only with limiting enzyme. With (poly dI-dC)·(poly dI-dC) at 28 μg per ml and one unit of enzyme, almost 100% inhibition of transfer is seen at 140 mM NaCl. However, with a high ratio of enzyme to DNA, even 200 mM NaCl has a relatively slight inhibitory action on either *de novo* or maintenance methylation (Fig. 4). Thus, although salt can inhibit interaction of the enzyme with DNA, this is not complete and is overcome if sufficient enzyme molecules are present to ensure that each DNA molecule is always associated with an enzyme molecule. In fact, throughout an incubation, the reaction becomes progressively less sensitive to inhibition by 200 mM NaCl such that, after 90 min, further methylation of (poly dI-dC)·(poly dI-dC) is essentially resistant to inhibition implying that all unmethylated DNA is enzyme bound (Fig. 4b). This is consistent with the conclusions of the gel retardation studies referred to above.

Another factor that has been considered important in methylation studies is the size of the substrate DNA. We have studied the accepting ability of (poly dI-dC)·(poly dI-dC) molecules ranging in size from about 20 bp up to 600 bp (generated by either DNase I or *Cfo*I digestion) and find only minor changes in accepting ability.

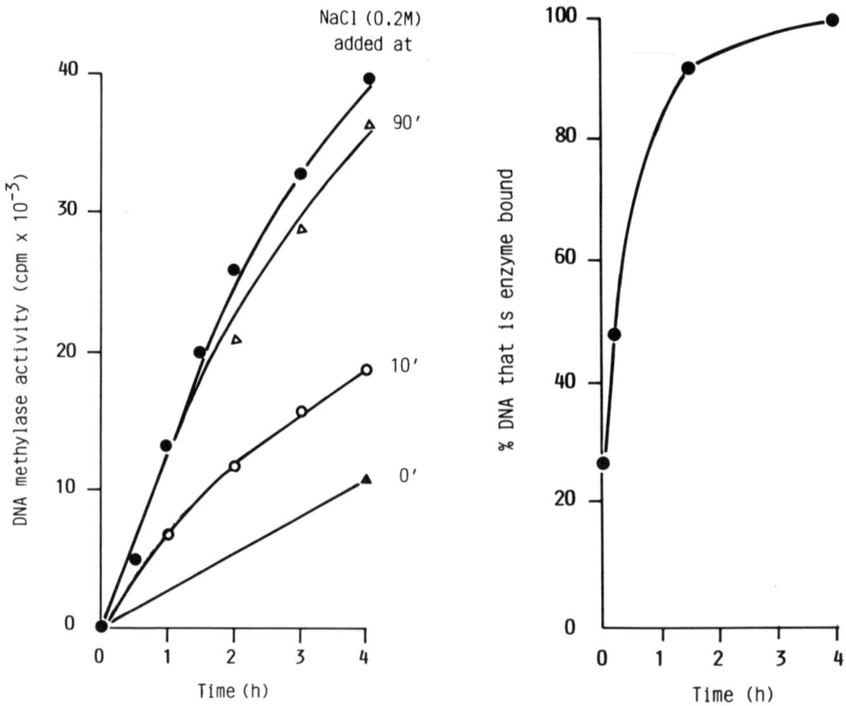

Figure 4. Formation of DNA/enzyme complex. Part A indicates the effect of adding NaCl (200 mM) at various times during an incubation of 0.1 μg (poly dI-dC)·(poly dI-dC) with mouse DNA methylase (18 μg protein). The values for part B are derived from the relative incorporation at 4 h.

2.4 The time course of de novo and maintenance methylation

Methylation of (poly dI-dC)·(poly dI-dC) will initially produce a hemimethylated product. A detailed study of the time course of methylation of (poly dI-dC)·(poly dI-dC) and the hemimethylated analogue (which contains both unmethylated CI and hemimethylated CI dinucleotide pairs) shows that the reaction with the former starts off slowly whereas, with the partially hemimethylated substrate, there is a very rapid initial reaction (Fig. 5). This rapid initial burst of methylation with the partially hemimethylated substrate can be attributed largely to the completion of methylation at the hemimethylated sites which is followed by the slower *de novo* methylation.

Looking at Figure 5 in more detail, the solid circles show the time course of addition of methyl groups to the 143 pmoles of cytosine in 0.1 μg (poly dI-dC)·(poly dI-dC). Incorporation starts off slowly (a) and then accelerates (b). We interpret this to mean that the slow reaction represents methylation of single cytosines (2–3 pmol of the

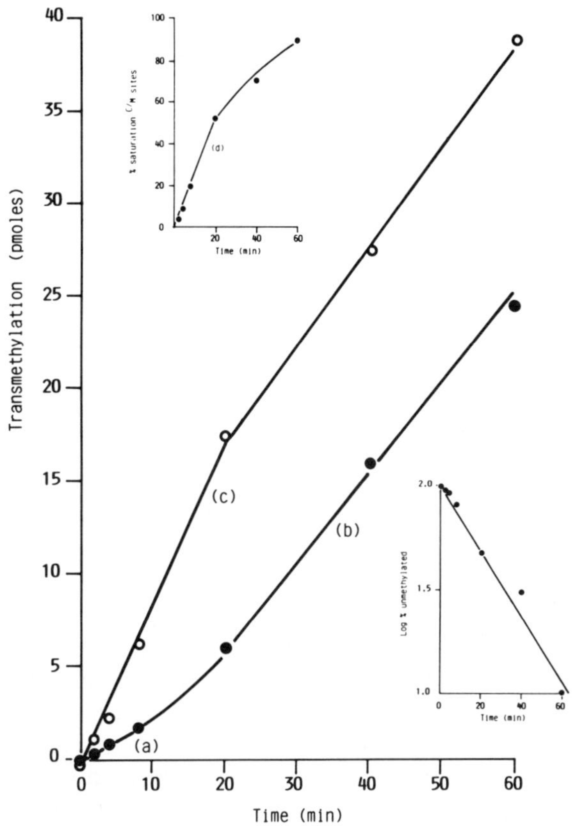

Figure 5. Time course of incubation of mouse DNA methylase (18 mg protein) with 0.1 mg of unmethylated (solid circles) or hemimethylated (open circles) (poly dI-dC)·(poly dI-dC). For an explanation of the two inserts see text.

143 pmol in the first 10 min) and only later does the rate increase as a proportion of hemimethylated sites arise. Thus the enzyme appears not to methylate both strands of an unmethylated duplex at the same time; rather the two reactions are independent. Were methylation of the two strands to occur essentially simultaneously, it is likely that the incorporation would be linear with time. The 0.1 μg of hemimethylated (poly dI-dC)·(poly dI-dM) contains 80 pmol of cytosines as unmethylated pairs and 27 pmol of cytosine in hemimethylated form. If we assume that the 80 pmol is methylated in parallel with the 143 pmol in the unmethylated substrate we can subtract these values from those obtained for the transmethylation of the partially hemimethylated substrate which gives us the time course of methylation of the hemimethylated sites, shown in the upper insert in Figure 5. Clearly, maintenance methylation starts off quickly (d) and 50% of the sites are

saturated in 20 min whereafter the rate slows so that only 90% saturation is achieved in the hour. The shape of this curve is what would be predicted for an enzyme interacting with the substrate DNA as a prelude to each methyl transfer reaction (Fig. 5, lower insert).

Comparing initial rates one can get an estimate of the relative rate of *de novo* and maintenance methylation per cytosine (d/a) of about 17.

2.5 Distributive action of de novo methylation

If the rate of maintenance methylation is *very much greater* than that of *de novo* methylation, then the product of the latter reaction should be DNA with fully methylated CG dinucleotides. However, the results of Figure 5, discussed above, suggest that this is not the case and that there is an independent methylation of the two strands.

The use of (poly dI-dC)·(poly dI-dC) as substrate allows an investigation of the efficiency of the methylase to act *de novo* at all available sites. If, on incubation with this substrate, the enzyme travels along the DNA methylating every cytosine, then long lengths of the methylated (radioactive) substrate should become resistant to cleavage by *Cfo*I (cleaves unmethylated GCGC or ICIC sites but not IMIC sites) even though the bulk of the DNA remains sensitive. However, this is not the case. Electrophoresis and fluorography shows that the fragment size of *Cfo*I-treated DNA, either methylated and unmethylated, is less than 15 bp. These results indicate that the enzyme either dissociates from the DNA after each methyltransfer or, more likely, that it slides past many cytosines in potential substrate sites without catalysing methyltransfer.

It is clear from these studies that for both maintenance and *de novo* methylation the enzyme interacts independently with the DNA substrate for each methyltransfer. This does not exclude a sliding mechanism but it does imply that if such a mechanism prevails it fails to direct the enzyme to each contiguous site on the DNA. The situation appears different for the pea enzyme which interacts very strongly and rapidly with hemimethylated CNG sequences (Houlston et al., 1992) and this may ensure that methylation of these sequences is much more strictly maintained than for CG sequences in plants or animals.

2.6 Sequence preference and poison sequences. CpG islands

Weissbach's group (Bolden et al., 1985; 1986; Ward et al., 1987) have provided evidence that the partially purified DNA methylases from mouse erythroleukaemia cells and HeLa cells show intrinsic selection of the cytosines that they will methylate and the rate at which methyltransfer occurs. They have shown that the enzymes preferentially methylate cytosines in a CG-rich environment but the opposite conclusion has been reached by Carotti et al. (1989); and Szyf et al. (1990) have found a region in a CpG island that protects that island from methylation *in vivo*. All these effects could be mediated by protein factors present in the DNA methylase preparation and in the cell nucleus and may not be a property of the pure enzyme.

We have shown that the mouse ascites DNA methylase is able to methylate particular oligonucleotide molecules to completion (Saluz et al., unpublished results) but we have not found it possible to methylate a population of DNA molecules to completion. This problem is greater when using tritiated AdoMet at low concentrations and this points to product inhibition by accumulating AdoHcy though attempts to remove this using adenosine deaminase have been unsuccessful. In addition, some of the explanation for incomplete methylation lies in the inactivation of the enzyme by covalent binding to substrate molecules discussed above. Methylation of a plasmid DNA to about 60%, followed by Maxam Gilbert sequencing (Bryans et al., 1992) has failed to indicate significant methylation of the GC boxes in the SV40 promoter region despite the good accepting activity of the oligonucleotide SVpro-ds referred to above. Similar methylation of the plasmid by the bacterial methylase M*SssI* completely saturates all sites studied. These results might point to the presence of a poison sequence in this plasmid and this is currently under investigation.

2.7 Inhibitors of de novo methylation

If the DNA methylase is capable, albeit inefficiently, of *de novo* methylation, what prevents all CG sequences (and CNG sequences in plants) from being methylated? Progression towards complete methylation should be particularly apparent the greater the time since division as the opportunity for *de novo* methylation would be greater.

Clearly, there is a window immediately after replication when the levels of DNA methylase are high and when the chance of methylation is greatest. This is essential in order to efficiently maintain methylation patterns. On the other hand, as cells enter G_0 the activity of the enzyme falls though there is still evidence for activity even in unstimulated peripheral lymphocytes and brain cells which have not undergone division for weeks or months. Activity is readily assayed in liver nuclear extracts though the size of the enzyme as measured on Western blots is much reduced. Although methylation is required following repair of damaged DNA, there is no evidence for methylation of preexisting DNA in nondividing cells. What prevents such action?

One might suppose that inhibitors of methylation fall into two classes; the general and the specific. Chromatin structure may play an integral role in preventing access of DNA methylase to the DNA at times other than S-phase. We have already shown (Davis et al., 1986) that histones inhibit DNA methylase activity and this effect has been narrowed down by Caiafa et al. (1991) to histone H1. This histone, which also interferes with transcription (Laybourn and Kadonaga, 1991), binds to the nucleosomal DNA at linker regions and serves to condense the chromatin into a tight solenoidal configuration that is largely inaccessible to other proteins. Such heterochromatin would be refractory to methylation as well as to transcription. This is confirmed by the finding that the endogenous, tightly-bound DNA methylase present in isolated mouse cell nuclei carries out delayed, maintenance methylation of newly synthesized DNA (Davis et al., 1985). It is only when enzyme is added exogenously to such nuclei that any *de novo* methylation occurs and this is at the more accessible nucleosomal linker regions (Davis et al., 1986). Such studies indicate the presence of a general inhibitor of *de novo* methylation that is present in nuclei. To what extent such an inhibitor could act at regions of transcriptionally active chromatin, deficient in histone H1, is not known and such regions could be potential targets for *de novo* methylation.

Some *de novo* methylation may occur during S-phase, though if it occurs on only one strand of DNA it may be of limited significance. At certain sites, however, the presence of even a single methyl group could be a problem and special mechanisms may exist to prevent *de novo* methylase activity at such sites. Such mechanisms may involve the participation of sequence-specific DNA binding proteins.

The majority of CG dinucleotides (apart from those in repetitive DNA) are in genes that are frequently transcribed (e.g. ribosomal and transfer RNA genes) or in the CpG-islands that form a characteristic feature of many promoter/proximal regions. Other CG dinucleotides occur only sparsely and seldom has the importance of their methylation status (nor their presence) been shown. Binding of transcription factors to promoter regions may inhibit methylation. Although this has not been shown in any single instance, it is the implicit assumption in models of gene regulation that propose a competition on newly replicated DNA between the various factors and enzymes that might recognise specific sequences on the DNA.

We have been looking for protein factors, other than histones, that might interfere with DNA methylase action. As substrates we have been using the unmethylated and hemimethylated SVpro-ds oligonucleotide duplexes where the preference for the hemimethylated site is already 31-fold (Table 1). We have looked in nuclear extracts for factors that selectively inhibit *de novo* methylation. Table 2 shows that a hypotonic buffer wash of crude nuclei from Krebs II ascites tumour cells (S_0) increases this discrimination a further sixfold by strongly inhibiting the *de novo* reaction. A subsequent wash in buffer containing 0.2 M NaCl

Table 2. Inhibition of *de novo* methylation. The table shows the results of two separate experiments in which SV40pro-ds DNA (either 0.1 or 1.0 μg of unmethylated or hemimethylated) was incubated with 18 μg of mouse ascites DNA methylase for one hour in the presence of nuclear extracts. S_0 was a zero salt extract of nuclei and S_2 a subsequent extract with buffer containing 0.2 M NaCl. In the first experiment a combined extract was used. Incorporation with S_0 or S_2 alone has been subtracted. Results are given as cpm.

DNA	control	$+S_0+S_2$	% inhibition
0.1 μg CG/CG	55	0	100
0.1 μg CG/MG	1,995	700	65
Main/*de novo*	36	∞	
1.0 μG CG/CG	975	35	96
1.0 μg CG/MG	13,405	6,030	55
Main/*de novo*	14	172	

DNA	control	$+S_0$	$+S_2$	% inhibition by S_0
1.0 μg CG/CG	3,320	150	960	95
1.0 μg CG/MG	16,400	9,590	14,730	42
Main/*de novo*	5	64	15	

(S_3) (or higher concentration of NaCl – not shown) does not contain significant inhibitory material and, in fact, is the starting point for our DNA methylase purification. We are at present fractionating the hypotonic buffer extract in order that we might determine the nature of the inhibitory component. However, this is not the extract where we would expect to find transcription factors nor other proteins that might interact ionically with DNA and this leads us to suggest that here there is a protein that interacts directly with the methyltransferase. Proteins that affect the DNA or AdoMet directly are unlikely to be involved as the inhibition is restricted to the *de novo* reaction.

On incubating the mouse DNA methylase with nuclear fractions S_0 or S_2, there is no change in the position or intensity of bands which react to anti DNA methylase antibody on Western blotting and so we have no evidence for covalent modification of the enzyme by factors in S_0 or S_2. Surprisingly, however, it is clear that S_0, while showing little DNA methylase activity, does show evidence of significant amounts of DNA methylase protein; and the 0.2 M NaCl nuclear extract is much more active when the nuclei are first given a hypotonic wash than when this step is omitted.

2.8 Location of activities on the enzyme molecule

The mouse DNA methylase is a large protein which is particularly prone to proteolysis both *in vivo* and *in vitro*. We are attempting to localise functional regions on the enzyme in an attempt to locate the domains that are essential for the limitation of *de novo* methylase activity. A number of "affinity" labels are being developed to identify important regions of the enzyme.

DNA methylase is inhibited by —SH reactive chemicals, such as N-ethylmaleimide (NEM). This is believed to be the result of the binding of the NEM to a cysteine residue in the active site. We have incubated ^{14}C-labelled NEM with the enzyme and then subjected it to partial proteolysis in the hope that we would be able to consistently locate the fragment carrying the active site. However, in the initial enzyme preparation there are lower molecular weight species that react strongly with NEM (Fig. 6) and, although we are convinced that these are fragments of DNA methylase, this tends to complicate the picture. A similar pattern of bands is seen with ^{14}C-NEM labelled enzyme immunoprecipitated by an antibody raised against the purified enzyme (Fig. 6, lane 4). It is clear that a series of higher molecular weight species of between 140 and 195 kDa bind NEM and that, on extensive trypsinization, binding is seen strongly with five fragments of 29 to

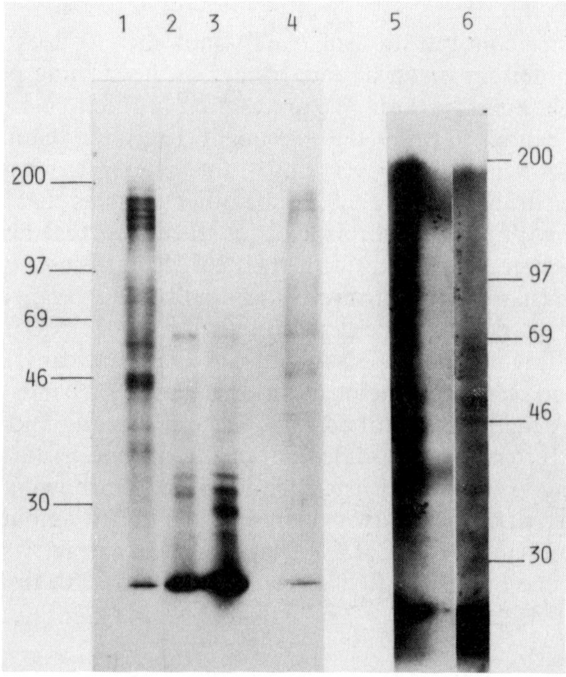

Figure 6. Covalent binding of ^{14}C-labelled N-ethylmaleimide (lanes 1–4) or ^{3}H AdoMet (lanes 5 and 6) to mouse DNA methylase. Enzyme was incubated or cross-linked to the radiolabelled molecule as described in the methods section and then separated on an SDS polyacrylamide gel either immediately (lanes 1 and 5) or after partial trypsinization (lanes 2, 3 and 6) or immunoprecipitation with an antimethylase antiserum (lane 4). The figure shows the resulting fluorograph with the position of size markers (in kDa) down the sides.

37 kDa (Fig. 6, lane 3). Intermediate fragments that had bound NEM prior to proteolysis are of 95, 85, 73, 66, 55 and 48–50 kDa. Assuming that the —SH group at the active site is the only one that is accessible to NEM in the native enzyme, one possible explanation for these results is that the active site to which the NEM binds is within 29 kDa (about 290 amino acids) of one (the C-terminal) end of the enzyme. In this case, all the fragments could arise by removal of increasing amounts of amino-terminal protein. A more likely explanation is that fragments are being removed from both sides of the active site, though the 55 kDa fragment could contain the C-terminus. If this is the situation, then the tryptic fragments could arise by cleavage at the run of lys-gly residues that occurs at about 100 amino acids before the presumptive active site. There is another run of lysines about 600 amino acids from the start of the protein and cleavage at this site would generate a fragment of around 100 kDa. A fragment of this size is typically found in liver nuclear extracts and the presence of such a fragment has been associated with high levels of *de novo* methylase activity.

Binding of AdoMet is also expected to occur at the active site of the methylase and we have used ultraviolet light to crosslink ^{3}H-labelled AdoMet to the enzyme (Fig. 6, lanes 5 and 6). As well as the high molecular weight enzyme, fragments of 33–35 kDa also crosslink to AdoMet. Although it is impossible to be certain that the various proteolytic fragments line up exactly, it is likely that the same set of fragments is labelling with each of these reagents.

Interaction with DNA would also be expected to occur at the active site of the enzyme and we have described above experiments in which covalently linked enzyme-DNA complexes form (Fig. 3). As well as high molecular weight, DNA-bound enzyme molecules, proteolytic fragments that bind DNA include those of 70–100 kDa together with bands at 46 and 50 kDa. In some cases, labelled fragments of 33 and 35 kDa are distinguishable from the background radiation.

The major "affinity" reagents we are investigating are antibodies raised to peptides corresponding to different regions of the deduced amino acid sequence of the enzyme. This approach requires that the nucleotide sequence is accurate and that no frame shifts are present which is apparently not the case for the published sequence. Tim Bestor was kind enough to let us have a revised sequence but there is still no guarantee of 100% accuracy. Nonetheless, we have raised five antipeptide antibodies and others will be available soon.

More detailed results of these studies with peptide antibodies will be presented elsewhere, but preliminary findings show that the antibodies to peptides towards the C-terminus do not react with a band at 29 kDa (confirming this is an internal peptide) but that they react strongly with a band at 44 kDa that is not illuminated with an antibody to the active site (about 500 amino acids from the C-terminus). This band is a likely candidate for the C-terminal fragment. It is noteworthy that all the antipeptide antibodies that we have raised to regions of the enzyme towards the C-terminus react with the set of high molecular weight bands. This confirms the proposal that these arise by sequential processing of the N-terminal region of the enzyme. We have not yet any evidence, however, that this is what causes the increase in *de novo* methylase activity as there is concomitant processing of the enzyme at other sites.

3 Discussion

It is clear that only one gene for DNA methylase is present in cells and that all types of methylation are carried out by the products of this gene. This applies to both animals and plants (Houlston et al., 1992) and, although different proteolytic fragments of the enzyme may show a different ratio of maintenance to *de novo* activity, Zucker et al. (1985) were at pains to show that one species of enzyme was capable of both activities. In tissues such as placenta and liver, it appears that only lower molecular weight forms of the enzyme are present and this may increase the relative *de novo* methylase activity, though the significance of such changes to the cell is far from clear.

It is also probable that a single species of enzyme is able to carry out methylation of cytosines in CG dinucleotide and CNG trinucleotide (and possibly other) sequences: the balance betweeen the two specificities being different in the enzymes from the plant and animal kingdoms.

It is very difficult to extrapolate *in vitro* data for the ratio of maintenance to *de novo* activity to the *in vivo* situation. Results presented in this chapter suggest that, using initial rates of the *in vitro* reaction, this ratio may be around 15–30, but previous estimates (Adams, 1990) have suggested that the ratio may be as high as 400. Whichever is nearer the mark, it is apparent that *de novo* methylation can be catalysed by the purified enzyme. Pfeifer et al. (1990b) have estimated that the efficiency of *de novo* methylation of an unselected gene on the human X-chromosome in a human/hamster cell line is about 5% compared with an efficiency of maintenance methylation of 99.9%; yet *de novo* methylation in preimplantation embryos is clearly very much more efficient. A 5% efficiency of *de novo* methylation acting at random would be sufficient to bring about almost complete methylation of a gene over a large number of generations but to achieve efficient methylation over two or three cell division cycles, as appears to be the situation in the early embryo, would require the 100-fold greater activity found in mouse eggs and early embryos (Monk et al., 1991). In the presence of such high DNA methylase activity, the cell must have a mechanism of comparable efficiency for the prevention of methylation at specific sites.

Furthermore, methylation at non-CG sequences can be maintained where no evidence for symmetry exists (Toth et al., 1990) and this, together with the results we have presented for covalent interactions of the enzyme with different oligonucleotide duplexes, may indicate that the enzyme is very relaxed in its specificity; methylating any cytosine that is not protected in some way.

Although the presence of histone H1 can inhibit DNA methylation, this may be important only in so far that this histone will interfere with further methylation of inactive genes occurring long after replication (delayed methylation) in cells whose pattern of gene expression is established. As with transcription (Laybourn and Kadonaga, 1991), it may be important to inhibit the bulk of basal methylase activity in order to regulate the sites at which the enzyme may act and this could likewise be achieved by histone H1. Whether or not transcription factors can interfere with methylation has not been established but we have presented some evidence that proteins are present in nuclear extracts that can interact with the DNA methylase to reduce its activity on unmethylated regions of DNA. As such it is pertinent to ask questions about the mode of interaction of the DNA methylase and its substrate.

The mouse DNA methylase appears to act in a distributive manner *in vitro* and this is consistent with the kinetics of methylation observed *in vivo* by Pfeifer et al. (1990b) and the finding of hemimethylated CG dinucleotides occurring in a stable form *in vivo*. The enzyme that is active *in vivo* is very tightly associated with the chromatin and is not released by high salt treatment. Whether this involves a covalent binding of the enzyme to DNA is not known but the enzyme can form such a linkage, independent of the presence of AdoMet, and covalent interaction has been shown to be involved in the mechanism of methyltransfer (Santi et al., 1983). In addition, the enzyme binds loosely to unmethylated and hemimethylated DNA. The binding to hemimethylated CNG sequences is much stronger for the plant enzyme (Houlston et al., 1992). Gel retardation analysis has been used to discriminate between the fully methylated products of reaction that are protein free and the enzyme bound substrate that has yet to accept a methyl group. These observations indicate that the enzyme interacts with the DNA in a number of discrete steps culminating in dissociation of the complex when the DNA is fully methylated.

Such conclusions are readily arrived at for DNA substrates containing one or two sites of action but, in the *in vivo* situation, CG dinucleotides occur either in a well dispersed manner or clustered in islands. In the former case the situation may well not be much different from that observed *in vitro* with short oligonucleotides. Methylation of CpG islands may represent a special case and Carotti et al. (1989) have proposed that the DNA methylase works only inefficiently at such sites. Furthermore, a 'poison' sequence has been found associated with one such site (Szyf et al., 1990) that protects adjacent sequences from *de novo* methylation in early embryonic cells. The binding of a protein to such a sequence in certain genes in an allele-specific manner may be sufficient to protect these regions from the high methylase activity present in the egg and early embryo.

Acknowledgements. We would like to acknowledge financial support from the Medical Research Council, the Science and Education Research Council and the Wellcome Trust and also the support of Prof. Houslay and the University of Glasgow.

Achten, S., Behn-Krappa, A., Jücker, M., Sprengel, J., Hölker, I., Schmitz, B., Tesch, H., Diehl, V., and Doerfler, W. (1991) Cancer Res. *51*, 3702–3709.

Adams, R. L. P. (1990a) DNA methylation: the effect of minor bases on DNA-protein interactions. Biochem. J. *265*, 309–320.

Adams, R. L. P. (1990b) Cell culture for biochemists. Elsevier, Amsterdam.

Adams, R. L. P., Burdon, R. H., McKinnon, K., and Rinaldi, A. (1983) Simulation of *de novo* methylation following limited proteolysis of mouse ascites DNA methylase. FEBS Lett. *163*, 194–198.

Adams, R. L. P., Gardiner, K., Rinaldi, A., Bryans, M., McGarvey, M., and Burdon, R. H. (1986) Mouse ascites DNA methylase: characterisation of size, proteolytic breakdown and nucleotide recognition. Biochim. Biophys. Acta *868*, 9–16.

Adams, R. L. P., Hill, J., McGarvey, M., and Rinaldi, A. (1989) Mouse DNA methylase; intracellular location and degradation. Cell Biophys. *15*, 113–126.

Antequera, F., Boyes, J., and Bird, A. (1990) High levels of *de novo* methylation and altered chromatin structure at CpG islands in cell lines. Cell *62*, 503–514.

Behn-Krappa, A., Hölker, I., Sandaradura de Silva, U., and Doerfler, W. (1991) Patterns of DNA methylation are indistinguishable in different individuals over a wide range of human DNA sequences. Genomics *11*, 1–7.

Belanger, F. C., and Hepburn, A. G. (1990) The evolution of CpNpG methylation in plants. J. Mol. Evol. *30*, 26–35.

Bestor, T., Laudano, A., Mattaliano, R., and Ingram, V. (1988) Cloning and sequencing of a cDNA encoding DNA methyltransferase of mouse cells. J. Mol. Biol. *203*, 971–983.

Bestor, T., and Ingram, V. (1985) Growth dependent expression of multiple species of DNA methyltransferase in murine erythroleukaemia cells. Proc. Natl. Acad. Sci. USA *82*, 2674–2678.

Bolden, A. H., Nalin, C. M., Ward, C. A., Poonian, M. S., McComas, W. W., and Weissbach, A. (1985) DNA methylation: sequences flanking C-G pairs modulate the specificity of the human DNA methylase. Nucl. Acids Res. *13*, 3479–3494.

Bolden, A. H., Nalin, C. M., Ward, C. A., Poonian, M. S., and Weissbach, A. (1986) Primary DNA sequence determines sites of maintenance and *de novo* methylation by mammalian DNA methyltransferases. Mol. Cell. Biol. *6*, 1135–1140.

Bryans, M., Kass, S., Seivwright, C. and Adams, R. L. P. (1992) Vector methylation inhibits transcription from the SV40 early promoter. FEBS Letts. in press.

Caiafa, P., Reale, A., Allegra, P., Rispoli, M., D'Erme, M., and Strom, R. (1991) Histones and DNA methylation in mammalian chromatin-differential inhibition of *in vitro* methylation by histone H1. Biochim. Biophys. Acta *1090*, 38–42.

Campbell, J. L., and Kleckner, N. (1990) *E. coli oriC* and the *dnaA* gene promoter are sequestered from *dam* methyltransferase following the passage of the chromosomal replication fork. Cell *62*, 967–979.

Carotti, D., Palitti, F., Mastrantonio, S., Rispoli, M., Strom, R., Amato, A., Campagnari, F., and Whitehead, E. P. (1986) Substrate preferences of human placental DNA methyltransferase investigated with synthetic polydeoxynucleotides. Biochim. Biophys. Acta *866*, 135–143.

Carotti, D., Palitti, F., Lavia, P., and Strom, R. (1989) *In vitro* methylation of CpG islands. Nucl. Acids Res. *17*, 9219–9229.

Cedar, H., and Razin, A. (1990) DNA methylation and development. Biochim. Biophys. Acta *1049*, 1–8.

Chaillet, J. R., Vogt, T. F., Beier, D. R., and Leder, P. (1991) Parental-specific methylation of an imprinted transgene is established during gametogenesis and progressively changes during embryogenesis. Cell *66*, 77–83.

Christy, B., and Scangos, G. (1982) Expression of transferred thymidine kinase genes is controlled by methylation. Proc. Natl. Acad. Sci. USA *79*, 6299–6303.

Davis, R. L., Fuhrer-Krusi, S., and Kucherlapati, R. S. (1982) Modulation of transfected gene expression mediated by changes in chromatin structure. Cell *31*, 521–529.

Davis, T., Rinaldi, A., Clark, L., and Adams, R. L. P. (1986) Methylation of chromatin *in vitro*. Biochim. Biophys. Acta *866*, 233–241.

Davis, T., Kirk, D., Rinaldi, A., Burdon, R. H., and Adams, R. L. P. (1985) Delayed methylation and the matrix bound DNA methylase. Biochem. Biophys. Res. Comm. *126*, 678–684.

Doerfler, W., Toth, M., Kochanek, S., Achten, S., Freisem-Rabien, U., Behn-Krappa, A., and Orend, G. (1990) Eukaryotic DNA methylation: facts and problems. FEBS Lett. *268*, 329–333.

Frank, D., Keshet, I., Shani, M., Levine, A., Razin, A., and Cedar, H. (1991) Demethylation of CpG islands in embryonic cells. Nature (London) *351*, 239–241.

Giordano, M., Mattachini, M. E., Cella, R., and Pedrali-Noy, G. (1991) Purification and properties of a novel DNA methyltransferase from cultured rice cells. Biochem. Biophys. Res. Comm. *177*, 711–719.

Hepburn, P. A., and Tisdale, M. J. (1991) Importance of the O^6 position of guanine residues in the binding of DNA methylase to DNA. Biochim. Biophys. Acta *1088*, 341–344.

Holliday, R., and Pugh, J. E. (1975) DNA modification mechanisms and gene activity during development. Science *187*, 226–232.

Houlston, C. E., Lindsay, H., and Adams, R. L. P. (1992) DNA substrate specificity of pea DNA methylase (in preparation).

Howlett, S. K., and Reik, W. (1991) Methylation levels of maternal and paternal genomes during preimplantation development. Development *113*, 119–127.

Jähner, D., Stuhlmann, H., Stewart, C. L., Harbers, K., Löhler, J., Simon, I., and Jaenisch, R. (1982) *De novo* methylation and expression of retroviral genomes during mouse embryogenesis. Nature (London) *298*, 623–628.

Jones, P. A., Wolkowicz, M. J., Rideot, W. M., Gonzales, F. A., Marziasz, C. M., Coetzee, G. A., and Tapscott, S. J. (1990) *De novo* methylation of the *MyoD1* CpG island during the establishment of immortal cell lines. Proc. Natl. Acad. Sci. USA *87*, 6117–6121.

Kruczek, I., and Doerfler, W. (1982) The unmethylated state of the promoter/leader and 5′-regions of integrated adenovirus genes correlates with gene expression. EMBO J. *1*, 409–414.

Laybourn, P. J., and Kadonaga, J. T. (1991) Role of nucleosomal cores and histone H1 in regulation of transcription by RNA polymerase II. Science *254*, 238–245.

Lyon, M. F. (1991) The quest for the X-inactivation centre. Trends Genet. *7*, 69–70.

Migeon, B. R. (1990) Insights into X chromosome inactivation from studies of species variation, DNA methylation and replication, and vice versa. Genet. Res. Camb. *56*, 91–98.

Monk, M., Adams, R. L. P., and Rinaldi, A. (1991) Decrease in DNA methylase activity during preimplantation development in the mouse. Development *112*, 189–192.

Monk, M., Boubelik, M., and Lehnert, S. (1987) Temporal and regional changes in DNA methylation in the embryonic, extraembryonic and germ cell lineages during mouse embryo development. Development *99*, 371–382.

Orend, G., Kuhlmann, I., and Doerfler, W. (1991) Spreading of DNA methylation across integrated foreign (adenovirus type 12) genomes in mammalian cells. J. Virol. *65*, 4301–4308.

Palitti, F., Carotti, D., Grünwald, S., Rispoli, M., Whitehead, E. P., Salerno, C., Strom, R., and Drahovsky, D. (1987) Inactivation of *de novo* DNA methyltransferase activity by high concentrations of double-stranded DNA. Biochim. Biophys. Acta *919*, 292–296.

Pellicer, A., Robins, D., Wold, B., Sweet, R., Jackson, J., Lowry, I., Roberts, J. M., Sim, G. K., Silverton, S., and Axel, R. (1980) Altered genotype and phenotype by DNA-mediated gene transfer. Science *209*, 1414–1422.

Pfeifer, G. P., Spiess, E., Grünwald, S., Boehm, T. L. J., and Drahovsky, D. (1985a) Mouse DNA-cytosine-5-methyltransferase: sequence specificity of the methylation reaction and electron microscopy of enzyme-DNA complexes. EMBO J. *4*, 2879–2884.

Pfeifer, G. P., Grünwald, S., Palitti, F., Kaul, S., Boehm, T. L. J., Hirth, H.-P., and Drahovsky, D. (1985b) Purification and characterisation of mammalian DNA methyltransferases by use of monoclonal antibodies. J. Biol. Chem. *260*, 13787–13793.

Pfeifer, G. P., Tanguay, R. L., Steigerwald, S. D., and Riggs, A. D. (1990a) *In vivo* footprint and methylation analysis by PCR-aided genomic sequencing: comparison of active and inactive X chromosomal DNA at the CpG island and promoter of human *PGK-1*. Genes Develop. *4*, 1277–1287.

Pfeifer, G. P., Steigerwald, S. D., Hansen, R. S., Gartler, S. M., and Riggs, A. D. (1990b) Polymerase chain reaction-aided genomic sequencing of an X chromosome-linked CpG island: methylation patterns suggest clonal inheritance, CpG site autonomy, and an explanation of activity state stability. Proc. Natl. Acad. Sci. USA *87*, 8252–8256.

Rubin and Modrich (1977) *Eco*RI methylase: physical and catalytic properties of the homogeneous enzyme. J. Biol. Chem. *252*, 7265–7272.

Sambrook, J., Fritsch, E. F., and Maniatis, T. (1989) Molecular Cloning, 2nd edn. Cold Spring Harbor Laboratory Press.

Santi, D. V., Garrett, C. E., and Barr, P. J. (1983) On the mechanisms of inhibition of DNA-cytosine methyltransferases by cytosine analogs. Cell *33*, 9–10.

Santi, D. V., Norment, A., and Garrett, C. E. (1984) Covalent bond formation between a DNA-cytosine methyltransferase and DNA containing 5-azacytosine. Proc. Natl. Acad. Sci. USA *81*, 6993–6997.

Simon, D., Grunert, F., Acken, U. V., Döring, H. P., and Kröger, H. (1978) DNA-methylase from regenerating rat liver: purification and characterisation. Nucleic Acids Res. *5*, 2153–2167.

Simon, D., Grunert, F., Kroger, H., and Grassman, A. (1980) Dinucleotide specificity of rat liver DNA methylase. Eur. J. Cell Biol. *22*, 33.

Smith, S. S., Kan, J. L. C., Baker, D. J., Kaplan, B. E., and Dembek, P. (1991) Recognition of unusual DNA structures by human DNA(cytosine-5)methyltransferase. J. Mol. Biol. *217*, 39–51.

Stein, R., Razin, A., and Cedar, H. (1982) *In vitro* methylation of the hamster adenosine phosphoribosyltransferase gene inhibits its expression in mouse cells. Proc. Natl. Acad. Sci. USA *79*, 3418–3422.

Szyf, M., Tanigawa, G., and McArthy, P. L. (1990) A DNA signal from the *Thy-1* gene defines *de novo* methylation patterns in embryonic stem cells. Mol. Cell. Biol. *10*, 4396–4400.

Szyf, M., Bozovic, U., and Tanigawa, G. (1991) Growth regulation of mouse DNA methyltransferase gene-expression. J. Biol. Chem. *266*, 10027–10030.

Theiss, G., Schleicher, R., Schimpff-Weiland, G., and Follmann, H. (1987) DNA methylation in wheat: purification and properties of DNA methyltransferase. Eur. J. Biochem. *167*, 89–96.

Toth, M., Müller, U., and Doerfler, W. (1990) Establishment of *de novo* DNA methylation patterns. J. Mol. Biol. *214*, 673–683.

Vaccaro, M., Pawlak, A., and Jost, J.-P. (1990) Positive and negative regulatory elements of chicken vitellogenin II gene characterized by *in vitro* transcription competition assays in a homologous system. Proc. Natl. Acad. Sci. USA *87*, 3047–3051.

Ward, C., Bolden, A., Nalin, C. M., and Weissbach, A. (1987) *In vitro* methylation of the 5′-flanking regions of the mouse β-globin gene. J. Biol. Chem. *262*, 11057–11063.

Wigler, M., Levy, D., and Perucho, M. (1981) The somatic replication of DNA methylation. Cell *24*, 33–40.

Woodcock, D. M., Crowther, P. J., and Diver, W. P. (1987) The majority of methylated deoxycytidines in human DNA are not in the CpG dinucleotide. Biochem. Biophys. Res. Commun. *145*, 888–894.

Yesufu, H. M. I., Hanley, A., Rinaldi, A., and Adams, R. L. P. (1991) DNA methylase from *Pisum sativum*. Biochem. J. *273*, 469–475.

Zucker, K. E., Riggs, A. D., and Smith, S. S. (1985) Purification of human DNA(cytosine-5-)-methyltransferase. J. Cell. Biochem. *29*, 337–349.

DNA Methylation: Molecular Biology and Biological Significance
ed. by J. P. Jost & H. P. Saluz

Effect of DNA methylation on the binding of vertebrate and plant proteins to DNA

Melanie Ehrlich[a] and Kenneth C. Ehrlich[b]

[a]*Department of Biochemistry, Tulane Medical School, 1430 Tulane Avenue, New Orleans, LA 70112, and* [b]*Southern Regional Research Center, New Orleans, LA 70179, USA*

1 Introduction

Many studies have shown correlations between increased gene expression and decreased DNA methylation (e.g., Sullivan et al., 1989; Piva et al., 1989; Shinar et al., 1989; Jahroudi et al., 1990; Burbelo et al., 1990). The demethylation associated with gene expression has been found either in localized sites or in extensive regions. When the tissue-specific demethylation is localized, it may be in the upstream regulatory regions, at the 5′ end of the gene, in an intron, or in the 3′ flanking region of the gene (Jump et al., 1987; Broad et al., 1989; Lamson and Stockdale, 1989; Umeno et al., 1988; Sullivan et al., 1989). Sometimes the localized demethylation occurring before or concomitantly with the activation of transcription is followed by a more generalized demethylation of the gene region (Umeno et al., 1988; Sullivan et al., 1989; Blackman and Koshland, 1985; Toth et al., 1989). In other cases, all the detected demethylation occurred before or concurrently with the onset of transcription (Lamson and Stockdale, 1989; Benvenisty et al., 1985; Saluz et al., 1988). However, for some genes, the demethylation associated with the activation of transcription was detected only after the turn on of the gene (Lock et al., 1987; Enver et al., 1988). In the latter cases, DNA methylation may just help keep a switched-off gene region silent. Another possibility is that the demethylation of critical mCpGs that preceded the turn on of transcription may have been missed because only a small fraction of CpG sites in the gene region (usually only 5′-CCGG-3′ [HpaII] or 5′-GCGC-3′ [HhaI] sites) were studied. DNA sequences methylated *de novo* after the turn on of transcription might have been subject to secondary methylation changes due to the spreading of new DNA methylation patterns (Toth et al., 1989). The DNA probe used in these studies may not have revealed methylation changes in an important part of the gene (such as part of the promoter, an intragenic silencer, or a distal enhancer).

Besides numerous studies documenting correlations between the changes in vertebrate DNA methylation and transcription, DNA methylation levels have been manipulated leading to increases in expression (upon demethylation) or decreasees in expression (upon increased methylation) of specific genes. Many gene activation experiments involve the treatment of the cells with 5-azacytidine (azaC), a strong inhibitor of DNA methyltransferase. Among the genes thus activated were those involved in muscle, thyroid, or erythroleukemia cell differentiation (Creusot et al., 1982; Tapscott et al., 1988; Avvedimento et al., 1989), silent retroviral genes in intact mice (Jaenisch et al., 1985), and the phosphoenolpyruvate carboxykinase gene in rat fetuses (Benvenisty et al., 1985). In one such study, DNA isolated from azaC-induced muscle cells, but not from the progenitor fibroblasts, was shown to genetically transform a fibroblastoid mouse cell line to muscle cells. This result indicates that the induction of differentiation by azaC was due to a *covalent* change in the DNA, most probably demethylation (Lassar et al., 1986). Similarly, spontaneous conversion of hamster cell lines to a thymidine kinase-negative (TK^-) phenotype was often associated with hypermethylation of a CCGG site within 20 bps of the putative cap site and the demethylation of this site was found in untreated TK^+ revertants (Dobrovic et al., 1988). Treatment with azaC led to an $\sim 10^5$-fold increase in TK^+ revertants (Harris, 1982). Changes in DNA methylation have also been strongly implicated in mammalian X chromosome inactivation in studies involving either azaC treatment of cells or the transfection of DNA prepared from such treated cells (Jones et al., 1982; Venolia et al., Grant and Chapman, 1988).

A number of studies show that *in vitro* methylation of genes often greatly inhibits their expression upon subsequent transfection into vertebrate cells or in the generation of transgenic mice (Guntaka et al., 1987; Shinar et al., 1989). *In vitro* methylation of an α-actin gene construct led to the inhibition of its expression upon transfection into fibroblasts where it stayed methylated but not in myoblasts where it underwent partial demethylation (Yisraeli et al., 1986). Site-specific methylation of the Ha-ras-1 gene promoter was shown to largely silence the expression of the adjacent CAT reporter gene whereas methylation of the CAT gene was without effect in transient expression assays in CV-1 cells (Rachal et al., 1989). *De novo* methylation of the HIV LTR silenced the CAT transcription from a chimeric DNA construct (Bednarik et al., 1990).

How DNA methylation controls transcription of certain vertebrate genes is unclear. As discussed in this chapter, evidence suggests that a physiologically important discrimination between 5-mC residues and C residues at certain sites in vertebrate DNA involves differential binding of positive or negative *trans*-acting transcription factors. Just as bacteria have exploited adenine methylation at 5′-GATC-3′ sequences for many functions (directing mismatch repair, fine-tuning control of gene expression, controlling transposition, facilitating and determining the timing of the initiation of chromosome replication; Roberts et al., 1985; Messer and Noyer-Weidner, 1988; Barras and Marinus, 1989; Bakker and Smith, 1989; Blyn et al., 1990; Louarn et al., 1990), so it is likely that the much more highly conserved vertebrate CpG methylation plays critical roles in the regulation of chromosome activity. Only some of the bacterial species studied (Brooks and Roberts, 1982; Ehrlich et al., 1985) display this bacterial-type methylation at the N^7 position of the A residue at the GATC sequence (*dam* methylation) and yet it plays the above-mentioned multiple roles in regulating DNA function in addition to participating in restriction-modification systems. In contrast, 5-methylation of C residues has been observed in all vertebrates and vascular plants so far studied and even its location in CpG dinucleotide pairs in the former and CpG plus CpNpG sequences in the latter is conserved (Vanyushin et al., 1973; Ehrlich and Wang, 1981; Wagner and Capesius, 1981; Gruenbaum et al., 1981).

Given the fact that the DNA methylation could be easily lost during the course of evolution by mutation of DNA methyltransferase genes, genomic 5-mC residues should confer important selective advantages to these vertebrates and higher plants, which invariably contain such modification. A selective advantage is all the more likely, given the genetically programmed tissue-specific differences in 5-mC levels involving up to tens of millions more 5-mC residues per diploid cell (Ehrlich et al., 1982; Gama-Sosa et al., 1983). Probably, only a small percentage of these differentiation-linked changes in DNA methylation (Gama-Sosa et al., 1983) are involved in transcription control. Most of the rest may be due to overshooting by the *de novo* methylation machinery which is probably not narrowly targeted to the DNA sequences whose methylation status is physiologically relevant. The determinants of which DNA sequences undergo these changes in methylation status is poorly understood. Further obscuring the role of DNA methylation in vertebrates, some of this tissue-specificity in DNA methylation patterns could be important mostly during embryogenesis, as may be the case for sperm-specific and placenta-specific hypomethylation seen in a class of human DNA sequences that often are rich in CpG dinucleotides (Zhang et al., 1985 and 1987).

Not only does *E. coli dam* methylation establish a proven precedent for multiple functions for the genetically programmed DNA methylation not associated with restriction, but also, detailed studies of DNA·protein interactions involving recognition of the 5-methyl group of thymine residues argue for a role of 5-mC in protein interactions *in vivo*. The 5-methyl group of both T and 5-mC residues is in the outer major groove of B-DNA and can make critical contacts with sequence-specific DNA-binding proteins (Seeman et al., 1976; Goeddel et al., 1978; Ivarie, 1987). It is, therefore, highly probable that analogous mechanisms involving sequence-specific DNA-binding proteins are used by vertebrates and higher plants to distinguish DNA sequences containing 5-mC from analogous unmethylated sequences. As discussed below and summarized in Tables 1 and 2, there is evidence that numerous proteins with this kind of specificity function in the control of transcription in these types of higher eucaryotes. Other eucaryotes not containing 5-mC in their DNA (Ehrlich and Wang, 1981) probably use other mechanisms for transcription control. This is not surprising given the multiple transcription control proteins and pathways that are involved in regulating the expression of eucaryotic genes.

Table 1. Eucaryotic DNA methylation-inhibited DNA-binding proteins and their specific DNA-binding sites

Protein	Binding site[1]	Reference
Vertebrate proteins		
E2F	TTTTCG$\overset{*}{C}$GC	Kovesdi et al., 1987
EBP-80	ATCTG$\overset{*}{C}$GCATATGCC	Falzon and Kuff, 1991
AP-2[2]	CCCGC$\overset{*}{C}$GGCG[2]	Comb and Goodman, 1990
AP-2	CTC$\overset{*}{C}$GGGG(C/T)TG	Hermann et al., 1989
MLTF	GGCCA$\overset{*}{C}$GTGACC	Watt and Malloy, 1988
Ah receptor	TTG$\overset{*}{C}$GTG	Shen and Whitlock, 1989
c-Myc/Myn	GACCA$\overset{*}{C}$GTGGTC	Prendergast et al., 1991
CREB/ATF	TGA$\overset{*}{C}$GTCA	Iguchi and Schaffner, 1990; Inamdar et al., 1991
CREB/ATF	CTG$\overset{*}{C}$GTCA	Becker et al., 1987; Weih et al., 1991
Plant proteins		
CG1	CA$\overset{*}{C}$GTG	Staiger et al., 1989
TnpA	$\overset{*}{C}\overset{*}{C}$GACACTCTTA	Gierl et al., 1988
MIB-1	TGA$\overset{*}{C}$GTCA	Inamdar et al., 1991
HBP/ASF-1/HSBF	CCA$\overset{*}{C}$GTCA	Inamdar et al., 1991
GBF	ACA$\overset{*}{C}$GTGG	Inamdar et al., 1991

[1]Binding sites are written 5′ to 3′ and given for only one strand. They are based upon consensus sequence, DNase I footprinting, or dimethyl sulfate interference data and in some cases represent only the approximate boundaries of the recognition site.
[2]Evidence suggests that AP-2 is binding to this DNA region. The sequence is repeated once in this region and the site of methylation that inhibits binding is shown.

Table 2. Proteins with specificity for methylated DNA

Protein	Binding site[1]	Reference
Vertebrate sequence-specific DNA-binding proteins		
MDBP-1	ATMGTCAMGGMGAT GTMGCCAMGGMGAT ATTGTCAMGGTGAC GCTGTCAMGGMGAC GCMGTCATGGMGCC ATTCCCAMGGCTAC	Wang et al., 1986; Zhang et al., 1990
Vertebrate sequence non-specific DNA-binding proteins		
MDBP-2	≥ 30 bp duplex with 1 5-mC	Saluz et al., 1988; Jost et al., 1991 Jost et al., 1992
MeCP	≥ 100 bp duplexes with ≥ 15 5-mC's	Meehan et al., 1989
Plant sequence non-specific DNA-binding protein		
DBP-m	oligonucleotide duplexes with ≥ 1 5-mC	Zhang et al., 1989a

[1]Sequences are given for one strand in the 5′→3′ direction. Binding was usually tested with bifilarly methylated CpG sites. M, 5-methylcytosine.

2 Inhibition of the binding of sequence-specific DNA-binding proteins to methylated viral DNA sequences

A number of sequence-specific DNA-binding proteins have been shown to have their binding activity to animal viral DNA sequences strongly decreased by CpG methylation (Table 1). The human DNA-binding protein, E2F, was one of the first proteins shown to bind much more weakly to a DNA recognition site when it was 5-methylated at CpG dinucleotides. E2F is involved in the control of adenoviral early gene transcription and recognizes sites in the adenovirus 5 (Ad5) E2 promoter and E1A enhancer (Kovesdi et al., 1987). An E2F binding site in the E1A enhancer overlaps a recognition site for the restriction endonuclease HhaI. Methylation at the CpG dinucleotide of this site in an E1A enhancer subfragment was carried out with HhaI DNA methyltransferase. The methylation abrogated the ability of this site to compete for binding to E2F.

In this study, Kovesdi and coworkers (1987) correlated the effects on protein-binding and on transcription. HhaI methylase was used to methylate a reporter plasmid carrying the mouse β-globin gene and its immediate upstream region with or without an 83-bp DNA fragment containing the E1A enhancer's E2F binding site. Reporter gene expression was compared after transient transfection of HeLa cells with *in vitro* methylated or unmethylated plasmid. There was no effect on the mouse β-globin gene transcription upon methylation by HhaI methylase when the transcription of this gene was driven only by its own promoter region. However, the upstream introduction of the 83-bp DNA fragment containing the E1A enhancer's E2F site that overlaps its single HhaI site resulted in ~8-fold increase in mouse β-globin gene transcription when the transfecting DNA was unmethylated and only an ~2-fold increase when it had been methylated with HhaI methylase.

Analysis of transcription *in vitro* revealed that methylation, by HhaI methylase, of a viral recognition site for a human DNA-binding protein, EBP-80, interfaces with the transcription-promoting activity of that site (Falzon and Kuff, 1991). The tested binding site was in the long terminal repeat (LTR) of an intracisternal A-particle DNA. When affinity-purified EBP-80 was added to the HeLa whole-cell extracts used for *in vitro* transcription, a higher level of transcription was obtained and a greater difference was observed in the LTR-driven transcription between methylated and unmethylated templates. This inhibition induced by the methylation of CpG was seen with covalently closed circular templates but not with linear templates. Analogous methylation of an oligonucleotide duplex containing the EBP-80 site also decreases the binding of affinity-purified EBP-80. Methylation of both strands (bifilar methylation) of the oligonucleotide duplex ligand was achieved by using 5-methylated derivatives of cytidine phosphoramidites for its synthesis.

Similarly, methylation of a few HpaII sites in the adenovirus type 2 (Ad2) late E2A promoter and immediate downstream region inhibits both *in vitro* transcription driven by that promoter and the formation of specific DNA·protein complexes (Hermann et al., 1989; Dobrzanski et al., 1988). A recent study suggests that one or two AP-2 sites (see proenkephalin promoter, below) at position +6 and +24 may be involved in the methylation-mediated down-regulation of transcription (Hermann and Doerfler, 1991). As for the above-mentioned EBP80 site in an LTR, site-specific methylation of the late E2A promoter region (at positions −215, +6, and +24 by HpaII methylase) strongly inhibits *in vitro* transcription but only when circular templates are used (Dobrzanski et al., 1988). Similar results were obtained upon methylation of a −13 or −52 site in the major late promoter (Dobrzanski et al., 1988). A comparison of *in vivo* dimethyl sulfate and DNase I genomic footprinting of the late E2A upstream promoter region in several related cell lines with different levels of methylation showed a correlation between methylation and inhibition of specific factor binding to DNA *in vivo* (Toth et al., 1990).

Methylation of the CpG in the binding site for the transcription factor MLTF in the Ad2 major late promoter is also implicated in negative transcription control. Such site-directed methylation inhibits *in vitro* transcription driven by that promoter more than 5-fold while methylation of an adjacent CpG has no effect (Watt and Malloy, 1988).

In this study, linearized templates were used for *in vitro* transcription. Methylation at the MLTF site decreases the binding by affinity-purified MLTF more than 15-fold as well as causing a much more rapid dissociation of the specific MLTF·DNA complex.

It has been proposed that methylation of adenoviral, herpesviral, and animal retroviral DNA sequences may have an important role to play in the latent stages of the viral life cycle (Doerfler, 1991; Tasseron-de Jong et al., 1989; Zhang et al., 1989b; Desrosiers et al., 1979; Breznik et al., 1984). A recent study led Bednarik et al. (1990) to suggest that methylation of the HIV long terminal repeat (LTR) helps to keep the proviral DNA repressed until the methylation-associated inhibition is overridden. This might happen, for example, by activation of a signal transduction pathway.

An uncharacterized protein in nuclear extracts of cultured human cells was shown to bind much less to a radiolabeled HIV LTR DNA fragment methylated at its two HpaII sites than to the corresponding unmethylated DNA fragment (Bednarik et al., 1990). A competition experiment with an excess of the unmethylated or methylated DNA fragment indicates that the binding is specific. The methylation of these two HpaII sites reduces the expression of the HIV *tat* gene or the *E. coli cat* gene driven by this LTR transfected into a human T-cell or simian cell line. Furthermore, HpaII methylation greatly inhibits the synthesis of proviral DNA upon transfection into a human adenocarcinoma cell line.

However, the lost *tat* gene expression in the construct with a methylated HIV LTR could be regained by treating the transfected T-cell line with phorbol ester and the loss of this LTR-driven *cat* gene expression due to methylation could be prevented by cotransfection of the simian cell line with an unmethylated HIV LTR/*tat* expression vector (Bednarik et al., 1990). Similarly, the inhibition of activity of the Ad2 E2A promoter by methylation at its three HpaII sites was overcome, at least partly, by transactivation with the Ad2 E1A protein or by *cis* activation by the IE1 enhancer from the human cytomegalovirus (CMV; Knust et al., 1989). Also, transient treatment with the DNA methylation inhibitor azaC reversed the effects of methylation in Ad2-transformed cell lines (Knust et al., 1989).

3 Inhibition of the binding of sequence-specific DNA-binding proteins to methylated mammalian DNA sequences

AP-2 is an inducible transcription regulatory factor that binds to a specific site in the human proenkephalin gene promoter (Comb and Goodman, 1990). Binding to this site is largely inhibited by CpG methylation within the binding site as detected by DNase I footprinting assays with affinity-purified AP-2 (Comb and Goodman, 1990). HpaII methylase was used to catalyze this methylation because this sequence overlaps a HpaII site. This methylation inhibits transient and stable, cAMP-induced or basal expression of a downstream reporter gene in human or monkey cell lines, both at the RNA and protein levels. By comparing a series of DNA constructs, the HpaII site mediating this effect on transcription was localized in the AP-2 site of the promoter.

Another DNA-binding protein whose recognition of DNA sequences is inhibited by CpG methylation is the Ah receptor protein (Shen and Whitlock, 1989). This protein resides intracellularly, binds to certain aromatic hydrocarbons, and subsequently increases the synthesis of cytochrome P-450IA1. Using nuclear extracts from dioxin-induced or uninduced cells, Shen and Whitlock showed by electrophoretic mobility shift assays that methylation of an oligonucleotide duplex containing the Ah receptor's DNA-binding site abolishes the formation of specific DNA·protein complexes. When the reporter constructs for transient transfection were prepared using the methylated or the unmethylated oligonucleotide duplexes ligated upstream of a viral promoter, the site-directed methylation was shown to strongly suppress the dioxin-responsiveness of reporter gene expression. Also, an oncoprotein, c-Myc and a heterodimer of c-Myc and a newly identified protein, Myn, have been shown to have their sequence-specific DNA-binding activity strongly inhibited by CpG methylation. However, the effect of this protein on the transcription of the methylated or unmethylated DNA templates have not yet been reported (Prendergast et al., 1991).

One example of the sequence-specific DNA-binding proteins that contains a CpG dinucleotide in its consensus sequence is the cyclic AMP response element binding protein (CREB or ATF), which belongs to a large family of DNA-binding proteins having the same or similar DNA sequence specificity (Hai et al., 1989; Hurst et al., 1990; Liu et al., 1988). Iguchi-Ariga and Schaffner (1989) showed that a radiolabeled oligonucleotide duplex containing a consensus sequence for CREB-type transcription factors formed specific complexes with proteins in crude nuclear extracts of HeLa cells. The protein-DNA complex could be competed by an excess of unlabeled unmethylated CREB ligand but not by a similar excess of the analogous bifilarly methylated ligand.

With the methylated radiolabeled CREB-specific ligand, fewer and much weaker bands of DNA·protein complexes were obtained than with the analogous unmethylated radiolabeled oligonucleotide duplex (Iguchi-Ariga and Schaffner, 1989). Furthermore, *in vitro* transcription and transient transfection assays in two mammalian cell lines indicated that a chimeric β-globin/CREB site promoter fragment whose two CREB-specific sites were methylated before the ligation to a reporter DNA construct had very much less promoter activity than the analogous unmethylated construct.

By electrophoretic mobility shift assays, we have confirmed that bifilar methylation of an oligonucleotide containing the CREB consensus sequence greatly decreases the formation of specific complexes upon incubation with HeLa cell nuclear extract, especially when low concentrations of extract are used (Inamdar et al., 1991; Ehrlich and Ehrlich, unpubl. data). Similarly, a nuclear extract prepared from HL60 cells, a promyelocytic leukemic cell line, gave less than 0.4% as much binding to an oligonucleotide duplex containing a bifilarly methylated CREB consensus sequence than to the analogous unmethylated nucleotide duplex. Compared to the unmethylated CREB consensus sequence oligonucleotide duplex, a corresponding duplex with a base transition from CpG to TpG/CpA in the middle of the CREB site (instead of 5-mCpG/5-mCpG) showed only a 40% reduction in binding of CREB-type proteins from the HL60 nuclear extract.

We also demonstrated that hemimethylation inhibited the binding less than the bifilar methylation (Inamdar et al., 1991; unpubl. data). Similar effects of methylation of an oligonucleotide duplex containing a nonconsensus CREB site on the binding of purified rat CREB were reported by Weih et al. (1991). Weih and coworkers (1991) examined the binding of proteins *in vivo* to two regions upstream of the rat tyrosine aminotransferase (TAT) gene including a cAMP-inducible enhancer that contains a CREB site. By genomic sequencing, it had previously been demonstrated that several cell lines which do not express the TAT gene are hypermethylated at two DNase I-hypersensitive sites upstream of the gene while an expressing cell line is hypomethylated in these upstream sequences (Becker et al., 1987). Furthermore, *in vivo* dimethyl sulfate protection experiments and *in vitro* DNase I footprinting experiments indicated that undefined sequence-specific nuclear proteins had their binding to upstream TAT gene sequences inhibited by DNA methylation. In this promoter region, correlations between DNA hypomethylation and expression were found in rat tissues as well as in cultured cells. In nonexpressing rat tissues, this region was hypermethylated at these sites while it was hypomethylated in expressing liver (Becker et al., 1987). These differentially methylated regions are DNase I hypersensitive in TAT-expressing but not in non-expressing cell populations. AzaC treatment of the cell lines that do not express TAT resulted in the loss of detectable methylation at the examined upstream sites, including the CREB site (Weih et al., 1991). However, as detected by dimethylsulfate experiments on intact cells either induced for cAMP formation or not, azaC treatment did not lead to the binding of proteins to the upstream region of the expressed TAT gene. As Weih et al. note, although DNA demethylation does not lead to interaction of protein factors with TAT gene sequences, cytosine methylation could eventually lead to the establishment of an inert chromatin configuration that may no longer require this methylation to be maintained.

Correlations between DNase I hypersensitivity, DNA hypomethylation, and gene transcription have been described for a number of other vertebrate genes (e.g., Levy-Wilson and Fortier, 1989; Winter et al., 1990; Lubbert et al., 1991). In some cases, it appears that the changes in DNA methylation occur before activation of gene expression indicating that, in these cases, demethylation was not simply a consequence of transcription (Winter et al., 1990; Lubbert et al., 1991).

Faber et al. (1991) examined the methylation in a DNA region containing multiple DNase I hypersensitive sites upstream of the rat phosphoenolpyruvate carboxykinase (PEPCK) gene. A cell line expressing the PEPCK gene and two non-expressing cell lines were studied by genomic sequencing and by DNase I protection experiments (*in vivo* DNase footprinting). The expressing cell line was unmethylated at all CpG sites from the cap site to 505 bps upstream. One of the non-expressing lines was methylated at about half of these CpGs and the other, at only one of these CpGs, which was centered over a site recognized by a CREB-type protein. AzaC treatment of the latter, nonexpressing cell line led to demethylation of this CREB site but did not lead to the synthesis of detectable PEPCK mRNA. It also did not result in detectable *in vivo* binding of protein factors as assayed by *in vivo* DNase footprinting. Such binding of proteins to the promoter region of the PEPCK gene was observed only in the PEPCK-expressing cell line. Therefore, in this study too, demethylation of a promoter site in a nonexpressing cell line did not suffice to allow detectable binding of specific proteins to the promoter. It should be noted that cultured cell lines may give misleading results in studies of DNA methylation and gene expression because of their usually low and frequently changing level of DNA methylation (Shmookler-Reis and Goldstein, 1982; Ehrlich et al., 1982) as well as their aneuploidy and absence of normal systemic and tissue-specific cell interactions.

4 DNA-binding proteins whose recognition of CpG-containing sites in mammalian DNA is not inhibited by cytosine methylation

The mammalian transcription factor Sp1 binds to specific DNA sites with a highly conserved CpG dinucleotide in its consensus sequence, namely, 5′-G/TGGGCGGPuPuPy-3′ (Kadonaga et al., 1986). Because the majority of CpGs in vertebrate DNA are methylated (Ehrlich and Wang, 1981) and because Sp1 binding sites (GC boxes) are often found in promoter regions where they are implicated in the control of transcription (Courey and Tijan, 1988; Evans et al., 1988), it was of interest to determine whether methylation of the conserved CpG dinucleotide pair in such sequences would inhibit the binding of the respective proteins.

Site-directed methylation of the CpG on both strands of an oligonucleotide duplex containing an Sp1 recognition site gave little (Ben-Hattar et al., 1989) or no (Holler et al., 1988; Harrington et al., 1988) change in the extent of specific DNA·protein complex formation when HeLa cell nuclear extracts were used as the source of Sp1. Similarly, transcription *in vitro* and *in vivo* showed that CpG methylation of the Sp1 binding site did not detectably affect the ability of human Sp1 to help drive the transcription from a Sp1-responsive promoter (Holler et al., 1988). Apparently unrelated to the Sp1 binding effects was the finding that single-site methylation of an Sp1 site in a truncated herpes virus thymidine kinase promoter resulted in a large decrease in downstream transcription from an M13-type recombinant after introduction into *Xenopus laevis* oocytes (Ben-Hattar and Jiricny, 1988). *In vitro* binding studies showed that the oocyte Sp1-like DNA-binding protein was inhibited only slightly by the methylation (Ben-Hattar and Jiricny, 1988). Similarly, CpG methylation did not affect the *in vitro* binding, of an uncharacterized protein that binds to a CpG-containing region in the human phosphoglycerate kinase promoter (Yang et al., 1988).

5 Sequence-specific DNA-binding proteins from plants whose binding to DNA is inhibited by DNA methylation

Vascular plants have about 4–8 times higher levels of 5-mC in their DNA than do mammals (Ehrlich and Wang, 1981; Wagner et al., 1981). Among the studies implicating plant DNA methylation in control of DNA functions are those showing that the binding of several sequence-specific DNA-binding proteins from plants is inhibited by the methylation of CpG or CpNpG at their recognition sites (Gierl et al., 1988; Staiger et al., 1989; Inamdar et al., 1991). The chalcone synthase gene of snapdragon (*Antirrhinum majus*) contains the palindromic motif 5′-CACGTG-3′ within 150 bps of the transcription start site. As tested by gel mobility shift assays an oligonucleotide containing this sequence was not recognized by the nuclear protein CG-1 when methylated at the CpG dinucleotide and the hemimethylated DNA bound with reduced efficiency compared to that unmethylated DNA (Staiger et al., 1989). Similar results were found for chalcone synthase genes in other plants. Staiger et al. proposed that methylation of the chalcone synthase gene at the CpG dinucleotide within the binding motif for CG-1 inhibits this gene's expression during specific stages of plant development and in tissues lacking UV light-responsiveness (Staiger et al., 1989).

Proteins that we described in nuclear extracts from several types of plants recognize the mammalian cAMP response element motif 5′-TGACGTCA-3′ (CREB/ATF binding consensus sequence) and, like CG-1, are sensitive to ligand DNA cytosine methylation. These proteins (called MIB-1, for methylation-inhibited binding protein), like CREB in mammalian cells, have greatly reduced binding capacity to specific DNA ligands when the central CpG dinucleotide is replaced by 5-mCpG (Inamdar et al., 1991). Similarly, we observed that the related DNA motif, 5′-CCACGTCA-3′, recognized by other specific DNA-binding proteins (Mikami et al., 1987; Mikami et al., 1989; Katagiri et al., 1989; Inamdar et al., 1991), showed upon methylation of the CpG site a decreased binding activity for these proteins. Methylation of the central CpG of an 5′-ACACGTGG-3′ motif that forms complexes with a plant sequence-specific DNA-binding protein (Giuliano et al., 1988) also decreased the affinity for this protein (Inamdar et al., 1991). Several different plant cDNAs coding for proteins binding specifically to the latter oligonucleotide duplexes were cloned. Their deduced amino acid sequences include a basic region adjacent to a leucine zipper motif (bZIP region; Katagiri et al., 1989; Tabata et al., 1989). These proteins are distinct from MIB-1 and do not have the same DNA sequence-specificity as CREB (Inamdar et al., 1991).

We recently cloned a broad bean cDNA that expressed in bacteria a protein with CREB-like (MIB-1-type) DNA sequence-specificity (Ehrlich et al., 1992). This recombinant protein, whose deduced amino acid sequence displays the typical bZIP motif characteristic of the DNA-binding domain of mammalian CREB/ATF family of proteins (Hoeffler et al., 1988) as well as other DNA-binding proteins (Landschultz et al., 1988; Gentz et al., 1989). The protein expressed in bacteria shows the same inhibition of binding upon methylation of the CREB consensus sequence as seen from MIB-1.

A number of studies implicate DNA methylation in suppression of transposition of plant transposable elements (Cone et al., 1986; Schwartz et al., 1986; Banks et al., 1988; Schwartz, 1989; Dennis et al., 1990). Gierl et al. (1988) isolated a protein, TnpA, encoded by the maize (*Zea mays*) transposon, En-1. TnpA, which recognizes the motif, 5′-CCGACACTCTTA-3′, present in 14 copies at the two termini of En-1, shows strongly reduced binding when the overlapping CpG and CpNpG sequences at the 5′ end are bifilarly methylated. Hemimethylation also largely reduces the binding although not to as great an extent.

Because six of the copies of this TnpA-binding motif are in the putative promoter region for the *tnpA* gene, it has been proposed that TnpA is part of an autoregulatory cycle in which this protein prevents methylation at its recognition motifs in its own promoter. This, in turn, might allow its binding to this promoter and the consequent expression of the *tnpA* gene. It was further proposed that if TnpA levels are depleted, then the element would be subject to methylation, and subsequent production of TnpA could be permanently lost (Gierl et al., 1988). Such a balance between protein-binding to unmethylated sites preventing *de novo* methylation may be involved in gene control during development at other plant and animal genes.

6 DNA-binding proteins whose binding to DNA is greatly stimulated by DNA methylation

In this chapter, the previous descriptions of proteins that discriminate between methylated and unmethylated DNA sequences involved proteins whose binding to DNA is inhibited by DNA methylation. Just as there are some restriction endonucleases in bacteria, albeit a minority, that recognize their substrate sequences only if they are appropriately methylated (Noyer-Weidner et al., 1986; Raleigh and Wilson, 1986; Nelson and McClelland, 1989), so there are some eucaryotic DNA-binding proteins that are specific for methylated DNA sequences. There are two classes of DNA-binding proteins which bind much better to DNA sequences that contain 5-mC than to analogous unmethylated DNA sequences, namely, sequence-specific DNA-binding proteins and proteins responding to 5-mC levels in DNA but without sequence-specificity.

The first methylation-specific, sequence-nonspecific DNA-binding protein to be identified was a plant protein. This activity, DBP-m, was found in nuclear extracts from peas (Zhang et al., 1989a), wheat germ, and soybean leaves (K. Ehrlich, unpubl. results). DBP-m recognizes a wide variety of arbitrarily chosen 22- to 35-bp oligodeoxynucleotide duplex sequences when they contain 5-mC residues and binds better when there are more than three per DNA duplex (Zhang et al., 1989a). In mobility shift assays, even a single 5-mC residue per oligonucleotide duplex was observed to confer DBP-m binding activity. A variety of unmethylated duplexes, notably including T-rich duplexes, showed no detectable binding under the same assay conditions.

In nuclear extracts from mammalian cells, a DNA-binding protein that bound better to 5-mC-containing DNA sequences was also found (Meehan et al., 1989). However, unlike DBP-m, this protein, MeCP, bound only to duplexes that were at least about 100 bps long and that contained an extremely high density of 5-mC residues. It has not been reported whether MeCP, like DBP-m (Zhang et al., 1989a), binds to 5-mC-rich DNA sequences but not to analogous T-rich sequences. The mol percent of (percentage of bases as) 5-mC in noncancerous mammalian DNAs ranges from ~0.77 to 1.10 depending on the tissue of origin as well as the species in question (Gama-Sosa et al., 1983). These values reflect both an approximately four-fold CpG suppression and the absence of methylation in a minor, but considerable, fraction of the genomic CpGs (Ehrlich and Wang, 1981). The standard ligand used by Meehan and coworkers to detect MeCP in mobility shift assays contained 20 mol percent 5-mC. With a 135-bp duplex containing 5 mol percent 5-mC, Meehan and coworkers (1989) did not obtain MeCP·DNA complexes. In contrast, Boyes and Bird (1991) showed, in mobility shift assays, that a human α-globin gene construct methylated at approximately 1% of its C residues competed with the 135-bp duplex for binding to MeCP while the unmethylated construct did not. Also, in transient transfection assays, this methylation inhibited expression of the α-globin gene driven by its own promoter.

Another protein that binds much better to 5-mCpG-containing DNA sequences than to the corresponding unmethylated DNA sequences has been described by Jost et al. We will give a short description of the protein for which a physiologically relevant change in methylation of a DNA binding site has been studied in detail. MDBP-2 (methylated DNA binding protein 2) was isolated by Jost and coworkers (Pawlak et al., 1991; Jost et al., 1991) from liver nuclear extracts of roosters or hens. It is present in a much higher concentration in roosters, and immature chicks than in egg-laying hens. Of these three groups, only in the latter animals is the vitellogenin II gene expressed and hypomethylated at a CpG site 10 bps downstream of the transcription initiation point (position +10). Genomic sequencing revealed this correlation between expression of this gene and demethylation of this site in contrast to the constitutive hypomethylation of a CpG site at position −52 (Saluz et al., 1988).

The demethylation of the position +10 site in rooster liver and the activation of transcription of the vitellogenin gene is elicited by a single injection of estradiol (Jost et al., 1991). This demethylation is first seen at 4 h after estradiol treatment and reaches a maximum 10–18 days after this treatment. The onset of this demethylation precedes appreciable vitellogenin mRNA synthesis although maximum demethylation of this site lags behind maximum transcription of this gene. The MDBP-2 binding activity decreases greatly just before considerable vitellogenin mRNA is detected.

Evidence from *in vitro* transcription studies indicates that MDBP-2 is a repressor of transcription of the vitellogenin gene (Pawlak et al., 1991). Jost et al. (1991) propose that a major decrease in MDBP-2 repressor binding activity may be the main factor turning on vitellogenin II gene transcription after estradiol treatment with demethylation of the position +10 site playing a lesser role. However, as noted by Jost and coworkers (1991), complete demethylation of this site, and possibly other sites, could facilitate the secondary activation of the gene as a memory effect. The demethylation of the 5-mCpG dinucleotide pair at the MDBP-2 site may be important to prevent high affinity binding of low levels of MDBP-2 during the prolonged activation of the gene. Only the single CpG dinucleotide pair within the MDBP-2 binding site need be methylated to cause a dramatic increase in binding although some binding can be detected even in the absence of DNA methylation. This residual binding may be of physiological importance in hens during their pause in egg-laying (Jost et al., 1991).

The only known vertebrate DNA-binding protein that is sequence-specific and that has DNA recognition sites that require cytosine methylation for appreciable binding is MDBP-1. MDBP-1 (previously known as MDBP) was found in nuclear extracts of all studied mammalian tissues and cell cultures, including an undifferentiated embryonal carcinoma cell line (Supakar et al., 1988) and was also seen in chicken liver extracts (P. S. Supakar and M. Ehrlich, unpubl. results) so that it is probably a ubiquitous vertebrate protein. This is in contrast to MDBP-2, which was found in avian liver but not in HeLa (human) cells (Pawlak et al., 1991). MDBP-1 was not found in extracts of cultured mosquito cells and so may be associated only with cells with vertebrate-type DNA methylation (Supakar et al., 1988).

MDBP-1 was first identified as a human nuclear protein that bound better to human DNA enriched in 5-mC residues *in vitro* compared to otherwise identical DNA whose 5-mC content was decreased (Huang et al., 1984). The initial sites identified as specific for MDBP-1 were six sequences in pBR322 and M13mp8 replicative form DNA that require cytosine methylation for binding (Wang et al., 1986). We subsequently found sites in phage λ DNA, mammalian viral DNAs, and mammalian DNA that do not require methylation for binding (Zhang et al., 1989b; Zhang et al., 1990a and b; Ehrlich and Ehrlich, 1990; Zhang et al., 1991). By DNase I footprinting, dimethyl sulfate interference, and proteolytic clipping bandshift assays, the protein that binds to these sites devoid of 5-mC residues was shown to be the same one that binds to other sites only when they are methylated at their CpG dinucleotides. The cytosine methylation-independent MDBP-1 binding sites have the equivalent of 5-mC $\rightarrow$ T substitutions relative to the methylation-dependent MDBP binding sites. All identified MDBP-1 sites resemble the 14- or 13-bp long consensus sequence for binding of this protein. The methylation-dependent sites have 1–3 5-mCpG dinucleotide pairs and bind better when more of these CpG are methylated. Symmetrical (bifilar) methylation generally is better than hemimethylation, which is, in turn, better than no methylation. A bifilarly methylated MDBP-1 site (methylation-dependent) can bind up to 100 times better than the unmethylated site (Khan et al., 1988).

A methylation-independent MDBP-1 site is located in the hepatitis B virus (HBV) enhancer I and two are found in a human cytomegalovirus (CMV) enhancer and a polyomavirus enhancer (Zhang et al., 1990a; Zhang et al., 1991). Transient transfection assays involving the HBV enhancer/reporter gene constructs indicate that the MDBP-1 site of the enhancer (also known as an EF-C of EP site) is important for the activation of transcription (Ostapchuk et al., 1989; Dikstein et al., 1990;

Zhang et al., 1990a; Garcia et al., 1991; C. K. Asiedu and M. Ehrlich et al., unpubl. results). However, the methylation-independent MDBP-1 sites in the CMV enhancer and one 5 bps sequence after the start point of transcription may play a negative role in the control of transcription (X.-Y. Zhang and M. Ehrlich, unpubl. results).

We identified methylation-responsive MDBP sites in mammalian DNA, which bound MDBP-1, under standard conditions, approximately 5–20 times better when their CpG dinucleotides were methylated (Zhang et al., 1990a). Four out of nine sites are located only about 30 to 110 bps downstream from the transcription start points. Several of these sites are located in genes whose CpG methylation is implicated in negative transcription control. We have proposed that MDBP-1 negatively controls expression of certain genes from the methylation-responsive MDBP-1 sites and that it also controls negatively or positively the transcription of other genes from methylation-independent MDBP-1 binding sites (Zhang et al., 1990a). During evolution, methylation-independent MDBP-1 sites may have arisen from methylation-dependent ones by spontaneous deamination of 5-mC residues (Ehrlich et al., 1986).

In the case of the human apolipoprotein(a) gene we found more than 30 MDBP-1 sites and evidence suggests that the number of these sites, which include methylation-dependent as well as methylation-independent MDBP-1 binding sequences, varies from allele to allele. The MDBP-1 site is 9 bps from an AP-2 site (Ehrlich and Ehrlich, 1990; C. K. Asiedu and Ehrlich, M., unpubl. data) and each of these resides within a repeated unit present within this internally repetitious gene. We have proposed that the number of copies of this repeat and, thereby, of the MDBP-1 and AP-2 sites within the gene are involved in negative transcription control (Ehrlich and Ehrlich, 1990). If so, this could help explain the otherwise enigmatic inverse relationship between the molecular mass of apolipoprotein(a) and plasma levels of this glycoprotein.

Interestingly, even though MDBP-1 sites have hyphenated dyad symmetry, methylation of only one strand or the other at a methylation-responsive DNA site often gives very different extents of complex formation with MDBP-1 (Zhang et al., 1990a). A similar finding was reported for MDBP-2. Therefore, as shown in Figure 1, upon replicative DNA synthesis, the two progeny chromosomes could have different fates in terms of binding of MDBP-type factors of which there may be many more yet to be discovered. Although usually both copies of the replicated site will eventually become bifilarly methylated, competition between sequence-specific DNA-binding proteins, histones, and other chromatin proteins for binding to DNA immediately after its replication may determine the long-term organization of chromatin around that sequence. For certain types of cells during certain stages

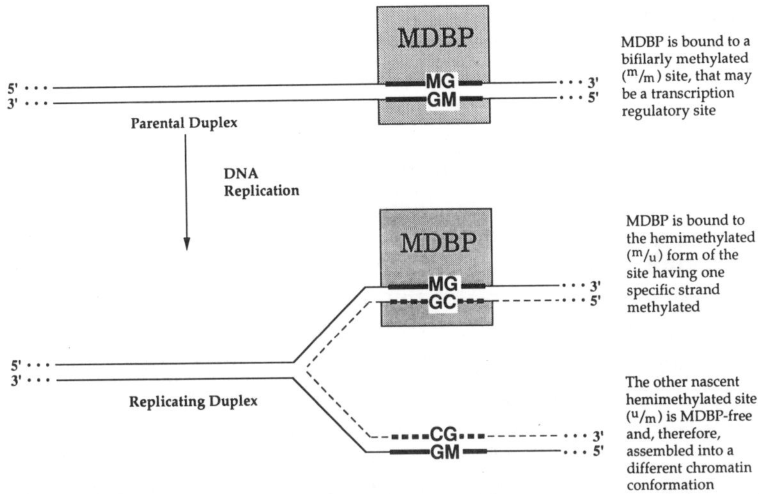

Figure 1. A cartoon depicting a model for how differential binding of an MDBP-like activity to newly replicated methylation-dependent DNA sites may help give rise to phenotypically different progeny cells. Just one 5-mCpG (MG) dinucleotide pair is shown in the parental duplex's MDBP recognition site (thick line) but there can be up to three such dinucleotide pairs per MDBP-1 recognition site. For simplicity, the MDBP protein is depicted as a rectangle. The newly replicated strand is shown as a dashed line. The hemimethylated m/u progeny duplex site has one strand methylated and the hemimethylated u/m progeny duplex site has the other strand methylated. Depending upon which strand in a hemimethylated site is methylated, there can be different probabilities that the MDBP will bind to the site (Zhang, et al., 1990a: Pawlak et al., 1991). Thereby, if these sites are involved in transcription regulation and if the cells contain proteins necessary for tissue-specific transcription, different extents of expression of the gene associated with this MDBP site could occur from the two progeny chromosomes. This could help to obtain phenotypic differences between progeny cells if one happens to receive two of the homologues that were of the m/u type at this locus and the other progeny cell gets the homologues that were of the corresponding u/m type.

of differentiation, this may allow the gene to be expressed in one of the two progeny cells but not in the other. Differential binding of MDBP-type proteins could thereby help one parental cell differentiate into two phenotypically different progeny cells.

Some of the DNA-binding proteins whose recognition of their specific DNA regulatory sites is inhibited by DNA methylation might also bind unequally to the two hemimethylated versions of the same sequence with opposite orientations for the methylated strand. If so, then this model coupling DNA replication, DNA methylation, and differentiation could also pertain to them. In accord with this model, hemimethylation of 5′-GATC-3′ sites associated with DNA replication appears to activate IS10 transposition in *E. coli* but does so unequally depending upon which of the DNA strands is methylated (Roberts et al., 1985).

7 Conclusions

Numerous reports have provided increasingly convincing evidence that cytosine methylation or demethylation associated with differentiation can control vertebrate transcription. Consistent with such a role for DNA methylation, there are many examples of sequence-specific DNA-binding proteins whose binding to DNA is strongly influenced by the state of CpG methylation of their cognate sites. Numerous DNA-binding proteins have recently been shown to bind much less to methylated forms of their recognition sites than to the unmethylated analogs.

Furthermore, there are proteins that bind much better to specific DNA sequences when these sites are methylated. Sequence non-specific proteins with a specificity for 5-mC-rich regions have also been described. DNA-binding proteins responsive to DNA methylation have been identified in higher plants as well as in vertebrates. That a variety of cytosine-methylation responsive DNA-binding proteins is found in vertebrates and higher plants, all of which appear to display genomic cytosine methylation, indicates the importance of protein-mediated effects of DNA methylation on transcription in these organisms.

Acknowledgments. We thank Drs. Xian-Yang Zhang and Clement Asiedu for critical reading of this manuscript. The research from our labs was supported in part by grants GM33999, DMB8614448, and 85-CRCR-1-1751 from the National Institutes of Health, National Science Foundation, and Department of Agriculture, respectively.

Avvedimento, E. V., Obici, S., Sanchez, M., Gallo, A., Musti, A., and Gottesman, M. E. (1989) Reactivation of thyroglobulin gene expression in transformed thyroid cells by 5-azacytidine. Cell *58*, 1135–1142.

Bakker, A., and Smith, D. W. (1989) Methylation of GATC sites is required for precise timing between rounds of DNA replication in *Escherichia coli*. J. Bacteriol. *171*, 5738–5742.

Banks, J. A., Masson, P., and Fedoroff, N. (1988) Molecular mechanisms in the developmental regulation of the maize suppressor-mutator transposable element. Genes Dev. *2*, 1364–1380.

Barras, F., and Marinus, M. G. (1989) The great GATC: DNA methylation in *E. coli*. *Trends Genet.* *5*, 139–143.

Becker, P. B., Ruppert, S., and Schütz, G. (1987) Genomic footprinting reveals cell type-specific DNA binding of ubiquitous factors. Cell *51*, 435–443.

Bednarik, D. P., Cook, J. A., and Pitha, P. M. (1990) Inactivation of the HIV LTR by DNA CpG methylation: evidence for a role in latency. EMBO J. *9*, 1157–1164.

Ben-Hattar, J., Beard, P., and Jiricny, J. (1989) Cytosine methylation in CTF and Spl recognition sites of an HSV tk promoter: effects on transcription *in vivo* and on factor binding *in vitro*. Nucl. Acids Res. *17*, 10179–10190.

Benvenisty, N., Szyf, M., Mencher, D., Razin, A., and Reshef, L. (1985) Tissue-specific hypomethylation and expression of rat phosphoenolpyruvate carboxykinase gene induced by *in vivo* treatment of fetuses and neonates with 5-azacytidine. Biochemistry *24*, 5015–5019.

Benvenisty, N., Mencher, D., Meyuhas, O., Razin, A., and Reshef, L. (1985) Sequential changes in DNA methylation patterns of the rat phosphoenolpyruvate carboxykinase gene during development. Proc. Natl. Acad. Sci. USA *82*, 267–271.

Blackman, M. A., and Koshland, M. E. (1985) Specific 5′ and 3′ regions of the μ-chain gene are undermethylated at distinct stages of B-cell differentiation. Proc. Natl. Acad. Sci. USA *82*, 3809–3813.

Blyn, L. B., Braaten, B. A., and Low, D. A. (1990) Regulation of *pap* pilin phase variation by a mechanism involving differential Dam methylation states. EMBO J. *9*, 4045–4054.

Boyes, J., and Bird, A. (1991) DNA methylation inhibits transcription indirectly via a methyl-CpG binding protein. Cell *64*, 1123–1134.

Breznik, T., Traina-Dorge, V., Gama-Sosa, M., Gehrke, C. W., Ehrlich, M., Medina, Butel, J. S., and Cohen, J. C. (1984) Mouse mammary tumor virus DNA methylation: tissue-specific variation. Virology *136*, 69–77.

Broad, P. M., Symes, A. J., Thakker, R. V., and Craig, R. K. (1989) Structure and methylation of the human calcitonin/αCGRP gene. Nucl. Acids Res. *17*, 6999–7011.

Brooks, J. E., and Roberts, R. (1982). Modification profiles of bacterial genomes. Nucl. Acids Res. *10*, 913–934.

Burbelo, P., Horikoshi, S., and Yamada, Y. (1990) DNA methylation and collagen IV gene expression in F9 teratocarcinoma cells. J. Biol. Chem. *265*, 4839–4843.

Comb, M., and Goodman, H. M. (1990) CpG methylation inhibits proenkephalin gene expression and binding of the transcription factor AP-2. Nucl. Acids Res. *18*, 3975–3982.

Cone, K. C., Burr, F. A., and Burr, B. (1986) Molecular analysis of the maize anthocyanin regulatory locus *C1*. Proc. Natl. Acad. Sci. USA *83*, 9631–9635.

Courey, A. J., and Tjian, R. (1988) Analysis of Sp1 *in vivo* reveals multiple transcriptional domains, including a novel glutamine-rich activation motif. Cell *55*, 887–898.

Creusot, G., Acs, G., and Christman, J. K. (1982) Inhibition of DNA methyltransferase and induction of friend erythroleukemia cell differentiation by 5-azacytidine and 5-aza-2′-deoxycytidine. J. Biol. Chem. *257*, 2041–2048.

Dennis, E. S., and Brettell, R. I. S. (1990) DNA methylation of maize transposable elements is correlated with activity. Phil. Trans. R. Soc. Lond. *326*, 217–229.

Desrosiers, R. C., Mulder, C., and Fleckenstein, B. (1979) Methylation of herpesvirus *saimiri* DNA in lymphoid tumor cell lines. Proc. Natl. Acad. Sci. USA *76*, 3839–3843.

Dikstein, R., Faktor, O., Ben-Levy, R., and Shaul, Y. (1990) Functional organization of the hepatitis B virus enhancer. Mol. Cell. Biol. *10*, 3683–3689.

Dobrovic, A., Gareau, J. L., Ouellette, G., and Bradley, W. E. (1988) DNA methylation and genetic inactivation at thymidine kinase locus: two different mechanisms for silencing autosomal genes. Som. Cell Mol. Genet. *14*, 55–68.

Dobrzanski, P., Hoeveler, A., and Doerfler, W. (1988) Inactivation by sequence-specific methylations of adenovirus promoters in a cell-free transcription system. J. Virol. *62*, 3941–3946.

Doerfler, W. (1991) Patterns of DNA methylation – evolutionary vestiges of foreign DNA inactivation as a host defense mechanism. Biol. Chem. Hoppe-Seyler *372*, 557–564.

Ehrlich, K. C., Carey, J. W., and Ehrlich, M. (1992) A broad bean DNA clone encoding a protein resembling mammalian CREB in its sequence specificity and DNA methylation-sensitivity. Gene *117*, 169–178.

Ehrlich, K. C., and Ehrlich, M. (1990) Highly repeated sites in the apolipoprotein(a) gene recognized by methylated DNA-binding protein, a sequence-specific DNA-binding protein. Mol. Cell. Biol. *10*, 4957–4960.

Ehrlich, M., Gama-Sosa, M. A., Carreira, L. H., Ljungdahl, L. G., Kuo, D. C., and Gehrke, C. W. (1985) DNA methylation in thermophilic bacteria: N^4-methylcytosine, and N^6-methyladenine. Nucl. Acids Res. *13*, 1399–1412.

Ehrlich, M., Gama-Sosa, M. A., Huang, L.-H., Midgett, R. M., Kuo, K. C., McCune, R. A., and Gehrke, C. (1982). Amount and distribution of 5-methylcytosine in human DNA from different types of tissues or cells. Nucl. Acids Res. *10*, 2709–2721.

Ehrlich, M., Norris, K. F., Wang, R. Y.-H., Kuo, D. C., and Gehrke, C. W. (1986) DNA cytosine methylation and heat-induced deamination. Bioscience Rep. *6*, 387–393.

Ehrlich, M., and Wang, R. Y.-H. (1981) 5-methylcytosine in eukaryotic DNA. Science *212*, 1350–1357.

Enver, T., Zhang, J., Papayannopoulou, T., and Stamatoyannopoulos, G. (1988) DNA methylation: a secondary event in globin gene switching? Genes Devel. *2*, 698–706.

Evans, T., DeChiara, T., and Efstratiadis, A. (1988) A promoter of the rat insulin-like growth factor II gene consists of minimal control elements. J. Mol. Biol. *199*, 61–81.

Faber, S., Ip, T., Granner, D., and Chalkley, R. (1991) The interplay of ubiquitous DNA-binding factors, availability of binding sites in the chromatin, and DNA methylation in the differential regulation of phosphoenolpyruvate carboxykinase gene expression. Nucl. Acids Res. *19*, 4681–4688.

Falzon, M., and Kuff, E. L. (1991) Binding of the transcription factor EBP-80 mediates the methylation response of an intracisternal A-particle long terminal repeat promoter. Mol. Cell. Biol. *11*, 117–125.

Gama-Sosa, M. A., Midgett, M., Slagel, V. A., Githens, S., Kuo, K. C., Gehrke, C. W., and Ehrlich, M. (1983) Tissue-specific differences in DNA methylation in various mammals. Biochim. Biophys. Acta *740*, 212–219.

Gama-Sosa, M. A., Wang, R. Y.-H., Kuo, K. C., Gehrke, C. W., and Ehrlich, M. (1983) The 5-methylcytosine content of highly repeated sequences in human DNA. Nucl. Acids Res. *11*, 3087–3095.

Garcia, A. D., Ostapchuk, P., and Hearing, P. (1991) Methylation-dependent and -independent DNA binding of nuclear factor EF-C. Virology *182*, 857-860.

Gentz, R., Rauscher, III, F. J., Abate, C., and Curran, T. (1989) Parallel association of Fos and Jun leucine zippers juxtaposes DNA binding domains. Science *243*, 1695–1699.

Gierl, A., Lütticke, S., and Saedler, H. (1988) TnpA product encoded by the transposable element En-1 of Zea mays is a DNA binding protein. EMBO J. *7*, 4045–4053.

Giuliano, G., Pichersky, E., Malik, M. P., Scolnik, P. A., and Cashmore, A. R. (1988) An evolutionarily conserved protein binding sequence upstream of a plant light-regulated gene. Proc. Natl. Acad. Sci. USA *85*, 7089–7093.

Goeddel, D. V., Yansura, D. G., and Caruthers, M. H. (1978) How *lac* repressor recognizes *lac* operator. Proc. Natl. Acad. Sci. USA *75*, 3578–3582.

Grant, S. G., and Chapman, V. M. (1988) Mechanisms of X-chromosome regulation. Annu. Rev. Genet. *22*, 199–233.

Gruenbaum, Y., Naveh-Many, T., Cedar, H., and Razin, A. (1981) Sequence specificity of methylation in higher plant DNA. Nature *292*, 860–862.

Guntaka, R. V., Gowda, S., Wagner, H., and Simon, D. (1987) Methylation of the enhancer region of avian sarcoma virus long terminal repeat suppresses transcription. FEBS Lett. *221*, 332–336.

Hai, T., Liu, F., Coukos, W., and Green, M. (1989) Transcription factor ATF cDNA clones: an extensive family of leucine zipper protein able to selectively form DNA-binding hetero-dimers. Genes Devel. *3*, 2083–2090.

Harrington, M. A., Jones, P. A., Imagawa, M., and Karin, M. (1988) Cytosine methylation does not affect binding of transcription factor Sp1. Proc. Natl. Acad. Sci. USA *85*, 2066–2070.

Harris, M. (1982) Induction of thymidine kinase in enzyme-deficient Chinese hamster cells. Cell *29*, 483–492.

Hermann, R., Hoeveler, A., and Doerfler, W. (1989) Sequence-specific methylation in a downstream region of the late E2A promoter of adenovirus type 2 DNA prevents protein binding. J. Mol. Biol. *210*, 411–415.

Hermann, R., and Doerfler, W. (1991) Interference with protein binding at AP2 sites by sequence-specific methylation in the late E2A promoter of adenovirus type 2 DNA. FEBS Lett. *281*, 191–195.

Hoeffler, J. P., Meyer, T. E., Yun, Y., Jameson, J. L., and Habener, J. F. (1988) Cyclic AMP-responsive DNA-binding protein: structure based on a cloned placental cDNA. Science *242*, 1430–1433.

Höller, M., Westin, G., Jiricny, J., and Schaffner, W. (1988) Sp1 transcription factor binds DNA and activates transcription even when the binding site is CpG methylated. Genes Dev. *2*, 1127–1135.

Hong, J. C., Nagao, R. T., and Key, J. L. (1987) Characterization and sequence analysis of a developmentally regulated putative cell wall protein gene isolated from soybean. J. Biol. Chem. *262*, 8367–8376.

Huang, L.-H., Wang, R., Gama-Sosa, M. A., Shenoy, S., and Ehrlich M. (1984) A protein from human placental nuclei binds preferentially to 5-methylcytosine-rich DNA. Nature *308*, 293–295.

Hurst, H. C., Masson, N., Hones, N. C., and Lee, K. A. W. (1990) The cellular transcription factor CREB corresponds to activating transcription factor 47 (ATF-47) and forms complexes with a group of polypeptides related to ATF-43. Mol. Cell. Biol. *10*, 6192–6203.

Iguchi-Ariga, S. M., and Schaffner, W. (1989) CpG methylation of the cAMP-responsive enhancer/promoter sequence TGACGTCA abolishes specific factor binding as well as transcriptional activation. Genes Dev. *3*, 612–619.

Inamdar, N. M., Ehrlich, K. C., and Ehrlich, M. (1991) CpG methylation inhibits binding of several sequence-specific DNA-binding proteins from pea, wheat, soybean and cauliflower. Plant Mol. Biol. *17*, 111–123.

Ivarie, R. (1987) Thymine methyls and DNA-protein interactions. Nucl. Acids Res. *15*, 9975–9983.

Jaenisch, R., Schnieke, A., and Harbers, K. (1985) Treatment of mice with 5-azacytidine efficiently activates silent retroviral genomes in different tissues. Proc. Natl. Acad. Sci. USA *82*, 1451–1455.

Jahroudi, N., Foster, R., Price-Haughey, J., Beitel, G., and Gedamu, L. (1990) Cell-type specific and differential regulation of the human metallothionein genes. J. Biol. Chem. *265*, 6506–6511.

Jones, P. A., Taylor, S. M., Mohandas, T., and Shapiro, L. J. (1982) Cell cycle-specific reactivation of an inactive X-chromosome locus by 5-azadeoxycytidine. Proc. Natl. Acad. Sci. USA *79*, 1215–1219.

Jost, J.-P., Saluz, H.-P., and Pawlak, A. (1991) Estradiol down regulates the binding activity of an avian vitellogenin gene repressor (MDBP-2) and triggers a gradual demethylation of the m^5CpG pair of its DNA binding site. Nucl. Acids Res. *19*, 5771–5775.

Jost J. P., and Hofsteenge, J. (1992) The repressor MDBP-2 is a member of the Histone H1 family that binds preferentially to methylated DNA *in vitro* and *in vivo*. Proc. Natl. Acad. Sci. USA (in press).

Jump, D. B., Wong, N. C. W., and Oppenheimer, J. H. (1987) Chromatin structure and methylation state of a thyroid hormone-responsive gene in rat liver. J. Biol. Chem. *262*, 778–784.

Kadonaga, J. T., Jones, K. A., and Tjian, R. (1986) Promotor-specific activation of RNA polymerase II transcription by Sp1. Trends Biol. Sci. *11*, 20–23.

Katagiri, F., Lam, E., and Chua, N.-H. (1989) Two tobacco DNA-binding proteins with homology to the nuclear factor CREB. Nature *340*, 727–730.

Khan, R., Zhang, X.-Y., Supakar, P. C., Ehrlich, K. C., and Ehrlich, M. (1988) Human methylated DNA-binding protein: Determinants of a pBR322 recognition site. J. Biol. Chem. *263*, 14374–14383.

Knust, B., Brüggemann, U., and Doerfler, W. (1989) Reactivation of a methylation-silenced gene in adenovirus-transformed cells by 5-azacytidine or by E1A trans activation. J. Virol. *63*, 3519–3524.

Kovesdi, I., Reichel, R., and Nevins, J. R. (1987) Role of an adenovirus E2 promoter binding factor in E1A-mediated coordinate gene control. Proc. Natl. Acad. Sci. USA *84*, 2180–2184.

Lamson, G., and Stockdale, F. E. (1989) Developmental and muscle-specific changes in methylation of the myosin light chain LC1f and LC3f promoters during avian myogenesis. Devel. Biol. *132*, 62–68.

Landschulz, W. H., Johnson, P. F., and McKnight, S. L. (1988) The leucine zipper: a hypothetical structure common to a new class of DNA binding proteins. Science *240*, 1759–1764.

Lasser, A. B., Paterson, B. M., and Weintraub, H. (1986) Transfection of a DNA locus that mediates the conversion of 10T1/2 fibroplasts to myoblasts. Cell *47*, 649–656.

Levy-Wilson, B., and Fortier, C. (1989) Tissue-specific undermethylation of DNA sequences at the 5′ end of the human apolipoprotein B gene. J. Biol. Chem. *264*, 9891–9896.

Liu, Y.-S., and Green, M. (1988) Interaction of a common cellular transcription factor, ATF, with regulatory elements in both Ela- and cyclic AMP-inducible promotors. Proc. Natl. Acad. Sci. USA *85*, 3396–3400.

Lock, L. F., Takagi, N., and Martin, G. R. (1987) Methylation of the *hprt* gene on the inactive X occurs after chromosome inactivation. Cell *48*, 39–46.

Louarn, J., Francois, V., and Louarn, J.-M. (1990) Chromosome replication pattern in *dam* mutants of *Escherichia coli*. Mol. Gen. Genet. *221*, 291–294.

Lübbert, M., Miller, C. W., and Koeffler, H. P. (1991) Changes of DNA methylation and chromatin structure in the human myeloperoxidase gene during myeloid differentiation. Blood *78*, 345–356.

Meehan, R. R., Lewis, J. D., McKay, S., Kleiner, E. L., and Bird, A. P. (1989) Identification of a mammalian protein that binds specifically to DNA containing methylated CpGs. Cell *58*, 499–507.

Messer, W., and Noyer-Weidner, M. (1988). Timing and targeting: the biological functions of *dam* methylation in *E. coli*. Cell *54*, 735–737.

Mikami, K., Tabata, T., Kawata, T., Nakayama, T., and Iwabuchi, M. (1987) Nuclear protein(s) binding to the conserved DNA hexameric sequence postulated to regulate transcription of wheat histone genes. FEBS Lett. *223*, 273–278.

Mikami, K., Takase, H., Tabata, T., and Iwabuchi, M. (1989) Multiplicity of the DNA-binding protein HBP-1 specific to the conserved hexameric sequence ACGTCA in various plant gene promoters. FEBS Lett. *256*, 67–70.

Nelson, M., and McClelland, M. (1989) Effect of site-specific methylation on DNA modification methyltransferases and restriction endonucleases. Nucl. Acids Res. *17*, 389–415.

Noyer-Weidner, M., Diaz, R., and Reiners, L. (1986) Cytosine-specific DNA modification interferes with plasmid establishment in *Escherichia coli* K12: involvement of *rgl* B. Mol. Gen. Genet. *205*, 469–475.

Ostapchuk, P., Scheirle, G., and Hearing, P. (1989) Binding of nuclear EF-C to a functional domain of the hepatitis B virus enhancer region. Mol. Cell. Biol. *9*, 2787–2797.

Pawlak, A., Bryans, M., and Jost, J.-P. (1991) An avian 40 KDa nucleoprotein binds preferentially to a promoter sequence containing one single pair of methylated CpG. Nucl. Acids Res. *19*, 1029–1034.

Piva, R., Kumar, L. V., Hanau, S., Maestri, I., Rimondi, A. P., Pansini, S. F., Mollica, G., Chambon, P., and del Senno, L. (1989) The methylation pattern in the 5′ end of the human estrogen receptor gene is tissue specific and related to the degree of gene expression. Biochem. Intern. *19*, 267–275.

Prendergast, G. C., Lawe, D., and Ziff, E. B. (1991) Association of Myn, the murine homolog of Max, with c-Myc stimulates methylation-sensitive DNA binding and Ras cotransformation. Cell *65*, 395–407.

Rachal, M. J., Yoo, H., Becker, F. F., and Lapeyre, J.-N. (1989) *In vitro* DNA cytosine methylation of *cis*-regulatory elements modulates c-Ha-ras promoter activity *in vivo*. Nucl. Acids Res. *17*, 5135–5147.

Raleigh, E. A., and Wilson, G. (1986) *Escherichia coli* K-12 restricts DNA containing 5-methylcytosine. Proc. Natl. Acad. Sci. USA *83*, 9070–9074.

Roberts, D., Hoopes, B. C., McClure, W. R., and Kleckner, N. (1985) IS10 transposition is regulated by DNA adenine methylation. Cell *43*, 117–130.

Saluz, H. P., Feavers, I. M., Jiricny, F. J., and Jost, J. P. (1988) Genomic sequencing and *in vivo* footprinting of an expression-specific DNase I-hypersensitive site of avian vitellogenin II promoter reveal a demethylation of a mCpG and a change in specific interactions of proteins with DNA. Proc. Natl. Acad. Sci. USA *85*, 6697–6700.

Schwartz, D. (1989) Gene-controlled cystosine demethylation in the promotor region of the *Ac* transposable element. Proc. Natl. Acad. Sci. USA *86*, 2789–2793.

Schwartz, D., and Dennis, E. (1986) Transposase activity of the *Ac* controlling element in maize is regulated by its degree of methylation. Mol. Gen. Genet. *205*, 476–482.

Seeman, N. C., Rosenberg, J. M., and Rich, A. (1976) Sequence-specific recognition of double helical nucleic acids by proteins. Proc. Natl. Acad. Sci. USA *73*, 804–808.

Shen, E. S., and Whitlock, J. P. Jr. (1989) The potential role of DNA methylation in the response to 2,3,7,8-tetrachlorodibenzo-p-dioxin. J. Biol. Chem. *264*, 17754–17758.

Shinar, D., Yoffe, O., Shani, M., and Yaffe, D. (1989) Regulated expression of muscle-specific genes introduced into mouse embryonal stem cells: inverse correlation with DNA methylation. Differentiation *41*, 116–126.

Shmookler-Reis, R. J., and Goldstein, S. (1982) Variability of DNA methylation patterns during serial passage of human diploid fibroblasts. Proc. Natl. Acad. Sci. USA *79*, 3949–3953.

Staiger, D., Kaulen, H., and Schell, J. (1989) A CACGTG motif of the *Antirrhinum majus* chalcone synthase promoter is recognized by an evolutionarily conserved nuclear protein. Proc. Natl. Acad. Sci. USA *86*, 6930–6934.

Sullivan, C. H., Norman, J. T., Borras, T., and Grainger, R. M. (1989) Developmental regulation of hypomethylation of δ-crystallin genes in chicken embryo lens cells. Mol. Cell. Biol. *9*, 3132–3135.

Supakar, P. C., Weist, D., Zhang, D., Inamdar, X.-Y., Khan, R., Ehrlich, K. C., and Ehrlich, M. (1988) Methylated DNA-binding protein in various mammalian cell types. Nucl. Acids Res. *16*, 8029–8044.

Tabata, T., Takase, H., Takayama, S., Mikami, K., Nakatsuka, A., Kawata, T., Nakayana, T., and Iwabuchi, M. (1989) A protein that binds to a *cis*-acting element of wheat histone genes has a leucine zipper motif. Science *245*, 965–966.

Tapscott, S. J., Davis, R. L., Thayer, M. J., Cheng, P. F., Weintraub, H., and Lassar, A. B. (1988) MyoD1: a nuclear phosphoprotein requiring a Myc homology region to convert fibroblasts to myoblasts. Science *242*, 405–411.

Tasseron-de Jong, J. G., den Dulk, H., van de Putte, P., and Giphart-Gassler, M. (1989) *De novo* methylation as major event in the inactivation of transfected herpesvirus thymidine kinase genes in human cells. Biochim. Biophys. Acta *1007*, 215–223.

Toth, M., Müller, U., and Doerfler, W. (1990) Establishment of *de novo* DNA methylation patterns. J Mol. Biol. *214*, 673–683.

Toth, M., Lichtenberg, U., and Doerfler, W. (1989) Genomic sequencing reveals a 5-methylcytosine-free domain in active promoters and the spreading of preimposed methylation patterns. Proc. Natl. Acad. Sci. USA *86*, 3728–3732.

Umeno, M., Song, B. J., Kozak, C., Gelboin, H. V., and Gonzalez, F. J. (1988) The rat P450IIE1 gene: complete intron and exon sequence, chromosome mapping, and correlation of developmental expression with specific 5′ cytosine demethylation. J. Biol. Chem. *263*, 4956–4962.

Vanyushin, B. F., Mazin, A. L., Vasilyev, V. K., and Belozersky, A. N. (1973) The content of 5-methylcytosine in animal DNA: the species and tissue specificity. Biochim. Biophys. Acta *299*, 397–403.

Venolia, L., Gartler, S. M., Wassman, E. R., Yen, P., Mohandas, T., and Shapiro, L. J. (1982) Transformation with DNA from 5-azacytidine-reactivated X chromosomes. Proc. Natl. Acad. Sci. USA *79*, 2352–2354.

Wagner, I., and Capesius, I. (1981) Determination of 5-methylcytosine from plant DNA by high-performance liquid chromatography. Biochim. Biophys. Acta. *654*, 52–56.

Wang, R.-H., Zhang, X.-Y., Khan, R., Zhou, Y., Huang, L.-H., and Ehrlich, M. (1986) Methylated DNA-binding protein from human placenta recognizes specific methylated sites on several prokaryotic DNAs. Nucl. Acids Res. *14*, 9843–9860.

Watt, F., and Molloy, P. L. (1988) Cytosine methylation prevents binding to DNA of a HeLa cell transcription factor required for optimal expression of the adenovirus major late promoter. Genes Dev. *2*, 1136–1143.

Weih, F., Nitsch, D., Reik, A., Schütz, G., and Becker, P. B. (1991) Analysis of CpG methylation and genomic footprinting at the tyrosine aminotransferase gene: DNA methylation alone is not sufficient to prevent protein binding *in vivo*. EMBO J. *10*, 2559–2567.

Winter, H., Rentrop, M., Nischt, R., and Schweizer, J. (1990) Tissue-specific expression of murine keratin K13 in internal stratified squamous epithelia and its aberrant expression during two-stage mouse skin carcinogenesis is associated with the methylation state of a distinct CpG site in the remote 5-flanking region of the gene. Differentiation *43*, 105–114.

Yamamoto, K. K., Gonzalez, G. A., Menzel, P., Rivier, J., and Montminy, M. R. (1990) Characterization of a bipartite activator domain in transcription factor CREB. Cell *60*, 611–617.

Yang, T. P., Singer-Sam, J., Flores, J. C., and Riggs, A. D. (1988) DNA binding factors for the CpG-rich island containing the promoter of the human X-linked PGK gene. Somat. Cell Molec. Gen. *14*, 461–472.

Yisraeli, J., Aldelstein, R. S., Melloul, D., Nudel, U., Yaffe, D., and Cedar, H. (1986) Muscle-specific activation of a methylated chimeric actin gene. Cell *46*, 409–416.

Zhang, D., Ehrlich, K. C., Supakar, P. C., and Ehrlich, M. (1989a) A plant DNA-binding protein that recognizes 5-methylcytosine residues. Mol. Cell Biol. *9*, 1351–1356.

Zhang, X.-Y., Asiedu, C. K., Supakar, P. C., Khan, R., Ehrlich, K. C., and Ehrlich, M. (1990a) Binding sites in mammalian genes and viral gene regulatory regions recognized by methylated DNA-binding protein. Nucl. Acids Res. *18*, 6253–6260.

Zhang, X.-Y., Inamdar, N. M., Supakar, P. C., Wu, K., Ehrlich, K. C., and Ehrlich, M. (1991) Three MDBP sites in the immediate-early enhancer-promoter region of human cytomegalovirus. Virology *182*, 865–869.

Zhang, X.-Y., Lofln, P. T., Gehrke, C. W., Andrews, P. A., and Ehrlich, M. (1987) Hypermethylation of human DNA sequences in embryonal carcinoma cells and somatic tissues but not in sperm. Nucl. Acids Res. *15*, 9429–9449.

Zhang, X.-Y., Supakar, P. C., Khan, R., Ehrlich, K. C., and Ehrlich, M. (1989b) Related sites in human and herpes DNA recognized by methylated DNA-binding protein from human placenta. Nucl. Acids Res. *17*, 1459–1474.

Zhang, X.-Y., Supakar, P. C., Wu, K., Ehrlich, K. C., and Ehrlich, M. (1990b) An MDBP site in the first intron of the human c-*myc* gene. Cancer Res. *50*, 6865–6869.

Zhang, X.-Y., Wang, R. Y.-H., and Ehrlich, M. (1985) Human DNA sequences exhibiting gamete-specific hypomethylation. Nucl. Acids Res. *13*, 4837–4851.

DNA Methylation: Molecular Biology and Biological Significance
ed. by J. P. Jost & H. P. Saluz

CpG Islands

Francisco Antequera and Adrian Bird

Institute of Cell and Molecular Biology, University of Edinburgh, Kings Buildings, Edinburgh EH9 3JR, Scotland

1 Introduction

DNA methylation is a conspicuous feature of vertebrate genomes. Approximately 4% of total cytosines are methylated, representing about 5×10^7 5-methylcytosine (5-mC) residues per diploid nucleus. All 5-mC is present in the dinucleotide CpG, although only 70 to 80% of the potentially methylatable sites are actually in a methylated form. Most studies on the relative distribution of methylated cytosines in DNA have relied on the ability of some restriction enzymes to distinguish between the methylated and unmethylated version of the same sequence. This approach has provided information about the methylation patterns at specific genes in different tissues during development and has revealed that the vertebrate genome can be divided into two distinct compartments: the CpG island fraction, representing 1 to 2% of the total genome and containing all the sites that are consistently nonmethylated at all stages of development; and the remaining 98% containing all methylated sites.

Early work on DNA methylation showed that regions that were sensitive to nucleases and enriched in actively transcribed sequences were hypomethylated relative to bulk DNA (Razin and Cedar, 1977; Naveh-Many and Cedar, 1981). These studies did not focus on specific sequences, but they showed that 5-mC was nonrandomly distributed in chromatin. More relevant to the discovery of the CpG islands was the observation of the asymmetrical distribution of CpG dinucleotides across some gene sequences. McClelland and Ivarie (1982) studied the distribution of CpGs in 15 randomly selected mammalian genes and found that, on average, the frequency of CpGs at the 5′ end was significantly higher than at the 3′ end. At the time it was difficult to realize that the "averaging" strategy would place in the same category genes with and without CpG islands, and therefore the CpG-richness of the former would be masked by the scarcity of CpGs in the latter. Results reported by Tykocinski and Max (1984) overcame this problem by concentrating on individual genes. They showed striking clustering of CpGs across the polymorphic exons of some MHC class I and II genes and the 5′ end of other vertebrate genes.

At the same time, Cooper et al. (1983) tested the possible existence of unmethylated CpG clusters in vertebrate genomes. The experiments were prompted by the finding that a number of invertebrates (Bird et al., 1979; Bird and Taggart, 1980) had been shown to have their genomes divided between methylated and nonmethylated compartments. Similar observations were also made in some species of fungi (Whittaker and Hardman, 1980; Antequera et al., 1984). Digestion of vertebrate DNA with methylation-sensitive endonucleases, such as HpaII (CCGG) and HhaI (GCGC), resulted in extremely poor cleavage due to the high levels of CpG methylation. As a result, staining of restriction fragments following electrophoresis on agarose gels did not reveal an obvious nonmethylated fraction (Bird and Taggart, 1980). When, however, the HpaII restriction fragments of a number of vertebrate DNAs were end-labelled before electrophoresis, Cooper et al. (1983) detected a fraction representing 1–2% of the total genome that was very rich in clustered, nonmethylated HpaII sites. This HpaII Tiny Fragment (HTF) fraction was present in all tissues examined, including sperm.

Further characterization of randomly selected small HpaII fragments of this fraction (Bird et al., 1985), revealed that they were derived from genomic clusters approximately 1-kb long. The clusters showed a significantly higher G + C content (65%) than average bulk DNA (40%) and were non-methylated at all testable sites. These properties were identical to those of DNA sequence domains at the 5′ end of certain genes; notably the chicken alpha-2(I) collagen gene (McKeon et al., 1982), the mouse DHFR and hamster APRT genes (Stein et al., 1983), and the human HPRT gene (Yen et al., 1984). Such nonmethylated and G + C-rich domains would contribute to the HTF fraction. Since a significant fraction of total genomic DNA had these properties, it was suggested that many genes (about 30,000) are associated with HTF sequences.

It was apparent quite early that this situation did not apply to all genes, since several highly tissue-specific genes showed no evidence of HTF-like sequences (see Fig. 1). In these cases CpG frequency at the beginning of the genes was not higher than the average for bulk DNA and the entire gene was heavily methylated in sperm and in other non-expressing cell types. Examples of genes in this category are the chicken ovalbumin gene (Mandel and Chambon, 1979) and the human β-globin gene (van der Ploeg and Flavell, 1980).

2 Structure of CpG islands

CpG islands can be identified by their lack of methylation, elevated CpG frequency and G + C-rich base composition (see Fig. 1). The

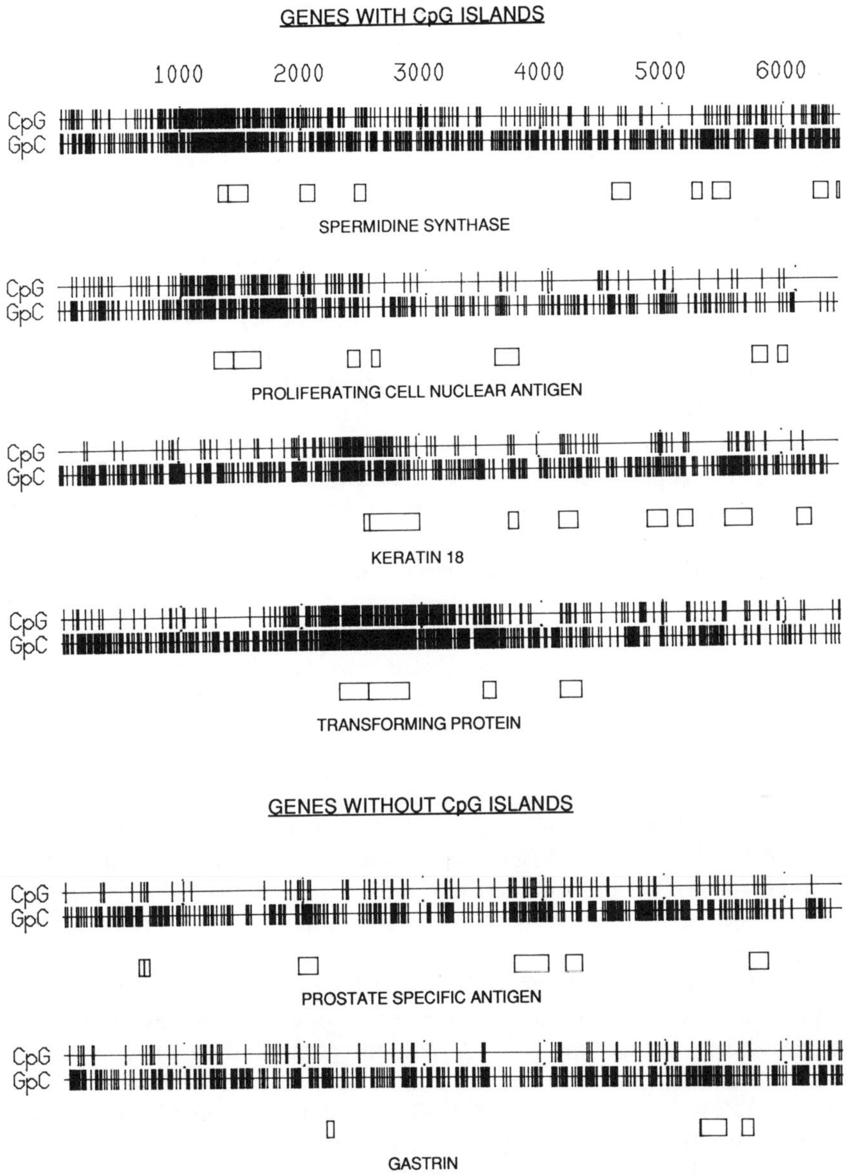

Figure 1. Maps showing the frequencies of CpG and GpC across six human genes. Exons are indicated by open boxes. A vertical line within an exon signifies the translation start site. Each occurrence of CpG or GpC is marked by a vertical on the respective map. The CpG map shows that each of the upper four genes has a cluster of CpGs over several hundred base pairs at the 5′ end of the gene. The densities of GpC and CpG in these regions are about the same, and all CpGs are expected to be non-methylated. These are the CpG islands. Outside the islands CpG is under-represented compared to GpC, and is predominantly methylated. The two lower genes are highly tissue-specific genes that lack CpG islands. They are uniformly CpG deficient, and most probably methylated, across the entire transcribed region.

frequency ratio of the dinucleotides CpG/GpC at CpG islands is about 1, compared to bulk genomic DNA where this ratio is 0.2. This five-fold deficiency of CpG in bulk DNA is a direct consequence of methylation at non-island CpGs. Unmodified cytosine is prone to deamination, giving uracil which is removed by a uracil-glycosylase (Lindahl, 1982). Deamination of 5-mC, however, yields thymine, converting a methyl-CpG dinucleotide into TpG. Despite the existence of a mechanism that corrects the mismatch in favour of the original cytosine with more than 90% efficiency (Brown and Jiricny, 1987; Wiebauer and Jiricny, 1990), it is clear that mutation of methyl-CpG to TpG is a relatively frequent mutational event. For example, the CpG deficiency of various genomes is matched by a proportional TpG/CpA excess (Bird, 1980). Since CpG islands have been free of methylation, presumably for millions of years, they have avoided the loss of CpG that has afflicted the majority of the genome. The five-fold deficiency of CpG probably represents an equilibrium between the accelerated loss of CpGs due to deamination and their creation due to point mutations at other sequences (Sved and Bird, 1990). A relatively small contribution to the total number of CpGs is probably due to the maintenance of some of them in critical coding positions.

Lack of methylation at CpG islands has been deduced mainly from the use of methyl-sensitive restriction enzymes. Since no methylation was detectable at testable sites for HpaII or HhaI, it was suggested that CpGs in general were non-methylated in these regions (Bird et al., 1985). This was shown to be correct when all CpGs across particular regions could be analysed by genomic sequencing methods. The 5′ CpG island at the human phosphoglycerate kinase 1 gene is non-methylated at all 61 CpGs over a region of about 500 bp (Pfeifer et al., 1990). Similar results were found over a 950-bp region at the 5′ CpG island at the 5′ end of the maize alcohol dehydrogenase 1 gene (Nicke et al., 1986).

In addition to the absence of methylation, and consequent lack of CpG suppression, another notable feature of mammalian and avian CpG islands is their high G + C content compared to bulk DNA. In general, islands are between 55% and 70% G + C, whereas bulk genomic DNA is on average 40% G + C. An elevated G + C content contributes significantly to the CpG-richness of islands, since the frequency of CpGs is increased more than 2.5 fold by raising the base composition from 40% to 60% G + C. Not all vertebrates, however, have CpG islands that are G + C-rich. Cross et al. (1991) have shown that the HTF fraction in fish genomes is much less prominent than in mammals. Comparison of two genes (metallothionein I and β-actin) with their mammalian counterparts revealed that all possessed a non-methylated region at their 5′ ends, but the G + C content of the fish genes was the same as the surrounding DNA. The difficulty in detecting the HTF fraction was therefore due to a reduced frequency of HpaII sites in fish CpG islands. In other respects (lack of methylation and undiminished frequency of CpG) the fish islands resemble their mammalian counterparts. The importance of these results is that they dissociate the two main characteristics of CpG islands, showing that lack of methylation and G + C richness are independent properties.

CpG islands are easily distinguishable at the level of purified naked DNA. Are they also distinctive at the level of chromatin conformation? To answer this question it was necessary to separate CpG island chromatin from the remaining 98% of the genome. The approach chosen by Tazi and Bird (1980) was based on the previous observation that restriction endonucleases that recognise CpG will release from nuclei oligonucleosomes that are exclusively derived from CpG islands (Antequera et al., 1989). These can be separated from the remaining undigested chromatin and analysed biochemically. Tazi and Bird (1990) found that the CpG island nucleosomes contain histones H3 and H4 in a highly acetylated form. Histone H1 is greatly reduced compared to bulk chromatin (or is less tightly bound to acetylated histones) and there is a nucleosome-free gap within most CpG islands. On either side of the gap, nucleosomal spacing is similar to that of bulk chromatin. These results indicate that CpG islands have the characteristics of active chromatin (see Elgin, 1990; and Turner, 1991 for recent reviews) which is in line with the fact that most of them include the promoter of housekeeping genes. The interesting question of how the lack of methylation, DNase I sensitivity and histone acetylation relate to each other is still largely unanswered.

The coincidence of CpG islands with promoters fits well with their altered chromatin structure, though it does not explain their relatively large size and often asymmetrical position at the promoter (see Fig. 1). Could it be that the distinctive CpG-rich sequences have evolved in order to simplify the nuclear genome? This could be achieved by making the relevant region of the gene available for the transcription machinery, while leaving the rest in a conformation less likely to interact with transcription factors (Bird, 1987). Support for this idea is provided by the observation that only organisms with very large genomes, consisting largely of non-coding DNA, (e.g. vertebrates and some plants) have CpG islands (Antequera and Bird, 1988). Organisms with smaller genomes (e.g. the sea urchin, *Arabidopsis* and some fungi) have a small fraction of the genome in a methylated form. In other organisms widely used in molecular biology (e.g. *Saccharomyces cerevisiae*, *Schizosaccharomyces pombe*, *Drosophila* and *Caenorhabditis elegans*) 5-mC has not been found so far. In a sense, most or all of the genome is CpG island-like (no methylation, no CpG deficiency) in organisms of this kind.

While it is possible that CpG islands have evolved in response to selective pressure, it is equally possible that they are a passive evolutionary consequence of the presence of promoters at these sites. Consistent failure to methylate any region of DNA in the germline would automatically lead to CpG island-like properties at that sequence, since CpGs would not be methylated and would not therefore be lost (see below).

3 CpG Islands and genes

Much of the interest in CpG islands is due to their association with genes. Two relevant questions are: which genes are associated with CpG islands? and are all CpG islands associated with genes? All housekeeping genes studied so far have CpG islands. A plausible hypothesis is that they remain nonmethylated because the continuous presence of transcription factors and preinitiation complexes at the 5′ end of housekeeping genes prevents the access by the DNA-methyltransferase (Bird, 1986). The actual passage of the RNA polymerase along the gene does not seem to be affected by methylation, because the body of the gene remains methylated even under conditions of active transcription. This model, however, is insufficient to explain the presence of *bona fide* CpG islands downstream from the promoter of some genes (see for example, Tykocinski and Max, 1984; Shemer et al., 1990). Nor can it easily account for the persistence of nonmethylated CpG islands at tissue-specific genes in non-expressing tissues (e.g. the human alpha-globin gene, the mouse MyoD gene and a number of brain-specific genes).

Are all CpG islands associated with genes? A direct test of this possibility showed that CpG islands randomly selected from a genomic library by the presence of sites for rare-cutting restriction enzymes were indeed associated with polyadenylated transcripts (Lindsay and Bird, 1987). Further confirmation derives from attempts to map the position of genes onto very large regions of DNA by identifying CpG islands. In nearly all cases, CpG islands are associated with transcripts. If all CpG islands are associated with genes, it should be possible to find genes in the genome by finding CpG islands. The approach has proved successful in the identification of genes in loci associated with human disease (e.g. the WAGR syndrome; Bonetta et al., 1990) or for finding new genes in already well-characterized loci (e.g. the H-2K region of the mouse MHC; Abe et al., 1988).

These mapping studies have also illustrated that the spacing between CpG islands in the genome is variable. The average distance has been estimated to be around 100 kb (Brown and Bird, 1986) which fits well with the nine CpG islands mapped across the 850-kb region in the Wilms tumour locus (Bonetta et al., 1990). However, variation about the average is large. Six CpG islands were found in a 170-kb region at the H-2K region of the mouse MHC (Abe et al., 1988), and four mapped in only 32 kb at the mouse Surfeit locus (Huxley and Fried,

1990). At the other extreme, the Duschenne muscular dystrophy locus showed no evidence of CpG islands over a distance of several megabases (Burmeister et al., 1988). The suitability and uses of different restriction enzymes to detect genes in mammalian genome based on the properties of the CpG islands has been recently reviewed (Bickmore and Bird, 1992).

4 CpG Islands as promoters

CpG islands contain the promoters of housekeeping genes. Very often they lack canonical TATA boxes, and transcription is initiated at multiple sites (Dynan, 1986). One extreme example is the mouse c-Myb gene which displays 13 different initiation sites spread along a region 900-bp long (Watson, 1988). It has been shown that the TFIID complex, which contains the TATA box-binding protein (TBP) and many TBP-associated factors (TAFs), is required for transcription from both TATA-containing and TATA-less promoters (Pugh and Tjian, 1991). In the case of TATA-less promoters, the TFIID complex is positioned onto the DNA template through protein-protein interactions with promoter-bound factors that normally act as activators, such as Sp1 (Pugh and Tjian, 1991). It may therefore be significant that most CpG islands contain potential Sp1 binding sites. These results suggest that different genes are transcribed by different TFIID complexes whose TAFs interact with the particular arrangement of transcription factors bound to each promoter (Sharp, 1992).

CpG islands are about 1-kb long and they contain multiple binding sites for transcription factors (Somma et al., 1991a; Pfeifer and Riggs, 1991). Is there any relation between the distribution of promoter elements and the size of the CpG island? A number of detailed deletion studied have revealed that housekeeping promoters can support transcription even after removal of most of the nonmethylated GC-rich sequences that define the CpG island. Some specific examples include the human HPRT (Melton et al., 1986), the mouse Surf1 and Surf2 (Lennard and Fried, 1991) and HTF9 genes (Somma et al., 1991a). These observations have led Somma et al. (1991a) to suggest that the promoters of housekeeping genes are composed of multiple, independent and functionally redundant elements. The heterogeneity of transcriptional initiation sites may be due to different combinations of factors acting in different nuclei. The attractive feature of such an organization is the flexibility that it confers on the promoter. Transcription is guaranteed in different cell types that have distinct populations of transcription factors. Somma et al. (1991b) have provided support for this idea by finding that different transcription factors bind to the mouse HTF9 promoter *in vitro* depending on the origin of the nuclear extract. Furthermore, they have shown by transient expression analysis that transcription from the same promoter is driven by different factors in different cell types.

5 *De novo* methylation of CpG islands

5.1 The inactive X chromosome

A fundamental feature of CpG islands is that they are nonmethylated in all tissues, including sperm. Artificial methylation of CpG island promoters shows that lack of methylation is essential for expression of the associated gene (Stein et al., 1982). Surprisingly, CpG islands at tissue-specific genes show no expression-related changes in methylation: they are nonmethylated in all tissues, regardless of the activity of the gene. There is one case, however, where gene repression is accompanied by island methylation during development. This exceptional case involves the CpG islands of the inactive X chromosome in female eutherian mammals.

Early in development, one of two X chromosomes is inactivated at random in every cell of the embryo to compensate for the differing genetic dosage between males and females. At the cytological level, the inactive X condenses to form the Barr body which is distinguished by its extensive heterochromatization and late replication during S phase relative to the active X. This pattern is somatically inherited and remains stable during the entire life of the organism. Well-characterized examples of X-linked CpG islands which become methylated upon X inactivation include the human HPRT gene (Yen et al., 1984; Wolf et al., 1984), the G6PD gene (Toniolo et al., 1988) and the PGK1 gene (Pfeifer et al., 1990). A recent analysis of 32 randomly selected CpG islands from the distal long arm of the human X chromosome shows a strict correlation between inactivity and methylation, suggesting that this situation probably applies to most CpG island genes in the inactive X (Tribioli et al., 1992). Significantly, the CpG island of the MIC2 gene, which maps to the pseudoautosomal region of the human chromosome X and therefore escapes inactivation, is nonmethylated in both active and inactive X chromosomes and in the Y chromosome (Goodfellow et al., 1988).

Direct evidence for the importance of DNA methylation in maintaining the inactive state is provided by the reactivation of the X-linked human HPRT gene with 5-azacytidine, a nucleoside analog that inhibits the DNA methyltransferase (Mohandas et al., 1981; Yen et al., 1984; Wolf et al., 1984). Purified DNA from the inactive X chromosome is unable to transform $HPRT^-$ cells to an $HPRT^+$ phenotype unless derived from cells that were treated with 5-azacytidine (Venolia et al., 1982).

It should be stressed that methylation appears not to be the primary mechanism of X inactivation, since transcriptional silencing of the mouse HPRT gene takes place some days before methylation of the CpG island (Lock et al., 1987). Moreover, in somatic cells of marsupials, extra-embryonic tissues of the mouse, and premeiotic germ cells of the mouse, X-inactivation is not accompanied by methylation of CpG islands (Kaslow and Migeon, 1987; Kratzer et al., 1983). In two of these cases (marsupials and extra-embryonic tissues) inactivation is unstable. Since X-inactivation in somatic cells of eutherian mammals is highly stable, it seems very likely that methylation locks-in repression, leading to stable maintenance of the inactive state.

5.2 The mechanism of methylation-mediated repression

A number of studies have suggested that the repression caused by *de novo* methylation of CpG islands involves changes in chromatin structure. The changes are not obvious when assayed by susceptibility to cleavage with DNase I (Yang and Caskey, 1987), though more subtle differences can be observed by using ligation-mediated PCR techniques (Pfeifer and Riggs, 1991). Profound alterations can be detected, however, by using restriction endonucleases that contain the dinucleotide CpG in their cleavage site. A number of studies in various laboratories have shown that methylated CpG islands are completely insensitive to MspI in intact nuclei (Wolf and Migeon, 1985; Hansen et al., 1988; De Bustros et al., 1988; Antequera et al., 1989). This enzyme, which cleaves the sequence CCGG, is indifferent to the presence of CpG methylation when purified DNA is the substrate. In nuclei, however, MspI is unable to cut any of the multiple sites in the methylated mouse HPRT island, even in large excess of enzyme. Non-methylated sites on the active X chromosome can be cleaved. Moreover, AluI, a restriction enzyme whose cleavage site (AGCT) does not contain CpG, cuts the methylated (inactive X) and nonmethylated (active X) HPRT islands equally well (Antequera et al., 1989). These results indicate that DNA methylation affects chromatin by specific interaction of methylated CpGs with nuclear components that protect them against nuclease attack. Nuclease sensitivity of sequences flanking the methylated CpGs is relatively mildly affected. Protection of CpGs is most likely mediated by methyl-CpG binding proteins (Meehan et al., 1989; Lewis et al., 1992).

5.3 Fragile X syndrome

De novo methylation of a CpG island has recently been dramatically linked to a disease phenotype in the case of the fragile-X syndrome.

This syndrome is the most common inherited form of mental retardation in males. Following isolation of the fragile-X locus, it was noted that affected individuals displayed unstable length and variable methylation in a closely linked CpG island (Kremer et al., 1991; Heitz et al., 1991; Vincent et al., 1991; Bell et al., 1991). The variability was traced to a tandemly repeated CGG trinucleotide whose copy number dramatically increased in individuals suffering from the disease. It is not yet clear whether the syndrome is a direct result of the increased number of repeats, or if it is due to methylation of the normally unmethylated CpG island. We know, however, that the associated gene, FMR1, is silent in affected individuals (Pieretti et al., 1991). Since methylation of CpG islands is known to inactivate many genes, it is tempting to suggest that *de novo* methylation is directly responsible for transcriptional shutdown of the gene. *De novo* methylation itself might then be due to the abrupt expansion of the CGG block, though as yet no mechanism-linking rearrangement to *de novo* methylation is known.

5.4 Methylation of CpG islands in cultured cells

A recent systematic study of the incidence of the *de novo* methylation of CpG islands in cell lines has shown that it is strikingly high (Antequera et al., 1990). In two widely used mouse cell lines, L and NIH 3T3, over half of the islands are methylated, while the corresponding sequences in mouse tissues remained nonmethylated. Chromatin analysis showed that methylated CpG islands adopt the MspI resistant conformation indistinguishable from the one found in the inactive X chromosome (Antequera et al., 1990). Assuming that there are about 30,000 CpG islands per haploid mouse genome (Bird et al., 1985), and that all of them are gene-associated, these results imply that over 15,000 genes have been inactivated by methylation in the L and NIH 3T3 cell lines. If this deduction is correct, genes that are essential for cell viability should have been spared. Consistent with this idea, it was found that both cell lines had inactivated the same subset of CpG islands, suggesting that the *de novo* methylation process has reached a "plateau" at which CpG islands of all the nonessential genes have been methylated. More direct support for this view came from analyzing the methylation status of a few specific genes. The CpG islands of genes essential for the survival of the cell (e.g. Triose phosphate isomerase) were not found methylated in any cell line. In contrast, the CpG islands of the tissue-specific genes which are not thought to be essential for survival in tissue culture (e.g. mouse Thy1, MHC Class I and human alpha globin), were frequently methylated.

Not all cell lines show evidence of CpG island methylation. None of the embryonal carcinoma cells that were tested in the study of Antequera et al. (1990) showed evidence of CpG island methylation. These embryonic cells, though not pluripotent, are considered as a model of the very early mouse stages of development. The F9 line was established at approximately the same time as other somatic cells included in the study, thereby excluding the possibility that there had been insufficient time for extensive CpG island methylation to occur. Paradoxically, these cells have been reported to have a much higher *de novo* methylase activity than somatic cells (Jahner and Jaenisch, 1984).

The results suggest that genes are subject to random *methylation associated gene inactivation* (MAGI) in culture. MAGI may have different phenotypic consequences depending on the affected gene. For example, inactivation of many tissue-specific genes (like those mentioned above) is probably neutral. In other cases, inactivation of genes involved in the biosynthesis or processing of various metabolites will give rise to auxotrophic variants that will not be selected against as long as the relevant metabolite is supplied in the culture medium (see Holliday, 1987 for review). Finally, in some cases there might be a selective advantage in the inactivation of genes that exert a negative control on cell division or that would drive the cell into a terminal differentiation pathway, such as the mouse MyoD1 gene (Jones et al., 1990). These results do not mean that the mechanism of *de novo* methylation can distinguish between different CpG islands. MAGI acts at random. It is selection which determines that essential genes are spared.

Epigenetic changes leading to heritable gene inactivation have been called "epimutations" because they do not involve alterations in the DNA sequence (Holliday, 1987). Several examples have been detected in cultured cells due to the reversibility of the defect upon treatment with the demethylating agent 5-azacytidine. Examination of the reactivated genes shows that most or all have CpG islands. Since 5-azacytidine causes general demethylation of DNA, genes that have been inactivated by MAGI will be expressed again. Analysis of the methylation status of the gene before and after treatment with the drug shows a strict correlation between demethylation and expression. This has sometimes been interpreted as evidence for a role for DNA methylation in the normal control of gene expression. This view is not justified, however, as in all cases studied so far, the island concerned is never methylated in normal cells of the organism. Thus the altered methylation pattern is confined to cultured cells, and does not appear to occur during development of the organism from which those cells are derived.

Although MAGI has not been detected in normal body cells, it is possible that a low level of CpG island methylation can occur, for example during aging. Even if one out of one hundred cells had suffered a *de novo* methylation event at a particular island, it would escape detection by the normal methods (methyl-sensitive restriction enzymes). It is, however, possible to detect MAGI in the organism when the *de novo* methylation of a CpG island contributes to the appearance of a tumour phenotype. Tumour suppressor genes would be the most obvious candidates to produce a phenotype upon inactivation, and in fact a few such cases have been reported. Analysis of sporadic retinoblastoma tumours has shown that in 1 out of 21 (Greger et al., 1989) and 5 our of 56 cases (Sakai et al., 1991) the CpG island associated with the retinoblastoma gene has been *de novo* methylated. In these cases, the retinoblastoma island shows the expected nonmethylated pattern in normal tissues from the same patients. The CpG island of the calcitonin gene is also methylated in a number of human cancers (De Bustros et al., 1988). Though this gene does not qualify as a tumour suppressor, it is localized in the short arm of chromosome 11 in a region where abnormal hypermethylation has been detected in some tumours, and two tumour suppressor genes have been mapped nearby. Altogether, the results are consistent with the idea that MAGI of the kind described in cell lines could be involved in some cases of tumourigenesis.

6 Origin and maintenance of CpG islands

The existence of CpG islands depends upon their lack of methylation in germ cells. Germ cells are not mortal in the same sense as somatic cells, because they are set aside very early during development and later on pass into the next generation through the gametes. Whatever happens to the CpG islands in somatic cells, important though it might be for the organism, will not be transmitted to the offspring. When CpG islands do become methylated in the germ line, the CpGs are gradually lost by mutation. This can be seen in the methylated human alpha globin pseudogene, which has remained GC-rich in base composition, but which has lost a high proportion of its CpGs to TpG and its complement CpA (Bird et al., 1987). How do CpG islands normally avoid this fate and remain free of methylation? The simple possibility that CpG island sequences are not substrates for the methyltransferase is untenable as they do in fact become methylated on the inactive X chromosome and on autosomes in cultured somatic cells. Moreover, CpG islands are substrates for the mammalian DNA methyltransferase *in vitro*, albeit with a reduced efficiency (Carrotti et al., 1989). It is conceivable, however, that special mechanisms exist in germ cells that prevent *de novo* methylation of CpG islands. Two kinds of mechanism can be envisaged. Either CpG islands could adopt a conformation that would make them refractory to the DNA-methyltransferase, or, alternatively, they could be subject to *de novo* methylation, but would be kept methylation-free by an opposing activity that removes 5-methylcytosine. A number of recent studies suggest that both alternatives could be operative.

Szyf et al. (1990) have reported on the existence of a 214-base pair region near the promoter of the mouse Thy-1 gene that can act as a signal to prevent *de novo* methylation in mouse embryonic stem (ES) cells. This signal is claimed to be portable because it can protect heterologous flanking sequences. These results are an extension of a previous study by Kolsto et al. (1986) which reported the fate of a nonmethylated hybrid mouse-human construct containing the mouse Thy-1 CpG island in transgenic mice. They found that the island region was maintained free of methylation in all tissues regardless of expression, while the flanking regions became *de novo* methylated. Similar results have been obtained by Shemer et al. (1990) for the human apolipoprotein A1 gene. The 3′ island of this gene remains free of methylation in all tissues of transgenic mice carrying this gene regardless of its expression. Taken together, these results suggest that the information necessary to protect CpG islands from methylation is intrinsic to them and independent of the chromosomal context. This situation may apply in all cells (somatic and germline), or it may only apply to very early stages of development, after which the pattern is passively maintained by the maintenance DNA-methyltransferase.

Evidence for a mechanism capable of replacing 5-methylcytosine with cytosine has been reported (Razin et al., 1986). It is thought to be responsible for the genome-wide transient demethylation in absence of DNA replication during differentiation of mouse Friend erythroleukemia cells. Support for a more specific CpG demethylating mechanism operative during early embryogenesis has been obtained by Frank et al. (1990). The evidence derives from experiments showing demethylation of the hamster APRT gene. The gene was methylated *in vitro* at HpaII and HhaI sites before introduction into embryonic cells and transgenic mice. Demethylation took place only at the CpG island region but not at the body of the gene. No demethylation occurred in somatic cell lines. Again the process seems to depend on an enzymatic activity and to operate very quickly upon the introduction of the DNA into the cells. Demethylation was detected within 48 h after transfection in absence of DNA replication. Unfortunately it has not yet proved possible to detect the demethylation activity *in vitro*. This would be an important step forward, as the characterization of the demethylating mechansims is likely to provide an important key to understanding the origin and maintenance of CpG islands during evolution.

Acknowledgements. F.A. is supported by a Fellowship from the Imperial Cancer Research Fund.

Abe, K., Wei, J., Wei, F., Hsu, Y., Ueara, H., Artzt, K., and Bennett, D. (1988) Searching for coding sequences in the mammalian genome: The H-2K region of the mouse MHC is replete with genes expressed in embryos. EMBO J. *7*, 3441–3449.

Antequera, F., Tamame, M., Villanueva, J., and Santos, T. (1984) DNA methylation in the fungi. J. Biol. Chem. *259*, 8033–8036.

Antequera, F., and Bird, A. (1988) Unmethylated CpG islands associated with genes in higher plant DNA. EMBO J. *7*, 2295–2299.

Antequera, F., Macleod, D., and Bird, A. (1989) Specific protection of methylated CpGs in mammalian nuclei. Cell *58*, 509–517.

Antequera, F., Boyes, J., and Bird, A. (1990) High levels of de novo methylation and altered chromatin structure at CpG islands in cell lines. Cell *62*, 503–514.

Bell, M., et al. (1991) Physical mapping across the fragile X: hypermethylation and clinical expression of the fragile X syndrome. Cell *64*, 861–866.

Bickmore, W., and Bird, A. (1992) The use of restriction endonucleases to detect and isolate genes from mammalian cells. Meth. Enzymol. *216*, 224–244.

Bird, A. (1980) DNA methylation and the frequency of CpG in animal DNA. Nucl. Acids Res. *8*, 1499–1504.

Bird, A. (1986) CpG-rich islands and the function of DNA methylation. Nature *321*, 209–213.

Bird, A. (1987) CpG islands as gene markers in the vertebrate nucleus. Trends Genet. *3*, 342–347.

Bird, A., and Taggart, M. (1980) Variable patterns of DNA and rDNA methylation in animals. Nucl. Acids Res. *8*, 1485–1497.

Bird, A., Taggart, M., and Smith, B. (1979) Methylated and unmethylated DNA compartments in the sea urchin genome. Cell *17*, 889–901.

Bird, A., Taggart, M., Frommer, M., Miller, O., and Macleod, D. (1985) A fraction of the mouse genome that is derived from islands of nonmethylated, CpG-rich DNA. Cell *40*, 91–99.

Bird, A., Taggart, M., Nichols, D., and Higgs, D. (1987) Non-methylated CpG-rich islands at the human alpha-globin locus: implications for evolution of the alpha-globin pseudogene. EMBO J. *6*, 999–1004.

Bonneta, L., Kuehn, S., Huang, A., Law, D., Kalikin, L., Koi, M., Reeve, A., Brownstein, B., Yeger, H., Williams, B., and Feinberg, A. (1990) Wilms tumor locus on 11p13 defined by multiple CpG island-associated transcripts. Science *250*, 994–997.

Brown, W., and Bird, A. (1986) Long-range restriction site mapping of mammalian genomic DNA. Nature *322*, 477–481.

Brown, T., and Jiricny, J. (1987) A specific mismatch repair event protects mammalian cells from loss of 5-methylcytosine. Cell *50*, 945–950.

Burmeister, M., Monaco, A., Gillard, E., van Ommen, G., Affara, N., Fergusen-Smith, M., Kunkel, L., and Lehrach, H. (1988) A 10 megabase physical map of human Xp21, including the Duschene muscular dystrophy gene. Genomics *2*, 189–202.

Carrotti, D., Palitti, F., Lavia, P., and Strom, R. (1989) In vitro methylation of CpG islands. Nucl. Acids Res. *17*, 9219–9229.

Cooper, D., Taggart, M., and Bird, A. (1983) Unmethylated domains in vertebrate DNA. Nucl. Acids Res. *11*, 647–658.

Cross, S., Kovarik, P., Schmidke, J., and Bird, A. (1991) Non-methylated islands in fish genomes are GC-poor. Nucl. Acids Res. *19*, 1469–1474.

De Bustros, A., Nelkin, B., Silverman, A., Ehrlich, G., Poiesz, B., and Baylin, S. (1988) The short arm of chromosome 11 is a "hot spot" for hypermethylation in human neoplasia. Proc. Natl. Acad. Sci. USA *85*, 5693–5697.

Dynan, W. (1986) Promoters of housekeeping genes. Trends Genet. *2*, 196–197.

Elgin, S. (1990) Chromatin structure and gene activity. Curr. Opin. Cell Biol. *2*, 437–445.

Frank, D., Keshet, I., Shani, M., Levine, A., Razin, A., and Cedar, H. (1991) Demethylation CpG islands in embryonic cells. Nature *351*, 239–241.

Goodfellow, P., Mondello, C., Darling, S., Pym, B., Little, P., and Goodfellow, P. (1988) Absence of methylation of a CpG-rich region at the 5′ end of the MIC2 gene on the active X, the inactive X, and the Y chromosome. Proc. Natl. Acad. Sci. USA *85*, 5605–5609.

Greger, V., Passarge, E., Hopping, W., Messmer, E., and Horsthemke, B. (1989) Epigenetic changes may contribute to the formation and spontaneous regression of retinoblastoma. Hum. Genet. *83*, 155–158.

Hansen, R., Ellis, N., and Gartler, S. (1988) Demethylation of specific sites in the 5′ region of the inactive X-linked human phosphoglycerate kinase gene correlates with the appearance of nuclease sensitivity and gene expression. Mol. Cell. Biol. *8*, 4692–4699.

Heitz, D. et al. (1991) Isolation of sequences that span the fragile X and identification of a fragile X-related CpG island. Science *251*, 1263–1239.

Holliday, R. (1987) The inheritance of epigenetic defects. Science *238*, 163–170.

Huxley, C., and Fried, M. (1990) The mouse Surfeit locus contains a cluster of six genes associated with four CpG-rich islands in 32 kilobases of genomic DNA. Mol. Cell. Biol. *10*, 605–614.

Jahner, D., and Jaenisch, R. (1984) DNA methylation in early mammalian development, in: DNA methylation. Biochemistry and biological significance. pp. 189–219. Eds A. Razin, H. Cedar, and A. Riggs. Springer-Verlag, Heidelberg/Berlin.

Jones, P., Wolkowicz, M., Rideout, W., Gonzales, F., Marziasz, C., Coetzee, G., and Tapscott, S. (1990) De novo methylation of the MyoD1 CpG island during the establishment of immortal cell lines. Proc. Natl. Acad. Sci. USA, *87*, 6117–6121.

Kaslow, D., and Migeon, B. (1987) DNA methylation stabilizes X chromosome inactivation in eutherians but not in marsupials: evidence for multistep maintenance of mammalian X dosage compensation. Proc. Natl. Acad. Sci. USA *84*, 6210–6214.

Kolsto, A., Kollias, G., Giguere, V., Isobe, K., Prydz, H., and Grosveld, F. (1986) The maintenance of methylation-free islands in transgenic mice. Nucl. Acids Res. *14*, 9667–9678.

Kratzer, P., Chapman, V., Lambert, H., Evans, R., and Liskay, R. (1983) Differences in the DNA of the inactive X chromosomes of fetal and extra embryonic tissues of mice. Cell *33*, 37–42.

Kremer, E. et al. (1991) Mapping of instability at the fragile X to a trinucleotide repeat sequence p(CCG)n. Science *252*, 1711–1714.

Lennard, A., and Fried, M. (1991) The bidirectional promoter of the divergently transcribed mouse Surf-1 and Surf-2 genes. Mol. Cell. Biol. *11*, 1282–1294.

Lewis, J., Meehan, R., Henzel, W., Maurer-Fogey, I., Jeppesen, P., Klein, F., and Bird, A. (1992) Purification, sequence and cellular localisation of a novel chromosomal protein that binds to methylated DNA. Cell *69*, 905–914.

Lindahl, T. (1982) DNA repair enzymes. Annu. Rev. Biochem. *51*, 67–87.

Lindsay, S., and Bird, A. (1987) Use of restriction enzymes to detect potential gene sequences in mammalian DNA. Nature *327*, 336–338.

Lock, L., Takagi, N., and Martin, G. (1987) Methylation of the HPRT gene on the inactive X occurs after chromosome inactivation. Cell *48*, 39–46.

Mandel, J., and Chambon, P. (1979) DNA methylation: organ specific variations in the methylation pattern within and around ovalbumin and other chicken genes. Nucl. Acids Res. *7*, 2081–2103.

McClelland, M., and Ivarie, R. (1982) Asymmetrical distribution of CpG in an "average" mammalian gene. Nucl. Acids Res. *10*, 7865–7877.

McKeon, C., Ohkubo, H., Pastan, I., and Crombrugghe, B. (1982) Unusual methylation pattern of the alpha-2(I) collagen. Cell *29*, 203–210.

Meehan, R., Lewis, J., McKay, S., Kleiner, E., and Bird, A. (1989) Identification of a mammalian protein that binds specifically to DNA containing methylated CpGs. Cell *58*, 499–507.

Melton, D., McEwan, C., and McKie, A. (1986) Expression of the mouse HPRT gene: Deletional analysis of the promoter region of an X-chromosome linked housekeeping gene. Cell *44*, 319–328.

Mohandas, T., Sparkes, R., and Shapiro, L. (1981) Reactivation of an inactive human X chromosome: Evidence for X inactivation by DNA methylation. Science *211*, 393–396.

Naveh-Many, T., and Cedar, H. (1981) Active gene sequences are undermethylated. Proc. Natl. Acad. Sci. USA *78*, 4246–4250.

Nicke, H., Bowen, B., Ferl, R., and Gilbert, W. (1986) Detection of cytosine methylation in the maize alcohol dehydrogenase gene by genomic sequence. Nature *319*, 243–246.

Pfeifer, G., Tanguay, R., Steigerwald, S., and Riggs, A. (1990) In vivo footprint and methylation analysis by PCR-aided genomic sequencing: Comparison of active and inactive X chromosomal DNA at the CpG island and promoter of human PGK-1. Genes Devel. *4*, 1277–1287.

Pfeifer, G., and Riggs, A. (1991) Chromatin differences between active and inactive X chromosomes revealed by genomic footprinting of permeabilized cells using DNase I and ligation-mediated PCR. Genes Devel. *5*, 1102–1113.

Pieretti, M., Zhang, F., Fu, Y., Warren, S., Oostra, B., Caskey, C. T., and Nelson, D. (1991) Absence of expression of the FMR1 gene in fragile X syndrome. Cell *66*, 817–822.

Pugh, B., and Tjian, R. (1991) Transcription from a TATA-less promoter requires a multisubunit TFIID complex. Genes Devel. *5*, 1935–1945.
Razin, A., and Cedar, H. (1977) Distribution of 5-methylcytosine in chromatin. Proc. Natl. Acad. Sci. USA *74*, 2725–2728.
Razin, A., Szyf, M., Kafri, T., Roll, M., Giloh, H., Scarpa, S., Carotti, D., and Cantoni, G. (1986) Replacement of 5-methylcytosine: A possible mechanism for transient DNA demethylation during differentiation. Proc. Natl. Acad. Sci. USA *83*, 2827–2831.
Sakai, T., Toguchida, J., Ohtani, N., Yandell, D., Rapaport, J., and Dryja, T. (1991) Allele-specific hypermethylation of the retinoblastoma tumor suppressor gene. Am. J. Hum. Genet. *48*, 880–888.
Sharp, P. (1992) TATA-binding protein is a classless factor. Cell *68*, 819–821.
Shemer, R., Walsh, A., Eisenberg, S., Breslow, J., and Razin, A. (1990) Tissue-specific methylation patterns and expression of the human apolipoprotein AI gene. J. Biol. Chem. *265*, 1010–1015.
Somma, P., Pisano, C., and Lavia, P. (1991a) The housekeeping promoter from the mouse CpG island HTF9 contains multiple protein-binding elements that are functionally redundant. Nucl. Acids Res. *19*, 2817–2824.
Somma, P., Gambino, I., and Lavia, P. (1991b) Transcription factors binding to the mouse HTF9 housekeeping promoter differ between cell types. Nucl. Acids Res. *19*, 4451–4458.
Stein, R., Razin, A., and Cedar, H. (1982) In vitro methylation of the hamster adenine phosphoribosyltransferase gene inhibits its expression in mouse L cells. Proc. Natl. Acad. Sci. USA *79*, 3418–3422.
Stein, R., Sciaky-Gallili, N., Razin, A., and Cedar, H. (1983) Pattern of methylation of two genes coding for housekeeping functions. Proc. Natl. Acad. Sci. USA *80*, 2422–2426.
Sved, J., and Bird, A. (1990) The expected equilibrium of the CpG dinucleotide in vertebrate genomes under a mutational model. Proc. Natl. Acad. Sci. USA *87*, 4692–4696.
Szyf, M., Tanigawa, G., and McCarthy, P. (1990) A DNA signal from the Thy-1 gene defines de novo methylation patterns in embryonic stem cells. Mol. Cell. Biol. *10*, 4396–4400.
Tazi, J., and Bird, A. (1990) Alternative chromatin structure at CpG islands. Cell *60*, 909–920.
Toniolo, D., Martini, G., Migeon, B., and Dono, R. (1988) Expression of G6PD locus on the human X chromosome is associated with demethylation of three CpG islands within 100 kb of DNA. EMBO J. *7*, 401–406.
Tribioli, C., Tamanini, F., Patrosso, C., Milanesi, L., Villa, A., Pergolizzi, R., Maestrini, E., Rivella, S., Bione, S., Mancini, M., Vezzoni, P., and Toniolo, D., (1992) Methylation and sequence analysis around Eagi sites: identification of 28 new CpG islands in XQ24–XQ28. Nucl. Acids Res. *20*, 727–733.
Turner, B. (1991) Histone acetylation and control of gene expression. J. Cell Sci. *99*, 13–20.
Tykocinski, M., and Max, E. (1984) CG dinucleotide clusters in MHC genes and in 5′ demethylated genes. Nucl. Acids Res. *12*, 4385–4396.
Van der Ploeg, L., and Flavell, R. (1980) DNA methylation in the human gamma-delta-beta-globin locus in erythroid and nonerythroid tissues. Cell *19*, 947–958.
Venolia, L., Gartler, S., Wassman, E., Yen, P., Mohandas, T., and Shapiro, L. (1982) Transformation with DNA from 5-azacytidine-reactivated X chromosomes. Proc. Natl. Acad. Sci. USA *79*, 2352–2354.
Vincent, A., Heitz, D., Petit, C., Kretz, C., Oberle, I., and Mandel, J. L. (1991) Abnormal pattern detected in fragile X patients by pulsed field gel electrophoresis. Nature *349*, 624–626.
Watson, R. (1988) A transcriptional arrest mechanism involved in controlling constitutive levels of mouse c-myb mRNA. Oncogene *2*, 267–272.
Whittaker, P., and Hardman, N. (1980) Methylation of nuclear DNA in Physarum polycephalum. Biochem. J. *191*, 859–862.
Wiebauer, K., and Jiricny, J. (1990) Mismatch-specific thymine DNA glycosylase and DNA polymerase B mediate the correction of G.T. mispairs in nuclear extracts from human cells. Proc. Natl. Acad. Sci. USA *87*, 5842–5845.
Wolf, S., Jolly, D., Lunnen, K., Friedman, T., and Migeon, B. (1984) Methylation of the hypoxanthine phosphoribosyltransferase locus on the human X chromosome: Implications for X-chromosome inactivation. Proc. Natl. Acad. Sci. USA *81*, 2806–2810.
Wolf, S., and Migeon, B. (1985) Clusters of CpG dinucleotides implicated by nuclease hypersensitivity as control elements of housekeeping genes. Nature *314*, 467–469.

Yang, T., and Caskey, T. (1987) Nuclease sensitivity of the mouse HPRT gene promoter region: Differential sensitivity on the active and inactive X chromosomes. Mol. Cell. Biol. *7*, 2994–2998.

Yen, P., Patel, P., Chinault, A., Mohandas, T., and Shapiro, L. (1984) Differential methylation of hypoxanthine phosphoribosyltransferase genes on active and inactive human X chromosomes. Proc. Natl. Acad. Sci. USA *81*, 1759–1763.

DNA Methylation: Molecular Biology and Biological Significance
ed. by J. P. Jost & H. P. Saluz

DNA methyltransferases and DNA site-specific endonucleases encoded by chlorella viruses

Michael Nelson, Yanping Zhang and James L. Van Etten

Department of Plant Pathology, University of Nebraska, Lincoln, NE 68583-0722, USA

1 Introduction

Large polyhedral (diameter of 150 to 190 nm) dsDNA-containing (> 300 kbp) viruses which infect certain unicellular, eukaryotic, chlorella-like green algae are common in fresh water collected throughout the world (Van Etten et al., 1985; Schuster et al., 1986; Zhang et al., 1988; Reisser et al., 1988; Yamada et al., 1991). The hosts for these lytic chlorella viruses are exsymbiotic *Chlorella* strains NC64A and Pbi, originally isolated from the protozoan *Paramecium bursaria.* Chlorella viruses, which can be produced in large quantities, are the first viruses infecting a photosynthetic eukaryotic organism which can be plaque assayed (Van Etten et al., 1983) and have been given family status with the name Phycodnaviridae (Francki et al., 1991). A comprehensive review on the chlorella viruses has recently been published (Van Etten et al., 1991).

One novel property of the chlorella viruses is that their DNAs contain methylated nucleotides (Van Etten et al., 1985a; Schuster et al., 1986; Zhang et al., 1988; Reisser et al., 1988). Each viral DNA contains 5-methylcytosine (5-mC), ranging in concentration from 0.1% to 47% of the total cytosines; however, the viral DNAs do not contain 4-methylcytosine. Many of the viral DNAs also contain N^6-methyladenine (6-mA) at levels as high as 37% of the total adenines. The finding of sequence-specific methylation led to the discovery that infection of host chlorella results in synthesis of viral encoded DNA methyltransferases and DNA site-specific (restriction) endonucleases (Xia and Van Etten, 1986; Xia et al., 1986; 1986a; 1987; 1987a; 1988; Zhang et al., 1992a).

This chapter will briefly describe the general properties and life cycle of the prototype chlorella virus PBCV-1 (*Paramecium bursaria* chlorella virus 1) and then focus on the molecular properties, possible biological functions, and evolutionary origins of the viral encoded DNA methyltransferases and site-specific endonucleases. Three recent discoveries will be discussed. First, the viruses can be grouped into DNA methylation phenotypes by their resistance/sensitivity to restriction endonucleases. Second, a change in the extraction and assay conditions reveals that most chlorella virus genomes encode multiple restriction/modification systems. Third, at least some of the chlorella viruses also encode nonfunctional DNA methyltransferase genes, i.e. pseudogenes. We suspect that the viruses also contain nonfunctional site-specific endonuclease genes. Sequence comparison between natural chlorella virus functional genes and pseudogenes will identify critical amino acids required for restriction/modification enzyme activity.

2 Properties of chlorella virus PBCV-1

PBCV-1, the prototype chlorella virus, is easy to purify and 80 to 105 A_{260} units (8 to 10 mg of PBCV-1), equivalent to 1 to 2×10^{12} plague forming units of purified PBCV-1, are obtained per liter of infected cell lysate. The virus is a large (175 to 190 nm in diameter) polyhedron with a multilaminate shell. PBCV-1 sediments at about 2300 S and has an estimated size of 1×10^{3} Md (Yonker et al., 1985). The virus contains about 64% protein, 21 to 25% dsDNA, and 5 to 10% lipid (Van Etten et al., 1983a; Skrdla et al., 1984). PBCV-1 contains at least 50 structural proteins ranging in apparent size from 10 to more than 200 kd. Four proteins, including the major capsid protein (Vp54), are located on the virion surface (Skrdla et al., 1984).

Protein Vp54 comprises about 40% of the total PBCV-1 protein and is one of three virion glycoproteins. The glycan portion of Vp54 is located on the external surface of the virus and probably accounts for the resistance of PBCV-1 to most proteases. Four of the virus proteins, including Vp54 are specifically labeled with myristic acid suggesting they are myristylated. Six virus proteins are labeled with $^{32}PO_4$ indicating they are phosphoproteins (Que, Lee and Van Etten, unpublished data).

The lipid component of PBCV-1, which is located beneath the outer glycoprotein shell, is required for infectivity, since organic solvents inactivate the virus. However, PBCV-1 infectivity is not affected by neutral detergents (Skrdla et al., 1984).

PBCV-1 contains a large dsDNA genome with 40% G + C content; 5-mC comprises 1.9% of the total cytosines and 6-mA comprises 1.5% of the total adenines (Van Etten et al., 1985a). The size of the PBCV-1 genome, 333 kbp, was derived from summing restriction fragment sizes (Girton and Van Etten, 1987) and then verified by pulsed field electrophoresis (Rohozinski et al., 1989). The PBCV-1 genome is a linear, nonpermuted molecule with covalently closed hairpin ends (Rohozinski et al., 1989), whose termini (not including the hairpin) are identical inverted repeats of at least 2185 bases (Strasser et al., 1991). The

Table 1. Classification of 37 *Chlorella* NC64A and 5 *Chlorella* Pbi viruses based on serology, restriction patterns, and methylated base content.[a]

Virus	Plaque size (mm)	Reaction with antibody to PBCV-1	Reaction with antibody to NY-2A	Restriction group[b]	Level of methylation 5-mC[c]	Level of methylation 6-mA[d]
NC64A viruses						
NE-8D	3	Yes	No	A	0.44	nd[e]
NYb-1	3	Yes	No	A	1.60	nd
CA-4B	3	Yes	No	A	0.12	nd
AL-1A	3	No	Partial	A	0.45	nd
NY-2C	3	No	Partial	A	0.39	nd
NC-1D	3	No	Partial	A	0.33	nd
PBCV-1	3	Yes	No	C	1.86	1.5
NC-1C	3	Yes	No	C	1.72	1.6
CA-1A	3	Yes	No	B	10.0	nd
CA-2A	3	Yes	No	B	—	—
IL-2A	3	Yes	No	B	9.4	nd
IL-2B	3	Yes	No	B	10.9	nd
IL-3A	3	Yes	No	B	9.7	nd
IL-3D	3	Yes	No	B	12.6	nd
SC-1A	1	Yes	No	D	1.94	7.3
SC-1B	1	Yes	No	D	2.04	7.5
NC-1A	3	Yes	No	D	7.1	7.3
NE-8A	3	Yes	No	E	14.3	8.1
Al-2C	3	Yes	No	E	—	—
MA-1E	3	Yes	No	E	14.9	8.1
NY-2F	3	Yes	No	E	14.6	8.1
CA-1D	3	Yes	No	E	—	—
NC-1B	3	Yes	No	E	13.4	8.2
NYs-1	1	No	Yes	F	47.5	11.3
IL-5-2s1	1	No	Yes	G	45.0	16.1
AL-2A	1	No	Yes	G	35.8	14.6
MA-1D	1	No	Yes	G	47.0	16.7
NY-2B	1	No	Yes	G	36.5	16.2
CA-4A	1	No	Yes	H	39.8	19.6
NY-2A	0.5	No	Yes	I	44.9	37.0
XZ-3A	1	Yes	No	J	12.8	2.2
SH-6A	1	Yes	No	K	12.6	10.3
BJ-2C	1	Yes	No	K	12,8	11.5
XZ-6E	1	Yes	No	E	21.2	15.2
XZ-4C	1	No	Yes	G	46.7	20.8
XZ-5C	1	No	Yes	H	42.7	27.9
XZ-4A	1	No	Yes	H	44.1	28.3

Table 1. contd

Virus	Plaque size (mm)	Reaction with antibody to PBCV-1	Reaction with antibody to NY-2A	Restriction group[b]	Level of methylation 5-mC[c]	Level of methylation 6-mA[d]
Pbi viruses						
CVA-1	1	No	No	M	43.1	nd[e]
CVB-1	1	No	No	N	42.7	17.7
CVG-1	1	No	No	L	19.2	nd
CVM-1	1	No	No	F	41.9	10.1
CVR-1	1	No	No	L	14.2	nd

[a]The separation is based on at least one of the following criteria: plaque size, reaction with four viral antisera, sensitivity of DNA to 13 restriction endonucleases, and percentage of 5-mC and 6-mA in genomic DNA.
[b]Viral DNAs were tested for susceptibility to the 13 restriction endonucleases *Bg*III, *Eco*RI, *Bam*HI, *Sal*I, *Xho*I, *Bcl*I, *Pst*I, *Hind*III, *Sst*I, *Mbo*I, *Dpn*I, *Msp*I, and *Hap*II. The DNAs were grouped according to the resistance of the DNAs to these enzymes: group A: *Dpn*I; group B: *Hind*III, *Sst*I, and *Dpn*I; group C: *Bc*II and *Mbo*I; group D: *Sal*I, *Xho*I, *Bcl*I, *Pst*I, and *Mbo*I; group E: *Sal*I, *Xho*I, *Bcl*I, *Pst*I, *Hind*III, *Sst*I, and *Mbo*I; group F: *Bam*HI, *Bcl*I, *Pst*I, *Hind*III, *Sst*I, *Mbo*I, *Msp*I and *Hpa*II; group G: *Bam*HI, *Sal*I, *Xho*I, *Bcl*I, *Pst*I, *Hind*III, *Sst*I, *Mbo*I, *Msp*I, and *Hpa*II; group H: *Eco*RI, *Bam*HI, *Sal*I, *Xho*I, *Bcl*I, *Pst*I, *Hind*III, *Sst*I, *Mbo*I, *Msp*I, and *Hpa*II; group I: *Bgl*II, *Eco*RI, *Bam*HI, *Sal*I, *Xho*I, *Bcl*I, *Pst*I, *Hind*III, *Sst*I, *Mbo*I, *Msp*I, and *Hpa*II; group J: *Bcl*I, *Hind*III, *Sst*I, and *Mbo*I; group K: *Bcl*I, *Pst*I, *Hind*III, *Sst*I, and *Mbo*I; group L: *Sal*I and *Dpn*I; group M: *Bam*HI, *Sal*I, *Hind*III, *Sst*I, *Dpn*I, *Msp*I, and *Hpa*II; group N: *Eco*RI, *Bam*HI, *Sal*I, *Bcl*I, *Pst*I, *Hind*III, *Sst*I, *Mbo*I, *Msp*I, and *Hpa*II
[c]Percentage of 5-mC per C plus 5-mC plus deoxyuridine.
[d]Percentage of 6-mA per A plus 6-mA plus deoxyinosine.
[e]nd, not detected.

remainder of the PBCV-1 genome contains primarily unique DNA and has the potential to code for several hundred proteins (Girton and Van Etten, 1987). PBCV-1 and the chlorella virus group, therefore, are among the largest and most complex viruses known.

3 Replication of PBCV-1

PBCV-1 attaches rapidly, specifically, and irreversibly to cell walls, but not to protoplasts, of its host *Chlorella* NC64A (Meints et al., 1984; 1988). The virus always attaches to the host wall via a hexagonal vertex and dissolves the wall at the point of attachment. Viral DNA enters the cell, leaving an empty capsid on the cell surface. Thus the PBCV-1 infection process resembles bacteriophage infection.

The destination of infecting DNA and the intracellular site of PBCV-1 transcription and DNA replication are unknown. However, the recent finding of a short 101 bp intron in the DNA polymerase gene of PBCV-1, with splice junction sequences characteristic of eukaryotic nuclear-spliced mRNA introns, suggests that some nuclear processing of viral mRNA occurs prior to virus DNA synthesis (Grabherr et al., 1992). Early PBCV-1 transcripts and translation products (translation occurs on cytoplasmic ribosomes) can be detected within 10 min post infection (p.i.) (Schuster et al., 1986a). PBCV-1 DNA synthesis and late virus transcription begin about 1 h p.i. (Van Etten et al., 1984; Schuster et al., 1986a). Host nuclear and chloroplast DNAs are degraded beginning about 1 h p.i. The total DNA level in the cell increases four to ten-fold by 4 h p.i. (Van Etten et al., 1984).

The PBCV-1 capsid assembly, first observed about 2 h p.i., followed by DNA packaging, occurs in cytoplasmic regions called virus assembly centers (Meints et al., 1986). At 4 to 5 h p.i. infected cells contain many filled virus particles which are distributed throughout the cytoplasm. PBCV-1 release, which involves cell wall lysis, begins at 4 h p.i. and is complete at 8 to 10 h p.i. The burst size is typically 200 to 350 plaque forming units/cell. Mechanical disruption of cells releases infectious virus 30 to 50 min prior to spontaneous lysis (Van Etten et al., 1983a). Consequently, PBCV-1 is completely assembled inside the host and does not acquire a membrane by budding through the host membrane.

4 Additional *Chlorella* NC64A viruses

Hundreds of plaque-forming viruses, all of which infect *Chlorella* NC64A (called NC64A viruses), have been isolated from fresh water collected throughout the continental United States (Van Etten et al., 1985; Schuster et al., 1986), China (Zhang et al., 1988), and Japan (Yamada et al., 1991). Like PBCV-1, each of these viruses is a 150–190 nm diameter polyhedron that contains many structural proteins, a large dsDNA genome (> 300 kb) with 40% G + C content, and is chloroform sensitive. Some viral DNAs hybridize extensively with PBCV-1 DNA, while others hybridize poorly (Schuster et al., 1986; Zhang et al., 1988). Highly conserved genes such as the DNA polymerase gene, from poorly hybridizing viruses such as PBCV-1 and NY-2A, show highly conserved amino acid sequence identity (ca. 90%), although the corresponding DNA sequences are only 76% identical (Grabherr et al., 1992).

Thirty-seven of the NC64A viruses have been partially characterized and grouped into 16 classes according to plaque size, antiserum sensitivity, DNA restriction patterns, sensitivity of the DNAs to restriction endonuclease, and the nature and abundance of methylated bases (Table 1). DNA from each of the viruses contains 5-mC in amounts varying from 0.12% to 47.5% of the total cytosine. In addition, 25 of these 37 viral DNAs contain 6-mA in amounts ranging from 1.45% to 37% of the total adenine. Viruses in these 16 classes can be placed into ten DNA methylation phenotypes by the sensitivity/resistance of their DNAs to over 50 DNA restriction endonucleases (Table 2).

5 Viruses infecting *Chlorella* Pbi

Plaque-forming viruses that infect *Chlorella* strain Pbi (called Pbi viruses) have been isolated from European fresh water (Reisser et al., 1986; 1988; 1988a). The Pbi viruses, like the NC64A viruses, are large polyhedra with diameters of 140 to 150 nm, are chloroform sensitive, have many structural proteins, and have large dsDNA genomes of at least 300 kb that contain methylated bases (Table 1). However, the Pbi viruses are serologically distinct from the NC64A viruses; their DNAs hybridize poorly to NC64A virus DNAs and have a higher G + C (46%) content than the NC64A virus DNAs (40% G + C). Pbi viruses neither infect nor attach to *Chlorella* NC64A and the NC64A viruses neither infect nor attach to *Chlorella* Pbi (Reisser et al., 1991). We have used resistance/sensitivity to over 50 restriction endonucleases to define five DNA methylation phenotypes among five Pbi viruses (Table 3). Three of these Pbi virus DNA methylation phenotypes are the same as those of the NC64A viruses, whereas two DNA modification phenotypes are unique to the Pbi viruses.

6 DNA methyltransferases and DNA site-specific endonucleases in PBCV-1 infected cells

Given that chlorella virus DNAs contain 5-mC and 6-mA, it is not surprising that these viruses encode DNA methyltransferases. However, the finding of virus-encoded DNA site-specific endonucleases was unexpected. The following observations led to the discovery of the first virus-encoded DNA methyltransferase: (i) virus PBCV-1 DNA contained 6-mA in the sequence G^{m}ATC, whereas the chlorella host DNA did not (Van Etten et al., 1985a). (ii) PBCV-1 infected cell extracts contained an adenine DNA methyltransferase activity, named M.*C*viAI, that methylated GATC sequences (Xia and Van Etten, 1986). (iii) M.*C*viAI activity appeared about 30 min p.i. and its activity increased until 3 to 4 h p.i., which indicated that M.*C*viAI is a virus early gene product (Xia and Van Etten, 1986).

PBCV-1 infection also led to the appearance of a DNA site-specific endonuclease, R.*C*viAI, which recognizes the same GATC target sequence as M.*C*viAI (Xia et al., 1986). R.*C*viAI cleaves DNA 5′ to G, does not cleave DNAs containing G^{m}ATC sequences, and is unaffected

Table 2. DNA modification phenotypes of some representative *Chlorella* NC64A viruses. The ten phenotypes were determined by the resistance/sensitivity of the DNAs to over 50 restriction endonucleases. The modification listed as X is an undefined base but is probably not N.

											Level of methylation	
Virus	^{m}CC	RG^{m}CY	^{m}CGR	TCG^{m}A	G^{m}ATC	G^{m}ANTC	R^{m}AX	GT^{m}AC	TGC^{m}A	C^{m}ATG	5-mC	6-mA
CA-4B	−	−	−	−	−	−	−	−	−	−	0.12	nd
AL-1A	−	−	−	−	−	−	−	−	−	−	0.45	nd
PBCV-1	−	−	−	−	+	−	−	−	−	+	1.9	1.5
IL-3A	−	+	−	−	−	−	−	−	−	−	9.7	nd
SC-1A	−	−	−	+	+	−	−	?	+	+	1.94	7.3
NC-1A	−	−	−	+	+	+	−	−	+	+	7.1	7.3
NE-8A	−	+	−	+	+	+	?	?	+	+	14.3	8.1
NYs-1	+	+	+	−	+	+	−	−	+	+	47.5	11.3
AL-2A	+	+	?	+	+	+	−	−	+	+	35.8	14.6
MA-1D	+	+	?	+	+	+	−	−	+	+	47.0	16.7
CA-4A	+	+	+	+	+	?	+	+	+	+	39.8	19.6
NY-2A	+	+	+	+	+	−	+	+	+	+	44.9	37.0
XZ-3A	−	+	?	−	+	−	−	−	−	+	12.8	2.2
BJ-2C	−	+	−	−	+	+	−	−	+	+	12.8	11.5
XZ-6E	−	+	+	+	+	?	+	+	+	+	21.2	15.2
XZ-4C	+	+	+	+	+	?	+	+	+	−	46.7	20.8
XZ-4A	+	+	+	+	+	?	+	+	+	+	44.1	28.3

Table 3. DNA modification phenotypes of 5 *Chlorella* Pbi viruses. The five phenotypes were determined by the resistance/sensitivity of the DNAs to over 50 restriction endonucleases.

										Level of methylation	
PBi Viruses	^{m}CC	$R^{m}CY$	$(^{G}_{T})T^{m}C$	$TCG^{m}A$	$G^{m}ATC$	$R^{m}AX$	$GT^{m}AC$	$TGC^{m}A$	$C^{m}ATG$	5-mC	6-mA
CVR-1	–	–	–	–	–	–	–	–	–	14.2	nd
CVG-1	–	–	+	–	–	–	–	–	–	19.2	nd
CVA-1	+	+	–	–	–	–	–	–	–	43.1	nd
CVM-1	+	+	+	–	+	–	–	–	–	41.9	10.1
CVB-1	+	+	+	–	+	–	–	+	–	42.7	17.7

by 5-mC methylation. Therefore, R.*C*viAI is an isochizomer of the bacterial restriction endonucleases *Mbo*I and *Nde*II. Like bacterial type II restriction endonucleases, R.*C*viAI activity requires Mg^{++} but is unaffected by ATP or S-adenosylmethionine (SAM). *In vitro*, R.*C*viAI cleaves *Chlorella* NC64A DNA, but not PBCV-1 DNA, because host DNA contains unmodified GATC sequences, whereas PBCV-1 DNA contains G^{m}ATC sequences. R.*C*viAI endonuclease activity is first detected between 30 and 60 min p.i., and its appearance coincides with the onset of host DNA degradation.

At the time M.*C*viAI and R.*C*viAI were discovered, several observations indicated that these restriction/modification genes were virus-encoded. (i) Both enzyme activities first appeared 30 to 60 min after PBCV-1 infection and the appearance of both enzyme activities required *de novo* protein synthesis (Xia et al., 1986; Xia and Van Etten, 1986). (ii) Both enzyme activities appeared in UV-irradiated chlorella cells inoculated with PBCV-1, where presumably only virus encoded genes were expressed (Xia et al., 1986; Xia and Van Etten, 1986). (iii) Infection of *Chlorella* NC64A by other viruses led to the synthesis of several different DNA methyltransferases and site-specific endonucleases (see below). More recently, four DNA methyltransferase genes and one endonuclease gene have been cloned from viral genomic DNAs, which conclusively establishes that the DNA methyltransferases are virus-encoded (see below).

Recently a second, less abundant, DNA methyltransferase (named M.*C*viAII) and its cognate site-specific endonuclease (named R.*C*viAII) were discovered in PBCV-1 infected cells (Zhang et al., 1992a). M.*C*viAII methylates adenine in CATG sequences. R.*C*viAII, like the bacterial restriction endonuclease *Nla*III, recognizes the sequence CATG and does not cleave C^{m}ATG sequences. However, unlike *Nla*III, which cleaves 3′ to the G (Qiang and Schildkraut, 1986) and does not cleave ^{5m}CATG sequences, R.*C*viAII cleaves between the C and A and is unaffected by ^{5m}CATG methylation.

The M.*C*viAII/R.*C*viAII system was not detected at the time the M.*C*viAI/R.*C*viAI system was discovered, because both M.*C*viAII and R.*C*viAII are unstable *in vitro* and are extremely sensitive to chloride ion, pH, and temperature conditions. R.*C*viAII has a higher pH optimum (ca. 8.5) and prefers lower temperature (23°C) and lower salt concentrations (< 20 mM K^{+} or Na^{+}) than most bacterial DNA restriction endonucleases. M.*C*viAII, like M.*C*viAI, is efficiently expressed *in vivo* because PBCV-1 DNA is resistant to both R.*C*viAI and R.*C*viAII endonucleases, as well as other 6-mA-sensitive restriction endonucleases which overlap GATC and CATG sequences.

In addition to the two adenine-specific restriction/modification systems, virus PBCV-1 also codes for at least one cytosine DNA methyltransferase, named M.*C*viAIII. M.*C*viAIII, has been partially purified from PBCV-1 infected cells; although the exact sequence specificity of M.*Cvi*AIII has not been determined, it is a short four bp G + C rich sequence (Zhang et al., 1992). A site-specific endonuclease, corresponding to M.*Cvi*AIII, has not been detected in PBCV-1 infected cells even with a variety of extraction and assay conditions (Nelson, unpublished results).

Although PBCV-1 is a relatively simple virus in terms of its methylated base content (1.9% 5-mC and 1.5% 6-mA), it encodes two functional DNA site-specific endonucleases, two functional adenine DNA methyltransferases, one functional cytosine DNA methyltransferase, and at least one cytosine DNA methyltransferase pseudogene (see below). We suspect that additional nonfunctional DNA methyltransferase genes and possibly DNA site-specific endonuclease genes are present in the PBCV-1 genome. The six known restriction/modification genes, including the pseudogene, comprise at least 2–3% of the 333 kbp PBCV-1 genome. Chlorella virus restriction/modification systems are therefore among the most complex restriction/modification loci known. Furthermore, as described below, many other chlorella viruses appear to contain even more restriction/modification systems.

7 Additional viral DNA modification phenotypes, methyltransferases and site-specific endonucleases

Each of the *Chlorella* NC64A virus genomes contains 5-mC, and 25 viral DNAs also contain 6-mA (Table 1). The modified sequences in most of these viral DNAs have not been defined. However, based on the sensitivity/resistance of the viral DNAs to about 50 methylation-sensitive restriction endonucleases (Nelson and McClelland, 1991), we have identified at least ten sequence-specific modification phenotypes in the 37 NC64A viruses (Table 2) and five modification phenotypes in five Pbi viruses (Table 3). At least some of the NC64A virus DNAs contain methylated bases at sites not listed in Table 2. For example, the 5-mC-modified site in virus PBCV-1 DNA is unknown. Likewise, NC-1A virus has 7.1% 5-mC at unknown sequences. DNA methyltransferase activities have been isolated from one or more chlorella virus infected cells that correspond to nine of the ten NC64A virus modification phenotypes.

Several DNA site-specific endonucleases have also been isolated from NC64A virus-infected cells (Table 4). Many of these enzymes were isolated prior to our recent discovery that some chlorella virus endonucleases are very sensitive to salt, especially chloride ion, and prefer lower temperature and higher pH than bacterial restriction endonucleases. We suspect that additional activities will be found when virus-infected cells are re-examined using newly developed chloride-free, low ionic strength buffers and assay conditions. The following paragraphs will briefly describe our current knowledge of the restriction/modification systems from five NC64A viruses; these viruses – Il-3A, NC-1A, XZ-6E, NYs-1, and NY-2A – are arranged in order of increasing level of methylation.

7.1 Virus IL-3A

Virus IL-3A DNA contains 9.7% 5-mC and no detectable 6-mA (Table 1). The virus encodes at least one type II DNA site-specific endonuclease, named R.*C*viJI. R.*C*viJI cleaves PuG/CPy, but not PuG^{m}CPy, sequences (Xia et al., 1987a). R.*C*viJI is the first enzyme to recognize the sequence PuGCPy and on a statistical basis is the first site-specific endonuclease to recognize a three base pair sequence. Consequently R.*C*viJI cleaves DNA frequently, with an average size of 64 bases, making it ideal for fingerprinting short DNA fragments such as polymerase chain reaction products. Addition of ATP and S-adenosylmethionine to R.*C*viJI assays reduces the specificity of R.*C*viJI to two bases (GC) (Xia et al., 1987). This property is being exploited to develop a rapid shotgun cloning system for large-scale DNA sequencing (Fitzgerald et al., 1992).

The cognate methyltransferase, M.*C*viJI [methylates (A/G)GC(T/C/G) sequences], has been cloned and sequenced (Shields et al., 1990). Three spontaneously derived, 5-azacytidine resistant [5-azacytidine is a mechanism based inhibitor of cytosine methylation (Jones, 1984)], deletion mutants of virus IL-3A were isolated which had lost both M.*C*viJI and R.*C*viJI activities. DNA from these deletion mutants contains 1.6% 5-mC, whereas wild-type IL-3A DNA contains 9.7% 5-mC (Burbank et al., 1990). The growth cycle and burst size of the mutant and parent viruses were identical. We conclude that M.*C*viJI and R.*C*viJI activities are not essential for virus growth and that virus IL-3A codes for a second cytosine methyltransferase of unknown specificity.

7.2 Virus NC-1A

Virus NC-1A DNA contains 7.1% 5-mC and 7.3% 6-mA. Cells infected with NC-1A contain a type II GANTC site-specific endonuclease named R.*C*viBI. R.*C*viBI, like its bacterial restriction endonuclease isoschizomers *Hin*fI and *Hha*II, cleaves DNA between G and A and methylation of the A prevents cleavage (Xia et al., 1986a). NC-1A infected cells also contain the cognate adenine methyltransferase (named M.*C*viBI). Based on the sensitivity/resistance of NC-1A DNA to restriction endonucleases we predict that NC-1A also encodes four additional adenine methyltransferases that modify the sequences TCG^{m}A, G^{m}ATC, TGC^{m}A, and C^{m}ATG (Table 2). Two of these activities, M.*C*viBII (methylates G^{m}ATC sequences) and M.*C*viBIII (methylates TCG^{m}A sequences) have been isolated from NC-1A infected cells (Xia and Van Etten, unpublished data). Since NC-1A DNA contains 7.1% 5-mC, it also must encode one or more 5-mC methyltransferases. It is not known if these additional DNA methyltransferases have cognate site-specific endonucleases. However, as a minimum, NC-1A codes for one site-specific endonuclease, five adenine methyltransferases, and at least one cytosine methyltransferase.

M.*C*viBIII, which has been cloned and sequenced (see below), is not required for virus NC-1A replication. A mutant of NC-1A which has the M.*C*viBIII gene deleted also replicated in the host.

7.3 Virus XZ-6E

Virus XZ-6E DNA contains 21.2% 5-mC and 15.2% 6-mA. Extracts of virus XZ-6E infected cells contain at least two abundant type II site-specific endonucleases (Nelson, Traylor, Xia, Roy and Van Etten, manuscript in preparation). One of these enzymes, named R.*C*viRI, is the first site-specific endonuclease to recognize the sequence TGCA. R.*C*viRI cleaves between the G and C and is inhibited by either 6-mA or 5-mC methylation. The second enzyme, R.*C*viRII, recognizes the sequence GTAC and is a heteroschizomer of the bacterial endonuclease *Rsa*I. However, unlike *Rsa*I, which cleaves between the T and A, R.*C*viRII cleaves between the G and T. R.*C*viRII does not cleave GT^{m}AC sequences. The two cognate adenine methyltransferases, M.*C*viRI (TGC^{m}A) and M.*C*viRII (GT^{m}AC) have been partially purified from XZ-6E infected cells; M.*C*viRI has also been cloned and sequenced (see below). In addition to these two restriction/modification systems, we predict that virus XZ-6E also encodes at least two cytosine methyltransferases that recognize RG^{m}CY and ^{m}CGR sequences and four more adenine methyltransferases that recognize G^{m}ATC, R^{m}AX, TCG^{m}A, and C^{m}ATG sequences (Table 2).

7.4 Virus NYs-1

Virus NYs-1 DNA contains 47.5% 5-mC and 11.3% 6-mA. Cells infected with NYs-1 contain a base-specific nicking enzyme (named NYs-1 endo) that cleaves 5′ to 5′-CC-3′ sequences and does not cleave the opposite strand 5′-GG-3′ sequences (Xia et al., 1988). The enzyme cleaves 5′-C^{m}C-3′ sequences but not 5′-^{m}CC-3′ sequences. Because NYs-1 endo only cleaves one strand of the duplex, it does not produce a completely stable digestion pattern.

Since NYs-1 DNA is resistant to NYs-1 endo, the virus must encode a cytosine methyltransferase that methylates the sequence ^{m}CC. Based on the resistance of NYs-1 DNA to other restriction endonucleases, we predict the virus also codes for two cytosine methyltransferases that recognize the sequences RG^{m}CY and ^{m}CGR and four adenine methyltransferases that recognize the sequences G^{m}ATC, G^{m}ANTC, TGC^{m}A, and C^{m}ATG (Table 2).

7.5 Virus NY-2A

Virus NY-2A, which has the most heavily methylated genome, (44.9% 5-mC and 37% 6-mA) induces at least two site-specific endonucleases. One of these enzymes, named R.*C*viQI (previously named R.*C*viII), is identical to R.*C*viRII and cleaves G/TAC sequences (Xia et al., 1987a).

The other NY-2A encoded site-specific endonuclease (named NY-2A endo) is a base specific nicking enzyme which cleaves Pu/AG sequences but not Pu^{m}AG sequences (Nelson, Ropp, Xia and Van Etten, manuscript in prep.). Like NYs-1 endo, NY-2A endo only cleaves one strand of duplex DNA and does not produce a completely stable digestion pattern. In contrast to most restriction endonucleases, which function as dimers, preliminary experiments suggest that NY-2A endo functions as a 50-kD monomer.

DNA methyltransferases from NY-2A infected cells are presently under study. Two adenine methylated activities have been detected (GT^{m}AC and Pu^{m}APu) (Nelson, unpublished data), but at least seven more methyltransferases can be predicted from NY-2A DNA modification phenotypes (Table 2): Three cytosine methyltransferases that modify the sequences ^{m}CC, RG^{m}CY, and ^{m}CGR and four adenine methyltransferases that modify TCG^{m}A, G^{m}ATC, TGC^{m}A, and C^{m}ATG. Therefore, as a minimum, NY-2A codes for two site-specific endonucleases and nine different DNA methyltransferases. We suspect that several additional salt/pH sensitive endonucleases are present in NY-2A infected cells.

7.6 Pbi viruses

European Pbi virus genomes, like the NC64A viruses, also contain various levels of 5-mC and 6-mA (Table 1), indicating that they also encode DNA methyltransferases. Three of the Pbi viral encoded DNA methyltransferases modify the same sequences as some of the NC64A virus methyltransferases, whereas the other two cytosine methyltransferases modify different sequences, R^{m}CY and (G/T)T^{m}C (Table 3). Preliminary experiments indicate that infected cells of *Chlorella* Pbi, like *Chlorella* NC64A, also contain DNA site-specific endonucleases (Danko and Van Etten, unpublished data).

In conclusion, the virus infected chlorella cells are a rich source of DNA methyltransferase and DNA site-specific endonucleases, especially enzymes that recognize 2 to 4 base pair sequences. The low level of methylated bases in some viruses, e.g. CA-4B has 0.12% 5-mC, suggests that some of the viruses may also encode methyltransferases that recognize 5 or 6 base pair sequences. Some chlorella virus encoded enzymes differ from known bacterial enzymes with respect to their (i) sequence specificities, (ii) cleavage sites, and/or (iii) sensitivities to site-specific methylation. Furthermore, two site-specific, methylation sensitive, DNA nicking (cleave one strand) enzymes represent a new class of endonucleases and await further characterization.

The presence of virus encoded DNA methyltransferases and DNA site-specific endonucleases leads to two obvious questions. (i) What is the biological function of these enzymes? (ii) What is the evolutionary origin of these enzymes? Experiments designed to address these two questions are discussed in the next two sections.

8 Function of virus-encoded DNA site-specific endonucleases and DNA methyltransferases

Traditionally it is believed that bacterial restriction/modification systems confer cellular resistance to foreign DNAs and DNA viruses (e.g. Wilson and Murray, 1991). In fact, the name "restriction" refers to the role of site-specific endonucleases in excluding foreign phage and plasmid DNA. In bacterial cells, DNA methyltransferases serve to prevent self-digestion of bacterial DNA by self-protective modification.

The biological function of the chlorella virus encoded restriction/modification systems are presently unknown. We have considered three functions for the chlorella virus restriction/modification systems, which are not necessarily mutually exclusive. (i) The site-specific endonucleases help degrade host DNA; resultant deoxynucleotides could then be recycled into viral DNA. Methylation of nascent virus DNA by the cognate methyltransferase protects it from "self-digestion". (ii) The site-specific endonucleases are responsible for inhibiting host DNA synthesis. (iii) The site-specific endonucleases prevent infection of the host cell by a second virus. It should be noted, however, that most of the experiments to test these hypotheses were conducted prior to our recent discovery that multiple DNA site-specific endonucleases can be detected in many virus-infected cells using milder extraction and assay conditions.

Several observations are consistent with the first hypothesis. (i) Host nuclear and chloroplast DNAs, but not virus DNA, are digested by the virus-encoded site-specific endonuclease(s) *in vitro*. (ii) *In vivo* degradation of host nuclear and chloroplast DNA coincides with the appearance of DNA site-specific endonuclease activity. (iii) Initiation of virus DNA synthesis *in vivo* occurs after the appearance of DNA methyltransferase activity (Xia et al., 1986a; Xia and Van Etten, 1986).

The isolation of three deletion mutants of virus IL-3A, each of which had lost their DNA methyltransferase (M.*C*viJI) and site-specific endonuclease (R.*C*viJI) activities, allowed us to test the host DNA degradation hypothesis directly. If R.*C*viJI activity was essential for host DNA degradation, nuclear and/or chloroplast DNA should be preserved or at least degraded more slowly in cells infected with the mutants than in cells infected with wild type IL-3A. However, both nuclear and chloroplast DNAs decreased at nearly identical rates following infection with each of the four viruses (Burbank et al., 1990). Thus we concluded that R.*C*viJI activity was not essential for host DNA degradation.

It should also be noted that the total DNA in the cell increases four to ten fold by 4 h after PBCV-1 infection (Van Etten et al., 1984). Even if recycling of nucleotides from host DNA to virus DNA occurs, salvaged nucleotides could only supply a small portion of the nucleotides required for virus DNA synthesis. The remainder of the nucleotides must be synthesized *de novo* after virus infection.

The second hypothesis, that the endonucleases are responsible for the rapid inhibition of host DNA synthesis following virus infection (Van Etten et al., 1984), arises from the recent discovery that a small amount of site-specific endonuclease activity can be isolated from purified, protease treated virions (Grabherr, Traylor and Van Etten, unpublished results). It is not known if this enzyme activity is specifically or accidentally packaged into the virions. However, if one or more endonucleases are specifically packaged into the virion and transported into the cell during infection, then these enzyme(s) could begin to degrade host DNA almost immediately after infection, leading to the rapid cessation of host DNA synthesis. If packaged virion endonuclease activity is involved in inhibiting host DNA synthesis, then one would predict that host DNA synthesis should continue in chlorella cells infected with the M.*C*viJI/R.*C*viJI deficient mutants of virus IL-3A relatively longer than in cells infected with wildtype IL-3A virus. However, host DNA synthesis stopped immediately after infection with both mutant and wildtype viruses (Grabherr and Van Etten, unpublished results).

To determine whether viral endonuclease(s) are involved in excluding infection of a cell by a second virus, chlorella cells were dually inoculated with different viruses at a high multiplicity of infection and plaques arising from infective centers were distinguished by immunoblotting (Chase et al., 1989). These experiments produced several conclusions. (i) At least 90% of the plaques resulting from single cells inoculated with two viruses contained only one of the viruses. Thus the chlorella viruses, like bacteriophages (e.g. reviewed in Doermann, 1983), exclude one another. (ii) Infection of the alga by one virus did not prevent attachment of a second virus to the host. (iii) Cells inoculated with one virus 30 min before inoculation with a second virus preferentially replicated the first virus. Thus the exclusion mechanism is probably triggered within 30 to 45 min after infection. (iv) A faster-growing virus does not necessarily dominate in a dual inoculation since virus SC-1B (replicates in 8 to 15 h) competed very well with virus NY-2C (replicates in 4 to 9 h) even though it replicated much slower. (v) Some viruses dominated in certain combinations. For example, IL-3A dominated NY-2C and the PBCV-1 serotype EPA-1. However, the dominance could not be predicted from the site-specific endonucleases. For example, virus PBCV-1 encodes the site-specific endonucleases R.*C*viAI and R.*C*viAII and virus NC-1A encodes R.*C*viBI. *In vitro*, R.*C*viBI digests PBCV-1 DNA but R.*C*viAI and R.*C*viAII do not digest NC-1A DNA. Consequently if the site-specific endonucleases mediate exclusion, NC-1A should dominate in a mixed infection. Such a dominant infection did not occur; about 40% of the infective centers contained only NC-1A, whereas 60% contained only PBCV-1. Although it is possible that PBCV-1 and NC-1A code for additional site-specific endonucleases which influence these results, we have also found viral exclusion in cells inoculated with the isogenic viruses PBCV-1 and EPA-1, which have identical restriction/modification systems. The finding that exclusion occurs in cells inoculated with isogenic viruses probably explains the low level of recombination (1 to 2%) between temperature sensitive mutants of PBCV-1 (Tessman, 1985). In summary, exclusion in the chlorella viruses appears to be independent of the presence of functional restriction/modification systems.

The complexity of chlorella virus DNA methyltransferases and site-specific endonucleases and their widespread geographic distribution implies that these enzymes confer an evolutionary advantage to these viruses in their native environments. Unfortunately, very little is known about the natural history of chlorella viruses, or is it known whether they replicate exclusively in exsymbiotic chlorella cells or have other natural host(s). We have conducted one simple experiment to determine whether viral restriction/modification enzymes confer an advantage to the virus in the laboratory. A culture of *Chlorella* NC64A was inoculated with a 1:1 mixture of virus IL-3A and a IL-3A mutant deficient in R.*C*viJI and M.*C*viJI activities (the two viruses grow at identical rates). A fresh culture of *Chlorella* NC64A was inoculated with a portion of the lysate. After five serial transfers, the lysate was titered. The mutant made up the majority of the plaques (Nietfeldt and Van Etten, unpublished data). Thus R.*C*viJI and M.*C*viJI activities conferred no selective advantage to the virus under these laboratory conditions.

Therefore, the biological function(s) of the chlorella virus encoded DNA methyltransferases and DNA site-specific endonucleases is unknown. Since spontaneous deletion mutants of virus NC-1A (missing the M.*C*viBIII gene) and virus IL-3A (missing the M.*C*viJI and R.*C*viJI genes) have been isolated and these mutants are viable, at least

Table 4. DNA methyltransferases and DNA site-specific endonucleases purified from virus infected *Chlorella* NC64A cells.

Virus	DNA methyl-transferase	Sequence	DNA site-specific endonuclease	Sequence[a]
PBCV-1	M.*C*viAI	G^{m}ATC	*C*viAI	/GATC
	M.*C*viAII[b]	C^{m}ATG	*C*viAII[b]	C/ATG
NC-1A	M.*C*viBI	G^{m}ANTC	*C*viBI	G/ANTC
	M.*C*viBII	G^{m}ATC		
	M.*C*viBIII[b]	TCG^{m}A		
NY-2A	M.*C*viQI	GT^{m}AC	*C*viQI	G/TAC
	M.*C*viQII	(Pu)^{m}A(Pu)	NY-2A endo	Pu/AG[c]
IL-3A	M.*C*viJI[b]	(G/A)G^{m}C(T/C/G)	*C*viJI	PuG/CPy
XZ-6E	M.*C*viRI[b]	TGC^{m}A	*C*viRI	TG/CA
	M.*C*viRII	GT^{m}AC	*C*viRII	G/TAC
NYs-1		^{m}CC	NYs-1 endo	/CC[c]

[a]The slash indicates the cleavage site; [b]Enzymes that have been cloned and sequenced; [c]Only cleaves one strand of dsDNA.

some restriction/modification genes are not essential for virus replication in the laboratory. On the other hand, if a virus contains a functional cytosine DNA methyltransferase, then expression of the gene is apparently necessary for virus growth because 5-azacytidine inhibits virus replication. There is a direct correlation between increasing viral 5-mC levels and the sensitivity of virus replication to 5-azacytidine (Burbank et al., 1990). We have also noticed that the most heavily methylated viruses typically form the smallest plaques (Burbank, unpublished results).

9 Sequence of virus-encoded DNA methyltransferases

The fact that the chlorella virus-encoded DNA methyltransferases and DNA site-specific endonucleases resemble bacterial type II restriction/modification enzymes suggests a common ancestry. One way to analyze the chlorella virus and bacterial enzymes is to compare their predicted amino acid sequences. Over 80 bacterial cytosine and adenine DNA methyltransferases have been cloned in *E. coli* (Wilson and Murray, 1991). Most of these DNA methyltransferase genes have been cloned using a strategy first proposed by Mann et al. (1978) and refined by Lunnen et al. (1988). A gene library is screened for expression of a particular methyltransferase by digesting recombinant plasmid DNAs with the cognate restriction endonuclease prior to transforming *E. coli*. Plasmids expressing a given DNA methyltransferase activity are resistant to the cognate restriction endonuclease and are isolated as modification-proficient transformants.

We have used this strategy to clone four (three adenine and one cytosine) chlorella virus DNA methyltransferase genes in *E. coli* (Table 4). All four DNA methyltransferase genes are expressed in both orientations in pUC or pBluescript vectors, suggesting transcription from a virus promoter. The three adenine methyltransferases, M.*Cvi*BIII [methylates TCG^{m}A (Narva et al., 1987)] from virus NC-1A, M.*Cvi*RI [methylates TGC^{m}A (Stefan et al., 1991) from virus XZ-6E, and M.*Cvi*AII [methylates C^{m}ATG (Zhang et al., 1992a)] from virus PBCV-1 contain 377, 379, and 326 predicted amino acids, respectively. These three virus enzymes, which are the first adenine methyltransferases to be cloned from a eukaryotic system, contain two amino acid motifs which are characteristic of bacterial adenine methyltransferases: (i) an N-terminal (D/E)x_3Gx(G/C) sequence which is probably the SAM binding site (Klimasauskas et al., 1989) and (ii) a (D/N)PP(Y/F) sequence that is probably the aminomethyl transfer site (Klimasauskas et al., 1989).

Comparison of the three viral adenine methyltransferases to each other and to bacterial adenine methyltransferases revealed that M.*Cvi*BIII and M.*Cvi*RI are most similar to each other. Forty percent of the amino acids are identical and the predicted size of the two proteins are very similar (Stefan et al., 1991). M.*Cvi*BIII is also similar to the bacterial isoschizomer M.*Taq*I (39% of the amino acids are identical) and M.*Pae*R7, whose recognition sequence (CTCGAG) contains the subset TCGA (Narva et al., 1988).

M.*Cvi*AII (CATG) has extensive similarity to its bacterial heteroschizomer M.*Nla*III (Labbe et al., 1990) in which 36% of the amino acids are identical (Zhang et al., 1992a). M.*Cvi*AII also has 36% amino acid identity with the portion of M.*Fok*I which methylates the GGATG sequence (Kita et al., 1989; Sugisaki et al., 1989).

At present catalytic and target recognition domains in adenine DNA methyltransferases are poorly defined. The M.*Cvi*RI (TGCA) and M.*Cvi*BIII (TCGA) adenine methyltransferases differ in their target recognition sequences by their two central base pairs. Because these two proteins are about the same size and share extensive amino acid homology, we are trying to identify the target recognition domain in these two adenine methyltransferases by constructing gene fusions between M.*Cvi*RI and M.*Cvi*BIII genes.

The virus IL-3A gene encoding the cytosine methyltransferase M.*Cvi*JI [methylates (G/A)G^{m}C(T/C/G) sequences] has also been cloned and sequenced (Shields et al., 1990). The M.*Cvi*JI amino acid sequence, which is 367 amino acids long, has no obvious similarity to the chlorella virus adenine methyltransferases. This is not surprising since the primary amino acid sequence of bacterial adenine and cytosine methyltransferases also differ substantially (e.g. Slatko et al., 1987; Posfai et al., 1989; but see Lauster, 1989 and Klimasauskas et al., 1989 for some evidence of similarity).

However, M.*C*viJI shares amino acid sequence motifs with bacterial cytosine methyltransferases. Posfai et al. (1989) reported that bacterial cytosine methyltransferases contain ten conserved amino acid motifs which are in the same order in each enzyme. The M.*C*viJI gene contains amino acid sequences which resemble each of the bacterial motifs. Motif IV (11 of 12 amino acids) is the most highly conserved motif in M.*C*viJI and the least conserved sequences are the three motifs [motif VIII (5 of 13 amino acids), motif IX (3 of 9 amino acids) and motif X (4 of 11 amino acids)] toward the C-terminus of the protein. Of the 21 invariant amino acids found in the 13 bacterial enzymes (Posfai et al., 1989), M.*C*viJI contains 17.

One difference between M.*C*viJI and bacterial cytosine methyltransferases is the spacing between motif VIII and motif IX. The bacterial enzymes contain a long variable region of 80 to more than 250 amino acids between these two motifs; this region contains the target recognition domain (TRD) (Posfai et al., 1989; Lauster et al., 1989; Klimasauskas et al., 1991). However, M.*C*viJI motifs VIII and IX are immediately adjacent to one another; thus the M.*C*viJI TRD is located elsewhere in the protein. The most likely candidate for the M.*C*viJI TRD is the 68-amino-acid stretch between motifs IX and X (Zhang et al., 1992).

Bacterial DNA methyltransferase genes are always located near their cognate DNA restriction endonuclease gene, although the spacing and relative orientation of the two genes vary (Wilson and Murray, 1991). The R.*C*viAII endonuclease gene from virus PBCV-1 was recently cloned and sequenced (Zhang et al., 1992a); this is the first site-specific endonuclease gene to be cloned and sequenced from a chlorella virus. The M.*C*viAII methyltransferase and R.*C*viAII endonuclease genes, which are located between map positions 125 and 130 on the PBCV-1 physical map, are tandemly arranged head-to-tail such that the TAA termination codon of the M.*C*viAII gene overlaps the ATG translation start site of the R.*C*viAII gene.

Spontaneously derived deletion mutants of virus IL-3A have also been isolated which have simultaneously lost both M.*C*viJI and R.*C*viJI activities (Burbank et al., 1990). Assuming these mutants arose from a single deletion, this finding suggests that the M.*C*viJI/R.*C*viJI virus genes are located near one another.

In conclusion, two types of evidence suggest that the chlorella virus modification/restriction genes have a common ancestry with bacterial genes: (i) The primary amino acid sequences between the chlorella viruses and bacterial adenine and cytosine DNA methyltransferases are similar, and (ii) the arrangement of the restriction/modification genes in bacteria resemble those of the chlorella viruses.

10 Virus encoded pseudogenes

We routinely hybridize cloned DNA methyltransferase genes from one virus to all other chlorella virus DNAs to see if this gene is present in other viruses. Expression of a DNA methyltransferase gene, or a gene with a similar function in the other viruses, is determined by testing the sensitivity of the viral DNAs to restriction endonuclease digestion. We suspected that nonfunctional DNA methyltransferases could be identified in some of the virus DNAs if a DNA methyltransferase gene hybridized strongly to a nonexpressing virus DNA.

Such an experiment was conducted with the viral cytosine methyltransferase M.*C*viJI [methylates PuG^{m}C(T/C/G) sequences] (Zhang et al., 1992). As expected, the M.*C*viJI gene hybridized to DNA from many NC64A viruses whose DNAs were resistant to the restriction endonuclease R.*C*viJI. However, M.*C*viJI also strongly hybridized to DNA from virus PBCV-1, which does not express a functional M.*C*viJI gene because its DNA is cleaved by R.*C*viJI, *Alu*I and *Hae*III (Shields et al., 1990). Subsequent cloning and sequencing the region from PBCV-1 that hybridized to M.*C*viJI revealed an open reading frame (ORF), called P17-ORF4, which differed by 8 amino acids from M.*C*viJI. P17-ORF4 is located at the right terminus of the PBCV-1 physical map immediately upstream of the inverted terminal repeat. P17-ORF4 is not transcribed and thus it is a nonfunctional gene (pseudogene).

Gene fusions between P17-ORF4 and M.*C*viJI and site-directed point mutations established that changing Gln^{188} to Lys^{188} completely abolished M.*C*viJI activity in *E. coli* (Zhang et al., 1992). Conversely, changing Lys^{188} in P17-ORF4 to Gln^{188} resulted in M.*C*viJI activity. The other seven amino acids did not affect M.*C*viJI activity. As mentioned above, bacterial cytosine methyltransferases contain ten conserved amino acid motifs. Posfai et al. (1989) identified twenty-one invariant amino acids in the thirteen bacterial enzymes, of which seventeen residues were conserved in M.*C*viJI. Of the eight amino acids which differed between M.*C*viJI and the P17-ORF4 pseudogene, one occurred in an "invariant" position: Glu^{167} was changed to a Lys^{167}. However, this change had no obvious effect on M.*C*viJI methyltransferase activity in *E. coli*. The amino acid change that abolished M.*C*viJI activity occurred outside the ten conserved motifs. Speculation on how the Gln^{188} to Lys^{188} change makes the enzyme inactive are described in the next section.

The presence of a M.*C*viJI DNA methyltransferase pseudogene in virus PBCV-1 is not an isolated occurrence; the M.*C*viJI gene also hybridized strongly to DNA from two other chlorella viruses, SC-1A and SC-1B, which do not express the M.*Cvi*JI modification (Shields et al., 1990). We also suspect that the chlorella viruses contain DNA site-specific endonuclease pseudogenes. Thus the chlorella virus systems are a unique and versatile source of naturally occurring functional and non-functional restriction/modification mutants. Comparing nonfunctional and functional gene sequences from different viruses provides a powerful tool for determining critical amino acids necessary for enzyme activity. Site-directed amino acid changes can then be focused on selected DNA binding or catalytic regions.

11 Secondary structure predictions of M.*Cvi*JI based on thymidylate synthetase

The M.*Cvi*JI methyltransferase is one of over fifty known enzymes which catalyze sequence-specific reductive methylation of the [5,6] double bond of cytosine residues. Other laboratories (Lauster et al., 1989; Wilke et al., 1988; Klimasauskas et al., 1989; Posfai et al., 1989) have identified conserved blocks of primary amino acid sequence in several cytosine methyltransferases, including a highly conserved catalytic Pro-Cys motif found in M.*Cvi*JI methyltransferase at amino acids 72-73 (Fig. 1).

This Pro-Cys motif is also conserved in other methyltransferase enzymes such as M.*Hha*I cytosine DNA methyltransferase (Santi et al., 1983; Wu and Santi, 1987; Ho et al., 1991), dCMP hydroxymethylase (Santi et al., 1983) and thymidylate synthetase (Wu and Santi, 1987; Santi et al., 1974), all of which catalyze reductive methylation of pyrimidine bases. Thymidylate synthetase from *Lactobacillus casei* is the best-studied example of an enzyme which catalyses reductive [5,6] pyrimidine methylation. Structure-function correlations, based on X-ray crystallography studies, indicate that the Pro-Cys sequence is the active site for covalent attachment of thymidylate synthetase to the 6-pyrimidine ring position of cytosine via a proposed Michael addition (Santi et al., 1974; Hardy et al., 1987).

The X-ray crystal structure of *L. casei* thymidylate synthetase reveals a mostly parallel beta sheet arrangement, in which the conserved Pro-Cys residues lie within one of five approximately planar beta sheets.

We have interpreted the eight amino acids changes in the nonfunctional M.*Cvi*JI methyltransferase in terms of a molecular model, based on the known X-ray structure of thymidylate synthetase. This interpretation relies on the following observations: (i) *L. casei* thymidylate synthetase and M.*Cvi*JI methyltransferase have a common sequence [PED(I/V)$x_{6\text{-}7}$$\underline{PC}$] spanning the thymidylate synthetase active site (Hardy et al., 1987). (ii) The underlined Pro-Cys motif is conserved in

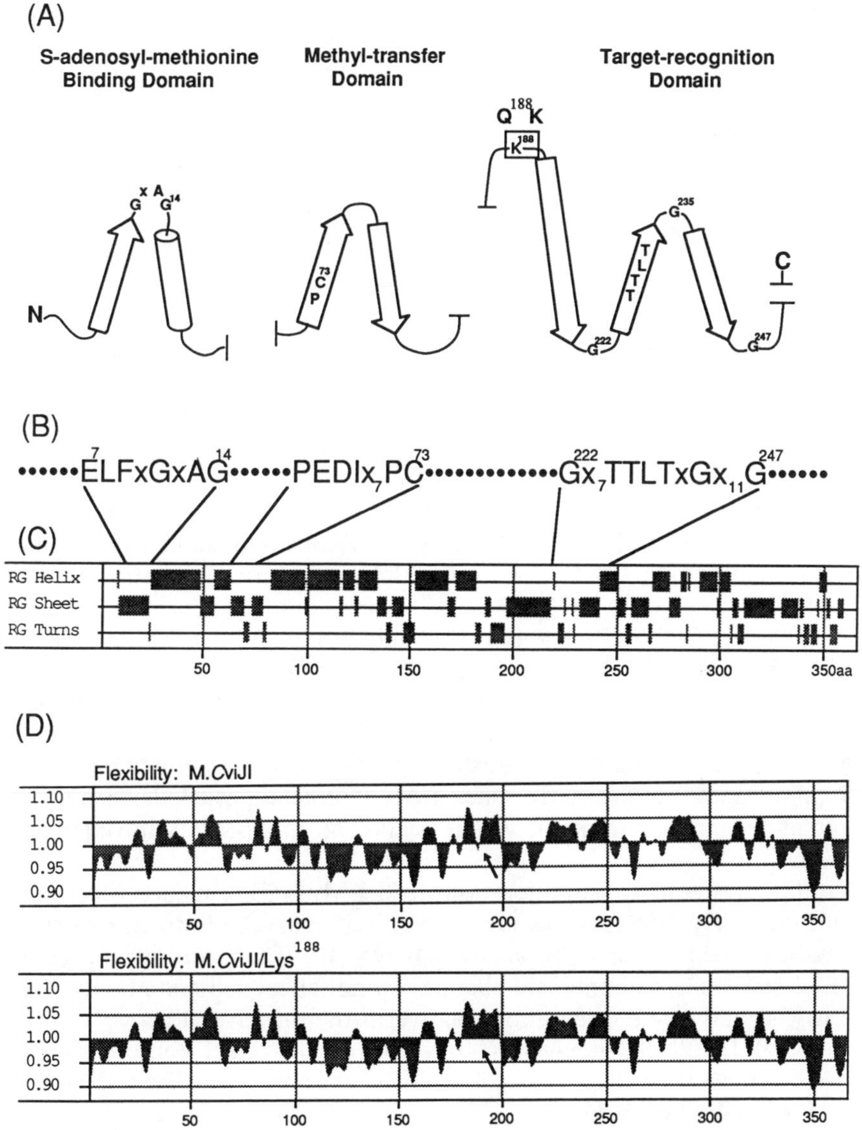

Figure 1. (A-B) Model of M.*Cvi*JI methyltransferase showing the SAM binding site (E^7LFxGxAG), the catalytic site ($P^{61}EDIx_7PC$), and the putative 26-amino-acid target recognition site ($G^{222}x_7TTLTxGx_{11}G$). (C) The predicted protein secondary structure of M.*Cvi*JI using Robson-Garnier algorithms. An identical profile was predicted for the non-functional M.*Cvi*JI/Lys^{188} methyltransferase. (D) Predicted protein flexibility comparison of M.*Cvi*JI and M.*Cvi*JI/Lys^{188}. Note: replacing Gln^{188} with Lys^{188} alters the predicted flexibility (arrow) of the protein.

all 5-mC DNA methyltransferases (Klimasauskas et al., 1989; Posfai et al., 1989). (iii) Thymidylate synthetase and 5-mC DNA methyltransferases share a common enzyme reaction mechanism (Santi et al., 1983; Wu and Santi, 1987), although the former enzyme uses tetrahydrofolate rather than SAM as a methyl doner. Therefore, it is likely that pyrimidine reductive methyltransferases have common superstructural elements, i.e., an active site Pro-Cys which lies at the stem of a beta-turn-beta structure.

We determined that the Robson-Garnier algorithm accurately predicted a secondary structure for the active site of thymidylate synthetase, the beta turn-beta sheet MALP*PC*H region, which agreed with its known high resolution X-ray crystal structure (Hardy et al., 1987). In contrast, the Chou-Fasman algorithm (Chou and Fasman, 1978) did not accurately predict extensive beta sheet/turn structures of thymidylate synthetase. Therefore, we assumed that the Robson-Garnier algorithm might accurately predict DNA cytosine methyltransferase secondary structures, for which no X-ray crystal data is available. The proposed SAM-binding site (E^7LFxGxAG), the catalytic-site (Pro^{72}-Cys), and the putative target recognition domain ($G^{222}x_7$TTLTxG$x_{11}G^{247}$) of M.*C*viJI are shown in Fig. 1. The SAM binding site is proposed to be a typical Rossmann dinucleotide fold (Rossmann et al., 1975), a beta-turn-alpha structure. The critical Gln^{188} to Lys^{188} change (Q^{188}K) in M.*C*viJI occurs in a predicted proline-rich beta turn (Fig. 1A-C), between the catalytic site and the putative TRD. This Q^{188}K substitution would be expected to alter the flexibility of this beta turn. This prediction is supported by a "flexibility" plot, (Karplus and Schulz, 1985) of M.*C*viJI and M.*C*viJI/Lys^{188} (Fig. 1D). Therefore, we propose that the Q^{188}K substitution may sterically hinder the SAM binding and Pro-Cys active sites from interacting with the proposed target recognition domain, which begins 34 amino acids downstream from Gln^{188}.

The seven-amino acid alterations in the M.*C*viJI sequence which did not adversely affect expression of M.*C*viJI in *E. coli*, do not disrupt predicted beta sheet or alpha helical regions. Therefore, as expected, these seven silent amino acid changes do not affect substrate binding or Pro-Cys active site orientation relative to the putative TRD.

12 Concluding remarks

The chlorella viruses are a rich source of DNA methyltransferases and DNA site-specific endonucleases which typically recognize short (2 to 4 bp) sequences. Most chlorella viruses encode three to nine different modification/restriction enzymes, making them among the most complex restriction/modification systems described to date. Many of these enzymes specify DNA methylation or cleavage reactions different from those of known bacterial enzymes. Furthermore, at least some of these viruses encode nonfunctional DNA methyltransferases (pseudogenes) and possibly nonfunctional DNA site-specific endonucleases. Sequence-comparisons between cloned nonfunctional genes and functional genes should indicate critical regions in the proteins which are required for enzyme activity and/or DNA binding.

Acknowledgments. The research in our laboratory was supported by grants from the Department of Energy and the National Institutes of Health.

Burbank, D. E., Shields, S. L., Schuster, A. M., and Van Etten, J. L. (1990) 5-Azacytidine resistant mutants of *Chlorella* virus IL-3A. Virology *176*, 311–315.

Chase, T. E., Nelson, J. A., Burbank, D. E., and Van Etten, J. L. (1989) Mutual exclusion occurs in a *Chlorella*-like green alga inoculated with two viruses. J. Gen. Virol. *70*, 1829–1836.

Chou, P. Y., and Fasman, G. D. (1978) Prediction of the secondary structure of proteins from their amino acid sequence. Adv. Enzymol. *47*, 45–148.

Doermann, A. H. (1983) Introduction to the early years of bacteriophage T4, in: Bacteriophage T4, pp. 1–7. Eds C. K. Mathews, E. M. Kutter, G. Mosig and P. B. Berget. American Society for Microbiology, Washington, D.C.

Fitzgerald, M. C., Skowron, P., Van Etten, J. L., Smith, L. M., and Mead D. A. (1992) Rapid shotgun cloning utilizing the two base recognition endonuclease *C*viJI. Nucl. Acids Res. *20*, 3753–3762.

Francki, R. I. B., Fauguet, C. M., Knudson, D. L., and Brown, F. (1991) Classification and nomenclature of viruses, in: Arch. Virol. suppl. 2. Springer-Verlag, Vienna.

Garnier, J., Osgnthorpe, D. J., and Robson, B. (1978) Analysis of the accuracy and implication of simple methods for predicting the secondary structure of globular proteins. J. Mol. Biol. *120*, 97–120.

Girton, L. E., and Van Etten, J. L. (1987) Restriction site map of the *Chlorella* virus PBCV-1 genome. Plant Mol. Biol. *9*, 247–257.

Grabherr, R., Strasser, P., and Van Etten, J. L. (1992) The DNA polymerase gene from *Chlorella* viruses PCBV-1 and NY-2A contains an intron with nuclear splicing sequences. Virology *188*, 721–731.

Hardy, L. W., Finer-Moore, J. S., Montfort, W. R., Jones, M. O., Santi, D. V., and Stroud, R. M. (1987) Atomic structure of thymidylate synthase: Target for rational drug design. Science *235*, 448–455.

Ho, D. K., Wu, J. C., Santi, D. V., and Floss, H. G. (1991) Stereochemical studies of the C-methylation of deoxycytidine catalyzed by *Hha*I methylase and the N-methylation of deoxyadenosine catalyzed by *Eco*RI methyltransferase. Archs. Biochem. Biophys. *284*, 264–269.

Jones, P. A. (1984) Gene activation by 5-azacytidine, in: DNA Methylation: Biochemistry and Biological Significance, pp. 165–187. Eds A. Razin, H. Cedar and A. D. Riggs. Springer-Verlag, Berlin.

Karplus, P. A., and Schulz, G. E. (1985) Prediction of chain flexibility in proteins. A tool for the selection of peptide antigens. Naturwissenschaften *72*, 212–213.

Kita, K., Kotani, H., Sugisaki, H., and Takanami, M. (1989) The *Fok*I restriction-modification system. I. Organization and nucleotide sequences of the restriction and modification genes. J. Biol. Chem. *264*, 5751–5756.

Klimasauskas, S., Nelson, J. L., and Roberts, R. J. (1991) The sequence specificity domain of cytosine-C5 methylases. Nucl. Acids Res. *19*, 6183–6190.

Klimasauskas, S., Timinskas, A., Menkevicius, S., Butkiene, D., Butkus, V., and Janulaitis, A. (1989) Sequence motifs characteristic of DNA [cytosine-N4] methyltransferases: Similarity to adenine and cytosine-C5 DNA-methylases. Nucl. Acids Res. *17*, 9823–9832.

Labbe, D., Holtke, H. J., and Lau, P. C. K. (1990) Cloning and characterization of two tandemly arranged DNA methyltransferase genes of *Neisseria lactamica*: An adenine-specific M.*Nla*III and a cytosine-type methyltransferase. Mol. Gen. Genet. *224*, 101–110.

Lauster, R. (1989) Evolution of type II DNA methyltransferases: A gene duplication model. J. Mol. Biol. *206*, 313–321.

Lauster, R., Trautner, T. A., and Noyer-Weidner, M. (1989) Cytosine-specific type II DNA methyltransferases: A conserved enzyme core with variable target-recognizing domains. J. Mol. Biol. *206*, 305–312.

Lunnen, K. D., Barsomian, J. M., Camp, R. R., Card, C. O., Chen, S.-Z., Croft, R., Looney, M. C., Meda, M. M., Moran, L. S., Nwankwo, D. O., Slatko, B. E., Van Cott, E. M., and Wilson, G. G. (1989) Cloning type-II restriction and modification genes. Gene *74*, 25–32.

Mann, M. B., Rao, R. N., and Smith, H. O. (1978) Cloning of restriction and modification genes in *E. coli*: The *Hha*II system from *Haemophilus haemolyticus*. Gene *3*, 97–112.

Meints, R. H., Burbank, D. E., Van Etten, J. L., and Lamport D. T. A. (1988) Properties of the *Chlorella* receptor for the virus PBCV-1. Virology *164*, 15–21.

Meints, R. H., Lee, K., Burbank, D. E., and Van Etten, J. L. (1984) Infection of a *Chlorella*-like alga with the virus, PBCV-1: Ultrastructural studies. Virology *138*, 341–346.

Meints, R. H., Lee, K., and Van Etten, J. L. (1986) Assembly site of the virus PBCV-1 in a *Chlorella*-like green alga: Ultrastructural studies. Virology *154*, 240–245.

Narva, K. E., Van Etten, J. L., Slatko, B. E., and Benner J. S., (1988) The amino acid sequence of the eukaryotic DNA [N^6-adenine]methyltransferase, M.*C*viBIII, has regions of similarity with the prokaryotic isoschizomer M.*Taq*I and other DNA [N^6-adenine] methyltransferases. Gene *74*, 253–259.

Narva, K. E., Wendell, D. L., Skrdla, M. P., and Van Etten, J. L. (1987) Molecular cloning and characterization of the gene encoding the DNA methyltransferase, M.*C*viBIII, from *Chlorella* virus NC-1A. Nucl. Acids Res. *15*, 9807–9823.

Nelson, M., and McClelland, M. (1991) Site-specific methylation: Effect on DNA modification methyltransferases and restriction endonucleases. Nucl. Acids Res. *19*, 2045–2071.

Posfai, J., Bhagwat, A. S., Posfai, G., and Roberts, R. J. (1989) Predictive motifs derived from cytosine methyltransferases. Nucl. Acids Res. *17*, 2421–2435.

Qiang, B.-Q., and Schildkraut, I. (1986) Two unique restriction endonucleases from *Neisseria lactamica*. Nucl. Acids Res. *14*, 1991–1999.

Resisser, W., Becker, B., and Klein, T. (1986) Studies on ultrastructure and host range of a *Chlorella* attacking virus. Protoplasma *135*, 162–165.

Reisser, W., Burbank, D. E., Meints, R. H., Becker, B., and Van Etten, J. L. (1991) Viruses distinguish symbiotic *Chlorella* spp. of *Paramecium bursaria*. Endocytobiosis Cell Res. *7*, 245–251.

Reisser, W., Burbank, D. E., Meints, S. M., Meints, R. H., Becker, B., and Van Etten, J. L. (1988) A comparison of viruses infecting two different *Chlorella*-like green algae. Virology *167*, 143–149.

Reisser, W., Klein, T., and Becker, B. (1988a) Studies on phycoviruses. I. On the ecology of viruses attaching Chlorellae exsymbiotic from an european strain of *Paramecium bursaria*. Arch. Hydrobiol. *111*, 575–583.

Rohozinski, J., Girton, L. E., and Van Etten, J. L. (1989) *Chlorella* viruses contain linear nonpermuted double stranded DNA genomes with covalently closed hairpin ends. Virology *168*, 363–369.

Rossmann, M. G., Liljas, A., and Banasak, L. J. (1975) in: The enzymes, vol. 11, pp. 61–102. Ed. P. D. Boyer. Academic Press, New York.

Santi, D. V., Garrett, C. E., and Barr, P. J. (1983) On the mechanism of inhibition of DNA-cytosine methyltransferases by cytosine analogs. Cell *33*, 9–10.

Santi, D. V., McHenry, C. S., and Sommer, H. (1974) Mechanism of interaction of thymidylate synthetase with 5-fluorodeoxyuridylate. Biochemistry *13*, 471–481.

Schuster, A. M., Burbank, D. E., Meister, B., Skrdla, M. P., Meints, R. H., Hattman, S., Swinton, D., and Van Etten, J. L. (1986) Characterization of viruses infecting a eukaryotic *Chlorella*-like green alga. Virology *150*, 170–177.

Schuster, A. M., Girton, L., Burbank, D. E., and Van Etten, J. L. (1986a) Infection of a *Chlorella*-like alga with the virus PBCV-1: Transcriptional studies. Virology *148*, 181–189.

Shields, S. L., Burbank, D. E., Grabherr, R., and Van Etten, J. L. (1990) Cloning and sequencing the cytosine methyltransferase gene M.*C*viJI from *Chlorella* virus IL-3A. Virology *176*, 16–24.

Skrdla, M. P., Burbank, D. E., Xia, Y., Meints, R. H., and Van Etten, J. L., (1984) Structural proteins and lipids in a virus, PBCV-1, which replicates in a *Chlorella*-like alga. Virology *135*, 308–315.

Slatko, B. E., Benner, J. S., Jager-Quinton, T., Moran, L. S., Simcox, T. G. Van Cott, E. M., and Wilson, G. G. (1987) Cloning, sequencing and expression of the *Taq*I restriction-modification system. Nucl. Acids Res. *15*, 9781–9796.

Stefan, C., Xia, Y., and Van Etten, J. L. (1991) Molecular cloning and characterization of the gene encoding the adenine methyltransferase M.*Cvi*RI from *Chlorella* virus XZ-6E. Nucl. Acids Res. *19*, 307–311.

Strasser, P., Zhang, Y., Rohozinski, J., and Van Etten, J. L. (1991) The termini of the *Chlorella* virus PBCV-1 genome are identical 2.2-kbp inverted repeats. Virology *180*, 763–769.

Sugisaki, H., Kita, K., and Takanami, M. (1989) The *Fok*I restriction-modification system. II. The presence of two domains in *Fok*I methylase responsible for modification of different DNA strands. J. Biol. Chem. *264*, 5757–5761.

Tessman, I. (1985) Genetic recombination of the DNA plant virus PBCV-1 in a *Chlorella*-like alga. Virology *145*, 319–322.

Van Etten, J. L., Burbank, D. E., Joshi, J., and Meints, R. H. (1984) DNA synthesis in a *Chlorella*-like alga following infection with the virus PBCV-1. Virology *134*, 443–449.

Van Etten, J. L., Burbank, D. E., Kuczmarski, D., and Meints, R. H. (1983) Virus infection of culturable *Chlorella*-like algae and development of a plaque assay. Science *219*, 994–996.

Van Etten, J. L., Burbank, D. E., Schuster, A. M., and Meints, R. H. (1985) Lytic viruses infecting a *Chlorella*-like alga. Virology *140*, 135–143.

Van Etten, J. L., Burbank, D. E., Xia, Y., and Meints, R. H. (1983a) Growth cycle of a virus, PBCV-1, that infects *Chlorella*-like algae. Virology *126*, 117–125.

Van Etten, J. L., Lane, L. C., and Meints, R. H. (1991) Viruses and viruslike particles of eukaryotic algae. Microbiol. Revs. *55*, 586–620.

Van Etten, J. L., Schuster, A. M., Girton, L., Burbank, D. E., Swinton, D., and Hattman, S. (1985a) DNA methylation of viruses infecting a eukaryotic *Chlorella*-like green alga. Nucl. Acids Res. *13*, 3471–3478.

Wilke, K., Rauhut, E., Noyer-Weidner, M., Lauster, R., Pawlek, B., Behrens, B., and Trautner, T. A. (1988) Sequential order of target-recognizing domains in multispecific DNA-methyltransferases. EMBO J. *7*, 2601–2609.

Wilson, G. G., and Murray, N. E. (1991) Restriction and modification systems. Annu. Rev. Genet. *25*, 585–627.

Wu, J. C., and Santi, D. V. (1987) Kinetic and catalytic mechanism of *Hha*I methyltransferase. J. Biol. Chem. *262*, 4778–4786.

Xia, Y., Burbank, D. E., Uher, L., Rabussay, D., and Van Etten, J. L. (1986) Restriction endonuclease activity induced by PBCV-1 virus infection of a *Chlorella*-like green alga. Mol. Cell. Biol. *6*, 1430–1439.

Xia, Y., Burbank, D. E., Uher, L., Rabussay, D., and Van Etten, J. L. (1987) IL-3A virus infection of a *Chlorella*-like green alga induces a DNA restriction endonuclease with novel sequence specificity. Nucl. Acids Res. *15*, 6075–6090.

Xia, Y., Burbank, D. E., and Van Etten, J. L. (1986a) Restriction endonuclease activity induced by NC-1A virus infection of a *Chlorella*-like green alga. Nucl. Acids Res. *14*, 6017–6030.

Xia, Y., Morgan, R., Schildkraut, I., and Van Etten, J. L. (1988) A site-specific single strand endonuclease activity induced by NYs-1 virus infection of a *Chlorella*-like green alga. Nucl. Acids Res. *16*, 9477–9487.

Xia, Y., Narva, K. E., and Van Etten, J. L. (1987a) The cleavage site of the *Rsa*I isoschizomer, *Cvi*II, is G/TAC. Nucl. Acids Res. *15*, 10063.

Xia, Y., and Van Etten, J. L. (1986) DNA methyltransferase induced by PBCV-1 virus infection of a *Chlorella*-like green alga. Mol. Cell. Biol. *6*, 1440–1445.

Yamada, T., Higashiyama, T., and Fukuda, T. (1991) Screening of natural waters for viruses which infect *Chlorella* cells. Appl. Environ. Microbiol. *57*, 3433–3437.

Yonker, C. R., Caldwell, K. D., Giddings, J. C., and Van Etten, J. L. (1985) Physical characterization of PBCV virus by sedimentation field flow fractionation. J. Virol. Meth. *11*, 145–160.

Zhang, Y., Burbank, D. E., and Van Etten, J. L. (1988) *Chlorella* viruses isolated in China. Appl. Environ. Microbiol. *54*, 2170–2173.

Zhang, Y., Nelson, M., Nietfeldt, J. W., Burbank, D. E., and Van Etten, J. L. (1992a) Characterization of chlorella virus PBCV-1 *Cvi*AII restriction and modification system. (submitted for publication).

Zhang, Y., Nelson, M., and Van Etten, J. L. (1992) A single amino acid change restores DNA cytosine methyltransferase activity in a cloned chlorella virus pseudogene. Nucl. Acids Res. *20*, 1637–1642.

DNA Methylation: Molecular Biology and Biological Significance
ed. by J. P. Jost & H. P. Saluz

Control of DNA methylation in fungi

Eric U. Selker

Institute of Molecular Biology, University of Oregon, Eugene, Oregon 97403, USA

Fungi represent an evolutionarily diverse set of eukaryotic organisms including, for example, mushrooms, slime molds, true molds and yeasts. Methylation of cytosines in DNA has been observed in many but not all fungi (Table 1). Considering the diversity of eukaryotic organisms with DNA methylation, it is possible that this modification serves distinct roles in higher and lower eukaryotes and even distinct roles in different fungi. The control of methylation may also differ in different eukaryotes. Why study methylation in fungi? First, studying methylation in fungi may reveal novel biological phenomena. The discovery of repeat-induced point mutation (RIP) occurred in this way (Selker et al., 1987b; Selker, 1990b). Second, fungi offer favorable systems to address general questions having to do with DNA methylation. Some fungi are extremely amenable to both genetic and molecular approaches, and may provide the easiest, fastest and cheapest means to understand the basics of the control and function of DNA methylation in eukaryotes. Since studies of DNA methylation in fungi are still in their infancy, in this chapter I will point out some attractions of studying methylation in fungi such as the ascomycete *Neurospora crassa*, the organism with which I am most familiar.

In many fungi, including *Phycomyces blakesleeanus* (Antequera et al., 1984; 1985), *Sporotrichum dimorphosporum* (Antequera et al., 1984), *Coprinus cinereus* (Zolan and Pukkila, 1985), and *Schizophyllum commune* (Buckner et al., 1988), 5-mC appears exclusively, or principally, in CpG dinucleotides. Appreciable non-CpG methylation has been observed in some fungi, however, including *Neurospora* (Selker and Stevens, 1985), *Phycomyces* (Antequera et al., 1985), and *Physarum polycephalum* (Peoples et al., 1985). Some modulation of the gross level of methylation during development has been seen in *Phycomyces* (Antequera, 1985), *Phymatotrichum* (Jupe et al., 1986), *Candida* (Russell et al., 1987a), *Neurospora* (Russell et al., 1987b), *Mucor* (Cano et al., 1988), and *Schizophyllum* (Buckner et al., 1988). Antequera et al. (1985) found evidence of a gradual increase in the fraction of methylation at CpG during development in *Phycomyces*. Information from *Physarum*

Table 1. Fungi with reported DNA methylation*

Fungus	Principal references
Physarum polycephalum	Evans and Evans, 1970; Evans et al., 1973; Peoples et al., 1985
Neurospora crassa	Bull and Wootton, 1984; Selker and Stevens, 1985; Russell et al., 1987b; Selker et al., 1987a; 1987b; Selker and Garrett, 1988; Selker et al., 1990a; Cambareri et al., 1991
Phycomyces blakesleeanus	Antequera et al., 1984; 1985
Sporotrichum dimorphosporum	Antequera et al., 1984
Coprinus cinereus	Zolan and Pukkila, 1985; 1986
Phymatotrichum omnivorum	Jupe et al., 1986
Candida albicans	Russell et al., 1987a
Schizophyllum commune	Buckner et al., 1988
Ascobolus immersus	Goyon and Faugeron, 1989; Rossignol and Picard, 1991
Mucor rouxii	Cano et al., 1988
Cochliobolus heterostrophus	Keller et al., 1991

*A number of fungi have been reported to lack DNA methylation including *Saccharomyces cerevisiae* (Proffit et al., 1984) and *Aspergillus nidulans* (Tamame et al., 1983). Although it seems likely that many fungi, including these, lack methylation, it is virtually impossible to be sure.

indicates that methylation of DNA is not always limited to one stage in the mitotic cycle (Evans and Evans, 1970) and that DNA may accumulate methylation though several cell cycles (Evans et al., 1973). It will be interesting to learn if CpG and non-CpG methylation occur at different stages in the mitotic cycle, and if they result from different mechanisms.

The finding of non-CpG methylation in fungi opened new routes to understand control of DNA methylation in eukaryotes. In animal cells, the observation that almost all 5-mC is located in CpG dinucleotides led to the elegant model of Riggs (1975) and Holliday and Pugh (1975) for "inheritance" of methylation patterns and suggested that the methylation normally observed in animal cells is largely "maintenance methylation". Results of many studies suggested that animal cells can indeed efficiently maintain methylation patterns (see Wigler et al., 1981 and Harland, 1982, for example) and imply that "*de novo* methylation" is rather rare in animal cells. We were excited to find evidence of non-CpG methylation in *Neurospora* because of the possibility that this reflected a *de novo* methylation activity amenable to study. Methylated regions stand out in *Neurospora* because most of the genome appears devoid of methylation. We tested the possibility that the scarce patches of methylation contain some sort of signal for *de novo* methylation by assaying restoration of methylation to such patches after they were passaged through *Escherichia coli* and then introduced back into the fungus. Specific methylation was observed (Selker et al., 1987a; Cambareri et al., 1991). For example, the 1.6-kb $\zeta - \eta$ region, a diverged tandem duplication, became methylated *de novo* when placed either at its native chromosomal location or at a variety of arbitrary ectopic sites. In contrast, control Neurospora sequences that are normally unmethylated in the genome did not become methylated in transformation experiments. We are testing segments of the $\zeta-\eta$ region and segments of other methylated sequences of the *Neurospora* genome to gain insight into what distinguishes sequences that are prone to *de novo* methylation from those that are not.

The discovery of RIP, a process that riddles duplicated sequences with G:C to A:T mutations at a specific stage in the sexual phase of the *Neurospora* life cycle, provided us with important clues concerning the control of DNA methylation. The mutations resulting from RIP frequently render the affected sequence a substrate for *de novo* methylation even when the sequence is engineered to be unique in the genome (Cambareri et al., 1989; 1991; Selker, 1991). The $\zeta-\eta$ region and some other methylated regions of the genome appear to be relics of RIP (Grayburn and Selker, 1989; E. Selker, P. Garrett, D. Yen-Fritz, J. Stevens, unpublished). It is important to note that the methylation resulting from RIP: (1) can result from a change in base sequence of 2% or less (M. Singer and E. Selker, unpublished), (2) extends roughly the entire length of the mutated sequence, including through short sections (e.g. 100 bp) devoid of mutations (Cambareri et al., 1989), (3) can affect most, or all, cytosines in the region (E. Selker, V. Miao, D. Yen-Fritz, unpublished), and (4) is variable among cells in a population, resulting in incomplete modification at most if not all sites (Selker and Stevens, 1985). This and other information allowed us to eliminate many possible models that might otherwise account for the distribution of DNA methylation. Nevertheless, many possibilities remain. In one class of models, methylation is the default state, i.e., methylation occurs when a sequence loses something that prevents methylation. In one such model, the collapsed chromatin model, sequences are kept unmethylated by the indirect action of sequence-specific DNA binding proteins (Selker, 1990a). In another class of models, *lack* of methylation is the default state and methylation reflects action of a positive signal. A special sequence or structural feature of DNA, for example, might invite methylation. An understanding of the control of DNA methylation in *Neurospora* awaits the results of additional experiments. What is encouraging, however, is that the tools are now in place to move forward towards elucidating the mechanism of *de novo* methylation in this organism.

While *Neurospora* offers a nearly ideal system for studying *de novo* methylation, other fungi may give us new insights into maintenance methylation. As mentioned above, 5-mC is primarily found at CpG dinucleotides in some fungi, in keeping with the classical model for maintenance methylation. However, in *Ascobolus immersus*, an ascomycete that shows evidence of maintenance methylation, methylation is not limited to CpG dinucleotides (Goyon and Faugeron, 1989). Curiously, the evidence of maintenance methylation came from discovery of MIP (methylation induced premeiotically), a process similar to RIP (Rossignol and Picard, 1991). Like RIP, MIP acts in the period between fertilization and nuclear fusion and inactivates duplicated genes. Unlike RIP, however, MIP does not appear to result in mutations. Rather, inactivation seems to be solely due to DNA methylation. Genes inactivated by MIP can revert in vegetative cells, especially after treatment with 5-azacytidine. The implication is that the methylation established during the sexual phase of the life cycle is propagated in the vegetative phase by a maintenance mechanism. Assuming that this methylation, like that in *Neurospora*, is not limited to symmetrical sites, the mechanism of this maintenance methylation must be different from that thought to operate in higher cells. Propagation of methylated stretches might reflect the action of a highly cooperative methyltransferase. Alternatively, propagation of methylation might reflect the stability of alternate structural or physiological states of chromatin. I have previously presented one such model (Selker, 1990a). A feature of maintenance methylation models that do not rely on perfect copying at symmetrically opposed cytosines is that they are compatible with spreading and/or contraction of methylation tracts, such as observed in *Coprinus cinereus* (Zolan and Pukkila, 1985; 1986). The non-traditional models also accommodate observed heterogeneity of methylation patterns (i.e. incomplete methylation), whereas the classical copying mechanism does not. Considering that changes in the size of methylation tracts and incomplete methylation have been observed in both lower and higher eukaryotes, insight into maintenance methylation in fungi may be of general interest.

Besides their possible use in illuminating control of methylation, fungi promise to lead us to a fuller understanding of the function(s) of DNA methylation. In particular, it seems likely that characterization of mutants defective in DNA methylation will be fruitful. We have embarked on this approach.

Acknowledgments. I thank Jette Foss, Vivian Miao, Chris Roberts, Jeanne Selker and Mike Singer for comments on the manuscript and past and present members of my laboratory for stimulating discussions. The work from my laboratory was supported by National Science Foundation Grant DCB 9019036 and Public Health Services grant GM-35690 from the National Institutes of Health. The article was written during the tenure of an Established Investigatorship of the American Heart Association.

Antequera, F., Tamame, M., Villanueva, J. R., and Santos, T. (1984) DNA methylation in the fungi. J. Biol. Chem. *259*, 8033–8036.

Antequera, F., Tamame, M., Villanueva, J. R., and Santos, T. (1985) Developmental modulation of DNA methylation in the fungus *Phycomyces blakesleeanus*. Nucl. Acids Res. *13*, 6545–6558.

Buckner, B., Novotny, C. P., and Ullrich, R. C. (1988) Developmental regulation of the methylation of the ribosomal DNA in the basidiomycete fungus *Schizophyllum commune*. Curr. Genet. *14*, 105–111.

Bull, J., and Wootton, J. C. (1984) Heavily methylated amplified DNA in transformants of *Neurospora crassa*. Nature *310*, 701–704.

Cambareri, E. B., Singer, M. J., and Selker, E. U. (1991) Recurrence of repeat-induced point mutation (RIP) in *Neurospora crassa*. Genetics *127*, 699–710.

Cano, C., Herrera-Estrella, L., and Ruiz-Herrera, J. (1988) DNA methylation and polyamines in regulation of development of the fungus *Mucor rouxii*. J. Bacteriol. *170*, 5946–5948.

Evans, H. H., Evans, T. E., and Littman, S. (1973) Methylation of parental and progeny DNA strands in *Physarum polycephalum*. J. Mol. Biol. *74*, 563–572.

Evans, H. H., and Evans, T. E. (1970) Methylation of the deoxyribonucleic acid of *Physarum polycephalum* at various periods during the mitotic cycle. J. Biol. Chem. *245*, 6436–6441.

Goyon, C., and Faugeron, G. (1989) Targeted transformation of *Ascobolus immersus* and *de novo* methylation of the resulting duplicated DNA sequences. Mol. Cell. Biol. *9*, 2818–2827.

Grayburn, W. S., and Selker, E. U. (1989) A natural case of RIP: Degeneration of DNA sequence in an ancestral tandem duplication. Mol. Cell. Biol. *9*, 4416–4421.

Harland, R. (1982) Inheritance of DNA methylation in microinjected eggs of *Xenopus laevis*. Proc. Natl. Acad. Sci. USA *79*, 2323–2327.

Holliday, R., and Pugh, J. E. (1975) DNA modification mechanisms and gene activity during development. Science *187*, 226–232.

Jupe, E. R., Magill, J. M., and Magill, C. W. (1986) Stage-specific DNA methylation in a fungal plant pathogen. J. Bacteriol. *165*, 420–423.

Keller, N. P., Bergstrom, G. C., and Yoder, O. C. (1991) Mitotic stability of transforming DNA is determined by its chromosomal configuration in the fungus *Cochliobolus heterostrophus*. Curr. Genet. *19*, 227–233.

Peoples, V., Whittaker, P., Pearston, D., and Hardman, N. (1985) Structural organization of a hypermethylated nuclear DNA component in *Physarum polycephalum*. J. Gen. Microbiol. *131*, 1157–1165.

Proffitt, J. H., Davie, J. R., Swinton, D., and Hattman, S. (1984) 5-methylcytosine is not detectable in *Saccharomyces cerevisiae* DNA. Mol. Cell. Biol. *4*, 985–988.

Riggs, A. (1975) X inactivation, differentiation, and DNA methylation. Cytogenet. Cell Genet. *14*, 9–25.

Rossignol, J.-L., and Picard, M. (1991) *Ascobolus immersus* and *Podospora anserina*: sex recombination, silencing, and death, in: More gene manipulations in fungi, pp. 266–290. Eds J. W. Bennett and L. Lasure. Academic Press, New York.

Russell, P. J., Rodland, K. D., Rachlin, E. M., and McCloskey, J. A. (1987a) Different levels of DNA methylation in yeast and mycelial forms of *Candida albicans*. J. Bacteriol. *169*, 4393–4395.

Russell, P. J., Welsch, J. A., Rachlin, E. M., and McCloskey, J. A. (1987b) Differential DNA methylation during the vegetative life cycle of *Neurospora crassa*. J. Bacteriol. *169*, 2902–2905.

Selker, E. U. (1990a) DNA methylation and chromatin structure: A view from below. TIBS *15*, 103–107.

Selker, E. U. (1990b) Premeiotic instability of repeated sequences in *Neurospora crassa*. Ann. Rev. Gen. *24*, 579–613.

Selker, E. U. (1991) Repeat-induction point mutation (RIP) and DNA methylation, in: More gene manipulations in fungi, pp. 258–265. Eds J. W. Bennett and L. Lasure. Academic Press, New York.

Selker, E. U., Cambareri, E. B., Jensen, B. C., and Haack, K. R. (1987b) Rearrangement of duplicated DNA in specialized cells of Neurospora. Cell *51*, 741–752.

Selker, E. U., and Garrett, P. W. (1988) DNA sequence duplications trigger gene inactivation in *Neurospora crassa*. Proc. Natl. Acad. Sci. USA *85*, 6870–6874.

Selker, E. U., Jensen, B. C., and Richardson, G. A. (1987a) A portable signal causing faithful DNA methylation *de novo* in Neurospora crassa. Science *238*, 48–53.

Selker, E. U., and Stevens, J. N. (1985) Methylation at asymmetric sites is associated with numerous transition mutations. Proc. Natl. Acad. Sci. USA *82*, 8114–8118.

Tamame, M., Antequera, F., Villanueva, J. R., and Santos, T. (1983) High-frequency conversion to a "fluffy" developmental phenotype in *Aspergillus* spp. by 5-azacytidine treatment: evidence for involvement of a single nuclear gene. Mol. Cell. Biol. *3*, 2287–2297.

Wigler, M., Levy, D., and Perucho, M. (1981) The somatic replication of DNA methylation. Cell *24*, 33–40.

Zolan, M. I., and Pukkila, P. J. (1985) DNA methylation in *Coprinus cinereus*, in: Molecular Genetics of Filamentous Fungi, pp. 333–344. Ed. W. Timberlake. A.R. Liss, Inc., New York.

Zolan, M. I., and Pukkila, P. J. (1986) Inheritance of DNA methylation in *Coprinus cinereus*. Mol. Cell. Biol. *6*, 195–200.

DNA Methylation: Molecular Biology and Biological Significance
ed. by J. P. Jost & H. P. Saluz

The role of DNA methylation in the regulation of plant gene expression

E. J. Finnegan, R. I. S. Brettell and E. S. Dennis

CSIRO, Division of Plant Industry, PO Box 1600, Canberra, ACT, 2601 Australia

1 Introduction

The most common modification of DNA in plant cells is methylation of cytosine residues at carbon 5. In contrast to mammalian cells in which 3–8% of cytosine residues are methylated (Shapiro, 1975), in plants up to 30% of cytosine residues are modified (Adams and Burdon, 1985). There is considerable inter-species variation in the level of cytosine methylation, ranging from 4.6% in *Arabidopsis thaliana* (Leutwiler et al., 1984), which has a small genome with relatively little highly repeated DNA, to 33% in rye, *Secale cereale* (Thomas and Sherratt, 1956). The difference in the extent of methylation, between plants and animals, is due to two factors. The CG dinucleotide, which is methylated to about the same extent (70–80%) in plants and animals, occurs more frequently in plant DNA. In addition plant DNA is methylated in CNG trinucleotides where N can be any base (Gruenbaum et al., 1981). The CG dinucleotide has symmetrical cytosine residues in the two DNA strands and it has been observed that, when modified, both cytosines are methylated (Bird, 1978; Cedar et al., 1979). It is this symmetry that allows the pattern of methylation to be maintained through DNA replication. Newly replicated DNA is hemimethylated with methylation at specific sites on the parental strand. Methylation of the new strand at the unmethylated cytosine of a hemimethylated CG dinucleotide restores the original pattern of methylation; this is termed maintenance methylation (Holliday and Pugh, 1975; Riggs, 1975; Razin and Riggs, 1980). The CNG motif also has strand symmetry suggesting that methylation of this motif will probably be maintained through replication by the same mechanism.

The frequency of 5-mC, the conservation of methylation patterns through cycles of DNA replication and the ability of 5-mC to interfere with normal protein DNA interactions led to the hypothesis that cytosine methylation might influence gene expression (Holliday and Pugh, 1975; Riggs, 1975). Consequently interest has focused on obtaining direct evidence for the role of methylation in gene expression. The role of methylation in regulation of expression in animal cells has now been well documented (reviewed recently by Razin and Riggs, 1991). We discuss briefly some methods for investigating methylation then general features of methylation with particular reference to plant DNA methylation. We then review the evidence suggesting that DNA methylation affects gene expression in plants and consider a possible role for methylation in plant gene regulation.

2 Methods for studying DNA methylation

While changes in the gross level of methylation can be measured by determining the frequency of 5-mC in total DNA, this will not detect changes in methylation at specific sites. The methylation status of specific cytosine residues can be assessed by using methylation-sensitive restriction enzymes to cleave DNA isolated from a number of different tissues. Any change in the size of fragments generated can be detected by Southern hybridization using the appropriate probe. The methylation status of some motifs, for example CC(A/T)GG, can be examined in detail by using isoschizomers ApyI, EcoRII and BstNI, all of which recognize this sequence but differ in their ability to cleave methylated DNA. While BstNI cleaves DNA whether or not the internal cytosine residue is methylated, EcoRII cleaves only when this cytosine is unmethylated. ApyI cleaves only when the internal cytosine is methylated and the external cytosine is unmethylated ($C^{m}C$(A/T)GG). Similarly, the sequence CCGG is cleaved by the enzyme HpaII only when both cytosines are unmethylated. The enzyme MspI will cleave DNA at this site either when the internal cytosine is methylated ($C^{m}CGG$) or if neither cytosine is modified (Kessler et al., 1985). Unless the presence of a restriction site can be confirmed by cleavage with a methylation-insensitive enzyme the failure of a methylation-sensitive restriction enzyme to cleave DNA may reflect loss of the site by mutation rather than methylation. The methylation status of sites that do not lie within a restriction enzyme recognition site will not be examined in an analysis of this type.

Methylation at cytosines which are not part of a restriction site can be detected by genomic sequencing which distinguishes between modified and unmodified cytosines because of the nature of the chemical reactions (Church and Gilbert, 1984). Although this technique is technically more difficult than an analysis based on restriction enzymes and Southern hybridization it allows all cytosine residues in each strand to be examined. A refinement of this technique, based on PCR amplification of fragments generated by chemical cleavage of DNA at unmodified cytosine residues, has both simplified and increased the sensitivity of the original method (Saluz and Jost, 1989; Pfeifer et al., 1989). Another modification of genomic sequencing provides a means of positive identification of methylated cytosine residues. Under the appropriate conditions, treatment of genomic DNA with bisulphite converts unmethylated cytosines to uracil, while 5-mC is not affected. The DNA is then amplified by PCR and sequenced. Uracil will be replaced by thymine during PCR amplification, therefore the remaining cytosine residues in the sequence obtained must correspond to a methylated cytosine in the starting DNA (Frommer et al., 1992) (see also chapter by Saluz and Jost, pp. 11–26 of this book).

DNA demethylation can be induced by treatment with 5-azacytidine (5-azaC) or 5-azadeoxycytidine (5-azadeoxyC). This is incorporated into DNA during replication or repair and is believed to inhibit methylation by binding the DNA methylase (reviewed in Taylor et al., 1984). Although demethylation of DNA is not the only result of 5-azaC treatment the reactivation of genes following exposure to 5-azaC has frequently been ascribed to demethylation of specific cytosines in the vicinity of the gene in question. At least in some cases a correlation between reactivation and demethylation of a gene has been observed (Groudine et al., 1981; Niwa and Sugahara, 1981; Weber et al., 1990).

3 Properties of DNA methylation

The distribution of 5-mC in the plant genome is far from random. For example unique sequences which comprise approximately 38% of the cotton genome contain only 4% of methylated cytosine residues (Guseinov et al., 1975). Furthermore, even though approximately 80% of cytosines in CG dinucleotides are modified, a number of plants have a fraction of unmethylated DNA with closely spaced sites for restriction enzymes that include CG in their recognition sequence, for example HpaII (CCGG). Digestion of genomic DNA with HpaII released a prominent low molecular weight fraction that could be detected, after end labelling, on agarose gels. These fragments ranged in size from ~25 bp to ~250 bp and resembled the HpaII tiny fragments (HTF) characteristic of CG islands in vertebrate genomes (Cooper et al., 1983). The islands described in vertebrates are CG rich, are about 500–2,000 bp in length and are often associated with the 5′ region of housekeeping genes sometimes extending into the first few exons. They are typically unmethylated in a wide range of tissues whether or not the associated gene is transcribed (Bird, 1986). Closer scrutiny of the known sequence for several maize genes revealed that there are clustered CG motifs associated with the *A1* (dihydroflavonol 4-reductase), *Adh1* (alcohol dehydrogenase) and *Sh1* (sucrose synthase) genes. Investigation of the methylation status of these CG clusters showed that sites within the coding region of the *A1* gene were unmethylated while sites about 2 kb upstream were methylated. The promoter of this gene does not contain sites for the methylation-sensitive restriction enzymes used in this analysis and as a result the methylation state was not examined. The 5′ CG rich regions of both *Adh1* and *Sh1* were not methylated nor were the neighbouring coding sequences (Antequera and Bird, 1988). The CG cluster associated with the maize *Adh1* gene remained unmethylated in leaf tissue where it is not transcribed (Nick et al., 1986). Thus the CG clusters associated with plant genes closely resemble CG islands in vertebrates. Islands may have arisen independently in plants and vertebrates because there is no known common ancestor, with CG islands, linking the two groups (Antequera and Bird, 1988).

Methylation of cytosine residues alters the structure of chromatin; in general DNaseI sensitive chromatin has a lower level of DNA methylation than that of bulk chromatin. This correlation is protein dependent, because when naked DNA was isolated from chromatin, sensitivity to DNaseI was no longer correlated with the level of methylation (Klaas and Amasino, 1989). Nucleosomes isolated from chromatin after digestion with methylation sensitive restriction enzymes had a lower level of histone H1 and histones H3 and H4 were more highly acetylated than those in bulk chromatin (Lewis and Bird, 1991). Acetylation of histones H3 and H4 has been associated with active chromatin (Hebbes et al., 1988) suggesting that hypomethylated DNA in chromatin was active. Furthermore, DNaseI-sensitive DNA was highly represented in cDNA corresponding to mRNA isolated from the same tissue. This indicates that DNaseI-sensitive DNA was both hypomethylated and actively transcribed (Spiker et al., 1983). Although hypomethylation is associated with active transcription it is not known whether methylation prevents transcription or whether DNA becomes methylated because it is not transcribed.

Methylation may affect the temporal regulation of DNA replication during the cell cycle (reviewed in Lewis and Bird, 1991). A large proportion of plant DNA is highly repeated; this repetitive DNA is normally heavily methylated (Guseinov et al., 1975), transcriptionally inactive and is replicated late in S phase.

Another consequence of cytosine methylation is to increase the frequency of spontaneous mutation. Deamination of cytosine produces uracil (Shapiro and Klein, 1966) which can be detected and removed from DNA by the enzyme uracil-specific N-glycosylase (Lindahl, 1974). In contrast, 5-mC converts to thymidine following deamination; as thymidine is a normal component of DNA the mismatched T-G pair could be repaired either way in the absence of a directional repair system. The observation that the frequency of CG dinucleotides in the genomes of both plants and animals is lower than predicted from the known base composition and that the TG frequency is higher than expected could be explained by the conversion of CG into TG, following deamination of 5-mC (Adams and Eason, 1984; McClelland, 1983; Bird, 1986). In contrast, in plants the frequency of CNG trinucleotides does not appear to be reduced, as might be expected, by deaminational loss in methylated DNA. It has been proposed that a mechanism to correct for deamination of 5-mC evolved prior to the occurrence of methylation in CNG trinucleotides (McCelland, 1983). The relationship between cytosine methylation and the low frequency of CG dinucleotides does not hold for ribosomal genes which are heavily methylated and yet have normal levels of CG (Adams and Eason, 1984; Delseny et al., 1984; Kaufman et al., 1987). This suggests that there may be a mechanism to correct deamination loss of methylated cytosines in at least some cell compartments (Hepburn et al., 1987). Alternatively, as deamination occurs only on single stranded DNA, methylated cytosine residues that lie in GC rich DNA may be protected from deamination by the stability of the duplex (Adams and Eason, 1984).

The primary effect of methylating cytosine residues is to modify the interaction of DNA with a wide range of DNA binding proteins. It has been shown that replacement of cytosine with 5-mC in the alternating polynucleotide poly(dG.dC)·poly(dG.dC) facilitated and stabilized the transition from the right-handed B form to the left-handed Z DNA (Behe and Felsenfeld, 1981; Behe et al., 1981; Klysik et al., 1983). The Z form of DNA seems to prevent formation of nucleosomes in contrast to B DNA which is stabilized by formation of nucleosome core particles (Nickol et al., 1982). Thus methylation can affect both secondary structure of DNA and tertiary structure of the chromatin. Interaction of DNA with transcription factors is important in the regulation of gene expression. DNA methylation can affect the accessibility of promoter sequences to transcription factors either directly, if methylation is within the recognition sequence, or indirectly, as a consequence of chromatin structure.

DNA methylation is catalyzed by the enzyme DNA methyltransferase or DNA methylase, which transfers methyl groups to cytosine residues from S-adenosyl methionine (Adams and Burdon, 1983). It appears that at least in mouse, both maintenance and *de novo* methylation (methylation of previously unmodified cytosine residues) are catalyzed by the same enzyme (Bestor and Ingram, 1983), but it is not known what governs the specificity of the *de novo* reaction. Methyltransferases have been purified from pea shoots, wheat embryos and cultured rice cells (Yesufu et al., 1991; Theiss et al., 1987; Giordano et al., 1991). A single enzyme activity with a molecular weight of approximately 160 kDa was isolated from pea (Adams et al., 1990); this is comparable in size to the methylases described in mammalian cells (Bestor et al., 1988; Adams et al., 1990). The enzymes purified from both wheat and rice were of lower molecular weight (50–55 kDa) and, because the pea methylase was unstable, it has been suggested that these may be the breakdown products of *in vivo* or *in vitro* proteolysis (Theiss et al., 1987; Giordano et al., 1991; Adams et al., 1990). There was some evidence for *de novo* methylation by the pea methylase (Adams et al., 1990) but hemimethylated DNA, the substrate for the maintenance reaction, was the preferred template for the enzymes from both pea and rice (Adams et al., 1990; Giordano et al., 1991). The pea methylase modified cytosine residues in all four dinucleotides (CA, CC, CG and CT), which suggests that it is capable of both CG and CNG methylation. The level of CG methylation was higher than that of the other dinucleotides; this may reflect methylation of CG in addition to CGG (Yesufu et al., 1991). Thus the purified pea methylase has the capacity to methylate cytosines in different sequence motifs and can act on both hemimethylated and unmodified DNA. However the methylase activity described in pea may not be the product of a single gene. Genomic Southern hybridization using a homologous probe coding for a portion of a methylase gene indicated that there are three copies of this gene in *Arabidopsis* (Finnegan, unpublished); it is not known whether these genes differ in target specificity, or whether they catalyse either *de novo* or maintenance methylation reactions.

4 The effect of *in vitro* DNA methylation on gene expression and binding of plant proteins

Direct evidence for the inhibitory effect of cytosine methylation on transcription in plants comes from experiments in which *in vitro* methylated DNA was introduced into protoplasts of either tobacco or petunia (Hershkovitz et al., 1989; Weber and Graessmann, 1989). These experiments indicate that to inhibit gene expression the level of methylation must exceed a certain threshold and/or cover specific sites. Methylation of a chimaeric 35SGUS reporter gene, with the bacterial HpaII methylase, prior to introduction into tobacco protoplasts did not inactivate the GUS gene. There are only nine HpaII recognition sites in the GUS coding sequence and no sites within the putative DNA binding domains of the 35S promoter. Therefore, it is likely that sites essential for gene activity were not methylated (Prat et al., 1989; Benfey et al., 1989). In contrast, when every cytosine of the coding strand was replaced with 5-methylcytosine the 35SGUS gene was not expressed in tobacco protoplasts. Methylation of the non-coding strand also resulted in strong inhibition of GUS activity (Weber and Graessmann, 1989). Similar results were obtained when either the coding or non-coding strand of the 35SCAT gene was methylated *in vitro* then introduced into petunia protoplasts (Hershkovitz et al., 1989). In plants, where methylation of cytosine residues is normally limited to CG dinucleotides and

CNG trinucleotides, DNA with one strand in which every cytosine is methylated is a non-physiological template. While it is hard to assess what effect comprehensive hemimethylation will have on secondary structure, it is clear that it does inhibit gene expression in protoplasts. It is not known whether complete cytosine methylation of one strand is sufficient to prevent transcription or whether the plant DNA methylase methylates the other strand of this hemimethylated plasmid DNA. In some cases hemimethylation does not inhibit gene expression. For example, in avian cells, activation of the vitellogenin II gene precedes demethylation of the non-coding strand suggesting that physiological hemimethylation of this strand does not inhibit transcription of this gene (Saluz et al., 1986).

The activity and methylation status of the integrated 35SGUS gene was examined in stably transformed tobacco callus derived from protoplasts transfected with DNA methylated, *in vitro*, on one strand. Methylation-sensitive restriction enzymes were used to digest DNA isolated from a number of independently transformed callus cultures. In every culture examined, the 35SGUS gene was hypermethylated indicating that methylation of the non-methylated strand had occurred prior to the first round of replication. While methylation introduced *in vitro* at CG and CNG motifs was retained, methylation at other sites was not inherited (Weber et al., 1990). In general, hypermethylation of the 35SGUS gene was associated with suppression of GUS activity suggesting that methylation at all CG and CNG motifs inhibits transcription. In the majority of lines the 35SGUS gene was reactivated following treatment of calli or protoplasts with the demethylating agent, 5-azaC. The gene remained active and demethylated in plants regenerated from callus that had been treated with 5-azaC (Weber et al., 1990). In this analysis the methylation status of the regions essential for 35S promoter activity was not examined, however it has been assumed that the overall level of methylation of the 35SGUS gene is indicative of methylation of the promoter. These observations support the hypothesis that transgene expression was inhibited by methylation and that demethylation induced by 5-azaC treatment was sufficient to activate the 35SGUS gene.

Methylation can interfere with the interaction between specific DNA sequences and DNA binding proteins that recognize the sequence in its unmodified form. For example, the restriction-modification systems of prokaryotes are based on site specific methylation of endogenous DNA which prevents cleavage by the corresponding restriction endonuclease (McClelland, 1981; McClelland, 1982; Kessler et al., 1985; Nelson and McClelland, 1987). Similarly, binding of promoter or enhancer specific transcription factors from mammalian cells can be inhibited by *in vitro* CG methylation of the sequences to which they bind (Comb and Goodman, 1990; Iguchi-Ariga and Schaffner, 1989; Kovesdi et al., 1987; Shen and Whitlock, 1989). The protein encoded by the most abundant transcript of the maize transposable element *Spm* (*En1*), *tnpA*, has been shown to bind to a 12-bp sequence motif that is repeated several times in the subterminal repetitive region of *Spm*. Binding of this protein was inhibited by methylation of cytosine residues present in the CCG trinucleotide on one strand and the corresponding CG dinucleotide of the complementary strand (Gierl et al., 1988). The exact role of the *tnpA* protein in *Spm* excision has not been elucidated; however inhibition of *tnpA*-DNA binding by methylation of its target site is consistent with the transpositional inactivity of methylated *Spm* elements (Banks *et al*., 1988; see section 8.2). Another plant DNA binding protein, CG-1, has been identified in tobacco nuclear extracts. This protein binds a sequence containing a CACGTG motif in the promoter of *Antirrhinum* chalcone synthase; binding was inhibited by *in vitro* methylation of the target sequence (Staiger et al., 1989). Other sequence specific DNA binding proteins have been identified in nuclear extracts of pea, wheat, cauliflower and soybean using gel retardation assays with probes containing sequences from the upstream regulatory regions of some plant genes (Inamdar et al., 1991). Methylation of a single cytosine residue in the target sequences used to identify these proteins was sufficient to reduce their binding *in vitro* (Inamdar et al., 1991). As *in vitro* methylation of the target sequence can directly inhibit the binding of a number of plant DNA binding proteins, it is likely that *in vivo* binding will also be inhibited by methylation.

Proteins that bind specifically to methylated DNA have been found in nuclear extracts of mammalian and avian cells. DNA binding proteins that have both methylation-specific and sequence-specific binding have been described in human and chick cells (Wang et al., 1986; Saluz et al., 1988). In contrast a methyl-binding protein, MeCP, identified in mouse cells binds preferentially to DNA containing 15 or more 5-mCG dinucleotides but has no other sequence specificity (Meehan et al., 1989). Binding of MeCP inhibited transcription from methylated promoters both *in vitro* and *in vivo* (Boyes and Bird, 1991). A DNA binding protein, MDBP, with binding properties similar to MeCP has been identified in pea nuclear extracts (Zhang et al., 1989). Thus DNA methylation could indirectly inhibit transcription as binding of methylation specific binding proteins could block the binding of transcription factors.

Given that DNA methylation has the capacity, at least *in vitro*, to disrupt the normal interactions between DNA and proteins that recognize a sequence in an unmethylated state, methylation of plant DNA could affect gene expression *in vivo*. Furthermore, the correlation between inhibition of *tnpA* binding by *in vitro* methylation, and the *in vivo* loss of *Spm* transposition with increased methylation of the element suggests that methylation could indeed play a role in gene regulation in plants.

5 DNA methylation of genes regulated in a tissue-specific or inducible manner

5.1 Tissue-specific genes

The methylation status of a number of genes and multigene families exhibiting a tissue-specific pattern of expression has been examined, both in the tissue in which the genes were expressed and in other plant organs where the genes were quiescent. When assayed by digestion with methylation-sensitive restriction enzymes there was a clear correlation between active transcription and the demethylation of specific cytosine residues, for some genes or gene families. For other genes there was no detectable change in methylation pattern when DNA isolated from tissues in which the gene was expressed was compared to DNA from tissues in which the gene was not expressed. Even within a particular class of genes in a single plant, for example the seed storage proteins of *Phaseolus vulgaris*, the occurrence of tissue-specific demethylation was not universal (Riggs and Chrispeels, 1990). Although for some genes there was a strong correlation between demethylation of a particular residue and gene expression, neither the causal relationship nor the mechanism for specific demethylation are clear. Methylation of specific cytosine residues may inactivate the gene by preventing the binding of transcription factors; in this event demethylation would precede activation. Alternatively, a sequence may be unmethylated because the binding of transcription factors prevents *de novo* or maintenance methylation. Thus, transcription itself may give specificity to demethylation.

5.1.1 Seed storage proteins. The seed storage protein genes of maize are organized in clusters on different chromosomes and constitute multigene families; transcription of these genes is specifically activated in developing endosperm (Soave and Salamini, 1984). An analysis using methylation-sensitive restriction enzymes showed that sequences encoding zein and glutelin gene families were hypomethylated in endosperm compared to tissues in which these genes were not expressed (Bianchi and Viotti, 1988; Spena et al., 1983). A fine mapping analysis of the zein zE/19.31 cluster defined a number of sites, in both coding and noncoding sequences, that were fully methylated in shoot, root, unfertilized ear and embryo but were either unmethylated or partially methylated in endosperm (Bianchi and Viotti, 1988). Mutation of *opaque-2*, a transcription factor regulating heavy chain zein synthesis in maize, drastically reduced the levels of M1 heavy chain zein transcripts (Langridge et al., 1982). Methylation of the M1 zein family in endosperm of an *opaque-2* mutant line was identical to wildtype maize, indicating that even in the absence of the *opaque-2* protein, there was endosperm specific demethylation of these genes (Bianchi and Viotti, 1988). This observation may point to demethylation prior to transcription rather than as a consequence of active gene expression. Consistent with this notion is the observation that endosperm-specific demethylation was established by 8 days after pollination (dap) when zein and glutelin transcripts were barely detectable. By 20 dap zein transcripts make up about 15% of the cellular mRNA. Thus demethylation occurred prior to the massive DNA amplification between 10–12 to 14–18 dap, that marked the end of cell division and differentiation of this tissue (Bianchi and Viotti, 1988).

Tissue-specific demethylation has also been correlated with endosperm-specific expression of the high molecular weight glutenins (HMW glutenins) of hexaploid wheat (Flavell and O'Dell, 1990). Restriction analysis of DNA isolated from seed and leaf tissue from the same plants showed specific demethylation of sites in seed DNA. The location of these sites with respect to the coding region of the HMW glutenin genes has not been defined.

Tissue-specific demethylation of sites 5′ to one of the seed storage proteins was observed in cotyledon tissue of *Phaseolus vulgaris*. Even though the sites at which demethylation occurs have been located with respect to the coding region of a gene that is expressed only in cotyledon tissue, no functional role in transcription has been ascribed to these sites. During embryogenesis in *P. vulgaris*, the cotyledons synthesize the seed storage proteins phytohaemagglutinin (PHA) and β-phaseolin. Phytohaemagglutinin is made up of two subunits PHA-E and PHA-L, which in this species are encoded by closely linked genes, *lec*1 and *lec*2, respectively (Hoffman and Donaldson, 1985). Tissue-specific demethylation was observed at two restriction sites (MspI and PstI) which lie in the intergenic region of the PHA locus, 600 bp and 400 bp respectively, upsteam of the *lec*1 gene (Riggs and Chrispeels, 1990). In contrast some sites in the region spanning the PHA locus were unmethylated in all tissues while other sites were uniformly methylated. Demethylation in cotyledon DNA was very specific; cleavage of the CCGG motif by MspI, but not by the isoschizomer HpaII, indicated that only the first cytosine of this motif had been demethylated. Although cotyledon-specific nuclear proteins bound the region 5′ to the *lec*2 gene, no factors bound the region, upstream to *lec*1, that spanned the sites that were demethylated in this tissue (Riggs et al., 1989).

No cotyledon-specific demethylation of β-phaseolin was seen when this locus was examined using a large number of methylation-sensitive restriction enzymes even though it shows the same tissue-specific pattern of expression as PHA (Riggs and Chrispeels, 1990). This example is not unique amongst seed storage proteins of legumes. No differences in methylation around the soybean glycinin genes, G1 and G2, nor the genes encoding the less prevalent seed proteins, lectin and Kunitz trypsin inhibitor, were detected in DNA isolated from developing seeds, post-germination cotyledons or leaf tissue (Walling et al., 1986). However, seed protein genes were undermethylated when compared to average soybean DNA (Walling et al., 1986).

Thus tissue-specific demethylation is correlated with the expression of some but not all seed storage protein genes, but no functional role has been assigned to the residues demethylated. Specific demethylation of most residues may not have been detected because of the limitations inherent in restriction enzyme based assays. If methylation is important in tissue-specific regulation of these genes it is not the only factor involved.

5.1.2 Photosynthetic proteins. There is some evidence for tissue-specific, developmentally regulated changes in methylation of the genes encoding photosynthetic enzymes of C4 plants. However, as the results obtained by different groups working in this area are conflicting the true picture is hard to discern.

In mature leaves of C4 plants two distinct cell types, bundle-sheath (BS) and mesophyll (M), are required for photosynthesis. The photosynthetic enzymes function in one cell type or the other and, as a consequence, the photosynthetic reactions are partitioned between BS and M cells (reviewed by Hatch, 1987). In M cells atmospheric CO_2 is assimilated into C4 dicarboxylic acids; this requires the enzymes phosphoenolpyruvate carboxylase (PEPCase), malate dehydrogenase (MDH) and pyruvate phosphate dikinase (PPdK). Decarboxylation and the subsequent refixation of carbon occurs in BS cells where malic enzyme (ME), ribulose bisphosphate carboxylase (RuBPCase) and several other C3 cycle enzymes are located (reviewed by Hatch, 1987). Accumulation of PEPCase, MDH and ME proteins is both light dependent and cell specific and is regulated at the level of transcription (Langdale et al., 1987; Langdale et al., 1988). In contrast, PPdK accumulation is rarely cell specific although its photosynthetic activity is limited to M cells. RuBPCase accumulates specifically in BS cells in true leaves of C4 plants but in C3 photosynthetic organs (for example, maize husk leaves) and in etiolated leaves, the enzyme accumulates in both cell types (Langdale et al., 1988).

Expression and methylation status of the genes encoding the RuBPCase large and small subunits (LS, and SS), PEPCase and PPdK proteins were investigated in isolated bundle sheath and mesophyll cells during the transition from etiolated to fully green tissue (Ngernprasirtsiri et al., 1989). The dimorphic chloroplasts that are characteristic of BS and M cells differentiate during greening. Cell-specific expression of these C4 photosynthetic enzymes and increased DNA methylation paralleled chloroplast differentiation. The observed changes in methylation, in the vicinity of each gene examined, were cell type specific such that increased methylation correlated with the loss of transcript. For example, methylation in the vicinity of the genes encoding large and small subunits of RuBPCase was detected in M cells of greening and fully green tissue but not in BS cells where the enzyme accumulates (Ngernprasirtsiri et al., 1989). This suggests that selective methylation could be involved in differential gene expression.

In direct contrast Langdale et al. (1991) did not detect any change in methylation that correlated with tissue-specific expression of either the large or small subunit of RuBPCase. This conflict is difficult to reconcile particularly as the same combination of probe and restriction enzymes was used to investigate methylation of the LS locus (Langdale et al., 1991; Ngernprasirtsiri et al., 1989). This difference could arise because different lines of maize were used in the two studies. However if true, this implies that while methylation may play a part in tissue-specific regulation of the large subunit, it is not universally important in different lines of maize.

Ngernprasirtsiri et al. (1989) observed a correlation between increased methylation and decreased expression of the PEPCase, PPdK, LS and SS genes during differentiation. The enzymes PEPCase, PPdK and the SS of RuBPCase are encoded by multigene families and, from the data presented, it is not possible to locate the methylated site(s) with respect to the coding sequence or to a member of the gene family. Therefore it is not possible to ascribe transcriptional inactivity of a particular member of these multigene families to methylation.

A more convincing example in which specific demethylation has been correlated with tissue-specific, light-regulated expression is provided by the PEPCase gene of maize (Langdale et al., 1991). In this case demethylation of a single PvuII site located approximately 3.3 kb upstream of the transcription start of the PEPCase gene correlated with accumulation of PEPCase mRNA during greening. Demethylation at this site was restricted to mesophyll cells, in which the PEPCase gene was expressed. The PvuII recognition sequence is CAGCTG suggesting that methylation of CNG trinucleotides, specifically ^{m}CTG, as well as CG dinucleotides may be important in gene regulation. As yet this site has no known functional role. A BstNI/EcoRII site located about 2.1 kb upstream was methylated in all tissues examined while most other sites assayed were unmethylated. Close examination of the sequence of the PEPCase gene indicates that there is a CG island spanning bases −160 to +800, in which potential methylation sites tested were unmethylated. The presence of a CG island in a tightly regulated (rather than a housekeeping) gene is not unique in plants as the anaerobically inducible maize *Adh1* gene is located within a CG island (see below).

5.2 Inducible genes

Some genes which have no detectable change in methylation during the transition from an inactive to an active state do undergo an alteration in chromatin structure. This can be monitored by changes in DNaseI hypersensitivity and/or the appearance of inducible DNaseI sensitive sites.

The multigene family encoding the small subunit of the ribulose-1,5 bisphosphate carboxylase/oxygenase (*rbc*S genes) of *Pisum sativum* are expressed in a tissue-specific manner in response to light (Coruzzi et al., 1984; Fluhr and Chua, 1986; Fluhr et al., 1986). DNaseI hypersensitivity of the *rbc*S genes was investigated in tissue in which these genes are inactive (roots), potentially active (leaves grown in the dark) and active (leaves grown in the light). A number of DNaseI-hypersensitive sites mapped to both the 5′ upstream regions and the coding region in all tissues. An additional site, centred around −190 was seen only in light-induced leaves (Görz et al., 1988). The DNaseI sensitivity of constitutive sites differed between tissues and, in general, sensitivity was increased in light grown material. No changes in methylation were detected within the promoter region of the *rbc*S genes during the transition from an inactive to an active state and all sites tested were unmethylated (Görz et al., 1988). The sensitivity to DNaseI seen in all tissues tested may reflect the potential to be activated rather than the actual activity state (Görz et al., 1988).

The maize *Adh1* gene shows both developmentally regulated tissue-specific expression and is induced in response to a range of environmental stimuli (reviewed by Freeling and Bennett, 1985). There are a number of constitutive DNaseI-sensitive sites in the maize *Adh1* gene. Under anaerobic stress, which activates the gene, additional DNaseI-sensitive sites appeared and the constitutive sites showed increased sensitivity (Paul et al., 1987). The methylation status of the *Adh1* gene was examined by genomic sequencing of mature leaf DNA. The 5′ end of the coding region and at least 600 bp upstream from the translation start were unmethylated, despite being comparatively rich in potential sites for methylation (Nick et al., 1986). This region includes all sequences required for anaerobic induction and pollen specific expression of *Adh1* (Walker et al., 1987; Shimamoto, pers. comm.). Two methylated cytosines were detected about 950 bp upstream of the translation start; the region between 650–900 bp was not sequenced but as the cytosines flanking this region were uniformly unmethylated, it was predicted that this region was also unmethylated (Nick et al., 1986). As *Adh1* is never expressed in mature leaf tissue, it appears that methylation of the DNA 950 bp 5′ to the translation start is neither a requirement for nor a consequence of the gene inactivity. A second analysis, using methylation sensitive restriction enzymes, showed that DNA flanking the *Adh1* gene was unmethylated for about 1 kb upstream of the first exon (Antequera and Bird, 1988; Walbot and Warren, 1990). It is not known whether the two methylated cytosines seen in leaf DNA (Nick et al., 1986) remained methylated in tissues in which this gene is expressed because they do not lie within the recognition sites of the restriction enzymes used.

A CG island spans the 5′ upstream region of the *Adh1* gene and extends into the coding region (Hepburn et al., 1987). The first methylated cytosine 5′ to the gene (Nick et al., 1986) lies outside the region defined as an island (Hepburn et al., 1987) while the CG island itself is unmethylated, even in mature leaves which never express the gene. Perhaps *Adh1* is maintained in a potentially active (unmethylated?) state because it can be expressed in response to different environmental stimuli in the absence of DNA replication. As there is no clear evidence for active demethylation in plants this may be an essential feature of genes, the expression of which is not confined to a discrete developmental stage, and which must therefore be accessible in the absence of replication.

5.3 Changes in methylation in response to environmental stimuli

Although for some genes, for example maize *Adh1* and pea *rbc*S, induction by enviromental stimuli is not correlated with a change in methylation status, changes in methylation do occur in response to environmental factors. The gross level of DNA methylation decreased in cotton plants infected with the fungus *Verticillium dahliae* (Guseinov et al., 1975). Methylation of unique sequences was not significantly affected while cytosine methylation in highly repeated DNA was decreased about three fold.

Another process in which changes in methylation have been implicated is vernalization which is the promotion of flowering by cold (Lang, 1965). Several characteristics of vernalization are consistent with the response being mediated by changes in DNA modification. For example, non-dividing cells do not respond to cold treatment indicating that at least cell division and probably DNA replication are essential for cells to be receptive to cold (Wellensiek, 1964). The response is not mediated by a diffusible product because vernalization occurs only when the cells giving rise to the floral meristem are subjected to cold (Schwabe, 1954; Metzger, 1988). Each generation requires vernalization, that is, the response to cold is not transmitted to the progeny, suggesting that the programme is reset during gamete formation (Lang, 1965). The vernalization response of late flowing ecotypes of *Arabidopsis* was mimicked by 5-azaC treatment of imbibed seed, although in some cases the reduction in the time to flower was not as great as that caused by extended cold treatment (Burn pers. comm.). This suggests that demethylation of specific genes involved in floral initiation may be the cause of early flowering following 5-azaC treatment and in vernalization. This hypothesis can only be tested definitively when the genes involved have been identified and cloned.

5.4 Conclusions

A major limitation of most of the analyses discussed is that the detection of cytosine methylation is restricted to those residues that lie within the recognition sequence of a restriction enzyme that is sensitive to methylation. Specific changes in methylation will not be detected if the residue concerned does not lie within the recognition sites of the enzymes used. Genomic sequencing or modifications of this technique allow a comprehensive survey of 5-mC residues to be made. Another limitation is the extent of the analysis; it is not known how far upstream of the translation start should be included or indeed whether sites within the coding region, the introns or the 3′ end of a gene should be examined. Tissue-specific demethylation was detected 3.3 kb upstream of the maize PEPCase gene (Langdale et al., 1991). While this distance seems large it should be noted that there is a tissue-specific enhancer 4.8 kb upstream of the rat PEPCase gene (Ip et al., 1990). Extensive genomic sequencing of tightly regulated genes is needed to determine whether the alteration in methylation patterns which has been correlated with changes in activity of some genes, is a widespread phenomenon.

While the observation of tissue-specific changes in methylation is in itself interesting, the relevance of these changes to gene regulation cannot be assessed unless the residues involved are put into a functional context. While methylation of a single cytosine residue within the target site has been shown to inhibit interaction with the corresponding DNA binding protein (Inamdar et al., 1991), methylation of residues outside protein binding sites may have no effect on gene activity. Even if a methylated residue lies within the recognition sequence for a transcription factor it may not affect gene expression. For example, binding of the mammalian transcription factor Sp1 and activation of transcription by this protein were insensitive to methylation of its target site (Höller et al., 1988). Thus to ascribe a role in gene regulation to a particular methylated cytosine residue, both its location with respect to protein binding sites and the effect of methylation on binding of transcription factors must be determined.

Methylation of DNA can alter the chromatin structure (reviewed by Lewis and Bird, 1991) and so it is possible that methylation of residues not involved in DNA-protein interactions may indirectly affect gene expression in this way. In many of the examples cited the location of the sites that show altered methylation is unknown and as a result the significance of these changes for gene regulation remains unclear.

6 Methylation and ribosomal gene expression

The nucleolar organizer (NOR) of plants is a complex locus consisting of many tandemly repeated copies (from 100–10,000) of a DNA unit encoding the ribosomal RNA. A single ribosomal repeat unit is 7–15 kb and encodes 25S-28S, 18S and 5.8S RNA as well as sequences for control, initiation and termination of transcription which are located in a non-transcribed spacer (NTS) region (Long and Dawid, 1980). Although the ribosomal repeat units may differ in length (Gerlach and Bedbrook, 1979; Oono and Sugiura, 1980), sequence (Maggini and Carmona, 1981) and DNA modification (Gerlach and Bedbrook, 1979), in general there is more similarity between repeat units within a locus than between repeats at different nucleolar organizer loci of the same plant. Not all ribosomal genes are transcribed, and the viability of aneuploid wheat lines that have major deletions of rDNA indicates that there are excess rRNA genes (Flavell et al., 1988). The condensed heterochromatin adjacent to the NOR contains inactive rRNA genes while the active or potentially active genes are within the nucleolus (Appels, 1983; reviewed by Flavell et al., 1986). Hypomethylation of rDNA has been correlated with active rDNA units (Flavell et al., 1988; Kaufman et al., 1987); at best these findings are correlative because it is not possible to associate any one gene with its product. However, as the hypomethylated sites lie within the NTS upstream of the transcription start this suggests that the promoter is unmethylated in transcribed genes.

The relative activity of an array of ribosomal RNA genes can be assessed by determining the relative volume of the nucleolus formed at the locus, and by the relative length of the constriction at the NOR during metaphase (Martini and Flavell, 1985). The NTS region of wheat rDNA contains a number of repeated units, called the A repeats which are 135 or 136 bp in length, and span the region −180 to −1964 with respect to the transcription start (Appels and Dvorak, 1982; Barker et al., 1988). In wheat the rDNA repeat unit of the dominant (more active) locus had more A repeats in the NTS suggesting that the number of these repeats is important in determining gene activity (Flavell et al., 1990). Similarly, the dominant rDNA locus contained the majority of genes which were unmethylated at both the HpaII site located at −165 and HhaI sites in the A repeats. Even though all A repeats contain HhaI sites only a few of these sites were unmethylated and these were frequently in the A repeats located towards the 3′ end of the NTS, adjacent to the unmethylated HpaII site. Other HhaI and HpaII sites within the NTS remained methylated (Flavell et al., 1988; Flavell et al., 1990). As all A repeat elements are essentially identical the variation in methylation status implies that it is not the primary sequence that determines which cytosine residues will be methylated.

Introduction of the NOR from *Aegilops umbellulata* into a Chinese Spring wheat background suppressed expression from the wheat NOR (Martini et al., 1982; Martini and Flavell, 1985). A concomitant increase in methylation was observed at the wheat NOR loci while many of the *A. umbellulata* rDNA genes remained unmethylated in the intergenic region (Flavell et al., 1988). The changes in methylation observed at any one of the NOR loci in different genetic backgrounds, for example in the presence of dominant NOR or in aneuploid lines, suggests that methylation of the intergenic regions in rRNA genes at a locus is highly regulated and inversely correlated with the relative activity of the locus (Flavell et al., 1990).

A model that relates ribosomal gene activity to the presence of specific unmethylated cytosine residues within the NTS of active genes has been proposed (Flavell, 1989). In this model, binding of a protein to specific sites within the NTS region prevents methylation at these sites. If this protein is limiting then only a subset of genes will be unmethylated at these sites. Genes with more copies of the A repeat compete more successfully for the limiting protein and thus are more likely to be unmethylated. Genes with unmethylated sites will be included in the nucleolus while their methylated counterparts become condensed into heterochromatin and are unavailable for transcription. It is envisaged that competition for transcription complexes within the nucleolus may also occur (Flavell, 1989).

Demethylation of HpaII sites within the NTS region of pea rDNA has been correlated with a two-fold increase in rRNA synthesis induced by light (Gallagher and Ellis, 1982; Kaufman et al., 1987). Two major length classes of rDNA (L and S) are found in pea; the two forms differ in the number of 100 bp repeats contained within the NTS (Watson et al., 1987) and are located at different NORs (Ellis et al., 1984). The L variant showed the same pattern of DNaseI-hypersensitive sites in the repeated region of the NTS in both light and dark grown material. In contrast, in light grown material, the S variant had additional hypersensitive sites in the distal half of the NTS (Kaufman et al., 1987). Furthermore, the location of DNaseI-hypersensitive sites within the NTS corresponded closely to HpaII/MspI sites in this region (Kaufman et al., 1987). The light induced change in chromatin structure, indicated by the appearance of new DNaseI-sensitive sites, correlated with demethylation of HpaII sites within the NTS. These changes were primarily in the S variant (Kaufman et al., 1987). Changes in sensitivity to DNaseI and HpaII of the S variant may reflect an inducible component of gene expression or an increase in the number of these genes that are transcribed under these conditions (Kaufman et al., 1987). Changes in both methylation status and DNaseI sensitivity of the S variant were observed during organ development suggesting that rRNA genes at this locus are also developmentally regulated (Watson et al., 1987).

Differential methylation of rDNA repeats has also been observed in pumpkin (Siegel and Kolacz, 1983), radish (Delseny et al., 1984) and flax (Ellis et al., 1983) where the site involved mapped 5′ to the transcription start. These observations suggest that differential methylation of rDNA genes may be ubiquitous in higher plants. While there is good evidence that the activity of rDNA is inversely correlated to the level of methylation, this may be a secondary event reflecting activity rather than the primary factor controlling gene expression.

7 Methylation and gene expression in organelles

Methylation of cytosine residues in CNG trinucleotides has been observed both in mitochondria isolated from a cytoplasmic male-sterile line of maize and its cytoplasmic revertant (Zabala, pers comm.). However, as little is known about methylation of plant mitochondrial DNA, the discussion will be confined to the role of methylation in controlling gene expression in plastids, the only organelle unique to plants.

Plastids exist in different forms, each with a distinct function, the best known of which is the chloroplast of green tissue. In response to light, chloroplasts differentiate from proplastids present in meristematic cells (Mohr, 1984; Tobin and Silverthorne, 1985). Both proplastids and chloroplasts can differentiate into specialized plastids which have other functions in nonphotosynthetic organs, for example, chromoplasts in flowers and ripe fruits or amyloplasts in tubers and roots (Kirk and Tilney-Bassett, 1978; Thomson and Whatley, 1980). The complete sequence of the plastid genome of both tobacco and the liverwort, *Marchantia polymorpha*, has been determined (Shinozaki et al., 1986; Ohyama et al., 1986). Many plastid encoded genes are required for photosynthesis, which occurs only in the chloroplasts of green tissue, and much interest has centred on the regulation of these genes. While the protein products of many photosynthetic genes are not normally present in chromoplasts or amyloplasts, it is not clear whether gene expression is regulated at the transcriptional level (Mullet and Klein, 1987; Kobayashi et al., 1990) or whether post-transcriptional controls, such as differential mRNA stability (Mullet and Klein, 1987; Deng and Gruissem, 1987), potential to be translated (Klein and Mullet, 1986; Kuntz et al., 1989), or protein stability, are involved (Gruissem, 1989). Therefore the question of methylation and its role in differential gene expression in plastids of higher plants remains a controversial one.

Methylation of plastid DNA was first observed in chloroplasts of *Chlamydomonas* where maternal chloroplast DNA was extensively methylated soon after the formation of the zygote (Burton et al., 1979). Paternal chloroplast DNA which remained unmethylated was degraded after zygote formation (Burton et al., 1979). It has been suggested that selective methylation is part of a modification-restriction system, analogous to bacterial restriction-modification systems, that is the molecular basis for maternal inheritance of chloroplast genes in *Chlamydomonas* (Sager et al., 1984).

Methylation of DNA from nonphotosynthetic plastids has been observed in cultured sycamore cells (Ngernprasirtsiri et al., 1988a), ripe tomato fruits (Ngernprasirtsiri et al., 1988b; Kobayashi et al., 1990) and in the differentiated chloroplasts of maize mesophyll cells (Ngernprasirtsiri et al., 1989). HPLC analysis of amyloplast DNA from a white heterotrophic cell line of sycamore indicated that amyloplast DNA contained a range of methylated bases in addition to approximately 5% 5-mC. Chloroplast DNA of an autotrophic line, a mutant of the original heterotrophic line, had a low level of 3-mC but no 5-mC. Methylation of EcoRII sites in the vicinity of DNA encoding genes for the large subunit of RuBPCase (*rbc*L), the α, β and ε subunits of chloroplast coupling factor1 (*atp*A, *atp*B, *atp*E), apoprotein of P700 (*psa*A) and ribosomal protein S4 (*rps*4) was detected in Southern analyses of amyloplast DNA (Ngernprasirtsiri et al., 1988a). Following treatment with 5-azaC the level of 5-mC decreased and the *rbc*L gene was reactivated (Ngernprasirtsiri and Akazawa, 1990). This is consistent with the hypothesis that methylation of the *rbc*L gene inhibited transcription.

Methylation was not detected in amyloplasts of spinach roots (Deng and Gruissem, 1988), nor in chromoplasts of *Capsicum annuum* (Gounaris and Price quoted by van Grinsven and Kool, 1988) or daffodil (Hansmann, 1987). Unlike many differentiated plastids which retain the ability to re-differentiate into chloroplasts (Thomson and Whatley, 1980), amyloplasts of the heterotrophic sycamore line were unable to differentiate into functional chloroplasts upon illumination of the tissue (Ngernprasirtsiri et al., 1988a). Therefore it is possible that methylation of DNA in these plastids represents a terminally differentiated state.

A comparative analysis of chloroplasts in green tomato fruit and chromoplasts of red fruits indicated that only chromoplast DNA contained methylated bases. Methylation of plastid DNA increased during differentiation of chloroplasts into chromoplasts. A comparison of HpaII and MspI digests indicated that HpaII sites were methylated in chromoplast DNA in the vicinity of a number of genes examined (*rbc*L, *atp*B, *atp*E, *psa*A and *rsp*4) (Ngernprasirtsiri et al., 1988b). These genes were not expressed in the chromoplast but transcripts of *rbc*L, *atp*B and *atp*E were present at high levels in chloroplasts of green fruits. As no hybridization data were presented for green fruit chloroplasts it is impossible to ascertain whether modification at these sites was specific to chromoplasts. Nuclear run-on experiments indicated that regulation of gene expression was at the level of transcription. Furthermore, modification of the plastid DNA was implicated in gene control because genes not expressed in chromoplasts were not transcribed from chromoplast DNA *in vitro*. In contrast, genes expressed in chloroplasts were transcribed from chloroplast DNA in the same *in vitro* transcription assay (Kobayashi et al., 1990).

In direct contrast, Marano and Carrillo (1991) found no evidence for DNA methylation in ripe tomato fruit chromoplasts, neither by HPLC analysis of hydrolysed DNA nor in Southern analyses using methylation-sensitive enzymes. Different cultivars (*cv.*) of tomato were used by the two groups, and it is worth noting that the chloroplast genome of *cv.* Firstmore, used in the study of Ngernprasirtsiri et al. (1988b) is approximately 15 kb smaller than the plastid genome of *cv.* Platense which was studied by Marano and Carrillo (1991). Restriction and isoschizomer analyses of four other cultivars gave results similar to those obtained with *cv.* Platense (Marano and Carrillo, 1991). It seems unlikely that different cultivars of tomato have evolved different mechanisms for gene regulation and further work is needed to clarify this enigma.

Methylation in the vicinity of the gene encoding the large subunit (*rbc*L) of RuBPCase enzyme has also been observed in mesophyll (M) cells of green maize leaves (Ngernprasirtsiri et al., 1989). No methylation was detected in chloroplasts expressing this gene, for example, in M cells from etiolated seedlings or bundle sheath (BS) cells at any developmental stage. It appears that methylation at this site occurred during the functional differentiation of M cell chloroplasts which do not have RuBPCase activity (Ngernprasirtsiri et al., 1989). This observation was not confirmed by other workers (Langdale et al., 1991).

Methylation has also been implicated in regulation of plastid DNA replication (Gauly and Kossel, 1989). Methylation was detected, in some copies of the plastid genome, at two restriction enzyme sites located in the ribosomal spacer region of maize plastid DNA. Methylation was tissue-specific; both sites were partially methylated in embryo while one site showed methylation in some plastids of root and endosperm. Neither site was methylated in leaf chloroplast DNA (Gauly and Kossel, 1989). The role of this tissue-specific methylation is unknown, but as an origin of replication has been located in the corresponding spacer region of rDNA of pea chloroplasts (Meeker et al., 1988), it is possible that methylation of this region is involved in plastid DNA replication rather than gene expression.

Methylation at some sites in plastid DNA is clearly not involved in gene regulation. In a recent study, pea leaf chloroplast DNA was shown to contain 0.08% 5-mC (Ohta et al., 1991). A detailed analysis of restriction enzyme recognition sites at known locations within 3 genes, *rbc*L, *atp*B and *psb*DC (the genes for the D2 and 43 kDa proteins of photosystem II), using methylation sensitive and insensitive isoschizomers indicated that a number of sites were partially methylated in both chloroplast and amyloplast DNA. These sites were more frequently methylated in chloroplast DNA (Ohta et al., 1991). These genes are highly transcribed in chloroplasts, but not in amyloplasts suggesting that methylation at these sites is not important for regulation of gene expression.

While methylation of plastid DNA has been observed by some workers its role in gene regulation remains controversial. In some studies cited the methylated sites have not been mapped with respect to the probe used (Ngernprasirtsiri et al., 1988a; Ngernprasirtsiri et al., 1988b; Kobayashi et al., 1990) and so it is a blind assumption to ascribe the inactivity of a particular gene to methylation. Furthermore methylation at some sites has no role in regulating gene expression (Ohta et al., 1991). The comparatively high level of methylation seen in amyloplasts of a heterotrophic sycamore cell line (Ngernprasirtsiri et al., 1988a) and chromoplasts of ripe tomato fruits (Ngernprasirtsiri et al., 1988b; Kobayashi et al., 1990) may be characteristic of a terminally differentiated state. If this is the case it appears that not all tomato cultivars inactivate chromoplast genes in the same way (Marano and Carrillo, 1991). Plastids that retain the ability to redifferentiate to other plastid types may reflect this competence by the low level of DNA methylation.

8 Methylation and transposable elements

Some of the best evidence that methylation of plant DNA has a regulatory role comes from studies on maize transposable elements. The maize controlling elements are arguably the best characterized of all plant transposable elements. This is in part because some of the genes expressed in the maize kernel can be inactivated with little or no effect on viability and yet can cause dramatic phenotypic changes. These genes include those affecting biosynthesis of starch and storage proteins of the endosperm and those involved in production of the red and purple anthocyanin pigments of the aleurone. The maize cob is, therefore, almost ideal for studying the activity of a transposable element by observing changes in phenotype caused by insertion or excision of an element. An understanding of transposition at the molecular level has been facilitated by the cloning and sequencing of several transposable elements and the identification of transcripts that they encode. Changes in the activity states of three maize transposable elements, *Ac*, *Spm* and *Mu* have been studied indirectly using pigment and storage protein genes of maize kernels as visual markers for activity.

8.1 Activator (Ac)

Reversible inactivation of the *Ac* transposable element, inserted in the *waxy* locus (*wx-m7*) was first described by McClintock (1964). In the active state the *Ac* element was able to excise and to transactivate excision of non-autonomous *Ds* elements inserted at other locations in the genome. In the inactive state, *Ac* did not excise nor was it able to catalyse excision of *Ds* elements. A molecular analysis of the *Ac* element at *wx-m7* showed that *Ac* in the inactive state was the same size and had the same general restriction pattern as the active *Ac*, with the important exception that an internal PvuII fragment was present only in the active *Ac*. Cloning of the inactive form of *Ac wx-m7* showed that both PvuII sites were still present suggesting that the PvuII sites in genomic DNA were protected by methylation (Dellaporta and Chomet, 1985). EcoRII and SstII sites were also shown to be methylated only in the inactive *Ac* (Chomet et al., 1987).

Similar results, correlating inactivation with methylation, were obtained when another copy of *Ac*, that in *wx-m9*, inserted at a different site in the *waxy* gene, was characterized (Schwartz and Dennis, 1986). The fully active *Ac*, the inactive *Ac* (*wx-m9 Ds-cy*) as well as several active revertant alleles were cloned and analysed by restriction endonuclease mapping. No differences between the alleles were detected with restriction enzymes that cut frequently indicating that the elements did not differ in size by insertion or deletion of more than approximately 10 bp (Schwartz and Dennis, 1986). When genomic DNA was cut with methylation sensitive enzymes and probed with internal fragments of the *Ac* element, there was a clear correlation between *Ac* activity and hypomethylation of the region upstream of the initiation codon of the transposase protein. The *Ac* element in *wx-m9* contains 13 HpaII sites, three of which are in the 3′ untranslated region of the transposase gene while the remaining 10 lie in the 5′ region. In the active element, all 5′ HpaII sites were unmethylated, while in the inactive *wx-m9 Ds-cy* all these sites were methylated. In revertants to full activity, the majority of 5′ HpaII sites were unmethylated (Schwartz and Dennis, 1986). There are two PvuII sites in the *Ac* element located in the coding region and in intron IV of the gene, respectively. These were unmethylated in the active *Ac*, methylated in the inactive allele and remained methylated in the revertants. Thus there was a correlation between activity of *Ac* and lack of methylation in the 5′ region but not in the coding region of the gene. The presence of the single *Ac* transcript has been correlated both with activity and with lack of methylation of the 5′ region (Kunze et al., 1988; Dennis and Brettell, 1990). This suggests that transposition of *Ac* is regulated indirectly by methylation which inhibits transcription of the transposase.

Although reversible inactivation of the *Ac* element has been correlated with methylation of transposable element sequences, methylation did not extend into the flanking *waxy* DNA (Schwartz and Dennis, 1986). The inactive *wx-m9 Ds-cy* reverted spontaneously to full activity at a frequency of approximately 1% and the reversion frequency was enhanced by the presence of other active *Ac* elements in the genome (Schwartz and Dennis, 1986). There is evidence to suggest that in addition to the presence of an active element another, unlinked gene is involved in reactivation of cryptic *Ac* elements (Schwartz, 1989).

8.2 Suppressor-mutator (Spm)

Methylation has also been correlated with lack of activity of a second transposable element system in maize. The *Spm* element can exist in one of three heritable forms, stably active, stably inactive or cryptic and programmable, which can be distinguished by the extent of methylation of cytosine residues surrounding the start of the major transcript (Banks et al., 1988). Hypomethylation of the upstream control region (UCR) was characteristic of active *Spm* elements; cryptic or inactive elements were extensively methylated in both the UCR and in the GC rich first exon known as the downstream control region (DCR). The programmable form of the element showed variable levels of methylation, but was characteristically methylated at only a subset of sites within the DCR (Banks et al., 1988; Fedoroff et al., 1989). The presence of an active element promoted the demethylation and transcriptional activation of an inactive programmable element but not of an inactive cryptic element (Fedoroff et al., 1989; Fedoroff, 1989).

As observed for the *Ac* element, the maize DNA flanking an *Spm* element that was inactivated by methylation, remained unmethylated (Fedoroff and Banks, 1988). Some cycling *Spm* elements were inactivated more frequently in progeny derived from upper ears and tassels than in progeny from lower ears and tillers. This suggests that the general increase in DNA methylation that occurs during maize development is reflected by increased methylation of the programmable form of *Spm* (Banks and Fedoroff, 1989).

8.3 Mutator (Mu)

A third example in which transposable element activity has been correlated with methylation is the Robertson's mutator system (*Mu*). There are at least nine classes of *Mu* elements each with similar ~220-bp terminal inverted repeats but with different internal sequences (reviewed by Chomet et al., 1991). None of these elements is capable of autonomous transposition. *Mu* transposition is controlled by an 11-kb element, *MuR*1, which presumably encodes the transposase that catalyses transposition of all other classes of *mutator* element (Chomet et al., 1991).

The non-autonomous *Mu* elements have been studied in great detail particularly the element in the *bz2-mu* allele (Walbot et al., 1988). Loss of variegation in the bronze phenotype indicated loss of *mutator* excision activity, which correlated with increased DNA methylation of the *mutator* element (Chandler and Walbot, 1986; Bennetzen et al., 1988). Direct estimates of the levels of cytosine methylation showed that on average only 41% of cytosine residues in mobile *Mu*1 elements were methylated; this compares with bulk DNA in which 65% of cytosine residues were methylated, and inactive *Mu* lines where on average 72% of cytosine residues in the element were methylated (Walbot pers. comm.).

In general methylation was restricted to the *Mu* element and the flanking DNA was not methylated (Bennetzen et al., 1988). In one *bz2-mu* line six sites proximal to the *Mu* element insertion and one site distal showed partial methylation (Walbot, pers. comm.). Loss of excision of the *Mu*1 element at the *bz2-mu* locus was associated with coordinate methylation of all *Mu*1 elements scattered throughout the genome (Bennetzen, 1987; Chandler and Walbot, 1986; Martienssen et al., 1990). Simultaneous loss of *MuR*1 (the putative autonomous *Mu* element) and methylation of the non autonomous *Mu*1 elements suggests that the loss of *MuR*1 is associated with *Mu*1 methylation (Chomet et al., 1991). Methylation of *Mu*1 elements may be inhibited by the binding, to these elements, of the protein product(s) of *MuR*1.

Insertion of a *Mu*1 element close to the 5′ untranslated region of the *hcf* gene, in the mutant *hcf*106, prevented normal assembly of the photosynthetic electron transport complex. The mutant *hcf*106 phenotype was suppressed when *Mu*1 elements were inactive; that is when *Mu*1 was methylated, plants with the *hcf*106 mutation were identical to the wild type. Suppression of the mutant phenotype may be due to the lack of transposase bound to *Mu*1 allowing transcription of the *hcf* gene. The mutant phenotype reappeared in progeny of plants crossed to a line containing active *Mu*1 elements suggesting that inactive *Mu*1 elements were reactivated in the presence of active elements, and by implication, of *MuR*1 (Martienssen et al., 1990).

8.4 Activation of transposable elements by passage through tissue culture

Passage through tissue culture was first implicated in the reactivation of transposable elements when *Ac* activity was detected in maize plants regenerated from cultures derived from explants that contained no active *Ac* elements (Peschke et al., 1987). Activation of *Spm* has also been observed following passage through tissue culture (Peschke and Phillips, 1991). A link between the activation of previously silent transposable elements following tissue culture and demethylation of transposable element sequences has now been demonstrated in one case.

Maize plants were regenerated from tissue cultures initiated from embryos that carried an inactive *Ac* in *wx-m9 Ds-cy*. These plants showed a high frequency of *Ac* activity which was transmitted to their progeny. There was a good correlation between *Ac* activity and demethylation of HpaII sites in the region 5′ to the *Ac* transposase gene. These sites remained methylated in regenerated plants in which the element was inactive (Dennis and Brettell, 1990; Brettell and Dennis, 1991). In a parallel study with tissue-culture-derived lines of maize, Peschke et al. (1991) also identified new *Ac* activity. However they were unable to demonstrate a clear correlation between *Ac* activity and hypomethylation of sequences homologous to *Ac*. This analysis was complicated by the presence of multiple copies of sequences homologous to *Ac*.

8.5 Conclusions

Although so far the discussion has been limited to *Ac*, *Spm* and *Mu* the correlation between methylation of transposable elements and loss of activity is not restricted to maize elements. When the *Tam*3 element from *Antirrhinum* was introduced into tobacco there was an initial burst of transposition, after which the element was inactive. The inactive element was methylated at its ends while tobacco sequences flanking the element remained unmethylated. Imbibition of tobacco seeds in the presence of 5-azaC resulted in some reactivation, suggesting that methylation was inhibiting the transposition of *Tam*3 in transgenic tobacco (Martin et al., 1989).

Thus the correlation between loss of activity and methylation of the element is a feature common to all plant transposable elements that have been examined in detail. Inactivation of autonomous elements, for example *Ac* and *Spm*, is marked by methylation of the 5′ end of the transposase gene. Other sites may be methylated in both the active and inactive state (Schwartz and Dennis, 1986; Banks et al., 1988). Both the inactive programmable *Spm* and the inactive state of cycling *Ac* elements, for example *Ac* in *wx-m9 Ds-cy*, show an increased frequency of activation in the presence of an active form of the corresponding element (Schwartz and Dennis, 1986; Fedoroff and Banks, 1988; Fedoroff, 1989). Co-ordinate methylation of the non-autonomous *Mu* elements is concomitant with the loss of the autonomous element *MuR*1. Conversely *Mu* activation and loss of methylation occurred in the presence of *MuR*1 (Chomet et al., 1991). It is possible that for each of these elements the transposase can prevent methylation by binding to the element DNA.

In general, the pattern of methylation and activity state of transposable elements is stably inherited through meiosis, with only a low frequency of reversion to an active state. Reversion is more likely to occur if an element passes through the female rather than the male gamete (Fedoroff and Banks, 1988; Schwartz, 1988). Passage through tissue culture frequently results in reactivation of an inactive transposable element (Peschke et al., 1987; Dennis and Brettell, 1990; Brettell and Dennis, 1991; Peschke et al., 1991; Peschke and Phillips, 1991). This may result from overall changes in methylation that occur during culture rather than from specific demethylation of transposable elements (see Section 9).

The observation that methylation associated with an inactive transposable element does not extend into flanking plant DNA is a surprising one (Schwartz and Dennis, 1986; Fedoroff and Banks, 1988; Bennetzen et al., 1988). One possibility is that the plant methylase specifically recognizes and methylates foreign DNA. However, as it is likely that transposable elements have been an integral part of the plant genome for a long time, it is difficult to perceive them as foreign. Transposable elements are frequently present as dispersed repeats, with up to 50–100 copies per cell (Sutton et al., 1984). The high copy number or the disruption of plant chromatin structure caused by transposition may be factors that identify transposable elements as targets for methylation. It is possible that plants have evolved a mechanism to control the transpositional activity of transposable elements to maintain the relative stability of their genome.

9 Tissue culture and methylation

Alterations in phenotype have been recognized in plants exposed to extreme environmental stresses (Walbot and Cullis, 1985) and are commonly observed in plants regenerated from cultured cells and tissues. The genetic variation found among plants regenerated from tissue culture, termed somaclonal variation (Larkin and Scowcroft, 1981), has been described for a wide range of species. It has become evident that there is a diversity of genetic and epigenetic changes which may not be the result of a single causal mechanism. In some cases the alterations are clearly due to a major genomic trauma that for example produces an abnormal configuration of chromosomes (D'Amato, 1985), whereas in other cases more subtle changes have been identified (Larkin, 1987). There are an increasing number of instances in which both stable and reversible genetic alterations can be linked to changes in methylation.

Plants regenerated from tissue culture commonly have reduced stature and abnormal morphology. This is particularly true of determinate species such as maize which have a strict developmental programme. While many of the abnormalities will disappear in the following generation (Brettell et al., 1980; Novak et al., 1988) others show stable inheritance. Rice dwarf 'mutants', regenerated from callus, persisted through eight self-pollinated generations, but disappeared when outcrossed to control plants or to the progeny of normal plants derived from tissue culture (Oono, 1985). This type of genetic variation is reminiscent to that observed following the treatment of germinating rice seeds with demethylating agents, 5-azaC or 5-azadeoxyC (Sano et al., 1990). The reduced level of methylation of genomic DNA and the reduction of stem length were correlated, were heritable, and a causal link was suggested.

The hormone composition of tissue culture medium can affect the level of DNA methylation in cultured cells. Methylation in carrot cell cultures undergoing somatic embryogenesis increased when cultures were exposed to high levels of auxin (Lo Schiavo et al., 1989). In contrast, on medium containing cytokinin, proliferative carrot cultures showed reduced levels of methylation, whereas in cells cultured without cytokinin there was a tendency towards increased methylation at certain sites (Arnholdt-Schmitt et al., 1991). Decreased methylation was also observed in satellite DNA of melon callus tissue compared to hypercotyl (Grisvard, 1985) and in 5S RNA genes in soybean suspension cultures compared to those genes in intact plants (Quemada et al., 1987). Re-methylation of 5S DNA occurred infrequently in some cell lines.

Growth in culture does not always result in detectable changes in methylation. Morrish and Vasil (1989) found no significant differences in levels of methylation between DNA isolated from callus tissue (embryogenic and non-embryogenic) and DNA from leaf tissue of *Pennisetum purpureum*.

Methylation has been implicated in the loss of betacyanin production by cultured red beet cells. The ratio of auxin to cytokinin in the culture medium was an important factor controlling the biosynthesis of betaxanthin and betacyanin pigments in cell lines of red beet exhibiting a range of colour phenotypes from white/green through yellow and orange to red and deep violet. Treatment of yellow cells with 5-azaC stimulated the appearance of betacyanin pigments suggesting that methylation may inhibit production of these pigments. However, 5-azaC induction was blocked by either 2,4-dichlorophenoxyacetic acid, an auxin, or 3-methoxybenzamide, an inhibitor of poly(ADP-ribose)polymerase, suggesting that demethylation alone may not be sufficient to induce betacyanin biosynthesis (Girod and Zryd, 1991).

Attempts to correlate changes in methylation, resulting from passage through tissue culture, with the appearance of phenotypic abnormalities seen in plants regenerated from culture have been largely unsuccessful. In one study approximately 16% of plants regenerated from maize tissue cultures showed altered patterns in HpaII digestion suggesting that gross changes in methylation had occurred. Changes in the methylation status of the *waxy* gene were demonstrated but these were not correlated with phenotypic abnormalities (Brown, 1989). In a second study, changes in methylation were detected in single copy sequences in 34% of families, progeny of plants regenerated from embryo-derived cultures of the maize inbred line A188. No changes were detected in 15 on cultured control plants (Phillips et al., 1991).

Restriction fragment length polymorphisms for methylation sensitive enzymes were also detected in DNA isolated from progeny of rice plants regenerated from embryogenic callus (Müller et al., 1990) or protoplast cultures (Brown et al., 1990). However a similar level of polymorphism was detected when DNA was digested with either methylation-sensitive or methylation-insensitive enzymes. This suggests that alterations associated with methylation were not the major cause of these polymorphisms (Brown et al., 1990).

In conclusion, alterations in methylation patterns can occur in cultured plant cells, with a general tendency towards demethylation of cytosine residues. Altered patterns of DNA methylation may result from imbalances between the activity of the enzymes involved in maintenance methylation and DNA replication and cell division, or could be related to other events that occur in cultured cells, such as chromosome breakage and repair (Peschke et al., 1991). Under certain conditions it has been shown that the altered patterns can be transmitted to the progenies of plants regenerated from the cultured cells. Nevertheless, it should not be construed that changes in methylation status are the sole cause of somaclonal variation observed among regenerant plants. Further research on replication and methylation of DNA in cultured plant cells may clarify the relationship between the multiplicity of genetic events occurring as a result of tissue culture.

10 The role of DNA methylation in the inactivation of introduced genes in transgenic plants

The variability observed in the expression of transgenes between independent transformants is frequently ascribed to differences in the number of copies of the transgene in each transformant and/or to position effect, that is, the effect on transgene expression of plant sequences flanking the site of integration. While both copy number (Stockhaus et al., 1987) and position of the integrated gene(s) (Jones et al., 1985; Eckes et al., 1985; Nagy et al., 1985) can play a role in gene expression in transgenic plants, other explanations have been invoked to account for non-Mendelian inheritance of introduced genes (Deroles and Gardner, 1988a; Deroles and Gardner, 1988b; Matzke et al., 1989; Scheid et al., 1991) and for the observation that some transgenic plants do not express all genes integrated (Amasino et al., 1984; Peerebolte et al., 1986; Matzke et al., 1989).

Methylation of integrated genes was first observed in *Agrobacterium tumefaciens* induced tumour lines that contained multiple copies of the T-DNA (Gelvin et al., 1983). A possible role for methylation in the suppression of T-DNA genes was indicated by the demonstration that silent genes could be reactivated by treatment with 5-azaC (Hepburn et al., 1983). This treatment resulted in both demethylation and increased transcription of the integrated gene suggesting that methylation inhibited expression (Hepburn et al., 1983). This correlation between methylation and repression of genes encoded by the T-DNA has been verified in many other studies (Amasino et al., 1984; Peerebolte et al., 1986; John and Amasino, 1989; Klaas et al., 1989; Hobbs et al., 1990); in each case multiple copies of the T-DNA were integrated into the plant genome (Gelvin et al., 1983; Hepburn et al., 1983; Amasino et al., 1984; Peerebolte et al., 1986). Extensive demethylation of T-DNA sequences with concomitant reactivation of T-DNA genes has been observed following spontaneous reactivation of an inactive *ipt* oncogene (John and Amasino, 1989) or after 5-azaC treatment (Amasino et al., 1984; Peerebolte et al., 1986; John and Amasino, 1989; Klaas et al., 1989). Grafting and redifferentiation resulting from extreme conditions in culture have been associated with reactivation of previously silent T-DNA genes (van Slogteren et al., 1984; Peerebolte et al., 1986). Here too, reactivation was associated with changes in methylation of the T-DNA but these were more quantitative than qualitative (Peerebolte et al., 1986).

Methylation of sites within the promoter has been correlated with inactivation of a chimaeric 35S*A1* gene in transgenic petunia. Transformation of a white flowering mutant of *Petunia hybrida* with this gene resulted in the production of pelargonidin derivatives manifested by flower colours ranging from brick red through variegated to white (Meyer et al., 1987; Linn et al., 1990). Most plants that produced white flowers had multiple copies of the *A1* gene integrated at a number of sites, while uniform pigmentation of the whole flower generally occurred in plants which carried a single intact copy of the gene. Two sites within the 35S*A1* chimaeric gene, located at −315 and +11 respectively, were methylated in DNA from plants with white flowers. However methylation at −315 alone did not prevent expression of the *A1* gene. Although the sites examined in this study do not lie within the putative protein binding domains (−90 to +8 and −343 to −90) for the 35S promoter (Benfey et al., 1989; Prat et al., 1989), methylation at −315 and +11 could be indicative of the general level of methylation of the entire promoter (Linn et al., 1990).

Initial observations of T-DNA methylation were made in plants that carried multiple copies of the same T-DNA. Methylation has now been implicated in the suppression of genes encoded on an integrated T-DNA following transformation with a second T-DNA encoding different selectable and screening markers (Matzke et al., 1989; Matzke and Matzke, 1990; Matzke and Matzke, 1991). Inactivation and methylation of a site in the promoters of the genes encoding markers on T-DNA-I occurred only in the presence of T-DNA-II; expression was restored when back-crossing or self-fertilization allowed the segregation of the two T-DNAs. Partial or complete demethylation of the promoters linked to genes on T-DNA-I was observed in progeny in which expression was restored. Although both selectable and screening markers were unique, there was considerable homology between the two T-DNA segments integrated in the doubly transformed plants. This included the *nos* promoter that controlled expression of the *npt*II, *nos* (both on T-DNA-I) and *ocs* (on T-DNA-II) genes as well as approximately 2 kb and 1 kb flanking the left and right T-DNA borders, respectively. It has been proposed that suppression results from the competition for binding to nuclear sites with fixed locations by the homologous regions on each T-DNA (Matzke and Matzke, 1990). The relative locations of the two T-DNA insertion sites must be important for this competition because in the majority (85%) of double transformants no interaction between the T-DNAs was observed (Matzke and Matzke, 1990).

The term co-suppression has been coined to describe a phenomenon in which the expression of both introduced and endogenous copies of a gene was co-ordinately suppressed. Although methylation of the inactivated genes has not been demonstrated, methylation of sequences required for active transcription of these genes could account for this phenomenon. Co-suppression was first described when an endogenous chalcone synthase gene was introduced into a deep violet flowering line of *Petunia hybrida*. Chalcone synthase (CHS), one of the enzymes of the anthocyanin biosynthetic pathway, is encoded by a multigene family, one member of which is highly expressed in petals (Koes et al., 1986; Koes et al., 1987; Koes et al., 1989). Surprisingly, no transgenic plants had flowers that were visibly darker than the untransformed control and up to 42% of the transgenic plants produced completely white flowers and/or flowers with white or pale sectors on a normally pigmented background (Mol et al., 1989; Napoli et al., 1990; van der Krol et al., 1990). Over-expression of chalcone synthase occurred in normally pigmented floral tissue of transgenic plants with one or more copies of the introduced gene. The level of CHS mRNA was about 50-fold lower in white or pale sectors of flowers although the developmental regulation of the endogenous CHS gene was unchanged. Homology between the promoters of the endogenous and introduced genes was not essential for co-suppression. Somatic reversion, which occurred at low frequency, was associated with increased steady-state levels of transcripts from both the endogenous and introduced copies of the CHS gene. The expression of other genes involved in anthocyanin biosynthesis was not affected by the introduction of a CHS gene.

Co-suppression has now been described for petunia dihydroflavonol-4-reductase (van der Krol et al., 1990), another gene in the pigment biosynthesis pathway, as well as the potato starch synthase (Visser et al., quoted in Jorgensen, 1991) and tomato fruit polygalacturonase gene (Smith et al., 1990). However, reduction of endogenous gene expression does not always follow the introduction of additional copies of the sense gene, for example, the introduction of a full-length cDNA for peroxidase resulted in increased levels of peroxidase in transgenic tobacco (Lagrimini et al., 1990).

Co-suppression is similar to pre-meiotic inactivation of duplicated genes in the filamentous fungi *Neurospora crassa* (Selker et al., 1987) and *Ascobolus immersus* (Goyon and Faugeron, 1989; Faugeron et al., 1990). In contrast to co-suppression, methylation of the inactivated genes has been demonstrated in pre-meiotic inactivation. In general both copies of duplicated genes in *Neurospora* were irreversibly inactivated at a pre-meiotic stage of the sexual cycle. Inactivation was associated with high rates of point mutation and cytosine methylation (Cambareri et al., 1989). Unlike *Neurospora*, the methylation and inactivation of duplicated genes in *Ascobolus* was reversible, and was not associated with increased mutation. Premeiotic pairing of repeated sequences prior to inactivation has been proposed to account for the selective methylation of both linked and unlinked copies in *Ascobolus* (Faugeron et al., 1989). By analogy, methylation of both introduced and endogenous genes could explain the variable and reversible nature of co-suppression in plants (Napoli et al., 1990).

Alternatively, co-suppression may result from the inactivation of transcripts, from both introduced and endogenous genes, by antisense RNA. This could be generated by transcription of the non-coding (antisense) strand by read-through from a convergently transcribed gene in the T-DNA (Grierson et al., 1991). Interference between sense and antisense transcription might be expected if this model is correct. It is worth noting that the 35S promoter which is driving the sense transcript in these constructs, is a stronger promoter in transgenic tobacco, sugarbeet, oilseed rape and *Arabidopsis* than the converging *nos* promoter (Sanders et al., 1987; Harpster et al., 1988; Haring et al., 1991). These hypotheses could be differentiated by looking for increased methylation in the promoter regions of inactivated genes and for antisense transcripts in the transgenic plants in which co-suppression is observed.

Some of the phenomena described above have parallels in natural plant systems. The suppression of genes encoded on one T-DNA by a second T-DNA is similar to paramutation which has been defined as an interaction *in trans* between two alleles at a locus; one allele causes a directed heritable and reversible change at the other (paramutable) allele (reviewed by Brink, 1973). This change in the paramutable allele is not permanent and is lost following segregation of the alleles. Similarly expression of T-DNA-I was restored after segregation away from T-DNA-II (Matzke et al., 1989). Methylation has been associated with the suppression of genes on T-DNA-I and could equally account for the changes that occur at the paramutable allele in paramutation. Paramutation at certain anthocyanin loci in maize and *Antirrhinum* has been well characterized (Brink, 1973; Harrison and Carpenter, 1973) and in many cases resulted in non-clonal patterns of pigmentation similar to those observed in co-suppression in petunia (Napoli et al., 1990; van der Krol et al., 1990).

In conclusion plants can, in some instances, recognize and inactivate introduced DNA; this could be construed as a defence mechanism to protect the plant against potentially deleterious effects of the expression of genes encoded on "foreign" DNA. Although the introduction of specific genes into plant cells in the laboratory is a very recent event, other, more natural means to introduce foreign DNA into plant cells have existed for millenia. Infection with *Agrobacterium* species is associated with concomitant transfer and integration of bacterial DNA into the plant genome. Transposition of mobile elements, which are extant in many species, disrupts the integrity of the plant genome by relocation of the element from one position to another. Inactivation of both integrated T-DNA and transposable elements is well documented (Gelvin et al., 1983; Hepburn et al., 1983; Chandler and Walbot, 1986; Schwartz and Dennis, 1986; Chomet et al., 1987; Fedoroff and Banks, 1988) suggesting that plants are well adapted to guard against events that have the potential to perturb the norm. Whether methylation is the primary event in the inactivation, or whether it occurs as a consequence of transcriptional inactivity of T-DNA or transposase genes is unknown (similar phenomenon is observed by vertebrates).

The mechanism by which plants recognize DNA as foreign has not been determined. Inactivation usually occurs when multiple copies of introduced DNA sequences are integrated, either at one chromosomal location or when dispersed throughout the genome. Inactivation of single copy inserts is observed, but at lower frequency. In addition, in all cases there is homology between the inactivated/methylated DNA sequences. The extent of homology may be important because the suppression of T-DNA-I was weaker when caused by a truncated T-DNA-II which lacks about 2 kb of homologous sequence, than by an intact T-DNA-II (Matzke and Matzke, 1991). However, as the relative location of the T-DNA sequences is also important, it is not clear whether the reduction in suppression was due to reduced homology or the relative positions of the T-DNAs. These two features, multi-copy sequences and a degree of homology, are also characteristic of the disruptive elements to which plants are already adapted, that is transposable elements and integrated T-DNA.

Whatever the mechanism, the inactivation of introduced genes has important implications for the maintenance of novel phenotypes created by the expression of transgenes.

11 Conclusions

The evidence presented in this review suggests that DNA methylation is important in gene regulation in plants. Methylation of DNA *in vitro* inhibits transcription, both transiently in protoplasts (Hershkovitz et al., 1989; Weber and Graessmann et al., 1989), and when the DNA is stably integrated in regenerated plants (Weber et al., 1990). Furthermore some plant DNA binding proteins no longer recognize their target site *in vitro* if one or more cytosine residues within the site are methylated (Gierl et al., 1988; Staiger et al., 1989; Inamdar et al., 1991). As some of these proteins are likely to be transcription factors, *in vivo* methylation of their binding sites may inhibit transcription. Methylation also affects chromatin structure; this is reflected by changes in the sensitivity of intact chromatin to DNaseI such that sensitive chromatin is generally associated with hypomethylated DNA (Klaas and Amasino, 1989). Transposable elements frequently insert into unmethylated DNA suggesting that the chromatin structure of methylated DNA is unsuitable as a target for transposition (Schmidt et al., 1987). Methylation could suppress transcription directly by disrupting the binding of transcription factors, or indirectly because it makes a gene inaccessible for transcription due to binding of methyl binding proteins or changes in chromatin structure.

The expression of at least some plant genes is correlated with specific, developmentally regulated changes in methylation in the vicinity of the gene (Bianchi and Viotti, 1988; Spena et al., 1983; Flavell and O'Dell, 1990; Riggs and Chrispeels, 1990; Langdale et al., 1990; Flavell et al., 1990; Kaufman et al., 1987). As a very limited number of the potential methylation sites can be assayed using methylation-sensitive restriction enzymes, the failure to detect specific changes in methylation in other genes (Riggs and Chrispeels, 1990; Walling et al., 1986) does not necessarily imply that methylation is not involved in regulation. The extent to which changes in methylation are associated with gene regulation will only be revealed by genomic sequencing of extensive regions around genes showing developmentally regulated expression.

Changes in methylation patterns occur during normal plant development, for example the methylation of programmable *Spm* elements is not constant throughout the life cycle (Fedoroff et al., 1989). Environmental stimuli can induce changes in methylation, for example, the overall level of methylation in cotton plants is altered by fungal infection (Guseinov et al., 1975). Changes in the methylation status of genes involved in floral initiation has been implicated in the vernalization response of late flowering ecotypes of *Arabidopsis* (Burn pers. comm.).

Genes introduced into plants, either by *Agrobacterium* or by more direct methods, are frequently subject to methylation (Gelvin et al.,

1983; Hepburn et al., 1983; Amasino et al., 1984; Peerebolte et al., 1986; John and Amasino, 1989; Klaas et al., 1989; Hobbs et al., 1990; Matzke et al., 1989; Linn et al., 1990). Inactivation of these genes is correlated with methylation, but whether methylation is the causative agent in inactivation or is secondary to transcriptional inactivation is unknown. If methylation is not the primary event in the inactivation of introduced DNA then it is not obvious how inactivation is mediated. One possible scenario is that competition for transcription factors could occur; if the introduced promoter has a lower affinity for these factors than the endogenous promoter elements, then the introduced DNA will not be transcribed. Methylation of the inactive DNA could then follow.

Plants contain highly repeated sequences within their genome, many of which are heavily methylated (Guseinov et al., 1975). In most instances introduced genes which became methylated were present in multiple copies, or at least shared extensive regions of homology. Therefore inactivation of introduced DNA may simply reflect a natural process used by plants to deal with repetitive DNA, rather than a novel response to foreign DNA. Perhaps plants have evolved a mechanism to recognize and methylate repeated sequences, either at a single location or when dispersed. On the other hand, plants have multigene families that are not methylated suggesting that the presence of multiple copies of a sequence does not always lead to methylation. Alternatively disruption of the normal chromatin structure by the introduction of DNA not present in the homologue may be a signal for methylation. As plant DNA sequences are also specifically methylated, hemizygosity cannot be the only signal for methylation. The factors regulating methylation, both the sequence and developmental specificity, are unknown.

In vivo, not all cytosines residues in CG or CNG motifs are methylated; specificity may come from the primary DNA sequence or residues may be protected from methylation by bound proteins or secondary structure. *In vitro*, the purified pea methylase can catalyse both *de novo* and maintenance methylation and can methylate cytosine residues in each of the four dinucleotides (CA, CC, CG and CT) (Yesufu *et al.*, 1991). The sequence specificity of this purified methylase activity has not been determined. Further studies with the purified protein and future work with cloned methylase genes should provide more information about factors that regulate reaction specificity and methylase expression.

The field of methylation remains a challenging and important area for research as many fundamental questions remain unanswered. One important aspect of methylation for agriculture is the inactivation of introduced genes in transgenic plants, as this has far reaching implications for the stability of novel phenotypes, such as insect or herbicide resistance, that are created by genetic engineering.

Acknowledgements. The authors thank R. M. Amasino, J. Burn, P. T. H. Brown, N. V. Fedoroff, M. A. Matzke, R. L. Phillips, K. Shimamoto, V. Walbot, H. Weber and G. Zabala for providing information for this review. We also thank L. B. Farrell, P. J. Kerr and J. D. Metzger for critically reading the manuscript and L. B. Thorpe for assistance in preparation of the manuscript.

Adams, R. L. P., Bryans, M., Rinaldi, A., Smart, A., and Yesufu, H. M. I. (1990) Eukaryotic DNA methylases and their use for *in vitro* methylation. Phil Trans. R. Soc. Lond. B. *326*, 189–198.

Adams, R. L. P., and Burdon, R. H. (1983) in: Enzymes of nucleic acid synthesis and processing, pp. 119–144. Ed. S. T. Jacob. CRC Press, Boca Raton, FL.

Adams, R. L. P., and Burdon, R. H. (1985) Molecular Biology of DNA Methylation. Springer Verlag, New York.

Adams, R. L. P., and Eason, R. (1984) Increased G + C content of DNA stabilizes methyl CpG dinucleotides. Nucl. Acids Res. *12*, 5869–5877.

Amasino, R. M., Powell, A. L. T., and Gordon, M. P. (1984) Changes in T-DNA methylation and expression are associated with phenotypic variation and plant regeneration in a crown gall tumor line. Mol. Gen. Genet. *197*, 437–446.

Antequera, F., and Bird, A. P. (1988) Unmethylated CpG islands associated with genes in higher plant DNA. EMBO J. *7*, 2295–2299.

Appels, R. (1983) Chromosome structure in cereals: the analysis of regions containing repeated sequence DNA and its applications to the detection of alien chromosomes introduced into wheat, in: Genetic Engineering in Plants, pp. 229–256. Eds. T. Kosuge, C. P. Meredith and A. Hollaender. Plenum Publishing Corporation, New York.

Appels, R., and Dvorak, J. (1982) The wheat ribosomal DNA spacer region: its structure and variation in populations and among species. Theor. Appl. Genet. *63*, 337–348.

Arnholdt-Schmitt, B., Holzapfel, B., Schillinger, A., and Neumann, K.-H. (1991) Variable methylation and differential replication of genomic DNA in cultured carrot root explants during growth induction as influenced by hormonal treatments. Theor. Appl. Genet. *82*, 283–288.

Banks, J. A., and Fedoroff, N. (1989) Patterns of developmental and heritable change in methylation of the *Suppressor-mutator* transposable element. Devel. Genet. *10*, 425–437.

Banks, J. A., Masson, P., and Fedoroff, N. (1988) Molecular mechanisms in the developmental regulation of the maize *Suppressor-mutator* transposable element. Genes Devel. *2*, 1364–1380.

Barker, R. F., Harberd, N. P., Jarvis, M. G., and Flavell, R. B. (1988) Structure and evolution of the intergenic region in a ribosomal DNA repeat unit of wheat. J. Mol. Biol. *201*, 1–17.

Behe, M., and Felsenfeld, G. (1981) Effects of methylation on a synthetic polynucleotide: the B-Z transition in poly(dG-m^{5}dC).poly(dG-m^{5}dC). Proc. Natl. Acad. Sci. USA *78*, 1619–1623.

Behe, M., Zimmerman, S., and Felsenfeld, G. (1981) Changes in the helical repeat of poly(dG-m^{5}dC).poly(dG-m^{5}dC) and poly(dG-dC).poly(dG-dC) associated with the B-Z transition. Nature *293*, 233–235.

Benfey, P. N., Ren, L., and Chua, N.-H. (1989) The CaMV 35S enhancer contains at least two domains which can confer different developmental and tissue-specific expression patterns. EMBO J. *8*, 2195–2202.

Bennetzen, J. L. (1987) Covalent DNA modification and the regulation of *Mutator* element transposition in maize. Mol. Gen. Genet. *208*, 45–51.

Bennetzen, J. L., Brown, W. E., and Springer, P. S. (1988) The state of DNA modification within the flanking maize transposable elements, in: Plant transposable elements, pp. 237–250. Ed. O. Nelson. Plenum Press, New York.

Bestor, T., Laudano, A., Mattaliano, R., and Ingram, V. (1988) Cloning and sequencing of a cDNA encoding DNA methyltransferase of mouse cells. J. Mol. Biol. *203*, 971–983.

Bestor, T. H., and Ingram, V. M. (1983) Two DNA methyltransferases from murine erythroleukaemia cells: Purification, sequence specificity, and mode of interaction with DNA. Proc. Natl. Acad. Sci. USA *80*, 5559–5563.

Bianchi, M. W., and Viotti, A. (1988) DNA methylation and tissue-specific transcription of storage protein genes of maize. Plant Mol. Biol. *11*, 203–214.

Bird, A. P. (1978) Use of restriction enzymes to study eukaryotic DNA methylation: II the symmetry of methylated sites supports semi-conservative copying of the methylation pattern. J. Molec. Biol. *118*, 49–60.

Bird, A. P. (1986) CpG-rich islands and the function of DNA methylation. Nature *321*, 209–213.

Boyes, J., and Bird, A. (1991) DNA methylation inhibits transcription indirectly via a methyl-CpG binding protein. Cell *64*, 1123–1134.

Brettell, R. I. S., and Dennis, E. S. (1991) Reactivation of a silent *Ac* following tissue culture is associated with heritable changes in its methylation pattern. Mol. Gen. Genet. *229*, 365–372.

Brettell, R. I. S., Thomas, E., and Ingram, D. S. (1980) Reversion of Texas male-sterile cytoplasm maize in culture to give fertile, T-toxin resistant plants. Theor. Appl. Genet. *58*, 55–58.

Brink, R. A. (1973) Paramutation. Ann. Rev. Genet. *7*, 129–152.

Brown, P. T. H. (1989) DNA methylation in plants and its role in tissue culture. Genome *31*, 717–729.

Brown, P. T. H., Kyozuka, J., Sukekiyo, Y., Kimura, Y., Shimamoto, K., and Lörz, H. (1990) Molecular changes in protoplast-derived rice plants. Mol. Gen. Genet. *223*, 324–328.

Burton, W. G., Grabowy, C. T., and Sager, R. (1979) Role of methylation in the modification and restriction of chloroplast DNA in *Chlamydomonas*. Proc. Natl. Acad. Sci. USA *76*, 1390–1394.

Cambareri, E. B., Jensen, B. C., Schabtach, E., and Selker, E. V. (1989) Repeat-induced G-C to A-T mutations in *Neurospora*. Science *244*, 1571–1578.

Cedar, H., Solage, A., Glaser, G., and Razin, A. (1979) Direct detection of methylated cytosine in DNA by use of the restriction enzyme MspI. Nucl. Acids Res. *6*, 2125–2132.

Chandler, V. L., and Walbot, V. (1986) DNA modification of a maize transposable element correlates with loss of activity. Proc. Natl. Acad. Sci. USA *83*, 1767–1771.

Chomet, P., Lisch, D., Hardeman, K. J., Chandler, V. L., and Freeling, M. (1991) Identification of a regulatory transposon that controls the mutator transposable element system in maize, Genetics *129*, 261–270.

Chomet, P. S., Wessler, S., and Dellaporta, S. L. (1987) Inactivation of the maize transposable element *Activator* (*Ac*) is associated with its DNA modification. EMBO J. *6*, 295–302.

Church, G. M., and Gilbert, W. (1984) Genomic sequencing. Proc. Natl. Acad. Sci. USA *81*, 1991–1995.

Comb, M., and Goodman, H. M. (1990) CpG methylation inhibits proenkephalin gene expression and binding of the transcription factor AP-2. Nucl. Acids Res. *18*, 3975–3982.

Cooper, D. N., Taggart, M. H., and Bird, A. P. (1983) Unmethylated domains in vertebrate DNA. Nucl. Acids Res. *11*, 647–657.

Coruzzi, G., Broglie, R., Edwards, C., and Chua, N.-H. (1984) Tissue specific and light regulated expression of a pea nuclear gene encoding the small subunit of ribulose-1,5-bisphosphate carboxylase. EMBO J. *3*, 1671–1679.

D'Amato, F. (1985) Cytogenetics of plant cell and tissue cultures and their regenerates. C. R. C. Crit. Rev. Plant Sci. *3*, 73–112.

Dellaporta, S. L., and Chomet, P. S. (1985) The activation of maize controlling elements, in: Genetic Flux in Plants, pp. 170–217. Eds. B. Hohn and E. S. Dennis. Springer-Verlag, Berlin/Heidelberg.

Delseny, M., Laroche, M., and Penon, P. (1984) Methylation pattern of radish (*Raphanus sativus*) nuclear ribosomal RNA genes. Plant Physiol. *76*, 627–632.

Deng, X.-W., and Gruissem, W. (1987) Control of plastid gene expression during development: the limited role of transcriptional regulation. Cell *49*, 379–387.

Deng, X.-W., and Gruissem, W. (1988) Constitutive transcription and regulation of gene expression in non-phtotsynthetic plastids of higher plants. EMBO J. *7*, 3301–3308.

Dennis, E. S., and Brettell, R. I. S. (1990) DNA methylation of maize transposable elements is correlated with activity. Phil. Trans. R. Soc. Lond. B. *326*, 217–229.

Deroles, S. C., and Gardner, R. C. (1988a) Expression and inheritance of kanamycin resistance in a large number of transgenic petunias generated by *Agrobacterium*-mediated transformation. Plant Molec. Biol. *11*, 355–364.

Deroles, S. C., and Gardner, R. C. (1988b) Analysis of the T-DNA structure in a large number of transgenic petunias generated by *Agrobacterium*-mediated transformation. Plant Molec. Biol. *11*, 365–377.

Eckes, P., Schell, J., and Willmitzer, L. (1985) Organ-specific expression of three leaf/stem specific cDNAs from potato is regulated by light and correlated with chloroplast development. Molec. Gen. Genet. *199*, 216–221.

Ellis, T. H. N., Davies, D. R., Castleton, J. A., and Bedford, I. D. (1984) The organization and genetics of rDNA length variants in peas. Chromosoma *91*, 74–81.

Ellis, T. H. N., Goldsborough, P. B., and Castleton, J. A. (1983) Transcription and methylation of flax rDNA. Nucl. Acids Res. *11*, 3047–3064.

Faugeron, G., Rhounim, L., and Rossignol, J. L. (1990) How does the cell count the number of ectopic copies of a gene in the premeiotic inactivation process acting in *Ascobolus immersus*? Genetics *124*, 585–591.

Fedoroff, N., Masson, P., and Banks, J. A. (1989) Mutations, epimutations, and the developmental programming of the maize *Suppressor-mutator* transposable element. BioEssays *10*, 139–144.

Fedoroff, N. V. (1989) The heritable activation of *cryptic Suppressor-mutator* elements by an active element. Genetics *121*, 591–608.

Fedoroff, N. V., and Banks, J. A. (1988) Is the *Suppressor-mutator* element controlled by a basic developmental regulatory mechanism? Genetics *120*, 559–577.

Flavell, R. B. (1989) Variation in structure and expression of ribosomal DNA loci in wheat. Genome *31*, 963–968.

Flavell, R. B., and O'Dell, M. (1990) Variation and inheritance of cytosine methylation patterns in wheat at the high molecular weight glutenin and ribosomal RNA gene loci. Development *S*, 15–20.

Flavell, R. B., O'Dell, M., and Thompson, W. F. (1988) Regulation of cytosine methylation in ribosomal DNA and nucleolus organizer expression in wheat. J. Mol. Biol. *204*, 523–534.

Flavell, R. B., O'Dell, M., Thompson, W. F., Vincentz, M., Sardana, R., and Barker, R. F. (1986) The differential expression of ribosomal RNA genes. Phil. Trans. R. Soc. Lond. B. *314*, 385–397.

Flavell, R. B., Sardana, R., Jackson, S., and O'Dell, M. (1990) The molecular basis of variation affecting gene expression: evidence from studies on the ribosomal RNA gene loci of wheat, in: Gene Manipulation in Plant Improvement II, pp. 419–430. Ed. J. P. Gustafson. Plenum Press, New York.

Fluhr, R., and Chua, N.-H. (1986) Developmental regulation of 2 genes encoding ribulose-bisphosphate carboxylase small subunit in peas and transgenic petunia plants: phytochrome response and blue light induction. Proc. Natl. Acad. Sci. USA *83*, 2358–2362.

Fluhr, R., Moses, P., Morelli, G., Coruzzi, G., and Chua, N.-H. (1986) Expression dynamics of the pea rbcS multigene family and organ distribution of the transcripts. EMBO J. *5*, 2063–2071.

Freeling, M., and Bennett, D. C. (1985) Maize *Adh*1. Ann. Rev. Genet. *19*, 297–323.

Frommer, M., McDonald, L. E., Millar, D. S., Collis, C. M., Watt, F., Grigg, G. W., Molloy, P. L., and Paul, C. L. (1992) A genomic sequencing protocol which yields a positive display of 5-methylcytosine residues in individual DNA strands. Proc. Natl. Acad. Sci. USA *89*, 1827–1831.

Gallagher, T. F., and Ellis, R. J. (1982) Light-stimulated transcription of genes for two chloroplast polypeptides in isolated pea leaf nuclei. EMBO J. *1*, 1493–1498.

Gauly, A., and Kossel, H. (1989) Evidence for tissue-specific cytosine-methylation of plastid DNA from *Zea mays*. Curr. Gen. *15*, 371–376.

Gelvin, S. B., Karcher, S. J., and DiRita, V. J. (1983) Methylation of the T-DNA in *Agrobacterium tumefaciens* and in several crown gall tumors. Nucl. Acids Res. *11*, 159–174.

Gerlach, W. L., and Bedbrook, J. R. (1979) Cloning and characterization of ribosomal RNA genes from wheat and barley. Nucl. Acids Res. *7*, 1869–1885.

Gierl, A., Lütticke, S., and Saedler, H. (1988) *TnpA* product encoded by the transposable element En-1 of *Zea mays* is a DNA binding protein. EMBO J. *7*, 4045–4053.

Giordano, M., Mattachini, M. E., Cella, R., and Pedrali-Noy, G. (1991) Purification and properties of a novel DNA methyltransferase from cultured rice cells. Biochem. Biophys. Res. Comm. *177*, 711–719.

Girod, P.-A., and Zryd, J.-P. (1991) Secondary metabolism in cultured red beet (*Beta vulgaris* L.) cells: differential regulation of betaxanthin and betacyanin biosynthesis. Plant Cell Tiss. Organ Cult. *25*, 1–12.

Görz, A., Schäfer, W., Hirasawa, E., and Kahl, G. (1988) Constitutive and light-induced DNAseI hypersensitive sites in the rbcS genes of pea (*Pisum sativum*). Plant Molec. Biol. *11*, 561–573.

Goyon, C., and Faugeron, G. (1989) Targeted transformation of *Ascobolus immersus* and de novo methylation of the resulting duplicated DNA sequences. Mol. Cell. Biol. *9*, 2818–2827.

Grierson, D., Fray, R. G., Hamilton, A. J., Smith, C. J. S., and Watson, C. F. (1991) Does co-suppression of sense genes in transgenic plants involve antisense RNA? Tibtech. *9*, 122–123.

Grisvard, J. (1985) Different methylation pattern of melon satellite DNA sequences in hypercotyl and callus tissues. Plant Science *39*, 189–193.

Groudine, M., Eisenman, R., and Weintraub, H. (1981) Chromatin structure of endogenous retroviral genes and activation by an inhibitor of DNA methylation. Nature *292*, 311–317.

Gruenbaum, Y., Naveh-Many, T., Cedar, H., and Razin, A. (1981) Sequence specificity of methylation in higher plant DNA. Nature *292*, 860–862.

Gruissem, W. (1989) Chloroplasts gene expression: how plants turn their plastids on. Cell *56*, 161–170.

Guseinov, V. A., Kiryanov, G. I., and Vanyushin, B. F. (1975) Intragenome distribution of 5-methylcytosine in DNA of healthy and wilt-infected cotton plants (*Gossypium hirsutum* L.). Mol. Biol. Rep. *2*, 59–63.

Hansmann, P. (1987) Daffodil chromoplast DNA: comparison with chloroplast DNA, physical map and gene localization. Z. Naturforsch. *42C*, 118–122.

Haring, M. A., Rommens, C. M. T., John, J., Nijkamp, J. J., and Hille, J. (1991) The use of transgenic plants to understand transposition mechanisms and to develop transposon tagging strategies. Plant Mol. Biol. *16*, 449–461.

Harpster, M. H., Townsend, J. A., Jones, J. D. G., Bedbrook, J., and Dunsmuir, P. (1988) Relative strengths of the 35S cauliflower mosaic virus, 1′, 2′, and nopaline synthase promoters in transformed tobacco sugarbeet and oilseed rape callus tissue. Mol. Gen. Genet. *212*, 182–190.

Harrison, B. J., and Carpenter, R. (1973) A comparison of the instabilities at the *Nivea* and *Pallida* loci in *Antirrhinum majus*. Heredity *31*, 309–323.

Hatch, M. D. (1987) C4 photosynthesis: a unique blend of modified biochemistry, anatomy and ultrastructure. Biochim. Biophys. Acta *895*, 81–106.

Hebbes, T. R., Thorne, A. W., and Crane-Robinson, C. (1988) A direct link between core histone acetylation and transcriptionally active chromatin. EMBO J. *7*, 1395–1402.

Hepburn, A. G., Belanger, F. C., and Mattheis, J. R. (1987) DNA methylation in plants. Devel. Genet. *8*, 475–493.

Hepburn, A. G., Clarke, L. E., Pearson, L., and White, J. (1983) The role of cytosine methylation in the control of nopaline synthase gene expression in a plant tumor. J. Mol. Appl. Genet. *2*, 315–329.

Hershkovitz, M., Gruenbaum, Y., Zakai, N., and Loyter, A. (1989) Gene transfer in plant protoplasts. Inhibition of gene activity by cytosine methylation and expression of single stranded DNA constructs. FEBS Lett. *253*, 167–172.

Hobbs, S. L. A., Kpodar, P., and DeLong, C. M. O. (1990) The effect of T-DNA copy number, position and methylation on reporter gene expression in tobacco transformants. Plant Mol. Biol. *15*, 851–864.

Hoffman, L. M., and Donaldson, D. D. (1985) Characterization of two *Phaseolus vulgaris* phytohemagglutinin genes closely linked on the chromosome. EMBO J. *4*, 883–889.

Höller, M., Westin, G., Jiricny, J., and Schaffner, W. (1988) Sp1 transcription factor binds DNA and activates transcription even when the binding site is CpG methylated. Genes Devel. *2*, 1127–1135.

Holliday, R., and Pugh, J. E. (1975) DNA modification mechanisms and gene activity during development. Science *187*, 226–232.

Iguchi-Ariga, S. M. M., and Schaffner, W. (1989) CpG methylation of the cAMP responsive enhancer/promoter sequence TGACGTCA abolishes specific factor binding as well as transcriptional activation. Genes Devel. *3*, 612–619.

Inamdar, N. M., Ehrlich, K. C., and Ehrlich, M. (1991) CpG methylation inhibits binding of several sequence-specific DNA-binding proteins from pea, wheat, soybean and cauliflower. Plant Molec. Biol. *17*, 111–123.

Ip, Y. T., Poon, D., Stone, D., Granner, D. K., and Chalkely, R. (1990) Interaction of a liver-specific factor with an enhancer 4.8 kilobases upstream of the phosphoenolpyruvate carboxykinase gene. Mol. Cell. Biol. *10*, 3770–3787.

John, M. C., and Amasino, R. M. (1989) Extensive changes in DNA methylation patterns accompany activation of a silent T-DNA *ipt* gene in *Agrobacteriun tumefaciens*-transformed plant cells. Mol. Cell. Biol. *9*, 4298–4303.

Jones, J. D., Dunsmuir, P., and Bedbrook, J. (1985) High level expression of introduced chaemeric genes in regenerated transformed plants. EMBO J. *4*, 2411–2418.

Jorgensen, R. (1991) Altered gene expression in plants due to *trans* interactions between homologous genes. Tibtech. *8*, 340–344.

Kaufman, L. S., Watson, J. C., and Thompson, W. F. (1987) Light-regulated changes in DNasel hypersensitive sites in the rRNA genes of *Pisum sativum*. Proc. Natl. Acad. Sci. USA *84*, 1550–1554.

Kessler, C., Neumaier, P. S., and Wolf, W. (1985) Recognition sequences of restriction endonucleases and methylases – a review. Gene *33*, 1–102.

Kirk, J. T. O., and Tilney-Bassett, R. A. E. (1978) The plastids: their chemistry, structure, growth and inheritance. Elsevier/Nth Holland Biomedical Press.

Klaas, M., and Amasino, R. M. (1989) DNA methylation is reduced in DNaseI-sensitive regions of plant chromatin. Plant Physiol. *91*, 451–454.

Klaas, M., John, M. C., Crowell, D. N., and Amasino, R. M. (1989) Rapid induction of genomic demethylation and T-DNA gene expression in plant cells by 5-azacytidine derivatives. Plant Mol. Biol. *12*, 413–423.

Klein, R. R., and Mullet, J. E. (1986) Regulation of chloroplast-encoded chlorophyll-binding protein translation during higher plant chloroplast biogenesis. J. Biolog. Chem. *261*, 11138–11145.

Klysik, J., Stridivant, S. M., Singleton, C. K., Zacharias, W., and Wells, R. D. (1983) Effects of 5 cytosine methylation on the B-Z transition in DNA restriction fragments and recombinant plasmids. J. Mol. Biol. *168*, 51–71.

Kobayashi, H., Ngernprasirtsiri, J., and Akazawa, T. (1990) Transcriptional regulation and DNA methylation in plastids during transitional conversion of chloroplasts to chromoplasts. EMBO J. *9*, 307–313.

Koes, R. E., Spelt, C. E., and Mol, J. N. M. (1989) The chalcone synthase multigene family of *Petunia hybrida* (V30): differential, light-regulated expression during flower development and UV light induction. Plant Molec. Biol. *12*, 213–225.

Koes, R. E., Spelt, C. E., Mol, J. N. M., and Gerats, A. G. M. (1987) The chalcone synthase multigene family of *Petunia hybrida* (V30): sequence homology, chromosomal localization and evolutionary aspects. Plant Molec. Biol. *10*, 159–169.

Koes, R. E., Spelt, C. E., Reif, H. J., van den Elzen, P. J. M., Veltkamp, E., and Mol, J. N. M. (1986) Floral tissues of *Petunia hybrida* (V30) express only one member of the chalcone synthase multigene family. Nucl. Acids Res. *14*, 5229–5239.

Kovesdi, I., Rerchel, R., and Nevins, J. R. (1987) Role of an adenovirus *E*2 promoter binding factor in E1A-mediated co-ordinate gene control. Proc. Natl. Acad. Sci. USA *84*, 2180–2184.

Kuntz, M., Evrard, J.-L., d'Harlingue, A., Weil, J.-H., and Camara, B. (1989) Expression of plastid and nuclear genes during chromoplast differentiation in bell pepper (*Capsicum annuum*) and sunflower (*Helianthus annuus*). Mol. Gen. Genet. *216*, 156–163.

Kunze, R., Starlinger, P., and Schwartz, D. (1988) DNA methylation of the maize transposable element *Ac* interferes with its transcription. Mol. Gen. Genet. *214*, 325–327.

Lagrimini, L. M., Bedford, S., and Rothstein, S. (1990) Peroxidase-induced wilting in transgenic tobacco plants. Plant Cell. *2*, 7–18.

Lang, A. (1965) Physiology of flower initiation, in: Encyclopedia of Plant Physiology, pp. 1489–1536. Ed. W. Ruhland. Springer Verlag, Berlin.

Langdale, J. A., Metzler, M. C., and Nelson, T. (1987) The *Argentia* mutation delays normal development of photosynthetic cell-type in *Zea mays*. Dev. Biol. *122*, 243–255.

Langdale, J. A., Taylor, W. C., and Nelson, T. (1991) Cell-specific accumulation of maize phosphoenolpyruvate carboxylase is correlated with demethylation at a specific site >3 kb upstream of the gene. Mol. Gen. Genet. *225*, 49–55.

Langdale, J. A., Zelitch, I., Miller, E., and Nelson, T. (1988) Cell position and light influence C4 versus C3 pattern of photosynthetic gene expression in maize. EMBO J. *7*, 3643–3651.

Langridge, P., Pintor-Toro, J. A., and Feix, G. (1982) Transcriptional effects of the opaque-2 mutation of *Zea mays* L. Planta *156*, 166–170.

Larkin, P. J. (1987) Somaclonal variation: history, method, and meaning. Iowa State J. Res. *61*, 393–434.

Larkin, P. J., and Scowcroft, W. R. (1981) Somaclonal variation – a novel source of variability from cell cultures for plant improvement. Theor. Appl. Genet. *60*, 197–214.

Leutwiler, L. S., Hough-Evans, B. R., and Meyerowitz, E. M. (1984) The DNA of *Arabidopsis thaliana*. Mol. Gen. Genet. *194*, 15–23.

Lewis, J., and Bird, A. (1991) DNA methylation and chromatin structure. FEBS Lett. *285*, 155–159.

Lindahl, T. (1974) An N-glycosidase from *Escherichia coli* that releases free uracil from DNA containing deaminated cytosine residues. Proc. Natl. Acad. Sci. *71*, 3649–3653.

Linn, F., Heidmann, I., Saedler, H., and Meyer, P. (1990) Epigenetic changes in the expression of the maize A1 gene in *Petunia hybrida*: role of numbers of integrated gene copies and state of methylation. Mol. Gen. Genet. *222*, 329–336.

Long, E. O., and Dawid, I. B. (1980) Repeated genes in eukaryotes. Ann. Rev. Biochem. *49*, 727–764.

Lo Schiavo, F., Pitto, L., Giuliano, G., Torti, G., Nuti-Ronchi, V., Marazziti, D., Vergara, R., Orselli, S., and Terzi, M. (1989) DNA methylation of embryogenic carrot cell cultures and its variations as caused by mutation, differentiation, hormones and hypomethylating drugs. Theor. Appl. Genet. *77*, 325–331.

Maggini, F., and Carmona, M. J. (1981) Sequence heterogeneity of the ribosomal DNA in *Allium cepa* (*Liliaceae*). Protoplasma *108*, 163–167.

Marano, M. R., and Carrillo, N. (1991) Chromoplast formation during tomato fruit ripening. No evidence for plastid DNA methylation. Plant Molec. Biol. *16*, 11–19.

Martienssen, R., Barkan, A., Taylor, W. C., and Freeling, M. (1990) Somatically heritable switches in the DNA modification of *Mu* transposable elements monitored with a suppressible mutant in maize. Genes Devel. *4*, 331–343.

Martin, C., Prescott, A., Lister, C., and MacKay, S (1989) Activity of the transposon *Tam3* in *Antirrhinum* and tobacco: possible role of DNA methylation. EMBO J. *9*, 997–1004.

Martini, G., and Flavell, R. (1985) The control of nucleolus volume in wheat, a genetic study at three developmental stages. Heredity *54*, 111–120.

Martini, G., O'Dell, M., and Flavell, R. B. (1982) Partial inactivation of wheat nucleolus organizers by the nucleolus organizer chromosomes from *Aegilops umbellulata*. Chromosoma *84*, 687–700.

Matzke, M. A., and Matzke, A. J. M. (1990) Gene interactions and epigenetic variation in transgenic plants. Devel. Genet. *11*, 214–223.

Matzke, M. A., and Matzke, A. J. M. (1991) Differential inactivation and methylation of a transgene in plants by two suppressor loci containing homologous sequences. Plant Mol. Biol. *16*, 821–830.

Matzke, M. A., Primig, M., Trnovsky, J., and Matzke, A. J. M. (1989) Reversible methylation and inactivation of marker genes in sequentially transformed tobacco plants. EMBO J, *8*, 643–649.

McClelland, M. (1981) The effect of sequence specific DNA methylation on restriction endonuclease cleavage. Nucl. Acids Res. *9*, 5859–5866.

McClelland, M. (1982) The effect of site specific methylation on restriction endonuclease cleavage (update). Nucl. Acids Res. *11*, r169–r173.

McClelland, M. (1983) The frequency and distribution of methylatable DNA sequences in leguminous plant protein coding genes. J. Mol. Evol. *19*, 346–354.

McClintock, B. (1964) Aspects of gene regulation in maize. Carnegie Inst. Wash. Yearbook *62*, 486–493.

Meehan, R. R., Lewis, J. D., McKay, S., Kleiner, E. L., and Bird, A. P. (1989) Identification of a mammalian protein that binds specifically to DNA containing methylated CpGs. Cell *58*, 499–507.

Meeker, R., Nielsen, B., and Tewari, K. K. (1988) Localization of replication origins in pea chloroplast DNA. Molec. Cell. Biol. *8*, 1216–1223.

Metzger, J. D. (1988) Localization of the site of perception of thermoinductive temperatures in *Thlaspi arvense* L. Plant Physiol. *88*, 424–428.

Meyer, P., Heidmann, I., Forkmann, G., and Saedler, H. (1987) A new petunia flower colour generated by transformation of a mutant with a maize gene. Nature *330*, 677–678.

Mohr, H. (1984) Phytochrome and chloroplast development, in: Chloroplast Biogenesis, pp. 305–347. Eds. N. R. Baker and T. Barber. Elsevier Science Publishers, Amsterdam.

Mol, J. N. M., Stuitje, A. R., and van der Krol, A. (1989) Genetic manipulation of floral pigmentation genes. Plant Mol. Biol. *13*, 287–294.

Morrish, F. M., and Vasil, I. K. (1989) DNA methylation and embryogenic competence in leaves and callus of napiergrass (*Pennisetum purpureum* Schum.). Plant Physiol. *90*, 37–40.

Müller, E., Brown, P. T. H., Hartke, S., and Lörz, H. (1990) DNA variation in tissue culture-derived rice plants. Theor. Appl. Genet. *80*, 673–679.

Mullet, J. E., and Klein, R. R. (1987) Transcription and RNA stability are important determinants of higher plant chloroplast RNA levels. EMBO J. *6*, 1571–1579.

Nagy, F., Morelli, G., Fraley, R. T., Rogers, S. G., and Chua, N.-H. (1985) Photoregulated expression of a pea *rbc*S gene in leaves of transgenic plants. EMBO J. *4*, 3063-3068.

Napoli, C., Lemieux, C., and Jorgensen, R. (1990) Introduction of a chimeric chalcone synthase gene into petunia results in reversible co-suppression of homologous genes *in trans*. Plant Cell. *2*, 279–289.

Nelson, M., and McClelland, M. (1987) The effect of site-specific methylation on restriction modification enzymes. Nucl. Acids Res. (suppl). *15*, r219–r230.

Ngernprasirtsiri, J., and Akazawa, T. (1990) Modulation of DNA methylation and gene expression in cultured sycamore cells treated by hypomethylating base analog. Eur. J. Biochem. *194*, 513–520.

Ngernprasirtsiri, J., Chollet, R., Kobayashi, H., Sugiama, T., and Akazawa, T. (1989) DNA methylation and the differential expression of C4 photosynthesis genes in mesophyll and bundle sheath cells of greening maize leaves. J. Biol. Chem. *264*, 8241–8248.

Ngernprasirtsiri, J., Kobayashi, H., and Akazawa, T. (1988a) DNA Methylation as a mechanism of transcriptional regulation in nonphotosynthetic plastids in plant cells. Proc. Natl. Acad Sci. USA *85*, 4750–4754.

Ngernprasirtsiri, J., Kobayashi, H., and Akazawa, T. (1988b) DNA methylation occurred around lowly expressed genes of plastid DNA during tomato fruit development. Plant Physiol. *88*, 16–20.

Nick, H., Bowen, B., Ferl, R. J., and Gilbert, W. (1986) Detection of cytosine methylation in the maize alcohol dehydrogenase gene by genomic sequencing. Nature *319*, 243–246.

Nickol, J., Behe, M., and Felsenfeld, G. (1982) Effect of the B-Z transition in poly (dG-m^{5}dC).poly(dG-m^{5}dC) on nucleosome formation. Proc. Natl. Acad. Sci. USA *79*, 1771–1775.

Niwa, O., and Sugahara, T. (1981) 5-aza-cytidine induction of mouse endogenous type C virus and suppression of DNA methylation. Proc. Natl. Acad. Sci. USA *78*, 6290–6294.

Novak, F. J., Daskalov, S., Brunner, H., Nesticky, M., Afza, R., Dolezelova, M., Lucretti, S., Herichova, A., and Hermelin, T. (1988) Somatic embryogenesis in maize and comparison of genetic variability induced by gamma radiation and tissue culture techniques. Plant Breeding *101*, 66–79.

Ohta, N., Sato, N., Kawano, S., and Kuroiwa, T. (1991) Methylation of DNA in the chloroplasts and amyloplasts of the pea, *Pisum sativum*. Plant Sci. *78*, 33–42.

Ohyama, K., Fukuzawa, H., Kohchi, T., Shirai, H., Sano, T., Sano, S., Umesono, K., Shiki, Y., Takeuchi, M., Chang, Z., Aota, S., Inokuchi, H., and Ozeki, H. (1986) Chloroplast gene organization deduced from complete sequence of liverwort *Marchantia polymorpha* chloroplast DNA. Nature *322*, 572–574.

Oono, K. (1985) Putative homozygous mutations in regenerated plants of rice. Mol. Gen. Genet. *198*, 377–384.

Oono, K., and Sugiura, M. (1980) Heterogeneity of the ribosomal RNA gene cluster in rice. Chromosoma *76*, 85–89.

Paul, A.-L., Vasil, V., Vasil, I. K., and Ferl, R. J. (1987) Constitutive and anaerobically induced DNase-I-hypersensitive sites in the 5′ region of the maize *Adh1* gene. Proc. Natl. Acad. Sci. USA *84*, 799–803.

Peerbolte, R., Leenhouts, K., Hooykaas-van Slogteren, G. M. S., Wullems, G. J., and Schilperoort, R. A. (1986) Clones from a shooty tobacco crown gall tumor II: irregular T-DNA structures and organization, T-DNA methylation and conditional expression of opine genes. Plant Mol. Biol. *7*, 285–299.

Peschke, V. M., and Phillips, R. L. (1991) Activation of the maize transposable element *Suppressor-mutator (Spm)* in tissue culture. Theor. Appl. Genet. *81*, 90–97.

Peschke, V. M., Phillips, R. L., and Gengenbach, B. G. (1987) Discovery of transposable element activity among progeny of tissue-culture derived maize plants. Science *238*, 804–807.

Peschke, V. M., Phillips, R. L., and Gengenbach, B. G. (1991) Genetic and molecular analysis of tissue-culture-derived *Ac* elements. Theor. Appl. Genet. *82*, 121–129.

Pfeifer, G. P., Steigerwald, S. D., Mueller, P. R., Wold, B., and Riggs, A. D. (1989) Genomic sequencing and methylation analysis by ligation mediated PCR. Science *246*, 810–813.

Phillips, R. L., Plunkett, D. J., and Kaeppler, S. M. (1991) Novel approaches to the induction of genetic variation and plant breeding implications, in: Breeding plants for the 1990's North Carolina State Univ. Symp. 10–14 March 1991. C.A.B. Intl., Raleigh, NC.

Prat, S., Willmitzer, L., and Sanchez-Serrano, J. J. (1989) Nuclear proteins binding to a cauliflower mosaic virus 35S truncated promoter. Mol. Gen. Genet. *217*, 209–214.

Quemada, H., Roth, E. J., and Lark, K. G. (1987) Changes in methylation of tissue cultured soybean cells detected by digestion with the restriction enzymes HpaII and MspI. Plant Cell Rep. *6*, 63–66.

Razin, A., and Cedar, H. (1991) DNA methylation and gene-expression. Microbiol. Rev. *55*, 451–458.

Razin, A., and Riggs, A. D. (1980) DNA methylation and gene function. Science *210*, 604–610.

Riggs, A. D. (1975) X inactivation, differentiation and DNA methylation. Cytogenet. Cell Genet. *14*, 9–25.

Riggs, C. D., and Chrispeels, M. J. (1990) The expression of phytohemagglutinin genes in *Phaseolus vulgaris* is associated with organ-specific DNA methylation patterns. Plant Molec. Biol. *14*, 629–632.

Riggs, C. D., Voeller, T. A., and Chrispeels, M. J. (1989) Cotyledon nuclear proteins bind to DNA fragments harboring regulatory elements of phytohemagglutinin genes. Plant Cell. *1*, 609–621.

Sager, R., Sano, H., and Grabowy, C. T. (1984) Control of maternal inheritance by DNA methylation in *Chlamydomonas*, in: Curr. Top. Microbiol. Immunol., pp. 157–172. Ed. T. A. Trautner. Springer-Verlag, Berlin.

Saluz, H.-P., Jiricny, J., and Jost, J.-P. (1986) Genomic sequencing reveals a positive correlation between the kinetics of strand-specific DNA demethylation of the overlapping estradiol/glucocorticoid-receptor binding sites and the rate of avian vitellogenin mRNA synthesis. Proc. Natl. Acad. Sci. USA *83*, 7167–7171.

Saluz, H.-P., and Jost, J.-P. (1989) A simple high-resolution procedure to study DNA methylation and *in vivo* DNA-protein interactions on a single-copy gene level in higher eukaryotes. Proc. Natl. Acad. Sci. USA *86*, 2602–2606.

Saluz, H.-P., Feavers, I. M., Jiricny, J., and Jost, J.-P. (1988) Genomic sequencing and *in vivo* footprinting of an expression-specific DNaseI-hypersensitive site of an avian vitellogenin II promoter reveal a demethylation of a mCpG and a change in specific interactions of proteins with DNA. Proc. Natl. Acad. Sci. USA *85*, 6697–6700.

Sanders, P. R., Winter, J. A., Barnason, A. R., Rogers, S. G., and Fraley, R. T. (1987) Comparison of the cauliflower mosaic 35S and nopaline synthase promoters in transgenic plants. Nucl. Acids Res. *15*, 1543–1558.

Sano, H., Kamada, I., Youssefian, S., Katsumi, M., and Wabiko, H. (1990) A single treatment of rice seedlings with 5-azacytidine induces heritable dwarfism and undermethylation of genomic DNA. Mol. Gen. Genet. *220*, 441–447.

Scheid, O. M., Paszkowski, J., and Potrykus, I. (1991) Reversible inactivation of a transgene in *Arabidopsis thaliana*. Mol. Gen. Genet. *228*, 104–112.

Schmidt, R. J., Burr, F. A., and Burr, B. (1987) Transposon tagging and molecular analysis of the maize regulatory locus *opaque*-2. Science *238*, 960–963.

Schwabe, W. W. (1954) Factors controlling flowering in the chrysanthemum. J. Exp. Biol. *5*, 389–400.

Schwartz, D. (1988) Comparison of methylation of the male- and female-derived *wx-m9 Ds-cy* allele in endosperm and sporophyte, in: Plant Transposable Elements, pp. 351–354. Ed. O. Nelson. Plenum Press, New York.

Schwartz, D. (1989) Gene-controlled cytosine demethylation in the promoter region of the *Ac* transposable element in maize. Proc. Natl. Acad. Sci. USA *86*, 2789–2793.

Schwartz, D., and Dennis, E. (1986) Transposase activity of the *Ac* controlling element in maize is regulated by its degree of methylation. Mol. Gen. Genet. *205*, 476–482.

Selker, E. U., Cambareri, E. B., Jensen, B. C., and Haack, K. R. (1987) Rearrangement of duplicated DNA in specialized cells of *Neurospora*. Cell *51*, 741–752.

Shapiro, H. S. (1975) Content of 6-methylaminopurine and 5-methylcytosine in DNA, in: Handbook of Biochemistry, Selected Data for Molecular Biology: Nucleic Acids. Ed. G. D. Fasman. CRC Press, Boca Raton, FL.

Shapiro, R., and Klein, R. S. (1966) The deamination of cytidine and cytosine by acidic buffer solutions: mutagenic implications. Biochem. *5*, 2358–2362.
Shen, E. S., and Whitlock, J. P. (1989) The potential role of DNA methylation in the response to 2,3,7,8-tetrachlorodibenzo-*p*-dioxin. J. Biol. Chem. *264*, 17754–17758.
Shinozaki, K., Ohme, M., Tanaka, M., Wakasugi, T., Hayshida, N., Matsubayasha, T., Zaita, N., Chunwongse, J., Obokata, J., Yamaguchi-Shinozaki, K., Ohto, C., Torazawa, K., Meng, B. Y., Sugita, M., Deno, H., Kamogashira, T., Yamada, K., Kusuda, J., Takaiwa, F., Kata, A., Todoh, N., Shimada, H., and Sugiura, M. (1986) The complete sequence of the tobacco chloroplast genome: its gene organization and expression. EMBO J. *5*, 2043–2049.
Siegel, A., and Kolacz, K. (1983) Heterogeneity of pumpkin ribosomal DNA. Plant Physiol. *72*, 166–171.
Smith, C. J. S., Watson, C. F., Bird, C. R., Ray, J., Schuch, W., and Grierson, D. (1990) Expression of a truncated tomato polygalacturonase gene inhibits expression of the endogenous gene in transgenic plants. Mol. Gen. Genet. *224*, 477–481.
Soave, C., and Salamini, F. (1984) Organization and regulation of zein genes in maize endosperm. Phil. Trans. R. Soc. Lond. B. *304*, 341–347.
Spena, A., Viotti, A., and Pirrotta, V. (1983) Two adjacent genomic zein sequences: structure, organization and tissue specific expression. J. Mol. Biol. *169*, 799–811.
Spiker, S., Murray, M. G., and Thompson, W. F. (1983) DNaseI sensitivity of transcriptionally active genes in intact nuclei and isolated chromatin of plants. Proc. Natl. Acad. Sci. USA *80*, 815–819.
Staiger, D., Kauleen, H., and Schell, J. (1989) A CACGTG motif of the *Antirrhinum majus* chalcone synthase promoter is recognized by an evolutionary conserved nuclear protein. Proc. Natl. Acad. Sci. USA *86*, 6930–6934.
Stockhaus, J., Eckes, P., Blau, A., Schell, J., and Willmitzer, L. (1987) Organ-specific and dosage-dependent expression of a leaf/stem specific gene from potato after tagging and transfer into potato and tobacco plants. Nucl. Acids Res. *15*, 3479–3491.
Sutton, W. D., Gerlach, W. L., Schwartz, D., and Peacock, W. J. (1984) Molecular analysis of *Ds* controlling element mutations at the *Adh* locus of maize. Science *223*, 1265–1268.
Taylor, S. M., Constantinides, P. A., and Jones, P. A. (1984) 5-Azacytidine, DNA methylation, and differentiation, in: Curr. Top. in Microbiol. Immunol., pp. 115–127. Ed. T. A. Trautner. Springer-Verlag, Berlin/Heidelberg.
Theiss, G., Schleicher, R., Schimpff-Weiland, R., and Follman, H. (1987) DNA methylation in wheat. Eur. J. Biochem. *167*, 89–96.
Thomas, A. J., and Sherratt, H. S. A. (1956) The isolation of nucleic acid fractions from plant leaves and their purine and pyrimidine composition. Biochem. J. *62*, 1–4.
Thomson, W. W., and Whatley, J. M. (1980) Development of nongreen plastids. Ann. Rev. Plant Physiol. *31*, 375–394.
Tobin, E. M., and Silverthorne, J. (1985) Light regulation of gene expression in higher plants. Ann. Rev. Plant Physiol. *36*, 569–593.
van der Krol, A. R., Mur, L. A., Beld, M., Mol, J. N. M., and Stuitje, A. R. (1990) Flavonoid genes in petunia: addition of a limited number of gene copies may lead to a suppression of gene expression. Plant Cell. *2*, 291–299.
van Grinsven, M. Q. J. M., and Kool, A. J. (1988) Plastid gene regulation during development: an intriguing complexity of mechanisms. Plant Mol. Biol. Rep. *6*, 213–239.
van Slogteren, G. M. S., Hooykaas, P. J. J., and Schilperoort, R. A. (1984) Silent T-DNA genes in plant lines transformed by *Agrobacterium tumefaciens* are activated by grafting and by 5-azacytidine treatment. Plant Mol. Biol. *3*, 333–336.
Walbot, V., Britt, A. B., Luehrsen, K., McLaughlin, M., and Warren, C. (1988) Regulation of *Mutator* activities in maize, in: Plant Transposable Elements, pp. 121–136. Ed. O. Nelson. Plenum Press, New York.
Walbot, V., and Cullis, C. A. (1985) Rapid genomic exchange in plants. Ann. Rev. Plant Physiol. *36*, 367–396.
Walbot, V., and Warren, C. (1990) DNA methylation in the *alcohol dehydrogenase*-1 gene of maize. Plant Mol. Biol. *15*, 121–125.
Walker, J. C., Howard, E. A., Dennis, E. S., and Peacock, W. J. (1987) DNA sequences required for anaerobic expression of the maize alcohol dehydrogenase1 gene. Proc. Natl. Acad. Sci. USA *84*, 6624–6628.
Walling, L., Drews, G. N., and Goldberg, R. B. (1986) Transcriptional and post-transcriptional regulation of soybean seed protein mRNA levels. Proc. Natl. Acad. Sci. USA *83*, 2123–2127.

Wang, R. Y.-H., Zhang, X.-Y., and Erhlich, M. (1986) A human DNA-binding protein is methylation-specific and sequence-specific. Nucl. Acids Res. *14*, 1599–1614.

Watson, J. C., Kaufman, L. S., and Thompson, W. F. (1987) Developmental regulation of cytosine methylation in the nuclear ribosomal RNA genes of *Pisum sativum*. J. Mol. Biol. *193*, 15–26.

Weber, H., and Graessmann, A. (1989) Biological activity of hemimethylated and single-stranded DNA after direct gene transfer into tobacco protoplasts. FEBS Lett. *253*, 163–166.

Weber, H., Zeichmann, C., and Graessmann, A. (1990) *In vitro* DNA methylation inhibits gene expression in transgenic tobacco. EMBO J. *9*, 4409–4415.

Wellensiek, S. J. (1964) Dividing cells as the prerequisite for vernalization. Plant Physiol. *39*, 832–835.

Yesufu, H. M. I., Hanley, A., Rinaldi, A., and Adams, R. L. P. (1991) DNA methylase from *Pisum sativum*. Biochem. J. *273*, 469–475.

Zhang, D., Ehrlich, K. C., Supakar, P. C., and Erhlich, M. (1989) A plant DNA-binding that recognizes 5-methylcytosine residues. Molec. Cell. Biol. *9*, 1351–1356.

DNA Methylation: Molecular Biology and Biological Significance
ed. by J. P. Jost & H. P. Saluz

Patterns of *de novo* DNA methylation and promoter inhibition: Studies on the adenovirus and the human genomes

Walter Doerfler

Institut für Genetik, Universität zu Köln, Weyertal 121, D-5000 Köln 41, Germany

1 The problem

Mammalian DNA, in particular human DNA, is characterized by the existence of very specific, apparently inter-individually strictly conserved patterns of DNA methylation. These patterns are represented by the probably unique distribution of the modified nucleotide 5-methyl-deoxycytidine (5-mC) in many different sequences of mammalian DNA. Most likely, 5-mC is the only modified nucleotide in mammalian DNA, at least the one present in measurable quantity. On the basis of the occurrence of established patterns of 5-mC in mammalian genomes, it is justified to consider 5-mC a fifth, functionally significant nucleotide. We can state in rather general terms that 5-mC in a specific sequence motif can lead to the modulation of DNA-protein interactions, however, we cannot predict from the sequence environment of an individual 5-mC residue whether and what type of modulation will ensue. As for DNA-protein interactions in general, we do not yet know the code, the dictionary, that governs these interactions, even with non-5-mC-containing sequences. It can be safely predicted that DNA-protein interactions lie at the core of many biological mechanisms. Therefore, it is more than a bias when researchers working on DNA methylation ascribe a crucial role to 5-mC, one that is equivalent, though different in nature, to that of the other four nucleotides in DNA. The stability of patterns in the distribution of 5-mC across parts of the genome points to a function of these patterns in the structural and functional organization of the genome. This notion has, of course, not yet been proven, hence many of us are working to unravel in a step-by-step manner the structural and functional intricacies of this fascinating genetic signal.

The regulation of gene expression has been studied in many different laboratories and under various aspects. A multitude of transcription factors has been recognized which have to interact with specific promoter and upstream sequences to guarantee the regulated transcription of eukaryotic genes. In my laboratory, studies on the role of DNA methylation in integrated adenovirus genomes were initiated in 1977/78. Since then, a very considerable body of evidence has been collected from work on the adenovirus and several other systems to support the notion that sequence-specific promoter methylations can lead to the inactivation or inhibition of promoter function. This inactivation is probably long-term. It is not unconditional but can be modulated by other factors, as expected in a highly complex regulatory circuit. A great deal of experimental efforts will be required to elucidate the biochemical mechanism by which DNA methylation interferes with promoter functions. It is likely that the interference with DNA-protein interactions is at the core of this mechanism.

Patterns of DNA methylation in existing genomes can also be viewed under the aspect of evolution. Practically no information is available on how patterns of DNA methylation might have changed in the course of evolution. An attainable, however not yet accomplished goal in research on these problems would be the determination of methylation patterns for a gene or segment of DNA whose sequence has been highly conserved in phylogeny in the DNA of different species.

There is evidence, at least for some parts of the genome, that during the early stages of mammalian development demethylation and specific remethylation events take place. The mechanisms of *de novo* methylation following the loss of 5-mC residues are not understood. What signals does the DNA-methyltransferase system of the host recognize in the *de novo* generation of methylation patterns in foreign DNA or in endogenous DNA during development?

We started to investigate adenovirus (Ad) DNA and in particular the DNA of Ad-transformed mammalian cells with respect to DNA methylation as early as 1975/76. More recently, we have extended these investigations on the distribution of 5-mC to various parts of the human genome with the intent to determine what patterns of DNA methylation exist in this complex genome.

2 DNA methylation – a putative host defense mechanism against the expression of foreign DNA

As will be discussed in the following sections, foreign DNA that has become integrated into a preexisting mammalian genome is frequently *de novo* methylated in specific patterns, and different segments of the integrated foreign DNA can be methylated to very different extents. It has also been observed that foreign genes permanently fixed as transgenes in established cell lines or in transgenic animals or plants can become extensively *de novo* methylated and – probably as a consequence – inactivated. The apparently quite selective mechanism of *de novo* methylation of foreign DNA is not understood. Are sequence elements in the foreign DNA, the cellular site of foreign DNA insertion or the genetics of the host's DNA-methyltransferase system decisive factors in determining if and to what extent foreign DNA sequences will be modified?

In prokaryotic organisms, foreign DNA introduced into the cell by virus infection or by transfection is frequently specifically fragmented by the action of restriction endonucleases. In this way, the foreign (viral) DNA can be prohibited from replicating, and the integrity of the cell is salvaged. The genome of the prokaryote itself is protected from the action of its own restriction endonucleases by sequence-specific DNA-methyltransferases which methylate those cellular DNA sequences which would be recognized and cleaved by the host's restriction endonucleases. Methylation of specific DNA sequences often incapacitates the restriction endonucleases to recognize and cleave the modified recognition sequences. Thus, the host genome avoids being destroyed by its own protective enzymes. In the evolution of viral genomes, the specific recognition sequences for the host's protecting restriction endonucleases are eliminated by selective pressure, as much as sequence economy permits. Sequence adaptation and the evolution of new restrictase specificities appear to be the consequences of a protracted struggle for host protection versus viral survival. It is still not understood why the DNA-methyltransferase of the host does not also modify and thus protect the foreign viral genomes. Explanations may be sought in the rapid replication of the viral genomes and, possibly, in the compartmentalization of the DNA-methyltransferase on the host genome.

The genomes of phages specialized to replicate in eukaryotic green algae encode their own restriction endonucleases and DNA-methyltransferases (van Etten et al., 1985). In these instances, the invading foreign viral genome is endowed with the destructive and self-protective armatory otherwise used by prokaryotes.

Most higher eukaryotes lack restriction endonucleases as defense mechanisms but seem to have developed other protective devices against virus infections (e.g. interferon). Eukaryotes have, however, DNA-methyltransferase(s), remnants, as it were, of the prokaryotic defenses. I have proposed that these eukaryotic (mammalian) DNA-methyltransferases with an efficient potential for selective *de novo* methylation of integrated foreign DNA constitute a powerful defense mechanism against the activity of foreign genes which – if unchecked – might become quite detrimental to the function of the recipient cell (Doerfler, 1991b). The selectivity in foreign DNA methylation might be determined by the potential value of foreign genes for the functionality of the recipient cell. A potentially useful gene will not be methylated and inactivated. Neutral or harmful genes could become methylated. This screening process requires time and, in keeping with this line of reasoning, it can be observed that the *de novo* methylation of foreign DNA is a slowly progressing event (Kuhlmann and Doerfler, 1982; Orend et al., 1991). There is evidence that the insertion of foreign DNA can lead to changes in the extent of DNA methylation in the DNA sequences immediately adjacent to the cellular site of foreign DNA integration (Jähner and Jaenisch, 1985; Lichtenberg et al., 1988). It will be interesting to determine how far into the cellular genome these changes extend and what functional consequences – inactivations or reactivations – they might have on the activity of neighboring cellular genes.

Lastly, it will be important for those interested in gene therapy to understand in detail the mechanisms of *de novo* DNA methylation. In what conformation, at what site(s), or in what state of methylation will foreign genes have to be inserted into a pre-existing genome to escape from the host's own defense mechanisms which might eventually cause the inactivation of the therapeutic inserts and hence make them useless.

3 The selection of the experimental systems

3.1 Adenoviruses

In research on the molecular biology of eukaryotic, in particular of mammalian cells, adenoviruses (Ad) have served as a useful tool. I started to work with this system in 1966 and commenced work on Ad DNA methylation in 1975. Here, I shall present a brief survey of Ad biology as it can be related to problems of DNA methylation. We have utilized exclusively the human adenoviruses of type 2 (Ad2) and type 12 (Ad12). Ad can replicate in human cells growing in culture. The usually employed cell lines are KB cells derived from a carcinoma of the oral cavity or HeLa cells which originated from a cervical carcinoma. Primary human embryonic kidney cells are a very efficient permissive cell type for the replication of Ad. Hamster cells are permissive for Ad2, but absolutely non-permissive for Ad12, although Ad12 DNA enters the hamster cell nuclei where the early viral genes (see below) can be transcribed. In 1962, Trentin, Yabe, and Taylor (Trentin et al., 1962) reported the seminal discovery that the subcutaneous injection of small amounts of purified Ad12 into newborn hamsters leads to the generation of undifferentiated sarcomas at the site of injection. This finding stimulated a great deal of work on the molecular biology of adenoviruses which thus had proven themselves as tools not only for research on the molecular biology of mammalian cells but also on viral oncogenesis.

The Ad genome is a double-stranded linear DNA molecule with a terminal 55-kiloDalton (kDa) protein (tp) covalently attached at either

5′ terminus. An 88-kDa precursor of this protein, which is encoded by the viral genome, functions as initiator of the bipolar replacement synthesis of Ad DNA. The nucleotide sequence of Ad2 DNA comprising 35,937 nucleotide pairs has been determined and many of the viral functions have been physically mapped on the genome. This map is schematically depicted in Figure 1. There are four gene clusters, E1, E2, E3, and E4 which are transcribed early (E) in the viral replication cycle, i.e., prior to the onset of Ad DNA replication which begins 6–8 h postinfection of Ad2-infected HeLa or KB cells. Genes in the late L1 region are also transcribed early in infection. The transcription of the E2A and E2B regions is controlled by the early (E) or late (L) E2A promoter. The E2A region encodes the DNA binding protein essential for Ad DNA replication, the E2B region the tp and the Ad-specific DNA polymerase. The transcription of all late viral functions (L1 to L5), which mainly comprise the viral structural proteins, is governed by the major late promoter (MLP). In the course of processing the very long primary Ad transcription products, RNAs are spliced in a complex way, and to the 5′-ends of all late Ad RNAs an identical tripartite leader sequence is added on by splicing. RNA splicing in eukaryotes has been discovered in the adenovirus system.

The genes in the E1 region of the Ad genome are thought to play a role in Ad oncogenesis and transformation. One of the E1A encoded proteins, a 289 amino acid polypeptide, acts as a transactivator (Flint and Shenk, 1989) of all other viral and of some cellular genes (Nevins, 1987). The mechanism of transactivation is not completely understood. The 289 amino acid E1A protein likely interacts with cellular transcription factors and activates them directly or by binding and sequestering inhibitory elements (Shi et al., 1991). There are as many as 24 related mRNAs encoded in the E4 region, and these functions are somehow involved in effecting the switch from early to late gene expression and in the establishment of "replication factories" in the nuclei of infected cells. Genes located in the E3 region are apparently not required for viral replication in culture. There is evidence that the E3 gene products help the virus to overcome the immune response of the host. A 19-kDa protein encoded in the E3 region of Ad DNA can interact with class I gene products of the cellular major histocompatibility complex. The VA RNAs, encoded at around 30 map units on the Ad genome, are transcribed by the DNA-dependent RNA polymerase III, whereas all other Ad genes are controlled by RNA polymerase II of the host. The VA RNAs are about 160 nucleotides in length. VA RNAI is an effective activator of translation and can also counteract translation inhibition by cellular interferon-dependent mechanisms.

Events in Ad-infected cells have been analyzed in considerable detail, and only a brief summary of the main features of the system could be presented here. I have not referenced this section but should like to

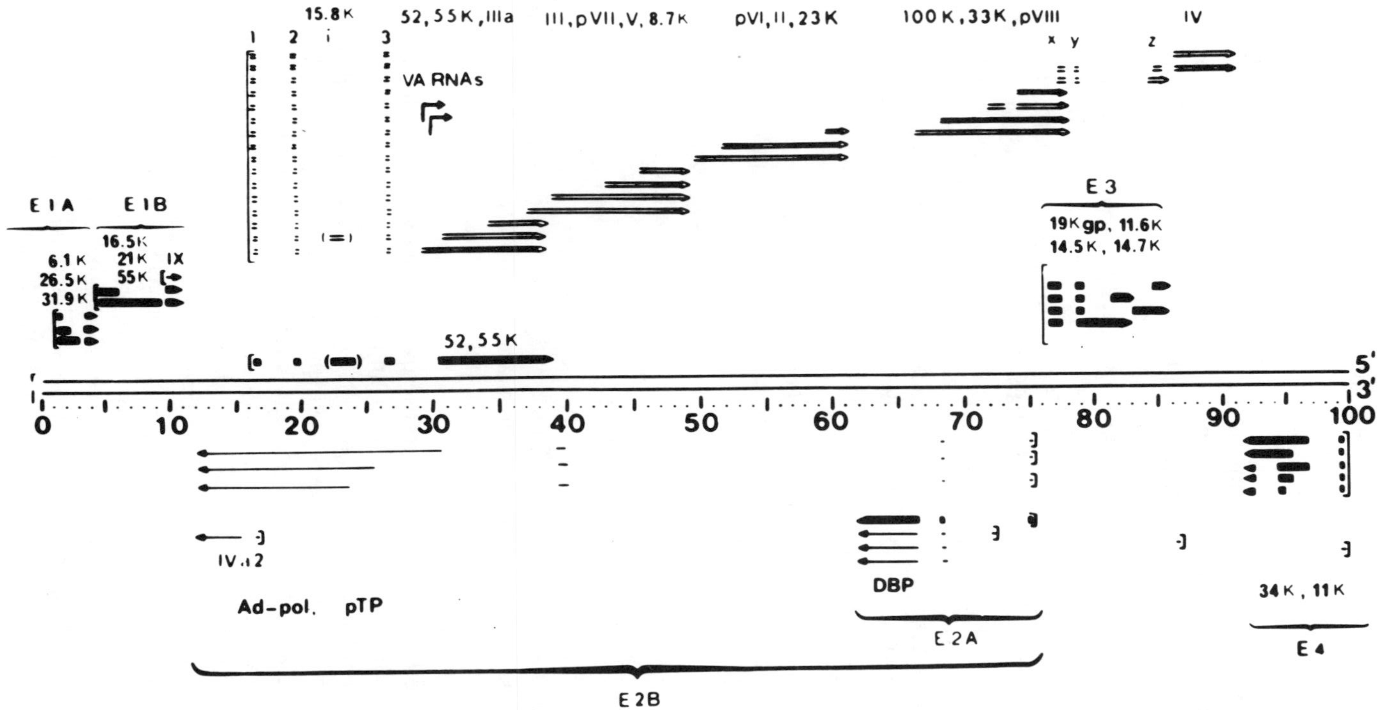

Figure 1. Functional map of the Ad2 genome. The double-stranded DNA molecule with the rightward (r) and leftward (l) transcribed strands and a fractional length scale are indicated in the center of the graph. The early (E) and late (L) transcription units are represented by arrows. The major gene products are presented by their molecular weights in kDaltons (K) or by the polypeptide number (Roman numerals) of the virion polypeptides. DBP – DNA binding protein; pTP – precursor terminal protein; Ad-pol. – Ad DNA polymerase. The tripartite leader sequences (1, 2, 3) spliced on to all groups of late messenger RNAs are also included. This map was taken from Akusjärvi et al. (1986).

direct the reader to textbooks on adenovirology for further information (Doerfler, ed., 1983, 1984; Ginsberg, ed., 1985).

Ad12 can induce tumors in newborn hamsters; Ad12 and Ad2 can transform a variety of rodent cells in culture to cells that have many properties of tumor cells. In Ad12-induced tumor cells and in Ad-transformed cells, the viral genome is integrated, frequently in multiple copies, into the host genome. Integration appears to occur preferentially at cellular sequences which are transcriptionally active. In hamster cells, which are non-permissive for Ad12, the entire Ad12 genome is in general integrated into the genomes of Ad12-induced tumor cells or Ad12-transformed cells. Persistence of the intact or nearly intact viral genome is not counterselected for, as it cannot replicate in hamster cells. In contrast, in Ad2-transformed hamster cells, the integrated viral genome usually carries smaller or larger deletions. This finding is consistent with the permissivity of hamster cells for Ad2 replication. I have proposed that persistence of intact Ad2 genomes in hamster cells would lead to viral replication and destruction of the cells; thus transformants could not arise (Doerfler, 1982; Doerfler et al., 1983), and persistence of intact Ad2 genomes is not likely to occur.

Studies on integrated Ad genomes (reviewed in Doerfler et al., 1983) have led us to a major effort in research on DNA methylation of foreign integrated DNA (Sutter et al., 1978; Kuhlmann and Doerfler, 1982, 1983; Kruczek and Doerfler, 1982; Orend et al., 1991). We recognized that methylation-sensitive restriction endonucleases failed to cleave integrated Ad12 genomes in hamster cells, whereas Ad12 virion DNA isolated from purified virions could be cut to completion by the same restrictases (Sutter et al., 1978; Sutter and Doerfler, 1979, 1980). The interest in mechanisms of viral DNA integration and transformation have thus opened unexpected avenues and committed us to investigations on *de novo* DNA methylation and its functional significance. Moreover, the discovery of inverse correlations between the extent of Ad DNA methylation and the transcriptional activities of integrated Ad12 genes (Ortin et al., 1976; Sutter and Doerfler, 1979, 1980; Vardimon et al., 1980) has motivated and helped to initiate a major project on the interrelationships between sequence-specific promoter methylation and gene activity. I have written several reviews on this complex subject which has grown into a major area of research since our original observations in the late seventies (Doerfler, 1981, 1983, 1984, 1989, 1990, 1991a; Doerfler et al., 1984, 1990).

As mentioned above, Ad12 infects hamster cells abortively, the viral DNA cannot replicate in these cells, and there is a block in late transcription, whereas many of the early viral genes can be transcribed. The Ad12 DNA reaches the hamster cell nucleus and can be integrated into the host genome (Doerfler, 1968, 1970). At least in part, the defect in the transcription of the late Ad12 genes in hamster cells can be explained by the presence of a mitigator element, a 32-nucleotide pair sequence located downstream of the major late promoter (MLP) of Ad12 DNA which prohibits the activity of the MLP in hamster cells or in intact hamster nuclei and dampens it in human cells (Weyer and Doerfler, 1985; Zock and Doerfler, 1990). Removal of the mitigator element facilitates the transcription of the late Ad12 genes in hamster cells. Surprisingly, in cell-free extracts of hamster nuclei, the MLP of Ad12 DNA can elicit transcription, hence an intact nuclear structure seems to be required for the mitigator to exert its negative influence on late Ad12 DNA transcription (Zock et al., 1993).

Our research on the abortive infection of hamster cells and the transformation of hamster cells by Ad12 (or Ad2) has been recently summarized in two reviews (Doerfler, 1991a, 1992). I have considered it important to investigate details of the interactions between Ad12 and hamster cells, since knowledge on the biology of this system might help to explain the mechanism of Ad12 oncogenesis.

3.2 Baculovirus – Autographa californica nuclear polyhedrosis virus (AcNPV)

Reviews on this virus system have been published (Doerfler, 1986; Luckow and Summers, 1988; Miller, 1988). I will not present details of this insect virus system here. DNA methylation does not appear to play a role in the replication cycle of this virus. In a limited set of experiments, we have utilized AcNPV promoters to demonstrate that their sequence-specific methylation can lead to inactivation even in insect cells (Knebel et al., 1985) in which the genetic signal 5-mC has as yet not been detected (e.g., Eick et al., 1983).

3.3 Patterns of DNA methylation in the human genome

In recent years, the elucidation of the biological significance of patterns of DNA methylation in a complex genome, like the human genome, has attracted our interest. These patterns might influence, if not determine, the organization and patterns of transcription in the human genome (Doerfler, 1990). Perhaps, patterns of DNA methylation can be used as indicators for gene activity. We have observed that the distribution of 5-mC residues in specific human genes is cell type specific and exhibits a surprisingly high interindividual concordance, at least in parts of the human genome (Kochanek et al., 1990, 1991; Behn-Krappa et al., 1991). We are currently extending these studies to other segments of the human genome. Since the mapping and potentially the determination of the complete nucleotide sequence of the human genome seems to have become the ambition of some molecular biologists, we may have to watch that the fifth nucleotide in the human genome receives appropriate attention. At the same time, as the nucleotide sequences of more extensive segments of the human genome become known, it might become more realistic to work successfully on the significance of 5-mC residues for the organization of the human genome.

4 DNA methylation in the viral replication cycle

In the DNAs of Ad2 or Ad12, modified nucleotides, like 5-mC, have not been detected by chemical analyses (Günthert et al., 1976; Eick et al., 1983) or by extensive restriction endonuclease analyses (Wienhues and Doerfler, 1985). The sensitivity of the methods employed does not permit one to rule out the presence of a very small number of 5-mC residues per molecule or in some of the viral molecules. Similarly, there is no evidence for the presence of methylated nucleotides in free viral DNA at any time in cells productively or abortively infected with Ad (Wienhues and Doerfler, 1985). The term "free viral DNA" is applied to DNA molecules in the nuclei of infected cells and in contrast to those viral DNA molecules that have become integrated into the host genome. It was conceivable that viral DNA methylation could be of importance in the shift from the early to the late viral transcription cycles. We have, therefore, initiated a study on the MLP of Ad2 DNA in virion DNA (from purified virions) and in intranuclear viral DNA at early and late times postinfection. Both restriction cleavage followed by Southern blot hybridization (Southern, 1975) and the genomic sequencing method (Church and Gilbert, 1984; Saluz and Jost, 1987) were applied as analytical tools. The results adduced so far do not suggest that the MLP of Ad2 DNA was methylated at any time after infection or that methylation of sequences in the MLP of Ad DNA was an important regulatory factor (Wienhues and Doerfler, 1985; C. Kremer, S. Kochanek, and W. Doerfler, unpublished results). This work is still in progress and, again, the possibility of a small subset of viral DNA molecules being methylated cannot be completely ruled out. The application of the genomic sequencing method was essential in this analysis, because one or a few cytidine residues in the MLP of Ad2 DNA could have been methylated, and their detection would have been impossible, unless these cytidine residues had been part of the recognition sequence of a methylation-sensitive restriction endonuclease. It is likely that DNA methylation does not play a role in the replication and transcription cycles of free Ad DNA. *Prima facie*, DNA methylation and adenovirus research would, therefore, have been very unlikely partners to meet had it not been for the permanent state of Ad DNA, i.e. those viral genomes integrated into the host cell's chromosome (see below).

5 Cellular DNA sequences in virion DNA and in the host genome: Determinants of *de novo* DNA methylation

In Ad12-infected human cells, we have discovered in Ad12 particles a subpopulation of hybrid molecules consisting of the 2081 left terminal nucleotides of Ad12 DNA linked to a large segment of human host cell DNA which harbors mainly, but not exclusively, repetitive DNA sequences. Analyses with the electron microscope have revealed that these molecules are apparently large mirror-image palindromes in which the right and left halves are symmetrically identical (Deuring et al., 1981; Deuring and Doerfler, 1983). These hybrid molecules have been termed symmetrical recombinants (SYREC). As they presumably carry the viral packaging signal (Hammarskjöld and Winberg, 1980) in the left terminal Ad12 DNA sequences, SYREC molecules are encapsidated into Ad12 particles and can thus be highly purified. We have demonstrated that the cellular DNA sequences in the virion-encapsidated SYREC molecules are not methylated at the 5′-CCGG-3′ (HpaII) sequences, whereas the same cellular DNA sequences in the genome of human KB cells are completely methylated (Deuring et al., 1981). These results directly demonstrate that the cellular SYREC DNA sequences are methylated when they are an endogenous part of the host chromosome, but apparently unmethylated when they replicate under the control of the adenovirus replication machinery. Thus, the nature of a nucleotide sequence by itself cannot be the sole decisive determinant influencing the extent of DNA methylation. Our results on the methylation status of the cellular SYREC sequences rather support the concept that a complex interplay of nuclear compartment localization and type of replication apparatus could be co-determinants of DNA methylation. As Ad DNA has its own esoteric DNA replication mechanism, in which initiation and polymerization processes are efficiently performed by virus-specific and viral genome-encoded proteins, Ad DNA freely persisting or replicating in the nucleus of infected cells does probably not get access to the cellular replication machinery. The DNA-methyltransferase system of the host is a likely component of the cellular replication apparatus. In this context, it makes sense that integrated Ad DNA as part of the host chromosome is subject to replication by the host machinery and in that course becomes *de novo* methylated in very distinct, probably cell-specific patterns (see below). Nevertheless, it is still puzzling that non-integrated Ad or SYREC DNA replicating in human cell nuclei does not become methylated at all, whereas the same cellular sequences in presumably the same nuclei are extensively methylated. We have shown that DNA-methyltransferase activities from uninfected and Ad-infected KB cell nuclei are indistinguishable (Doerfler et al., 1982). However, this rather crude analytical approach did not yet do full justice to the sophistication of the DNA-methyltransferase machinery in the host cell.

6 *De novo* methylation of integrated adenovirus DNA

In 1978, we reported that integrated Ad12 DNA molecules in the DNA of a number of Ad12-transformed cell lines could not be cleaved with the restriction endonuclease HpaII, whereas Ad12 virion DNA was readily cut to a large number of fragments (Sutter et al., 1978). We interpreted this finding at the time as evidence for the *de novo* methylation of Ad12 (foreign) DNA after its insertion into the host chromosome. In subsequent years, we pursued the process of *de novo* methylation in detail. I will summarize our major observations and refer the reader to glean details from the original publications.

6.1 Gradual establishment of de novo methylation patterns

Upon the integration of Ad12 DNA into the genome of Ad12-induced hamster tumor cells, the viral DNA is in general not immediately methylated. When the hamster tumor cells were explanted into culture, the integrated viral sequences were gradually methylated with increasing passage number in culture (Kuhlmann and Doerfler, 1982, 1983; Orend et al., 1991).

6.2 De novo methylation of transfected integrated DNA

De novo methylation of integrated Ad12 DNA occurring in a step-wise manner was also observed after co-transfection of hamster cells with Ad12 DNA together with the gene for neomycin phosphotransferase and selection for neomycin-resistant cells. In different cell lines thus established, methylation of Ad12 DNA increased with the number of generations following the transformation event (S. T. Tjia, G. Meyer zu Altenschildesche, and W. Doerfler, unpublished results). Thus, the eliciting of *de novo* DNA methylation of integrated foreign (Ad12) DNA was probably not primarily dependent on the transfer of cells from a hamster tumor in an animal into culture.

6.3 Specific patterns

The *de novo* methylation of integrated Ad12 DNA molecules is not a random process, but follows a certain pattern which is similar for different cloned Ad12-induced tumor cell lines (Kuhlmann and Doerfler, 1983; Orend et al., 1991).

6.4 Initiation of de novo methylation in internal parts of the foreign genome

The *de novo* methylation commences in the internal parts of the co-linearly integrated Ad12 DNA molecules. The sites of initiation of *de novo* methylation appear to be spread across a certain area and do not seem to be confined to a very narrow region. Methylation gradually spreads across large parts of the integrated Ad12 molecule. Surprisingly, multiple copies of Ad12 DNA follow a very similar, if not identical, pattern of *de novo* methylation (Sutter et al., 1978; Kruczek and Doerfler, 1982; Kuhlmann and Doerfler, 1983; Orend et al., 1991).

6.5 Exclusion of terminal parts and sites of junction

The terminal parts of the Ad12 genome, which are linked to cellular DNA, and the sites of linkage do not become methylated, at least not at their 5′-CCGG-3′ sequences (Orend et al., 1991). Again, this lack of methylation at the viral termini is common to multiple integrated Ad12 DNA molecules. These findings suggest that the nature of the integrated DNA sequences can at least influence the *de novo* pattern of DNA methylation. It is conceivable that the right and left termini of Ad12 DNA are not methylated and inactivated because their expression is functionally important and of survival value to the transformed cells.

6.6 Changes in DNA methylation in adjacent cellular DNA sequences

In one Ad12-induced hamster tumor cell line T1111/2 (Lichtenberg et al., 1987), we could demonstrate that at the cellular site of Ad12 DNA integration, in an area of about 1000 nucleotides of cellular DNA, the 5′-CCGG-3′ and 5′-GCGC-3′ sequences had lost the 5-mC residues after Ad12 DNA integration. In contrast, in non-Ad12-transformed hamster cells and at the equivalent to the integration site on the non-occupied chromosome in the Ad12-induced tumor cell line, the same sequences remained completely methylated (Lichtenberg et al., 1988). Thus, the integration of Ad12 (foreign) DNA can lead to changes in the methylation patterns of cellular DNA sequences close to the site of integration. It will be of interest to investigate how extensive an area of cellular DNA around the site of foreign DNA integration can be affected by these changes of DNA methylation. As a corollary to these investigations, we want to study to what extent these changes in the methylation of cellular DNA can cause alterations of cellular DNA (gene) transcription in Ad12-induced tumor cells or in Ad-transformed cells. A major study along these lines has been initiated (Rosahl and

Doerfler, 1992). Perhaps, these changes are important for the mechanisms of adenovirus transformation and oncogenesis (Doerfler, 1991a, 1992).

6.7 The spreading of DNA methylation

In the establishment of *de novo* patterns of DNA methylation, the gradual spreading of DNA methylation plays a significant role (Kuhlmann and Doerfler, 1983; Toth et al., 1989, 1990; Orend et al., 1991). We have used the indicator gene chloramphenicol acetyl transferase (CAT) coupled to the late E2A promoter. This construct was genomically integrated into hamster DNA in several cloned BHK21 cell lines (Müller and Doerfler, 1987). Prior to host genome integration, the construct had been enzymatically premethylated *in vitro* by HpaII DNA methyltransferase at its three 5′-CCGG-3′ sequences which reside in the promoter at nucleotide positions +24, +6, and −215, relative to the cap site at +1 of this promoter. Alternatively, the construct and the promoter were left unmethylated.

By using the genomic sequencing method, which allows the screening of each cytidine residue for the presence of 5-mC, we have demonstrated that DNA methylation gradually spreads from the enzymatically premethylated sites to neighboring nucleotides (Toth et al., 1989). This spreading progresses with increasing number of cell generations and can be detoured to some extent by proteins bound at certain sequences in the integrated promoter in living cells (Toth et al., 1990). The results of this spreading analysis are presented in Figure 2. In the experiments performed here, the non-premethylated promoter was not methylated during the time course of this analysis.

6.8 Methylation of non-CG sequences is the exception

It is also apparent from the data in Figure 2 that the *de novo* methylation of integrated foreign DNA pertains in most cases only to 5′-CG-3′ sequences. In the late E2A promoter of Ad2 DNA in cell line HE2, each one 5′-CA-3′ and one 5′-CT-3′ dinucleotide have been found methylated in two different passages of this cell line (Toth et al., 1990). In this cell line, all 5′-CG-3′ sequences were methylated in the late E2A promoter (Toth et al., 1990) which is inactive (Johansson et al., 1978; Vardimon et al., 1980; Esche, 1982). Since the 5′-CA-3′ and 5′-CT-3′ sequences are not palindromic, as 5′-CG-3′ is, it is difficult to understand how the methylated status of non-CG dinucleotides could be stably maintained. Inheritance of the 5-mC signal would be a problem, since the complementary sequences 3′-GT-5′ (to 5′-CA-3′) and 3′-GA-5′ (to 5′-CT-3′)

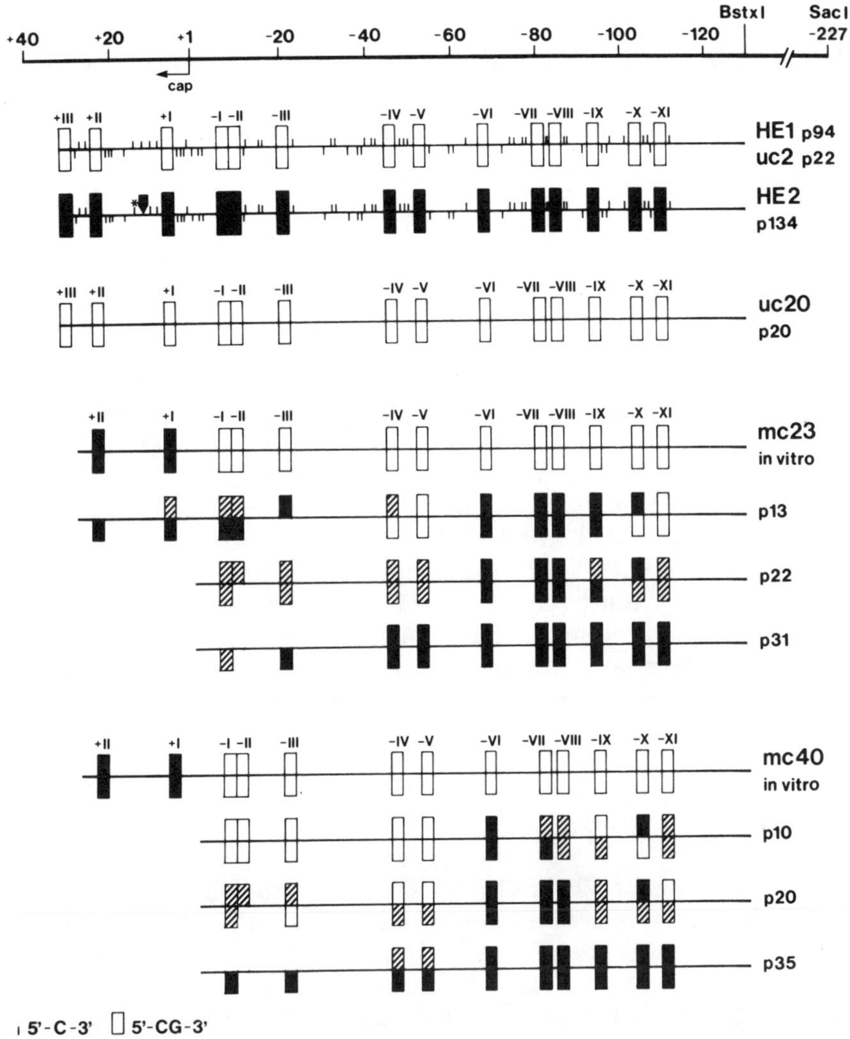

Figure 2. Summary of the genomic sequencing data from the late E2A promoter of Ad2 DNA. Fully and partially methylated or unmethylated 5′-CG-3′ dinucleotides in the late E2A promoter of Ad2 DNA are presented for the transformed cell lines HE1, HE2 (Vardimon and Doerfler, 1981), uc2, uc20, or for cell lines mc23 and mc40 (Müller and Doerfler, 1987) in different passages (p), as shown by genomic sequencing. The scale refers to nucleotide numbers in the late E2A promoter relative to the cap site (↲). The 5′-CCGG-3′ sequences are at nucleotides +6 (+I) and +24 (+II), which have been premethylated *in vitro* in the generation of cell lines mc23 and mc40 (Müller and Doerfler, 1987). Horizontal lines represent the late E2A promoter segment in individual cell lines. The 5′-CG-3′ sequences in this segment (+III to −XI) are represented by vertical bars: (□) unmethylated; (■) completely methylated; and (▨) 5′-CG-3′ sequences which are methylated in only some of the integrated promoter copies. The bars above the horizontal line designate 5′-CG-3′ dinucleotides in the top strand of the promoter sequence, the bars below the line represent the same dinucleotides in the bottom strand. In the E2A region of Ad2 DNA the bottom strand is the transcribed strand, the top strand the non-transcribed complement. This map has been taken from Toth et al. (1990).

cannot contain 5-mC as a potential signal for recognition by the DNA-methyltransferase system. In all the genomic sequencing work in human (Kochanek et al., 1990, 1991) and hamster DNA (Toth et al., 1989, 1990), which we have completed so far and which comprises in all >1000 nucleotides, the singular 5′-CA-3′ and 5′-CT-3′ sequences in cell line HE2 are the only non-CG located 5-mC residues we have been able to trace so far.

6.9 De novo methylation of foreign DNA in transgenic mice

We have also genomically integrated the aforementioned late E2A promoter-CAT gene construct in the methylated or in the unmethylated form into the genomes of transgenic mice (Lettmann et al., 1991). In general, the 5′-CCGG-3′ premethylated constructs remained methylated in the preimposed patterns in all organ systems investigated, except in the testes in which the constructs became at least partly demethylated in several founder and F1 (first generation beyond founder) animals. The non-methylated constructs remained largely unmethylated in one founder and its progeny. In other founder animals and their progeny, the non-premethylated construct was *de novo* methylated in the 5′-CCGG-3′ sequences, the only sequences investigated, in all organ systems analyzed (Lettmann et al., 1991).

6.10 General conclusions

Ad12 DNA (about 30 kbp) or the late E2A promoter of Ad2 DNA (about 160 bp) have been used as prototype foreign DNA molecules for studies on the mechanism of *de novo* methylation in mammalian cells. Very similar results were obtained irrespective of whether Ad12-induced tumor cell lines, Ad-transformed cell lines, cells transfected with Ad12 DNA in culture or mice rendered transgenic for the late E2A promoter of Ad2 DNA were used as the experimental system. Frequently, *de novo* methylation is characterized by the gradual spreading across a promoter sequence of about 160 nucleotides, as determined by genomic sequencing, or across an integrated viral genome of about 30 kilobase pairs, as analyzed by restriction and Southern blot hybridization analyses. The presence of a few preimposed 5-mC residues in a sequence might facilitate and speed up *de novo* methylation, is, however, not essential for the initiation of the process.

What are the decisive factors for the inception of *de novo* DNA methylation in integrated foreign (viral) DNA? Nucleotide sequence, the cellular site of foreign DNA integration, DNA-protein complexes, the structure of DNA, the genetics of the host's DNA-methyltransferase system and other unknown factors seem to determine singly or – perhaps more likely – in cooperation the kinetics and patterns of the *de novo* methylation of foreign (viral) DNA. *De novo* methylation can lead to the inactivation of foreign genes (Doerfler, 1981, 1983, 1991b). There is no evidence for the assumption that the data adduced with the adenovirus system would be valid only for this system. More likely, they are of general significance, although it will be important to extend this work to other types of foreign DNA molecules. For a number of reasons, it will be desirable to understand what parameters determine extent and speed of the methylation of foreign DNA. Frequently, foreign DNA integrated in transgenic cells in culture, animals or plants has been found to be inactivated by DNA methylation. Before one can hope to apply successfully concepts and techniques of gene therapy, the problems around *de novo* methylation of foreign DNA will have to be solved.

7 Differential viral gene expression in adenovirus-transformed or in Ad12-induced tumor cells: Inverse correlations to DNA methylation

The mechanism of transformation of mammalian cells by adenoviruses is not really understood. The chase for the decisive events continues. In my laboratory, we have tried to pursue less frequently discussed possibilities (Doerfler, 1992) and have mainly concentrated on the role of Ad DNA integration and alterations in methylation patterns in viral and in adjacent cellular DNA sequences. The long-established term of "insertional mutagenesis" can now be viewed in a different, so far not yet applied sense. In addition to the possibility that cellular genes at the site of foreign DNA integration can be directly destroyed or mutated, we have shown that the levels of DNA methylation in cellular DNA sequences surrounding the site of Ad DNA insertion can be altered (Lichtenberg et al., 1988). These changes could alter the transcriptional patterns of cellular DNA and thus have consequences similar to true mutations. Perhaps, one could designate these alterations epi-mutations. It will be a prime goal of our research to investigate how far these changes in cellular DNA methylation can extend from the site of insertion into the cellular genome.

In Ad12-induced hamster tumor cells or in Ad-transformed hamster cells, many of the early viral gene groups are transcribed, many of the late viral genes are permanently shut off (Ortin et al., 1976). There are, of course, exceptions to this "rule". In the Ad12-transformed hamster cell line HA12/7, e.g., the early region E3 is not transcribed and heavily methylated at 5′-CCGG-3′ sequences (Ortin et al., 1976; Kruczek and Doerfler, 1982). On the other hand, in several Ad12-induced hamster brain tumor cell lines some of the late viral gene groups can be expressed and they are hypomethylated (Ibelgaufts et al., 1980; Sutter and Doerfler, 1980). Viral replication has not been observed in any of these hamster tumor or transformed cell lines. Ad12 DNA replication in hamster cells has been shown to be completely blocked (for review, Doerfler, 1991a). Transcriptional details for all the early and late Ad12 genes in hamster cells have not yet been determined.

Using Ad12-transformed (Sutter et al., 1978; Sutter and Doerfler, 1979, 1980) or Ad2-transformed hamster cells (Vardimon et al., 1980) and analyzing the patterns of methylation in the different segments of the integrated Ad genomes, we discovered striking inverse correlations between the levels of viral gene transcription and the extent of DNA methylation at 5′-CCGG-3′ sequences in these Ad DNA segments. Active genes were undermethylated, inactive genes were extensively methylated. In a more detailed analysis, we refined this observation, in particular for the promoter and 5′ upstream regions of integrated Ad12 genes in several Ad12-transformed hamster cell lines (Kruczek and Doerfler, 1982). Concordant observations were reported from many laboratories for many viral and non-viral eukaryotic genes studied in many different systems (for reviews Doerfler, 1981, 1983, 1984; contributions in Razin, Cedar, and Riggs, eds., 1984).

In our own work, the late E2A promoter as a part of integrated Ad2 DNA in Ad2-transformed cell lines (Cook and Lewis, 1978; Vardimon and Doerfler, 1981) was studied in detail. In cell line HE1, this promoter is active (Johansson et al., 1978; Esche, 1982) and unmethylated at its three 5′-CCGG-3′ sequences (Vardimon et al., 1980), whereas in cell lines HE2 and HE3 this promoter does not function, and the same sequences are methylated in all copies of the integrated Ad2 genomes. Recently, we succeeded in refining this analysis by applying the genomic sequencing technique. In a promoter and downstream region of about 160 nucleotides within the late E2A promoter, all 14 5′-CG-3′ dinucleotide sequences are unmethylated in cell line HE1 and methylated in cell lines HE2 and HE3 (Toth et al., 1989, 1990) (Fig. 2).

On the basis of these early correlative results on DNA methylation and promoter activity, which suggested a possible causative role for sequence-specific promoter methylation in long-term promoter inactivation, I decided upon a detailed experimental analysis of this problem. It has proven a good choice to concentrate on work with one or a few viral promoters, for which inverse correlations had been established, and to test the working hypothesis of promoter methylation as a regulatory signal in different experimental systems. I selected the late E2A promoter of Ad2 DNA as our main experimental tool. This work was initiated and preceded by work on the E1A promoter of Ad12 DNA (Kruczek and Doerfler, 1983) and complemented by studies on the VAI gene of Ad2 DNA and the p10 promoter of AcNPV.

8 Sequence-specific promoter methylation and long-term gene inactivation: Experiments with viral promoters

8.1 The E1A promoter of Ad12 DNA

Our first experiments with the *in vitro* methylation of viral promoter sequences and the subsequent assessment of promoter inactivation were performed with the E1A promoter of Ad12 DNA (Kruczek and Doerfler, 1983). The E1A promoter-CAT gene construct was used in this work and consisted of the 525 left terminal nucleotide pairs of Ad12 DNA and the prokaryotic CAT gene which had just been established at that time as an activity indicator for eukaryotic promoter studies (Gorman et al., 1982). This construct was left unmethylated or was methylated at all 5′-CCGG-3′ (HpaII) or 5′-GCGC-3′ (HhaI) sequences. Upon the transfection of these constructs into mouse Ltk^- cells, the unmethylated promoter was active, the methylated constructs were non-functional (Kruczek and Doerfler, 1983). Since in the course of these experiments the entire construct was methylated, it could not be unequivocally decided which part of the methylated construct was responsible for promoter inactivation. In control experiments, the early SV40 promoter-CAT gene construct, in which the SV40 promoter was devoid of 5′-CCGG-3′ and 5′-GCGC-3′ sequences, did not lose its activity by *in vitro* methylation with these enzymes. This result was interpreted to point to the E1A promoter as the methylation-sensitive element in the construct. However, it could still be argued that the Ad12 E1A and early SV40 promoters differed in strengths, and the effects of methylation on the two different constructs could not be directly compared. We, therefore, devised a more subtle experimental approach using the late E2A promoter of Ad2 DNA (see below).

As a corollary to this study, we also investigated what effect the methylation of other sequences might have on the activity of the E1A promoter. A number of prokaryotic DNA-methyltransferases with different sequence specificities were employed in this study. The most striking and unexpected observation has been that the introduction of one N^6-methyldeoxyadenosine (N^6-mA) residue in the immediate vicinity of the left end of the E1A DNA segment (left terminal 525 nucleotide pairs) inhibited the E1A promoter. Thus, although N^6-mA is thought not to be present in mammalian DNA – a likely though not definitely proven notion – the transcriptional and control apparatus in human (HeLa) cells can nevertheless respond to this signal. The *in vitro* premethylation of AluI (5′-AGCT-3′) or MboI (5′-GATC-3′) sequences did not compromise E1A promoter function, whereas the methylation of HpaII (5′-CCGG-3′), HhaI (5′-GCGC-3′), or TaqI (5′-TCGA-3′) sequences inhibited the E1A promoter of Ad12 DNA in an E1A promoter-CAT gene construct (Kruczek and Doerfler, 1983; Knebel and Doerfler, 1986).

8.2 The late E2A promoter of Ad2 DNA

The experimental work with this promoter had the advantage that it could be premised on previous results on this promoter's activity in living cells in culture. On the other hand, it was known prior to initiating this project that the activity of the late E2A promoter of Ad2 DNA was dependent on and enhanced by the E1A transactivator of Ad DNA (for reviews Berk, 1986; Nevins, 1987; Flint and Shenk, 1989). The effect of 5′-CCGG-3′ sequence methylation on the activity of the late E2A promoter of Ad2 DNA was determined in transient expression systems, such as microinjection into oocytes of *Xenopus laevis*, or transfection into mammalian cells and *in vitro* transcription experiments. In addition, the effect of methylation was also assessed upon the genomic fixation of the unmethylated or the 5′-CCGG-3′ methylated late E2A-CAT gene construct in BHK21 hamster cells or in the different organs of transgenic mice.

The anatomy of the late E2A promoter with upstream and downstream sequences is shown in Fig. 3. For the interpretation of data described in this section, the three 5′-CCGG-3′ sequences are relevant which can be methylated by HpaII at the cytidine residues in promoter positions +24, +6, and −215, relative to the site of transcriptional initiation at nucleotide position +1.

8.2.1 Transient expression systems

8.2.1.1 Xenopus laevis oocytes. In these experiments, the BamHI-HindIII fragment of the E2A gene of Ad2 DNA (Fig. 3) was microinjected in the 5′-CCGG-3′ methylated or the unmethylated form into oocytes from *Xenopus laevis*. This fragment contains 14 HpaII sites. The methylated construct was inactive, the unmethylated construct was transcribed (Vardimon et al., 1981, 1982a). In a control experiment, the 5′-CCGG-3′ methylated E2A gene was coinjected with the unmethylated histone h22 gene from sea urchin. The histone gene of sea urchin was transcribed, the methylated E2A gene of Ad2 DNA was silenced indicating that inactivation of the E2A gene was not due to artifacts in these experiments. In the same experimental approach, 5′-GGCC-3′ methylation of the E2A gene of Ad2 DNA did not affect its expression upon microinjection into *Xenopus laevis* oocytes (Vardimon et al., 1982b).

In further studies, the late E2A promoter, the EcoRI-HindIII fragment (Fig. 3) with three 5′-CCGG-3′ sequences was excised from the BamHI-HindIII E2A fragment. Upon *in vitro* methylation at the three 5′-CCGG-3′ sequences, it was linked back to the unmethylated main body of the gene (BamHI-EcoRI fragment) with eleven unmodified 5′-CCGG-3′ sequences. Upon microinjection into oocytes of *Xenopus laevis*, this construct was not transcribed. The complementary construct

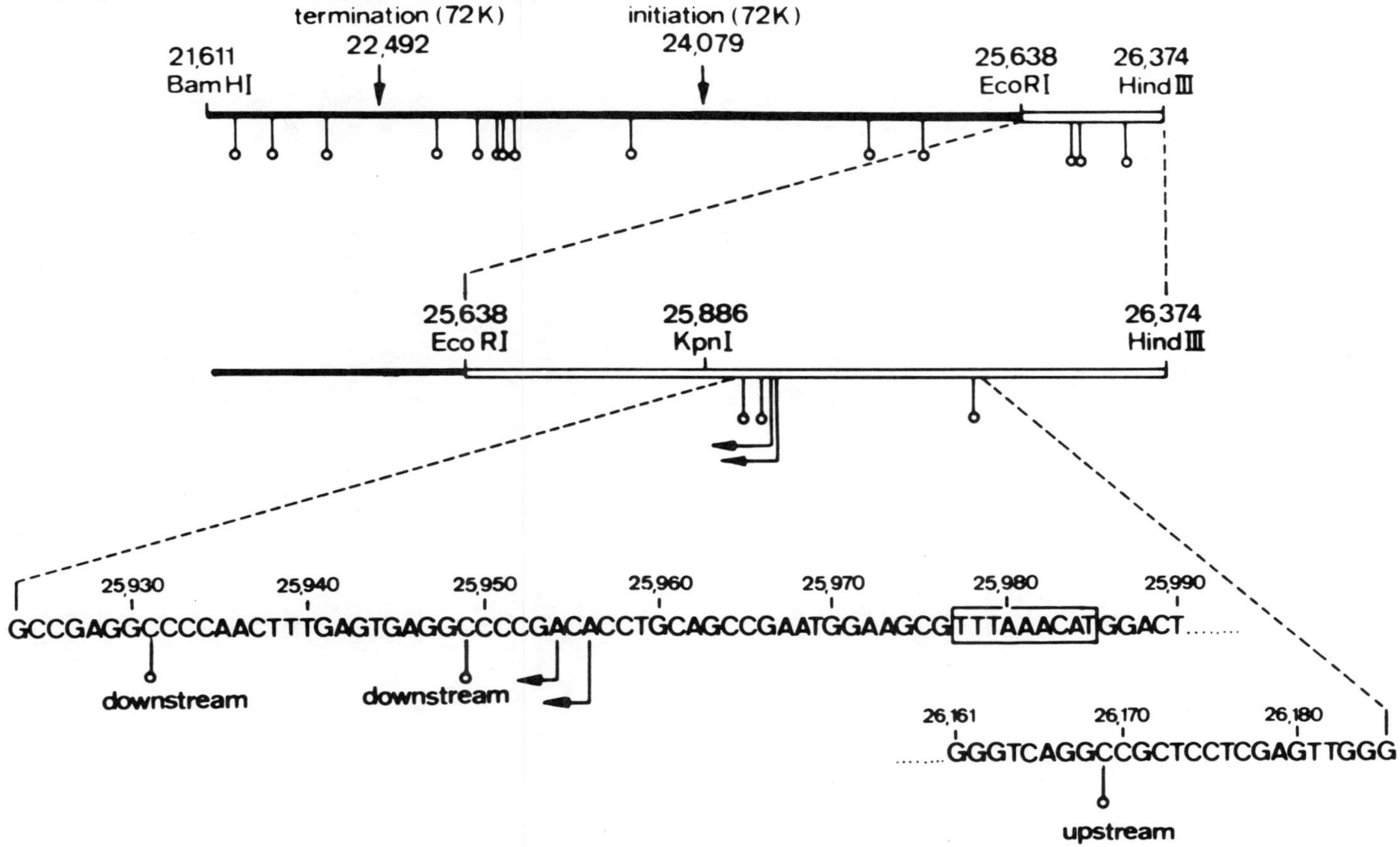

Figure 3. Map of the E2A subclone (BamHI-HindIII fragment) and maps of the promoter (EcoRI-HindIII) and gene (BamHI-EcoRI) fragments. Ad2 nucleotide numbers refer to the Ad2 DNA sequence as determined by Roberts et al. (1986). The DNA-binding protein (DBP = 72 kDalton protein) is read from right to left. ↓, Initiation and termination sites of translation of the 72 kDalton protein. ♁, designate 5′-CCGG-3′ (HpaII) sites. The E2A gene has a non-canonical TATA sequence (boxed). This map has been taken from Langner et al. (1984).

carrying the unmethylated promoter segment and the 5′-CCGG-3′ methylated sequences of the main body of the E2A gene was transcribed like the completely unmodified gene (Langner et al., 1984). These results indicate that, at least for the E2A gene of Ad2 DNA, the promoter region together with upstream and downstream elements contains sequences whose 5′-CCGG-3′ methylation leads to the inactivation or the strong inhibition of this gene. The methylation-sensitive sites are located at positions +24, +6, and −215. We do not know yet whether the methylation of other 5′-CG-3′ sequences in the late E2A promoter would also affect its functionality. Moreover, the results described demonstrate the inhibitory effect of methylation on a eukaryotic mammalian virus promoter in a rather heterologous system, as *Xenopus* oocytes are not the natural host for Ad. We have, therefore, turned to mammalian cells as an additional test system.

8.2.1.2 Mammalian cells. The late E2A promoter with upstream and downstream sequences (the KpnI-HindIII fragment, see Fig. 3) was fused to the prokaryotic CAT indicator gene and tested for its activity in the unmethylated or its 5′-CCGG-3′ methylated form after transfecting either of these constructs into human HeLa or hamster BHK21 cells. The results of CAT activity measurements confirmed the previously adduced observations. The methylated construct was inactive or strongly reduced in its activity, whereas the unmethylated DNA elicited CAT activity in extracts of transfected cells (Langner et al., 1986). In contrast, when human 293 cells or Ad-transformed hamster cells were used as recipients, which carried in an integrated form and constitutively expressed the left terminus of Ad2 or adenovirus type 5 (Ad5) DNA, the methylated construct was also active, although at a level somewhat lower than that of the unmethylated DNA which had attained higher activity than in HeLa or BHK21 cells due to transactivation (Langner et al., 1986; Weisshaar et al., 1988). It will be described below that Ad E1A functions can probably help to overcome the inhibitory function of promoter methylation (Weisshaar et al., 1988; Knust et al., 1989) by an as yet unknown mechanism.

8.2.1.3 In vitro transcription. Studies in an *in vitro* transcription system might eventually facilitate analyses on the biochemical mechanism by which sequence-specific promoter methylations could reduce the promoter function. The transcriptional activity of the unmethylated or of the 5′-CCGG-3′ methylated late E2A promoter was hence assessed in cell free nuclear extracts prepared from human HeLa cells. We found that three preconditions had to be fulfilled before differential activities of the two types of constructs could be demonstrated. (i) A circular template had to be used. (ii) Extracts of high protein concentration yielded the best results. (iii) For each extract prepared, the optimal template DNA concentration had to be empirically determined. Under this set of experimental conditions, the promoter methylated at the

5′-CCGG-3′ sequences was inactive or its activity was strongly decreased, whereas the unmodified template had full activity (Dobrzanski et al., 1988). In some of the experiments, the major late promoter (MLP) of Ad2 DNA in its unmethylated form was used as a second template and internal control together with the 5′-CCGG-3′ methylated late E2A promoter. The same extract that mediated MLP activity in the same test tube did not transcribe the 5′-CCGG-3′ methylated late E2A promoter. This finding ruled out unspecific inhibitory events in the nuclear extract, perhaps as a side effect of methylation on one of the templates.

The striking dependence on template concentration for the inhibitory effect of promoter methylation in cell free transcription extracts has been interpreted to signify the presence and importance of a factor(s) present in the extracts in limiting amounts. This factor(s) might be responsible for the system's reaction to the presence of 5-mC residues in the critical promoter or downstream positions (Dobrzanski et al., 1988). Similar results and conclusions for a number of different mammalian promoters have recently been published (Boyes and Bird, 1991).

In summary, all the data gleaned from different types of transient expression systems, i.e., microinjection into *Xenopus laevis* oocytes, transfection into mammalian cells, and *in vitro* transcription experiments in cell-free extracts from mammalian nuclei, attest to the inhibitory or completely inactivating effect on transcription of methylated sequences in the promoter and downstream region. For the work with the late E2A promoter it was striking that two of the sequences mediating inhibition upon methylation are located in the downstream region of the promoter, i.e., at nucleotide positions +24 and +6.

8.2.2 Genomic integration

These general conclusions on the regulatory effect of sequence-specific methylation on the late E2A promoter, and perhaps on many eukaryotic promoters, documented in transient expression systems had to be confirmed by studies in which the same promoter was genomically integrated. In these experiments, the late E2A promoter was again premethylated at the same three 5′-CCGG-3′ sequences in positions +24, +6, and −215 or was left unmethylated prior to integration into the genome of BHK21 hamster cells (Müller and Doerfler, 1987) or of transgenic mice (Lettmann et al., 1991). The results on the transcriptional block by 5′-CCGG-3′ methylation of the late E2A promoter gleaned from these systems are in good agreement with those obtained from studies under transient expression conditions.

8.2.2.1 Hamster cells. The *in vitro* 5′-CCGG-3′ premethylated or the unmethylated late E2A promoter-CAT gene construct was cotransfected with the unmethylated pSV2-neo construct into BHK21 hamster cells. A number of neomycin-resistant clones of stably E2A-CAT construct-transformed cell lines were thus established (Müller and Doerfler,

1987). The pSV2-neo construct carries the gene for neomycin-phosphotransferase under the control of the early SV40 promoter (Southern and Berg, 1982). The cell lines obtained contained multiple copies of the late E2A promoter-CAT gene construct in the unmethylated or, in general, stably methylated form. In some of the above cell lines, the construct lost 5-mC residues and became partly demethylated, at least in some of the integrated copies of the premethylated construct. When CAT activity was determined, the cell lines transformed with the unmethylated construct yielded active extracts. Cell lines, in which the premethylated constructs had remained methylated, produced extracts with no CAT activity. Partial loss of 5-mC residues in the course of transformed cell passages entailed reduced CAT activity in extracts prepared from these BHK21 cell lines (Müller and Doerfler, 1987).

8.2.2.2 Transgenic mice. Some of the results obtained with the late E2A promoter-CAT gene construct in several clones of transgenic mice have already been described in section 6.9 and dealt with the stability of or changes in methylation patterns, and with the *de novo* methylation of foreign DNA in transgenic mice (Lettmann et al., 1991). With respect to the inhibitory effect of sequence-specific methylation, the data with transgenic mice can be summarized as follows. From none of the organs in transgenic mice harboring the late E2A promoter-CAT gene construct methylated at the 5′-CCGG-3′ sequences could extracts be obtained that exhibited CAT activity. In that sense, these data were consistent with the previously drawn conclusions. However, in a number of transgenic founder or F1 animals, extracts of testes, where the constructs had at least partly lost 5′-CCGG-3′ methylation, showed CAT activity (Lettmann et al., 1991). Extracts from organs of transgenic animals that contained the unmethylated construct exhibited CAT activity in some but not all instances. Extracts from kidney, spleen, testis, and colon showed CAT activity, whereas extracts from other organs were inactive. Extracts from testis and colon had low activity levels. Since promoter activity is subject to the control by multiple factors and since the late E2A promoter is known to be dependent on transactivation, the finding of variable CAT activity is not surprising.

8.2.3 Conclusions

In this section, I have presented a detailed account of our work with Ad promoters in a number of different experimental systems that permit the assessment of the transcriptional activity of these promoters. The results in their entity are consistent with the interpretation that the sequence-specific methylation of promoter and/or downstream sequences leads to the inactivation of the promoter. The effect is most likely a causative one. The data presented so far, however, do not pertain to the mechanism by which the inhibitory or inactivating effect is elicited. There are numerous results in the literature on other eukaryotic systems that confirm the data obtained with viral promoters. This volume presents several relevant examples of such studies.

8.3 The VA gene of Ad DNA

Most genes of the Ad genome are transcribed in Ad-infected cells by the host's DNA-dependent-RNA polymerase II. In contrast, the ~160 nucleotide VAI RNA, a translational regulatory element in the Ad genome, is under the control of RNA polymerase III of the host cell. This biochemical property permitted us to use the VA gene with its internal control regions and to study the effect of 5′-CG-3′ methylations also on the activity of an RNA polymerase III-transcribed gene. We have utilized two experimental approaches, the transfection into mammalian cells and the *in vitro* transcription of the unmethylated or the methylated VA gene of Ad2 DNA in nuclear extracts from HeLa cells. The results can be summarized as follows. When all 5′-CCGG-3′ and 5′-GCGC-3′ sequences were methylated, transcription upon transfection into HeLa cells was not inhibited, whereas in a cell-free transcription system with HeLa nuclear extracts an inhibitory effect could be demonstrated (Jüttermann et al., 1991). When the VA construct was methylated in all 5′-CG-3′ sequences by the DNA methyltransferase from *Spiroplasma species* (Renbaum et al., 1990), transcription of the VA construct in both experimental systems was completely inhibited (Jüttermann et al., 1991). At the same time, it could be shown that 5′-CG-3′ methylation of three sequences in the internal control region A of the VAI gene interfered with protein binding. It is concluded that sequence-specific methylation can also inhibit the transcription of the VAI Ad gene that is transcribed by RNA polymerase III of the host.

8.4 The p10 promoter of AcNPV DNA

Although there is, as yet, no evidence for the occurrence of 5-mC in the DNA of lepidopteran insect cells, like *Spodoptera frugiperda* (S.f.) cells, which serve as host cells for the baculovirus AcNPV, we have tested the effect of 5-mC residues introduced into the p10 promoter of AcNPV DNA on the activity of this promoter in S.f. cells. The p10 promoter is one of the strong promoters in this viral DNA which controls the expression of a gene encoding a 10 kDalton protein. When one 5-mC residue was placed into the only 5′-CCGG-3′ site in the promoter and into two downstream 5′-CCGG-3′ sites of the p10 promoter, the activity of this promoter in driving CAT expression in *Spodoptera frugiperda* insect cells was reduced about 40-fold (Knebel et al., 1985). Obviously, the work with insect cells will have to be extended before general interpretations can be attempted. The data available, however, suggest that even insect cells, thought to be genuinely devoid of 5-mC as a genetic signal, are nevertheless capable of recognizing this signal and of responding to it with promoter inhibition. I want to reserve judgment towards the widely held view that insect DNA would completely lack 5-mC. As long as extensive insect DNA sequences have not been screened for the presence of 5-mC by a highly sensitive technique, like the genomic sequencing method, I consider this question as essentially unsolved. Work on Drosophila DNA, e.g., would be very important as this system holds such an important position in developmental biology.

8.5 Reversal of inhibition

Since promoter control in mammalian cells is subject to a rather complex combination of factors, which is only partly understood, it has not been surprising to find that promoter inhibition or inactivation by DNA methylation is not an unconditional event (Langner et al., 1986; Knebel-Mörsdorf et al., 1988; Weisshaar et al., 1988; Knust et al., 1989). Again working with the late E2A promoter of Ad2 DNA, we have been able to demonstrate that the inhibition of this promoter by methylating the 5′-CCGG-3′ sequences at nucleotide positions +24, +6, and −215 can be overcome by the E1A-encoded 289 amino acid protein of Ad DNA (Weisshaar et al., 1988; Knust et al., 1989) which is considered the prototype mammalian transactivator (Flint and Shenk, 1989). This cancellation of the inhibitory effect by promoter methylation is not accompanied by the loss of 5-mC residues from the late E2A promoter, at least not in both DNA complements. Loss of 5-mC from one complement is also unlikely since the available evidence suggests that the methylation of one strand of the late E2A promoter suffices to inactivate it (Doerfler, 1989; Hermann et al., 1989; U. Freisem-Rabien, and W. Doerfler, unpublished results). Reversal of inhibition by the paradigm E1A transactivator of Ad can be demonstrated both with the transactivating E1A gene in the genomically integrated position plus the 5′-CCGG-3′ methylated late E2A promoter in *trans* in a transfected plasmid construct (Weisshaar et al., 1988) and, conversely, with the methylated late E2A promoter genomically integrated plus the transactivating E1A gene in *trans* in a transfected plasmid (Knust et al., 1989). In the latter instance, experiments have been performed with the Ad2-transformed hamster cell line HE3 which carries major parts of the Ad2 genome including the endogenous E1A and E2A genes integrated into the host genome. It is puzzling that, although the endogenous E1A transactivator is apparently actively expressed in this cell line, it is unable to transactivate the *cis* located, fully 5′-CG-3′ methylated (Toth et al., 1989, 1990), endogenous E2A gene which can, however, be activated by transfecting the Ad2 E1A gene as a plasmid (Knust et al., 1989). This apparent discrepancy has not yet been resolved. It might point to the complicated compartmentalization problems in mammalian cells.

We have also been able to demonstrate that cotransfection of mammalian cells, with the E1A transactivating gene as the 13S cDNA in one plasmid and the 5′-CCGG-3′ methylated late E2A promoter in another, also shows the reversal effect on methylation inhibition (Weisshaar et al., 1988). Thus, it is likely that the 13S RNA-encoded, 289-amino acid E1A protein is responsible for the alleviation of the inhibition of the late E2A promoter by 5′-CCGG-3′ methylation.

Similar reversal effects on methylation inhibition in eukaryotic promoters have been described for an immediate early gene product of the iridovirus frog virus 3 (FV3) (Thompson et al., 1986) and for the transactivator tat of the human immunedeficiency virus (Bednarik et al., 1990).

In a different set of experiments, we have documented that the presence in the construct of the strong enhancer of an immediate early gene from the human cytomegalovirus DNA together with the 5′-CCGG-3′ methylated late E2A promoter cancels or largely compensates promoter inhibition (Knebel-Mörsdorf et al., 1988).

The results on these rather specific cancellation effects suggest that the inhibition of viral promoters by sequence-specific methylation is dependent on complex DNA protein interactions. These interactions can be decisively modulated by DNA methylation and counteracted by transactivators or by enhancer elements.

8.6 Studies on the mechanism of promoter inhibition by DNA methylation

As has been well documented for the prokaryotic restriction endonucleases, sequence-specific DNA methylation can lead to the interference with specific DNA-protein interactions and to the abrogation of restriction enzyme functions. It is, therefore, important to investigate transcription factor-promoter interactions for their sensitivities towards sequence-specific DNA methylations. I have proposed this mechanism as a possibility for the inhibition of promoters by DNA methylation in one of my first reviews on DNA methylation (Doerfler, 1983).

We have, therefore, continued our work on the late E2A promoter of Ad2 DNA and have used this sequence also for studies on protein-promoter interactions and on the effect of promoter methylations at the three aforementioned 5′-CCGG-3′ sequences, at nucleotide positions +26, +4, and −215. A detailed map of the late E2A promoter of Ad2 DNA is shown in Figure 4. When the ~1200 nucleotide PvuII-PvuII fragment of this Ad2 DNA segment was used in the unmethylated or in

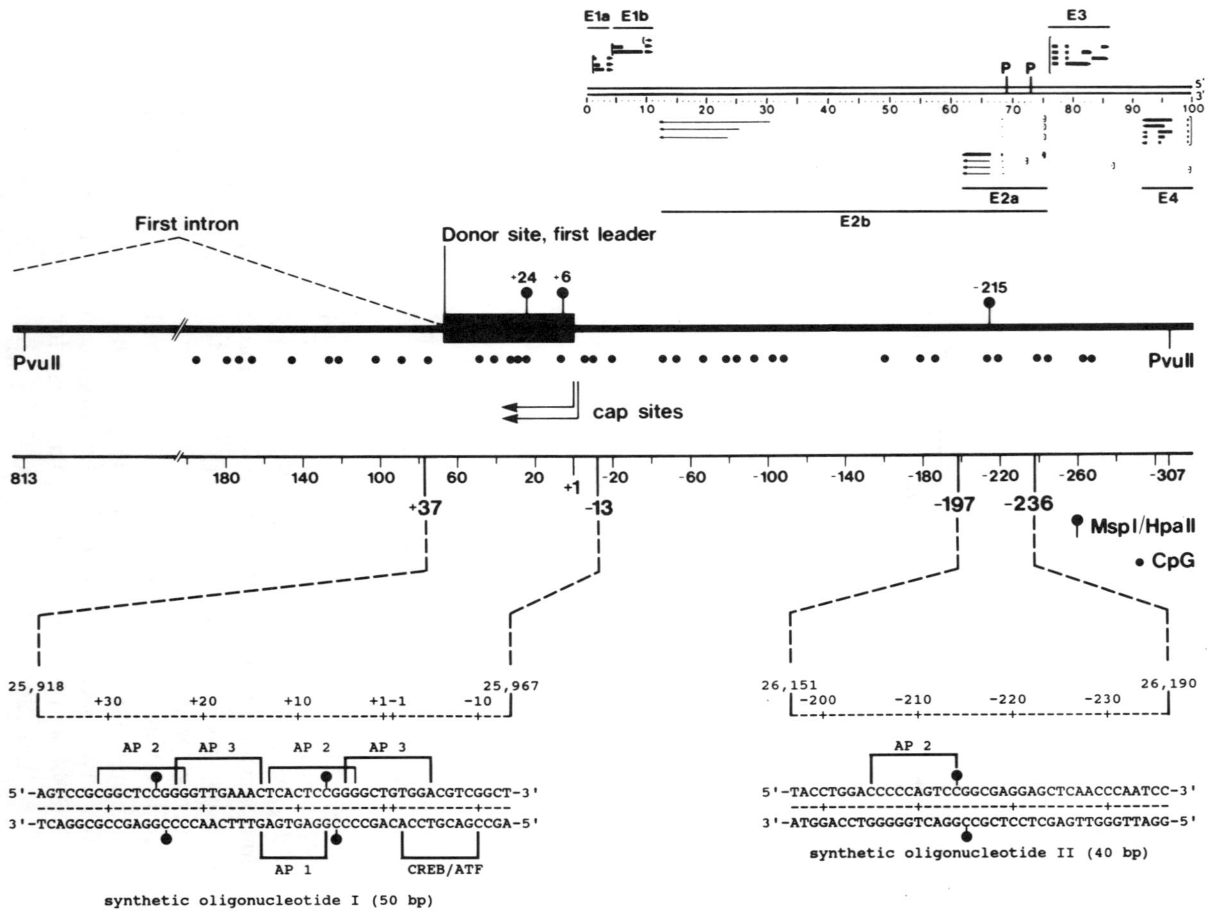
E1a E1b
E3
P P
E2a
E4
E2b
First intron
Donor site, first leader
+24 +6
-215
PvuII
PvuII
cap sites
813
180
140
100
60
20
+1
-20
-60
-100
-140
-180
-220
-260
-307
+37
-13
-197
-236
MspI/HpaII
CpG
25,918
25,967
+30 +20 +10 +1-1 -10
AP 2 AP 3 AP 2 AP 3
5'-AGTCCGCGGCTCCGGGGTTGAAACTCACTCCGGGGCTGTGGACGTCGGCT-3'
3'-TCAGGCGCCGAGGCCCCAACTTTGAGTGAGGCCCCGACACCTGCAGCCGA-5'
AP 1
CREB/ATF
synthetic oligonucleotide I (50 bp)
26,151
26,190
-200 -210 -220 -230
AP 2
5'-TACCTGGACCCCCAGTCCGGCGAGGAGCTCAACCCAATCC-3'
3'-ATGGACCTGGGGGTCAGGCCGCTCCTCGAGTTGGGTTAGG-5'
synthetic oligonucleotide II (40 bp)

the 5′-CCGG-3′ methylated form, differences in protein binding between the two different forms of this fragment could not be detected by the DNase I footprinting method (Hoeveler and Doerfler, 1987). It could not be ruled out that there were functionally significant differences in protein binding which could have remained undetectable with the relatively crude method of DNase I footprinting.

We subsequently resorted to using double-stranded synthetic oligodeoxyribonucleotides derived from the late E2A promoter sequence in gel retardation experiments. The 50 nucleotide long fragment from nucleotide positions −13 to +37 in this promoter (see Fig. 4) led to the formation of a high molecular mass DNA-protein complex and of several lower molecular mass DNA-protein complexes when incubated with partially purified nuclear extracts from HeLa cells. Proteins from these 0.42 M NaCl nuclear extracts were partially purified by $(NH_4)_2SO_4$ precipitation and heparin-Sepharose column chromatography. Formation of the high molecular mass complex was abrogated when the oligodeoxyribonucleotide was methylated in both or hemimethylated in the rightward transcribed or the leftward transcribed complement (Hermann et al., 1989; Hermann and Doerfler, 1991). These results provided evidence that the hemimethylation of the late E2A promoter might already compromise its function. The formation of the high molecular mass complex, which comprised at least four different proteins, could be competed by the unmethylated, but not by the 5′-CCGG-3′ methylated 50 base pair oligodeoxyribonucleotide. These data argued against the possible participation of a 5-mC-dependent protein in the formation of the high molecular mass complex. Similar results and a similar high molecular mass complex were obtained with a 73-base-pair-long oligodeoxyribonucleotide, which encompassed the nucleotide positions +24 and +6 in the downstream region of the late E2A promoter, but not with longer oligodeoxyribonucleotides (Hermann et al., 1989). Differences in complex formation between a methylated and a non-methylated 99 base pair oligodeoxyribonucleotide were much less distinct. The absence of differences in complex formation with longer DNA fragments cannot be easily explained at the present time. We have obtained encouraging results in experiments designed to purify more extensively by affinity chromato-

Figure 4. Map and factor binding sites in the late E2A promoter and its downstream region of Ad2 DNA. The insert (upper right) presents a map of the early regions of the Ad2 genome. The PvuII-PvuII fragment whose map is detailed in the central part of the graph is designated P-P in the Ad2 DNA map. In the PvuII-PvuII fragment, all 5′-CG-3′ dinucleotides (●) and the 5′-CCGG-3′ sequences (⚲) with the internal C at nucleotide positions +6, +24 and −215 relative to the downstream cap site have been indicated. The scale presents nucleotide positions upstream (−) and downstream (+) from this site. Nucleotide sequences akin to known factor binding sites have been bracketed in the bottom part of the scheme. This map has been taken from Hermann and Doerfler (1991).

graphy the proteins involved in the complex formation using the +37 to −13 oligodeoxyribonucleotide from the late E2A promoter. Further analytical work will be dependent on the availability of purified proteins that bind specifically to this downstream promoter segment. There is evidence that the transcription factor AP2 may play a role in the formation of the high molecular mass complex (Hermann and Doerfler, 1991). However, a definite interpretation of data will be difficult unless purified proteins become available for experimentation.

The 40 base pair synthetic oligodeoxyribonucleotide II straddling the −215 nucleotide upstream position in the late E2A promoter (Fig. 4) has also been used to study DNA-protein complex formation. Testing HeLa nuclear extracts, such complexes were formed but their generation was not abrogated by 5′-CCGG-3′ methylation at this site. It is, therefore, unlikely that methylation of the −215 position in the upstream region of the late E2A promoter is important in silencing this promoter. Of course, this conclusion must remain preliminary since, in this experimental approach, we are still working with rather short oligodeoxyribonucleotides and not with the intact promoter.

Recently, we have continued to use the *in vitro* transcription system to analyze further the role of DNA methylation in promoter silencing. Conditions were as described (Dobrzanski et al., 1988) and as summarized in section 8.2.1.3. The effect of inhibition of E2A promoter transcription by promoter methylation can be reproducibly demonstrated. This template activity can be inhibited by competition with increasing amounts of the unmethylated oligodeoxyribonucleotide comprising nucleotide positions −13 to +37, but not with the same DNA fragment in the 5′-CCGG-3′ methylated form. These results reason against, but do not exclude, the role of a 5-mC-dependent protein in the control of this promoter (R. Hermann and W. Doerfler, unpublished results).

The data available so far from this series of experiments render it likely that the interactions of specific proteins with downstream sequences in the promoter are at the core of the transcriptional control of the late E2A promoter. The methylation at positions +24 and +6 of the late E2A promoter compromises the formation of a DNA-protein complex, whereas another complex with the promoter segment around nucleotide position −215 is not sensitive to 5′-CCGG-3′ methylation. Of course, there are many additional 5′-CG-3′ sequences in other segments of the late E2A promoter whose methylation might affect promoter function and which have not been assayed so far. We recall that in the inactive late E2A promoter of integrated Ad2 DNA in cell line HE2 all 5′-CG-3′ sequences are methylated (Toth et al., 1989, 1990). Much more work will be required to elucidate the mechanism by which promoter methylation inhibits the late E2A promoter.

9 Human genes

9.1 Patterns of DNA methylation in complex genomes: Biological significance for the organization and expression of the human genome

During the last few years, our attention has been directed towards studies on DNA methylation patterns in the human genome. There are obvious medical implications that have rendered studies on human genes necessary and interesting, and there will be a great deal of sequence information available in the near future, due to worldwide efforts to map and even sequence the human genome. I consider it likely, and plan experiments with the working hypothesis in mind, that patterns of DNA methylation might be significant for the organization of a complex genome, like the human genome, and for the regulation of its expression. The term "organization of the genome" is obviously not strictly defined since we lack precise knowledge about the arrangement of DNA in the nucleus. It is, however, conceivable that patterns of 5-mC distribution in a complex genome relate to the intranuclear topology of DNA and perhaps to its interactions with the nuclear matrix. These interactions might also be dependent on DNA-protein complexes and might be modulated by specific patterns of DNA methylation.

These patterns could also serve as indicators of gene activity. Patterns of DNA methylation are cell type specific, and it will be a demanding task to determine the complexity of these patterns even for a few cell types. We, therefore, decided to seek data first which would support or disprove the notion of specific patterns of DNA methylation. One might compare the complexity of patterns of DNA methylation in the human genome to traffic regulation by traffic signals in a big city with a huge number of intersections. Of course, 5-mC residues might not only be equivalent to red traffic lights and stop transcriptional activity, but may have multiple additional functions. I have previously suggested that the actual patterns of 5-mC distribution represent the sum total of several interdigitating patterns with different functional significance (Doerfler, 1990). Whatever the complex functions of DNA methylation patterns will turn out to be, we have set out to determine these patterns as accurately as possible in a few segments of the human genome.

The limitations of the present technology suggest two independent approaches. (i) We have genomically sequenced small segments of DNA in the human genes for tumor necrosis factors (TNF)-α and -β (Kochanek et al., 1990, 1991). (ii) We have analyzed about 500 kbp of randomly selected human DNA by restriction and Southern blot hybridization analyses (Behn-Krappa et al., 1991). Further analyses are in progress on the human Alu sequences (Kochanek et al., 1993) and on the α-chain of the human gene for the interleukin 2 receptor (A. Behn-Krappa and W. Doerfler, unpublished results).

9.2 Interindividual concordance of patterns of DNA methylation in the human genome

The genomic sequencing method (Church and Gilbert, 1984) has been applied to determine the state of DNA methylation in the human genes for TNF-α and -β (Kochanek et al., 1990, 1991). DNA from human granulocytes, monocytes, T- and B-lymphocytes, from natural killer cells and from sperm has been analyzed. For comparison, the DNA from the human HeLa cell line has also been investigated. Methylation patterns proved to be highly cell type specific. The sequences studied, about 800 base pairs in TNF-α and about 200 base pairs in TNF-β, are methylated to different extents and in different patterns in the different populations of white blood cells and are rather fully methylated in sperm and in the continuous HeLa cell line. I will not repeat details of the published experimental findings (Kochanek et al., 1990; 1991). It is, however, striking that in 15 different DNA samples from human granulocytes, which were donated by 15 individuals of African, Asian or European origin, the four 5-mC residues found in the analyzed segment of the TNF-α gene were at exactly identical nucleotide positions (Fig. 5). Similar findings have been reported for nine different samples from human lymphocytes, again from 9 individuals of different ethnic origins. We have concluded that, at least in parts of the human genome, patterns of DNA methylation are highly concordant among individuals with different genetic backgrounds (Kochanek et al., 1990).

As an extension to these studies, though at a less precise level, we have also compared patterns of 5′-CCGG-3′ methylation in DNA samples of different individuals by restriction and Southern blot hybridization analyses. Although these methods afford a much lower resolution than the genomic sequencing method, the results of an investigation extended to about 500 kbp (0.02% of the human genome) of randomly selected human cellular DNA sequences demonstrate again interindividual concordance (Behn-Krappa et al., 1991). In these studies randomly selected cosmid clones of human DNA have been used as hybridization probes. Of course, we cannot rule out the possibility that there are segments in the huge human genome which might show interindividual variations in methylation patterns (Silva and White, 1988).

Patterns of DNA methylation can be viewed as indicators for several types of genetic activity – only one being transcriptional activity – and for genome organization. Different patterns would probably interdigi-

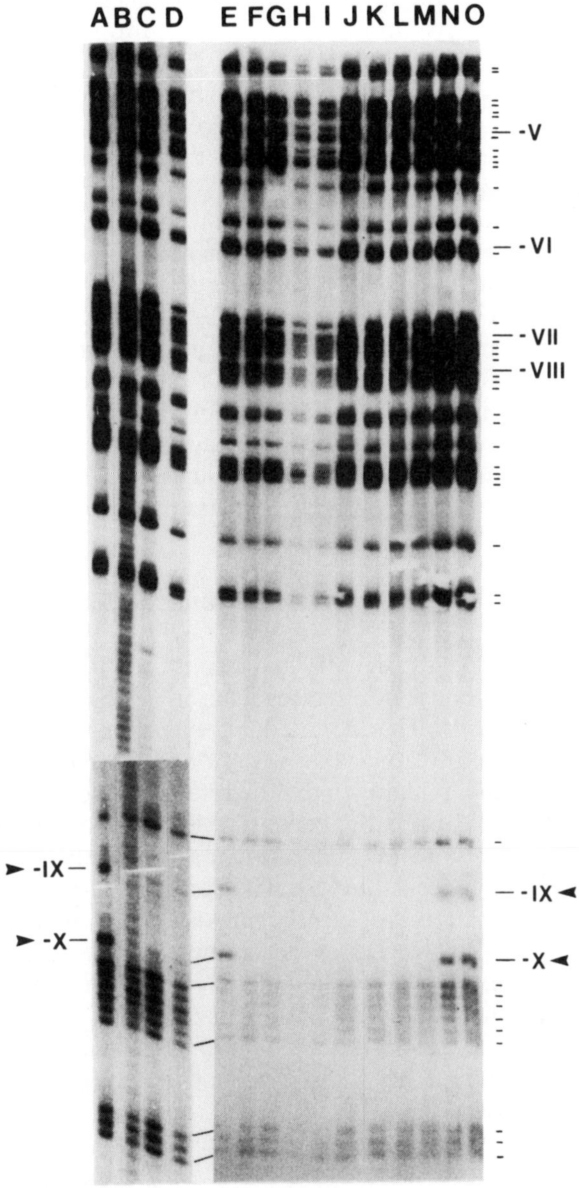

Figure 5. Genomic sequencing data from the TNF-α gene. Autoradiogram of a genomic sequencing gel: The top strand between 5′-CG-3′ positions -V to -X was sequenced. Lanes: A, E. N, O, plasmid controls; B, DNA from a Chinese individual; C and F-M, DNAs from Europeans; D, DNA from an African. The 5-mC residues in positions -IX, and -X are indicated by arrowheads. Note the absence of bands, i.e., presence of 5-mC residues in identical positions for all individuals investigated. This figure was taken from Kochanek et al. (1990). Roman numerals refer to 5′-CG-3′ sequences in the upstream sequence of the TNF-α promoter.

tate to yield an overall pattern. Depending on the functional states of different cell types in an intact organism or in cells kept in culture, patterns of DNA methylation would be expected to be quite different in the same genome segment in different cell types. It was, therefore, not surprising to find complete methylation of almost all 5′-CG-3′ sequences of the TNF-α and -β genes in the DNA from HeLa cells, whereas the same sequences were not methylated in the TNF-α gene in human lymphocytes or undermethylated in the TNF-β gene of the same cells (Kochanek et al., 1990). In contrast, some of the randomly selected human DNA sequences used in the aforementioned analyses of methylation patterns in human DNA (Behn-Krappa et al., 1991) were completely or almost completely 5′-CCGG-3′ methylated in human lymphocytes or granulocytes and strikingly undermethylated in the DNA from HeLa cells. It is important to recognize that patterns of DNA methylation in cultured cells do not have to reflect at all the patterns in individual cell types in an organism. Considering the vastly different transcriptional programs of cell lines and primary cell types, gross differences in methylation patterns have to be expected.

9.3 Alterations of patterns of DNA methylation in human malignancies

There is at present no satisfactory biochemical definition for a malignant cell. A huge body of literature has described a multitude of changes in tumor cells. This decade-long endeavor has not helped to recognize yet a generally applicable pattern of alterations. Each type of tumor could exhibit a highly individual set of biochemical changes, many of which would probably constitute secondary or tertiary effects that camouflage access to the primary causative events. It is a testable contention that tumor cells might differ from the normal precursor by alterations in the patterns of gene expression. Let us assume that a normal human lymphocyte expresses a complex set of genes in a highly regulated way depending on its state of activity. Is the same set of genes expressed in lymphoma or lymphatic leukemia cells? Considering the enormous number of human genes that would have to be assessed for activity changes, there is no realistic way at present to approach these fundamental problems.

In recent years, we have performed a number of experiments which were based on the premise that patterns of DNA methylation could be indicators for gene activity and for a certain organizational state of the entire genome. Consequently, alterations in gene activity or organizational status of the genome might be reflected in changes of patterns of DNA methylation. Obviously, exploratory experiments have to test the validity of these concepts. So far, we have used two types of systems and obtained the following results.

(i) Changes in patterns of DNA methylation at 5′-CCGG-3′ sequences in 17 different human genes or gene segments were investigated in the DNA of a number of established human lymphoma and leukemia cell lines in culture. Depending on cell line and gene segment studied, there were no changes in comparison to normal human lymphocytes and granulocytes or there were sometimes decreases or increases in DNA methylation (Achten et al., 1991). A clear-cut tendency could not be recognized, although part of the problem may have been that we had chosen to analyze the DNA from established cell lines. Taken at face value, these results suggest that changes in DNA methylation in tumor cell lines will be complicated, multidirectional and probably call for an approach in which gene for gene, promoter for promoter will have to be assessed. Current technology will render this type of experimentation very difficult if a major part of the genome had to be investigated.

(ii) We concentrated on the human TNF-α and -β genes (see above, Kochanek et al., 1990) in an analysis by the genomic sequencing method of the same DNA sequences that had been analyzed in human granulocytes and lymphocytes. We have recently extended these determinations to the DNA from different cases of acute or chronic myeloid or lymphatic leukemias or from non-Hodgkin lymphomas (Kochanek et al., 1991). Again, the data from cells of primary human normal or malignant origin with well-defined classifications demonstrated that in some cases of leukemias or lymphomas there were no alterations in the distribution of 5-mC residues, in others there were changes. One can only speculate about the reasons for these alterations. Could they be related to the DNA methylation patterns in precursors of cells which individual leukemias could have originated from? A great deal of work will still have to be done, perhaps most sensibly on early stages of malignancies. From our limited experience in this field, I would recommend to concentrate on the study in depth of methylation patterns in a few genes by using the genomic sequencing method which detects all 5-mC residues in a sequence. The analysis with restriction enzymes, like HpaII or HhaI, could help to localize areas of significant changes in patterns of DNA methylation. But this method by itself will not be sufficient.

Acknowledgments. Without the dedication, interest, and hard work of many of my pre- and postdoctoral colleagues this work could not have been performed. I have referenced their contributions in numerous publications. I am indebted to Petra Böhm for expert editorial work.

Research in the author's laboratory has been supported for many years by the Deutsche Forschungsgemeinschaft through Sonderforschungsbereich (SFB) 74-C1 until 1988 and through SFB 274-TP1 starting in 1989 to the present. I am particularly grateful for the long-term consideration of these projects by the Deutsche Forschungsgemeinschaft.

Achten, S., Behn-Krappa, A., Jücker, M., Sprengel, J., Hölker, I., Schmitz, B., Tesch, H., Diehl, V., and Doerfler, W. (1991) Patterns of DNA methylation in selected human genes in different Hodgkin's lymphoma and leukemia cell lines and in normal human lymphocytes. Cancer Res. *51*, 3702–3709.

Akusjärvi, G., Pettersson, U., and Roberts, R. J. (1986) Structure and function of the adenovirus-2 genome. Dev. Mol. Virol. *8*, 53–95.

Bednarik, D. P., Cook, J. A., and Pitha, P. M. (1990) Inactivation of the HIV LTR by DNA CpG methylation: evidence for a role in latency. EMBO J. *9*, 1157–1164.

Behn-Krappa, A., Hölker, I., Sandaradura de Silva, U., and Doerfler, W. (1991) Patterns of DNA methylation are indistinguishable in different individuals over a wide range of human DNA sequences. Genomics *11*, 1–7.

Berk, A. J. (1986) Adenovirus promoters and E1A transactivation. Annu. Rev. Genet. *20*, 45–79.

Boyes, J., and Bird, A. (1991) DNA methylation inhibits transcription indirectly via a methyl-CpG binding protein. Cell *64*, 1123–1134.

Church, G. M., and Gilbert, W. (1984) Genomic sequencing. Proc. Natl. Acad. Sci. USA *81*, 1991–1995.

Cook, J. L., and Lewis, A. M., Jr. (1979) Host response to adenovirus 2-transformed hamster embryo cells. Cancer Res. *39*, 1455–1461.

Deuring, R., and Doerfler, W. (1983) Proof of recombination between viral and cellular genomes in human KB cells productively infected by adenovirus type 12: structure of the junction site in a symmetric recombinant (SYREC). Gene *26*, 283–289.

Deuring, R., Klotz, G., and Doerfler, W. (1981) An unusual symmetric recombinant between adenovirus type 12 DNA and human cell DNA. Proc. Natl. Acad. Sci. USA *78*, 3142–3146.

Dobrzanski, P., Hoeveler, A., and Doerfler, W. (1988) Inactivation by sequence-specific methylations of adenovirus promoters in a cell-free transcription system. J. Virol. *62*, 3941–3946.

Doerfler, W. (1968) The fate of the DNA of adenovirus type 12 in baby hamster kidney cells. Proc. Natl. Acad. Sci. USA *60*, 636–643.

Doerfler, W. (1970) Integration of the deoxyribonucleic acid of adenovirus type 12 into the deoxyribonucleic acid of baby hamster kidney cells. J. Virol. *6*, 652–666.

Doerfler, W. (1981) DNA methylation – A regulatory signal in eukaryotic gene expression. J. gen. Virol. *57*, 1–20.

Doerfler, W. (1982) Uptake, fixation, and expression of foreign DNA in mammalian cells: The organization of integrated adenovirus DNA sequences. Curr. Top. Microbiol. Immunol. *101*, 127–194.

Doerfler, W. (1983) DNA methylation and gene activity. Annu. Rev. Biochem. *52*, 93–124.

Doerfler, W., ed. (1983) The Molecular Biology of Adenoviruses. Current Topics in Microbiology and Immunology, vols. 109 and 110. Springer Verlag, Heidelberg.

Doerfler, W., ed. (1984) The Molecular Biology of Adenoviruses. Current Topics in Microbiology and Immunology, vol. 111. Springer Verlag, Heidelberg.

Doerfler, W. (1984) DNA methylation: Role in viral transformation and persistence, in: Advances in Viral Oncology, vol. 4, pp. 217–247. Ed. G. Klein. Raven Press, New York.

Doerfler, W. (1986) Expression of the *Autographa californica* nuclear polyhedrosis virus genome in insect cells: homologous viral and heterologous vertebrate genes – The baculovirus vector system. Curr. Top. Microbiol. Immunol. *131*, 51–68.

Doerfler, W. (1989) Complexities in gene regulation by promoter methylation. Nucl. Acids Mol. Biol. *3*, 92–119.

Doerfler, W. (1990) The significance of DNA methylation patterns: Promoter inhibition by sequence-specific methylation is one functional consequence. Phil. Transact. Royal Soc., London B *326*, 253–265.

Doerfler, W. (1991a) The abortive infection and malignant transformation by adenoviruses: Integration of viral DNA and control of viral gene expression by specific patterns of DNA methylation. Adv. Virus Res. *39*, 89–128.

Doerfler, W. (1991b) Patterns of DNA methylation – evolutionary vestiges of foreign DNA inactivation as a host defense mechanism. A proposal. Biol. Chem. Hoppe-Seyler *372*, 557–564.

Doerfler, W. (1992) Transformation of cells by adenoviruses: less frequently discussed mechanisms, in: Malignant Transformation by DNA Viruses: Molecular Mechanisms. Eds W. Doerfler and P. Böhm, Verlag Chemie, Weinheim, pp. 87–109.

Doerfler, W., Kruczek, I., Eick, D., Vardimon, L., and Kron, B. (1982) DNA methylation and gene activity: The adenovirus system as a model. Cold Spring Harbor Symp. Quant. Biol. *47*, 593–603.

Doerfler, W., Gahlmann, R., Stabel, S., Deuring, R., Lichtenberg, U., Schulz, M., Eick, D., and Leisten, R. (1983) On the mechanism of recombination between adenoviral and cellular DNAs: The structure of junction sites. Curr. Top. Microbiol. Immunol. *109*, 193–228.

Doerfler, W., Langner, K.-D., Kruczek, I., Vardimon, L., and Renz, D. (1984) Specific promoter methylations cause gene inactivation, in: DNA Methylation, pp. 221–247. Eds A. Razin, H. Cedar, and A. D. Riggs. Springer Verlag, Heidelberg.

Doerfler, W., Toth, M., Kochanek, S., Achten, S., Freisem-Rabien, U., Behn-Krappa, A., and Orend, G. (1990) Eukaryotic DNA methylation: Facts and problems. FEBS Lett. *268*, 329–333.

Eick, D., Fritz, H.-J., and Doerfler, W. (1983) Quantitative determination of 5-methylcytosine in DNA by reverse-phase high-performance liquid chromatography. Analyt. Biochem. *135*, 165–171.

Esche, H., (1982) Viral gene products in adenovirus type 2-transformed hamster cells. J. Virol. *41*, 1076–1082.

Van Etten, J. L., Burbank, D. E., Schuster, A. M., and Meints, R. H. (1985) Lytic viruses infecting a Chlorella-like alga. Virology *140*, 135–143.

Flint, J., and Shenk, T. (1989) Adenovirus E1A protein paradigm viral transactivator. Annu. Rev. Genet. *23*, 141–161.

Ginsberg, H., ed. (1985) Adenoviruses. Plenum Press, New York.

Gorman, C. M., Moffat, L. F., and Howard, B. H. (1982) Recombinant genomes which express chloramphenicol acetyltransferase in mammalian cells. Mol. Cell. Biol. *2*, 1044–1051.

Günthert, U., Schweiger, M., Stupp, M., and Doerfler, W. (1976) DNA methylation in adenovirus, adenovirus-transformed cells, and host cells. Proc. Natl. Acad. Sci. USA *73*, 3923–3927.

Hammarskjöld, M.-L., and Winberg, G. (1980) Encapsidation of adenovirus 16 DNA is directed by a small DNA sequence at the left end of the genome. Cell *20*, 787–795.

Hermann, R., and Doerfler, W. (1991) Interference with protein binding at AP2 sites by sequence-specific methylation in the late E2A promoter of adenovirus type 2 DNA. FEBS Lett. *281*, 191–195.

Hermann, R., Hoeveler, A., and Doerfler, W. (1989) Sequence-specific methylation in a downstream region of the late E2A promoter of adenovirus type 2 DNA prevents protein binding. J. Mol. Biol. *210*, 411–415.

Hoeveler, A., and Doerfler, W. (1987) Specific factors binding to the E2A late promoter region of adenovirus type 2 DNA: No apparent effects of 5′-CCGG-3′ methylation. DNA *6*. 449–460.

Ibelgaufts, H., Doerfler, W., Scheidtmann, K. H., and Wechsler, W. (1980) Adenovirus type 12-induced rat tumor cells of neuroepithelial origin: Persistence and expression of the viral genome. J. Virol. *33*, 423–437.

Jähner, D., and Jaenisch, R. (1985) Retrovirus-induced *de novo* methylation of flanking host sequences correlates with gene inactivity. Nature *315*, 594–597.

Johansson, K., Persson, H., Lewis, A. M. Pettersson, U., Tibbetts, C., and Philipson, L. (1978) Viral DNA sequences and gene products in hamster cells transformed by adenovirus type 2. J. Virol. *27*, 628–639.

Jüttermann, R., Hosokawa, K., Kochanek, S., and Doerfler, W. (1991) Adenovirus type 2 VAI RNA transcription by polymerase III is blocked by sequence-specific methylation. J. Virol. *65*, 1735–1742.

Knebel, D., Lübbert, H., and Doerfler, W. (1985) The promoter of the late p10 gene in the insect nuclear polyhedrosis virus *Autographa californica*: activation by viral gene products and sensitivity to DNA methylation. EMBO J. *4*, 1301–1306.

Knebel, D., and Doerfler, W. (1986) N^6-Methyldeoxyadenosine residues at specific sites decrease the activity of the E1A promoter of adenovirus type 12 DNA. J. Mol. Biol. *189*, 371–375.

Knebel-Mörsdorf, D., Achten, S., Langner, K.-D., Rüger, R., Fleckenstein, B., and Doerfler, W. (1988) Reactivation of the methylation-inhibited late E2A promoter of adenovirus type 2 DNA by a strong enhancer of human cytomegalovirus. Virology *166*, 166–174.

Knust, B., Brüggemann, U., and Doerfler, W. (1989) Reactivation of a methylation-silenced gene in adenovirus-transformed cells by 5-azacytidine or by E1A transactivation. J. Virol. *63*, 3519–3524.

Kochanek, S., Toth, M., Dehmel, A., Renz, D., and Doerfler, W. (1990) Interindividual concordance of methylation profiles in human genes for tumor necrosis factors α and β. Proc. Natl. Acad. Sci. USA *87*, 8830–8834.

Kochanek, S., Radbruch, A., Tesch, H., Renz, D., and Doerfler, W. (1991) DNA methylation profiles in the human genes for tumor necrosis factors α and β in subpopulations of leukocytes and in leukemias. Proc. Natl. Acad. Sci. USA *88*, 5759–5763.

Kochanek, S., Renz, D., and Doerfler, W. (1993) DNA methylation in the Alu sequences of primary human cells. Submitted.

Kruczek, I., and Doerfler, W. (1982) The unmethylated state of the promoter/leader and 5′-regions of integrated adenovirus genes correlates with gene expression. EMBO J. *1*, 409–414.

Kruczek, I., and Doerfler, W. (1983) Expression of the chloramphenicol acetyltransferase gene in mammalian cells under the control of adenovirus type 12 promoters: Effect of promoter methylation on gene expression. Proc. Natl. Acad. Sci. USA *80*, 7586–7590.

Kuhlmann I., and Doerfler, W. (1982) Shifts in the extent and patterns of DNA methylation upon explantation and subcultivation of adenovirus type 12-induced hamster tumor cells. Virology *118*, 169–180.

Kuhlmann, I., and Doerfler, W. (1983) Loss of viral genomes from hamster tumor cells and nonrandom alterations in patterns of methylation of integrated adenovirus type 12 DNA. J. Virol. *47*, 631–636.

Langner, K.-D., Vardimon, L., Renz, D., and Doerfler, W. (1984) DNA methylation of three 5′ C-C-G-G 3′ sites in the promoter and 5′ region inactivates the E2a gene of adenovirus type 2. Proc. Natl. Acad. Sci. USA *81*, 2950–2954.

Langner, K.-D., Weyer, U., and Doerfler, W. (1986) *Trans* effect of the E1 region of adenoviruses on the expression of a prokaryotic gene in mammalian cells: Resistance to 5′-CCGG-3′ methylation. Proc. Natl. Acad. Sci. USA *83*, 1598-1602.

Lettmann, C., Schmitz, B., and Doerfler, W. (1991) Persistence or loss of preimposed methylation patterns and *de novo* methylation of foreign DNA integrated in transgenic mice. Nucl. Acids Res. *19*, 7131–7137.

Lichtenberg, U., Zock, C., and Doerfler, W. (1987) Insertion of adenovirus type 12 DNA in the vicinity of an intracisternal A particle genome in Syrian hamster tumor cells. J. Virol. *61*, 2719–2726.

Lichtenberg, U., Zock, C., and Doerfler, W. (1988) Integration of foreign DNA into mammalian genome can be associated with hypomethylation at site of insertion. Virus Res. *11*, 335–342.

Luckow, V. A., and Summers, M. D. (1988) Trends in the development of baculovirus expression vectors. BioTechnology *6*, 47–55.

Miller, L. K. (1988) Baculoviruses as gene expression vectors. Annu. Rev. Microbiol. *42*, 177–199.

Müller, U., and Doerfler, W. (1987) Fixation of the unmethylated or the 5′-CCGG-3′ methylated adenovirus late E2A promoter-cat gene construct in the genome of hamster cells: gene expression and stability of methylation patterns. J. Virol. *61*, 3710–3720.

Nevins, J. R. (1987) Regulation of early adenovirus gene expression. Microbiol. Rev. *51*, 419–430.

Orend, G., Kuhlmann, I., and Doerfler, W. (1991) The spreading of DNA methylation across integrated foreign (adenovirus type 12) genomes in mammalian cells. J. Virol. *65*, 4301–4308.

Ortin, J., Scheidtmann, K.-H., Greenberg, R., Westphal, M., and Doerfler, W. (1976) Transcription of the genome of adenovirus type 12. III. Maps of stable RNA from productively infected human cells and abortively infected and transformed hamster cells. J. Virol. *20*, 355–372.

Razin, A., Cedar, H., and Riggs, A. D., eds. (1984) DNA Methylation. Biochemistry and Biological Significance. Springer Verlag, New York, Berlin, Heidelberg, Tokyo.

Renbaum, P., Abrahamove, D., Fainsod, A., Wilson, G. G., Rottem, S., and Razin, A. (1990) Cloning, characterization, and expression in *Escherichia coli* of the gene coding for the CpG DNA methylase from *Spiroplasma* sp. strain MQ1(M·SssI). Nucl. Acids Res. *18*, 1145–1152.

Roberts, R. J., Akusjärvi, G., Aleström, P., Gelinas, R. E., Gingeras, T. R., Sciaky, D., and Pettersson, U. (1986) A consensus sequence for the adenovirus-2 genome. In: Adenovirus

DNA. The Viral Genome and its Expression, W. Doerfler, ed., pp. 1–51, Martinus Nijhoff Publ., Boston.

Rosahl, T., and Doerfler, W. (1992) Alterations in the levels of expression of specific cellular genes in adenovirus-infected and -transformed cells. Virus Res., in press.

Saluz, H. P., and Jost, J. P. (1987) A Laboratory Guide to Genomic Sequencing. Birkhäuser, Basel, Boston.

Shi, Y., Seto, E., Chang, L.-S., and Shenk, T. (1991) Transcriptional repression by YY1, a human GLI-Krüppel-related protein, and relief of repression by adenovirus E1A protein. Cell *67*, 377–388.

Silva, A. J., and White, R. (1988) Inheritance of allelic blueprints for methylation patterns. Cell *54*, 145–152.

Southern, E. M. (1975) Detection of specific sequences among DNA fragments separated by gel electrophoresis. J. Mol. Biol. *98*, 503–517.

Southern, P. J., and Berg, P. (1982) Transformation of mammalian cells to antibiotic resistance with a bacterial gene under control of the SV40 early region promoter. J. mol. appl. Genet. *1*, 327–341.

Sutter, D., and Doerfler, W. (1979) Methylation of integrated viral DNA sequences in hamster cells transformed by adenovirus 12. Cold Spring Harbor Symp. Quant. Biol. *44*, 565–568.

Sutter, D., and Doerfler, W. (1980) Methylation of integrated adenovirus type 12 DNA sequences in transformed cells is inversely correlated with viral gene expression. Proc. Natl. Acad. Sci. USA *77*, 253–256.

Sutter, D., Westphal, M., and Doerfler, W. (1978) Patterns of integration of viral DNA sequences in the genomes of adenovirus type 12-transformed hamster cells. Cell *14*, 569–585.

Thompson, J. P., Granoff, A., and Willis, D. B. (1986) Trans-activation of a methylated adenovirus promoter by a frog virus 3 protein. Proc. Natl. Acad. Sci. USA *83*, 7688–7692.

Toth, M., Lichtenberg, U., and Doerfler, W. (1989) Genomic sequencing reveals a 5-methylcytosine-free domain in active promoters and the spreading of preimposed methylation patterns. Proc. Natl. Acad. Sci. USA *86*, 3728–3732.

Toth, M., Müller, U., and Doerfler, W. (1990) Establishment of *de novo* DNA methylation patterns. Transcription factor binding and deoxycytidine methylation at CpG and non-CpG sequences in an integrated adenovirus promoter. J. Mol. Biol. *214*, 673–683.

Trentin, J. J., Yabe, Y., and Taylor, G. (1962) The quest for human cancer viruses. Science *137*, 835–841.

Vardimon, L., Neumann, R., Kuhlmann, I., Sutter, D., and Doerfler, W. (1980) DNA methylation and viral gene expression in adenovirus-transformed and -infected cells. Nucl. Acids Res. *8*, 2461–2473.

Vardimon, L., and Doerfler, W. (1981) Patterns of integration of viral DNA in adenovirus type 2-transformed hamster cells. J. Mol. Biol. *147*, 227–246.

Vardimon, L., Kuhlmann, I., Cedar, H., and Doerfler, W. (1981) Methylation of adenovirus genes in transformed cells and *in vitro*: influence on the regulation of gene expression. Eur. J. Cell Biol. *25*, 13–15.

Vardimon, L., Kressmann, A., Cedar, H., Maechler, M., and Doerfler, W. (1982a) Expression of a cloned adenovirus gene is inhibited by *in vitro* methylation. Proc. Natl. Acad. Sci. USA *79*, 1073–1077.

Vardimon, L., Günthert, U., and Doerfler, W. (1982b) *In vitro* methylation of the BsuRI (5′-GGCC-3′) sites in the E2a region of adenovirus type 2 DNA does not affect expression in *Xenopus laevis* oocytes. Mol. Cell. Biol. *2*, 1574–1580.

Weisshaar, B., Langner, K.-D., Jüttermann, R., Müller, U., Zock, C., Klimkait, T., and Doerfler, W. (1988) Reactivation of the methylation-inactivated late E2A promoter of adenovirus type 2 by E1A (13S) functions. J. Mol. Biol. *202*, 255–270.

Weyer, U., and Doerfler, W. (1985) Species dependence of the major late promoter in adenovirus type 12 DNA. EMBO J. *4*, 3015–3019.

Wienhues, U., and Doerfler, W. (1985) Lack of evidence for methylation of parental and newly synthesized adenovirus type 2 DNA in productive infections. J. Virol. *56*, 320–324.

Zock, C., and Doerfler, W. (1990) A mitigator sequence in the downstream region of the major late promoter of adenovirus type 12 DNA. EMBO J. *9*, 1615–1623.

Zock, C., Iselt, A., and Doerfler, W. (1993) A unique mitigator sequence and preserved nuclear topology determine the species specificity of the major late promoter in adenovirus type 12 DNA. Submitted.

DNA Methylation: Molecular Biology and Biological Significance
ed. by J. P. Jost & H. P. Saluz

DNA Methylation and retrovirus expression

Daniel P. Bednarik

Centers for Disease Control, National Center for Infectious Diseases, Division of Viral and Rickettsial Diseases, Retrovirus Diseases Branch, Molecular Genetics Section, 1600 Clifton Road, Mail Stop G-19, Atlanta, Georgia 30333, USA

1 Introduction

The methylation of cytosine in mammalian DNA has proven to be an enigma to scientists since its discovery more than 30 years ago. There are several apparent roles for the methylation of DNA in eukaryotic cells, the exceptions being *Drosophila* yeast. The absence of DNA methylation in the latter two examples suggests that the fundamental mechanisms of DNA metabolism can exist without modification of cytosine. Probably the most common phenomenon ascribed in DNA methylation is the control of gene activity at the level of transcription (Bird, 1986; Lindsay and Bird, 1988). Methylation of key cytosine moieties is mediated by the mammalian DNA cytosine-5-methyltransferase (MeTase), and is specific for the dinucleotide sequence CpG (Bestor et al., 1988; Doerfler, 1983; Santi et al., 1983; Smith et al., 1991). The CpG sequences are often found localized in clusters and are referred to as "CpG islands" (Bird, 1986). Gene promoters which occupy such islands are susceptible to inactivation, but the process which determines what genes are to be selectively inactivated is currently unknown. A key feature of the regulation of gene activity by hypermethylation of CpG sequences is the epigenetic manner by which the cell strictly maintains DNA methylation patterns. The term "epigenetic" was introduced to permit the addition of another level of information, separate from alteration of the actual gene sequence, to allow the stable inheritance of, in this case, gene activity patterns (Holliday, 1990). The breakdown in the relaying of epigenetic information would result in the disarray of gene expression and ultimately cellular dysfunction.

Retroviruses are positive-strand RNA viruses which occupy the family *retroviridae* (Fig. 1). Of these are the oncoviruses (*oncovirinae*), lentiviruses (*lentivirinae*), and spumaviruses (*spumavirinae*) all of which are, essentially, a movable genetic element or retrotransposon (Varmus, 1983; Varmus and Brown, 1990). Given the random nature by which

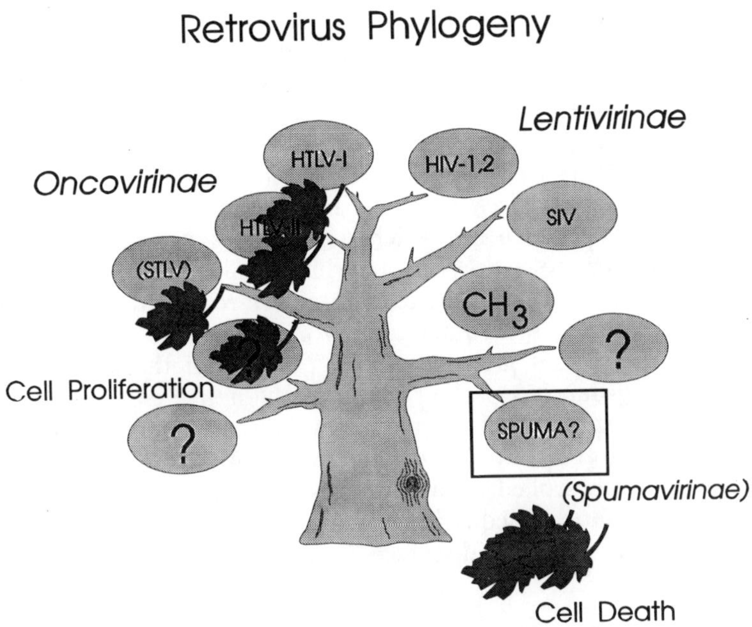

Figure 1. "Family Tree" of retrovirus phylogeny. The human oncoviruses (*oncovirinae*) are symbolized by the growing leaves which represent proliferation. The human lentiviruses (*lentivirinae*) which induce immunodeficiency disease are symbolized by falling leaves which represent death. The simian viruses (SIV/STLV) are included parenthetically while spumavirus (inside box) is not yet clearly associated with a human disease. The question marks represent the possibility of other unknown retroviruses which may be contributing elements to human diseases.

retroviruses integrate into the host cellular genome, they become susceptible to modification by DNA methylation depending upon the immediate DNA environment. The relationship of DNA methylation to retrovirus pathophysiology became apparent when Harbers et al. (1981) demonstrated that the murine leukemia virus long terminal repeats (MuLV) LTRs could become hypermethylated after integration into chromosomal DNA, with subsequent inactivation of viral transcription and expression. As such, the integrated provirus could be maintained in a latent, unexpressed state. This chapter will focus upon the role which DNA methylation plays on the expression of retroviruses utilizing HIV-1 as the primary model; the contribution of this process to the latency phenomenon will be examined in detail.

2 The mechanism of DNA methylation

The methylation of cytosine to form 5-methylcytosine in mammalian DNA is known to be the fundamental mechanism of gene expression control and/or recognition (Holliday, 1990; Doerfler, 1983, 1984; Keshet et al., 1986). The enzymatic reaction catalyzing the conversion to cytosine to 5-methylcytosine is carried out by a unique enzyme which, to date, has been very poorly characterized in the human system. the murine enzyme, however, has been studied by Bestor et al. (1988), and was characterized as a protein of ~180 kD in molecular mass. Little is known about the sequences which are recognized by this enzyme other than the CpG target sequence which is absolutely required for activity. The methyl donor molecule, *S*-adenosylmethionine (SAM), is required as a cofactor for catalysis of the reaction which results in the hydrolysis of SAM to *S*-adenosylhomocysteine (SAH), which the subsequent transfer of the methyl group to the 5 position of the cytidine ring. Recently, our group has partially purified the native enzyme from the T-cell line Jurkatt, and is characterizing a molecular clone encoding the human enzyme to gain a better understanding of how the expression of this protein is controlled, and how this enzyme is regulated during cellular differentiation (Bednarik, 1992). Much more extensive information regarding this topic is discussed elsewhere in this book.

3 Replicative life cycle of HIV

In this section we will review the retroviral life cycle; for the sake of brevity, only the main points will be addressed. Retroviruses require several crucial events to propagate in mammalian cells, the first of which is the binding to a cell-surface receptor molecule, usually a protein, followed by internalization. Once uncoated, the two copies of genomic mRNA are transported to the nucleus where reverse transcription and integration into the chromosomal DNA take place. Transcription of the provirus yields mRNA which is further spliced and processed for export and translation into viral proteins. Some of the full-length genomic mRNA is maintained for packaging and final assembly into the nucleoid body. This structure is released, via budding, into the supernatant encapsulated in a mature viral envelope. Several of the most important events, utilizing HIV as the model, are described below.

3.1 Infection

The receptor molecule which facilitates entry of the HIV particle has unequivocally been demonstrated to be the CD4 protein (for review see Bednarik and Folks, 1992). Although other cell surface molecules may exist to enhance or modulate virus entry, the CD4 protein is absolutely required.

3.2 Reverse transcription and provirus integration

Probably the most essential step in the life cycle of a retrovirus is the reverse transcription of viral genomic RNA (Bednarik and Folks, 1992; Goff, 1990). All of these reactions are catalyzed by the *pol*-encoded multifunctional enzyme reverse transcriptase (RT) (Gilboa et al., 1979). The human enzyme is composed of two subunits (p66 and p51), unlike the monomeric MuLV RT enzyme, which exist as a heterodimer (Davies et al., 1991). The p66-p66 homodimer is cleaved by the HIV protease via asymmetric processing to yield the p51 subunit. Mechanistically, RT transcribes RNA in the 3′ to 5′ direction, initiating from a primed RNA template to form a complementary DNA strand (cDNA). Simultaneously, the RNase H activity degrades the RNA template (Davies et al., 1991).

The apparent inability of retroviruses to replicate autonomously as episomes requires integration into the host chromosomal DNA to support stable maintenance in dividing cells. Once integrated, the provirus is transmitted, via DNA replication, as an integral element of the host genome. The integration of retroviral DNA is important for the transcription of new viral copies, the spliced mRNA of which is also translated into viral protein components, and the genomic message which is subsequently packaged into a mature virion.

The present working model of retroviral integration is based upon elegant work developed by Varmus and Brown (1987, 1990), Craigie et al. (1989, 1991) and, with respect to HIV, studies completed by Farnet and Haseltine (1990). Soon after retroviral infection, a double-stranded DNA molecule, which is terminally redundant, is synthesized by reverse transcriptase in the cytoplasm (Fig. 2). These redundant DNA sequences are better known as the long terminal repeats or LTRs. The preintegrative DNA species is almost certainly linear (Fujiwara and Craigie, 1989) and is a blunt-ended molecule which exists in the cytoplasm as a nucleoprotein complex. The 3′ termini of the linear DNA are processed, resulting in the loss of one or two nucleotides from either end. This event is dependent upon the integrase protein and exposes the 3′ OH groups which delineate the boundaries of the provirus. The nucleoprotein complex now enters the nucleus by a translocation-dependent mechanism. The actual integration of the provirus into the host chromosomal DNA is orchestrated into a series of events which are preceded by a staggered cut in the host target DNA. The broken phosphodiester bonds in the host DNA provide the energy for bond formation thereby facilitating the joining of both recessed ends. There is a DNA synthesis event, perhaps mediated by the viral reverse transcriptase or cellular enzymes, which repairs the gap between host and viral DNA, thereby simultaneously displacing the mismatched 5′ ends (Bushman and Craigie, 1991). Integration is completed by a final ligation step.

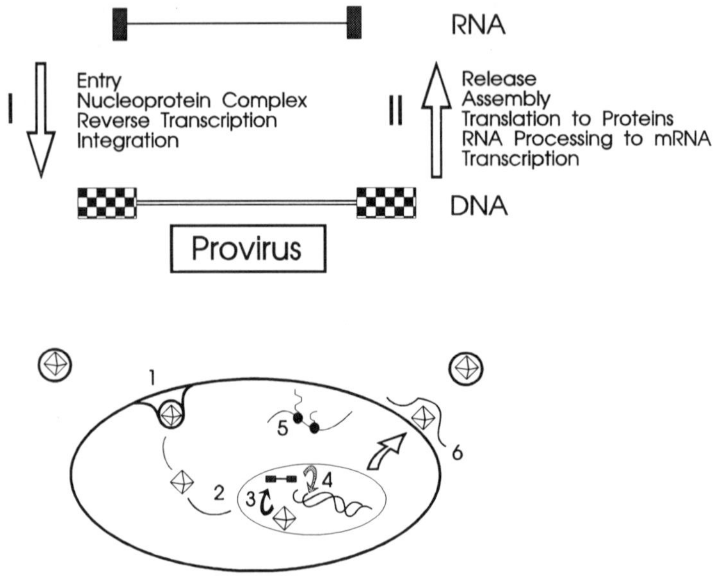

Figure 2. Simplified summary of processes involved in the life cycle of a retrovirus. (1) Virus entry is dependent upon the presence of a surface receptor molecule usually specific for an envelope protein domain. (2) Virus uncoating and translocation to the nucleus as an RNA:reverse transcriptase complex. (3) Reverse transcriptase synthesizes a complimentary DNA strand while simultaneously degrading the viral genomic RNA. (4) The double-standard proviral DNA exists as a nucleoprotein complex with integrase, which catalyzes random integration into the host genome. (5) The proviral DNA is transcribed to mRNA which is subsequently spliced, and translated into viral proteins. Proteins as assembled into a mature virion containing two copies of full-length genomic RNA, followed by budding into the supernatant (6).

3.3 Transcription

Figure 3 details the structure of the HIV LTR and the binding motifs of nuclear proteins. Initiation of transcription occurs at the cap site, and is dependent upon RNA polymerase II. Recently, much effort has been directed toward the precise localization of LTR motifs which respond *in trans* to cellular or viral transcriptional proteins. Upstream from the cap site are found a TATA element for the positional initiation of RNA polymerase II and three Sp1 motifs to which the binding of the corresponding proteins results in the augmentation of transcription (Majors, 1990; Jones et al., 1986, 1988). The HIV LTR responds to T-cell activation signals via a 11 base-pair repeat element with the sequence GGGGACTTTCC and is denoted as the NF-kB motif (Majors, 1990; Tong-Starksen et al., 1987). This family of proteins was originally demonstrated to be a constitutive B-cell transcription factor which was found to be induced in T-cells in response to activation/proliferation signals (Sen and Baltimore, 1986a, 1986b; Lenardo and Baltimore,

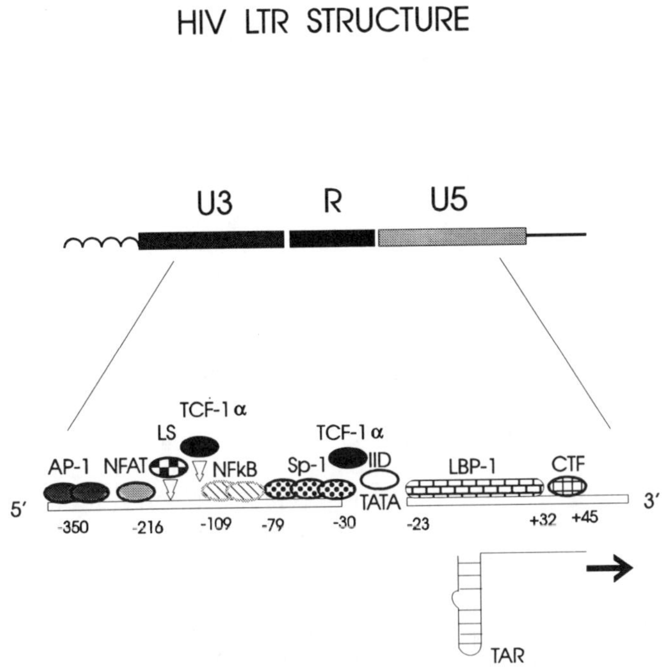

Figure 3. HIV-1 LTR Structure. The HIV LTR is divided into U3, R, and U5 regions and contains binding motifs specific for a plethora of nuclear proteins (Waterman et al., 1991). AP-1: activation protein; NFAT: nuclear factor AT-1; LS: lymphoid-specific protein, TCF-1a: T-cell-specific transcriptional activator; NF-kB: nuclear factor kappa-B; Sp1: transcription factor Sp1; TFIID: transcriptional initiation factor IID; LBP-1: latency binding protein; TAR: tat transacting response element.

1989). These proteins exist in the cytoplasm as a p50-p65 phosphoprotein complex bound to an inhibitor denoted as IkB. Upon dissociation of the inhibitor protein, the complex is translocated, as a heterodimer, to the nucleus where the p50 subunit mediates DNA-binding (Ruben et al., 1991; Nolan et al., 1991). Since the recent cloning of the NF-kB proteins, and their apparent relation to the *rel* and *dorsal* proteins, the understanding of the roles of these unique cellular factors in mediation of transcription has been greatly augmented (Kieran et al., 1990; Ghosh et al., 1990; Gilmore, 1990). However, the association of other cellular factors with the HIV LTR has given rise to a more complicated scenario.

It is the inhibition or modulation of retroviral transcription which has been most directly addressed by experiments which probe the mechanism(s) of DNA methylation-mediated transcriptional regulation. The retrovirus actually serves as a convenient model to study DNA methylation since the genome can be introduced into susceptible cells and can be monitored by a variety of experimental means. By far, the most frequent studies have focused upon the correlation of cytosine methylation with transcriptional activity such as described for the murine system (Jahner et al., 1982; Stewart et al., 1982). Treatment of cells in tissue culture with antagonistic agents such as 5-azacytidine (5AC) or bromodeoxyuridine (BUdR) resulted in the demethylation of the integrated provirus, and subsequent expression of infectious virions. The HIV, like other retroviruses, is dependent upon transcriptional activation *in trans* by the virally encoded tat gene (Sodroski et al., 1985). The HIV genome becomes transcriptionally competent after reverse transcription of the genomic RNA into the double-stranded provirus. The viral genome has been shown to exist as several unintegrated species (Bednarik and Folks, 1992; Besansky et al., 1991; Pauza and Galindo, 1989). These proviral forms can be linear, double LTR circular, or a single LTR circular species. The circular, unintegrated species are not precursors to integration, and the linear form only exists as a nucleoprotein complex which is the true precursor to integration (Fujiwara and Craigie, 1989). While circular, proviral DNA evidently do not integrate, they are readily transcribed, and can disseminate a spreading infection (Butera et al., 1991). There is no evidence, however, that the circular proviral forms become methylated, probably because these proviral species do not replicate autonomously in contrast to autonomously replicating genomes, such as Epstein-Barr virus (EBV), which are susceptible to methylation (Jannson et al., 1992). Indeed the *de novo* methylation of proviral DNA appears to be dependent upon the immediate chromosomal environment into which the retrovirus integrates, whether the methylation of CpG dinucleotides is the result of maintenance or *de novo* activities.

4 Retroviral LTRs and CpG islands

When one envisions a CpG island, or cluster of these dinucleotide sequences, a macromolecular structure such as the extensive region spanning part of the Fragile-X (Fmr1) gene certainly comes to mind (Annemieke et al., 1991). In this case, approximately 200 base-pairs of DNA may contain a high density of CpG dinucleotides in the unique sequence CGG, which are recognized by DNA MeTase during DNA replication. While retroviral LTRs are not long enough to fit into this category, they can integrate into such a region of the chromosome and, in effect, become part and parcel of the CpG island. This means that the signal(s) which influence the methylation of cytosine in this region will also affect the integrated provirus as well. This observation served as the basis of our rationale to investigate the mechanism(s) of latency during chronic HIV infection. The term latency evokes a variety of definitions, which depend upon views of viral persistence, and may take several

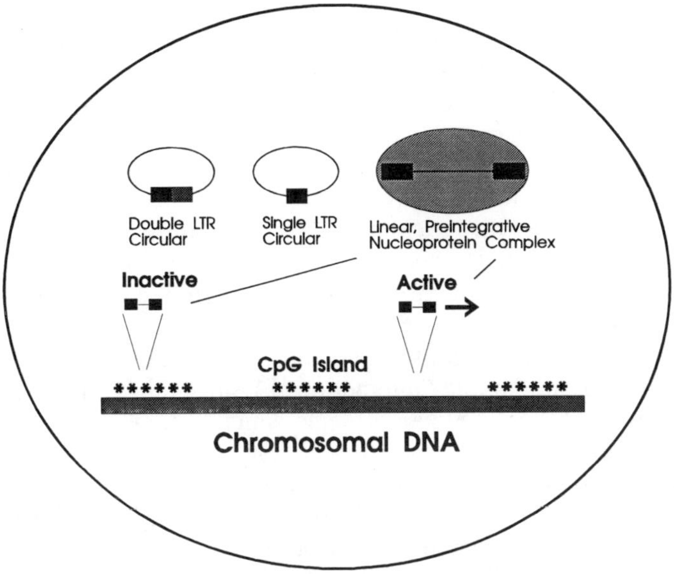

Figure 4. DNA methylation-mediated HIV latency is dependent upon the site of integration. The nuclear compartment contains primarily two forms of circular, unintegrated proviral DNA during acute infection. The double and single LTR species are not precursors to integration, but are transcriptionally active. The linear preintegrative species is depicted as the nucleoprotein complex. Integration of the provirus into a CpG island, shown by the asterisks, would be maintained as a hypermethylated provirus, and therefore transcriptionally inactive. In contrast, integration into actively transcribed, hypomethylated DNA would establish a transcriptionally active proviral species.

avenues of meaning. The word is derived from the Latin root "latens", which means "to lie hidden or concealed". Figure 4 illustrates an overview of the mechanism of methylation-mediated latency. The step which defines whether DNA methylation will modulate HIV transcription is dependent upon the site of viral integration. After integration, the HIV LTR contains several CpG dinucleotides which are plausible methylation sites. Prior to the transactivation of HIV transcription, the LTR must be activated *in cis* since the transactivator tat must first be expressed as a mature, spliced mRNA, and subsequently translated into active proteins.

5 DNA Methylation as an inactivator of retroviral transcription

5.1 Hypermethylation of retroviral LTRs as a mediator of viral latency

To map the region of the HIV LTR most sensitive to activation by deoxy-5-azacytidine (d5AC), permanent cell lines containing either the native LTR directing CAT expression (pU3RIIICAT) or various deletion mutations of this reported plasmid were established (Mosca et al., 1987). Figure 5a illustrates the overall structure of the HIV LTR and the deletion mutant LTRs are shown as those nucleotide positions established by Bal 31 nuclease digestion either upstream or downstream of the cap site. Panel B shows the relative activities of each deletion plasmid when transfected by electroporation into A3.01 T-cells, followed by treatment with 10 μM d5AC for 36 h. Inducibility of the LTR was greatly diminished when the nucleotide region $-57/-45$ was eliminated. Inducibility by d5AC was also severely compromised when the region spanning $+21/+80$ was deleted. These experiments suggested that the d5AC effect was not localized to a single location, but was due to a more complex interaction. Figure 5 illustrates the known binding sites for those cellular proteins which play a role in transcriptional regulation of HIV expression. These observations might be explained in one or several following ways. The HIV LTR is weakly active when expressed *in cis*, particularly in cells of lymphoid lineage. The reduction in CAT activity occurs in parallel with the loss of the core enhancer region which contains the binding sites for cellular transactivators and modulators such as NF-kB and Sp1. This region of the HIV LTR also contains CpG dinucleotides localized in or near the binding sites for these proteins. Hypermethylation of these sites may reduce transcriptional activity by precluding DNA-protein interaction. With respect to the Sp1 class of proteins, this is an unlikely event since methylation of these binding motifs *in vitro* does not prevent DNA-binding activity (Harrington et al., 1988). In contrast, the NF-kB motif contains a single core CpG, which our laboratory has shown to be selective for the inhibition of p50 DNA-binding when methylated (Bednarik et al., 1991); these experiments will be discussed in detail in Section 6. The progressive loss of d5AC inducibility may also be explained by the upregulation of cellular transactivators due to the demethylation of genes encoding these factors. The net result would be deletion of the corresponding binding motif with which the induced transactivator interacts. Since transiently-transfected plasmid DNA does not replicate autonomously, these effects may reflect the action of a *de novo* MeTase activity which would not be dependent upon DNA replication.

Deletion of the region spanning $+21/+80$ (Fig. 5A and B) resulted in a $\sim$70% loss of *cis*-level expression. this was somewhat surprising since this region has been defined as the *trans*-acting response region or TAR, which is required for action by tat (Rosen et al., 1985). The loss of these sequences may further impair the overall "processivity" of transcript initiation and progression (Marciniak and Sharp, 1991). Transcription of the HIV LTR has been shown to be very weakly processive *in cis* and very strongly processive *in trans* (Marciniak and Sharp, 1991); this

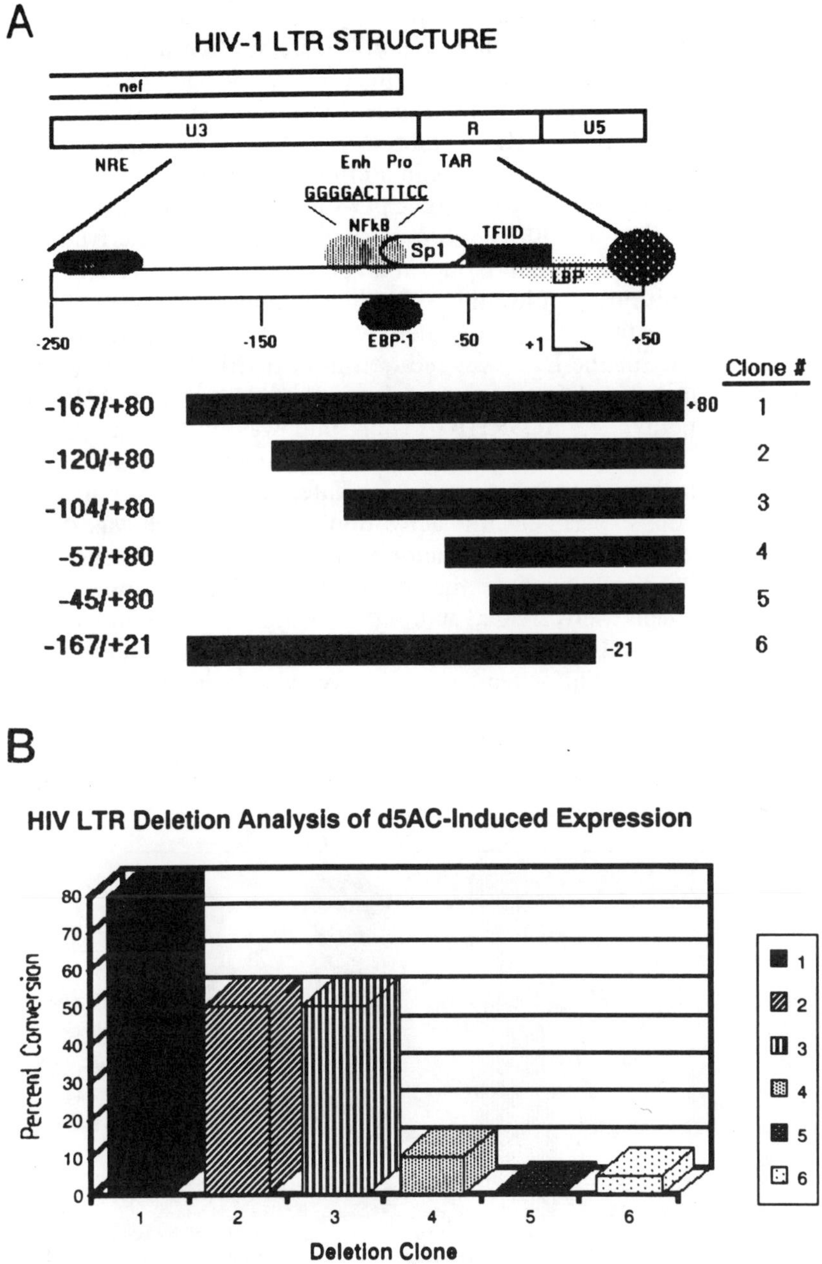

Figure 5. Deletion analysis of HIV LTR responsiveness to deoxy-5-azacytidine (d5AC) induction. Panel A: The overall structure of the LTR and the regions spanning each deletion mutant reporter plasmid. Panel B: Relative activities of chloramphenicol acetyltransferase (CAT) expressed by each deletion mutant clone 24 h after electroporation of DNA (10 μg) into A3.01 T-cells. Activity levels are expressed as percent conversion of ^{14}C-chloramphenicol to the acetylated forms.

suggested that transcriptional initiation was crippled upon deletion of this region, and thereby did not respond to d5AC treatment.

As a direct method to analyze the effect of site-specific methylation of the HIV LTR, we employed the transient electroporation of enzymatically methylated pU3RIIICAT into A3.01 T-cells (Bednarik et al., 1990). Experiments of this nature can address the *cis*-level activity of the LTR after selective methylation of CpG dinucleotides localized in the core enhancer. The pU3RIIICAT construct was originally derived from the pHCBX2 HIV molecular clone, and contains two Msp I/Hpa II sites located as positions −146 and −218 nucleotides respectively. When both of these sites were methylated with Hpa II methyltransferase, *cis*-level activity of the LTR was reduced to negligible levels (Fig. 6). In these experiments, CAT activity was determined 24 h after transfection. When the methylated pU3RIIICAT plasmid was co-transfected with pIIIextatIII, a plasmid directing tat expression under control of the HIV LTR, the block in transcription was obviated after 24 h. As a logical extension of this experiment, the activation state of the cell was studied to determine whether cellular factors, induced by the phorbol ester TPA, could similarly obviate the methylation-induced transcriptional block. When cells were treated wth TPA (50 ng/ml) immediately after transactivation, only a small increase in relative CAT activity was observed (Fig. 7). This effect was additive when cells were co-trans-

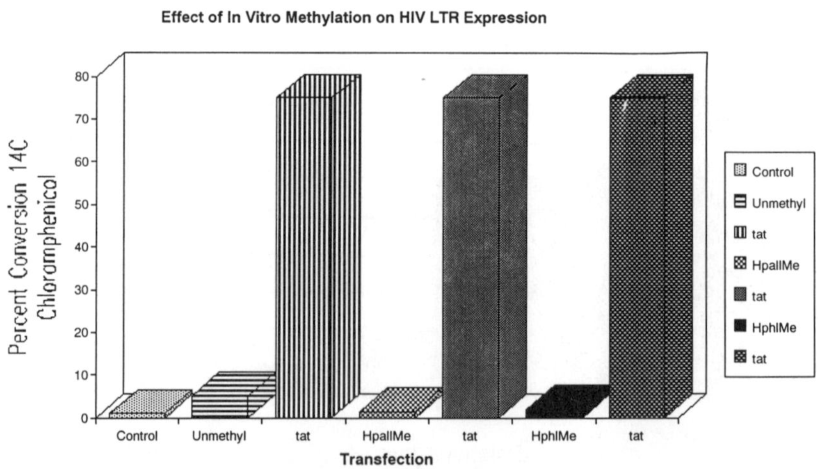

Figure 6. Inhibition of HIV LTR transactivation and expression by site-specific DNA methylation. The pU3RIIICAT reporter plasmid was enzymatically methylated with either Hpa II methyltransferase, or Hph I methyltransferase and electroporated into A3.01 cells in the presence or absence of plasmid DNA directing tat expression (pIIIextatIII) under control of the HIV LTR. The control is a mock transfection. Analysis of CAT activity was performed 24 h after transfection as described previously (Bednarik et al., 1990).

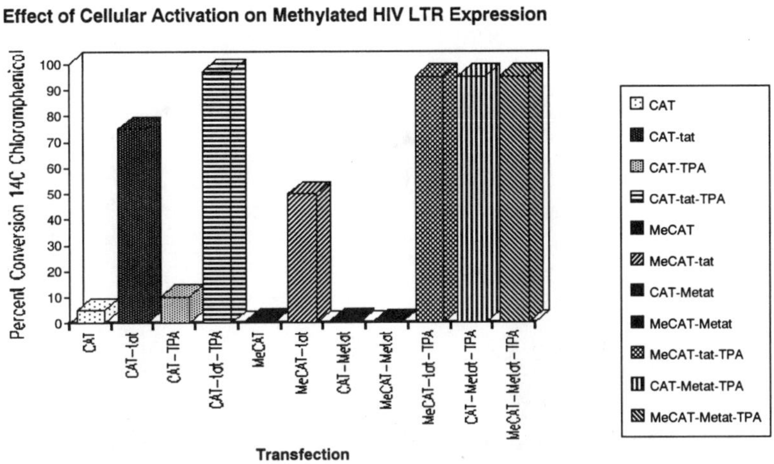

Figure 7. Methylation-mediated inhibition of HIV LTR expression is influenced by cellular factors. A3.01 cells were electroporated with pU3RIIICAT (10 μg) alone, or in combination with pIIIextatIII. In each case, the LTRs were either individually methylated *in vitro*, or both were methylated simultaneously in order to differentiate the effect of methylation of one plasmid DNA on the expression of the other. Methylated plasmid DNA is denoted by "Me". The effects of induced host cellular factors was determined by treatment with the phorbol ester TPA (50 ng/ml).

fected with pIIIextatIII. It was interesting to note in these experiments that the methylation of the LTR directing tat expression in the pIIIextatIII plasmid completely inhibited transacivation when co-transfected with pU3RIIICAT. The addition of TPA released the transcriptional block, presumably *in trans*, even when both LTRs directing CAT or tat epression were methylated suggesting that cellular factor(s) expressed by the activated T-cell can modulate this effect. Experiments designed to determine whether demethylation of the LTR was involved in TPA-mediated reactivation found no evidence of the loss of methylation at either Hpa II site (Bednarik et al., 1990).

5.2 Inhibition of tat-mediated transactivation by DNA methylation is time dependent

It is of interest to note that the apparent ability of tat to overcome the transcriptional block introduced by DNA methylation was almost identical to observations by Weisshaar et al. (1988) in that the transactivation of the methylated adenovirus promoter was facilitated by the E1A protein. Subsequent experiments performed by our laboratory were directed at the dissection of events which might explain this phenomenon. When the HIV LTR was again methylated at the two Hpa II

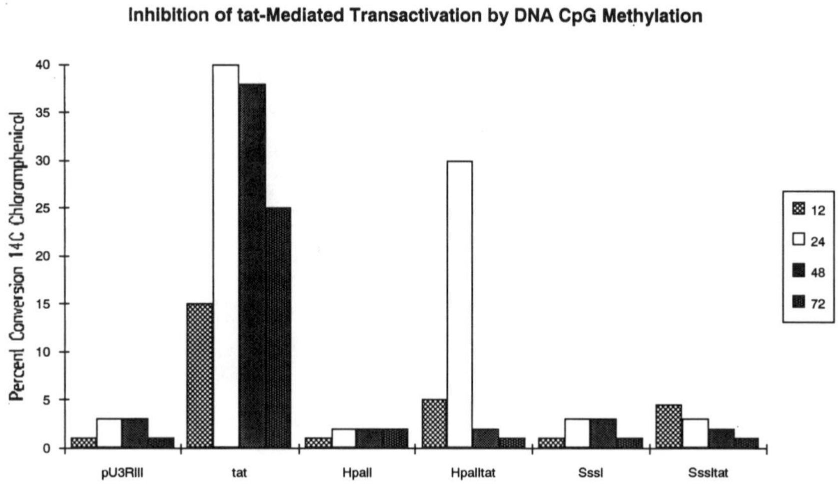

Figure 8. Inhibition of tat-mediated transactivation by DNA CpG methylation. Plasmid DNA (pU3RIIICAT) was methylated as described above with the exception that Sss I methyltransferase (CpG methylase, New England BioLabs) was employed to methylate all CpG dinucleotides. Plasmid DNA (10 μg) was electroporated in A3.01 cells (1×10^7 per ml) and cells were harvested at 12, 24, 48, and 72 h respectively. Expression of the methylated or unmethylated LTRs was reflected as percent conversion ^{14}C-chloramphenicol to the acetylated forms.

($-146/-218$) sites and co-transfected with pIIIextatIII, there was a rapid increase in CAT activity after 24 h as was previously observed (Bednarik et al., 1990). However, CAT activity decreased dramatically after 48 h even when tat was present (Fig. 8). The ability of tat to transactivate the methylated LTR decreases as a function of time, thereby indicating the occurrence of subsequent cellular event(s). This observation might be explained in the following way. The HIV LTR is initially methylated at two sites using Hpa II methyltransferase. Following transfection, the LTR may be repressed further by some cellular factor(s), or may become more extensively hypermethylated at other CpG dinucleotides as a result of the endogenous cellular *de novo* DNA MeTase recognizing the first two methyl groups as a signal. To directly test the latter scenario identical experiments were performed with the exception that Sss I methyltransferase was employed to methylate the HIV LTR. This enzyme recognizes all CpG dinucleotides regardless of their location; the net result is a completely methylated LTR. If the increased frequency of CpG methylation was involved with complete inhibition of tat-mediated transactivation, there should be no predicted initial increase in CAT activity. Co-transfection of the Sss I-methylated pU3RIIICAT with pIIIextatIII resulted in complete inhibition of transactivation (Fig. 8). Clearly, the density of methylated CpG dinucleo-

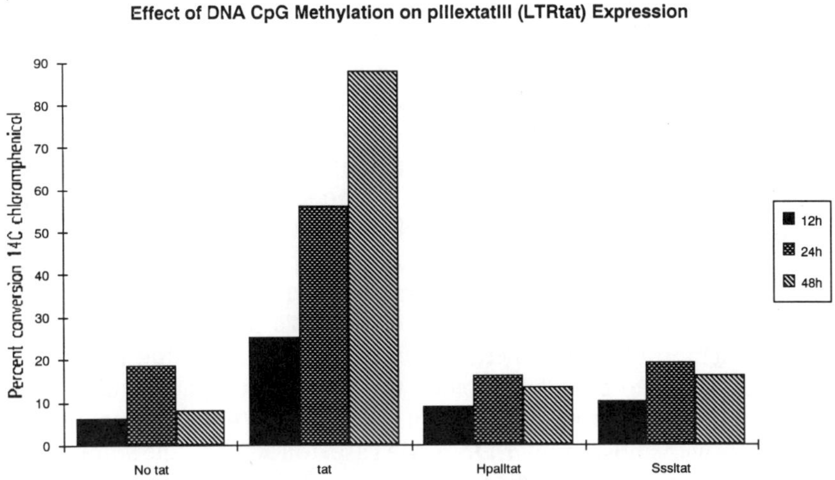

Figure 9. Effect of DNA CpG methylation on LTR-directed tat expression (pIIIextatIII). A3N92.2 cells containing a single stably integrated copy of pU3RIIICAT were electroporated with pIIIextatIII DNA which was either unmethylated, methylated with Hpa II methyltransferase/Sss I methyltransferase and subsequently assayed for CAT activity at the times indicated.

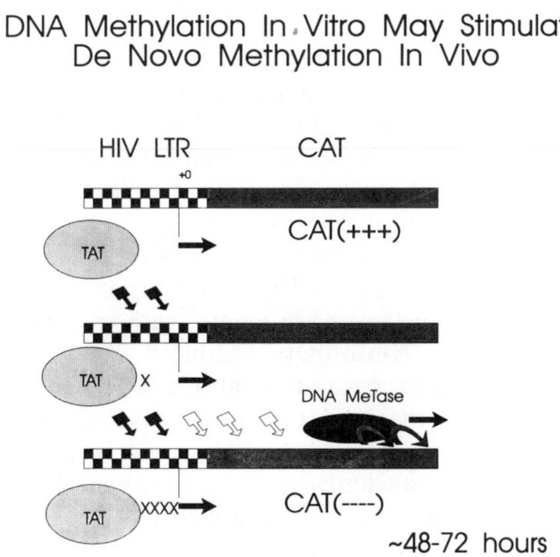

Figure 10. Does *in vitro* methylation stimulate the hypermethylation of CpG dinucleotides *in vitro*? Enzymatic methylation at one or two sites in the LTR is not sufficient to block transactivation by tat, however methylation at the first two sites (black arrows) is followed by recognition by a cellular *de novo* methyltransferase. Additional CpG dinucleotides are subsequently methylated which results in complete inhibition of tat activity.

tides plays a role in determining gene activity, and other laboratories have demonstrated the existence of a class of putative proteins which have DNA-binding affinity for hypermethylated DNA (Antequera *et al.*, 1989; Boyes and Bird, 1991, 1992). Such putative repressor molecules, denoted as MeCP for "methyl CpG inhibitory protein" may act at the transcriptional level by inhibiting transcript initiation (see chapter of Ehrlich and Ehrlich, pp. 145–168 of this book). To confirm that this observation was a time-dependent effect on tat action, A3.01 T-cells containing a single, stable integrated copy of pU3RIIICAT (A3N92.2) were electroporated with either unmethylated, HpaII-methylated, or SssI-methylated pIIIextatIII; CAT activity was monitored over 48 h. Figure 9 demonstrates the methylation of the LTR directing tat expression in sufficient to completely block transcription and accumulation of tat. A model depicting the recognition of the *in vitro* methylated HIV LTR by the host cellular DNA MeTase, followed by the stimulation of *de novo* methylation is shown in Figure 10.

5.3 Methylation of the HIV LTR does not inhibit transcription by RNA pol II in vitro

Previous studies have suggested that CpG methylation was inhibitory to RNA pol II-mediated transcription, but not to pol III-mediated transcription (Besser et al., 1990). In order to determine whether *cis*-level initiation of RNA pol II was affected by *in vitro* methylation of CpG sites, both methylated and unmethylated pU3RIIICAT was tested for expression by employing cell-free HeLa nuclear extracts (Marciniak and Sharp, 1991). Plasmid DNA (pU3RIIICAT) was methylated with either Hpa II or Sss I methylase and incubated with nuclear extracts. Figure 11 represents primer extension cDNA products resolved by denaturing polyacrylamide/urea gel electrophoresis. Both methylated reporter genes were transcribed with efficiencies equal to that of the unmethylated control. The major product was determined to be the expected 100 nucleotides in length, with the smaller products probably resulting from conformation-induced, reverse transcriptase pause sites. We have found these observations to be very exciting since these data suggest that the inhibitory effect of CpG methylation on transcription *in vivo* is uncoupled when assayed in partially purified nuclear extracts. We do not know as yet whether inhibition would be restored in a whole cell extract. Preparation of the nuclear extracts may allow the loss of important factor(s) which may normally recognize methylated DNA. Again, the MeCP proteins proposed by Boyes and Bird (1990, 1992), or similar inhibitory proteins might be involved. Experiments designed to dissect this system further are underway.

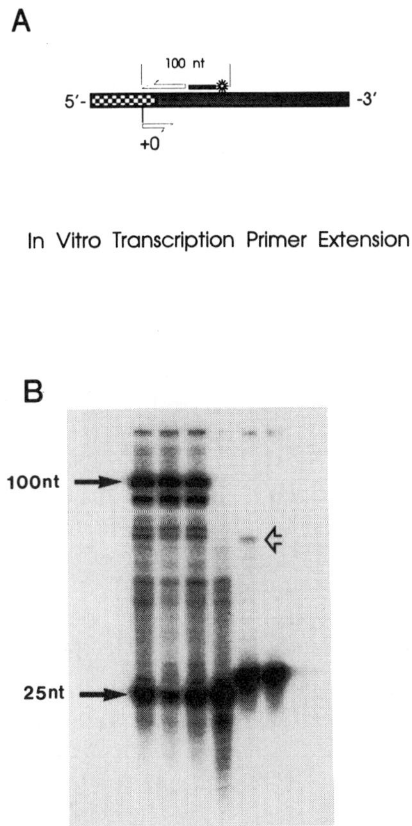

Figure 11. Methylation of the HIV LTR does not inhibit *in vitro* transcription by RNA polymerase II in HeLa nuclear extracts. Plasmid DNA (pU3RIIICAT; 100 ng) was methylated as described above and incubated at 37°C in cell-free HeLa nuclear extracts as prescribed by the vendor (Promega). Incubations were terminated after 30 minutes by phenol/chloroform extraction. Transcripts were annealed to a [^{32}P] end-labeled antisense oligonucleotide, and extended by incubation with MuLV reverse transcriptase. Panel A denotes the region of the plasmid DNA which encodes the corresponding mRNA. Primer extension products were resolved (panel B) by electrophoresis on a 20% polyacrylamide/7 molar urea gel. The gel was visualized by analysis on a Molecular Dynamics PhosphorImager. Lane 1: unmethylated pU3RIIICAT; lane 2: same as lane 1 except that the plasmid DNA was methylated with Hpa II methyltransferase; lane 3: same as lane 1 except that plasmid DNA was methylated with Sss I methyltransferase; lane 4: control reaction lacking plasmid DNA; lane 5: pAdeno plasmid positive control (open arrow); lane 6: same as lane 5 except that plasmid DNA was not included in the reaction.

6 Selective modulation of NF-kB by core CpG methylation

The regulation of gene expression is a complex process orchestrated by the binding of a spectrum of transcriptional proteins to defined DNA motifs localized in *cis*-acting promoter or enhancer elements (Mitchell and Tijan, 1989; Dynan, 1989; Bendarik et al., 1990). Nuclear factor kB

(NF-kB) participates in the control of both cellular and viral gene expression and is post-transcriptionally activated by phosphorylation of its inhibitor IkB, followed by subsequent translocation to the nucleus (Sen and Baltimore, 1986a and 1986b; Lenardo and Baltimore, 1989; Nabel and Baltimore, 1987; Nabel et al., 1988; Bielinska et al., 1989; Greene et al., 1989; Ruben et al., 1989).

Although NF-kB was first identified as a DNA-binding protein in the nucleus of B-cells (Sen and Baltimore, 1986b), more recent work has demonstrated the role of this factor in the modulation of retrovirus and DNA virus gene transcription, and in the activation of cytokine gene expression (Collart et al., 1990; Zhang et al., 1990; Hiscott et al., 1989; Sambucetti et al., 1989). The affinity-purified NF-kB has been shown to be a heteromeric complex which is composed of p50, which binds to DNA sometimes in association with p65, the latter of which is required for inactivation of NF-kB via IkB (Baeuerle and Baltimore, 1986a and 1986b). The p50 subunit which is initially synthesized as a precursor of $M_r \sim 105{,}000$ and is processed to the mature form of M_r 50,000, has been cloned and has been identified as a member of the *dorsal* and *rel* family of proteins (Ghosh et al., 1990; Kieran et al., 1990; Bours et al., 1990), in addition to the p65 subunit (Ruben et al., 1991; Nolan et al., 1991).

The binding of many nuclear transcription factors to defined DNA motifs is attenuated by the content of 5-methylcytosine localized near or at the site(s) of DNA attachment (Doerfler, 1983, 1984; Keshet et al., 1986; Murray and Grosveld, 1985). The mechanism by which methylated DNA modulates gene expression has recently been further characterized and is believed to involve two possible scenarios. A "direct model" in which promoter elements containing relatively few CpG sequences which are localized within sites of nuclear factor binding can directly inhibit protein-DNA association, such as the cyclic AMP responsive element (CREB), and two specific transcription factors found in HeLa cells (Doerfler et al., 1990; Iguchi-Ariga and Schaffner, 1989; Kovesdi et al., 1987; Watt and Molloy, 1988). Secondly, an "indirect model" in which sequences of DNA which contain a dense population of CpG dinucleotides, or CpG islands, may globally affect gene transcription by attracting the binding of an inhibitory protein which was recently described (Meehan et al., 1989; Boyes and Bird, 1991). This recently discovered methyl CpG inhibitory protein (MeCP) has been shown to have affinity for densely methylated DNA and is believed to displace the binding of transcriptional factors (Meehan et al., 1989; Boyes and Bird, 1991). The methylated CpG sequences are resistant to nonspecific nucleases (Solage and Cedar, 1978), and to endonucleases which domonstrate CpG specificity (Antequera et al., 1989, 1990). Evidence for the indirect model has been derived from experiments studying transcription of the γ-globin and herpes simplex thymidine kinase genes (Murray and Grosveld, 1987; Buschhausen et al., 1987).

Since the HIV LTR is regulated by several cellular DNA-binding proteins in addition to virus-encoded proteins, it was necessary to investigate the role(s) which DNA methylation played in the modulation of cellular factor/viral LTR DNA interaction. Indeed, methylation-sensitive sequence-specific DNA-binding by the c-myc basic region was recently shown, and the methylation of the c-myc motif may regulate the formation of the c-myc/myn complex *in vivo* (Prendergast and Ziff, 1991a and 1991b). It was therefore pertinent to the rationale of our studies to determine whether DNA-binding of cellular transcriptional factors are sensitive to methylation of CpG sequences located in their DNA-binding motifs. The role of NF-kB in the life cycle of HIV has been shown to bear strong physiological significance relevant to the activation of viral/cellular gene transcription (Gilmore, 1990; Lenardo and Baltimore, 1989), and was, therefore, an ideal candidate for these experiments.

6.1 Core CpG methylation in the NF-kB DNA-binding motif inhibits binding of the DNA:NF-kB protein complex

Analysis of protein binding was performed by employing the electrophoretic mobility shift assay (EMSA) initially utilizing crude nuclear extracts isolated from phorbol ester-activated Jurkat cells (Bendarik et al., 1990; Raj et al., 1983; Baldwin and Sharp, 1987). Incubation of the unmethylated, [32P]-labeled DNA probe with 1 μg of total nuclear extract resulted in the retardation of a DNA-protein complex corresponding to the p65/p50 heterodimer (Fig. 12). when labeled DNA containing the mutated NF-kB site was incubated with nuclear protein, no binding was observed. Incubation of the labeled, methylated NF-kB DNA probe with crude nuclear extract did not result in binding of the heterodimer complex. A secondary DNA-protein complex was consistently observed for the DNA probes incubated with the crude nuclear extracts, and this complex was non-specifically competed away by unlabeled Sp1, AP1, or mutant NF-kB competitor DNA (data not shown) as was observed previously (Hohmann et al., 1990).

When similar experiments employing the DNase I footprint protection assay were performed, protection of sequences corresponding to the NF-kB binding motifs (kB1/kB2) were significantly inhibited when the single CpG sequence located within both binding sites contained 5-methylcytosine (data not shown). Analysis of footprint autoradiographs suggested that the methylated NF-kB competitor DNA was a poor substrate since little competition of labeled probe DNA occurred. Interpretation of the data further suggested that both binding motifs become deprotected to the same degree when 5-methylcytosine is present at the number 10 position on the DNA probe. Since crude

DNA Probes (Coding Strand)

1. 5'-GGGACTTTCCGCTGGGGACTTTCC-3'
2. 5'-TCTACTTTCCGCTTCTGACTTTCC-3'
3. 5'-GGGACTTTCCGCTGGGGACTTTCC-3'
CH_3

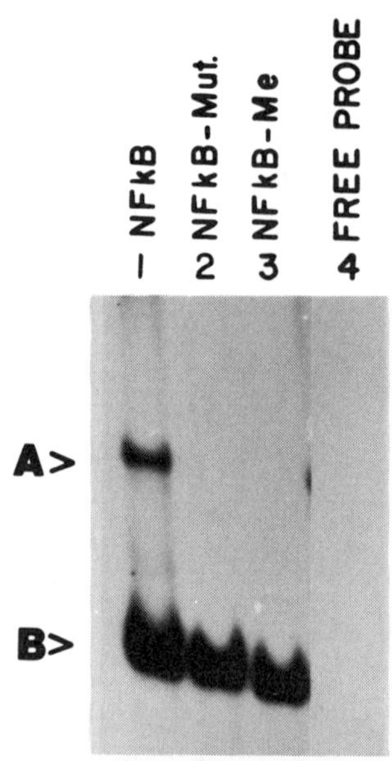

Figure 12. Methylation of cytosine at position number 10 in the sequence CpG inhibits NF-kB complex binding. EMSA DNA-binding analysis of crude nuclear extract NF-kB proteins utilizing either the native NF-kB motif (probe 1), mutant NF-kB motif (probe 2), or the CpG-methylated probe (probe 3). Lane 1: EMSA of the putative NF-kB p65:p50 heterodimer (arrow, A) complex on a 5–10%, 32 cm non-denaturing gradient polyacrylamide gel; lane 2: mutant NF-kB probe DNA; lane 3: methylated probe DNA; lane 4: free probe. Arrow B denotes the position of non-specific binding protein (unretarded probe not shown).

nuclear protein was employed in this experiment, additional protection of the probe DNA may have been due to non-specific binding activity. There was no evidence of another protein specifically binding to the methylated position on the DNA probe when crude nuclear extracts were employed.

6.2 Methylation of the NF-kB core CpG specifically inhibits the binding of bovine spleen or recombinant NF-kB p50 proteins

As an extension of the experiments described above, NF-kB proteins purified from bovine spleen were employed as a source of native protein (Lenardo *et al.*, 1988), and as a means of eliminating other non-specific DNA-binding factors which may interfere with the binding assay. Figure 13 demonstrated the specificity of complex formation as shown by competition with unlabeled NF-kB competitor DNA, but not by

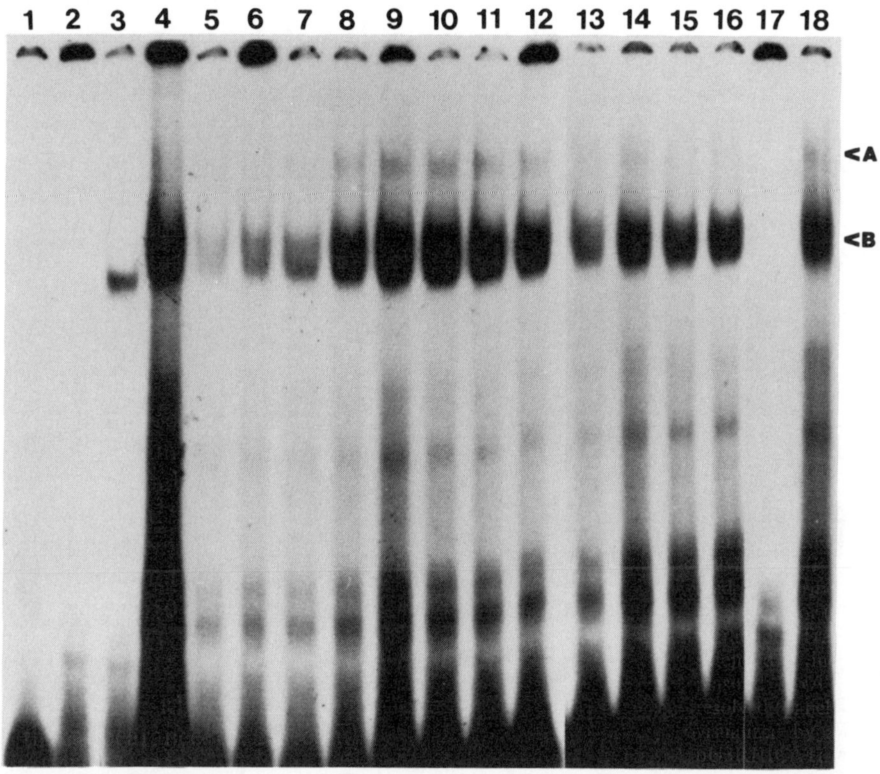

Figure 13. EMSA analysis of bovine spleen NF-kB DNA-binding to the methylated or unmethylated NF-kB DNA motifs. Labeled DNA probe (1 ng) was incubated at room temperature in the presence of 1 μg purified bovine spleen NF-kB protein for 20 min followed by electrophoresis on a 5% non-denaturing polyacrylamide gel. Lanes 1 and 2: Free native and mutant probe DNA respectively; lane 3: labeled NF-kB mutant probe; lane 4: labeled native NF-kB probe; lanes 5–8: native probe DNA incubated with 50, 10, 2.0, and 0.4 ng of unlabeled NF-kB competitor DNA; lanes 9–12: same as lanes 5–8 except that the unlabeled mutant DNA was employed as the competitor; lanes 13–16: same as lanes 5–8 except that unlabeled DNA containing the Sp1 binding motifs was employed as the competitor, lanes 17 and 18: incubation of bovine spleen NF-kB protein with the labeled, methylated and unmethylated DNA probes respectively. Arrows A and B denote the positions of NF-kB-specific DNA-protein complexes; the retarded band in lane 3 represents an apparently non-specific binding protein which associated with the mutant NF-kB probe.

competitor DNA containing the kB mutation, or by DNA containing the binding motifs specific for Spl transcriptional proteins. The bovine spleen NF-kB proteins displayed little or no affinity for the probe containing the methylated core CpG dinucleotide.

Purified native NF-kB protein consists of two major polypeptides, one of M_r 50,000 (p50) and the other of M_r 65,000 (p65). Of these two protein subunits, it is the p50 polypeptide which actually binds to DNA (Ghosh et al., 1990), while the p65 subunit is necessary for interaction with IkB and has less intrinsic DNA-binding capacity (Nolan et al., 1991). However, the p50/p65 heteromeric DNA-binding properties differ from those of p50 alone in that the heteromer preferentially binds to a less palindromic kB site such as the present in the HIV-1 enhancer (Baeuerle and Baltimore, 1988a, 1988b). Further characterization of single site CpG methylation on the interaction fo NF-kB protein with the methylated motif required recombinant p50, generated in *E. Coli*, to specifically interpret the inhibitory effect, and to determine if inhibition of DNA-binding adhered to the "direct" or "indirect" models. Figure 14A demonstrated the inhibition of p50 binding to the methylated CpG as compared to the unmethylated motif. The specificity of DNA binding is clearly demonstrated by complete competition of p50 by less than 20 ng of unlabeled competitor DNA. In order to determine whether the observed inhibition of DNA binding was due to the methylation of a single CpG and was not a consequence of experimental variability, DNA binding to the unmethylated NF-kB probe was quantitated by direct competition with various concentrations of the unlabeled, CpG-methylated NF-kB competitor DNA (Fig. 14B). While as little as 2 ng of unmethylated competitor DNA was sufficient to displace the binding of p50 protein to the labeled NF-kB probe, 50 ng of methylated competitor DNA did not completely compete the p50 binding, and at 10 ng of methylated competitor, essentially no competition occurred at all.

Figure 14. Effect of core CpG methylation upon recombinant p50 NF-kB subunit DNA-binding. Panel A: The effect of CpG methylation on the binding afinity of the NF-kB p50 subunit was determined by EMSA. Recombinant p50 protein (courtesy G. Nabel and C. Duckett) derived from the human molecular clone KBF1 (Kieran et al., 1990) was incubated with labeled probe DNA at a concentration of 25 ng p50 protein to 1 ng (15,000 cpm) DNA. Lane 1: free probe; lanes 2–4: incubation of recombinant p50 protein with the native NF-kB, methylated, or mutant probes respectively; lanes 5–7: same as lanes 2–4 except *E. Coli* BL21 Lys E extracts which do not contain the p50 protein were included as a negative protein control. Panel B: Competition analysis of p50 DNA-binding utilizing unmethylated or core CpG-methylated competitor DNA. Unmethylated and methylated NF-kB motifs were employed as competitor DNA-binding substrates in order to quantitate the affinity of each DNA for p50 protein. Lane 1: free unmethylated probe; lane 2: same as lane 1 except that 10 ng of p50 protein was included in the incubation reaction; lanes 3–6: incubation reactions were performed as in lane 2, however either 50, 10, 2.0, or 0.4 ng of serially diluted, unlabeled/unmethylated competitor DNA was included respectively; lanes 7–10: same as lanes 3–6 except that the unlabeled/methylated DNA was employed as the competitor; lanes 11–14: same as lanes 3–6 and 7–10 except that the kB mutant DNA was employed as the competitor.

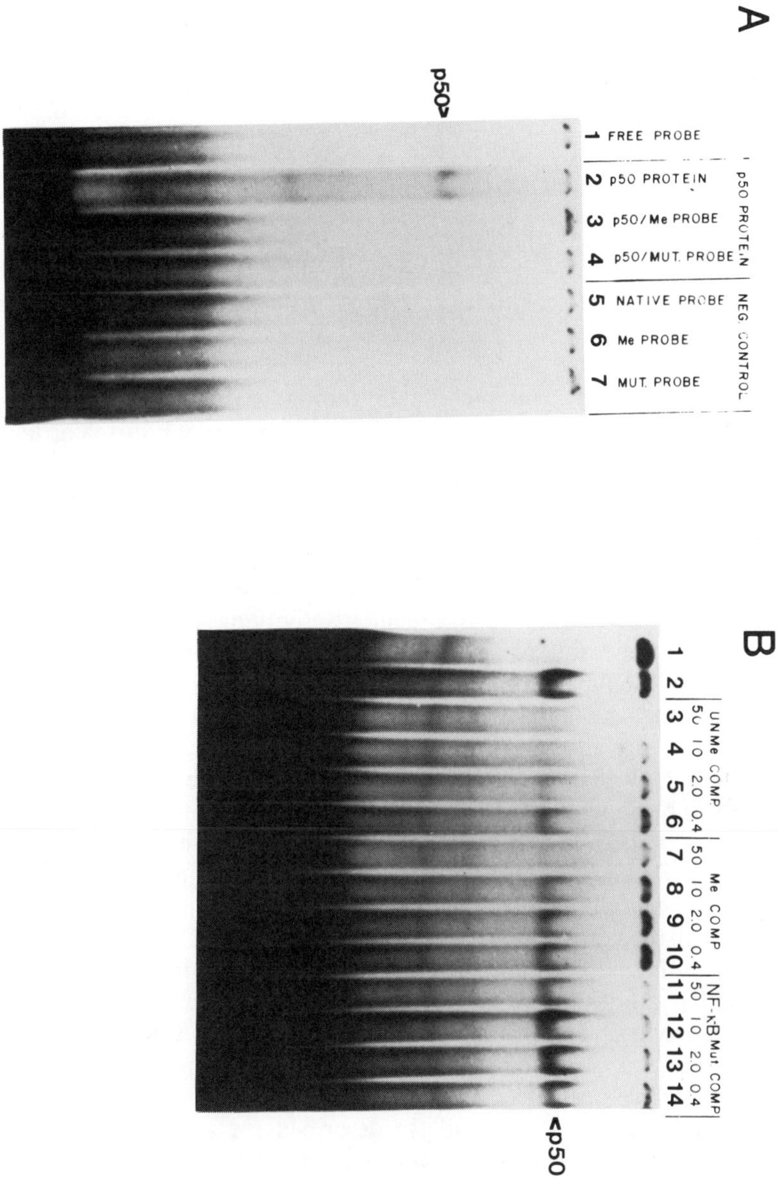
A
p50>
1 FREE PROBE
2 p50 PROTEIN
3 p50/Me PROBE
4 p50/MUT. PROBE
p50 PROTEIN
5 NATIVE PROBE
6 Me PROBE
7 MUT. PROBE
NEG. CONTROL
B
1 2 3 4 5 6 7 8 9 10 11 12 13 14
UNMe COMP
50 10 2.0 0.4
Me COMP
50 10 2.0 0.4
NF-κB Mut COMP
50 10 2.0 0.4
<p50

Complete competition of p50 has been shown to occur with at least 20 ng of competitor DNA (C. Duckett, personal communication), and in our hands was observed complete competition with as little as 10 ng of competitor DNA. These data collectively demonstrate a much lower binding affinity of p50 protein to the NF-kB motif when methylated at the single core CpG site.

6.3 Direct and indirect models for maintenance of HIV latency

It is becoming clear that the regulation of gene expression by CpG methylation is one which can be highly specific as has been demonstrated by the methylation sensitivity of c-myc basic region (Prendergrast and Ziff, 1991a). In this case, dimers of the chimeric protein, denoted as E6, specifically bind to an E box element with the sequence GGC-CACGTGACC. In a manner much analogous to the experiments described, the authors discovered that methylation of the E box central core CpG specifically inhibited the binding by E6, but not by other HLH proteins. In addition, methylation of the c-myc motif blocks the binding by c-myc, myn, and c-myc/myn which is the first report of an oncoprotein whose activity is regulated, in a direct manner, by DNA methylation (Prendergrast and Ziff, 1991b). Furthermore, the methylation of DNA leading to either cellular or viral gene inactivity has been shown to be at least partly the result of a direct mechanism, as is shown here by the methylation-mediated inhibition of NF-kB binding to its cognate motif, or by an indirect mechanism as shown by Boyes and Bird (1991). Both mechanisms are illustrated in Figure 15 as they would apply to the integrated HIV. The specificity of the type of inhibition which occurs appears to be dependent upon the transcriptional protein in question and the density of methylation in the region of the binding motif(s) (Boyes and Bird, 1991; Bednarik et al., 1990). Since the binding of p50 protein from *E. Coli* lysates was inhibited by methylation of the core CpG dinucleotide, the interaction of a methyl-CpG binding protein (MeCP) could not be a variable in these experiments since p50 was the only NF-kB-specific factor present. In addition, we never observed the appearance of a second protein(s) which preferably bound to methylated DNA when crude nuclear extracts were employed. Apparently, the regional density of methylated CpG sequences is a determining factor in order to facilitate the binding of the MeCP molecule(s) described for the "indirect" model (Boyes and Bird, 1991), as this is consistent with our observations.

These experiments have determined that NF-kB proteins are sensitive to DNA methylation. Since the expression of HIV in latently infected T-cells can be correlated with the activation state of the cell (Bednarik and Folks, 1992, the methylation of corresponding binding motifs in the HIV LTR may contribute to release of the virus from latency. It is interesting to note that NF-kB proteins are very similar to both *rel* and *dorsal* in both

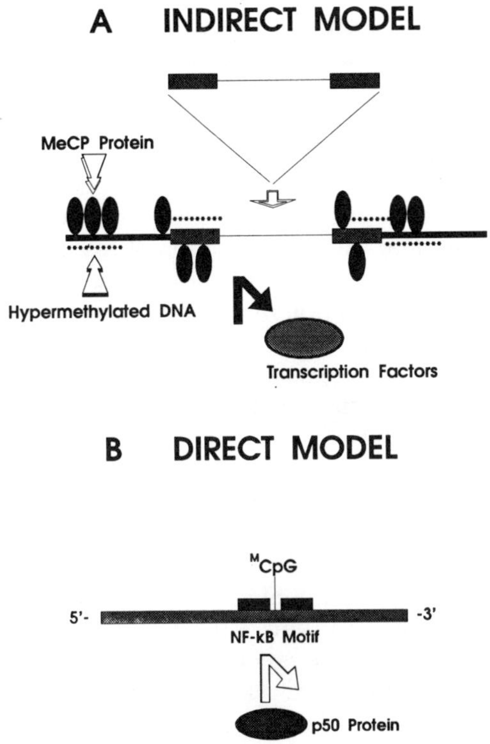

Figure 15. Possible models of DNA methylation-mediated inhibition of transcription factor binding. Panel A, Indirect Model: The inhibition of transcription factor binding to defined DNA motifs contained within the HIV LTRs is caused by the association of a putative methyl CpG-binding protein (MeCP; Boyes and Bird, 1991) which recognizes densely methylated DNA. The mechanism of transcriptional inhibition occurs via the displacement of transcription proteins by MeCP or related factors. Panel B, Direct Model: The binding of a specific transcription factor is obviated by the presence of methylated CpG(s) localized in close proximity to the site of protein-DNA attachment. This scheme is represented by the NF-kB motif repeats contained within the HIV core enhancer region. The core CpG is actually adjacent to the protein-DNA contact points.

structure and, like c-myc and myn, their possible roles in development and cellular differentiation. These results extend the list of proteins which are directly influenced by DNA methylation to include NF-kB.

7 DNA Methylation of human retroviruses; role(s) in human disease

7.1 DNA Methylation as a suppressive mechanism of human neoplasia or immune deficiency

The proviruses of retroviruses are the most sophisticated of transposable elements. Retroviruses are characterized as causing neoplasia because of their ability to insert into genes thereby inducing aberrant up- or down-regulation of gene products. Expression of retroviral gene products which are DNA-binding proteins or transactivators may act *in trans* upon other cellular genes, inducing the inappropriate accumulation of proteins which are themselves oncogenic. Many retroviruses express proteins which are the products of oncogenes. Obviously, the methylation of retroviral LTRs would maintain the virus in a latent state and should interfere with the induction of disease by obviating the expression of viral proteins. One would not expect DNA methylation to prevent the insertional disruption of a cellular gene by a retrovirus, but that it may prevent the insertional activation of a putative protooncogene. We do not as yet know exactly what, if any, sequences the cellular DNA MeTase recognizes other than the CpG dinucleotide, and there is no conclusive evidence which would lead us to believe the retroviral LTRs are preferential targets for DNA MeTase. Indeed, the methylation-mediated inactivation of retroviral transcription, and the obviation of disease may be totally dependent upon the site of integration.

The expression of HIV has been demonstrated to be intimately associated with the orchestration of virus-encoded proteins which modulate events such as transactivation, RNA processivity and splicing, etc. However, as previously demonstrated by the murine system, all of these HIV-encoded proteins bear no relevance to virus regulation if the LTR is methylated and virus expression is cryptically silent. We have previously suggested that the expression of low levels of tat would obviate the transcriptional block and cause a cascade of virus reactivation (Bednarik et al., 1990, 1991). In light of recent extensions of these experiments, the fully methylated HIV LTR is, surprisingly, not responsive to tat. Much less is known about the mechanisms by which retroviruses induce a wide spectrum of non-neoplastic disease, including anemia, arthritis, pneumonia, peripheral neuropathy, myelopathy, and encephalopathies.

The remaining questions really focus upon the true physiological relevance of DNA methylation and HIV latency in AIDS. Experiments to date have addressed this issue utilizing cell lines in which we do not really know how MeTase is regulated, if indeed its regulation is normal at all. We have now localized a population of individuals who are serologically negative for HIV infection, but remain positive by PCR (Jehuda-Cohen et al., 1990). There is the possibility that these persons harbor a defective provirus, or one in which a deletion has occurred. This observation is very similar to SIV-infected sooty mangabey monkeys which are seronegative for SIV but are PCR positive (Ansari et al., unpublished data). These foci of human and simian populations may prove to be a valuable resource to physiologically demonstrate DNA methylation-mediated HIV/SIV latency. The recent advances in ligation-mediated PCR (Pfeifer et al., 1989) are currently being employed to analyze this pool of individuals as to be the methylation status of the integrated provirus.

7.2 Future prospects

The mechanism by which DNA methylation recognizes and inactivates the expression of specific gene sequences remains largely unknown. The unique property of retroviruses is that they are valuable insertional "tags" or labels (Soriano et al., 1989). Since the retrovirus is a mobile set of genes, one can direct its insertion, *in vitro*, into a variety of target gene sequences. A hybrid virus containing a short CpG cluster is currently being constructed and tested for packaging efficiency. The artificial CpG repeats are arranged in the U3 region and will mimic a target for DNA MeTase. Similarly, LTR reporter constructs containing CGG repeats at variable distances upstream of the U3 region will be employed to test regional inactivation of the LTR via hypermethylation. The introduction of these hybrid proviral constructs into cells via transfection would permit the determination of those target sequences which allow viral expression, and those that do not. This method would be a direct approach to probe for possible DNA MeTase recognition sequences, thereby simulating the integration of the provirus into a region of DNA which is not normally transcriptionally active. For example, the CD8 gene localized in CD4+ cells is extensively hypermethylated in upstream regions and is not expressed (Bednarik et al., 1992). This cell lineage-dependent phenomenon absolutely requires specific recognition by the cellular DNA methylation machinery, probably during cell differentiation. One might choose this region of DNA as an *in vitro* target for retroviral insertion, and attempt to express the virus in CD4+ and non-CD4+ cells to determine whether it becomes a target for DNA methylation.

Since our initial finding that the HIV is sensitive to inactivation by DNA methylation, we have continued to employ it as a tool to further understand the mechanism of latency as well as the regulation of DNA methylation in the mammalian cell. Furthermore, the role of DNA methylation as a suppressor of retroviral-mediated disease is more likely to become apparent as we continue to investigate the progression of diseases which take relatively long periods of time to present themselves. The manifestation of HTLV-I induced HAM-TSP (Kaplan et al., 1991), or adult T-cell leukemia-lymphoma (ATLL; Dosaka et al., 1991) may require decades to develop. Similarly, many dementias and neuropathies may ultimately be the result of retroviral expression. During the aging process, there is a progressive loss of DNA methylation, and therefore epigenetic information. Retroviruses which were previously latent would become active, and disease would follow much in the same way that an erroneously expressed gene may generate malignancy. Only until the black box of DNA methylation is fully illuminated will the understanding of these disease states be fully realized.

Annemieke, J. M. H., Verkerk, M. P., Sutcliffe, J. S., Fu, Y-.H., et al. (1991) Identification of a gene (FMR-1) containing a CGG repeat coincident with a breakpoint cluster region exhibiting length variation in fragile X syndrome. Cell *65*, 905–914.

Antequera, F., Macleod, D., and Bird, A. P. (1990a) Specific protection of methylated CpGs in mammalian nuclei. Cell *58*, 509–517.

Antequera, F., Boyes, J., and Bird, A. P. (1990b) High levels of de novo methylation and altered chromatin structure at CpG islands in cell lines. Cell *62*, 503–514.

Baeuerle, P. A., and Baltimore, D. (1988a) Activation of DNA-binding activity in an apparently cytoplasmic precursor of the NF-kB transcription factor. Cell *53*, 211–217.

Baeuerle, P. A. and Baltimore, D. (1988b) IkB: A specific inhibitor of the NF-kB transcription factor. Science *242*, 540–546.

Baldwin, A. S., and Sharp, P. A. (1987) Binding of a nuclear factor to a regulatory sequence in the promoter of the mouse H-2KB class I major histocompatibility gene. Mol. Cell. Biol. *7*, 305–313.

Bednarik, D. P., and Folks, T. M. (1992a) Mechanisms of HIV-1 latency. AIDS *6*, 1–14.

Bednarik, D. P., Duckett, C., Kim, S. U., Perez, V. L., Griffis, K., Guenthner, P. C., and Folks, T. M. (1991) DNA CpG methylation inhibits binding of NK-kB proteins to the HIV-1 long terminal repeat cognate DNA motifs. New Biol. *3*, 969–976.

Bednarik, D. P., Guenthner, P. C., and Gutekunst, K. A. (1992b) Inactivation of the HIV LTR by a human T-cell DNA methyltransferase (submitted).

Bednarik, D. P., Cook, J. A., and Pitha, P. M. (1990) Inactivation of the HIV LTR by DNA CpG methylation: evidence for a role in latency. EMBO J. *9*, 1157–1164.

Bednarik, D. P., Mosca, J. D., and Raj, N. B. K. (1987) Methylation as a modulator of expression of the huma immunodeficiency virus. J. Virol. *61*, 1253–1257.

Besser, D., Gotz, F., Schulze-Forster, K., Wagner, H., Kroger, H., and Simon, D. (1990) DNA methylation inhibits transcription by RNA polymerase III of a tRNA gene, but not of a 5S rRNA gene. FEBS Lett. *269*, 358–362.

Bester, T., Laudano, A., Mattaliano, R., and Ingram, V. (1988) Cloning and sequencing of a cDNA encoding DNA methyltransferase of mouse cells. Mol. Biol. *203*, 971–983.

Besansky, N. J., Butera, S. T., Sinha, S., and Folks, T. M. (1991) Unintegrated human immunodeficiency virus type 1 DNA in chronically infected cells lines is not correlated with surface CD4 expression. J. Virol. *65*, 2695–2698.

Bielinska, A., Krasnow, S., and Nabel, G. J. (1989) NF-kB-mediated activation of the human immunodeficiency virus enhancer: site of transcriptional initiation is independent of the TATA box. J. Virol. *63*, 4097–4100.

Bird, A. P. (1986) CpG-rich and the function of DNA methylation. Nature (London) *321*, 209–213.

Bours, V., Villalobos, J., Burd, P. R., Kelley, K., and Siebenlist, U. (1990) Cloning of a mitogen-inducible gene encoding a kB DNA-binding protein with homology to the *rel* oncogene and to cell-cycle motifs. Nature (London) *348*, 76–80.

Boyes, J., and Bird, A. P. (1991) DNA methylation inhibits transcription indirectly via a methyl-CpG binding protein. Cell *64*, 1123–1134.

Boyes, J., and Bird, A. (1992) Repression of genes by DNA methylation depends on CpG density and promoter strength: evidence for involvement of a methyl-CpG binding protein. EMBO J. *11*, 327–333.

Brown, P. O. (1990) Integration of retroviral DNA in: Retroviruses, strategies of replication pp. 19–48. Eds R. Swanstrom and P. K. Vogt. Springer-Verlag, New York.

Buschhausen, G., Wittig, B., Graessmann, M., and Graessmann, A. (1987) Chromatin structure is required to block transcription of the methylated herpes simplex virus thymidine kinase gene. Proc. Natl. Acad. Sci. USA *84*, 1177–1181.

Butera, S. T., Perez, V. L., Besansky, N. J., Chan, W. C., Bei-Yu, W., Nabel, G. J., and Folks, T. M. (1991) Extrachromosomal human immunodeficiency virus type-1 DNA can initiate a spreading infection of HL-60 cells. J. Cell. Biochem. *45*, 366–373.

Collart, M. A., Baeuerle, P., and Vassalli, P. (1990) Regulation of tumor necrosis factor alpha transcription in macrophages: involvement of four kB-like motifs and of constitutive and inducible forms of NF-kB. Mol. Cell. Biol. *10*, 1498–1506.

Fujiwara, T., and Craigie, R. (1989) Integration of mini-retroviral DNA: a cell-free reaction for biochemical analysis of retroviral integration. Proc. Natl. Acad. Sci. USA *86*, 3065–3069.

Bushman, F. D., and Craigie, R. (1991) Activities of the human immunodeficiency virus (HIV) integration protein *in vitro*: specific cleavage and integration of HIV DNA. Proc. Natl. Acad. Sci. USA *88*, 1339–1343.

Davies, J. F., Zuana, H., Hostomsky, Z., Jordan, S. R., and Matthews, D. A. (1991) Crystal structure of the ribonuclease H domain of HIV-1 reverse transcriptase. Science *252*, 88–95.

Doerfler, W. (1983) DNA methylation and gene activity. Annu. Rev. Biochem. *52*, 93–124.

Doerfler, W. (1984) DNA methylation and its functional significance: studies on the adenovirus system. Curr. Top. Microbiol. Immunol. *108*, 79–98.

Doerfler, W., Toth, M., Kochanek, S., Achten, S., Freisem-Rabien, U., Behn-Krappa, A., and Orend, G. (1990) Eukaryotic DNA methylation: facts and problems. FEBS Lett. *268*, 329–333.

Dosaka, N., Tanaka, T., Miyachi, Y., Imamura, S., and Kakizuka, A. (1991) Examination of HTLV-I integration in the skin lesions of various types of adult T-cell leukemia (ATL): independent of cutaneous-type ATL confirmed by Southern blot analysis. J. Invest. Dermatol. *96*, 196–200.

Dynan, W. S. (1989) Modularity in promoters and enhancers. Cell *58*, 1–4.

Farnet, C. M., and Haseltine, W. A. (1990) Integration of human immunodeficiency virus type 1 into DNA *in vitro*. Proc. Natl. Acad. Sci. USA *87*, 4164–4168.

Ghosh, S., Gifford, A. M., Riviere, L. R., Tempst, P., Nolan, G. P., and Baltimore, D. (1990) Cloning of the p50 DNA binding subunit of NF-kB: homology to *rel* and *dorsal*. Cell *62*, 1019–1029.

Gilboa, E., Mitra, S. W., Goff, S., and Baltimore, D. (1979) A detailed model of reverse transcription and tests of crucial aspects. Cell *18*, 93–98.

Gilmore, T. D. (1990) NK-kB, KBF1, *dorsal* and *rel*ated matters. Cell *62*, 841–843.

Goff, S. (1990) Retroviral reverse transcriptase: synthesis, structure, function. J. Acquir. Immune Defic. Syndr. *3*, 817–825.

Greene, W. C., Bohnlein, E., and Ballard, D. W. (1989) HIV-1, HTLV-1 and normal T-cell growth: transcriptional strategies and surprises. Immunol. Today *10*, 272–278.

Gutekunst, K. A., Guenthner, P. C., and Bednarik, D. P. (1992) Transactivation of the methylated HIV-1 long terminal repeated by tat is inhibited in a time-dependent manner (submitted).

Harbers, K., Schieke, N., Stuhlmann, H., and Jahner, D. (1981) DNA methylation and gene expression: endogenous retroviral genome becomes infectious after molecular cloning. Proc. Natl. Acad. Sci. USA *78*, 7609–7613.

Harrington, M. A., Jones, P. A., Imagawa, M., and Karin, M. (1988) Cytosine methylation does not affect binding of transcription factor Sp1. Proc. Natl. Acad. Sci. USA *85*, 2066–2070.

Hohmann, H. P., Remy, R., Poschl, B., and van Loon, A. P. G. M. (1990) Tumor necrosis factors-a and -b bind to the same two types of tumor necrosis factor receptors and maximally activate the transcription factor NF-kB at low receptor occupancy and within minutes after receptor binding. J. Biol. Chem. *265*, 15183-15188.

Hiscott, J., Alper, D., Cohen, L., Leblanc, J. F., Sportza, L., Wong, A., and Xanthoudakis, S. (1989) Induction of human interferon gene expression is associated with a nuclear factor that interacts with the NF-kB site of the human immunodeficiency virus enhancer. J. Virol. *63*, 2557–2566.

Iguchi-Ariga, S. M. M., and Schaffner, W. (1989) CpG methylation of the cAMP responsive enhancer/promoter sequence TGACGTCA abolishes specific factor binding as well as transcriptional activation. Genes Dev. *3*, 612–619.

Jahner, D., Stuhlmann, H., Stewart, C. L., Harbers, K., Lohler, J., Simon, I., and Jaenisch, R. (1982) *De novo* methylation and expression of retroviral genomes during mouse embryogenesis. Nature *298*, 623–627.

Jansson, A., Masucci, M., and Rymo, L. (1992) Methylation of discrete sites within the enhancer region regulates the activity of the Epstein-Barr virus Bam HI W promoter in Burkitt lymphoma lines. J. Virol. *66*, 62–69.

Jehuda-Cohen, T., Slade, B., Powell, J. D., Villinger, F., De, B., Folks, T. M., McClure, H. M., Sell, K. W., and Ansari, A. A. (1990) Polyclonal B-cell activation reveals antibodies against human immunodeficiency virus type 1 (HIV-1) in HIV-1-seronegative individuals. Proc. Natl. Acad. Sci. USA *87*, 3972–3976.

Kaplan, J. E., Litchfield, B., Rouault, C., Lairmore, M. D., Luo, C. C., Williams, L., Brew, B. J., Price, R. W., Janssen, R., Stoneburner, R., Ou, C. Y., and Folks, T., De, B. (1991) HTLV-l-associated myelopathy associated with blood transfusion in the United States. Neurology *41*, 192–197.

Keshet, I., Yisraeli, J., and Cedar, H. (1985) Effect of regional DNA methylation on gene expression. Proc. Natl. Acad. Sci. USA *82*, 2560–2564.

Keshet, I., Leiman-Hurwitz, J., and Cedar, H. (1986) DNA methylation affects the formation of active chromatin. Cell *44*, 535–543.

Kieran, M., Blank, V., Logeat, F., Vandekerckhove, J., Lottspeich, F., Le Bail, O., Urban, M. B., Kourilsky, P., Baeuerle, P. A., and Israel, A. (1990) The DNA binding subunit of NF-kB is identical to factor KBF1 and homologous to the *rel* oncogene product. Cell *62*, 1007–1018.

Kovesdi, I., Reichel, R., and Nevins, J. R. (1987) Role of an adenovirus E2 promoter binding factor in E1A-mediated coordinate gene control. Proc. Natl. Acad. Sci. USA *84*, 2180–2184.

Lenardo, M. J., Kuang, A., Gifford, A., and Baltimore, D. (1988) NF-kB protein purification from bovine spleen: Nucleotide stimulation and binding site specificity. Proc. Natl. Acad. Sci. USA *85*, 8825–8829.

Lenardo, M. J., and Baltimore, D. (1989) NF-kB: a pleiotropic mediator of inducible and tissue-specific gene control. Cell *58*, 227–229.

Lindsay, S., and Bird, A. P. (1988) Use of restriction enzymes to detect potential gene sequences in DNA. Nature (London) *327*, 336–338.

Marciniak, R., and Sharp, P. (1991) HIV-1 tat protein promotes formation of more-processive elongation complexes. EMBO J. *10*, 4189–4196.

Meehan, R. R., Lewis, J. D., McKay, S., Kleiner, E. L., and Bird, A. P. (1989) Identification of a mammalian protein that binds specifically to DNA containing methylated CpGs. Cell *68*, 499–507.

Mikovitz, J. A., Raziuddin, Gonda, M., Ruta, M., et al. (1990) Negative regulation of human immune deficiency virus replication in monocytes. Distinction between restricted and latent expression in THP-1 cells. J. Exp. Med. *171*, 1705–1720.

Michell, P. J., and Tijan, R. (1989) IkB: A specific inhibitor of the NF-kB transcription factor. Science *245*, 371–378.

Majors, J. (1990) The structure and function of retroviral long terminal repeats, in: Retroviruses, Strategies of Replication, pp. 49–92. Eds R. Swanstrom and P. K. Vogt.

Mosca, J. D., Bednarik, D. P., Raj, N. B. K., Rosen, C. A., Sodroski, J. G., Haseltine, W. A., Hayward, G. S., and Pitha, P. M. (1987) Activation of human immunodeficiency virus by herpes virus infection: identification of a region within the long terminal repeat that responds to a trans-acting factor encoded by herpes simplex virus 1. Proc. Natl. Acad. Sci. USA *84*, 7408–7412.

Murray, E., and Grosveld, F. (1985) in: Biochemistry and Biology of DNA Methylation, pp. 157–176. Eds G. L. Cantoni and A. Razin. Alan R. Liss Inc., New York.

Murray, P. J., and Grosveld, F. (1987) Site specific demethylation in the promoter of human g globin gene does not alleviate methylation mediated suppression. EMBO J. *6*, 2329–2335.

Nabel, G. J., and Baltimore, D. (1987) An inducible transcription factor activates expression of the human immunodeficiency virus in T cells. Nature (London) *326*, 711–713.

Nabel, G. J., Rice, S. A., Knipe, D. M., and Baltimore, D. (1988) Alternative mechanisms for activation of human immunodeficiency virus enhancer in T cells. Nature (London) *239*, 1299–1302.

Nolan, G. P., Ghosh, S., Liou, H.-C., Tempst, P., and Baltimore, D. (1991) DNA binding and IkB inhibition of the cloned p65 subunit of NF-kB, a *rel*-related polypeptide. Cell *64*, 961–969.

Pauza, C. D., Galindo, J. E., and Richman, D. D. (1990) Reinfection results in accumulation of unintegrated viral DNA in cytopathic and persistent human immunodeficiency virus type 1 infection of Cem cells. J. Exp. Med. *172*, 1035–1042.

Prendergast, G. C., and Ziff, E. B. (1991a) Methylation-sensitive sequency-specific DNA binding by the c-myc basic region. Science *251*, 186–189.

Prendergast, G. C., Lawe, D., and Ziff, E. B. (1991b) Association of myn, the murine homology of max, with c-myc stimulates methylation-sensitive DNA binding and ras cotransformation. Cell *65*, 395–407.

Raj, N. B. K., and Pitha, P. M. (1983) Two levels and regulation of b-interferon gene expression in human cells. Proc. Natl. Acad. Sci, USA *80*, 3923–3927.

Ruben, S. M., Perkins, A., and Rosen, C. A. (1989) Activation of NF-kB by the HTLV-1 transactivator protein tax requires an additional factor present in lymphoid cells. New Biol. *1*, 275–284.

Ruben, S. M., Dillon, P. J., Schreck, R., Henkel, T., Chen, C.-H., Maher, M., Baeuerle, P. A., and Rosen, C. A. (1991) Isolation of a *rel*-related human cDNA that potentially encodes the 65-kD subunit of NF-kB. Science *251*, 1490–1493.

Sambucetti, L. C., Cherrington, J. M., Wilkinson, G. W. G., and Mocarski, E. S. (1989) NF-kB activation of the cytomegalovirus enhancer is mediated by a viral transactivator and by T cell simulation. EMBO J. *8*, 4251–4258.

Santi, D. V., Garret, C. E., and Barr, P. J. (1983) On the mechanism of inhibition of DNA cytosine methyltransferases by cytosine analogs. Cell *33*, 9–10.

Sen, R., and Baltimore, D. (1986a) Inducibility of k immunoglobulin enhancer-binding protein NF-kB by a posttranslational mechanism. Cell *47*, 921–928.

Sen, R., and Baltimore, D. (1986b) Multiple nuclear factors interact with the immunoglobulin enhancer sequences. Cell *46*, 705–716.

Smith, S. S., Kan, J. L. C., Baker, D. J., Kaplan, B. E., and Dembek, P. (1991) Recognition of unusual DNA structures by human DNA (cytosine-5) methyltransferase. J. Mol. Biol. *217*, 39–51.

Sodroski, J. G., Rosen, C., Wong-Staal, F., Salahuddin, S. K., Popovic, M., Arya, S., Gallo, R. C., and Haseltine, W. A. (1985) Trans-acting transcriptional activation of human T-cell leukemia virus type III long terminal repeat. Science *227*, 171–173.

Solage, A., and Cedar, H. (1978) Organization of 5-methylcytosine in chromosomal DNA. Biochemistry *17*, 2934–2938.

Soriano, P., Gridley, T., and Jaenisch, R. (1989) Retroviral tagging in mammalian development and genetics, in: Mobile DNA. Eds D. Berg and M. M. Howe, ASM Press, Washington, D.C.

Stewart, C. L., Stuhlmann, H., Jahner, D., and Jaenisch, R. (1982) *De novo* methylation, expression, and infectivity of retroviral genomes introduced into embryonal carcinoma cells. Proc. Natl. Acad. Sci. USA *79*, 4098–4102.

Tong-Starksen, S. E., Luciw, P. A., and Peterlin, B. M. (1987) Human iommunodeficiency virus long terminal repeat responds to a T-cell activation signal. Proc. Natl. Acad. Sci. USA *85*, 6845–6849.

Varmus, H. E. (1983) Retroviruses, in: Mobile Genetic Elements, pp. 411–503. Ed. J. Shapiro. Academic Press, Inc., New York.

Varmus, H. E., and Brown, P. (1990) Retroviruses, in: Mobile DNA, pp. 53–108. Eds D. E. Berg and M. M. Howe. ASM Press, Inc., Washington, D.C.

Watt, F., and Molloy, P. L. (1988) Cytosine methylation prevents binding to DNA of a HeLa cell transcription factor required for the optimal expression of the adenoviral major late promoter. Genes Dev. *2*, 1136–1143.

Waterman, M. L., Sheridan, P. L., Milocco, L. H., and Jones, K. (1991) Nuclear proteins implicated in HIV-1 transcriptional control, in: Genetic Structure and Regulation of HIV, pp. 391–403. Eds W. A. Haseltine and F. Wong-Staal. Raven Press, New York.

Weisshaar, B., Langner, K. D., Juttermann, R., Muller, U., Zock, C., Klimkait, T., and Doerfler, W. (1988) Reactivation of the methylation-inactivated late E2A promoter of adenovirus type 2 by E1A (13S) functions. J. Mol. Biol. *202*, 255–270.

Zhang, Y., Lin, J.-X., and Vilcek, J. (1990) Interleukin-6 induction by tumor necrosis factor and interleukin-1 in human fibroblasts involves activation of a nuclear factor binding a kB-like sequence. Mol. Cell. Biol. *10*, 3818–3823.

DNA Methylation: Molecular Biology and Biological Significance
ed. by J. P. Jost & H. P. Saluz

Effect of DNA methylation on dynamic properties of the helix and nuclear protein binding in the H-ras promoter

Mack J. Rachal, Paula Holton, and Jean-Numa Lapeyre

Department of Molecular Pathology, The University of Texas M.D. Anderson Cancer Center, 1515 Holcombe Boulevard, Houston, Texas 77030, USA

1 Introduction

The regulatory role of cytosine methylation in the genome is incompletely understood despite considerable evidence that it is often associated with the repression of specific gene expression (Doerfler, 1983; Cedar, 1988). The emerging data on distribution of the CpG dinucleotide, which is the major target for cytosine methylation, is thought to reflect a basic type of organization of regulatory sequences in the genome. This dinucleotide is severely underrepresented in bulk vertebrate DNA; but it occurs at the frequency predicted by base composition in CpG islands (Bird, 1986). It is estimated that from 15% to 30% of known genes contain CpG islands. Although they can be found in the 3′ flanking region and in the transcriptional unit of associated genes, a large proportion occur in the 5′ flanking regions which overlap with promoter regions (Gardiner-Garden and Frommer, 1987). Since CpG island sequences are potentially targets for methylation, it is not clear why some of these regions cannot be methylated in the cell (Lock et al., 1986; Dobkin et al., 1989; Szyf et al., 1990) and if, functionally, cytosine methylation of CpG islands in promoter region inactivates transcription.

The H-ras promoter occurs in a dense CpG island region in both the human and mouse genome (Gardiner-Garden and Frommer, 1987; Brown et al., 1988). Besides symmetric CpG sites, this region features a preponderance of asymmetric CpG sites not accessible to restriction enzyme analysis. We have examined the activity of the H-ras gene promoter as a function of methylation in a transient expression assay and found that this promoter's activity is partially inhibited by methylation with bacterial methylases (MT HpaII and MT HhaI) but completely turned off after methylation with eukaryotic methylase from human placenta (Yoo et al., 1987) which methylated about 60% of the potential CpG sites, including a large number of asymmetric CpG sites and SpI-binding sites (Rachal et al., 1989).

The H-ras core promoter region spans XhoI to SmaI sites at coordinate positions 194 and 553 relative to the upstream BamHI site. This region contains 6 functional GC boxes and a degenerate CCAAT site for NFI/CTF binding (Ishii et al., 1986; Honkawa et al., 1987). We show through nitrocellulose filter binding assays and mobility shift assays, that the protein-DNA interactions are altered by cytosine methylation, and that an anomalous electrophoretic mobility of the DNA from in this region is observed after cytosine methylation with eukaryotic DNA methyltransferase, but not after methylation with bacterial methyltransferases.

2 Alteration in promoter DNA after methylation

The 359 bp XhoI-SmaI fragment from nucleotide positions 194 to 553, contains all the *cis*-acting elements necessary for the maximum promoter activity of the human c-H-ras (Ishii et al., 1986; Honkawa et al., 1987; Lowndes et al., 1989). It features 57 CpG dinucleotides clustered in that region and a GC content of 75%, which is highly indicative of a CpG island (Gardiner-Garden and Frommer, 1987). This fragment was enzymatically methylated with prokaryotic HpaII and HhaI methylases alone or to saturate all the available CpG sites by a method involving pretreatment with HpaII and HhaI methylase followed by incubation with purified human placental (HP) DNA methyltransferase (MTase) which effectively stimulates the eukartyotic enzyme (Rachal et al., 1989). On the basis of incorporated tritiated methyl groups, approximately 60%–70% of all CpGs present in the 359 bp fragment can be methylated. The insensitivity to cognate restriction endonucleases indicates that all the HpaII and HhaI sites were saturated and, thus, an additional 30–40% of total CpG sites, including asymmetric CpGs, were methylated by the mammalian enzyme. This represents a net increase in mass of approximately 75 methyl groups for each mole of 359 bp fragment which translates into a mass effect of approximately 1100 daltons (equivalent to 1.8 bp increased length per molecule). When these fragments were separated by electrophoresis on a 6% nondenaturing polyacrylamide gel, the HpaII and HhaI methylated fragment migrated with the same mobility; however, extensive methylation with mammalian DNA methyltransferase produced a slower migrating fragment (Fig. 1B). Based on the relative migration, the degree of retardation is equivalent to approximately a 20 bp change in mass (length), which is well in excess of the increase in mass contributed by cytosine methylation calibrated against a similarly GC-rich fragment from H-ras promoter (XhoI-MboI, 397 bp). The anomalous gel migration suggests

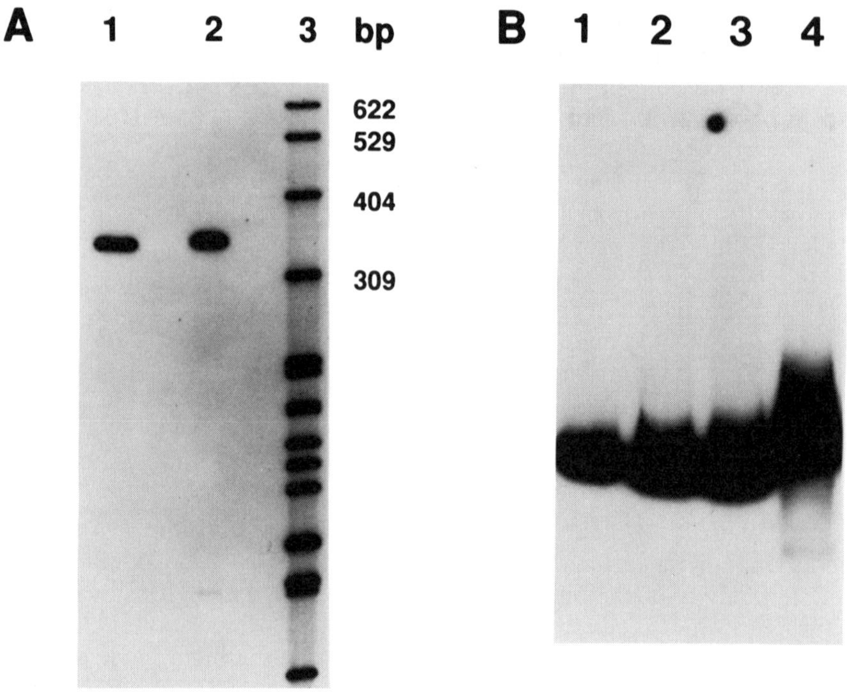

Figure 1. Effect of methylation on DNA mobility. A) Denaturing gel electrophoresis (7 M urea, 89 mM TBE). Lane 1, 359 bp unmethylated XhoI/SmaI H-ras promoter fragment; Lane 2, same fragment methylated with HhaI/HpaII and eukaryotic methylases; Lane 3, pBR 322 MspI molecular size markers. B) Non-denaturing gel electrophoresis. Lane 1, Unmethylated XhoI/SmaI H-ras promoter fragment; Lane 2, methylated only with Hha I/Hpa II methylase (B-MTase); Lane 3, methylated only with eukaryotic methylase (E-MTase); Lane 4, methylated with combination eukaryotic methylase and HhaI/HpaII methylase (E + B-MTase).

that a concerted change in helical conformation or dynamic property takes place after methylation as this effect is abolished when the helical structure is disrupted by denaturation (Fig. 1A). Here both unmethylated and methylated DNA strands migrate at their expected size with the methylated strand showing a very slight retardation, which is consistent with the additional mass of approximately 1.8 bp contributed by cytosine methylation. As a first step toward unraveling the basis of this phenomena, the double-stranded methylated and unmethylated fragments were separated on 2-D polyacrylamide gels containing chloroquine under conditions that resolve effects of helical bending or distortion (Ulanovsky et al., 1986; Mizuno, 1987). In the first dimension, the native gel was run under conditions that resolve effects of helical distortion or curvature, while in the second, chloroquine phosphate was added to abolish or minimize these effects. Electrophoresis of standards for unbent DNA (MspI pBR 322 fragments) are displayed on

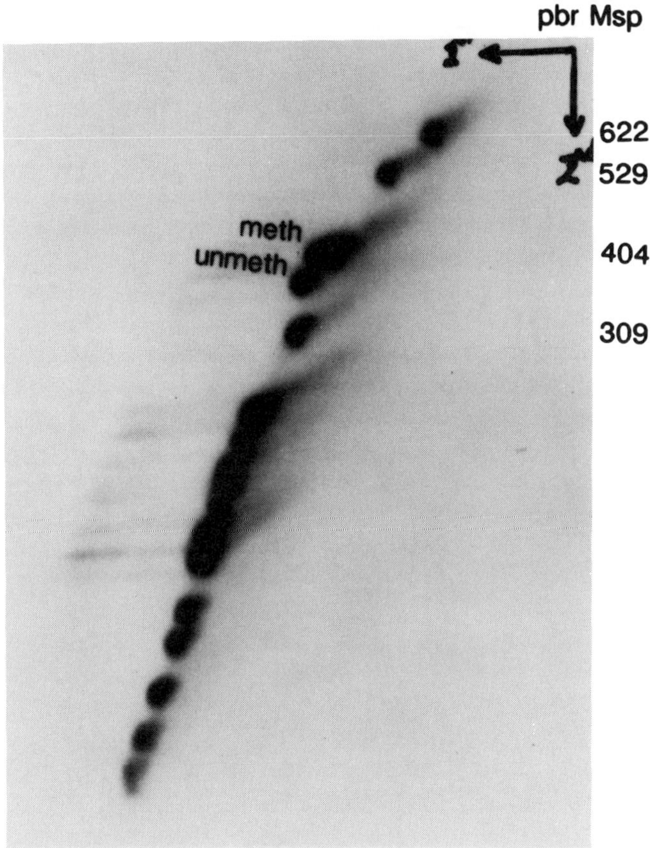

Figure 2. Two dimensional gel electrophoresis of methylated and unmethylated promoter. 1:1 mixture of methylated and unmethylated H-ras promoter were separated by electrophoresis on 6% polyacrylamide 1X TBE gel in the first dimension gel, then in the second dimension (downward) in the same buffer containing 50 μg/ml chloroquine phosphate. The diagonal spots represent pBR 322 MspI digested size markers. The methylated and unmethylated H-ras promoter fragments that migrate off-diagonal are labeled as indicated.

the main diagonal, along with a 1:1 mix of the unmethylated and methylated H-ras promoter fragment, indicated by labels in Figure 2. The unmethylated and methylated fragments are clearly resolved into two species with the methylated one showing the most retardation and off-diagonal position relative to the MspI pBR 322 standards. The migration in the second dimension also displays an anomalous migration as the two bands do not migrate at the same level. This suggests a conformational shift or alteration in helical dynamic property since it is not abolished by chloroquine intercalation as it is with bent or distorted molecules (Trifonov and Ulanovsky, 1988). To examine whether there is a basis for a sequence-dependent bending of the H-ras promoter, head-to-tail dimers of the XhoI-SmaI fragment were cloned in pUC 18 and cleaved with restriction endonuclease at unique internal sites to produce a monomer length set of cyclically permutated sequences (Wu and Crothers, 1984). As shown in Figure 3, the PCR amplified dimers

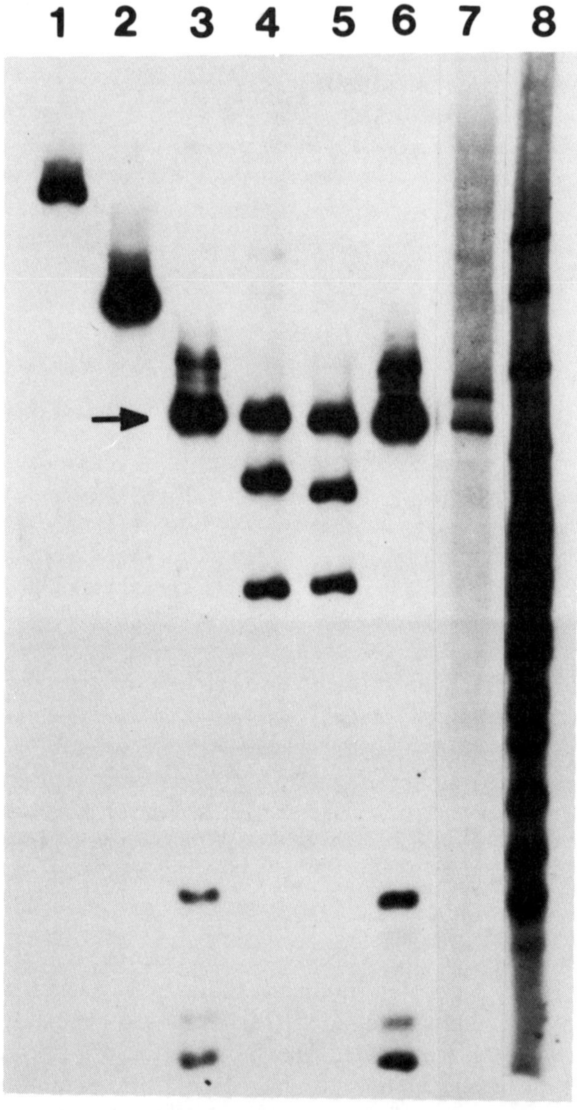

Figure 3. Cyclic permutation test for curvature in H-ras promoter. Lane 1, excised PCR amplified head-to-tail dimer of Xho/Sma fragment; Lane 2, PCR amplified XhoI-MboI monomer. Lanes 3 and 6, SmaI/TaqI cut Lane 2 PCR product giving SmaI/TaqI monomer length standard; Lane 4, DdeI cut PCR dimer; Lane 5, FspI curt PCR dimer; Lane 7, AluI cut PCR dimer; Lane 8, MspI cut pBR 322 length markers. Non-denaturing 6% polyacrylamide gel in 89 mM TBE, pH 8.3. Note SmaI/TaqI-sized DdeI, FspI, and AluI excised cyclic permuted fragments, all have the same mobility denoted by arrow.

when cut as positions 210, 325, and 414 with AluI, DdeI, and FspI, respectively, all yielded monomer length fragments having identical mobilities in 6% polyacrylamide gels as the XhoI-SmaI monomer suggesting that there is no intrinsic static curvature in this region. Further analyses by this method to determine whether anomalous mobility due to methylation could be correlated with a center of bending moment was not possible due to the inhibition by methylated cytosine of all the above used restriction enzymes and the paucity of small tracts of AT rich sequences in this region.

3 Nuclear protein binding

In order to address the effects of cytosine methylation on protein-DNA interactions, a partially purified HeLa nuclear extract enriched in several transcription factors including SpI was employed for binding studies. We used a nitrocellulose filter (NC) binding assay to investigate the stoichiometry of binding and relative binding affinity interaction in this region with unfractionated proteins. Saturation curves were generated by adding increasing amounts of partially purified nuclear extract to a fixed amount of labeled DNA fragment as shown in Figure 4A. There were significant differences in protein binding between the methylated and unmethylated promoter fragments. The stoichiometry of protein binding by the methylated fragment, is reduced for each given input and reaches a lower saturation plateau, which suggests fewer binding sites. However, the drawback of using a heterogeneous mixture of proteins for binding studies is that it is not feasible to derive a relevant equilibrium constant describing the interaction since each of the protein binding components would have to be characterized and any cooperative effects determined. The slopes, however, give an approximate measure of the ability of the proteins to bind to this DNA under our experimental conditions. From this we can estimate that, at saturation, there is approximately 60% less binding to the methylated fragment. Since the plateau may involve contributions of secondary nonspecific binding, a test for specificity involving competition experiments was performed to estimate the relative binding affinities (Fig. 4B). The labeled probe and nuclear extract were preincubated in binding buffer at the saturation level (input 2 μg extract/ng DNA), then increasing amounts of competitor DNA were added, and 15 min later the amount of complex formation assessed. Noncompetitor DNA will not interfere with the formation of specific DNA binding complexes, but will gradually decrease the amount of complex recovered due to a shift in equilibrium from nonspecific DNA interactions (Lin and Riggs, 1975) as shown for a 375 bp nonspecific competitor EcoRI/BamHI fragment from pBR 322. By comparing the point at which 50% of the total

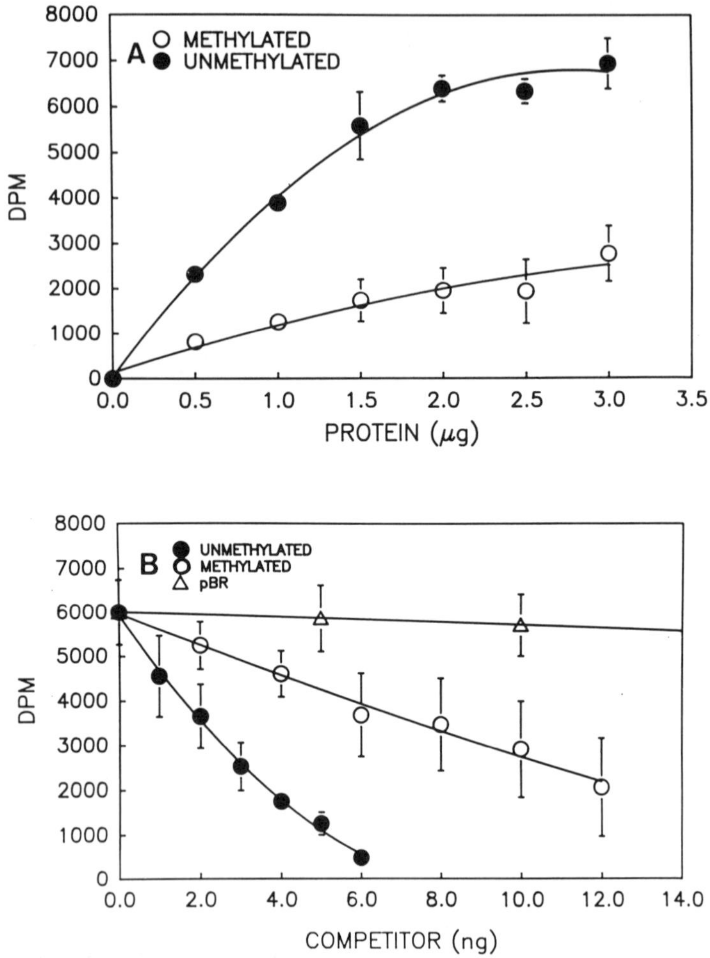

Figure 4. Nitrocellulose filter binding assays. A) Saturation curve employing 1 ng of with E + B-methylated H-ras promoter fragment with increasing amounts of partially purified HeLa nuclear extract denoted on abscissa, (●—●), unmethylated; (○—○), methylated. B) Competition analysis at saturation plateau (2.5 μg Hela nuclear extract) with various competitor DNAs, (●—●), unmethylated H-ras promoter DNA; (○—○), E + B methylated H-ras promoter DNA; or (△—△), 375 bp EcoRI/BamHI pBR 322 fragment (non-specific control).

protein DNA complexes are competed away, the relative binding affinities for competitor DNAs were estimated assuming protein dissociation rate constants, k, are on the order of 10^{-3} sec^{-1}, and a diffusion-limited associated constant K_a of at least 10^7, which is within the range for most specific DNA binding protein interaction (Lin and Riggs 1970; Lapeyre and Bekhor, 1976; Emerson et al., 1985). The results of the competition experiments show that the C50s for methylated compared to self-competition with an unmethylated fragment, occurs at a concentration of 8.7×10^{-10} M, and 2.5×10^{-10} M, respectively, which translates to a 3.5-fold increased relative affinity for the unmethylated H-ras promoter fragment over its methylated counterpart.

This effect on protein DNA interaction was also explored by gel mobility shift assay, where the specificity of protein-DNA interactions can be determined by a competition analysis to yield semi-quantitative information regarding the number of proteins bound (Fried and Crothers, 1981). To investigate whether DNA methylation affected the formation of these complexes in the gel retardation assays, both methylated and unmethylated H-ras fragments were incubated with increasing amounts of the partially purified HeLa nuclear extract. The methylated promoter produced a marked diminution of two bands associated with specific complex formation, (Fig. 5). This was observed in the range for stoichiometric input which reaches a plateau at 2 μg/ng DNA probe in the NC filter binding assays, and corroborates with the inhibition of binding data also derived from nitrocellulose filter assays.

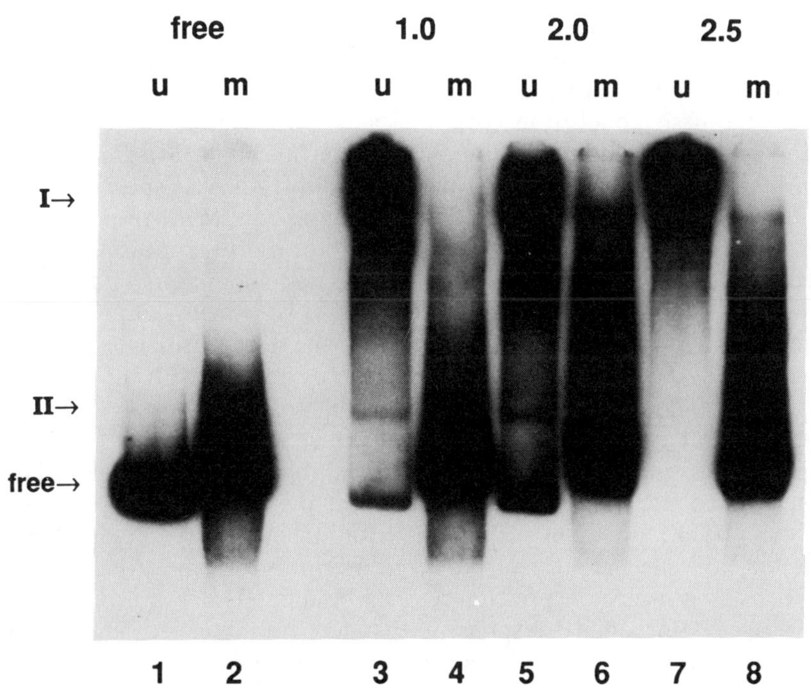

Figure 5. Complex formation of methylated versus unmethylated H-ras promoter fragments assayed by gel retardation. Lanes 1 and 2, free (minus protein extract) of unmethylated (U) and methylated (M) promoter, respectively; Lanes 3 and 4, 5 and 6, 7 and 8, denote 1.0, 2.0 and 2.5 μg of partially purified HeLa extract incubated with unmethylated or methylated promoter.

4 Discussion

How proteins interact with discrete DNA sequences to effect transcription and replication is just beginning to be elucidated particularly with respect to recognition and binding of their specific *cis*-acting sequences. One process capable of altering these interactions is the covalent modification of the genomic DNA by the methylase. In eukaryotic nuclei, this concerns mainly but not exclusively, the CpG dinucleotide sequence (Hubrich-Kuhner et al., 1989). The methyl group in 5-methylcytosine protrudes into the major groove of B-DNA helix and provides an additional constraint on sequence recognition. There are two basic models for how dC methylation can alter this process. The first involves steric hindrance by methyl groups and/or altered hydrophobic interactions that may disrupt finely tuned interactions which destabilize the specific binding of proteins to DNA. In the second model, methyl groups may serve as specific recognition points for stabilization of 5-mC-dependent binding proteins such as MDBP (Khan et al., 1988) and MeCP (Antequera et al., 1989; Meehan et al., 1989). Their interaction with methylated cytosines in promoter regions might repress promoter activity as suggested from recent experiments where promoter repression was alleviated by excess methylated competitor DNA (Boyes and Bird, 1991; Levine et al., 1991).

In the first model, transcriptional activity can be altered by disruption of specific binding at discrete sites. In this context, a small change in binding affinity induced by methylation of DNA at specific binding sites could exert large effects on gene expression. With respect to this mechanism, the binding of SpI to GC box motifs appears not to be affected (Harrington et al., 1988; Ben-Hattar and Jiricny, 1988). Hoveler and Doerfler (1987) have noted that *in vitro* methylation at HpaII and HhaI sites prevents transcription of the adenovirus late gene promoter; however, this effect could not be correlated with an inhibition of factor binding at the DNase I footprint level. In contrast, Becker et al. (1987) have reported that methylation of the promoter region of the rat tyrosine aminotransferase gene inhibited recognition of a HhaI methylated site by a transacting factor at the DNase I footprint level. Other specific examples include methylation of the Hpa II site located within an element which binds transcription factor AP-2 inhibits proenkephalin gene expression and its binding (Comb and Goodman, 1990), and the cAMP responsive enhancer/promoter element (Iguchi-Ariga and Schaffner, 1989). Kovesdi et al. (1987) and Watt and Molloy (1988) have shown transcriptional factors E2 and MLTF, respectively, have lost their binding and activation property when challenged with methylated adenovirus promoter DNA. As shown herein, a third possibility is that dC methylation effects the local conformation of DNA, which in turn is the key factor affecting protein binding and recognition. There is ample evidence for sequence-directed DNA bending affecting nucleosome assembly (Hsieh and Griffith, 1988; Kimura et al., 1989) and for activity of regions regulating transcriptional activity (Bossi and Smith, 1984; Gourse et al., 1986). In this context, methylation could play a role by either enhancing or depressing bending if it is in phase with the bending moment or if out of phase, affect other parameters, such as helix pitch and/or flexural properties.

The change in mobility observed in this study appears to be a shift that encompasses the whole sequence. No local unpaired base regions that would be expected to form S1 hypersensitive sites could be detected in either the methylated fragment or methylated pXM18 plasmid when analyzed for cruciform structures, B/Z junctions or H-DNA-containing regions (unpublished data). Since extensive methylation might produce a more drastic effect on the overall topology of the DNA fragment, it may possibly explain why it can affect overall binding. The separation of the methylated species from the unmethylated species on 2-dimensional +/− chloroquine and cold/heated gels (unpublished data) strongly suggests a regional shift in the DNA molecule by methylation that involves either a change in helical properties or conformation. This effect exhibits a threshold since it was not observed at 30% saturation of methylatable sites with HpaII and HhaI methylase. The data on experimental retardation on two-dimensional gels are suggestive of DNA-cytosine methylation affecting stable helix curvature in this region. This has been reported for the c-myc promoter, which is also extremely GC-rich in an unmethylated state (Kumar and Leffak, 1989). However, the anomalous migration was not abolished in the presence of chloroquine, ruling out a simple static bending model. In addition, the evidence that this is not related to curvature effects is that unlike methylation at N6 of adenine, which increases the wedge angle of adenine tracts (Diekmann, 1987), cytosine methylation at the 5 position, though possibly distorting the major groove slightly, is not known to increase curvature as determined experimentally with sequenced defined cytosine containing oligonucleotides (Diekmann, 1987). Our own studies on the mobility of ligated oligomers of methylated and unmethylated SpI consensus decanucleotide sequences in polyacrylamide gels used to determine curvature (unpublished results) were negative and suggest a very small effect, if any, of cytosine methylation on bending. On the other hand, dC methylation may have an effect on helical periodicity (Zacharias et al., 1988) or possibly on the dynamics of reptation through the gel by affecting helix flexibility. In the former case, cytosine methylation has been shown to increase the pitch of right-handed CpG containing DNA regions from 10.5 to 10.7 bp/turn, and when in left-handed form from 11.5 to 12.8 bp/turn. The multiple protein factors involved in the interaction with H-ras promoter are presently being characterized with respect to specific contact points in order to determine whether a specfic placement of methyl groups is sufficient to alter the interaction or whether only a regional effect on helix conformation/dynamic property can produce this effect in the H-ras promoter.

Acknowledgments. This investigation was supported by Grant No. BC650 from the American Cancer Society. The stipend for MGR was provided by funding from the Kennerly Foundation. The authors wish to thank Kamal Emami for technical assistance, and Marjorie Nortin and Betty Martz for preparation of the manuscript.

Antequera, F., Macleod, D., and Bird, A. P. (1989) Specific protection of methylated CpGs in mammalian nuclei. Cell *58*, 509–517.

Becker, P. B., Ruppert, S., and Schutz, G. (1987) Genomic footprinting reveals cell type-specific DNA binding of ubiquitous factors. Cell *51*, 435–443.

Ben-Hattar, J., and Jiricny, J. (1988) Methylation of single CpG dinucleotides within a promoter element of the herpes simplex virus tk gene reduces its transcription *in vivo*. Gene *65*, 219–227.

Ben-Hattar, J., Beard, P., and Jiricny, J. (1989) Cytosine methylation in CTF and Spl recognition sites of an HSV tk promoter: effects on transcription *in vivo* and on factor binding *in vitro*. Nucl. Acids Res. *17*, 10179–10190.

Bird, A. P. (1986) CpG-rich islands and the function of DNA methylation. Nature *321*, 209–213.

Bossi, L., and Smith, D. M. (1984) Conformational change in the DNA associated with an unusual promoter mutation in a tRNA operon of salmonella. Cell *39*, 643–652.

Borello, M. G., Pierotti, M. A., Bongarzone, I., Donghi, R., Mondellini, P., and Della Porta, G. (1987) DNA methylation affecting the transforming activity of the human Ha-ras oncogene. Cancer Res. *47*, 75–79.

Boyes, J., and Bird, A. (1991) DNA methylation inhibits transcription indirectly via a methyl-CpG binding protein. Cell *64*, 1123–1134.

Brown, K., Bailleul, B., Ramsden, M., Fee, F., Krumlauf, R., and Balmai, A. (1988) Isolation and characterization of the 5′ flanking region of the mouse C-Harvey-ras gene. Mol. Carcinogenesis *1*, 161–170.

Cedar, H. (1988) DNA methylation and gene activity. Cell *53*, 3–4.

Chandler, L. A., Ghazi, H., Jones, P. A., Boukamp, P., and Fusenig, N. E. (1987) Allele specific methylation of the human c-Ha-ras-l gene. Cell *50*, 711–717.

Chodosh, L. A., Baldwin, A. S., Carthew, R. W., and Sharp, P. A. (1988) Human CCAAT-binding proteins have heterologous subunits. Cell *53*, 11–24.

Comb, M., and Goodman, H. M. (1990) CpG methylation inhibits proenkephalin gene expression and binding of the transcription factor AP-2. Nucl. Acids Res. *18*, 3975–3982.

Diekmann, S. (1987) DNA methylation can enhance or induce DNA curvature. EMBO J. *6*, 4213–4217.

Dobkin, C., Ferrando, C., and Brown, W. T. (1987) PFGE of human DNA: 5-azacytidine improves restriction. Nucl. Acids Res. *15*, 3183.

Doerfler, W. (1983) DNA methylation and gene activity. Ann. Rev. Biochem. *52*, 93–124.

Dorn, A., Bollenkens, J., Staub, A., Benoist, C., and Matthis, D. (1987) A multiplicity of CCAAT box-binding proteins. Cell *50*, 863–872.

Emerson, B. M., Lewis, C. D., and Felsenfeld, G. (1985) Interaction of specific nuclear factors with the nuclease-hypersensitive region of the chicken adult β-globin gene: nature of the binding domain. Cell *41*, 21–30.

Fried, M., and Crothers, D. M. (1981) Equilibria and kinetics of lac repressor-operator interactions by polyacrylamide gel electrophoresis. Nucl. Acids Res. *9*, 6506–6525.

Gardiner-Garden, M., and Frommer, M. (1987) CpG islands in vertebrate genomes. J. Mol. Biol. *196*, 261–282.

Gourse, R. L., de Boer, H. A., and Nomura, M. (1986) DNA determinants of rRNA synthesis in *E. coli*: growth rate dependent regulation feedback inhibition, upstream activation, antitermination. Cell *44*, 197–205.

Harrington, M. A., Jones, P. A., Imagawa, M., and Karin, M. (1988) Cytosine methylation does not affect binding of transcription factor Spl. Proc. Natl. Acid. Sci. USA *85*, 2066–2070.

Höller, M., Westin, G., Jiricny, J., and Schaffner, W. (1988) Sp1 transcription factor binds DNA and activates transcription even when the binding site is CpG methylated. Genes & Develop. *2*, 1127–1135.

Honkawa, H., Masahashi, W., Hashimoto, S., and Hashimoto-Gotoh, T. (1987) Identification of the principal promoter sequence of the c-H-ras transforming oncogene: Deletion analysis of the 5′-flanking region by focus formation assay. Mol. Cell Biol. *7*, 2933–2940.

Hoeveler, A., and Doerfler, W. (1987) Specific factors binding to the late E2A promoter region of adenovirus type 2 DNA: No apparent effects of 5′-CCGG-3′ methylation. DNA *6*, 449–460.

Hsieh, C. H., and Griffith J. D. (1988) The terminus of SV40 DNA replication and transcription contains a sharp sequence-directed curve. Cell *52*, 535–544.

Hubrich-Kühner, K., Buhk, H. J., Wagner, H., Kröger, H., and Simon, D. (1989) Non-C-G recognition sequences of DNA cytosine-5-methyltransferase from rat liver. Biochem. Biophys. Res. Comm. *160*, 1175–1182.

Iguchi-Ariga, S. M. M., and Schaffner, W. (1989) CpG methylation of the cAMP-responsive enhancer/promoter sequence TGACGTCA abolishes specific factor binding as well as transcriptional activation. Genes & Develop. *3*, 612–619.

Ishii, S., Kadonaga, J. T., Tjian, R., Brady, J. N., Merlino, G. T., and Pastan, I. (1986) Binding of the Sp1 transcription factor by the human harvey ras 1 proto-oncogene promoter. Science *232*, 1410–1413.

Khan, R., Zhang, X. Y., Supakar, P. C., Ehrlich, K. C., and Ehrlich, M. (1988) Human methylated DNA-binding protein, determinants of a pBR 322 recognition site. J. Biol. Chem *263*, 14374–14383.

Kimura, T., Asai, T., Imai, M., and Takanami, M. (1989) Methylation strongly enhances DNA bending in the replication origin region of the *Escherichia coli* chromosome. Mol. Gen. Genet. *219*, 69–74.

Kimura, T., Takeya, T., and Takanami, M. (1989) Reconstitution of nucleosomes in vitro with a plasmid carrying the long terminal repeat of Moloney murine leukemia virus. Biochim. Biophys. Acta *1007*, 318–324.

Kolsto, A. B., Kollias, G., Giguere, V., Isobe, K. I., Prydz, H., and Grosveld, F. (1986) The maintenance of methylation-free islands in transgenic mice. Nucl. Acids Res. *14*, 9667–9678.

Kovesdi, I., Reichel, R., and Nevins, J. R. (1987) Role of an adenovirus E2 promoter binding factor in E1A-mediated coordinate gene control. Proc. Natl. Acad. Sci. USA *84*, 2180–2184.

Kumar, S., and Leffak, M. (1989) DNA topology of the ordered chromatin domain 5′ to the human c-myc gene. Nucl. Acids Res. *17*, 2819–2833.

Lapeyre, J. N., and Bekhor, I. (1976) Chromosomal protein interactions in chromatin and with DNA. J. Mol. Biol. *104*, 25–58.

Levine, A., Cantoni, G. L., and Razin, A. (1991) Inhibition of promoter activity by methylation: possible involvement of protein mediators. Proc. Natl. Acad. Sci. USA *88*, 6515–6518.

Lin, S. Y., and Riggs, A. D. (1975) The general affinity of lac repressor for *E. coli* DNA: implications for gene regulation in procaryotes and eucartyotes. Cell *4*, 107–111.

Lock, L. F., Melton, D. W., Caskey, C. T., and Martin, G. R. (1986) Methylation of the mouse hprt gene differs on the active and inactive X-chromosomes. Mol. Cell Biol. *6*, 914–924.

Lowndes, N. F., Paul, J., Wu, J., and Allan, M. (1989) C-Ha-ras gene bidirectional promoter expressed *in vitro*: location and regulation. Mol. Cell Biol. *9*, 3758–3770.

Meehan, R. R., Lewis, J. D., McKay, S., Kleiner, E. L., and Bird, A. P. (1989) Identification of a mammalian protein that binds specifically to DNA containing methylated CpGs. Cell *58*, 499–507.

Mizuno, T. (1987) Random cloning of bent DNA segments from *Escherichia coli* chromosome and primary characterization of their structures. Nucl. Acids Res. *15*, 6827–6841.

Rachal, M. J., Yoo, H., Becker, F. F., and Lapeyre, J. N. (1989) *In vitro* DNA cytosine methylation of cis-regulatory elements modulates c-Ha-ras promoter activity *in vivo*. Nucl. Acids Res. *17*, 5135–5147.

Ramsden, M., Cole, G., Smith, J., and Balmain, A. (1985) Differential methylation of the c-H-ras gene in normal mouse cells and during skin tumour progression. EMBO J. *4*, 1449–1454.

Riggs, A. D., Suzuki, H., and Bourgeois, S. (1970) *lac* Repressor-operator interaction. J. Mol. Biol. *48*, 67–83.

Szyf, M., Tanigawa, G., and McCarthy, P. L. (1990) A DNA signal from the Thy-1 gene defines *de novo* methylation patterns in embryonic stem cells. Mol. Cell Biol. *10*, 4396–4400.

Trifonov, E. N., and Ulanovsky, L. E. (1988) in: Unusual DNA structures, pp. 173–187. Eds R. D. Wells and S. C. Harvey. Springer, New York.

Ulanovsky, L., Bodner, M., Trifonov, E. N., and Choder, M. (1986) Curved DNA: design, synthesis, and circularization. Proc. Natl. Acad. Sci. USA *83*, 862–866.

Watt, F., and Molloy, P. L. (1988) Cytosine methylation prevents binding to DNA of a HeLa cell transcription factor required for optimal expression of the adenovirus major late promoter. Genes & Develop. *2*, 1136–1143.

Wu, H. M., and Crothers, D. M. (1984) The locus of sequence-directed and protein-induced DNA bending. Nature *308*, 509–513.

Yoo, H., Noshari, J., and Lapeyre, J. N. (1987) Subunit and functional size of human placental DNA methyltransferase involved in *de novo* and maintenance methylation. J. Biol. Chem. *262*, 8066–8070.

Zacharias, W., O'Connor, T. R., and Larson, J. E. (1988) Methylation of cytosine in the 5-position alters the structural and energetic properties of the supercoil-induced Z-helix and of B–Z junctions. Biochemistry *27*, 2970–2978.

DNA Methylation: Molecular Biology and Biological Significance
ed. by J. P. Jost & H. P. Saluz

DNA methylation and embryogenesis

Aharon Razin and Howard Cedar

Department of Cellular Biochemistry, Hebrew University, Hadassah Medical School, Jerusalem, Israel

1 Introduction

A large number of studies have documented the methylation patterns of individual genes and repeated sequences in different tissues of various mammalian species. This has revealed a consistent picture of DNA modification where each cell type has its characteristic pattern. Tissue-specific genes are undermethylated in their cell type of expression but fully modified in other cells, while housekeeping genes contain 5′ CpG islands which are constitutively unmethylated in all cells (see Yeivin and Razin, pp. 523–568). Experiments in tissue culture have clearly demonstrated that these somatic patterns are fixed and conservatively passed on from generation to generation (Stein et al., 1982; Wigler et al., 1991), but little is known about how this modification profile is established during early stages of embryo development.

It is most likely that a basic methylation profile is set up either during gametogenesis or at very early stages of embryonic development prior to organogenesis (Razin and Riggs, 1980). Since all tissue-specific gene sequences that have been analyzed in sperm are known to be fully methylated (Yisraeli and Szyf, 1984) it was reasonable to assume that the female germ line genome is also hypermethylated and that this pattern is maintained throughout early embryogenesis. The adult methylation state would then be produced by tissue-specific demethylation events associated with the activation of gene expression (Razin and Riggs, 1980). Testing this hypothesis required the analysis of methyl moieties in the germ line and early embryo, and until recently this could not be done, since the available analytical methods were not sensitive enough to detect specific sites in very small amounts of DNA. In fact, the quantities of DNA that can be isolated from mammalian oocytes or early embryos are in the picogram to nanogram range. To get around the problem of sensitivity several approaches were developed to assess the general methylation state of the genome in early embryos. Using uniformly labelled mouse embryo DNA as substrate for restriction enzyme digestion initially revealed that CCGG sites in the blastocyst are methylated to an extent comparable to that of adult tissues (Singer et al., 1977), and similar results were obtained in an analysis of rabbit embryo DNA (Manes and Menzel, 1981). In both of these cases, the methodology was not sensitive enough to detect even large differences in DNA methylation, and more recent results indicate that these conclusions were probably incorrect. In one critical study (Monk et al., 1987) DNA from various stages of development were digested with HpaII, end-labelled and subsequently size-analyzed by gel electrophoresis. This semi-quantitative assay revealed that oocyte DNA is substantially undermethylated in comparison to sperm. It also showed, for the first time, that the genome is extensively undermodified in initial early stage embryos and then remains that way in the blastula.

Recent advancements in the isolation of microquantities of DNA from mouse embryos and oocytes, together with the development of a highly sensitive PCR technology (Singer-Sam et al., 1990) now allow one to focus on the methylation status of individual sites at specific gene loci in the mouse genome (Kafri et al., 1992). These studies confirm the results of Monk et al. and provide additional insights into the dynamics of methylation during gametogenesis and embryogenesis.

2 Methylation status of the mature male and female germ cells

A considerable amount of data indicate that sperm DNA carries the basic bimodal pattern of methylation whereby most sites in specific genes are fully methylated (Yisraeli and Szyf, 1984; Razin et al., 1984) while CpG island sequences are unmodified. Despite this picture of hypermethylation, however, the overall level of DNA methylation in mouse sperm is relatively low. This is probably a result of the fact that a large portion of the CpG sites in the mouse derive from the major satellite (Solage and Cedar, 1978) and these sequences are considerably undermethylated in sperm (Sanford et al., 1984; Ponzetto-Zimmerman and Wolgemuth, 1984). On the other hand, other repetitive sequence families such as L1, MUP and IAP are actually methylated in mouse sperm (Sanford et al., 1978: see also Fig. 1).

The analysis of methylation in mature oocytes has been greatly limited by the difficulty in extracting sufficient quantities of DNA to perform Southern blot analysis on single copy DNA sequences. Studies focusing on the methylation status of minor and major satellite DNA revealed a pattern of partial methylation in the dictyate oocyte, and this is similar to that found in sperm (Sanford et al., 1984). Dispersed repetitive sequences such as MUP and IAP were found to be methylated in ovulated oocytes but at least one repetitive sequence, L1, is less methylated in the oocyte than in the sperm (Sanford et al., 1987; Howlett and Reik, 1991; see also Fig. 1). These findings are consis-

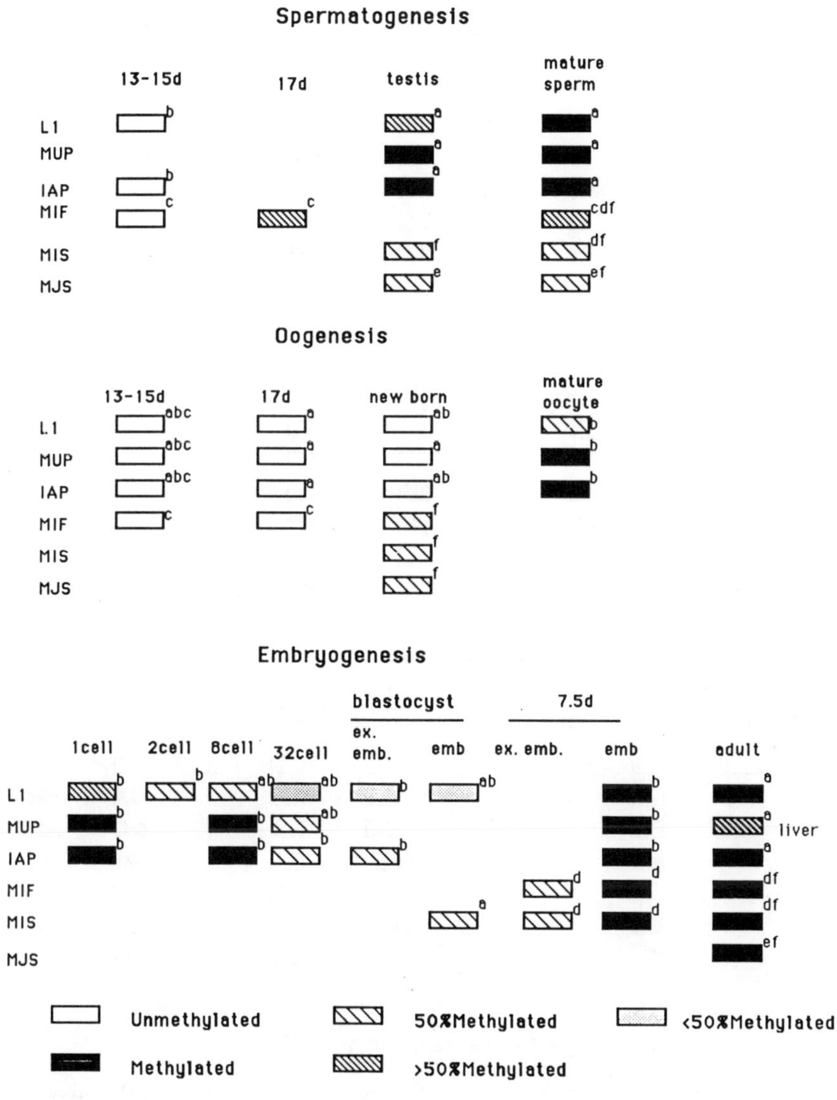

a Sanford et.al. 1987 ;b. Howlett and Reik 1991;c. Monk et. al. 1987

d. Chapman et. al. 1984; e. Ponzetto-Zimmerman and Wolgemuth 1984f. Sanford et. al. 1984

Figure 1. Methylation status of repetitive sequences at various stages during gametogenesis and embryogenesis. This is a compilation of published data obtained in various laboratories. It should be kept in mind that various analytical techniques were used in the different laboratories. The following abbreviations were used: L1 – Line 1 repetitive sequence: MUP – Major urinary protein: IAP – Intercisternal A particle: MIF – Mouse interspersed family: MIS – Minor satellite: MJS – Major satellite: 13-15d – 13–15 days post coitum: 17d – 17 days post coitum: ex.emb. – extraembryonic tissue: emb. – embryo proper.

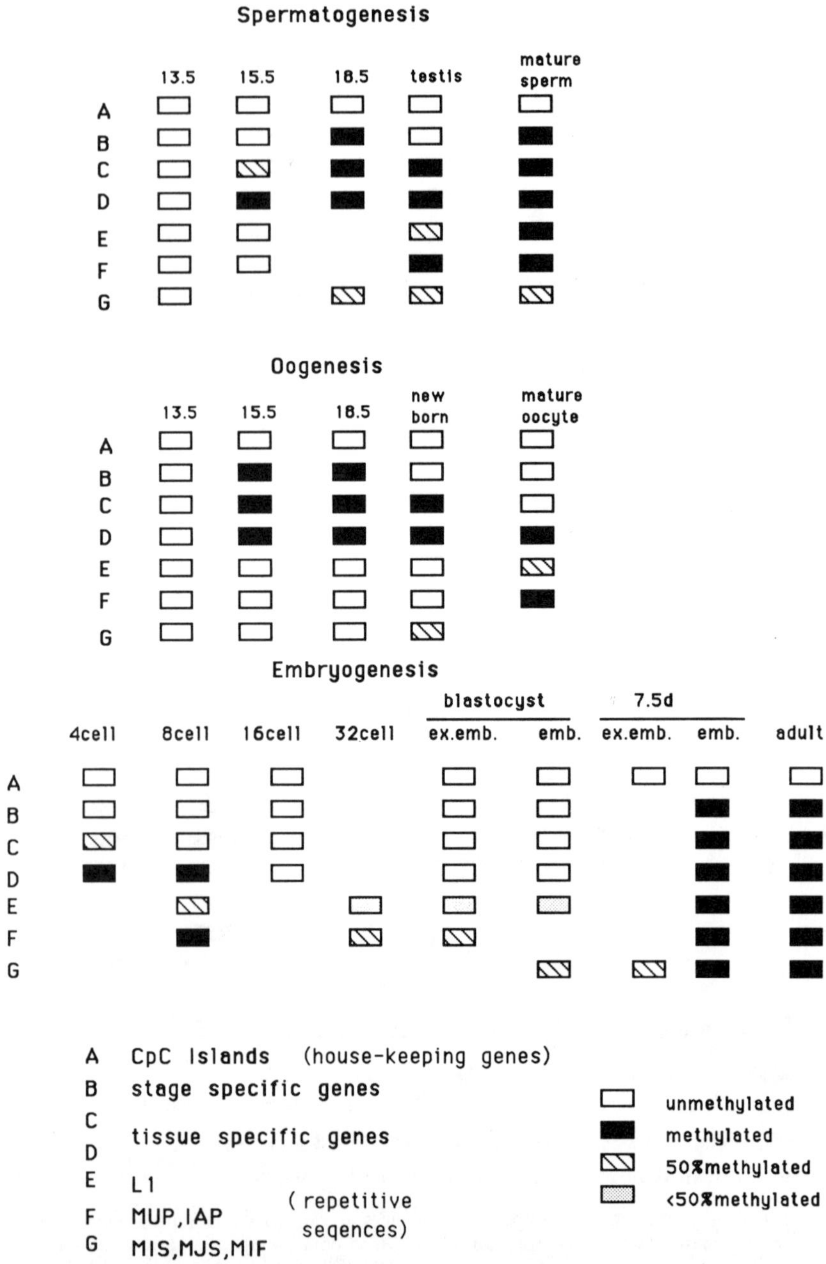

Figure 2. Changes in methylation patterns of seven paradigms of genomic sequences during development. Paradigms A–G are as listed at the bottom of the figure. Repetitive sequences are abbreviated as in legend to Fig. 1. 13.5, 15.5 and 18.5 are days post coitum: ex. emb. represents extraembryonic tissue and emb. represents embryo proper.

tent with the suggestion that the level of DNA methylation is lower in female germ cells than those of the male (Monk et al., 1987). Using PCR to assay individual sites in specific gene sequences it has now been shown unequivocally that the male and female germ line DNAs are differentially methylated (Kafri et al., 1992, see Fig. 2). Many sites are indeed modified in both sperm and oocyte, but other loci are unmethylated in the oocyte despite their full methylation in sperm. Since most of the genes tested in this study are not expressed in the germ cells or in the early embryo, the significance of this differential pattern is not yet clear (see Fig. 2).

3 Dynamic changes in methylation at the preimplantation stage

An overall view of the extent of global methylation at various stages of early embryonic development was obtained by Monk and coworkers (Monk et al., 1987). They concluded that the level of methylation in the 8-cell embryo was intermediate between the values observed in sperm and oocytes. However, the modification state in the blastula was even lower than that found in the oocyte. Although this data was only based on a semi-quantitative assay it strongly suggested that extensive demethylation of the genome takes place in the early embryo between the 8-cell and blastula stages. However, this study did not provide information on the mechanism and precise stage at which demethylation takes place, nor did it define the sequences that are subject to demethylation. Recently, the L1, MUP and IAP repetitive sequences were analyzed in further detail (Howlett and Reik, 1991). Both MUP and IAP were highly methylated in the zygote and remained that way until the 8-cell stage when these sequences began to undergo extensive demethylation. In the case of L1, the level of methylation in the sperm component became partially demethylated by the late 2-cell embryo and thus adopted a pattern similar to that found in the ovulated unfertilized egg (see Fig. 1). Methylation then remained at this intermediate level until after the 8-cell stage when further demodification occurred. This two step process of demethylation appears to be a general phenomenon, since single-copy sequences also behave in a similar manner. In the case of CpG loci which are modified in sperm DNA, but have already undergone demethylation in the oocyte, the paternal allele becomes unmethylated very soon after fertilization and prior to the 4-cell stage (Shemer et al., 1991b; Kafri et al., 1992). This represents a continuation of the process that begins during late oogenesis and is probably mediated by the same *trans*-acting factors. A second class of CpG sites includes those loci methylated both in the oocyte and sperm. These methyl groups are also removed from both alleles, but this takes place in a delayed manner during a small window of development between the 8- and 16-cell cleavage stages (see Fig. 2). The end result of these reactions is that a large number of individual sites in the genome are unmethylated in the morula and blastula, and this includes both the inner cell mass and the extra-embryonic tissues.

It is clear that not all CpG sites in the genome are unmethylated in the blastula. By employing a quantitative assay, it was recently reported that approximately 8% of the cytosines in TaqI sites and 14% in HpaII sites were methylated in DNA derived from 3.5 d embryos (Chaillet et al., 1991). While much of this methylation may be accounted for by the partial methylation of highly repetitive sequences (see Fig. 1) it is possible that other unique loci also remain modified at this stage. In one documented case, a multicopy imprinted transgene was shown to be differentially methylated on only one allele in the blastula (Chaillet et al., 1991). It is still too early to know whether this pattern is representative of the natural endogeneous imprinted genes, as well (see below).

The biological significance of this early embryonic demethylation is unknown. One possibility is that demodification of the genome provides a mechanism for removing differences in specific gene methylation patterns which emerge from the male and female germ lines. Alternatively this may represent part of a mechanism for reformatting the genome prior to the initiation of the normal program of embryonic development. Indeed, the timing of the initial demethylation appears to coincide with activation of zygotic transcription (McLaren, 1976; Clegg and Piko, 1983). DNA methylation is known to have profound effects on both chromatin structure and gene expression (Cedar and Razin, 1990). Thus, it is likely that this massive undermethylation, encompassing a major part of the genome, may influence the physiology of the embryo. Since DNA methylation represents one of the cellular mechanisms for the global repression of gene activity, it is certainly possible that these changes may bring about a partial derepression of genes which are normally only expressed in somatic tissues. A phenomenon of this nature has indeed been described in *Xenopus* embryos, where several genes, including MyoD, are constitutively expressed following the midblastula transition (Rupp and Weintraub, 1991). Although some uncertainty still remains with regard to what CpG sites remain methylated in the blastula it seems that general *de novo* methylation of the entire genome including repetitive sequences takes place in the embryo proper at some time around implantation (between 5.5 and 6.5 d p.c.) leaving only CpG island sequences unmethylated (Monk et al., 1987; Kafri et al., 1992; Shemer et al., 1991b). In the extraembryonic tissues, however, some repetitive sequences may retain their pregastrula state of partial methylation (Chapman et al., 1984; Fig. 1).

The mechanism of this demethylation reaction is not known, but most of the data support the existence of an active demodification process. Sites in the mouse ApoAI gene, for example, appear fully methylated at the 8-cell stage, but are completely unmethylated one division cycle later at the 16-cell stage (Shemer et al., 1991b). This abrupt demodification strongly suggests, but does not prove, that this is an active process which is not dependent on DNA replication, and this would be consistent with findings in other systems (Paroush et al., 1990; Razin et al., 1986; Wilks et al., 1984; Sullivan and Grainger, 1987). In one preliminary experiment, a methylated fragment of the human ApoAI gene was microinjected into the male pronucleus of a fertilized egg. This DNA was found to integrate into the genome and to undergo demethylation during the first stages of embryogenesis in a manner similar to that which occurs to endogenous genes (unpublished, this laboratory). First, the restriction enzyme DpnI was used to digest the non-integrated bacterially derived 6-methyl adenine. It was then possible to use MboI to trace the original integrated plasmid derived molecules and examine CpG methylation on these molecules. Such analysis clearly showed that this DNA also underwent demodification at CpG residues. Since this demethylation must have occurred on unreplicated strands, this represents direct proof for an active form of demethylation. In contrast to these results, Howlett and Reik (1991) recently showed that the demethylation of repeated sequences in the embryo could be inhibited by the drug aphidicolin. Their results thus suggest that either DNA replication or cell cycle progression are required for demodification. Obviously more direct experiments will be required to delineate this biochemical reaction.

4 Reconstruction of the somatic methylation pattern

There are many indications in the literature that implantation embryos may have the capacity to *de novo* methylate foreign DNA. Indeed, when preimplantation embryos were infected with retroviruses the viral genome underwent *de novo* methylation, and remained modified in the adult organism. In contrast, following infection of post implantation embryos the same retroviral sequences remained unmethylated (Jahner and Jaenisch, 1984). A similar *de novo* methylation activity is also observed when exogenous DNA sequences are inserted into transgenic mice. It is now clear that this process is part of a normal developmental pathway whereby the unmethylated genomic DNA of the blastula undergoes methylation just prior to gastrulation. This is not only true for the unique DNA in the genome (Shemer et al., 1991b; Kafri et al., 1992) but also for repetitive sequences such as LI, MUP, IAP (Howlett and Reik, 1991) MIF, and the major and minor satellites (Chapman et al., 1984). A careful look at the specificity of this *de novo* reaction shows that while most DNA becomes remodified at this stage, CpG island sequences remain unmethylated (see Fig. 2). This was first observed for island containing transgenes and was then confirmed by showing that endogenous CpG islands also remain unmethylated despite the massive wave of modification that occurs in the rest of the genome. The mechanism of this selective *de novo* methylation is as yet unknown, but recent experiments clearly show that the embryo does have some mechanism for distinguishing island DNA from the bulk DNA.

When methylated DNA is introduced into embryonic cells in culture, all non-island sequences retain their modification pattern, while CpG islands undergo specific demethylation (Frank et al., 1991; Shemer et al., 1991a). It is also clear that a similar island demethylating activity must exist *in vivo*, since *in vitro* methylated island DNA injected into the pronucleus of the fertilized egg lost its methyl moieties during embryogenesis as judged by the analysis of the resulting transgenic mice (Frank et al., 1991). It thus appears likely that a mechanism of this nature could serve to protect CpG island DNA from the wave of *de novo* modification which takes place prior to gastrulation (Kolsto et al., 1986; Szyf et al., 1990; Shemer et al., 1991a), and in this way the genome reconstructs its bimodal pattern which remains relatively fixed in somatic cells.

The exact timing of the methylation reaction has not yet been determined, but there are several indications that this occurs at about the time of implantation, probably between 5.5 and 6.5 days of development. Repeated and satellite sequences have been shown to be fully methylated by 7.5 days in the embryo proper and a large number of individual CpG residues which were unmodified in the blastula were shown to be methylated in the 6.5 d embryo. In preliminary experiments, it was demonstrated in this laboratory that a paternally derived multicopy transgene was still unmodified at 5.5 days and this suggests that the methylation reaction takes place subsequent to or concomitant with implantation. It has also been shown that the X-linked inactivated Pgk-1 gene undergoes modification at this same time (Singer-Sam et al., 1990). It is not yet known whether other genes on the inactive X chromosome became methylated at this stage and there are some indications that methylation of the Hprt gene may take place as a slow process which continues until day 10 (Lock et al., 1987).

The *de novo* methylation process is probably confined to the embryo proper and may not include the DNA in extraembryonic tissues. In the case of satellite DNA, microsurgery clearly demonstrated that these sequences remain partially undermethylated in extraembryonic cells (Fig. 1), and it is also known that both placenta and yolk sac DNA are generally hypomethylated throughout the life of the embryo (Razin et al., 1984).

5 Formation of DNA methylation patterns in the soma

The final methylation patterns of gene sequences in different adult tissues is established during late embryonic development and may even be delayed to adult life. Specific demethylations take place during the differentiation of mature cell types. Since DNA modification is known to mediate gene repression this scheme is consistent with its role during development. According to this model, essentially all tissue-specific genes are ubiquitously modified in every cell and, as such, are inaccessible to the transcription machinery. In contrast, housekeeping genes contain constitutively unmethylated 5′ CpG islands and this ensures that these sequences will be expressed in every cell. While this pattern is fairly constant throughout the organism, specific genes undergo demethylation in their cell type of expression and this is an integral part of their activation process.

Demethylation in differentiated tissues is highly selective and only takes place on the relevant specific genes of that cell type. This strongly implies that each event must be directed by cell type specific *cis* elements and *trans*-acting factors. Using DNA-mediated gene transfer in tissue culture it has been possible to study some of these demodification reactions in detail. One of the best documented cases is the muscle specific rat α actin gene. When this sequence is introduced/methylated into fibroblast cells by DNA-mediated gene transfer, modification inhibits its expression. In contrast, when the same methylated construct is inserted into L8 myoblasts, α actin is actively transcribed with approximately the same efficiency as the unmethylated form. In the same experiment these sequences were also found to have undergone demethylation (Yisraeli et al., 1986). Thus, as opposed to fibroblasts, these muscle-like cells have the ability to recognize the methylated α actin gene and direct both its activation and demethylation. These specific recognition events must, of course, be mediated by *cis*-acting sequences, and by applying reversed genetics it was possible to identify the precise DNA loci which are required to drive the demethylation events (Paroush et al., 1990). These sites are located in the upstream regulatory region of the gene and, while they are not directly required for RNA transcription, they undoubtedly play an important role in the developmental activation process (Paroush et al., 1990). While these preliminary experiments pinpointed the factors involved in demethylation, they could not determine the temporal order of the activation events. In order to assess whether demodification indeed precedes the initiation of transcription a methylated α actin gene was introduced into L8 cells by transient transfection. These experiments showed that demethylation occurs in a two-step process involving a hemimethylated intermediate. Transcription from the α actin promoter was observed only when both symmetrical methyl moieties were removed from the critical sites. Further evidence supporting this idea was obtained by altering the *cis*-acting elements required for the specific demodification reaction. Unlike the wild type α actin gene, these constructs were unable to undergo demethylation in L8 cells and thus remained transcriptionally silent (Paroush et al., 1990).

Although demethylation clearly preceded activation in this example this sequence of events may not be obligatory. In the developing mouse or rat liver, for example, several gene sequences have been shown to undergo demodification at about the time of birth despite the fact that these genes are already fully active several days earlier in the embryonic liver (Kunnath and Locker, 1983; Wilks et al., 1984; Benvenisty et al., 1984; Shemer et al., 1991b). In these cases it has not yet been shown that these particular modified sites are indeed involved in gene regulation, and further studies will be required to show that they indeed have biological function. Another reservation of these experiments is that DNA methylation was measured by restriction enzymes which do not distinguish between fully methylated and hemimethylated sites. In the case of the chicken vitellogenin gene, genomic sequencing revealed, for example, that this gene is actually hemidemethylated at critical loci prior to or comcomitant with its activation, suggesting that some form of partial demodification may in fact be a required first step (Saluz et al., 1986).

It had previously been assumed that demethylation reactions of this nature probably take place through a passive mechanism whereby DNA is replicated in the absence of maintenance methylation. Several studies have now provided clear cut proof that this process is actually an active one which does not require DNA replication. Firstly, in the developing liver of the chicken, gene-specific demethylation can occur even in the presence of DNA synthesis inhibitors (Wilks et al., 1984). In addition, it has been shown that demodification of the α actin gene can take place on marked DNA molecules which remain unreplicated (Paroush et al., 1990). It should be noted that while this is clearly an active process, the exact biochemical mechanism has not been elucidated. It is very likely that this involves either base or nucleotide substitution similar to that seen in the glycosylase reaction used to remove uridine residues from DNA. Biochemical studies on the rapid demethylation seen upon the induction of mouse erythroleukemia cells are also consistent with this model (Razin et al., 1986).

6 Gametogenesis

Both in males and females, the germ line is established from primordial cells which originate in the epiblast. These cells are first seen at 7 days in the yolk sac where they slowly accumulate in the region of the hind gut before making their way to the genital ridge at about 11 days p.c (McLaren, 1988). Strikingly, every DNA sequence that has been analyzed in primordial germ cells is unmethylated (see Fig. 2). This was originally shown using a semi-quantitative assay (Monk et al., 1987), but has now been confirmed for many individual sites in specific genes (Kafri et al., 1992) and for several repeated sequences (Sanford et al., 1987; Monk et al., 1987; Howlett and Reik, 1991). Furthermore, a similar phenomenon has been observed in human fetal ovaries (Driscoll and Migeon, 1990).

How primordial germ cells acquired their undermethylated state is, as yet, an open question. One possibility is that these cells escape from the wave of *de novo* methylation which takes place in the embryo proper prior to gastrulation at about 6 days of development. Alternatively, germ line DNA may undergo *de novo* methylation following blastulation along with the rest of the cell types in the organism and then be subject to an additional round of cell-specific demethylation. In any event, at least one transgene sequence is methylated at the blastula stage and unmethylated in primordial germ cells (Chaillet et al., 1991), suggesting that these cells retain the potential to demethylate genomic DNA. It was originally suggested that remethylation of the genome at later stages of gametogenesis takes place only in the male, while female germ line DNA might remain undermethylated throughout oogenesis (Monk et al., 1987). Recent experiments on individual CpG sites in a variety of genes, however, clearly show a wave of *de novo* methylation which begins at around 15 days and culminates at 18.5 days p.c. in both sexes (Kafri et al., 1992). In the sample of genes tested in this study, all non-CpG island sites were fully methylated and this suggests that gametic DNA also undergoes a rebuilding of its methylation pattern, similar to that which occurs in the rest of the embryo at the pregastrulation stage. This *de novo* methylation is not associated with the developmental events which occur during gametogenesis, since it takes place following the beginning of meiosis in the female but prior to meiotic stages in the testis.

These alterations in DNA modification are clearly general in nature and involve a large portion of the genome (see Fig. 2). In contrast, further changes in methylation patterns during later stages of spermatogenesis are quite gene specific. A good example is the Pgk-2 gene which is turned on premeiotically and remains transcriptionally active during a narrow window of testis development. In keeping with this expression pattern, at least two CpG sites in the 5′ region of the gene were shown to undergo specific demethylation sometime before the spermatogonial stage and this event takes place prior to activation (Ariel et al., 1991). The fact that other sites 3′ to the gene region remained modified suggests that his change is highly sequence specific and is thus probably related to transcription. This developmental methylation pattern may represent a common theme for genes which are expressed during spermatogenesis and at least one other gene (ApoAI) has been shown to undergo similar methylation changes (Kafri et al., 1992). In normal somatic tissues, specific genes undergo demethylation and then remain in that state throughout the lifetime of the organism. In contrast, both the Pgk-2 and ApoAI genes undergo a further remethylation step at late stages of spermatogenesis and are thus fully modified in mature sperm (Fig. 2). In a sense this represents a process of reorganization which is probably necessary to reconstruct the basic DNA methylation pattern prior to the transfer of the genome to the next generation. It should be noted that the large majority of testis-specific genes have CpG island sequences at their 5′ end. Since they are constitutively unmethylated, their methylation pattern does not have to be readjusted during late gametogenesis (Ariel et al., 1991).

Specific changes in DNA methylation also occur during oogenesis. A first round of demodification seems to take place just prior to birth and this involves the exact same sites as those which become unmethylated in the testis (Fig. 2). Since these genes are not known to be expressed in oocytes, it appears that this demethylation represents a gamete-specific pathway. Further demethylation events clearly take place during oocyte maturation and these involve a large number of sites in various gene sequences. While the rules governing this selective reaction are not known, this change does not appear to be related to gene expression and probably represents the initiation of the generalized demodification which occurs during early embryogenesis. It should be noted that during the entire process of oocyte maturation germ cells are locked within the second meiosis and this implies that demodification in those cells must occur by an active mechanism which does not involve DNA replication.

7 Genomic imprinting

Several endogenous genes have been shown to be subject to parental imprinting in the mouse (see corresponding chapter of this book). The Igf-II gene, for example, is turned on during embryonic development in many cell types, but only the paternal allele is active (DeChiara et al., 1991). In contrast, both the Igf II receptor (Barlow et al., 1991) and the H19 gene (Bartolomei et al., 1991) are exclusively transcribed from the maternal allele. The mechanism for marking the parental origin of these genes is unknown, but experiments in transgenic mice suggest that DNA methylation may be involved. In these cases, the methylation state of the transgene in adult tissues is defined by the parental source of the allele. For most of these exogenous sequences the maternal allele is methylated, while the paternally derived constructs are unmodified. The fact that the parental alleles are differentially methylated suggests that

DNA modification is associated with the imprinting process but does not prove that these methyl moieties provide the signals for marking the gene during gametogenesis. In keeping with this idea it would be expected that endogenous imprinted loci might be differentially methylated in the mature oocyte and sperm and that this state might be maintained throughout early development. Careful studies on one imprinted transgene support this general model (Chaillet et al., 1991). In this case, specific sequences on the IgMycCAT construct are methylated in mature oocytes and unmethylated in sperm DNA. This imprint was stably inherited and was detectable in the blastula, as well. It should be noted that the maternal allele remained methylated despite the massive general demethylation which occurs during these early divisions. This suggests that the differential methylation pattern in the gamete is not sufficient to direct the imprinting process and there must be an additional mechanism for recognizing imprinted sequences during early development. It is clear that further research on endogenous genes will be necessary to clarify the role of DNA modification in the imprinting process.

Ariel, M., McCarrey, J., and Cedar, H. (1991) Methylation patterns of testis-specific genes. Proc. Natl. Acad. Sci. USA *88*, 2317–2321.

Bartolomei, M. S., Sharon, Z., and Tilghman, S. M. (1991) Parental imprinting of the mouse H19 gene. Nature *351*, 153–155.

Barlow, D. P., Stoger, R., Herrmann, B. G., Saito, K., and Schweifer, N. (1991) The mouse insulin-like growth factor type-2 receptor is imprinted and closely linked to the Tme locus. Nature *349*, 84–87.

Benvenisty, N., Mencher, D., Meyuhas, O., Razin, A., and Reshef, L. (1985) Methylation of cytosolic PEPCK gene: Pattern associated with tissue specificity and development. Proc. Natl. Acad. Sci. USA *82*, 267–271.

Cedar, H., and Razin, A. (1990) DNA methylation and development. Biochim. Biophys. Acta *1049*, 1–8.

Chaillet, J. R., Vogt, T. F., Beier, D. R., and Leder, P. (1991) Parental-specific methylation of an imprinted transgene is established during gametogenesis and progressively changes during embryogenesis. Cell *66*, 77–83.

Chapman, V., Forrester, L., Sanford, J., Hastie, N., and Rossant, J. (1984) Cell lineage-specific undermethylation of mouse repetitive DNA. Nature *307*, 284–286.

Clegg, K. B., and Piko, L. (1983) Poly(A) length, cytoplasmic adenylation and synthesis of poly(A) + RNA in early mouse embryos. Dev. Biol. *95*, 331–341.

DeChiara, T. M., Robertson, E. J., and Efstratiadis, A. (1991) Parental imprinting of the mouse insulin-like growth factor II gene. Cell *64*, 849–859.

Driscoll, D. J., and Migeon, B. R. (1990) Evidence that human single copy genes which are methylated in male meiotic cells are unmethylated in female meiotic germ cells and fetal germ cells of both sexes. Somatic Cell Genet. *16*, 267–275.

Frank, D., Mintzer-Lichenstein, M., Paroush, Z., Bergman, Y., Shani, M., Razin, A., and Cedar, H. (1990) Demethylation of genes in animal cells. Phil. Trans. Royal Soc. *326*, 241–251.

Frank, D., Keshet, I., Shani, M., Levine, A., Razin, A., and Cedar, H. (1991) Demethylation of CpG islands in embryonic cells. Nature *351*, 239–241.

Howlett, S. K., and Reik, W. (1991) Methylation levels of maternal and paternal genomes during preimplantation development. Development *113*, 119–127.

Jahner, D., and Jaenisch, R. (1984) DNA methylation in early mammalian development, in: DNA Methylation: Biochemistry and Biological Significance, pp. 189–219. Eds A. Razin, H. Cedar and A. D. Riggs. Springer-Verlag, New York.

Kafri, T., Ariel, M., Brandeis, M., Shemer, R., Urven, L., McCarrey, J., Cedar, H., and Razin, A. (1992) Developmental pattern of gene specific DNA methylation in the mouse embryo and germ line. Genes and Develop. *6*, 705–714.

Kolsto, A. B., Kollias, G., Giguere, V., Isobe, K. I., Prydz, H., and Grosveld, F. (1986) The maintenance of methylation-free islands in transgenic mice. Nucl. Acids Res. *14*, 9667–9678.

Lock, L. F., Takagi, N., and Martin, G. R. (1987) Methylation of the Hprt gene on the inactive X occurs after chromosome inactivation. Cell *48*, 39–46.

Manes, C., and Menzel, P. (1981) Demethylation of CpG sites in DNA of early rabbit trophoblast. Nature *293*, 589–590.

McLaren, A. (1976) Genetics of the early mouse embryo. Ann. Rev. Genet. *10*, 361–388.

McLaren, A. (1988) The developmental history of female germ cells in mammals. Oxford Rev. Reprod. Biol. *10*, 162–179.

Monk, M., Boubelik, M., and Lehnert. S. (1987) Temporal and regional changes in DNA methylation in the embryonic, extraembryonic and germ cell lineages during mouse embryo development. Development *99*, 371–382.

Paroush, Z., Keshet I., Yisraeli, J., and Cedar, H. (1990) Dynamics of demethylation and activation of the α actin gene in myoblasts. Cell *63*, 1229–1237.

Ponzetto-Zimmerman, C., and Wolgemuth, D. J. (1984) Methylation of satellite sequences in mouse spermatogenic and somatic DNAs. Nucl. Acids Res. *12*, 2807–2822.

Razin, A., and Cedar, H. (1991) DNA methylation and gene expression. Microbiol. Rev. *55*, 451–458.

Razin, A., and Riggs, A. D. (1980) DNA methylation and gene regulation. Science *210*, 604–610.

Razin, A., Webb, C., Szyf, M., Yisraeli J., Rosenthal, A., Naveh-Many, T., Sciaky-Gallili, N., and Cedar, H. (1984) Variations in DNA methylation during mouse cell differentiation in vivo and in vitro. Proc. Natl. Acad. Sci. USA *81*, 2275–2279.

Razin, A., Szyf, M., Kafri, T., Roll, M., Giloh, H., Scarpa, S., Carotti, D., and Cantoni, G. L. (1986) Replacement of 5-methylcytosine by cytosine: a possible mechanism for transient DNA demethylation during differentiation. Proc. Natl. Acad. Sci. USA *83*, 2827–2831.

Rupp, R. A. W., and Weintraub, H. (1991) Ubiquitous MyoD transcription at the midblastula transition precedes induction-dependent MyoD expression in presumptive mesoderm of X. laevis. Cell *65*, 927–937.

Saluz, H. P., Jiricny, J., and Jost, J. P. (1986) Genomic sequencing reveals a positive correlation between the kinetics of strand specific DNA demethylation of the overlapping estratiol/glucocorticoid receptor binding sites and the rate of avian vitellogenin mRNA synthesis. Proc. Natl. Acad. Sci. USA *83*, 7167–7171.

Sanford, J., Forrester, L., Chapman, V., Chandley, A., and Hastie, N. (1984) Methylation patterns of repetitive DNA sequences in germ cells of Mus musculus. Nucl. Acids Res. *12*, 2823–2836.

Singer, J., Roberts-Ems, J., Luthardt, F. W., and Riggs, A. D. (1979) Methylation of DNA in mouse early embryos, teratocarcinoma cells and adult tissues of mouse and rabbit. Nucl. Acids Res. *7*, 2369–2385.

Singer-Sam, J., Lebon, J. M., Tanguay, R. L., and Riggs, A. D. (1990) A quantitative HpaII-PCR assay to measure methylation of DNA from a small number of cells. Nucl. Acids Res. *18*, 687–692.

Shemer, R., Eisenberg, S., Breslow, J. L., and Razin, A. (1991a) Methylation patterns of the human ApoAI-CIII-AIV gene cluster in adult and embryonic tissue suggest dynamic changes in methylation during development. J. Biol. Chem. *266*, 23676–23681.

Shemer, R., Kafri, T., O'Connell, A., Eisenberg, S., Breslow, J. L., and Razin, A. (1991b) Methylation changes in the ApoAI gene during embryonic development of the mouse. Proc. Natl. Acac. Sci. USA *88*, 11300–11304.

Solage, A., and Cedar, H. (1978) Organization of 5 methylcytosine in chromosomal DNA. Biochemistry *17*, 2934–2938.

Sullivan, C. H., and Grainger, R. M. (1987) δ Crystallin genes become hypomethylated in postmitotic lens cells during chicken development. Proc. Natl. Acad. Sci. USA *84*, 329–333.

Stein, R., Gruenbaum, Y., Pollack, Y., Razin, A., and Cedar, H. (1982) Clonal inheritance of the pattern of DNA methylation in mouse cells. Proc. Natl. Acad. Sci. USA *79*, 51–65.

Szyf, M., Tonigawa, G., and McCarthy, P. L. Jr. (1990) A DNA signal from the Thy-1 gene defines *de novo* methylation patterns in embryonic stem cells. Mol. Cell. Biol. *10*, 4396–4400.

Wigler, M., Levy, D., and Perucho, M. (1981) The somatic replication of DNA methylation. Cell *24*, 33–40.

Wilks, A. F., Seldran, M., and Jost, J. P. (1984) An estrogen-dependent demethylation at the 5′ end of the chicken vitellogenin gene is independent of DNA synthesis. Nucl. Acids Res. *12*, 1163–1177.

Yisraeli, J., and Szyf, M. (1984) Gene methylation patterns and expression, in: DNA Methylation: Biochemistry and Biological Significance, pp. 352–370. Eds A. Razin, H. Cedar and A. D. Riggs. Springer-Verlag, New York.

Yisraeli, J., Adelstein, R. S., Melloul, D., Nudel, U., Yaffe, D., and Cedar, H. (1986) Muscle-specific activation of a methylated chimeric actin gene. Cell *46*, 409–416.

DNA Methylation: Molecular Biology and Biological Significance
ed. by J. P. Jost & H. P. Saluz

X chromosome inactivation and DNA methylation

Judith Singer-Sam and Arthur D. Riggs

Beckman Research Institute of the City of Hope, Duarte, CA 91010, USA

1 Introduction

X chromosome inactivation (XCI) renders one of the two X chromosomes in female mammalian cells heterochromatin-like and genetically silent. This chromosome-wide phenomenon has long been considered a paradigm for the study of the effect of heterochromatinization and DNA methylation on gene expression in mammals (for reviews see Gartler and Riggs, 1983; Grant and Chapman, 1988; Lyon, 1988; Riggs, 1990b; Riggs and Pfeifer, 1992). The inactivation process is usually considered to consist of three components – initiation, spreading and maintenance. A role for DNA methylation is possible for all three of these processes, but most available information is relevant only to maintenance.

Three features of XCI are of particular interest: (i) XCI is one of the best examples of cell memory, the faithful maintenance of a determined chromosomal state in mitotically active cells. Once an X chromosome is inactivated by the epigenetic mechanism(s) of XCI, maintenance of the inactive state is very efficient – unless cells are treated with agents which inhibit DNA methylation, e.g. 5-azacytidine. (ii) The inactive X chromosome (the X^i) is in the same nucleus as the active X chromosome (the X^a), and yet housekeeping genes such as phosphoglycerate kinase (*PGK1*) and *HPRT* are actively transcribed only from the X^a. Therefore the X^i remains stably silent in spite of the presence of all the necessary transcriptions factors. What is the mechanism for this remarkable immunity to activation? (iii) XCI is a chromosome-wide but exclusively *cis* phenomenon, being limited to DNA that is in physical continuity with a locus called the X inactivation center, located at Xq13 on the human X chromosome (Brown et al., 1991b). X chromosomal DNA separated from the X inactivation center by translocation to autosomes is not inactivated. Inactivation often "spreads" into autosomal DNA translocated onto an X chromosome, but, on the other hand, the inactivation process sometimes can "skip over" the autosomal insert (Lyon et al., 1986; Mohandas et al., 1987).

Other recent reviews (Grünwald and Pfeifer, 1989; Holliday, 1989; Riggs, 1989; Cooney and Bradbury, 1990; Jones and Buckley, 1990; Riggs, 1990a; Riggs, 1990b; Lewis and Bird, 1991; Razin and Cedar, 1991) and other chapters in this book cover much of the available knowledge about methylation-mediated gene silencing of autosomal genes. With regard to X-linked genes and XCI, most reviews have included summaries of the methylation literature (Gartler and Riggs, 1983; Monk, 1986; Grant and Chapman, 1988; Lyon, 1988; Riggs, 1990b). In this chapter we will emphasize studies that have been reported since the last general reviews of X inactivation (Grant and Chapman, 1988; Riggs, 1990b), add to our knowledge of the role(s) of DNA methylation in mammalian gene control and chromatin structure, and give information on the changes in X chromosomal activity and methylation that occur during development. Additionally we will address two questions in detail that are relevant to methylation dynamics and development: (i) What is the rate of *failure* of methylation maintenance? (ii) What is the rate of *de novo* methylation, that is, the methylation of unmethylated sites?

2 An overview of X chromosome inactivation, with emphasis on development

Each sex has an inactivation-reactivation cycle in the germ line. During male meiosis the single X^a becomes condensed and transcriptionally inactive, associating with the Y chromosome to form the cytologically identifiable sex vesicle (Monesi et al., 1978). Female primordial germ cells migrating to the genital ridge contain an X^i, which then reactivates at about the time of the beginning of meiosis (Kratzer and Chapman, 1981; Gartler and Riggs, 1983; Grant and Chapman, 1988). Though the paternal X enters the egg inactive, soon after fertilization both X chromosomes in female embryos are transcriptionally active (reviewed in Grant and Chapman, 1988) and both replicate at the same time in the cell cycle (synchronous or isocyclic replication). From this both-active ground state, one of the two X chromosomes later becomes condensed during interphase, similar to heterochromatin, and genetically silent for most genes. Concomitant with these changes, the X^i begins to replicate at a different time in the cell cycle (asynchronous or allocyclic replication). For extraembryonic tissues, the X^i first replicates earlier than the autosomes and then switches to the late replication pattern seen in somatic cells (Sugawara et al., 1985).

XCI in the inner cell mass is generally random with respect to parental source of the X, although the probability of inactivation is affected by various alleles at the *Xce* locus (Johnston, 1981; Fowlis, 1991), which is very near to and may be identical to the mouse inactivation center (Cattanach, 1981). This random XCI occurs at about 6.5 days p.c., near the time of uterine implantation (McMahon and Monk, 1983). XCI occurs at an earlier time in cells which do not become part of the embryo; in the trophectoderm and primitive endoderm inactivation occurs at 3.5 and 4.5 days, respectively, when these tissues first show cytodifferentiation and departure from totipotency. In these exraembryonic tissues XCI is *not* random; instead the paternal X is preferentially inactivated (Takagi and Sasaki, 1975; Harper et al., 1981). Interestingly, preferential paternal XCI in marsupials occurs not only in extraembryonic tissues, but in all cells (VandeBerg et al., 1987); conversely in humans, at least one extraembryonic tissue, chorionic villi, shows random XCI (Mohandas et al., 1987).

Prior to fertilization, the chromatin states of the maternal X and paternal X are quite different since the paternal X is heterochromatic during meiosis (Monesi, 1965) and then becomes nucleosome-free in sperm. Both *PGK* and *HPRT* RNA levels are already reduced by the leptotene stage of male meiosis (Singer-Sam et al., 1990b). During maturation to sperm, most histone proteins are replaced by protamines. In spite of these differences, both the paternal X and maternal X become active very soon after fertilization. Evidence for an active X^P in female embryos comes from assay of the X-linked enzymes α-galactosidase and HPRT in single embryos (Adler et al., 1977; Epstein et al., 1978; Kratzer and Gartler, 1978). A unimodal distribution of the amount of enzyme activity is found until about the 8-cell stage ($2\frac{1}{2}$ day embryos), when a bimodal distribution starts to appear, suggesting that one X chromosome is active in male embryos, but that both X chromosomes are active in female embryos. Additionally, we have recently found (J. Singer-Sam, A. D. Riggs, and V. Chapman, unpublished) that transcripts of the paternally derived X-linked *Pkg-1* gene are present by the 2–4-cell stage. The other evidence that there are two active X chromosomes in early embryos is cytogenetic, based upon the isocyclic replication pattern of both maternal X and paternal X before XCI occurs (Takagi and Sasaki, 1975; Mukherjee, 1976).

3 X^a/X^i Differences other than methylation

There are several known differences between the X^a and the X^i. The X^i is condensed during interphase, like heterochromatin, and is cytologically recognizable as the Barr body in many species. Because the X^i is like constitutive heterochromatin but can become euchromatin, it is often called facultative heterochromatin. The X^i is generally seen near the nuclear membrane (Dyer et al., 1989; Manuelidis, 1990), but there is some variability in location, perhaps depending on the state of the cell (Borden and Manuelidis, 1988). Recently, Walker et al. (1991) found that the intertelomere distance of the X^a is 10-fold greater than that of the X^i. The two telomeres at each end of the X^i are at all phases of the cell cycle located within 1 μm of each other, probably indicating they are bound together, thus forcing a folded or looped organization of the X^i. This may partly explain why a characteristic bend near the X inactivation center is often seen in mitotic chromosome spreads (Van Dyke et al., 1986). Mitotic X^a and X^i chromosomes are morphologically very similar and stains do not distinguish them, with the exception of the Kanda staining procedure (Kanda, 1973; Rastan and Robertson, 1985) in which cells are incubated in hypotonic KCl prior to fixation and Giemsa staining. The molecular basis for the differential staining is not known.

Studies of nuclease sensitivity have revealed further differences between the X^a and X^i. Sites hypersensitive to DNase I or methylation-insensitive restriction enzymes such as Msp I are found on the X^a (Riley et al., 1986; Lin and Chinault, 1988; Antequera et al., 1989). Interestingly, the general sensitivity of the X^i to DNase I for both human *PGK1* and mouse *HPRT* is only slightly (about 2-fold), less than that of the X^a (Riley et al., 1986; Lin and Chinault, 1988; Pfeifer and Riggs, 1991). Recent DMS, DNase I, and UV photofootprinting experiments (Pfeifer et al., 1990b; Pfeifer et al., 1991; Pfeifer and Riggs, 1991) done with a new, improved method for genomic sequencing (Pfeifer et al., 1989) have, for the first time, given a picture of the X^a and the X^i at the human PGK locus at nucleotide level resolution. The promoter-containing CpG island of the *PGK1* gene on the X^a shows several clear DNase I footprints, indicative of protein factor binding. The X^i has none of these factors, but instead has two nucleosomes, or nucleosome-sized particles, positioned over much of the promoter; one of the nucleosomes is approximately centered over the transcription start region. The insensitivity of the X^i to Msp I can be explained by the observation that nucleosomes cover all of the CCGG sites. There is no footprint over the transcription start region of the X^i, but it may be of interest that only on the X^a is there increased UV photoreactivity just downstream of the major transcription start site (Pfeifer et al., 1992). These same studies also gave information on the methylation of this region, and this will be considered in the next and later sections.

4 DNA Methylation and maintenance of X chromosome inactivation

Historically, it was considerations of the maintenance of XCI that led to the concept that DNA sequences symmetrically methylated on both strands could provide a mechanism for maintaining epigenetic differences through DNA replication (Holliday and Pugh, 1975; Riggs, 1975). Numerous studies have established that DNA methylation is an important component of XCI, and, moreover, that methylated CpG sites are somatically heritable. The maintenance methylation mechanism is critically dependent on the postreplicational activity of an enzyme, called the maintenance DNA methyltransferase, which preferentially methylates hemimethylated DNA (one strand methylated) which is formed during DNA replication. The maintenance methylase system is reviewed in another chapter of this book.

Although methylation of CpG islands is now known to play a role in the maintenance of XCI, it is not known whether methylation of sites outside of CpG islands in involved. In brief summary, some of the well established, and well reviewed (Gartler and Riggs, 1983; Monk, 1986; Grant and Chapman, 1988) evidence implicating cytosine methylation in the maintenance of XCI is: (i) Genes on the X^i can be reactivated in some cell lines by agents, such as 5-azacytidine, which inhibit cytosine DNA methylation; (ii) DNA from cells treated with 5-azacytidine functions more efficiently in DNA mediated transfer of the gene for *HPRT*; and (iii) numerous studies have found that the CpG islands at the 5′ end of X-linked genes for housekeeping enzymes are highly methylated only on the X^i in somatic cells. Recent studies have shown that *in vitro* methylation of CpG islands inhibits *in vitro* transcription. The inhibition is an indirect effect that seems to be mediated by a protein, MeCP-1 that binds nonspecifically to methylated DNA (Boyes and Bird, 1991; Boyes and Bird, 1992).

Methylation differences between the X^a and X^i are clear but are confined primarily to X-linked CpG islands. In contrast to the CpGs clustered into islands, regions such as coding regions and exons are usually similarly methylated on the X^a and X^i (Wolf et al., 1984; Migeon et al., 1991). There are even some sites that are hypomethylated only on the X^i (Lindsay et al., 1985). In a recent cytological study, treatment of fixed mitotic chromosomes with Hpa II followed by nick translation was used to show that CCGG sites on the X chromosome are preferentially located in R bands, and that the X^i is more methylated than the X^a (Parantera and Ferraro, 1990). Since many CCGG sites are in islands, this result is not inconsistent with the methylation differential being mostly in the clustered CpGs.

It is well known that genes on the X^i can be reactivated by agents that inhibit DNA methylation. *HPRT* reactivation frequencies, for example, increase from less than 10^{-7} to as much as 10^{-2} (Grant and Chapman, 1988; Grant and Worton, 1989a; Grant and Worton, 1989b). Though the reactivation is stable for *HPRT* (Homman et al., 1987), it should be noted that reactivation is only piecemeal when somatic cells are the source of the X^i. For example, many clones that have reactivated human *HPRT* remain *PGK* or *G6PD* negative. This result suggests independent domains for X^i maintenance. Present results would be consistent with each looped domain or replicon being an independent unit with regard to maintenance, but the size of the reactivated region remains to be determined. Stable reactivation of an entire somatic cell X^i and conversion to early replication has only been seen upon fusion of thymocytes and embryonic carcinoma cells (EC cells) (Takagi et al., 1983; Takagi, 1988). Chorionic villi cells, which are of extraembryonic lineage, also completely reactivate (Migeon et al., 1986). In addition to stable reactivation of X-linked genes, 5-azacytidine can cause transient expression and transient conversion to early replication of the entire X chromosome (Jablonka et al., 1985; Hockey et al., 1989).

Embryonic carcinoma cells (EC cells) have been used to study XCI. McBurney and colleagues (Paterno et al., 1985; Hockey et al., 1989) have derived a number of female EC cell lines apparently arrested at various stages. One cell line, CM86AGM2, is not overtly differentiated but has an X^i that carries a distinguishable *Hprt* allele (Hockey et al., 1989). This cell line can be induced by retinoic acid to undergo differentiation to a wide variety of cell types. Prior to differentiation, treatment with 5-azacytidine efficiently reactivates *Hprt* expression from the X^i (10^{-7} spontaneous vs 10^{-2} posttreatment), with concomitant elimination of an observable late replicating X. If, however, the cells are induced to differentiate prior to 5-azacytidine treatment, reactivation of the X^i is not seen; the cells now respond more like adult, nontransformed somatic cells. The lack of reactivation is not due to a resistance to demethylation; both before and after differentiation, 5-azacytidine causes a similar transient reduction in methylation level from 3.6% to 1.8% 5-methylcytosine. These results have been interpreted to mean the XCI is at least a two-step process, but they are also consistent with a gradation of methylation, with partial methylation being reversible but full methylation being irreversible.

In cases where mammalian XCI is not random, DNA methylation may be involved much less or perhaps not at all. For example, DNA-mediated transfection experiments suggest that the X chromosome of sperm and the X^i in extraembryonic mouse tissue are not methylated in such a way as to prevent gene expression in transfected cells (Kratzer et al., 1983; Venolia et al., 1984). Additionally a key argument for an important role for methylation of CpG islands in XCI is the relative instability of XCI in marsupials, where little or no methylation of CpG islands has yet been seen either on the X^a or X^i. Most marsupial adult tissues show significant expression of paternal genes and this is thought to be a result of tissue-specific reactivation (VandeBerg et al., 1987). Migeon et al. (1989) found that cells grown in tissue culture from lung, kidney, cardiac and skeletal muscle from a female opossum (*Didelphis virginiana*), and heterozygous at the *G6PD* locus, expressed *G6PD* heterodimers, confirming that XCI is not stable. Similar results were found for the *HPRT* locus by measurement of reversion frequencies in cells whose only functional *HPRT* gene is paternally derived. Studies of the effect of aging also support the idea that DNA methylation plays a role in stabilizing the inactive state. For example, Wareham et al. found that the OTC gene, which does not contain a CpG-rich island, is reactivated with age in the mouse (Wareham et al., 1987) (as is the Mo coat color gene, which is located near the *Xce* locus (Brown and Rastan, 1988)). Human chorionic villi have a hypomethylated *HPRT*-associated CpG island and XCI in chorionic villi is unstable and reactivatable (Migeon et al., 1986). However, human X-linked genes known to be associated with heavily methylated CpG islands are not known to reactivate either with age or on passage in culture, although a larger survey of genes needs to be done. Migeon et al., looked carefully at the human *HPRT* locus and found no effect of aging (Migeon et al., 1988). The *G6PD*, *GdX* and *MIC2* loci show no detectable methylation changes with age (Pagani et al., 1990), although see Holliday (1991) for a discussion of this point.

5 Methylation changes in X-linked genes during development

Only two groups have studied methylation of single-copy, endogenous genes during early germ cell development. Driscoll and Migeon (1990) analyzed impure gonadal fractions from male and female fetuses 8–21 weeks of age by Southern blotting. They analyzed 57 Hpa II sites in six X-linked genes (*HPRT*, *G6PD*, *P3*, *PGK1*, *GLA* and *F9*) and one autosomal gene (*EPO*). Faint bands indicative of hypomethylation of a small percentage of the cell population were seen in both sexes of the youngest, 8-week-old, gonadal samples, and this was interpreted to mean that primordial, premeiotic germ cells are unmethylated in female as well as male gonia. Hypomethylation was seen both for clustered, island-located CpG dinucleotides and nonclustered CpGs in the body of the genes.

Singer-Sam et al. (1992) used PCR to assay pure germ cells individually micromanipulated by hand from male and female fetuses. In this study methylation was examined at a Hpa II site, H8, in the 5′ untranslated coding region of the human *PGK-1* gene, 21 bp downstream of the major transcription start site, at the 3′ edge of the heavily methylated CpG island. H8 was chosen for analysis because it might be a critical site for methylation-mediated silencing in that it shows a perfect inverse correlation between methylation and transcriptional activity. Site H8 was found to be unmethylated in germ cells of human female fetuses, even as early as 47 days, a time when germ cells are still about 90% premeiotic, and just beginning to migrate into the genital ridges (Baker, 1963). Thus this study on pure cells confirms the study of Driscoll et al. (1990), and it seems likely that premeiotic germ cells are very hypomethylated. Whether these sites are ever methylated in germ cells remains as an open question. It has been proposed (Monk, 1987) that the germ line is set aside early in development thereby avoiding methylation. In any case, if sites like H8 are never methylated in the female germ line, it would mean that methylation at these sites is not essential for XCI, at least at some stages of germ line development (Singer-Sam et al., 1992). Pachytene spermatocytes have inactivated the single X chromosome, but have hypomethylated CpG islands (Driscoll and Migeon, 1990), so here too inactivation can be accomplished without methylation of CpG islands.

Several autosomal and X-linked genes have been shown to have sperm-specific methylation patterns (Driscoll and Migeon, 1990; Ariel et al., 1991; Ghazi et al., 1992) indicative of a spermatogenesis-specific methylation and/or demethylation program. For example, male-specific DNA methylation was found at 4 of the 7 single-copy X-linked loci tested by Driscoll et al. (1990). In keeping with somewhat lower overall methylation levels is sperm (Gamma-Sossa et al., 1983), Driscoll et al. (1990) found the X-linked genes in sperm to be less methylated than somatic tissue, but only at a few sites. Several autosomal CpG islands have been shown to be unmethylated during spermatogenesis and in sperm (Ariel et al., 1991). Most nonclustered CpGs are highly methylated in sperm (Yisraeli and Szyf, 1984), and Pgk-2, which is a testis-specific autosomal gene, is unmethylated in spermatocytes and round spermatids, where it is expressed, and then it becomes methylated in mature sperm. Groudine and Conkin (1985) found sperm-specific methylation patterns in chicken testes and saw *de novo* methylation between the spermatogonial stage and first meiotic prophase. Mammalian spermatogonial cells, spermatocytes, and round spermatids do have high levels of DNA methyltransferase mRNA (Singer-Sam et al., 1990b). Taking all results together, it seems likely that premeiotic male germ cells are hypomethylated and that sperm-specific methylation is established largely by *de novo* methylation during spermatogenesis.

Less is known about the methylation program during female meiosis and maturation of oocytes. Transgene studies have shown that there can be an oocyte-specific methylation pattern (Chaillet et al., 1991), and in the majority of transgenes showing imprinting, the transgene in the embryo and adult is more highly methylated when inherited from the mother than from the father (Howlett and Reik, 1991). Studies on L1, MUP, and IAP repetitive sequences and end labelling of Hpa II sites led to the suggestion that the oocyte and preimplantation embryo are very hypomethylated relative to sperm and somatic cells (Monk et al., 1987; Sanford et al., 1987). However, a recent study by Howlett and Reik (1991) suggests a more complicated picture. They found that MUP and IAP sequences are undermethylated only in the immature oocytes; high methylation was found in both sperm and mature oocytes. L1 repeats, however, were highly methylated in sperm but only 40% methylated in mature eggs. Since the X^M and X^P are differentially imprinted and methylated, existing data is consistent with the possibility that parental imprinting information is written as DNA methylation patterns, but, of course, correlations do not rule out other possibilities.

It is now known that after fertilization the preimplantation mammalian embryo is a dynamic system with regard to methylation. For example, Frank et al. (1991) showed that normal development results in demethylation of CpG islands methylated *in vitro* and then injected into mouse oocytes. They also observed that in the F9 line of EC cells there was rapid demethylation of methylated islands, but that nonclustered CpG sites were not demethylated. The island-specific demethylation in F9 cells is rapid, being detectable within 48 h of transfection, prior to integration. Other non-embryonic cell lines such as L-cell fibroblasts and L8 myoblasts do not demethylate the *in vitro* methylated islands. Shemer et al. (1991) have shown that nonclustered Hpa II sites in or near the apolipoprotein AI gene are methylated in both egg and sperm, but are demethylated in the early embryo (8–16 cells) and then become remethylated during gastrulation. The general picture, which seems to apply for mouse satellite sequences and the majority of nonclustered CpG sites (Monk et al., 1987; Sanford et al., 1987; Howlett and Reik, 1991), is for these elements to either enter unmethylated or to become undermethylated in the early embryo. After reaching a relatively hypomethylated state by the late morula or early blastocyst stage, there must then be a period of active *de novo* methylation, because most centromeric satellites, L1 elements, viral LTRs (Jaenisch and Jähner, 1984), and nonclustered CpGs in single-copy genes are highly methylated in adult tissues. The exceptions are autosomal CpG islands which remain unmethylated (Bird, 1986; Lewis and Bird, 1991).

Studies on X-linked genes in the early embryo seem to be in agreement with the above picture, with the exception that CpG islands on the X^i do become methylated. In tissue culture, it is known that some autosomal CpG islands, such as that in the MyoD1 gene (Antequera et al., 1990; Jones et al., 1990) can become highly methylated, so neither X-linked nor autosomal islands are intrinsically immune to methylation. The question of what keeps CpG islands free of methylation remains to be answered. The size of the methylation free zone near CpG islands varies with development for both X-linked and autosomal islands. Migeon et al. (1991) studied methylation of 4 human X-linked genes (*PGK*, *G6PD*, *P3*, *GdX*, *HPRT*) and one autosomal gene (*DHFR*) at various developmental stages. For each gene, methylation of the 3′ end of the CpG islands was highest in 6–14 week embryos. Thus prior to 6 weeks some *de novo* methylation was occurring, but after 14 weeks of development a demethylation program then established the adult pattern. Ghazi et al. (1992) have also shown for several human CpG island-associated genes (*c-H-ras*, *c-Myc*, and *HPRT*) that regions of extensive methylation in sperm move closer to CpG islands in fetal and adult somatic cells.

Because of the difficulty of obtaining adequate material, only two studies have focused on DNA methylation during the time of XCI in mouse embryos. In the first of these, Lock et al. (1987) pooled many mouse embryos and found by Southern blotting that methylation at sites in the first intron of the *Hprt* gene takes place between 9.5 and 13.5 days p.c., well after the time of XCI. In the second study, Singer-Sam et al. (1990) used a PCR-based assay to examine DNA methylation in individual embryos. It was found that Hpa II site H-7 in the CpG-rich 5′ region of the X-linked *Pgk-1* gene is unmethylated prior to and after fertilization but then becomes methylated in female embryos between 5.5 and 6.5 days, roughly coincident with the time of XCI. Like adults, male embryos do not become methylated at this site. Surprisingly, analysis of dissected embryos showed that site H-7 is methylated in the non-randomly inactivating ectoplacental cone and extraembryonic ectoderm as well as the embryo proper (embryonic ectoderm). A recent analysis of XCI in differentiating EC cells included a study of the methylation of the mouse *Pgk-1* gene (Bartlett et al., 1991). In contrast to the *in vivo* situation where 4 Hpa II sites in the first intron of the mouse *Pgk-1* gene are methylated on the X^i, these sites are not methylated either before or after *in vitro* differentiation and the occurrence of XCI as assayed by late replication.

The above results have been interpreted to mean that XCI is a two-step process, with methylation not being involved in the first step (Lock et al., 1987; Bartlett et al., 1991). The conclusion about a two-step process is not likely to be wrong. However, it should be noted that since the key DNA elements for the first step have not been identified, none of these data address the role of methylation in the first step and several interpretations are possible. One model compatible with the data is that there is a hierarchy of sites. For example replication origins might be the primary sites and methylation could affect these first. Another testable model is that methylation begins at the X inactivation center and then slowly spreads down the chromosome. This model suggests that loci far from the X inactivation center would be methylated after loci near the center. Present results are compatible with this model because methylation of the *PGK* gene, which is very near the X inactivation center, occurs at 5.5–6 days p.c. (Singer-Sam et al., 1990a) whereas methylation of the distantly located *HPRT* gene takes place at 9–13 days (Lock et al., 1987). Study of additional genes at various distances from the X inactivation center is obviously needed.

6 The dynamics of DNA methylation

Programmed methylation changes have long been proposed as an important component of development, and since the first study of methylation in the mouse early embryo and teratocarcinoma cells (Singer et al., 1979), the prevailing model has been one of demethylation from a high methylation ground state. Proteins, sometimes called determinator proteins (Riggs and Jones, 1983; Riggs, 1989), were proposed to bind to methylated sites and either actively or passively cause conversion of the methylated state to an unmethylated state (Singer et al., 1979; Riggs, 1989). It is now known that the high-methylation ground state for the embryo proper is probably reached shortly after implantation (Howlett and Reik, 1991; Migeon et al., 1991).

Until recently, most information was on methylation loss during tissue culture passages or induced differentiation, and there was little information on the *de novo* gain of methylation at previously unmethylated sites. However, the only DNA methyltransferase so far identified in mammalian cells does have significant *de novo* activity *in vitro* (Pfeifer and Drahovsky, 1986). Moreover, it is now clear that *de novo* methylation does occur with significant frequency in some cell types (Szyf et al., 1989; Antequera et al., 1990; Jones et al., 1990; Migeon et al., 1991; Shemer et al., 1991). With regard to the loss of methylation, there is evidence for conversion of 5-methylcytosine to cytosine without DNA replication, thus indicating an active removal process (Razin et al., 1986; Sullivan and Grainger, 1986; Paroush et al., 1990). The mechanism of this active process is not yet known, but it could be, for example, a "repair"-type glycosylase.

Recent genomic sequencing studies have suggested that methylation of even the X^i is to a certain extent dynamic (Pfeifer et al., 1990a). Because of the laws of mass action, no biochemical system is all-or-none at the molecular level. Methylation maintenance must occasionally fail, and thus, theoretically, methylation must be dynamic to a certain extent, as illustrated by the scheme in Figure 1, which shows the methylation state (M, methylated; U, unmethylated) of a *specific* site in the DNA of a cell and its mitotic descendents. Two parameters are sufficient to describe the methylation level ($\hat{M}$) of a given site in a population of cells, and these are E_m, the efficiency for maintenance of the methylated state of that particular site, and E_d, the efficiency of *de novo* methylation of the unmethylated state of the site. In theory, E_m and E_d are independent variables that can be different depending on the sequence context, cell type, developmental history, cell environment, etc. Otto and Walbot (1990) assumed that methylation could change only during replication and derived a discontinuous, recursive equation describing methylation levels in a growing cell population. Pfeifer et al. (1990) derived a more general treatment (equations 1–4 in Fig. 1) in

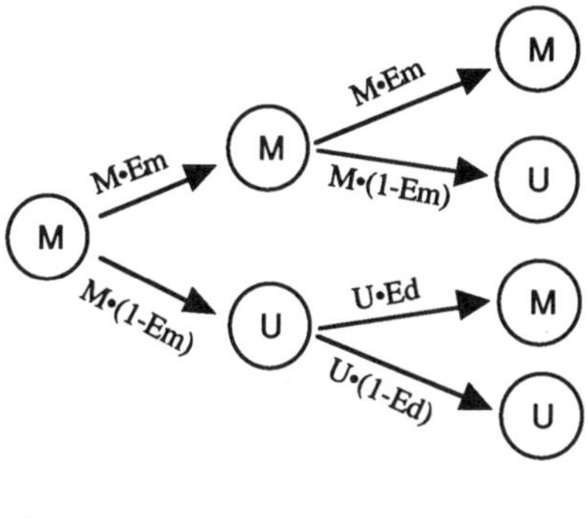

$$dM/dt = a \cdot Em \cdot M + a \cdot Ed \cdot U \quad [1]$$

$$dU/dt = a \cdot (1 - Ed) \cdot U + a \cdot (1 - Em) \cdot M \quad [2]$$

$$M^\wedge = M/Total = M/(M + U) \quad [3]$$

$$M^\wedge = Ed/(1 + Ed - Em) \quad [4]$$

Figure 1. Methylation levels at a specific site in a growing population of cells. Each circle represents a cell. M represents the methylated state of a specific CpG site in a specific gene, whereas U represents the unmethylated state of the same site. Since M is the starting state, U states arise in the population by a failure of maintenance. After some U states arise, they can be converted back to M states by *de novo* methylation. With time, a steady-state level of methylation of the site will be attained. The equations shown are from Pfeifer et al. (1990) and describe the fraction methylation, $M^\wedge$, as a function of two primary parameters – E_m, the efficiency of maintenance, and E_d, the efficiency of *de novo* methylation. It is assumed that the growth rate constant, a, is the same for M or U cells. The first three equations can be used to model $M^\wedge$ as a function of either time or population doublings (see Pfeifer et al., 1990a). For the special case where a steady state is reached and thus both dM/dt and dU/dt equal zero, $M^\wedge$ is given by equation 4, which can be used to estimate E_m and E_d as explained in the text.

which methylation loss may occur at any stage of the cell cycle, such as might occur by a "repair" type system removing methylated cytosines from the DNA.

Very few studies have been done that are appropriate for determining E_m or E_d, but the data of Pfeifer et al. (1990) can be used to make rough estimates for the *PGK* locus on the X^i. This study began when Hansen and Gartler (1988; 1990) treated briefly with 5-azacytidine a human-Chinese hamster cell line carrying an inactive human X chromosome. Clones were derived by selection only for reactivation of human *HPRT*, and then the *nonselected* human X-linked *PGK* gene was studied. All cytosines in a 450 bp region around the *PGK* promoter and transcription start were then analyzed for methylation by ligation-mediated genomic sequencing (Pfeifer et al., 1989). Representative results are shown in Figure 2. The region analyzed is a CpG island with 61 CpG sites. The genomic sequencing study revealed that: (i) prior to treatment with 5-azacytidine, 60 out of 61 of the CpG dinucleotides were fully

A. Before 5-Azacytidine

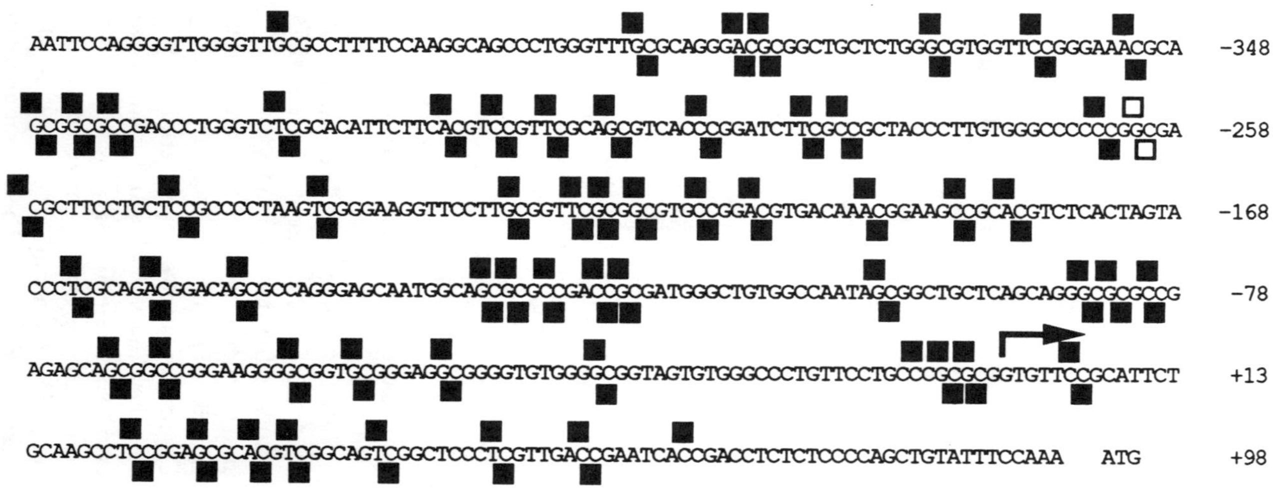

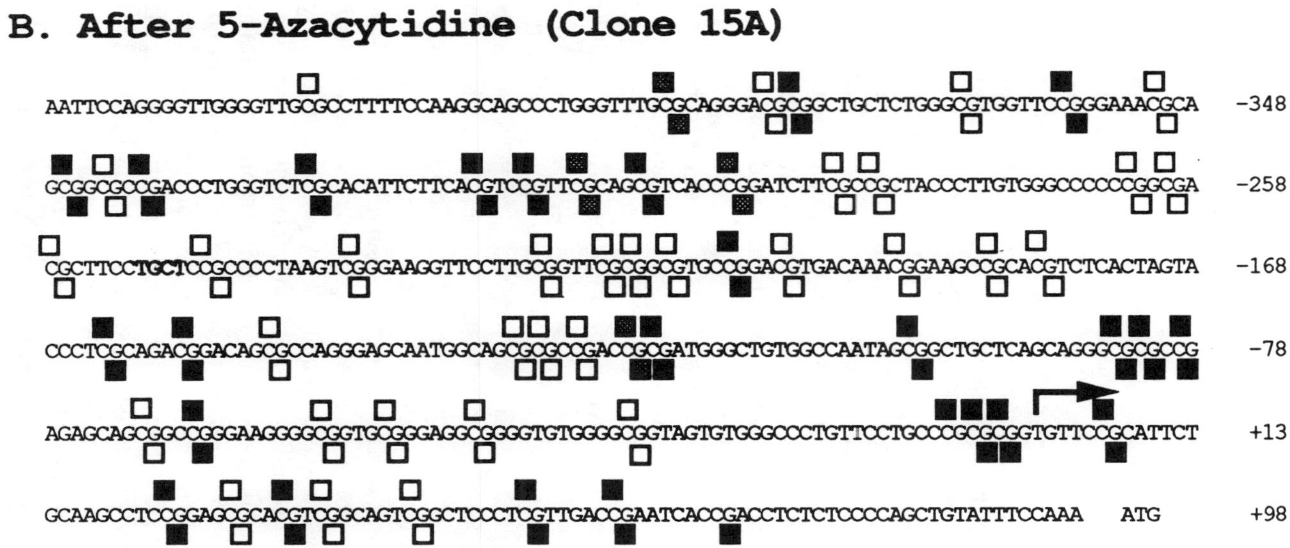

Figure 2. Methylation at the promoter of *PGK1* on an X^i before and after treatment with 5-azacytidine. (A) The methylation pattern prior to 5-azacytidine treatment. Black squares, methylated C; open squares, unmethylated C. (B) The methylation pattern in one clone derived after treatment with 5-azacytidine for one day. Shaded squares are partially methylated Cs. DNA was isolated from this clone after growth for at least 25 generations without any selection for PGK. Adapted from Pfeifer et al. (1990).

methylated; (ii) only CpGs are detectably methylated; (iii) a clone showing full reactivation for the expression of human PGK is fully unmethylated; (iv) clones not expressing human PGK remain methylated at about 50% of CpG sites, with a pattern of interspersed methylated (M) and unmethylated (U) sites different for each clone; (v) singles, defined as M-U-M or U-M-U, are common; and (vi) a few CpG sites are partially methylated.

The existence of frequent "singles" (M-U-M or U-M-U) at least *25 generations after cloning* is strong evidence that methylation is not highly cooperative in the clustered CpG sites in the upstream region of the *PGK* gene, either for "repair" methylation, which presumably takes place at the hemimethylated sites generated by 5-azacytidine treatment, or for *de novo* methylation during growth. For this reason, analysis of the data was made assuming considerable CpG site autonomy. The mosaic patterns seen were different for each clone, suggesting clonal inheritance, and subcloning and methylation-sensitive restriction analysis after 25–30 generations indicated that patterns were for the most part maintained. However, some *de novo* methylation could be detected in most of the subclones. These results establish that maintenance predominates in human-hamster hybrid cells, but, in addition, *de novo* methylation occurs at a significant rate.

Estimates of E_m and E_d were made as follows. A key experimental observation is that before 5-azacytidine treatment, full methylation was seen for all but one of the sites in the PGK promoter region. In genomic sequencing autoradiograms the loss of methylation gives a positive signal, so 20 percent demethylation at any given site during culture after cloning would have been easily apparent. Modeling analysis using the equations in Figure 1 indicates that an average E_m of 90% or less is clearly inconsistent with experimental results, since the methylation level, $M^\wedge$, would drop to 50% for each site by thirty generation (Pfeifer et al., 1990a). Thus E_m must be better than 90% for most of the sites. A rough estimate for E_d was derived from the observation that 10% of the measured CpG sites do, in fact, show partial methylation. The detection of partial methylation at a specific site by genomic sequencing is limited by background, probably due to random cleavage or termination during various steps of the procedure. For this reason, only sites at least 20% methylated in *both strands* were classified as partially methylated sites. To be this highly methylated, the *de novo* methylation event must have occurred in the first two or three divisions after cloning. It was found that 15 *de novo* events occurred in a potential of 150 sites during 2–3 generations; thus $E_d \approx 5\%$ methylation per site per generation for this human CpG island in Chinese hamster cells.

Having an estimate for E_d, a different approach to the minimum estimate for E_m was made by using equation 4 of Figure 1. The genomic sequencing data indicate that most sites are more than 80% methylated, that is, $M^\wedge \geq 0.8$. For $M^\wedge = 0.8$ and $E_d = 0.05$, $E_m = 0.9875$. This approach to the estimation of E_m is applicable for each site and thus the *average* maintenance efficiency for the entire region is at least 98.75%. An earlier study (Steigerwald et al., 1990) assayed two restriction sites in this CpG island and found them to be at least 98% methylated; thus by equation 4, $E_m = 99.9\%$. The Otto-Walbot treatment (Otto and Walbot, 1990) gives a similar estimate. In summary, analysis of the *PGK* promoter on the inactive X chromosome in human-Chinese hamster hybrid cells indicates that (i) mCpG is maintained with very good fidelity and (ii) *de novo* methylation occurs at a very significant rate. Some of the implications of these results will be discussed in the next section.

7 The maintenance of methylation and chromatin differentials

The maintenance parameters E_m and E_d determine whether, with time or population growth, a specific site becomes methylated, unmethylated, or partially methylated. It should be appreciated that stable partial methylation is not necesarily indicative of a mixture of cell types but could be merely a consequence of a particular set of E_m and E_d values. For example, if E_m is 0.90 and E_d is 0.10, the steady state level is 50% methylation.

The maintenance of the methylation differential between the X^a and X^i illustrates several points. The PGK promoter on the X^a is completely unmethylated, free of nucleosomes, and shows DMS (Pfeifer et al., 1990b), DNase I (Pfeifer and Riggs, 1991), and UV footprints (Pfeifer et al., 1991; Pfeifer et al., 1992) for eight putative transcription factors. The same sequence on the X^i is methylated at 60 out of 61 CpG sites, wrapped around two phased nucleosomes (or nucleosome-sized particles), and shows no footprints. Why does the X^i stay transcriptionally inactive? This question can be at least partially answered if we assume that high methylation of the promoter region precludes transcription. Given the experimentally determined estimates of E_m and E_d, a methylated PGK promoter will forever be methylated. In this sense, the maintenance of X inactive state can now be explained. Errors in methylation maintenance will not accumulate because they will be "repaired" by the *de novo* methylation that occurs at about 5% per CpG site per generation. A more difficult question and perhaps now the key question is: Why is the PGK promoter on the X^a not methylated? One reasonable possibility is that the binding of transcription factors would decrease E_m by interfering with maintenance methylase action during DNA replication, and/or decrease E_d by inhibiting *de novo* methylation. Another intriguing and experimentally testable idea is that the maintenance parameters in early S phase, when the X^a replicates, are different from the maintenance parameters in late S phase, when the X^i replicates.

8 Replication timing, methylation and X inactivation

As recently reviewed by Riggs and Pfeifer (1992), it has become increasingly likely that late replication (replication in the second half of S phase) is not just a trivial consequence of X inactivation, but rather is perhaps the ancestral and thus the most fundamental mechanism for X inactivation. Studies with 5-azacytidine have shown that changes in methylation are correlated with changes in replication timing (Shafer and Priest, 1984; Gregory et al., 1985; Jablonka et al., 1985; Schmidt et al., 1985). When an X^i is entirely reactivated, it becomes early replicating, and the piecemeal reactivation usually seen also is associated with regional changes in replication timing. Replication timing could bring about the temporal separation of the X^a from the X^i; in effect, the two X chromosomes would replicate in two different environments, and this could aid in the maintenance of differential methylation and chromatin structure. Methylation could also be involved in the control of replication origins. In both prokaryotes and eukaryotes there is evidence for this notion, which was first discussed by Taylor (1984). DNA modification is apparently used as part of the mechanism for controlling the timing of the reinitiation of *E. coli* replication (Landoulsi et al., 1990) and cytosine methylation may determine replication origin usage in *Physarum* ribosomal genes (Cooney et al., 1988). Mammalian replication origins have only recently begun to be identified, and so far no X-linked origins have been characterized. It will be of interest to see if there is a methylation differential between X^a and X^i origins.

9 Fragile X and DNA methylation

An emerging example of a correlation between abnormal genetic silence, late DNA replication, and methylation is seen at the fragile X site, located at Xq27.3. Fragile X mental retardation is the most common form of heritable mental retardation, with a frequency in males of about 5×10^{-3}. This syndrome has long been known to show a very unusual pattern of heredity, and has been a genetic puzzle as well as a significant medical problem. Normal males carrying a "premutation" pass the syndrome, not to their children, but to their grandchildren, more specifically, only to their daughter's children. It seems clear that passage of the premutation through the female germ line is necessary for manifestation of the syndrome. In 1987 Laird proposed an attractive explanation (Laird, 1987). The essence of his proposal was that the premutation prevents the Xq27.3 region from reactivating during oogenesis. He further suggested that the lack of reactivation was due to abnormally late replication for the Xq27.3 region and retention of the X^i methylation pattern. In support of the model, the region of the affected locus is indeed late replicating even on the X^a (Yu et al., 1990). Also there is now evidence for an inverse correlation between methylation of the fragile X locus and manifestation of the syndrome (Bell et al., 1991; Vincent et al., 1991). How methylation and late replication contribute to the general picture remains to be determined, but the nature of the premutation has recently been elucidated (Fu et al., 1991; Oberlé et al., 1991; Verkerk et al., 1991; Yu et al., 1991). Fu et al. (1991) have established that the fragile X syndrome results from mutations in a $(CGG)_n$ repeat found in the coding region of the FMR-1 gene, a gene that is expressed mainly in the brain. In the normal population, the CGG repeat number varies considerably but is usually close to 30. In affected individuals the repeat number is usually 200 or more. Premutations showing no phenotypic effect in fragile X families range in size from 52 to over 200 repeats. All alleles with greater than 52 repeats are unstable on passage through oogenesis, usually with an increase in repeat number. FMR-1 is not expressed in cells having large repeat numbers and this probably explains the disease. Large alleles are probably also mitotically unstable, because some affected individuals are somatic cell mosaics, with some cells having reduced repeat numbers and some expression of FMR-1. The methylation status of the repeat units is not yet known, but if mosaicism is taken into account, there is an excellent inverse correlation between FMR-1 transcription and the methylation of a CpG island located 250 bp upstream of the repeat (Bell et al., 1991; Fu et al., 1991; Vincent et al., 1991). Nothing more is yet known about the connection between late replication, methylation, and transcriptional silencing, but the fragile X system promises to be a good one for addressing these questions. Methylation stabilizes duplex DNA, so a reasonable and experimentally testable model, in keeping with Laird's earlier proposal, is that methylation of the premutation, such as may occur in the premeiotic X^i, would cause stalling of DNA replication and thus abnormally late completion of replication, with retention of the X^i chromatin state for the Xq27.3 replication domain. Retention of the methylated X^i state would insure future meiotic and mitotic instability.

10 The spreading of X chromosome inactivation and methylation

Because segments of the X chromosome that are physically separated from the X inactivation center fail to inactivate, XCI seems to be a positive process brought about by some event that travels along the chromosome (Lyon et al., 1986). Is this process methylation? DNA methyltransferase may be a processive enzyme, not freely transferring from one molecule to another, but rather "walking" down one DNA molecule, methylating as is goes (Bestor and Ingram, 1983; Drahovsky and Pfeifer, 1988). Methylation also has been reported to initiate in the approximate middle of integrated adenovirus genomes and to spread slowly to cover all but the ends of the virus upon repeated passages in cell culture (Orend et al., 1991). Integrated retroviruses often become methylated, and this methylation can extend into flanking cellular sequences (Jahner and Haenisch, 1985). For these reasons methylation should be, and has been, considered as a possible mechanism aiding or causing the spread of XCI. In this context it is interesting to note that the timing of methylation of the only two genes that have been studied at the time of initiation of XCI is compatible with the methylation spreading hypothesis.

The above mentioned results and ideas about processivity suggest that the methylation of one CpG might increase the probability of *de novo* methylation of adjacent CpGs. However, the actual evidence for methylation cooperation is minimal. The *in vitro* data indicating processivity of methyltransferase are not definitive because the concentration of DNA in the microenvironment of a DNA macromolecule was not taken into account for competition experiments. The general difficulty in interpreting experiments measuring enzyme transfer between DNA molecules has been recently discussed by Kornberg and Lorch (1991). The results seen for integrated adenoviruses (Orend et al., 1991) may be atypical because in the cases studied a cluster of 10 or more viral genomes integrates. These new "repeat" units may be subjected to methylation-mediated silencing, as repeated genes are in *Neurospora* (Selker and Garrett, 1988; Selker, 1990). Alternatively, temporary improper pairing of homologous sequences between the colinearly integrated repeat units may transiently generate unusual DNA structures that may be more susceptible to *de novo* methylation. *In vitro* studies have shown that *de novo* methylation activity is high on some mispaired or "loose" DNA structures (Smith et al., 1991). Some transgenes are methylated, especially upon transmission through the female germ line, but the propensity for methylation can be a property of the transgene (or transgene cluster), independent of its integration site and the methylation status of the integration site (Engler et al., 1991), so these data are not suggestive of the spread of methylation to (or from) the transgene to surrounding genomic DNA. The only direct *in vivo* data on methylation cooperativity of a single-copy genomic gene has been reported by Pfeifer et al. (1990a), and this study showed that at nucleotide-level-resolution methylation of one CpG does not greatly increase the probability of methylation of adjacent CpGs.

Because of the evidence against high cooperativity of methylation per se, the working hypothesis in this laboratory is that the initial inactivation of a chromosomal domain is not brought about by the spreading of methylation along the DNA stand. It has been proposed (Riggs, 1990a) that spreading of XCI could be from one silenced replication origin to adjacent origins, which will be nearby because of the looping of 30 nm chromatin fibers. In this model methylation of the origin need not be the primary event, but it could be that methylation stabilizes the late replication state of the origin. For autosomal genes and X-linked genes that escape inactivation, methylation of the origin may not occur and then these origins may be more likely to revert to early replication soon after the spreading of inactivation in the early embryo. There is one report indicating that the late replication of autosomal material translocated to a human X chromosome is not clonally stable in culture, but rather reverts to early replication (Schanz and Steinbach, 1989).

11 Future directions: Replication origins, *Xist* and the X inactivation center

It is very likely that none of the key sites for the initiation and spreading of XCI have yet been studied, and thus no current methylation data is relevant to anything but maintenance, with the possible exception of the studies showing that 5-azacytidine causes an advance in replication time of heterochromatin and portions of the X chromosome. There is reason to be optimistic that this situation may change however. As outlined above, replication origins are candidates for key sites for XCI and methods are now becoming available for the identification of mammalian replication origins (Handeli et al., 1989; Burhans et al., 1990). Over the next few years it should be possible to determine the methylation status of mammalian replication origins.

In another direction, it may soon be possible to study methylation at the X chromosome inactivation center, the single locus on the X chromosome at which XCI is believed to be initiated. If the locus (*Xce* in the mouse; XIC in human) is deleted, XCI does not occur (Cattanach and Papworth, 1981; Rastan and Robertson, 1985; Lyon et al., 1986). Recent fine mapping of the *Xce* locus (Brown et al., 1991a; Brown, 1991b) places it in a region that contains the genes *Pgk-1*, *Ccg-1*, and *Ta*. One gene, termed *XIST* in human and *Xist* in mouse, is located in the region of the X-inactivation center, and is expressed only from the inactive X chromosome in man and mouse (Borsani et al., 1991; Brockdorff et al., 1991; Brown et al., 1991a). Furthermore, the degree of expression seems greatest in those mouse strains which in interspecies crosses are most likely to donate the X chromosome which becomes inactivated. Originally the gene was isolated as a cDNA insert that bound to anti-STS antibody in a human expression library. Later both the mouse and human genes were found to map between *Ccg-1* and

Pgk-1. Despite its antigenic activity in the cDNA library from which it was originally isolated, the human cDNA contains no open reading frame (ORF) greater than 300 bp in length; similarly the moust *Xist* was found to have only one ORF, of about 500 bp. Northern blots suggest great heterogeneity in the size of the transcripts, with the largest being about 14.3 kb for the mouse and multiple splice sites for humans. A conserved region has been detected between man and mouse; however this does not correspond to the ORF regions.

What is the relationship of *Xist* to the center of X chromosome inactivation (*Xce*)? While results to date have shown a correlation of *Xist* expression with inactivation of X-linked genes *in cis* (Brockdorff et al., 1991), its exact role is not known (Brown, 1991a). It may, for example, provide a counting mechanism after one X chromosome is activated. Does the *Xist* transcript stabilize the inactive state or does it initiate and stabilize it? Does it exert its effect *in trans*? If it does play a role in XCI, this may give some clues as to how heterochromatin forms in general. As more is learned about the center of XCI and the *Xist* locus, it will become possible to do molecular studies to test the effect of DNA methylation on the initiation of XCI.

Adler, D. A., West, J. D., and Chapman, V. M. (1977) Expression of alpha-galactosidase in preimplantation mouse embryos. Nature *267*, 838–839.

Antequera, F., Macleod, D., and Bird, A. P. (1989) Specific protection of methylated CpGs in mammalian nuclei. Cell *58*, 509–517.

Antequera, F., Boyes, J., and Bird, A. (1990) High levels of *de novo* methylation and altered chromatin structure at CpG islands in cell lines. Cell *62*, 503–514.

Ariel, M., McCarrey, J., and Cedar, H. (1991) Methylation patterns of testis-specific genes. Proc. Natl. Acad. Sci. USA *88*, 2317–2321.

Baker, T. G. (1963) A quantitative and cytological study of germ cells in human ovaries. Proc. Roy. Soc. B. *158*, 417–433.

Bartlett, M. H., Adra, C. N., Park, J., Chapman, V. M., and McBurney, M. W. (1991) DNA methylation of two X chromosome genes in female somatic and embryonal carcinoma cells. Somatic Cell Mol. Genet. *17*, 35–47.

Bell, M. V., Hirst, M. C., Nakahori, Y., MacKinnon, R. N., Roche, A., Flint, T. J., Jacobs, P. A., Tommerup, N., Tranebjaerg, L., Froster-Iskenius, U., Kerr, B., Turner, G., Lindenbaum, R. H., Winter, R., Pembrey, M., Thibodeau, S., and Davies, K. E. (1991) Physical mapping across the fragile X: hypermethylation and clinical expression of the fragile X syndrome. Cell *64*, 861–866.

Bestor, T. H., and Ingram, V. M. (1983) Two DNA methyltransferases from murine erythroleukemia cells: purification, sequence specificity, and mode of interaction with DNA. Proc. Natl. Acad. Sci. USA *80*, 5559–5563.

Bird, A. P. (1986) CpG-rich islands and the function of DNA methylation. Nature *321*, 209–213.

Borden, J., and Manuelidis, L. (1988) Movement of the X chromosome in epilepsy. Science *242*, 1687–1691.

Borsani, G., Tonlorenzi, R., Simmler, M. C., Dandolo, L., Arnaud, D., Capra, V., Grompe, M., Pizzuti, A., Muzny, D., Lawrence, C., Willard, H. F., Avner, P., and Ballabio, A. (1991) Characterization of a murine gene expressed from the inactive X chromosome. Nature *351*, 325–329.

Boyes, J., and Bird, A. (1991) DNA methylation inhibits transcription indirectly via a Methyl-CpG Binding Protein. Cell *64*, 1123–1134.

Boyes, J., and Bird, A. (1992) Repression of genes by methylation depends on CpG density and promoter strength: evidence for involvement of a methyl-CpG binding protein. EMBO J. *11*, 327–333.

Brockdorff, N., Ashworth, A., Kay, G. F., Cooper, P., Smith, S., McCabe, V. M., Norris, D. P., Penny, G. D., Patel, D., and Rastan, S. (1991) Conservation of position and exclusive expression of mouse *Xist* from the inactive X chromosome. Nature *351*, 329–331.

Brown, S. D. M. (1991) *XIST* and the mapping of the X chromosome inactivation center. BioAssays *13*, 607–611.

Brown S., and Rastan, S. (1988) Age-related reactivation of an X-linked gene close to the inactivation centre in the mouse. Genet. Res. *52*, 1512–1514.

Brown, C. J., Ballabio, A., Rupert, J. L., Lafreniere, R. G., Grompe, M., Tonlorenzi, R., and Willard, H. F. (1991a) A gene from the region of the human X inactivation centre expressed exclusively from the inactive chromosome. Nature *349*, 38–44.

Brown, C. J., Lafreniere, R. G., Powers, V. E., Sebastio, G., Ballabio, A., Pettigrew, A. L., Ledbetter, D. H., Levy, E., Craig, I. W., and Huntington, H. F. (1991b) Localization of the X inactivation centre on the human X chromosome in Xq13. Nature *349*, 82–84.

Burhans, W. C., Vassilev, L. T., Caddle, M. S., Heintz, N. H., and DePamphilis, M. L. (1990) Identification of an origin of bidirectional DNA replication in mammalian chromosomes. Cell *62*, 955–965.

Cattanach, B. M., and Papworth, D. (1981) Controlling elements in the mouse V. Linkage tests with X-linked genes. Genet. Res. Camb. *38*, 57–70.

Chaillet, J. R., Vogt, T. F., Beier, D. R., and Leder, P. (1991) Parental-specific methylation of an imprinted transgene is established during gametogenesis and progressively changes during embryogenesis. Cell *66*, 77–83.

Cooney, C. A., and Bradbury, E. M. (1990) DNA methylation and chromosome organization in eucaryotes, in: The Eukaryotic Nucleus: Molecular Biochemistry and Macromolecular Assemblies, pp. 813–843. Eds. P. R. Strauss and S. H. Wilson. Telford Press, Caldwell, NJ, USA.

Cooney, C. A., Eykholt, R. L., and Bradbury, E. M. (1988) Methylation is coordinated on the putative replication origins of *Physarum* rDNA. J. Molec. Biol. *204*, 889–901.

Drahovsky, D., and Pfeifer, G. P. (1988) Enzymology of DNA methylation in mammalian cells, in: Architecture of Eukaryotic Genes, pp 435–445. Ed. G. Kahl. VCH Verlagsgesellschaft mbH, Weinheim.

Driscoll, D. J., and Migeon, B. R. (1990) Sex difference in methylation of single-copy genes in human meiotic germ cells: Implications for X chromosome inactivation, parental imprinting, and origin of CpG mutations. Somatic Cell Mol. Genet. *16*, 267–282.

Dyer, K. A., Canfield, T. K., and Gartler, S. M. (1989) Cytogenet. Cell Genet. *50*, 116.

Engler, P., Haasch, D., Pinkert, C. A., Doglio, L., Glymour, M., Brinster, R., and Storb, U. (1991) A strain-specific modifier on mouse chromosome 4 controls the methylation of independent transgene loci. Cell *65*, 939–947.

Epstein, C. J., Smith, S., Travis, B., and Tucker, G. (1978) Both X chromosomes function before visible X-chromosome inactivation in female mouse embryos. Nature *274*, 500–502.

Fowlis, D. J., Ansel, J. D., and Micklem, H. S. (1991) Further evidence for the importance of parental source of the *Xce* allele in X chromosome inactivation. Genet. Res. *58*, 63–65.

Frank, D., Keshet, I., Shani, M., Levine, A., Razin, A., and Cedar, H. (1991) Demethylation of CpG islands in embryonic cells. Nature *351*, 239–241.

Fu, Y., Kuhl, D. P. A., Pizzuti, A., Pieretti, M., Sutcliffe, J. S., Richards, S., Verkerk, A. J. M. H., Holden, J. J. A., Fenwick, R. G., Warren, S. T., Oostra, B. A., Nelson, D. L., and Caskey, C. T. (1991) Variation of the CGG repeat at the fragile X site in genetic instability: Resolution of the Sherman paradox. Cell *67*. 1047–1058.

Gamma-Sossa, M. A., Midgett, R. M., Slagel, V. A., Githens, S., Kuo, K. C., Gehrke, C. W., and Ehrlich, M. (1983) Tissue-specific differences in DNA methylation in various mammals. Biochim. Biophys. Acta *740*, 212–219.

Gartler, S. M., and Riggs, A. D. (1983) Mammalian X-chromosome inactivation. Ann. Rev. Genet. *17*, 155–190.

Ghazi, H., Gonzales, F., and Jones, P. A. (1991) Methylation of CpG-island-containing genes in human sperm, fetal and adult tissues. Gene *114*, 203–210.

Grant, S. G., and Chapman, V. M. (1988) Mechanisms of X-chromosome regulation. Ann. Rev. Genet. *22*, 199–233.

Grant, S. G., and Worton, R. G. (1989a) Activation of the *Hprt* gene on the inactive X chromosome in transformed diploid female Chinese hamster cells. J. Cell Sci. *92*, 723–732.
Grant, S. G., and Worton, R. G. (1989b) Differential activation of the hprt gene on the inactive X chromosome in primary and transformed Chinese hamster cells. Mol. Cell. Biol. *9*, 1635–1641.
Gregory, P., Greene, C., Shapira, E., and Wang, N. (1985) Alterations in the time of X chromosome replication induced by 5-azacytidine. Cytogenet. Cell Genet. *39*, 234–236.
Groudine, M., and Conkin, K. F. (1985) Chromatin structure and *de novo* methylation of sperm DNA: implications for activation of the paternal genome. Science *228*, 1061–1068.
Grünwald, S., and Pfeifer, G. P. (1989) Enzymatic DNA methylation. Progr. Clin. Biochem. Med. *9*, 61–103.
Handeli, S., Klar, A., Meuth, M., and Cedar, H. (1989) Mapping replication units in animal cells. Cell *57*, 909–920.
Hansen, R. S., and Gartler, S. M. (1990) 5-Azacytidine-induced reactivation of the human X chromosome-linked *PGK1* gene is associated with a large region of cytosine demethylation in the 5′ CpG island. Proc. Natl. Acad. Sci. USA *87*, 4147–4178.
Hansen, R. S., Ellis, N. A., and Gartler, S. M. (1988) Demethylation of specific sites in the 5′ region of the inactive X-linked human phosphoglycerate kinase gene correlates with the appearance of nuclease sensitivity and gene expression. Mol. Cell. Biol. *8*, 4692–4699.
Harper, M. I., Monk, M., and Fosten, M. (1981) Preferential paternal X-inactivation in extra-embryonic tissues of early mouse embryos. J. Embryol. exp. Morph. *67*, 127–135.
Hockey, A. J., Adra, C. N., and McBurney, M. W. (1989) Reactivation of *Hprt* on the inactive X chromosome with DNA demethylating agents. Som. Cell Mol. Genet. *15*, 421–434.
Holliday, R. (1989) A different kind of inheritance. Sci. Amer. *260*, 60–73.
Holliday, R. (1991) Is DNA methylation of X chromosome genes stable during aging? Somatic Cell Mol. Genet. *17*, 101–102.
Holliday, R., and Pugh, J. E. (1975) DNA modification methanisms and gene activity during development. Science *187*, 226–232.
Homman, N., Heuertz, S., and Hors, C. M. C. (1987) Time-dependence of X-gene reactivation induced by 5-azacytidine; possible progressive restructuring of chromatin. Exp. cell. Res. *172*, 481–486.
Howlett, S. K., and Reik, W. (1991) Methylation levels of maternal and paternal genomes during preimplantation development. Development *113*, 119–127.
Jablonka, E., Goitein, R., Marcus, M., and Cedar, H. (1985) DNA hypomethylation causes an increase in DNase I sensitivity and an advance in the time of replication of the entire inactive X chromosome. Chromosoma *93*, 152–156.
Jaenisch, R., and Jähner, D. (1984) Methylation, expression and chromosomal position of genes in mammals. Biochim. Biophys. Acta *782*, 1–9.
Jahner, D., and Jaenisch, R. (1985) Retrovirus-induced *de novo* methylation of flanking host sequences correlate with gene activity. Nature *315*, 594–597.
Johnston, P. G., and Cattanach, B. M. (1981) Controlling elements in the mouse IV. Evidence of non-random X-inactivation. Genet. Res. Camb. *37*, 151–160.
Jones, P. A., and Buckley, J. D. (1990) The role of DNA methylation in cancer. Adv. Cancer Res. *54*, 1–24.
Jones, P. A. Wolkowicz, M. J., Rideout, I. W. M., Gonzales, F. A., Marziasz, C. M., Coetzee, G. A., and Tapscott, S. J. (1990) *De novo* methylation of the MyoD1 CpG island during the establishment of immortal cell lines. Proc. Natl. Acad. Sci. USA *87*, 6117–6121.
Kanda, N. (1973) A new differential technique for staining the heteropycnotic X chromosome is female mice. Expl. Cell Res. *80*, 463–467.
Kornberg, R. D., and Lorch, Y. (1991) Irresistible force meets immovable object: transcription and the nucleosome. Cell *67*, 833–836.
Kratzer, P. G., and Chapman, V. M. (1981) X chromosome reactivation in oocytes of *Mus caroli*. Proc. Natl. Acad. Sci. USA *78*, 3093–3097.
Kratzer, P. G., and Gartler, S. M. (1978) HGPRT activity changes in preimplantation mouse embryos. Nature *274*, 503–504.
Kratzer, P. G., Chapman, V. M., Lambert, H., Evans, R. E., and Liskay, R. M. (1983) Differences in the DNA of the inactive X chromosomes of fetal and extraembryonic tissues of mice. Cell *33*, 37–42.

Laird, C. D. (1987) Proposed mechanism of inheritance and expression of the human fragile-X syndrome of mental retardation. Genetics *117*, 587–599.

Landoulsi, A., Malki, A., Kern, R., Kohiyama, M., and Hughes, P. (1990) The *E. coli* cell surface specifically prevents the initiation of DNA replication at oriC on hemimethylated DNA templates. Cell *63*, 1053–1060.

Lewis, J., and Bird, A. (1991) DNA methylation and chromatin structure. FEBS Lett. *285*, 155–159.

Lin, D., and Chinault, A. C. (1988) Comparative study of DNase I sensitivity at the X-linked human *HPRT* locus. Somat. Cell Mol. Genet. *14*, 261–272.

Lindsay, S., Monk, M., Holliday, R., Huschtscha, L., Davies, K. E., Riggs, A. D., and Flavell, R. A. (1985) Differences in methylation on the active and inactive human X chromosomes. Ann. Hum. Genet. *49*, 115–127.

Lock, L. F., Takagi, N., and Martin, G. R. (1987) Methylation of the *Hprt* gene on the inactive X occurs after chromosome inactivation. Cell *48*, 39–46.

Lyon, M. F. (1988) The William Allan Memorial Award address: X-chromosome inactivation and the location and expression of X-linked genes. Am. J. Hum. Genet. *42*, 8–16.

Lyon, M. F., Zenthon, J., Evans, E. P., Burtenshaw, M. D., Wareham, K. A., and Williams, E. D. (1986) Lack of inactivation of a mouse X-linked gene physically separated from the inactivation center. J. Embryol. Exp. Morph. *97*, 75–85.

Manuelidis, L. (1990) A view of interphase chromosomes. Science *250*, 1533–1540.

McMahon, A., and Monk, M. (1983) X-chromosome activity in female mouse embryos heterozygous for *Pgk-1* and Searle's translocation, T(X;16)16H. Genet. Res. Camb. *41*, 69–83.

Migeon, B. R., Schmidt, M., Axelman, J., and Cullen, C. R. (1986) Complete reactivation of X chromosomes from human chorionic vili with a switch to early DNA replication. Proc. Natl. Acad. Sci. USA *83*, 2182–2186.

Migeon, B. R., Axelman, J., and Beggs, A. H. (1988) Effect of ageing on reactivation of the human X-linked *HPRT* locus. Nature *335*, 93–96.

Migeon, B. R., de Beur, S. J., and Axelman, J. (1989) Frequent depression G6PD and *HPRT* on the marsupial inactive X chromosome associated with cell proliferation *in vitro*. Exp. Cell Res. *182*, 597–609.

Migeon, B. R., Holland, M. M., Driscoll, D. J., and Robinson, J. C. (1991) Programmed demethylation in CpG islands during human fetal development. Somatic Cell Mol. Genet. *17*, 159–168.

Mohandas, T., Geller, R. L., Yen, P. H., Rosendorff, J., Bernstein, R., Yoshida, A., and Shapiro, L. J. (1987) Cytogentic and molecular studies on a recombinant human X chromosome: implications for the spreading of X-chromosome inactivation. Proc. Natl. Acad. Sci. USA *84*, 4954–4958.

Monesi, V. (1965) Synthetic activities during spermatogenesis in the mouse. Exp. Cell Res. *39*, 197–224.

Monesi, V., Geremia, R., D'Agostino, A., and Boitani, C. (1978) Biochemistry of male germ cell differentiation in mammals: RNA synthesis in meiotic and postmeiotic cells, in: Current Topics in Developmental Biology, Vol. 12, pp. 11–36. Eds. A. A. Moscone and A. Monroy. Academic Press, New York.

Monk, M. (1986) Methylation and the X chromosome. BioEssays *4*, 204–208.

Monk, M. (1987) Memories of mother and father (genomic imprinting). Nature *328*, 203–204.

Monk, M., Boubelik, M., and Lehnert, S. (1987) Temporal and regional changes in DNA methylation in the embryonic, extraembryonic, and germ-cell lineages during mouse embryo development. Development *99*, 371–382.

Mukherjee, A. B. (1976) Cell cycle analysis and X-chromosome inactivation in the developing mouse. Proc. Natl. Acad. Sci. USA *73*, 1608–1611.

Oberlé, I., Rousseau, F., Heitz, D., Kretz, C., Devys, D., Hanauer, A., Boué, J., Bertheas, M. F., and Mandel, J. L. (1991) Instability of a 550-base pair DNA segment and abnormal methylation in fragile X syndrome. Science *252*, 1097–1102.

Orend, G., Kuhlmann, I., and Doerfler, W. (1991) Spreading of DNA methylation across integrated foreign (adenovirus type 12) genomes in mammalian cells. J. Virol. *65*, 4301–4308.

Otto, S. P., and Walbot, V. (1990) DNA methylation in eukaryotes; kinetics of demethylation and de novo methylation during the life cycle. Genetics *124*, 429–437.

Pagani, F., Toniolo, D., and Vergani, C. (1990) Stability of DNA methylation of X-chromosome genes during aging. Somat. Cell Mol. Genet. *16*, 101–103.

Paroush, Z., Keshet, I., Yisraeli, J., and Cedar, H. (1990) Dynamics of demethylation and activation of the alpha-actin gene in myoblasts. Cell *63*, 1229–1237.

Paterno, G. D., Adra, C. N., and McBurney, M. W. (1985) X chromosome reactivation in mouse embryonal carcinoma cells. Mol. Cell. Biol. *5*, 2705–2712.

Pfeifer, G. P., and Drahovsky, D. (1986) Preferential binding of DNA methyltransferase and increased *de novo* methylation of deoxyinosine containing DNA. FEBS Lett. *207*, 75–78.

Pfeifer, G. P., and Riggs, A. D. (1991) Chromatin differences between active and inactive X chromosomes revealed by genomic footprinting of permeabilized cells using DNase I and ligation-mediated PCR. Genes Dev. *5*, 1102–1113.

Pfeifer, G. P., Steigerwald, S. D., Mueller, P. R., Wold, B., and Riggs, A. D. (1989) Genomic sequencing and methylation analysis by ligation mediated PCR. Science *246*, 810–813.

Pfeifer, G. P., Steigerwald, S. D., Hansen, R. S., Gartler, S. M., and Riggs, A. D. (1990a) Polymerase chain reaction-aided genomic sequencing of an X chromosome-linked CpG island: Methylation patterns suggest clonal inheritance, CpG site autonomy, and an explanation of activity state stability. Proc. Natl. Acad. Sci. USA *87*, 8252–8256.

Pfeifer, G. P., Tanguay, R. L., Steigerwald, S. D., and Riggs, A. D. (1990b) *In vivo* footprint and methylation analysis by PCR-aided genomic sequencing: comparison of active and inactive X chromosomal DNA at the CpG island and promoter of human *PGK-1*. Genes Dev. *4*, 1277–1287.

Pfeifer, G. P., Drouin, R., Riggs, A. D., and Holmquist, G. P. (1991) *In vivo* mapping of a DNA adduct at nucleotide resolution: Detection of pyrimidine (6-4) pyrimidone photoproducts by ligation-mediated polymerase chain reaction. Proc. Natl. Acad. Sci. USA *88*, 1374–1378.

Pfeifer, G. P., Drouin, R., Riggs, A. D., and Holmquist, G. P. (1992) Binding of transcription factors creates hot spots for UV photoproducts *in vivo*. Mol. Cell Biol. *12*, 1798–1804.

Prantera, G., and Ferraro, M. (1990) Analysis of methylation and distribution of CpG sequences on human active and inactive X chromosomes by *in situ* nick translation. Chromosoma *99*, 18–23.

Rastan, S., and Robertson, E. J. (1985) X-chromosome deletions in embryo-derived (EK) cell lines associated with lack of X-chromosome inactivation. J. Embryol. Exp. Morph. *90*, 379–388.

Razin, A., and Cedar, H. (1991) DNA methylation and gene expression. Microbiol. Rev. *55*, 451–458.

Razin, A., Szyf, M., Kafri, T., Rolle, M., Giloh, H., Scarpa, S., Carotti, D., and Cantoni, G. L. (1986) Replacement of 5-methylcytosine by cytosine: a possible mechanism for transient DNA demethylation during differentiation. Proc. Natl. Acad. Sci. USA *83*, 2827–2831.

Riggs, A. D. (1975) X chromosome inactivation, differentiation and DNA methylation. Cytogenet. Cell Genet. *14*, 9–25.

Riggs, A. D. (1989) DNA methylation and cell memory. Cell Biophysics *15*, 1–13.

Riggs, A. D. (1990a) DNA methylation and late replication probably aid cell memory, and type I DNA reeling could aid chromosome folding and enhancer function. Phil. Trans. R. Soc. Lond. *326*, 285–297.

Riggs, A. D. (1990b) Marsupials and mechanisms of X chromosome inactivation. Austr. J. Zool. *37*, 419–441.

Riggs, A. D., and Jones, P. A. (1983) Methylcytosine, gene regulation, and cancer. Adv. Cancer Res. *40*, 1–30.

Riggs, A. D., and Pfeifer, G. P. (1992) X chromosome inactivation and cell memory. Trends Genet. *8*, 169–174.

Riley, D. E., Goldman, M. A., and Gartler, S. M. (1986) Chromatin structure of active and inactive human X-linked phosphoglycerate kinase gene. Som. Cell Mol. Genet. *12*, 73–80.

Sanford, J. P., Clark, H. J., Chapman, V. M., and Rossant, J. (1987) Differences in DNA methylation during oogenesis and spermatogenesis, and their persistence during early embryogenesis in the mouse. Genes Dev. *1*, 1039–1046.

Schanz, S., and Steinbach, P. (1989) Investigation of the "variable spreading" of X inactivation into a translocated autosome. Hum. Genet. *82*, 244–248.

Schmidt, M., Wolf, S. F., and Migeon, B. R. (1985) Evidence for a relationship between DNA methylation and DNA replication from studies of the 5-azacytidine-reactivated allocyclic X chromosome. Exp. Cell Res. *158*, 301–310.

Selker, E. U. (1990) DNA methylation and chromatin structure: A view from below. Trends Biochem. Sci. *15*, 103–107.

Selker, E. U., and Garrett, P. W. (1988) DNA sequence duplications trigger gene inactivation in *Neurospora crassa*. Proc. Natl. Acad. Sci. USA *85*, 6870–6874.

Shafer, D. A., and Priest, J. H. (1984) Reversal of DNA methylation with 5-azacytidine alters chromosome replication patterns in human lymphocyte and fibroblast cultures. Am. J. Hum. Genet. *36*, 534–545.

Shemer, R., Kafri, T., O'Connell, A., Eisenberg, S., Breslow, J. L., and Razin, A. (1991) Methylation changes in the apolipoprotein AI gene during embryonic development of the mouse. Proc. Natl. Acad. Sci. USA *88*, 11300–11304.

Singer, J., Roberts-Ems, J., Luthardt, F. W., and Riggs, A. D. (1979) Methylation of DNA in mouse early embryos, teratocarcinoma cells and adult tissues of mouse and rabbit. Nucl. Acids Res. *7*, 2369–2385.

Singer-Sam, J., Grant, M., LeBon, J. M., Okuyama, K., Chapman, V., Monk, M., and Riggs, A. D. (1990a) Use of a HpaII-polymerase chain reaction assay to study DNA methylation in the *Pgk-1* CpG island of mouse embryos at the time of X-chromosome inactivation. Mol. Cell. Biol. *10*, 4987–4989.

Singer-Sam, J., Robinson, M. O., Bellvé, A. R., Simon, M. I., and Riggs, A. D. (1990b) Measurement by quantitative PCR of changes in *HPRT*, *PGK-1*, *PGK-2*, APRT, MTase, and ZFY gene transcripts during mouse spermatogenesis. Nucl. Acids Res. *18*, 1255–1259.

Singer-Sam, J., Goldstein, L., Dai, A., Gartler, S. M., and Riggs, A. D. (1992) A potentially critical Hpa II site of the X chromosome-linked *PGK1* gene is unmethylated prior to the onset of meiosis of human oogenic cells. Proc. Natl. Acad. Sci. USA *89*, 1413–1417.

Smith, S. S., Kan, J., Baker, D. J., Kaplan, B. E., and Dembek, P. (1991) Recognition of unusual DNA structures by human DNA(cystosine-5)methyltransferase. J. Mol. Biol. *217*, 39–51.

Steigerwald, S. D., Pfeifer, G. P., and Riggs, A. D. (1990) Ligation-mediated PCR improves the sensitivity of methylation analysis by restriction enzymes and detection of specific DNA strand breaks. Nucl. Acids Res. *18*, 1435–1439.

Sugawara, O., Takagi, N., and Sasaki, M. (1985) Correlation between X-chromosome inactivation and cell differentiation in female preimplantation mouse embryos. Cytogenet. Cell Genet. *39*, 210–219.

Sullivan, C. H., and Grainger, R. M. (1986) Delta-crystallin genes become hypomethylated in postmitotic lens cells during chicken development. Proc. Natl. Acad. Sci. USA *83*, 329–333.

Szyf, M., Schimmer, B. P., and Seidman, J. G. (1989) Nucleotide-sequence-specific *de novo* methylation in a somatic murine cell line. Proc. Natl. Acad. Sci. USA *86*, 6853–6857.

Takagi, N. (1988) Requirement of mitoses for the reversal of X inactivation in cell hybrids between murine embryonal carcinoma cells and normal female thymocytes. Exp. Cell. Res. *175*, 363–375.

Takagi, N., and Sasaki, M. (1975) Preferential inactivation of the paternally derived X chromosome in the extraembryonic membranes of the mouse. Nature *256*, 640–642.

Takagi, N., Yoshida, M. A., Sugawara, O., and Sasaki, M. (1983) Reversal of X inactivation in female mouse somatic cells hybridized with murine teratocarcinoma stem cells in vitro. Cell *34*, 1053–1062.

Taylor, J. H. (1984) Origins of replication and gene regulation. Mol. Cell. Biochem. *61*, 99–109.

Van Dyke, D. L., Flejter, W. L., Worsham, M. J., Roberson, J. R. Higgins, J. V., Herr, H. M., Knuutila, S., Wang, N., Babu, V. R., and Weiss, L. (1986) A practical metaphase marker of the inactive X chromosome. Am. J. Human Genet. *39*, 88–95.

VandeBerg, T. L., Robinson, E. S., Samollow, P. G., and Johnson, P. G. (1987) X-linked gene expression and X-chromosome inactivation: Marsupials, mouse, and man compared, in: Isozymes: Current Topics in Biological and Medical Research, pp. 225–253. Ed. C. L. Markert. Alan R. Liss, New York.

Venolia, L., Cooper, D. W., O'Brien, D. A., Millette, C. F., and Gartler, S. M. (1984) Transformation of the *Hprt* gene with DNA from spermatogenic cells. Chromosoma *90*, 185–189.

Verkerk, A. J. M. H., Pieretti, M., Sutcliffe, J. S., Fu, Y.-H., Kuhl, D. P. A., Pizzuti, A., Reiner, O., Richards, S., Victoria, M. F., Zhang, F., Eussen, B. E., Ommen, G.-J. B. v., Blonden, L. A. J., Riggins, G. J., Chastain, J. L., Kunst, C. B., Galjaard, H., Caskey, C. T., Nelson, D. L., Oostra, B. A., and Warren, S. T. (1991) Identification of a gene (FMR-1) containing a CGG repeat coincident with a breakpoint cluster region exhibiting length variation in fragile X syndrome. Cell *65*, 905–914.

Vincent, A., Heitz, D., Petit, C., Kretz, C., Oberlé, I., and Mandel, J. L. (1991) Abnormal pattern detected in fragile-X patients by pulsed-field gel electrophoresis. Nature *349*, 624–626.

Walker, C. L., Cargile, C. B., Floy, K. M., Delannoy, M., and Migeon, B. R. (1991) The Barr body is a looped X chromosome formed by telomere association. Proc. Natl. Acad. Sci. USA *88*, 6191–6195.

Wareham, K. A., Lyon, M. F., Glenister, P. H., and Williams, E. D. (1987) Age related reactivation of an X-linked gene. Nature *327*, 725–727.

Wolf, S. F., Jolly, D. J., Lunnen, K. D., Friedmann, T., and Migeon, B. R. (1984) Methylation of the hypoxanthine phosphoribosyltransferase locus on the human X chromosome: implications for X-chromosome inactivation. Proc. Natl. Acad. Sci. USA *81*, 2806–2810.

Yisraeli, Y., and Szyf, M. (1984) Gene methylation patterns and expression, in: DNA Methylation: Biochemistry and Biological Significance, pp. 353–378. Eds A. Razin, H. Cedar and A. D. Riggs. Springer-Verlag, New York.

Yu, S., Pritchard, M., Kremer, E., Lynch, M., Nancarrow, J., Baker, E., Holman, K., Mulley, J. C., Warren, S. T., Schlessinger, D., Sutherland, G. R., and Richards, R. I. (1991) Fragile X genotype characterized by an unstable region of DNA. Science *252*, 1179–1181.

Yu, W. D., Wenger, S. L., and Steele, M. W. (1990) X chromosome imprinting in fragile X syndrome. Hum. Genet. *85*, 590–594.

DNA Methylation: Molecular Biology and Biological Significance
ed. by J. P. Jost & H. P. Saluz

DNA methylation: Its possible functional roles in developmental regulation of human globin gene families

Chien Chu and C.-K. James Shen

Department of Genetics, University of California, Davis, CA 95616, USA

1 Introduction

The eukaryotic globin gene family is one of the first gene systems whose regulatory mechanisms of tissue-specific transcription appeared to involve DNA methylation of specific CpG residues (reviewed in Shen, 1984). In this chapter, we review the various experiments carried out since 1984 on the relationship between DNA methylation and transcriptional regulation of different human globin genes during erythroid development. We hope the article complements several earlier reviews on similar topics (Bird, 1984; Karlsson and Nienhuis, 1985; Cedar, 1989; Weissbach et al., 1989).

The human α-like and β-like globin gene families are individually clustered on chromosome 16 and 11, respectively. Functional members of both gene clusters are arranged in the order of their developmental regulation: β, 5′-ε (embryonic) – ${}^{G}\gamma$ (fetal) – ${}^{A}\gamma$ (fetal) – δ (adult) – β (adult) – 3′; α, 5′-ζ(embryonic) – $\alpha 2$ (adult) – $\alpha 1$ (adult) – $\theta 1$ (fetal/adult) – 3′ (Fig. 1). The expression of ε, γ, and β globin genes are all controlled, at least in part, by a constitutive promoter consisting of the CACC box, the CCAAT box, and the TATA box, which serve as binding sites of various ubiquitous nuclear factors (Anagnou et al., 1986; de Boer et al., 1988; Yu et al., 1991). Further erythroid cell-specific transcriptional control is contributed by other upstream promoter elements such as the binding site of erythroid-specific factor GATA-1 (reviewed in Evans et al., 1990; Orkin, 1990). Tissue-specific and developmental stage-specific regulation of the human β-like and α-like globin clusters, on the other hand, appears to be modulated by the action of locus control regions, (LCR), which are erythroid-specific enhancers located at many kilobase (kb) upstream of each gene cluster (Tuan et al., 1985; Forrester et al., 1987; Grosveld et al., 1987; Tuan et al., 1989; Higgs et al., 1990; Fig. 1). The molecular mechanisms of how

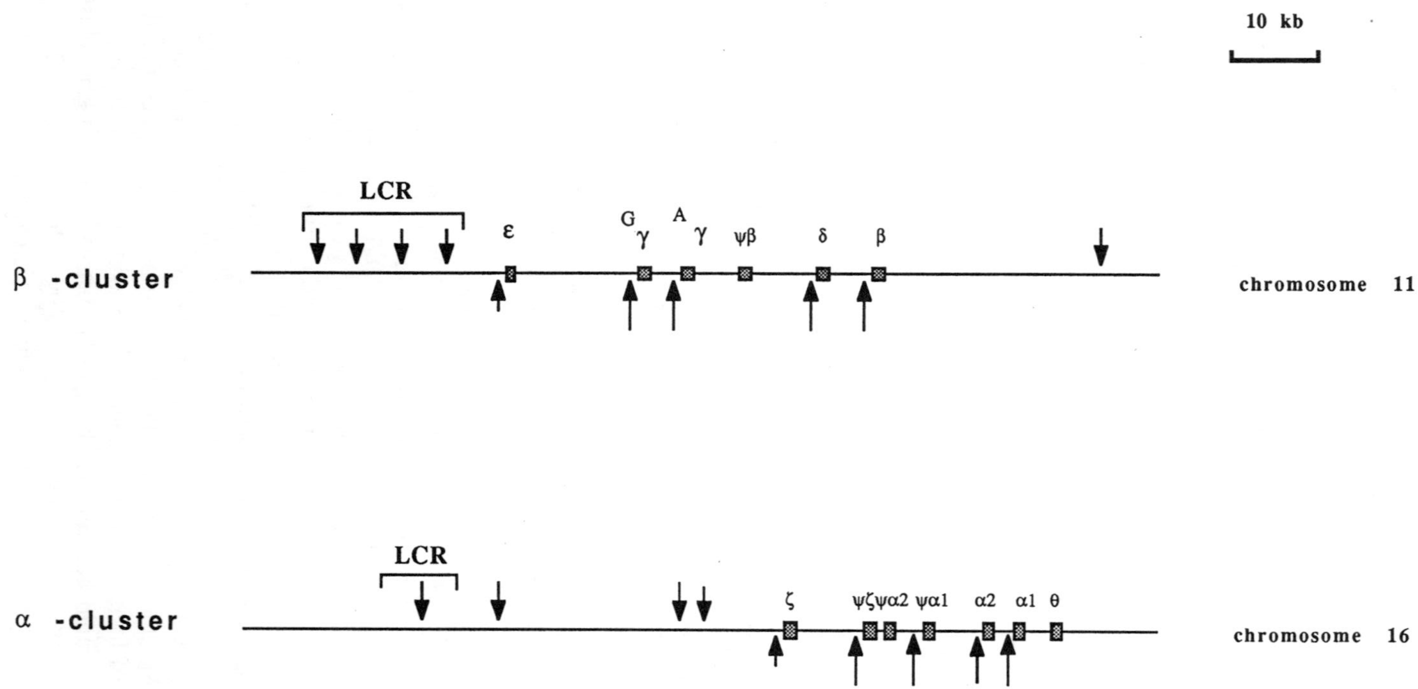

Figure 1. Linkage map of human α-like and β-like globin gene clusters. The transcription of all genes are from left to right. The vertical arrows indicate the positions of DNase I-hypersensitive sites that are present in expressing cells in the LCR enhancers (above the map) and in the globin upstream promoters (below the map). For further details and references, see text and Tables 1 and 2.

LCRs sequentially activate human globin genes during development are still not clear, but they could involve the competitive interaction between multiple nuclear factor-DNA complexes formed at the enhancer and the globin promoters, and/or the alteration of chromatin structure of specific domains of the two loci (reviewed in Evans et al., 1990; Orkin, 1990; Stamatoyannopoulos, 1991). Because of the relatively late discovery of the existence of these LCRs, many of the experiments described below were focused mainly on the analysis of the upstream promoters and gene bodies of different globins.

2 DNA Hypomethylation and promoter activities of human β-like globin genes

Earlier observations have clearly demonstrated the correlation between DNA hypomethylation and promoter activities of mammalian β globin gene clusters (van der Ploeg and Flavell, 1980; Shen and Maniatis, 1980). However, the hematopoietic tissues used in those studies, including fetal liver and bone marrow, were heterogeneous in cell composition; and no embryonic tissues were tested. Using the same approach, Mavilio et al. (1983) later did an extensive study of purified erythroid cells from embryonic yolk sac, 6–12-week fetal liver, and adult bone marrow. In this way, a more complete picture has been obtained in which hypomethylation occurred in the vicinities of ε, γ, and β globin genes correlated well with their expression in different hematopoietic tissues during development.

Oppenheim et al. (1985) have further developed a method to isolate immature erythroid cells, the orothochromatic normoblasts, from peripheral blood of patients with β-thalassemia major or other congenital hemolytic anemia after splenetomy. These normoblasts are at the last differentiation stage before releasing their nuclei and maturing into reticulocytes. Analysis of DNA methylation patterns of β-like globin gene cluster of this homogeneous population of nucleated erythroid cells of synchronized differentiation stage showed that two CpG sites flanking the unexpressed ε gene were methylated. However, the two CpG sites 5′ to the γ genes, which were expressed to a high level in these patients, were hypomethylated. These results are in agreement with the earlier reports (van der Ploeg and Flavell, 1980; Mavillio, et al., 1983). Most interestingly, the CpG site 3′ to the β gene was found to be completely unmethylated. Since the earlier study of Mavilio et al. (1983) revealed only partial hypomethylation of this site, Oppenheim et al. (1985) suggested that this hypomethylation occurred during the last erythroid cell division. They speculated that the onset of β-globin gene transcription may precede hypomethylation of this particular CpG site.

Perrine et al. (1988) studied the genomic DNA metylation of infants of diabetic mothers (IDM) who have a delayed fetal-to-adult globin switch. They analyzed the methylation patterns surrounding the active infant γ-globin genes, in comparison to those of the age-matched controls with a normal schedule of γ- to β-globin gene switching. Again, extensive hypomethylation was found to be associated with the still very active γ-globin gene expresssion in the IDM population.

All of the above studies support the notion that DNA hypomethylation *in vivo* correlates with globin gene activities in erythroid tissues. Methylation pattern analysis of human β-like globin gene clusters in several human erythroleukemic cell lines is in general agreement with the *in vivo* observations (Enver et al., 1988a; Enver et al., 1988b; see further discussion later).

3 Timing of DNA methylation and γ to β globin switch

It is now clear that human γ-globin genes are transcriptionally active as well as hypomethylated in the erythroid cells of fetal liver before the γ to β-globin gene switch, and they become silent and methylated in the erythroid cells of adult bone marrow after the switch. Is methylation one of the essential steps that silence γ globin genes during this switch? If it is, then one expects that methylation occurs either synchronously with or precedes the inactivation of the γ globin genes.

In humans, the developmental γ to β switch begins at around 24–30 weeks of gestation, and it is not complete until a few months after birth. Since it is difficult to obtain pure material to study this hematopoietic development, Stamatoyannopoulos and his colleagues have attempted this task by using somatic cell hybrids formed between mouse erythroleukemia cells and human erythroblasts (Enver et al., 1988b). These hybrid cells initially express human γ-globin but gradually switch to β-globin in culture at a rate roughly dependent on the developmental age of the fetus used. The cellular population during the γ to β switching is heterogeneous, containing cells not yet switched (γ^+), few cells in the actual process of switching ($\gamma^+\beta^+$), and cells already switched (β^+). Interestingly, by capturing and amplifying hybrid cells of particular phenotypes with cell sorter, Enver et al. (1988b) identified hybrids that no longer express γ-globin but still possess hypomethylated γ-globin genes. This data suggests that γ-gene inactivation can precede DNA methylation. Thus, methylation does not appear to be a primary mechanism for γ gene inactivation during human erythroid development.

However, in this study by Enver et al. (1988b) and in all other experiments described above, only those CpG sites recognized by C-methyl group-sensitive restriction enzymes, e.g. Hpa II, were analyzed. The CpG sites analyzed by Enver et al. (1988b) consist of only the ones at -53 and $+1805$, and these two sites may not play a functional role in γ to β switch (see further discussion later).

4 Does methylation inhibit expression of transfected γ and β globin genes?

Several elegant gene transfer experiments have been designed and carried out to test whether methylation *in vitro* affects transcription of human γ and β globin genes in transfected cell cultures. In an earlier study, Busslinger et al. (1983a) found that unmethylated human $^{A}\gamma$ globin gene was transcriptionally active after transfer into mouse L cells. However, if the 5′ portion, but not 3′ portion, of the gene (from -760 to $+100$) was heavily methylated by an enzymatic method at all C residues before transfer, it became transcriptionally inactive in mouse L cells. The beauty of this method, which was first designed by Stein et al. (1982), is that only methyl groups of CpG dinucleotides were maintained in the transfected cells after DNA replication.

Murray and Grosveld (1987) extended the above observation by the use of three different restriction fragments to obtain constructs that contain methylated C at all CpG sites downstream of -384, -210 or $+92$ of human γ-globin gene. It was found that methylation of CpGs downstream of $+92$ did not affect transcription. On the other hand, methylation downstream of -210 abolished γ expression. This suggests that methylation of the six CpG sites in between nucleotide positions -210 and $+92$, which are located at nucleotide positions -161, -53, -50, $+6$, $+18$ and $+49$, respectively, suppresses expression of the transfected human γ globin gene in the non-erythroid L cells.

Data from further site-direct mutagenesis experiments showed that simultaneous methylation of all six CpG sites in between nucleotide positions -210 and $+92$ was not required for the repression of transfected human globin gene expression *in vivo*. Also, presence of longer stretches of unmethylated DNA upstream of $+92$ prior to transfection conferred better expression of the human γ globin gene in mouse L cells (Murray and Grosveld, 1987).

As mentioned in previous sections, the human γ to β globin switch study of Enver et al. (1988b) only monitored the methylation status of CpG at -53 of the γ gene. It would be interesting if the other five CpG sites in between -210 and $+92$ could also be monitored in the hybrid cells used by Enver et al. (1988b). However, one should also bear in mind the difference of erythroid nature of the cell culture systems used by these two groups.

The effect of CpG methylation on human β globin gene expression *in vivo* has also been studied by the same *in vitro* methylation approach (Yisraeli et al., 1988). Methylation of all CpG sites within a 4.4 kb Pst I fragment containing the entire human β globin gene inhibited both basal transcription in mouse L cells, and hexamethylene-bisacetamide (HMBA) induced expression of the β gene in mouse erythroleukemic cells (MEL).

Partially methylated human β globin gene constructs were also used to transfect mouse L cells. Interestingly, unlike human γ globin gene (Busslinger et al., 1983; Murray and Grosveld, 1987), methylation of either the 5′ portion of the Pst I fragment, which contains mostly upstream region, or the 3′ portion of the Pst I fragment, which contains the β globin structural gene and its 3′ flanking DNA, inhibited β globin transcription. The inhibition by methylation of 3′ flanking DNA is in agreement with the observation that at least one CpG site in this region is unmethylated in adult erythroid cells (Van der Ploeg et al., 1980; Mavilio et al., 1983; Oppenheimer et al., 1985). Unfortunately, the partially methylated constructs were not tested in uninduced and induced MEL cells. This would be an interesting experiment since, MEL cells are of adult erythroid origin and the 3′ flanking DNA of human β globin gene has previously been shown to be involved in its induced expression in transfected MEL cells (Wright et al., 1984; Charnay et al., 1984).

5 Does de-methylation *in vivo* activate human β-like globin genes? The 5-azacytidine (5-azaC) experience

5-Azacytidine (5-azaC), as a base analog, is a potent demethylating agent (reviewed in Jones, 1984). γ-Globin gene expression can be stimulated transiently by 5-azaC in anemic baboons (DeSimone et al., 1982) and in the bone marrow cells of sickle cell anemia or β-thalassemia patients (Ley et al., 1982; Charache et al., 1983; Ley et al., 1983). Although non-random de-methylation of CpG sites occurred throughout the β-like globin gene cluster after 5-azaC treatment, only the γ globin gene expression, but not β, was increased. Furthermore, administering hydroxyurea and arabinosylcytosine (an inhibitor of DNA synthesis), both of which do not affect DNA methylation, also increases γ globin gene expression in anemic primates (Letvin et al., 1984; Papayannopoulou et al., 1984). It was suggested that 5-azaC, like other S phase-specific cytotoxic drugs, may act by recruiting a subpopulation of early erythroid progenitor cells whose progenies transcribe γ globin gene more efficiently and consequently synthesize more fetal hemoglobin. In addition to this mechanism, 5-azaC may also act directly on late erythroid cells to cause γ gene activation via demethylation.

The latter possibility was partially tested by Ley et al. (1984) who have studied the effect of 5-azaC on human β-like globin gene expression of M11-X, a mouse erythroleukemia cell line containing human chromosome 11. Since cellular sub-populations at different stages of erythroid differentiation do not exist in M11-X cells, the direct demethylation mechanisms of 5-azaC may play a more important role in gene activation in this system. Treatment of M11-X cells with HMBA alone, an inducing reagent of MEL and M11-X cells, 5-azaC alone, or 5-azaC followed by HMBA all induced mouse β and human β globin gene expression. However, only combined treatment with 5-azaC plus HMBA induced γ globin gene expression. None of these treatments could induce human ε globin gene expression. Methylated CpG dinucleotides of these treated M11-X cells were assayed by Ley et al. (1984). Transient decrease of genomic frequency of 5-methyl C was observed upon treatment with 5-azaC or (5-azaC + HMBA). However, induced demethylation of the human β-like globin locus following these two treatments persisted in culture.

The above experiments suggest that DNA demethylation of human β-like globin locus in M11-X cells treated with 5-azaC alone is not sufficient for γ globin activation. However, the combined use of HMBA, which causes terminal diffrentiation of M11-X cells, may induce the synthesis of trans-acting nuclear factors that are required for active transcription of the γ globin promoter. The inactivity of the similarly demethylated human ε globin gene in treated M11-X cells, on the other hand, could be due to a chromatin conformation unfavourable for transcription, or due to the lack of proper transcription factors. This simultaneous requirement of DNA demethylation, specific trans-acting factors, and/or active chromatin conformation for human β-like globin gene activation is independently supported by similar studies of mouse adult β and α globin gene expression in 5-azaC treated mouse cell cultures (Hsiao et al., 1984; Michalowsky and Jones, 1989), and of embryonic globin gene induction in chickens administered with 5-azaC and butyric acid (Burns et al., 1988).

6 DNA Methylation and expression of human α-like globin genes

The correlation between transcriptional regulation and methylation of the human α-like globin locus is much less clear. The α-like and β-like globin clusters differ in several aspects regarding their structures and developmental expression. The β-like globin cluster is relatively AT-rich and deficient in CpG sites. On the other hand, the human α globin locus is GC-rich. In particular, the adult $\alpha 2$ and $\alpha 1$ globin genes and their upstream regions up to 1 kb are highly enriched in CpG sites (reviewed in Bird et al., 1987) that are characteristic of the CpG islands in many house-keeping genes and several tissue-specific genes of mammals (Bird, 1986). In contrast to the β-like globin locus, there is only one switch, ζ to α, of the α-like globin expression during development.

This switch occurs early at 5-6 weeks of gestation whereas the adult β globin is only fully expressed in the final stage of development (Peschle et al., 1985).

The methylation patterns of the human α-like globin locus in different tissues or cell lines have been analyzed by the use of methyl-C sensitive restriction enzymes. Similar to other CpG island-containing genes, the $\alpha 2$ and $\alpha 1$ globin genes were found to be completely unmethylated in all non-erythroid and erythroid tissues tested including sperms (Bird et al., 1987; Antequera et al., 1990). This finding is supported by recent genomic sequencing analysis of different human tissue DNA (Kuang-Yu Hu and J. Shen, unpublished results). Although several of the human cells lines contained methylated α globin genes, this has been attributed to a *de novo* methylation mechanisms of non-essential CpG islands in permanent cell cultures (Antequera et al., 1990). In regions containing ζ, $\psi\zeta$, $\psi\alpha 2$ and $\psi\alpha 1$, on the other hand, the tested CpG sites are partially methylated *in vivo* and their extents of methylation vary among different cell types. Unlike the β-like globin locus (Table 1), there is no clear-cut relationship

Table 1. DNA methylation and chromatin structure of human β-like globin gene cluster

				Chromatin structure	
Cell types	Genes	Expression	Hypo-methylation	5′-DNase I hypersensitivity	Overall DNase I sensitivity
	ε	−	−	−	−
White blood cells	γ	−	−	−	−
	δ	−	+	−	−
	β	−	−	−	−
	ε	−	−	−	+
Fetal erythroid cells (late stages)	γ	+	+	+	+
	δ	+	+	+	+
	β	+	+	+	+
	ε	−	−	−	+
Adult erythroblasts	γ	−	−	−	+
	δ	+	+	+	+
	β	++	+	+	+
	ε	+	+	+	+
K562 cell line	γ	+	+	+	+
	δ	−	+	−	+
	β	−	+	−	+

The relationship among tissue-specific expression, DNA hypomethylation in the vicinities of the genes, and chromatin structure is tabulated for human β-like globin genes. The table has been synthesized from data of the following references: van der Ploeg et al., (1984); Oppenheim et al. (1985); Tuan et al. (1985); Arapinis et al. (1986); and Forrester et al. (1986). The information regarding methylation and chromatin structure of human β-like globin genes in early fetal erythroid cells and other non-erythroid tissues can be found in the references listed above. Note also that analysis of pure adult erythroblasts does reveal a good correlation between the degrees of overall DNase I sensitivity and the levels of expression of the β-like globin genes (Arapinis et al., 1986). No attempt was made to tabulate this correlation here.

Table 2. DNA methylation and chromatin structure of human α-like globin gene cluster

Cell types	Genes	Expression	Hypo-methylation	Chromatin structure: 5′-DNase I hypersensitivity	Chromatin structure: Overall DNase I sensitivity
White blood cells	ζ	−	+	−	−
and fetal brain	α	−	+	−	−
	ζ	+	ND	+	+
Fetal liver	α	+	ND	+	+
	ζ	+	+	+	ND
Adult bone marrow	α	++	+	+	ND
	ζ	+	+	+	+
K562 cell line	α	+	+	+	+

The relationship among tissue-specific expression, DNA hypomethylation in the vicinity of the gene, and chromatin structure is tabulated for the human α-like globin gene cluster. The α globin genes (α2 and α1) are un-methylated in all tissues tested including sperm. The "+" signs in the "hypo-methylation" column for ζ indicate that their 5′ ends are hypo-methylated in the specific types of cells relative to sperm, in which they are completely methylated. ND, not done. The data are derived from Yagi et al. (1986) and Bird et al. (1987).

between DNA methylation and expression of the human α-like globin genes in these different tissues and cell lines (Table 2).

The effects of *in vitro* methylation on the expression of human α globin genes have also been studied. Busslinger et al., (1983b) have found that α globin promoter activity was repressed in stably transfected cell cultures if the DNA was methylated prior to transfection. This observation was reproduced by Boyes and Bird (1991) in a transient expression assay. In this later series of experiments conducted by Bird and his colleagues (Meehan et al., 1989; Boyes and Bird, 1991; Boyes and Bird, 1992), a methyl CpG-binding protein, or MeCP-1, was also identified which may be responsible for transcriptional repression of a number of methylated eukaryotic genes in nuclear extracts or in transfected cell cultures. The immediate function of MeCP-1 is not clear as to the regulation of human adult α2 and α1 globin genes, since these two CpG island-containing genes are not methylated in erythroid as well as in non-erythroid tissues. However, it may play an important role in the silencing of individual human β-like globin genes during erythroid development (see further discussion later).

7 Possible mechanisms of transcriptional repression by DNA methylation

The developmental stage-specific expression of human β-like globin locus is thus closely associated with erythroid tissue-specific hypomethylation of DNA regions located within and flanking the different β-like globin genes. Results from different experiments carried out *in vitro* and *in vivo* on methylated DNA templates, as described in previous sections, are highly suggestive of a causative role of DNA methylation in repressing transcription of the β-like globin genes. Similar role of DNA methylation has also been suggested for other eukaryotic genes (see other chapters, this volume). How could DNA methylation repress transcription of β-like globin genes *in vivo*? We discuss below three possible pathways: (a) Inhibition of binding of specific transcription factors by DNA methylation; (2) Establishment of inactive chromatin structure by DNA methylation; (3) Repression of transcription by methyl CpG-binding protein(s).

7.1 Inhibition of binding of specific transcription factors by DNA methylation

There is accumulating evidence that CpG methylation may interfere with binding of transcription factors to specific DNA sequences, and consequently their function of transcriptional activation or repression (See chapter by Ehrlich and Ehrlich, pp. 145–168 of this book). For example, Becker et al. (1987) observed that methylation of the cloned tyrosine aminotransferase gene completely abolished the binding of at least one of the three transcriptional factors required for expression of this gene. It has also been shown that methylation of a single CpG centrally located in the recognition site of a transcription factor required for transcription activity of the adenovirus-2 major late promoter affects its DNA binding property (Watt and Molloy, 1988). Also, methylation of the sequence 5′-TGACGTCA-3′ of a cAMP responsive element of several promoters inhibited both the binding of transcriptional factors to this element and transcription activities *in vivo* of these promoters (Iguchi-Ariga and Schaffner, 1989).

A number of ubiquitous and erythroid-specific nuclear factors recognize specific DNA sequences,or motifs, in the upstream promoters of different human globin genes (Fig. 2). The experimental assays used to map these factor-binding sites include DNase I footprinting *in vitro*, gel mobility shift and, in some cases, genomic footprinting (for references, see legend of Fig. 2). Mutagenesis of the promoters have demonstrated that protein-DNA complexes formed at some of these motifs, e.g. the GATA-1 of γ globin promoter, the CACC promoter boxes of all three human β-like globin genes, etc. (Fig. 2), function to activate different β-like globin genes in transfected cell cultures. Despite of all these data, no CpG sites in the human β, γ, ε, or ξ promoter are located within the binding domains of the different nuclear factors (Fig. 2). Although CpG is part of the transcription factor Sp1-binding site 5′-CCCGCC-3′, and human $\alpha 2$ and $\alpha 1$ globin upstream promoters contain several of them (Shen et al., 1989;

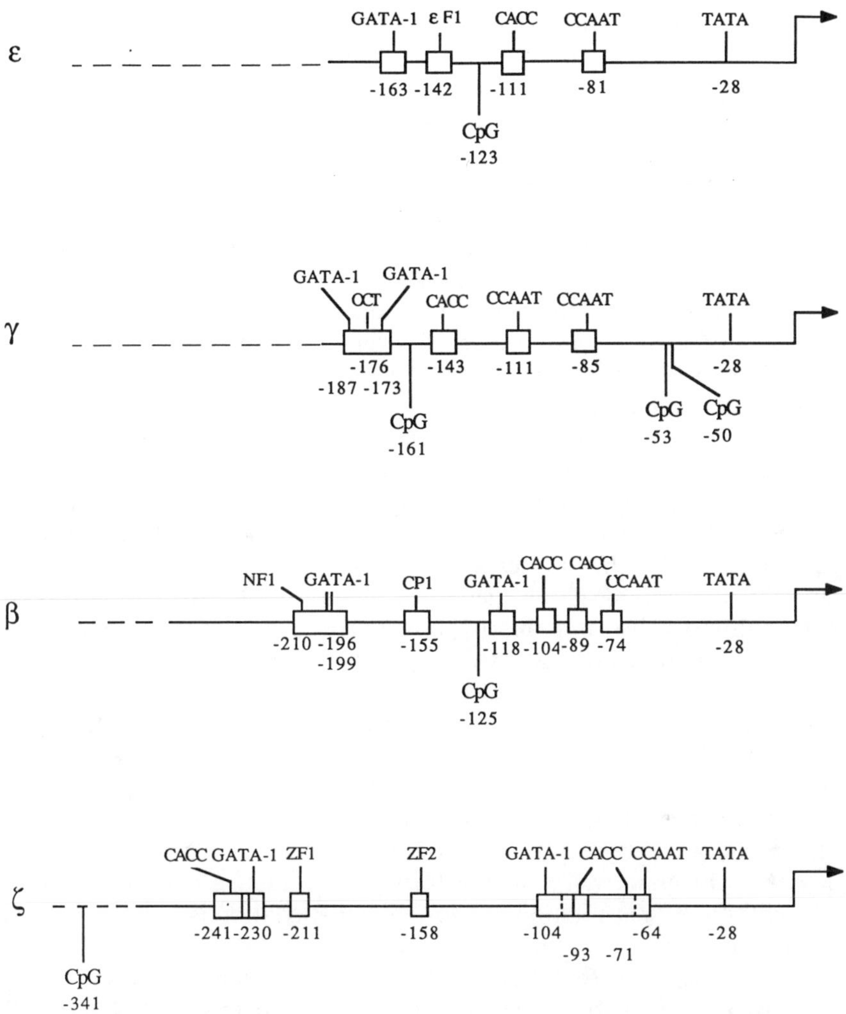

Figure 2. Relative locations of CpG sites and nuclear factor-binding sites in the upstream promoter regions of human ε, γ, β and ζ globin genes. The central nucleotides of each factor-binding motifs, and the positions of the CpG sites are individually indicated below each map. The factor-binding motifs have been mapped by protein-binding studies in nuclear extract and with partially purified factors: ε (Gong et al., 1991; Yu et al., 1991); γ (Anagnou et al., 1986; Gumucio et al., 1988; Superti-Furga et al., 1988; Catala et al., 1989; Lloyd et al., 1989; Martin et al., 1989; Nicholis et al., 1989); β (deBoer et al., 1988); ζ (Yu et al., 1990).

Whitelaw et al., 1989), CpG methylation of this hexanucleotide sequence does not affect Sp1 binding and its function (Harrington et al., 1988). Therefore, the involvement of methyl CpG-directed inhibition of DNA binding of specific transcription factors may be minimal in human globin gene regulation during erythroid development.

7.2 Establishment of inactive chromatin structure by DNA methylation

Altered chromatin structures are often associated with transcriptionally active genes as manifested by the overall DNase I sensitivity (Weintraub and Groudine, 1976), DNases I hypersensitivity in the upstream promoters and other transcription regulatory elements (Wu et al., 1979; reviewed by Elgin, 1988; Gross and Garrard, 1988), and sensitivity to micrococcal nuclease and restriction enzyme digestion *in vivo* (McGhee et al., 1981). The relationship among transcriptional activities of different human β-like globin genes, the extents of DNA methylation in the vicinities of these genes, and their chromatin structure as assayed by overall DNase I sensitivity and DNase I hypersensitivity of 5′ promoter regions are summarized in Table 1.

Except for human δ globin gene, several general rules are revealed by close examination of Table 1. 1) The chromatin structure of all five human β-like globin genes is more sensitive to DNase I digestion *in vivo* in erythroid cells than in non-erythroid cells, irrespective of the developmental stages of the former type of cells; 2) CpG sites surrounding actively transcribing globin genes are hypomethylated in the expressing erythroid cells, and this hypomethylation is always accompanied by both overall DNase I sensitivity and the presence of 5′-DNase I hypersensitive sites; 3) hypomethylated globin genes in erythroid cells always adopt an altered chromatin structure with an overall DNase I sensitivity, but they may or may not possess 5′-DNase I hypersensitive sites.

The data regarding methylation and chromatin structure of human ζ and α globin genes are summarized in Table 2. The transcription activities correlate well with the formation of active chromatin structure, although their relationship to DNA methylation is not obvious, as already mentioned in previous sections.

Whether CpG methylation leads to the formation of inactive chromatin structure has been elegantly tested by Keshet et al. (1986) in cell cultures. M13 plasmid constructs containing human β globin gene, as well as several other eukaryotic genes, were enzymatically methylated and then transfected into mouse L cells. It was demonstrated that unmethylated DNAs, after integration and stable propagation in the cell culture, were all DNase I-sensitive, and sometimes exhibited DNase I hypersensitivity and restriction enzyme sensitivity *in vivo*, depending on the sequence nature of the transfected plasmid. The extents of DNase I sensitivities of these un-methylated, transfected DNAs were similar to those of endogeneous, transcriptionally active mouse genes. However, *in vitro* methylation prior to transfection led to an "inactive" conformation of these different DNA sequences, including M13, in the stably transfected cells, as assayed by *in vivo* digestion with DNase I, restriction enzymes, or micrococcal nuclease. Their results are highly suggestive that CpG methylation *per se* is sufficient to cause stable formation of inactive or condensed chromatin structure. How could this happen? Experiments described below provide some interesting clues.

7.3 Repression of transcription by methyl CpG-binding proteins

The effect of DNA methylation on chromatin structure and transcription may involve a mediator capable of binding to methylated CpG sites, thus preventing the formation of active transcription complex. A methylated DNA-binding protein, or MDBP, has been identified and isolated from human cells (Khan et al., 1988 and references therein). MDBP has a higher binding affinity towards DNA containing methylated CpG residues than unmethylated DNA. However, because MDBP appeared to bind a specific set of DNA sequences (Khan et al., 1988) and because no experiments have been carried out to test its possible biological function, the role of MDBP in regulating expression of eukaryotic genes, including human globin family, cannot be speculated at this time.

A novel methyl-CpG binding protein, MeCP-1, has been identified and studied extensively by Bird and his colleagues. This protein was first isolated from nuclear extract of mouse liver cells, and later identified to exist in a number of different mammalian cells, by the use of various methylated oligonucleotides. MeCP-1 appeared to repress transcription of methylated DNA templates both *in vitro* and *in vivo*. As shown by Boyes and Bird (1991), transcription of four CpG island-containing genes in nuclear extract was completely repressed if the promoter DNAs were heavily methylated at CpG sites. Less extent of repression was seen when the level of CpG methylation was decreased by the use of different bacterial methyltransferases. Furthermore, the inhibition of transcription became increasingly obvious as the template concentration was lowered, or as more extracts were added. Also, addition of excess amounts of various methylated competitors restored transcription of methylated construct to the same level as that of the non-methylated genes. Methylated constructs containing human α globin gene were shown to bind MeCP-1 in HeLa extract by a bandshift assay. The methylated competitors, while competing away the inhibition, also competed for the bandshift activity of MeCP-1. Finally, very little inhibition was seen even at low concentrations of methylated constructs when nuclear extract from F9, a cell line deficient in MeCP-1 (Meehan et al., 1989), was used instread of HeLa extract. These observations together suggested that MeCP-1 functions as a mediator of transcriptional repression *in vitro* of genes containing methyl-CpG residues.

By the use of differently methylated gene constructs and methylated competitors in transient expression assay, Boyes and Bird (1991) further showed that MeCP-1 might also mediate indirect repression of transcription in living human or mouse cells via binding to the methylated template *in vivo*. In a recent study, it was demonstrated that transcription *in vivo* of human γ globin gene, which has a much lower density of CpG sites than human α globin gene, could also be inhibited if the gene was methylated prior to transfection (Boyes and Bird, 1992). Appropriate control experiments suggested that this inhibition also resulted from an indirect repression mechanism mediated by MeCP-1. Interestingly, the inhibition of transcription *in vivo* of genes with sparse density of methylated CpG residues, including completely methylated human γ globin gene, completely methylated mouse α globin gene, and partially methylated human α globin, could be overcome by the presence *in cis* of a viral enhancer (Boyes and Bird, 1992). Thus, unlike the scheme proposed for rat α-actin gene in which a precedent demethylation event is essential for transcriptional activation (Paroush et al., 1990), the MeCP-1 mediated repression of certain sparsely methylated genes may simply be abolished following the displacement of MeCP-1 by binding of appropriate transcription factor(s) at the promoter.

8 Summary and further comments

It has been known for a while that DNA hypermethylation is correlated with transcriptional inactivities of different members of eukaryotic β-like globin gene families in both erythroid and non-erythroid cells. Evidence accumulated during the past ten years are highly suggestive of a causative role of DNA methylation in switching-off of the human β-like globin genes. This inactivation mechanism may involve the formation of inactive chromatin conformation, which in turn could be mediated indirectly by nuclear proteins such as MeCP-1. Activation of the repressed β-like globin genes may be achieved by competitive binding of appropriate transcription factors, by a long-range interaction of the distal LCR enhancer with the promoters, or by a de-methylation event. Since the developmental regulation of human globin genes are likely modulated by a complex and mutually interactive array of molecular and cellular mechanisms, the above over-simplified scheme of DNA methylation-mediated regulation has to be carefully refined and further tested in conjunction with other regulatory pathways proposed (reviewed in Evans et al., 1990; Orkin, 1990; Stamatoyannopoulos, 1991).

Some further comments are listed in the following:

(1) Both human α-like and β-like globin clusters are controlled by the LCR enhancers. Whether methylation of these enhancers affects their chromatin structure and function has not been tested;

(2) Thus far, the *in vivo* methylated CpG sites have only been mapped by methyl C-sensitive restriction enzymes. Only 5–15% of CpG site could be detected by this method. The recently refined technique of genomic sequencing of different tissue DNAs should be applied to the analysis of human β- and α-like globin gene clusters, as has been initiated for the mouse β globin gene (Ward et al., 1990) and other eukaryotic genes (other chapters, this volume). Of particular interest is the mapping of methylated CpG sites of human globin loci in transgenic mice;
(3) All experiments testing the effects of DNA methylation on human globin gene transcription have used artificially methylated plasmid DNA constructs. It may be worthwhile to repeat some of these experiments by methylating specific CpGs according to the genomic maps of methylated CpG sites obtained from (2);
(4) The transfer of methylated plasmid constructs into erythroid cell lines of different developmental stages, and into mouse by transgenic technique should reveal interesting information;
(5) Caution should be taken in the interpretation of experimental results obtained from cell culture work. Both DNA methylation (Michalowsky and Jones, 1989; Antequera et al., 1990) and nuclear factor-binding (Reddy and Shen, unpublished) sometimes behave promiscously in cell lines;
(6) Knowledge of the relative timing of DNA de-methylation and onset of transcrtiption of particular human globin genes would have essential implication regarding the functional role of CpG methylation. The availability of pure human erythroblasts of different stages is highly limited at this time;
(7) Chromatin of the entire locus of either human α-like or β-like globin family is sensitive to DNase I in erythroid cells of different developmental stages. Thus, separate mechanism(s), other than DNA hypomethylation, is likely to modulate the formation of this altered chromatin structure;
(8) The evidence of MeCP-1-like proteins in repressing human globin genes is still indirect. The demonstration of cell type-specific binding *in vivo* of these proteins to specific globin gene region(s) would strengthen their proposed role;
(9) The human $\alpha 2$ and $\alpha 1$ globin genes contain CpG islands that are constitutively unmethylated. However, CpG sites outside of these CpG islands are differentially methylated in different tissues. These flanking CpG sites may be involved in the regulation of the two α globin genes by DNA methylation.

Acknowledgements. This project was supported by a U.S. Public Health grant (DK 29800). We thank Adrian Bird, Howard Cedar, and Melanie Ehrlich for sending us their preprints and reprints. Kuang-Yu Hu and Sekhar Reddy have generously communicated to us their unpublished results.

Antequera, F., Boyes, J., and Bird, A. (1990) High levels of *de novo* methylation and altered chromatin structure at CpG islands in cell lines. Cell *62*, 503–514.

Anagnou, N. P., Karlsson, S., Moulton, A. D., Keller, G., and Nienhuis, A. W. (1986) Promoter sequences required for function of human γ globin gene in erythroid cells. EMBO J. *5*, 121–126.

Arapinis, C., Elion, J., Labie, D., and Krishnamoorthy, R. (1986) Differences in DNase I sensitivity and methylation within the human β-globin gene domain and correlation with expression. Eur. J. Biochem. *156*, 123–129.

Becker, P. B., Ruppert, S., and Schutz, G. (1987) Genomic footprinting reveals cell type-specific DNA binding of ubiquitous factors. Cell *51*, 435–443.

Bird, A. P. (1984) DNA methylation – how important in gene control? Nature *307*, 503–504.

Bird, A. P. (1986) CpG-rich islands and the function of DNA methylation. Nature *321*, 209–213.

Bird, A. P., Taggart, M. H., Nicholls, R. D., and Higgs, D. R. (1987) Non-methylated CpG-rich islands at the human α-globin locus: implication for evolution of the α-globin pseudogene. EMBO J. *6*, 999–1004.

Boyes, J., and Bird, A. (1991) DNA methylation inhibits transcription indirectly via a methyl-CpG binding protein. Cell *64*, 1123–1134.

Boyes, J., and Bird, A. (1992) Repression of genes by DNA methylation depends on CpG density and promoter strength: evidence for involvement of a methyl-CpG binding protein. EMBO J. *11*, 327–333.

Burns, L. J., Glauber, J. G., and Ginder, G. D. (1988) Butyrate induces selective transcriptional activation of a hypomethylated embryonic globin gene in adult erythroid cells. Blood *72*, 1536–1542.

Busslinger, M., Hurst, J., and Flavell, R. A. (1983a) DNA methylation and regulation of globin gene expression. Cell *34*, 197–206.

Busslinger, M., and Flavell, R. A. (1983b) DNA methylation and the regulation of globin gene expression, in: Globin Gene Expression and Hematopoietic Differentiation, pp. 193–203. Eds G. Stamatoyannopoulos and A. Nienhuis. Alan R. Liss, New York.

Catala, F., deBoer, E., Habets, G., and Grosveld, F. (1989) Nuclear protein factors and erythroid transcription of the human Aγ globin gene. Nucleic Acids Res. *17*, 3811–3827.

Cedar, H. (1988) DNA methylation and gene activity. Cell *53*, 3–4.

Charache, S., Dover, G., Smith, K., Talbot, C. C., Moyer, M., and Boyer, S. (1983) Treatment of sickle cell anemia with 5-azacytidine results in increased fetal hemoglobin production and is associated with non-random hypomethylation of DNA around the γ-δ-β-globin gene complex. Proc. Natl. Acad. Sci. USA *80*, 4842–4846.

Charnay, P., Mellon, P., and Maniatis, T. (1985) Linker scanning mutagenesis of the 5′-flanking region of the mouse β-major-globin gene: sequence requirements for transcription in erythroid and nonerythroid cells. Mol. Cell. Biol. *5*, 1498–1511.

deBoer, E., Antoniou, M., Mignotte, V., Wall, L., and Grosveld, F. (1988) The human β globin promoter: Nuclear protein factors and erythroid specific induction of transcription. EMBO J. *7*, 4203–4212.

DeSimone, J., Heller, P., Hall, L., and Zwiero, D. (1982) 5-Azacytidine stimulates fetal hemoglobin synthesis in anemic baboons. Proc. Natl. Acad. Sci. USA *79*, 4428–4431.

Elgin, S. C. R. (1988) The formation and function of DNase I hypersensitive sites in the process of gene inactivation. J. Biol. Chem. *263*, 19259–19262.

Enver, T., Zhang, J.-W., Anagnou, N. P., Stamatoyannopoulos, G., and Papyannopoulou, T. (1988a) Developmental programs of human erythroleukemia cells: globin gene expression and methylation. Mol. Cell. Biol. *8*, 4917–4926.

Enver, T., Zhang, J.-W., Papyannopoulou, T., and Stamatoyannopoulos, G. (1988b) DNA methylation: a secondary event in globin gene switching? Genes Dev. *2*, 698–706.

Evans, T., Felsenfeld, G., and Reitman, M. (1990) Control of globin gene transcription. Annu. Rev. Cell Biol. *6*, 95–124.

Forrester, W. C., Thompson, C., Elder, J. T., and Groudine, M. (1986) A developmentally stable chromatin structure in the human β-globin gene cluster. Proc. Nat. Acad. Sci. USA *83*, 1359–1363.

Forrester, W. C., Takewaga, S., Papayannopoulou, T., Stamatoyannopoulos, G., and Groudine, M. (1987) Evidence for a locus activation region: The formation of developmentally stable hypersensitive sites in globin expressing hybrids. Nucl. Acids Res. *24*, 10,159–10,177.

Gong, R. H., Stern, J., and Dean, A. (1991) Transcriptional role of a conserved GATA-1 site in the human ε-globin gene promoter. Mol. Cell. Biol. *17*, 2558–2566.

Gross, D. S., and Garrard, W. T. (1988) Nuclease hypersensitive sites in chromatin. Annu. Rev. Biochem. *57*, 159–197.

Grosveld, F., Blom van Assendelft, G., Greaves, D. R., and Kollias, G. (1987) Position-independent high level expression of the human β-globin gene in transgenic mice. Cell *51*, 975–985.

Groudine, M., Kohwi-Shigematsu, T., Gelinas, R., Stamatoyannopoulos, G., and Papayannopoulou, T. (1983) Human fetal to adult hemoglobin switching: changes in chromatin structure of the β-globin gene locus. Proc. Natl. Acad. Sci. USA *80*, 7551–7555.

Gumucio, D. L., Rood, K. L., Gray, T. A., Riordan, M. F., Sartor, C. I., and Collins, F. S. (1988) Nuclear proteins that bind the human γ-globin gene promoter: Alterations in binding produced by point mutations associated with hereditary persistence of fetal hemoglobin. Mol. Cell. Biol. *8*, 5310–5322.

Harrington, M. A., Jones, P. A., Imagawa, M., and Karin, M. (1988) Cytosine methylation does not affect binding of transcription factor Sp1. Proc. Natl. Acad. Sci. USA *85*, 2066–2070.

Higgs, D. R., Wood, W. G., Jarman, A. P., Sharpe, J., Lida, J., Pretorius, I.-M., and Ayyub, H. (1990) A major positive regulatory region located far upstream of the human α-globin gene locus. Genes Dev. *4*, 1588–1601.

Hsiao, W.-L. W., Gattoni-Celli, S., Kirschmeier, P., and Weinstein, I. B. (1984) Effect of 5-azacytidine on methylation and expression of specific DNA sequences in C3H 10T 1/2 cells. Mol. Cell. Biol. *4*, 634–641.

Iguchi-Ariga, S., and Schaffner, W. (1989) CpG methylation of the cAMP-responsive enhancer/promoter sequence TGACGTCA abolishes specific factor binding as well as transcriptional activation. Genes Dev. *3*, 612–619.

Jones, P. A. (1984) Gene activation by 5-azacytidine, in: DNA Methylation: Biochemistry and Biological Significance, pp. 165–187. Eds A. Razin, H. Cedar and A. Riggs. Springer-Verlag, New York.

Karlsson, S., and Nienhuis A. (1985) Developmental regulation of human globin genes. Annu. Rev. Biochem. *54*, 1071–1108.

Keshet, I., Lieman-Hurwitz, J., and Cedar, H. (1986) DNA methylation affects the formation of active chromatin. Cell *44*, 535–543.

Khan, R., Zhang, X.-Y., Supakar, P. C., Ehrlich, K. C., and Ehrlich, M. (1988) Human methylated DNA-binding protein: determinants of a pBR322 recognition site. J. Biol. Chem. *263*, 14,374–14,383.

Letvin, N. L., Linch, D. C., Beardsley, G. P., McIntyre, K. W., and Nathan, D. G. (1984) Augmentation of fetal-hemoglobin production in anemic monkeys by hydroxyurea. N. Engl. J. Med. *310*, 869–873.

Ley, T. J., DeSimone, J., Anagnou, N. P., Keller, G., Humphries, R. K., Turner, P. H., Young, N. S., Heller, P., and Nienhuis, A. W. (1982) 5-Azacytidine selectively increases γ-globin synthesis in a patient with β^+-thalassemia. N. Engl. J. Med. *307*, 1469–1475.

Ley, T. J., DeSimone, J., Noguchi, C. T., Turner, P. H., Schechter, A. N., Heller, P., and Nienhuis, A. W. (1983) 5-Azacytidine increases γ-globin synthesis and reduces the proportion of dense cells in patients with sickle cell anemia. Blood *62*, 370–380.

Ley, T. J., Chiang, Y. L., Haidaris, D., Anagnou, N. P., Wilson, V. L., and Anderson, W. F. (1984) DNA methylation and regulation of the human β-globin-like genes in mouse erythroluekemia cells containing human chromosome 11. Proc. Natl. Acad. Sci. USA *81*, 6618–6622.

Lloyd, J. A., Lee, R. F., and Lingrel, J. B. (1989) Mutations in two regions of the Aγ globin canonical promoter affect gene expression. Nucl. Acids Res. *17*, 4339–4352.

Martin, D. I. K., Tsai, S. F., and Orkin, S. H. (1989) Increased γ-globin expression in a non-deletion HPFH mediated by an erythroid-specific DNA-binding factor. Nature *388*, 435–438.

Mavilio, F., Giampaolo, A., Care, A., Migliaccio, G., Calandrini, M., Russo, G., Pagliardi, G. L., Mastroberardino, G., Marinucci, M., and Peschle, C. (1983) Molecular mechanisms of human hemoglobin switching: selective under-methylation and expression of globin genes in embryonic, fetal and adult erythroblasts. Proc. Natl. Acad. Sci. USA *80*, 6907–6911.

McGhee, J. D., Wood, W. I., Dolan, J., Engel, J. D., and Felsenfeld, G. (1981) A 200 base pair region at the 5′ end of the chicken adult β-globin gene is accessible to nuclease digestion. Cell *27*, 45–55.

Meehan, R. R., Lewis, J. D., McKay, S., Kleiner, E. L., and Bird, A. P. (1989) Identification of a mammalian protein that binds specifically to DNA containing methylated CpGs. Cell *58*, 499–507.

Michalowsky, L. A., and Jones, P. A. (1989) Gene structure and transcription in mouse cells with extensively demethylated DNA. Mol. Cell. Biol. *9*, 885–892.

Murray, E. J., and Grosveld, F. (1987) Site specific demethylation in the promoter of human γ-globin gene does not alleviate methylation mediated suppression. EMBO J. *6*, 2329–2335.

Nicolis, S., Ronchi, A., Malgarette, N., Mantovani, R., Giglioni, B., and Ottolenghi, S. (1989) Increased erythroid-specific expression of a mutated HPFH γ-globin promoter requires the erythroid factor NF-E1. Nucl. Acids Res. *17*, 5509–5516.

Oppenheim, A., Katzir, Y., Fibach, E., Goldfarb, A., and Rachmilewitz, E. (1985) Hypomethylation of DNA derived from purified human erythroid cells correlates with gene activity of the β-globin cluster. Blood *66*, 1202–1207.

Orkin, S. H. (1990) Globin gene regulation and switching: circa 1990. Cell *63*, 665–672.

Papayannopoulou, T., Torrealba-de ron, A., Veith, R., Knitter, G., and Stamatoyannopoulos, G. (1984) Arabinosylcytosine induces fetal hemoglobin in baboons by perturbing cell differentiation kinetics. Science *224*, 617–618.

Paroush, A., Keshet, I., Yisraeli, J., and Cedar, H. (1990) Dynamics of demethylation and activation of the α-actin gene in myoblasts. Cell *63*, 1229–1237.

Perrine, S. P., Greene, M. F., Cohen, R. A., and Faller, D. V. (1988) A physiological delay in human fetal hemoglobin switching is associated with specific globin DNA hypomethylation. FEBS Lett. *228*, 139–143.

Peschle, C., Mavilio, F., Care, A., Migliaccio, G., Migliaccio, A. R., Salvo, G., Samoggia, P., Petli, S., Guerriero, R., Marinucci, M., Lazzaro, D., Russo, G., and Mastroberadino, G. (1985) Haemoglobin switching in human embryos: Asynchrony of ζ to α and ε to γ globin switches in primitive and definitive erythropoietic lineages. Nature *313*, 235–238.

Razin, A., Levine, A., Kafri, T., Agostini, S., Gomi, T., and Cantoni, G. L. (1988) Relationship between transient DNA hypomethylation and erythroid differentiation of murine erythroleukemia cells. Proc. Natl. Acad. Sci. USA *85*, 9003–9006.

Shen, C.-K. J. (1984) DNA methylation and developmental regulation of eukaryotic globin gene transcription, in: DNA methylation: Biochemistry and Biological Significance, pp. 249–268. Eds A. Razin, H. Cedar, and A. D. Riggs. Springer-Verlag, New York.

Shen, C.-K. J., and Maniatis, T. (1980) Tissue-specific DNA methylation in a cluster of rabbit β-like globin genes. Proc. Natl. Acad. Sci. USA *77*, 6634–6638.

Shen, C.-C., Bailey, A. D., Kim, S. H., Yu, C.-Y., Marks, J., Shaw, J.-P., Klisak, I., Sparkes, R. and Shen, C.-K. J. (1989) The human α2-α1-θ1 globin locus: some thoughts and recent studies of its evolution and regulation, in: Hemoglobin Switching. pp. 19–32. Eds G. Stamatoyannopoulos and A. Nienhuis. Alan R. Liss, New York.

Stamatoyannopoulos, G. (1991) Human hemoglobin switching. Science *252*, 383.

Stein, R., Gruenbaum, Y., Pollack, Y., Razin, A., and Cedar, H. (1982) Clonal inheritance of the pattern of DNA methylation in mouse cells. Proc. Natl. Acad. Sci. USA *79*, 61–65.

Superti-Furga, G., Barberis, A., Schaffner, G., and Busslinger, M. (1988) The −117 mutation in Greek HPFH affects the binding of three nuclear factors to the CCAAT region of the γ-globin gene. EMBO J. *7*, 3099–3107.

Tuan, D., Solomon, D. W., Qiliang, L., and London, I. M. (1985) The "β-like globin" gene domain in human erythroid cells. Proc. Natl. Acad. Sci. USA *82*, 6384–6388.

Tuan, D., Solomon, W., London, I. M., and Lee, D. P. (1989) An erythroid-specific, developmental-stage independent enhancer far upstream of human "β-like globin" genes. Proc. Natl. Acad. Sci. USA *86*, 2554–2558.

Van der Ploeg, L. H. T., and Flavell, R. A. (1980) DNA methylation in human γ-δ-β-globin locus in erythroid and non-erythroid tissues. Cell *19*, 947–958.

Ward, C., Bolden, A., and Weissbach, A. (1990) Genomic sequencing of the 5′-flanking region of the mouse β-globin major gene in expressing and nonexpressing mouse cells. J. Biol. Chem. *265*, 3030–3033.

Watt, F., and Molloy, P. L. (1988) Cytosine methylation prevents binding to DNA of a HeLa cell transcription factor required for optimal expression of the adenovirus major late promoter. Genes Dev. *2*, 1136–1143.

Weintraub, H., and Groudine, M. (1976) Chromosomal subunits in active genes have an altered conformation. Science *93*, 848–858.

Weissbach, A., Ward, C., and Bolden, A. (1989) Eukaryotic DNA methylation and gene expression. Curr. Top. Cell. Reg. *30*, 1–21.

Whitelaw, E., Hogben, P., Hanscombe, O., and Proudfoot, N. J. (1989) Transcriptional promiscuity of the human α-globin gene. Mol. Cell. Biol. *9*, 241–251.

Wright, S., Rosenthal, A., Flavell, R., and Grosveld, F. (1984) DNA sequences required for regulated expression of β-globin genes in murine erythroleukemia cells. Cell *38*, 265–273.

Wu, C., Bingham., P. M., Livak, K. J., Holmgren, R., and Elgin, S. C. R. (1979) The chromatin structure of specific genes: I. evidence for higher order domains of defined DNA sequences. Cell *16*, 797–806.

Yagi, M., Gelinas, R., Elder, J. T., Peretz, M., Papayannopoulou, T., Stamatoyannopoulos, G., and Groudine, M. (1986) Chromatin structure and developmental expression of the human α-globin cluster. Mol. Cell. Biol. *6*, 1108–1116.

Yisraeli, J., Frank, D., Razin, A., and Cedar, H. (1988) Effect of *in vitro* DNA methylation on β-globin gene expression. Proc. Natl. Acad. Sci. USA *85*, 4638–4642.

Yu, C.-Y., Chen, J., Lin, L. I., Tam, M., and Shen, C.-K. J. (1990) Cell type-specific protein-DNA interaction in the human ζ-globin upstream promoter region: Displacement of Sp1 by the erythroid cell-specific factor NF-E1. Mol. Cell. Biol. *10*, 282–294.

Yu, C.-Y., Motamed, K., Chen, J., Bailey, A. D., and Shen, C.-K. J. (1991) The CACC box upstream of human embryonic ε globin gene binds Sp1 and is a functional promoter element *in vivo* and *in vitro*. J. Biol. Chem. *266*, 8907–8915.

Zhu, J.-D., Allan, M., and Paul J. (1984) The chromatin structure of the human ε globin gene: Nuclease hypersensitive sites correlate with multiple initiation sites of transcription. Nucl. Acids Res. *12*, 9191–9204.

DNA Methylation: Molecular Biology and Biological Significance
ed. by J. P. Jost & H. P. Saluz

DNA Methylation, chromatin structure and the regulation of gene expression

M. Graessmann and A. Graessmann

Institut für Molekularbiologie und Biochemie der Freien, Universität Berlin, Arnimallee 22, 1 Berlin 33, Germany

1 Introduction

It is still not clear how eukaryotic cells regulate gene expression during differentiation and in the differentiated state. There is increasing experimental evidence that this requires a wide spectrum of different *cis*- and *trans*-acting elements (for a review, see Wasylyk, 1988). One of the *cis*-functional elements, crucial for gene activation, is the change in the DNA methylation pattern. In mammalian cells methylation occurs exclusively at the cytosine residue in the CpG dinucleotide sequence.

The current model proposes that undermethylation of the DNA within the promoter region is a prerequisite for gene activation. This hypothesis is based on the observation that some genes exhibit an inverse correlation between the extent of DNA methylation and gene activity (Waalwijk and Flavell, 1978). Furthermore, DNA demethylation induced by 5-azacytidine treatment eventually causes gene activation (Jones, 1985). The most direct evidence that DNA methylation can block gene expression has been obtained by gene transfer experiments (Wigler et al., 1981; Buschhausen et al., 1985).

However, the above model strongly oversimplifies the issue, because in mammalian cells, genes do not respond in a uniform manner to DNA methylation. Genes can be classified into at least three different categories: methylation-sensitive genes, where DNA hypermethylation can block gene expression (Wigler et al., 1981); methylation-insensitive genes, where DNA hypermethylation has no detectable effect on gene expression (Graessmann et al., 1983); and those genes where hypermethylation precedes gene activation (Tanaka et al., 1983).

A considerable obstacle is still the question of the mechanism by which DNA methylation prevents expression of the sensitive genes. Data obtained in different laboratories indicate that DNA methylation *per se* does not affect the function of the RNA polymerase II (Boyes and Bird, 1991). Therefore, inhibition of gene expression by DNA methylation should require, in addition, the interplay with *trans*-acting elements. Indeed, it has been shown that under certain *in vitro* conditions, binding of transcription factors can be inhibited by DNA methylation (Watt and Molloy, 1988; Kovesdi et al., 1987; see also chapter Ehrlich and Ehrlich, pp. 145–168 of this book). Since in mammalian cells the nuclear DNA is not naked, but packed around nucleosome cores to form chromatin, we have to ask what the functional correlation might be between DNA methylation and chromatin structure.

In this article we summarize what is known about DNA methylation, chromatin structure and the regulation of gene expression.

2 Relaxed DNA molecules are efficiently expressed before chromatin formation

Since most of the nuclear DNA in mammalian cells is present as chromatin, it was of interest to test whether free DNA is expressed with the same efficiency as chromatin, after the transfer into culture cells. As gene transfer method we used our microinjection technique, because it allows the transfer of a known number of test molecules into the cell compartment of choice (Graessmann and Graessmann, 1983). The test molecules were the SV40 (Simian virus) and the HSV-tk DNA (Herpes simplex virus I thymidine kinase gene).

In order to see when following injection, SV40 gene expression starts, 2–4 superhelical SV40 DNA molecules (form I) were microinjected into the nuclei of TC7 cells. At different times after DNA transfer the cells were fixed and stained for SV40 T-antigen and we found that T-antigen synthesis is detectable as early as 30 min after injection. Similar results were obtained after the injection of the HSV-tk DNA into tk-minus rat 2-cells analyzed by the ^{3}H thymidine incorporation into the cellular DNA. To follow the fate of the microinjected DNA, 100 cells were injected with the superhelical DNA (form I) and at various time points after the transfer, the DNA was reextracted and submitted to DNA blot analysis. We observed that immediately after intranuclear injection the entire DNA was converted into the relaxed circular DNA (form II). This conversion was mediated by the topoisomerase I (Graessmann et al., 1985). Generation of superhelical DNA, due to chromatin assembly, occurred about 8 h later. These results clearly demonstrated that free DNA without detectable torsional stress is efficiently expressed before being packed into minichromosomes. To test whether chromatin molecules are more efficiently expressed than the DNA, SV40 chromatin was isolated from virus-infected cells and HSV-tk DNA was reconstituted into chromatin *in vitro* with isolated histone octamers. Upon injection we found that the chromatin is expressed with the same efficiency as free DNA molecules both in terms of onset and rate of transcription (Buschhausen et al., 1987).

3 DNA Methylation does not block SV40 gene expression

As is the case for various mammalian genes, the CpG dinucleotide sequence – the potential methylation site in mammalian cells – is underrepresented in the SV40 genome. The SV40 DNA with 5243 bp has only 27 CpGs, and about 70% of them are located within the 300 bp regulatory region (early and late promoter, replication origin). This distribution suggested the possibility of an effect of DNA methylation on viral gene expression and DNA replication. In order to test this hypothesis, we methylated the SV40 DNA with rat liver methyltransferase and tested the methylation pattern by dinucleotide analysis (Graessmann et al., 1983). This enzyme methylated *in vitro* all CpGs of the SV40 DNA and also a significant portion of the cytosine residues within the CpA and CpT dinucleotide sequence, CpC methylation however was not detected (Table 1).

After *in vitro* methylation the SV40 DNA was microinjected into TC7 and rat 2 cells. Unexpectedly, the methylated SV40 DNA expressed the T- and V-antigen with the same efficiency as with the non-methylated DNA template. In addition, DNA methylation did not interfere with viral DNA replication (Table 2).

Table 1. *In vitro* methylation by rat liver methylase of 2′-deoxycytidine-containing dinucleotides in SV40

Dinucleotide	SV40 Total number	SV40 % methylated
C-G	27	106
C-A	423	21
C-T	399	3.5
C-C	245	0

Analysis of the dinucleotides of ^{3}H-methylated SV40.

Table 2. Biological activity of methylated and non-methylated SV40 DNA

Injection of two to four DNA molecules per cell	Antigen formation (%*) T	V	Virus production, % of injected cells+	Cell transformation, % of injected cells#
SV40 DNA I	100	100	40–60	20–30
SV40 DNA I methylated				
with rat liver methylase	100	100	40–60	20–30
with HpaII methylase	100	100	ND	ND

ND, not done. *T- and V-antigen formation was tested by the direct immunofluorescence technique 24 h after DNA injection. +Virus production was assayed by plaque formation. #Transformation was assayed by colony formation in soft agar.

Furthermore, the transformation efficiency was not impaired by DNA methylation. As with the non-methylated DNA, 20–30% of the recipient cells were transformed into permanently T-antigen positive cells and acquired the phenotype of fully transformed cells, thus excluding a long-term effect of DNA methylation on SV40 gene expression (Table 2).

These results clearly demonstrate that early SV40 gene expression and viral DNA replication cannot be blocked by DNA methylation. In *Xenopus oocytes*, T-antigen synthesis is also not blocked by DNA methylation (Götz et al., 1990). This means that SV40 represents a type of methylation-insensitive gene.

4 Inhibition of the HSV-tk gene by DNA methylation is indirect and requires chromatin formation

Since it was well documented that methylation of the HSV-tk gene by HpaII methyltransferase (M-HpaII) causes inhibition of gene expression (Wigler et al., 1981), we used this gene as a methylation-sensitive gene.

In the first set of experiments, the pHSV-106 DNA (BRL: HSV-tk gene inserted in the Bam H1 site of pBR 322) was methylated with the HpaII methyltransferase and injected into the nuclei of the thymidine kinase-negative rat 2 cells (200–400 molecules injected per cell). Directly after injection ^{3}H-thymidine was added to the culture medium and 24 h later cells were fixed and processed for autoradiography. Against our expectation, inhibition of tk gene expression was not observed. Figure 1a shows an autoradiogram of cells microinjected with the non-methylated DNA, and 1b cells injected with the methylated DNA. In both cases the number of thymidine kinase-positive cells and the intensity of thymidine incorporation were identical. To exclude the possibility that undermethylation of some DNA molecules, not detectable by Southern blotting, was the reason for the observed tk gene activity, a lower number of DNA molecules were injected. However, as with the non-methylated DNA, tk gene expression was also demonstrated, even after the transfer of only 2–4 methylated DNA molecules per cell (Table 3). This means that undermethylation was not the reason for tk gene expression. We can also exclude that demethylation of the injected DNA caused tk gene expression. To demonstrate this, 500 rat 2 cells were microinjected and the DNA was reextracted 8 h later; the methylation pattern was analyzed by HpaII and MspI restriction enzyme treatment and DNA blot analysis. As shown in Figure 2 the entire HSV-tk DNA was resistant to the HpaII endonuclease enzyme but was cleaved by the MspI enzyme.

Expression of the M-HpaII-methylated DNA was also demonstrated by the incorporation of ^{3}H-thymidine into DNA and by the nuclear

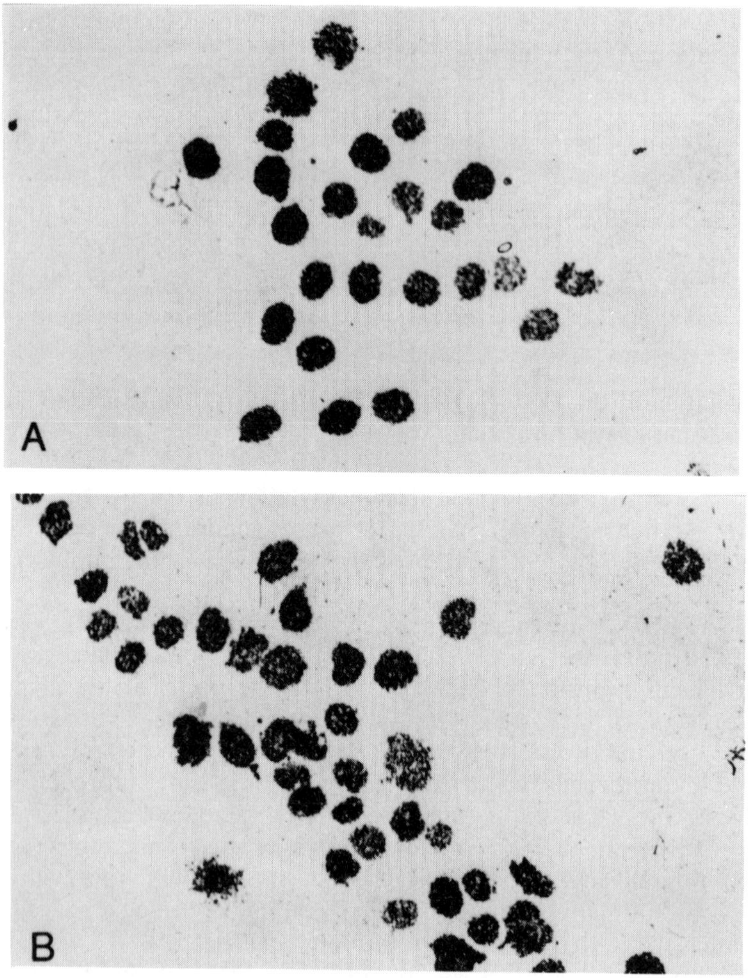

Figure 1. Autoradiogram of rat 2 cells after injection of: (A) HSV-tk DNA and (B) HSV-tk-CH3-DNA. ^{3}H-thymidine (1 μCi/ml) was added to the medium directly after injection.

Table 3. The biological activity of methylated and non-methylated HSV-tk DNA

	% of TK-positive cells after injection: number of DNA molecules injected/cell		
Material injected	200–400	20–40	2–4
pHSV-106	120–140	120–140	80–120
pHSV-106-CH_3	120–140	120–140	80–120

In all experiments the number of injected cells were counted as 100%. Rat 2 cells were labeled with ^{3}H-thymidine (1 μCi/ml) for 24 h directly after injection.

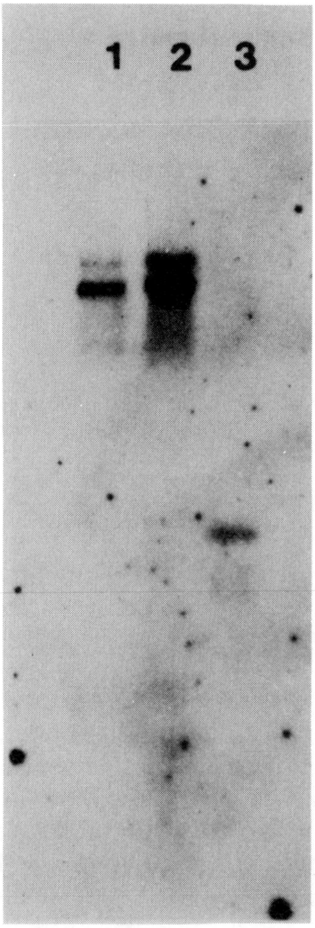

Figure 2. Methylated HSV-tk DNA reextracted from 500 rat 2 cells 8 h after injection. Isolated DNA was subdivided into three samples: (1) served as control, (2) treated with HpaII and (3) with MspI restriction endonucleases. Southern blot was hybridized with the nick translated ^{32}P-tk gene.

RNA "run" on transcription experiments after transfection of the recombinant DNA into rat 2 cells (Ca-phosphate coprecipitation). However, significant inhibition of the tk gene expression by DNA methylation was shown when ^{3}H-thymidine was added to the medium 48–72 h after injection of the DNA. While a continuous increase in the number of TK-positive cells was seen after injection of the non-methylated DNA, a sharp decrease in the number of TK positive cells occurred at this time in cells that received the methylated DNA. Cells which were incubated with thymidine 72–96 h after injection of the methylated DNA did not exhibit TK activity (Fig. 3) (Buschhausen et al., 1985).

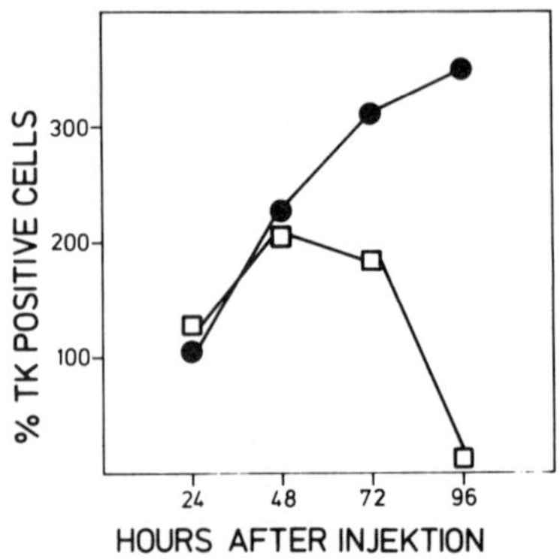

Figure 3. Time course of ^{3}H-thymidine incorporation into rat 2 cells after microinjection of the pHSV-106-CH3 and pHSV-106 DNAs. DNA was microinjected into the nuclei of rat 2 cells and ^{3}H-thymidine was added to the medium for 24 h intervals. The number of injected cells are counted as 100%. The data are the mean values from three independent injection experiments with 100 injected cells each.

To determine more precisely the time point of transition from methylation insensitivity to sensitivity, RNA was isolated from the cells at different times after DNA transfer and tested by hybridization. These experiments showed that expression of the methylated HSV-tk DNA is blocked about 8 h after microinjection (Fig. 4). This means that DNA methylation *per se* does not block the HSV-tk gene expression. What then does inhibit the transcription of the methylated HSV-tk DNA?

It has been shown that in some cases DNA methylation can prevent the binding of transcription factors (Watt and Molloy, 1988; Kovesdi et al., 1987; see also chapter Ehrlich & Ehrlich, pp. 145–168 of this book).

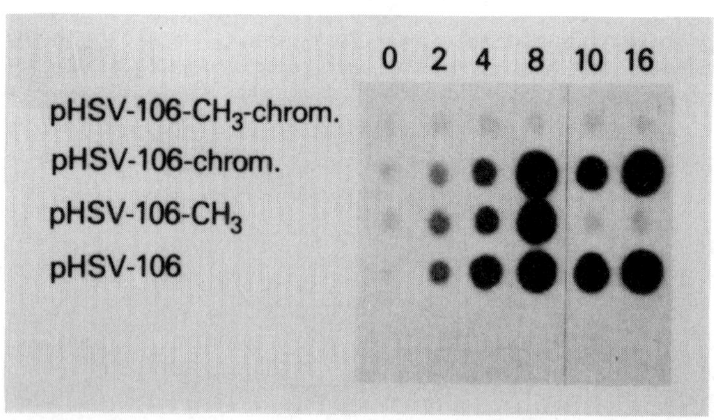

Figure 4. HSV-tk RNA dot blot. Total cellular RNA was isolated from rat 2 cells at various times, shown in hours, after microinjection of either methylated or mock-methylated HSV-tk DNA and chromatin.

But in this regard HSV-tk differs from other genes. Transcription factors, such as the Sp1 protein which is required for efficient HSV-tk gene expression, bind the methylated and non-methylated DNA *in vitro* (Ben-Hattar et al., 1988; Höller et al., 1988). Our results indicate that *in vivo* methylation does not block the binding of essential transcription factors to the HSV-tk DNA.

Furthermore, it is known that DNA methylation also facilitates the binding of proteins which may block gene transcription. For example, the MeCP-1 factor binds to the DNA when clusters of CGs are symmetrically methylated, but not to DNA with a low number of methylated sites. A specific DNA sequence is not required for the MeCP-1 binding. This factor has a cell type specificity: MeCP-1 is synthesized in HeLa cells but not in F9 cells (Boyes and Bird, 1991). It thus appears that different cell types may utilize different factors to inhibit gene expression. The MDBP factor, isolated from human placenta, also binds to hemimethylated DNA (Khan et al., 1988). Once again the fact that HSV-tk genes become methylation-sensitive only 8 h after injection argues against a direct function of these factors in rat 2 cells. Furthermore, inhibition of HSV-tk gene expression also occurs when the DNA is methylated in a non-physiological pattern.

The crucial question emerging from the experiments described above is: What happens to methylated HSV-tk DNA 8 h after injection when there is an inactivation of the tk-gene?

Among the various possibilities, one important step is the conversion of the injected naked DNA into chromatin.

To test this hypothesis, both the methylated and non-methylated HSV-tk "chromatin" were prepared. To obtain reasonable amounts of methylated chromatin, we first methylated the HSV-tk DNA with the HpaII methylase, and then reconstituted the chromatin out of DNA and purified histone octamers (Buschhausen et al., 1987). This procedure was chosen because minichromosomes are methylated very inefficiently *in vitro* by both the bacterial (e.g. M-HpaII) and the rat liver methyltransferase. This observation may imply a possible role of the nucleosomes in the *de novo* and maintenance methylation *in vivo*.

Significantly, as far we could judge, the *in vitro* reconstituted minichromosomes resemble the *in vivo* formed chromatin. Figure 5 shows an EM picture of *in vitro* reconstituted chromatin molecules. Under our reconstitution conditions (Buschhausen et al., 1987) no significant difference in the number of nucleosomes per DNA molecule and their possible phasing was observed between methylated and non-methylated chromatin.

Furthermore, *in vitro* reconstituted chromatin remained stable after injection into culture cells, without dissociation into free DNA and histone octamers. A normal nucleosomal ladder was obtained when the isolated nuclei were treated with DNaseI (pers. observ.).

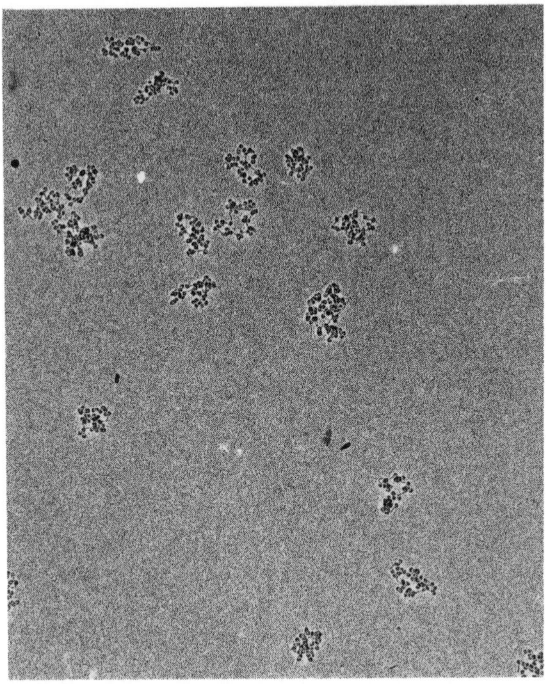

Figure 5. Electron micrographs of the *in vitro* reconstituted pHSV-tk chromatin.

In contrast to the non-methylated chromatin molecules the methylated chromatin molecules were immediately inactive after injection. As tested by thymidine incorporation or by RNA analysis the methylated chromatin remained permanently inactive (Table 4; Fig. 4). Moreover, the methylated DNA reextracted from the minichromosomes behaved as did the initial methylated DNA after injection (Table 4). These results clearly demonstrate that chromatin formation is of great importance for the inhibition of gene expression by DNA methylation.

Because methylated minichromosomes are inactive in an episomal state (Buschhausen et al., 1987) they do not require for their inactivation an integration into the higher order structure of the chromatin of host genome (e.g. 30 nm filaments).

The central question now is: What is the difference between the non-methylated and the methylated chromatin? This brings us back to the general question of what constitutes the difference between active and inactive chromatin. One feature frequently associated with active chromatin formation is the appearance of DNaseI-hypersensitive sites within or around the promoter region (Weintraub and Groudine, 1976). Although the nature of the DNase hypersensitivity is still not well understood, we investigated whether the methylated chromatin acquired

Table 4. TK activity in microinjected rat 2 cells

Material injected	^{3}H-thymidine incorporation at different times after injection 0–24 h	24–48 h	48–72 h
pHSV-106	120–140	>200	500*
pHSV-106-CH_3	120–140	160–180	>0.5
pHSV-106 chromatin	120–140	>200	500*
pHSV-106-CH_3 chromatin	0	0	0
pHSV-106 chromatin/ pHSV-106-CH_3 chromatin#	120–140	>200	500*
pHSV-106-CH_3 reextracted from pHSV-106-CH_3 chromatin^0	120–140	160–180	<0.5

After microinjection, cells were labeled with ^{3}H-thymidine (1 μCi/ml) for 24 h intervals at the time indicated and processed for autoradiography. The number of injected cells was counted as 100%. Data given are average values from five independent experiments with 50 injected cells each and are expressed as % TK positive cells.
*The number of positive cells (48–72 h after injection) was estimated and not counted.
#pHSV-106-CH_3 chromatin (0.01 mg/ml) was mixed before microinjection with mock-methylated chromatin (0.01 mg/ml) at 1:1 ratio (vol/vol).
^{0}pHSV-106-CH_3 DNA was reextracted from the reconstituted chromatin by NaDodSO4/phenol treatment.

DNase-hypersensitivity after injection. We therefore injected the methylated and the non-methylated chromatin into the nuclei of the rat 2 cells and treated the isolated nuclei with DNaseI. We found that 20 h after injection both the non-methylated and the methylated chromatin molecules contained a DNase-hypersensitive site within the promoter region. This hypersensitivity was not detected shortly after chromatin injection (pers. observ.). Keshet et al. (1986) described an inverse correlation between DNA methylation and DNase-hypersensitivity. However, these experiments were done with stable transformed cell lines many cell generations after DNA transfection. We can therefore conclude that the generation of DNaseI hypersensitivity does not suffice to allow expression of the HSV-tk (methylated) chromatin. The lack of hypersensitivity described by this group may, therefore, be more a consequence rather than the reason for gene inactivation.

5 Hemimethylated chromatin is inactive

We also investigated whether the methylation of only one DNA strand (hemimethylated DNA) may cause inhibition of gene expression. If so this would exclude the involvement of specific factors requiring symmetrical DNA methylation for the inactivation of the HSV-tk chromatin. To obtain hemimethylated molecules the HSV-tk gene was inserted into the replicative form of the M13 phage DNA and propagated in *E. coli*,

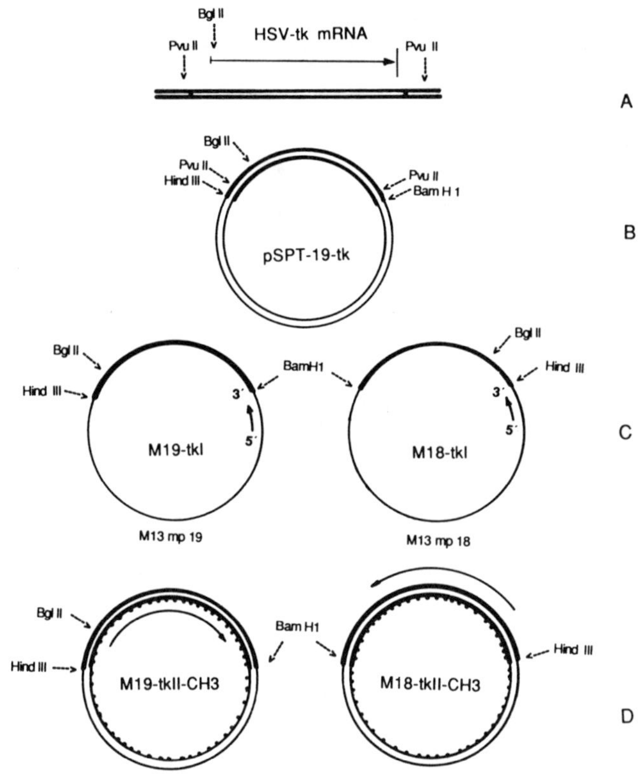

Figure 6. Schematic representation of *in vitro* replicated HSV-tk DNA. (A) The HSV-tk gene was isolated as a PvuII restriction fragment from plasmid pHSV-106. The transcriptional orientation is indicated by the arrow. (B) the PvuII fragment was inserted into the PvuII site of plasmid pSPT19 and propagated in *E. coli* as pSPT19-tk. (C) The tk gene was then isolated from pSPT19-tk as a HindIII-BamH1 DNA fragment and inserted into M13mp19 and M13mp18 RF DNA and propagated in *E. coli*. The single-stranded M19-tkI and M18-tkI was isolated from the bacteriophages. M19-tkI contains the non-coding strand and M18-tkI contains the coding strand of the HSV-tk gene. The location and orientation of the universal M13 primer are indicated. (D) The complementary DNA strands were synthesized *in vitro*. Dots on the inner circle of M19-tkII-CH3 and M18-tkII-CH3 represent 5-methyl-cytosine residues.

and then the single-stranded DNA was isolated from the purified phage particles. The cloning strategy is illustrated in Figure 6. The M19-tkI DNA contains the non-coding strand and the M18-tkI the coding strand of the HSV-tk gene. Both constructs were used for *in vitro* second-strand DNA synthesis, generating the newly synthesized DNA strand either in the methylated or in the non-methylated form. After chromatin formation *in vitro*, the hemimethylated molecules were injected into the rat 2 cells. We found that regardless of whether the coding or the non-coding strand was methylated the chromatin was always inactive.

We also methylated the pSPT19-tk DNA with HaeIII methylase. This bacterial enzyme modifies the DNA in a non-physiological fashion, methylating exclusively the first cytosine within the GGCC tetranucleotide sequence, a methylation pattern which cannot be maintained in rat 2 cells. In spite of this unusual methylation pattern the rat 2 cells did not allow expression of the M-HaeIII methylated chromatin for about 72 h after injection. Thereafter tk gene was expressed and a high

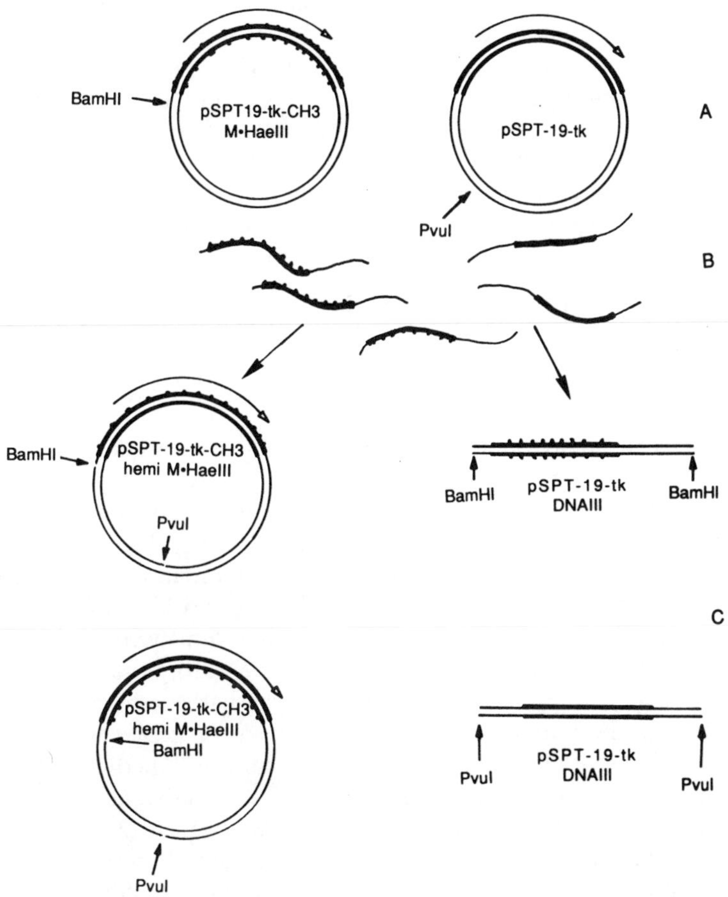

Figure 7. Generation of M-HaeIII hemimethylated HSV-tk DNA. (A) pSPT19-tk-CH3 DNA methylated by M-HaeIII enzyme and the non-methylated pSPT19-tk DNA. Heavy lines represent the HSV-tk gene inserted into the pSPT19 DNA. Dots represent the M-HaeIII sites within the HSV-tk gene. Transcriptional orientation is indicated by arrows. (B) The linearized single-stranded DNA. The pSPT19-tk-CH3 DNA was linearized with BamHI endonuclease and the pSPT19-tk DNA with the PvuI enzyme. Both linear double-stranded molecules were mixed in a 1:1 ratio, denatured by boiling for 10 min and renatured at 60°C. (C) The renatured DNA. The circular (form II) DNA represents the hemimethylated DNA (coding and non-coding DNA) and the linear (form III) DNA the fully methylated and the non-methylated DNA.

proportion of TK-positive cell clones were isolated from the injected cells (Deobagkar et al., 1990). Thus M-HaeIII methylation inactivates transiently the HSV-tk chromatin expression. To obtain M-HaeIII hemimethylated HSV-tk DNA, we followed the strategy illustrated in Figure 7. In short, the HSV-tk DNA was methylated with the HaeIII methylase, linearized by the Bam H1 restriction enzyme and mixed with the same amount of non-methylated HSV-tk DNA that had been linearized by the PvuI endonuclease. The DNA mixture was then boiled and allowed to renature. Under these reannealing conditions two kinds of double-stranded DNA molecules are formed: the circular (form II) and the linear (form III). The circular fraction contains hemimethylated DNA exclusively and the linear fraction both fully and non-methylated DNA. After microinjection of the hemimethylated chromatin the HSV-tk gene was inactive for about 48 h. The methylation of either the coding or the non-coding DNA strand caused HSV-tk chromatin inactivation. The reactivation after 48 h of incubation was linked to the generation of non-methylated chromatin molecules. These experiments clearly demonstrate that methylation of only one strand is sufficient to cause inactivation of the HSV-tk chromatin.

6 Symmetrical methylation occurs before DNA replication

So far it is not certain when, after injection, the hemimethylated molecules are converted into fully methylated molecules. It is assumed that in mammalian cells, *de novo* methylation and maintenance methylation are carried out by the same enzyme (Bestor and Ingram, 1983). The current hypothesis is that maintenance methylation occurs after DNA replication and that the enzyme is mainly active during the S and the G2 phase of the cell cycle. To investigate whether methylation occurs exclusively in the S and G2 phase 20–40 DNA molecules were injected into the nuclei of TK-minus cells grown in HAT-medium after the injection. Under these conditions only the TK-positive cells can grow. After the transfer of the non-methylated HSV-tk DNA, 25–30% of the recipient cells formed HAT-medium-resistant cell clones. In contrast, HAT-medium-resistant cell clones were not obtained after microinjection of either the hemimethylated M19-tkII-CH3 or the M18-tkII-CH3 DNA (Fig. 6, Table 5). This observation clearly demonstrates that the hemimethylated DNA was converted into symmetrical methylated molecules before DNA replication. If this were not so, the same number of HAT-medium positive cell clones would have been generated as after injection of the non-methylated DNA, because the first round of semi-conservative DNA replication would have generated non-methylated and hemimethylated DNA molecules, and hence HAT-medium-resis-

Table 5.

Material injected	HAT-medium-positive cell clones (% of injected cells)
M18-tkII	25–30
M18–tkII-CH_3	0
M18-tkII-CH_3 + 5-azaC*	20–25
M19-tkII	20–30
M19-tkII-CH_3	0
M19-tkII-CH_3 + 5-azaC*	20–25
pSPT-19-tk	25–30
pSPT-19-tk-CH_3	0
pSPT-19-tk-CH_3 + 5-azaC*	20–25

Rat 2 cells grown on glass slides were microinjected with 20–40 DNA molecules/cell and transferred into HAT-medium.
*After microinjection, 5-azacytidine (4 μM end concentration) was added for 20 h to the culture medium. Thereafter cells were transferred to HAT-medium.

tant cells. So far we have not tested whether all the potential CpG sites of the second DNA strand become methylated or not.

Since the HSV-tk constructs used in our experiments do not contain a replication origin specific for mammalian cells, replication of the trans-DNA requires integration into the host genome. It has not yet been determined when after injection the trans-DNA is integrated and replicated into the host DNA. To test whether this occurs during the first cell cycle after gene transfer, the demethylating agent 5-azacytidine was added for 20 h to the cells directly after injection. Trace amounts of 5-azacytidine were removed by washing the cells with DMEM-medium and the cells were further incubated in the HAT-medium. As shown in Table 5, the 5-azacytidine treatment caused growth of these cells in the HAT-medium. In order to get more precise information on the timing of the integration of the trans-DNA after injection, we used synchronized cells. To avoid chemical treatment for synchronization, shake-off cells (M-phase cells) were used in our experiments. At different times after the shake-off, the cells were microinjected and treated for two hours with 5-azacytidine (Fig. 8). The cells were then washed and transferred to the HAT-medium and the cell clones were counted. The earliest time point when the cells are accessible for microinjection is 2 h after the "shake off". Figure 8 shows that 10% of the cells injected 2 h after preparation and treated subsequently for 2 h with 5-azacytidine were converted into HAT-medium-resistant cells. The maximal number of HAT-medium-positive cell clones was obtained 4–6 h after the shake-off, when about 30% of the cells formed colonies in the HAT-medium. None of the cells which were injected and treated with 5-azacytidine 8 to 18 h after shake-off were converted into HAT-medium-resistant cells. HAT-medium-positive clones were again obtained 20–22 h after shake-off. If 5-azacytidine was omitted, none of the

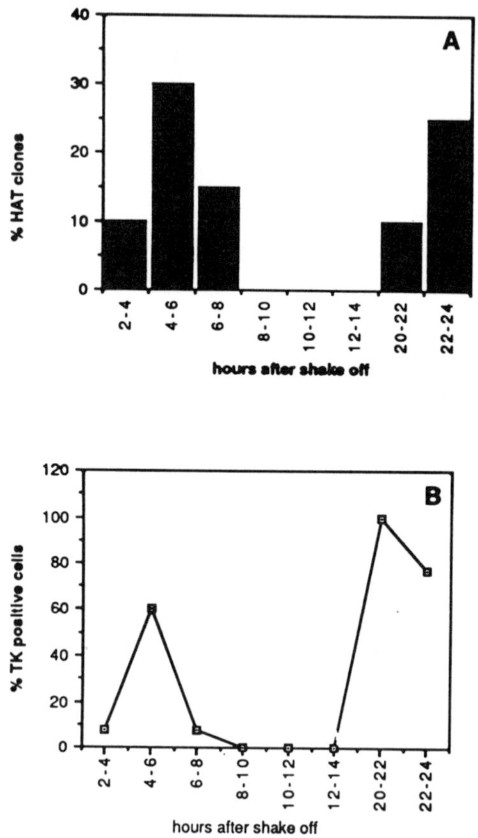

Figure 8. Shake-off cells were microinjected at different times after isolation as indicated and subsequently treated for a further 2 h with 5-azacytidine. Thereafter the cells were washed with DMEM medium and transferred into HAT-medium. (A) Shows the number of injected cells which grew out into HAT-medium-resistant cell clones. The number of injected cells were counted as 100%. (B) Shake-off cells were microinjected with the pSTP-19-tk at the time indicated. After injection, ^{3}H-thymidine (1 μCi/ml) was added to the culture medium for 2 h. Thereafter the cells were fixed and processed for autoradiography.

recipient cells grew in HAT-medium, regardless of the time after shake-off when microinjection occurred; however, positive clones were always obtained when 5-azacytidine remained with the cells for 20 h (data not shown). As shown by thymidine incorporation and autoradiography the injected, shake-off cells, enter into the S-phase as early as 2–4 h after the preparation (Fig. 8).

Reactivation of the methylated DNA by 5-azacytidine involves the inhibition of methyltransferase and possibly the incorporation of the cytosine analog into the DNA. After microinjection of the M-HpaII-methylated DNA (pSPT-19-tk-CH3) into the shake-off cells, followed by the 2-h 5-azacytidine treatment, reactivation occurred as it did after injection of the hemimethylated DNA (Sandberg et al., 1991). These results indicate that the 5-azacytidine inhibited methylation of DNA during the first round of replication, generating hybrid DNA molecules with one methylated DNA strand and leaving the second strand non-methylated. The subsequent replication cycle then caused demethylation of the second DNA strand and thus HSV-tk gene activation.

7 Methylation of single HpaII sites causes inactivation of the HSV-tk gene

The current hypothesis is that the change in the methylation pattern within the promoter region is crucial for the regulation of gene expression, which implies that transcription initiation is affected by DNA methylation. It has been demonstrated that the change in the methylation status of single sites within the promoter region can have a significant effect on gene transcription (see review by Dörfler, 1983).

We have demonstrated here 1) that the methylation of either the promoter or the coding region causes HSV-tk inactivation; and 2) that the methylation of a single site far down from the promoter blocks tk expression.

By second-strand DNA synthesis, it is possible to methylate selectively specific DNA segments. Using different DNA fragments as primer for *in vitro* DNA synthesis the desired DNA segments were methylated as illustrated in Figure 9. The effect of partial DNA methylation on tk gene expression was analyzed by autoradiography and colony formation in HAT-medium. These experiments have shown that methylation of either the promoter or the coding region is sufficient to cause tk inactivation. Furthermore, the first part of the HSV-tk coding region does not have to be methylated to cause gene inactivation. The HSV-tk gene was completely inactivated when the Sph1-HindIII DNA fragment, which hybridizes with the entire tk promoter and the first 499 nucleotides of the gene, was used as primer for the subsequent synthesis of methylated DNA (Fig. 9-(5)). Addition of 5-azacytidine to the medium reactivated the methylated DNA. However, when the DNA segment between the nucleotides 811-1309 was methylated, tk gene expression occurred as with the non-methylated DNA (Fig. 9-(6)).

We also investigated whether the methylation of single HpaII site either in the promoter or in the coding region may cause tk inactivation.

As shown in Figure 10 the HSV-tk gene contains 16 HpaII sites, one within the promoter region and 15 in the coding part. To test whether methylation of single HpaII sites leads to gene inactivation 16 primers (14 mer, HpaII-CH3 1-16) corresponding to the 16 HpaII sites were synthesized *in vitro* with the internal cytosine of the CCGG tetra-

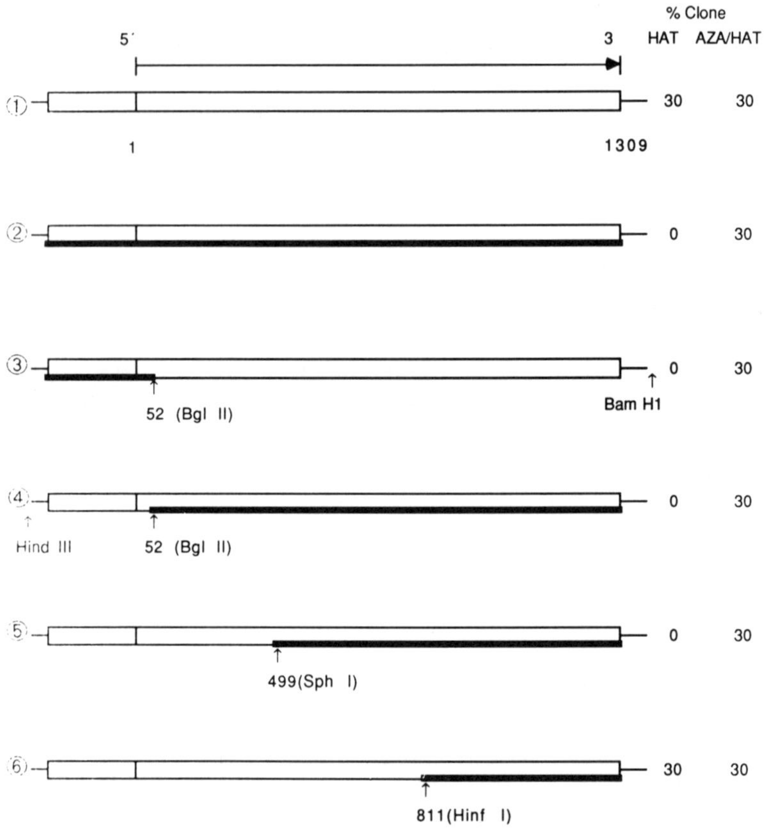

Figure 9. Partial methylation of the HSV-tk DNA by second strand synthesis *in vitro*. M19-tkI DNA was used as template for the second-strand synthesis and the black lines indicate the methylated part of the DNA. The M13 universal primer was used to obtain: 1) The second DNA strand in the non-methylated form (control); 2) in the methylated form; 3) HSV-tk BglII-BamH1 fragment for promoter methylation; 4) HSV-tk HindIII-BglII fragment for methylation of the coding region between nucleotides 52-1309; 5) HSV-tk SphI-HindIII fragment for methylation between nucleotides 499-1309 and 6) HSV-tk HinfI-HindIII fragment for methylation between nucleotides 811-1309. The efficiency of colony formation in HAT-medium is given in % of microinjected cells.

nucleotide sequence in the methylated form (Fig. 10) and these methylated oligonucleotides were used as primers for *in vitro* synthesis of the second strand of M19-tkI DNA. In this way 16 different constructs were obtained, each with only one methyl-cytosine residue. As summarized in Table 6 methylation of the HpaII sites 2–6 caused tk inactivation. After injection of these constructs no TK-positive cell clones were obtained. However, when 5-azacytidine was added for a 20-h period to the cells after injection, HAT-positive cell clones appeared as they had after injection of the non-methylated DNA. In contrast methylation of all the

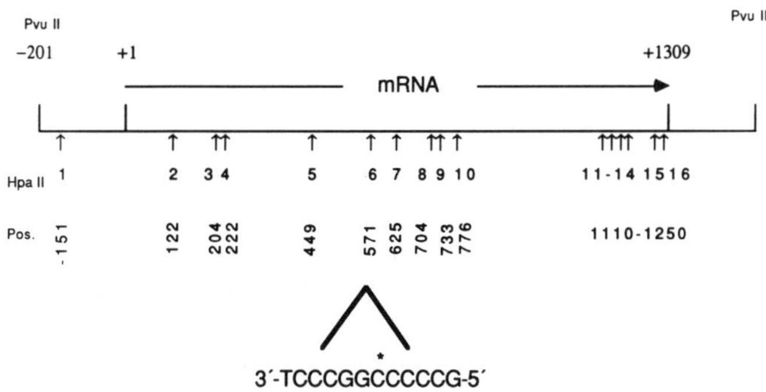

Figure 10. Methylation of single HpaII sites. The position of the 16 HpaII sites on the HSV-tk gene are indicated by arrows and the nucleotide number. The nucleotide sequence of the primer 6 is shown below.

Table 6. Inhibition of HSV-tk gene expression by methylation of single HpaII sites

Material injected	HAT-medium positive cell clones (% injected cells)	After 5-azacytidine* treatment
M19-tkII(1)-CH3	25–30	25–30
M19-tkII(2)-CH3	0	25–30
M19-tkII(3)-CH3	0	25–30
M19-tkII(4)-CH3	0	25–30
M19-tkII(5)-CH3	0	25–30
M19-tkII(6)-CH3	0	25–30
M19-tkII(7–16)-CH3	25–30	25–30

Rat 2 cells were microinjected with 20–40 DNA molecules/cell and transferred into HAT-medium. The number of injected cells were counted as 100%. HAT-positive cell clones were determined 10–14 days later.
*5-azacytidine (1–10 μM end concentration) was added for 20 h to the HAT-medium.

other sites as either single sites or in combination with each other failed to block HSV-tk gene expression.

8 Conclusion

Since in eukaryotic cells inactive and active genes (with the exception of, e.g., ribosomal genes; Conconi et al., 1989) exist as chromatin, and since chromatin formation is essential for the inhibition of the M-HpaII-methylated DNA, one has to ask what makes the difference between the methylated and the non-methylated chromatin and what is the fate of the nucleosomes during transcription? Up to now after *in vitro* reconstitution we have not observed a significant difference between methylated and non-methylated chromatin with regard to the number of nucleosomes and their positioning in the chromatin. However, the generation of DNaseI-hypersensitive sites observed after the injection of DNA indicates that the position of the nucleosomes may be changed in the recipient cells.

It is assumed that the histone octamers are not only of great importance for DNA packaging, but that the nucleosomes may also play an important role in the regulation of gene expression (Grunstein, 1990). However, the precise role of the nucleosomes in gene expression during cell differentiation and in differentiated cells remains a matter of debate. There is increasing evidence that key nucleosomes can prevent protein binding. It has been shown that a single nucleosome covering the promoter region suppresses the initiation of transcription *in vitro* (Lorch et al., 1987). Furthermore the displacement *in vivo* of key nucleosomes seems to be of importance for gene activation. For example, two nucleosomes flanking the promoter part of the yeast PHO5 gene are displaced before transcription initiation (Almer et al., 1986) and in SV40 virus-infected cells about 30% of the viral minichromosomes have a nucleosome-free promoter region, although only a small proportion of them is actively transcribed (see the review by Thoma, 1991).

If certain nucleosomes have to be displaced in order to mediate gene expression, the question must be raised as to what the signal for and the mechanism of the nucleosomal displacement might be. It is tempting to speculate that in some cases the change of the DNA methylation pattern may partly influence the nucleosome mobility. However, our microinjection experiments have clearly demonstrated that generation of DNaseI hypersensitivity, as an indication of the displacement or unfolding of nucleosomes, is not sufficient to support expression of M-HpaII-methylated chromatin. We have to ask therefore, whether methylation blocks the elongation of transcription. The hypothesis that methylation may also interfere with transcript elongation is further supported by our observation that methylation of single sites, up to 571 basepairs from the transcription initiation site, also blocks HSV-tk expression.

The inhibition of gene expression can be mediated by the methylation of different sites. This means that the cell has more options for achieving the inhibition of gene expression by DNA methylation. Furthermore, this suggests that there may be a variation in the DNA methylation pattern in cells in different organs.

Indeed, there is now increasing evidence that the elongation step of transcription is also important for the regulation of gene activation. This might well require a further modification of the RNA polymerase II, such as phosphorylation at the carboxyl terminal domain (Cisek and Corden, 1989; Lee and Greenleaf, 1989) and an interplay with additional transcription factors (e.g., TFII-S, Reines et al., 1989). Furthermore, it has become evident that during transcription the RNA polymerase II can pause or terminate prematurely (for a review see

Spencer and Groudine, 1990). This demonstrates the existence of down-stream regulatory elements modulating the gene expression in eukaryotic cells. The nature of these pausing points is not well understood. Efficient *in vivo* transcription elongation may require further transacting factors (e.g., antipausing factors) or/and a specific chromatin structure.

The possibility that nucleosomes can indeed block transcription elongation was elegantly documented by Rougvie and Lis (1988). In *Drosophila*, transcription initiation of the hsp 70 gene occurs in non-heat-shocked cells and transcription elongation occurs after nucleosome displacement. It is therefore tempting to speculate that under certain conditions DNA methylation may convert pausing sites into termination sites. This also would explain why single down-stream sites, when methylated, block gene expression. It is unlikely that this is due to a long-distance effect since the methylation-insensitive SV40 promoter does not overcome the methylation insensitivity of the HSV-tk coding region (data not shown).

In this context, it is quite possible that methylation-insensitive genes such as the SV40 T-antigen, do not contain critical sites that can be converted by methylation into stopping points. In the case where hypermethylation precedes gene activation DNA methylation may prevent nucleosomes from being placed in a critical position at either the promoter or the coding region of the gene, keeping it open for transcription initiation and elongation.

In conclusion, we have to assume that the change in the DNA methylation pattern as a regulatory element for gene expression acts at different levels by different mechanisms. This means that each individual gene has to be analyzed separately to learn what effect DNA methylation may have on its expression.

Acknowledgment. This work was supported by the Deutsche Forschungsgemeinschaft.

Almer, A., Rudolph, H., Hinnen, A., and Hörz, W. (1986) Removal of positioned nucleosomes from the yeast PHO5 promoter upon PHO5 induction releases additional upstream activating DNA elements. EMBO J. *5*, 2689–2696.

Ben-Hattar, J., and Jiricny, J. (1988) Methylation of single CpG dinucleotides within a promoter element of the herpes simplex virus tk gene reduces its transcription *in vivo*. Gene *65*, 219–227.

Bestor, T. H., and Ingram, V. M. (1983) Two DNA methyltransferases from murine erythroleukemia cells: purification, sequence specificity, and mode of interaction with DNA. Proc. Natl. Acad. Sci. USA *80*, 5559–5563.

Boyes, J., and Bird, A. (1991) DNA methylation inhibits transcription indirectly via a methyl-CpG binding protein. Cell *64*, 1123–1134.

Buschhausen, G., Graessmann, M., and Graessmann, A. (1985) Inhibition of herpes simplex thymidine kinase gene expression by DNA methylation is an indirect effect. Nucl. Acids Res. *13*, 5503–5513.

Buschhausen, G., Wittig, B., Graessmann, M., and Graessmann, A. (1987) Chromatin structure is required to block transcription of the methylated herpes simplex virus thymidine kinase gene. Proc. Natl. Acad. Sci. USA *84*, 1177–1181.

Cisek, L. J., and Corden, J. L. (1989) Phosphorylation of RNA polymerase by the murine homologue of the cell-cycle control protein cdc 2. Nature *339*, 679–684.
Conconi, A., Widmer, R. M., Koller, T., and Sogo, J. M. (1989) Two different chromatin structures coexist in ribosomal RNA genes throughout the cell cycle. Cell *57*, 753–761.
Deobagkar, D. D., Liebler, M., Graessmann, M., and Graessmann, A. (1990) Hemimethylation of DNA prevents chromatin expression. Proc. Natl. Acad. Sci. USA *87*, 1691–1695.
Dörfler, W. (1983) DNA methylation and gene activity. Ann. Rev. Biochem. *52*, 93–124.
Götz, F., Schulze-Forster, K., Wagner, H., Kröger, H., and Simon, D. (1990) Transcription inhibition of SV40 by *in vitro* DNA methylation. Biochim. Biophy. Acta *1087*, 323–329.
Graessmann, M., and Graessmann, A. (1983) Microinjection of tissue culture cells. Methods Enzymol. *101*, 482–492.
Graessmann, A., Bumke-Vogt, C., Buschhaussen, G., Bauer, M., and Graessmann, M. (1985) SV40 chromatin structure is not essential for viral gene expression. FEBS Lett. *179*, 41–45.
Graessmann, M., Graessmann, A., Wagner, H., Werner, E., and Simon, D. (1983) Complete DNA methylation does not prevent polyoma and simian virus 40 virus early gene expression. Proc. Natl. Acad. Sci. USA *80*, 6470–6474.
Grunstein, M. (1990) Nucleosomes: regulators of transcription. TIG *6*, 395–400.
Höller, M., Westin, G., Jiricny, J., and Schaffner, W. (1988) Spl transcription factor binds DNA and activates transcription even when the binding site is CpG methylated. Genes & Develop. *2*, 1127–1135.
Jones, P. A. (1985) Altering gene expression with 5-azacytidine. Cell *40*, 485–486.
Khan, R., Zhang, X. Y., Supakar, P. C., Ehrlich, K. C., and Ehrlich, M. (1988) Human methylated DNA-binding protein. Determinants of a pBR 322 recognition site. J. Biol. Chem. *263*, 14374–14383.
Keshet, I., Lieman-Hurwitz, J., and Cedar, H. (1986) DNA methylation affects the formation of active chromatin. Cell *44*, 535–543.
Kovesdi, I., Reichel, R., and Nevins, J. R. (1987) Role of an adenovirus E2 promoter binding factor in E1A-mediated coordinate gene control. Proc. Natl. Acad. Sci. USA *84*, 2180–2184.
Lee, J. M., and Greenleaf, A. L. (1989) A protein kinase that phosphorylates the C-terminal repeat domain of the largest subunit of RNA polymerase II. Proc. Natl. Acad. Sci. USA *86*, 3624–3628.
Lorch, Y., LaPointe, J. W., and Kornberg, R. D. (1987) Nucleosomes inhibit the initiation of transcription but allow chain elongation with the displacement of histones. Cell *49*, 203–210.
Rougvie, A. E., and Lis, J. T. (1988) The RNA polymerase II molecule at the 5′ end of the uninduced hsp 70 gene of D. melanogaster is transcriptionally engaged. Cell *54*, 795–804.
Reines, D., Chamberlin, M. J., and Kane, C. M. (1989) Transcription elongation factor SII (TFIIS) enables RNA polymerase II to elongate through a block to transcription in a human gene *in vitro*. J. Biol. Chem. *264*, 10799–10809.
Spencer, C. A., and Groudine, M. (1990) Transcription elongation and eukaryotic gene regulation. Oncogene *5*, 777–785.
Sandberg, G., Guhl, E., Graessmann, M., and Graessmann, A. (1991) After microinjection hemimethylated DNA is converted into symmetrically methylated DNA before DNA replication. FEBS Lett. *283*, 247–250.
Tanaka, K., Appella, E., and Jay, G. (1983) Developmental activation of the H-2K gene is correlated with an increase in DNA methylation. Cell *35*, 457–465.
Thoma, F. (1991) Structural changes in nucleosomes during transcription: strip, split or flip? TIG *7*, 175–177.
Waalwijk, C., and Flavell, R. A. (1978) MspI, an isoschizomer of HpaII which cleaves both unmethylated and methylated HpaII sites. Nucl. Acids Res. *5*, 3231–3236.
Wasylyk, B. (1988) Enhancers and transcription factors in the control of gene expression. Biochim. Biophy. Acta *951*, 17–35.
Watt, F., and Molloy, P. L. (1988) Cytosine methylation prevents binding to DNA of a HeLa cell transcription factor required for optimal expression of the adenovirus major late promoter. Genes & Develop. *2*, 1136–1143.
Wigler, M., Levy, D., and Perucho, M. (1981) The somatic replication of DNA methylation. Cell *24*, 33–40.
Weintraub, H., and Groudine, M. (1976) Chromosomal subunits in active genes have an altered conformation. Globin genes are digested by deoxyribonuclease I in red blood cell nuclei but not in fibroblast nuclei. Science *193*, 848–856.

DNA Methylation: Molecular Biology and Biological Significance
ed. by J. P. Jost & H. P. Saluz

Steroid hormone dependent changes in DNA methylation and its significance for the activation or silencing of specific genes

J. P. Jost[1] and H. P. Saluz[2]

[1]*Friedrich Miescher-Institut, P. O. Box 2543, CH-4002 Basel, Switzerland, and* [2]*IRBM, via Pontina KM 30.6, Pomezia, Rome, Italy*

1 Introduction

Steroid hormones are known to be directly involved in the differentiation of organs and tissues and to modulate the expression of specific genes. For the differentiation of specific tissues, methylation and demethylation of DNA may play an important role (Holliday, 1990), whereas for the modulation of gene expression specific steroid receptors rather than methylation of DNA have the key role.

In this chapter we first review two different systems where long-term deprivation or treatment of two different types of cells with steroid hormones influences the methylation and demethylation of the same gene. In our review of the third system we deal more directly with the specific interaction of nucleoproteins with the methylated DNA and the mechanism responsible for the site-specific demethylation of the avian vitellogenin gene.

2 Long-term treatment of mouse thymoma cells with glucocorticoids favors the selective demethylation of MMTV-LR DNA

The retrovirus, Mouse Mammary Tumor Virus (MMTV), which is responsible for mammary carcinomas in many inbred strains of mice, is transmitted either as a provirus integrated in the mouse genome, or as a milk borne-virus (Cohen, 1980 and references therein). There is usually no expression of the endogenous genetically transmitted MMTV proviral sequence in most strains of mice whereas the acquired milk-borne viral sequences are expressed. As shown by Cohen (1980), the main difference between the DNA of the endogenous and the milk-borne virus sequence is the state of cytosine methylation within CpG dinucleotides. The DNA of the silent endogenous provirus is methylated (Cohen, 1980), whereas expressed milk-borne virus genome is not methylated. It has also been shown that in certain tissues such as the liver and lymphoid cells, glucocorticoids regulate the expression of the integrated MMTV proviral DNA (Ringold et al., 1977; Stallcup et al., 1979).

Mermod et al. (1983) have analyzed in detail the control of MMTV-RNA synthesis by dexamethasone in wild-type T_1M_1, dexamethasone-resistant $T_1M_1^R$ and dexamethasone-supersensitive $T_1M_1^{SS}$ thymoma cell lines. They also studied by means of restriction-isochisomeric restriction endonucleases the state of methylation of several CpGs situated within or flanking the long terminal repeat (LTR). The basal level of MMTV RNA synthesis in the wild-type T_1M_1 and the glucocorticoid-hypersensitive cell line $T_1M_1^{SS}$ is extremely low, and the addition of glucocorticoids to the cells resulted within a very short time in a

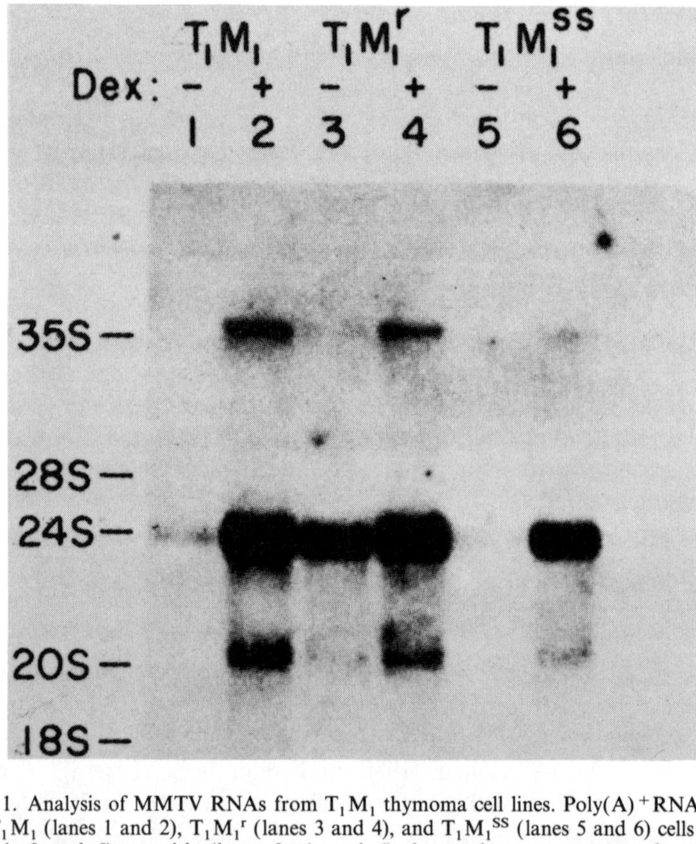

Figure 1. Analysis of MMTV RNAs from T_1M_1 thymoma cell lines. Poly(A)$^+$RNAs (5 μg) from T_1M_1 (lanes 1 and 2), $T_1M_1^r$ (lanes 3 and 4), and $T_1M_1^{SS}$ (lanes 5 and 6) cells without (lanes 1, 3 and 5) or with (lanes 2, 4, and 6) dexamethasone treatment for 6 h were electrophoresed in a 1.4% agarose gel. The RNAs were then transferred from the gel to nitrocellulose filters, hybridized with ^{32}P-labeled LTR and cloned envelope DNAs, and exposed for autoradiography for 5 days. The positions of the MMTV-specific RNAs are noted by their sedimentation constants (24S), which were determined by electrophoresis of the rRNAs in a parallel lane (data not shown) (Mermod et al., 1983). Reprinted with permission from Proc. Natl. Acad. Sci. USA *80* (1983) 110–114.

dramatic increase in MMTV RNA synthesis in the absence of DNA demethylation (Figs. 1 and 2). In sharp contrast, in the glucocorticoid-resistant cell line $T_1M_1^r$, the expression of MMTV RNA is constitutive. Addition of glucocorticoids to $T_1M_1^r$ cells had only a marginal effect on the synthesis of MMTV RNA (see Fig. 1), and the MMTV-LTR DNA was unmethylated (see Fig. 2). The resistant variant $T_1M_1^r$ was selected by continuous growth of the wild-type T_1M_1 over a period of several weeks in the presence of a high concentration (10 μM) of dexamethasone. The above results suggest that in the constitutive strain, $M_1T_1^r$, the demethylation of the mCpG occurred by a mechanism involving several cell divisions in the presence of the steroid hormone.

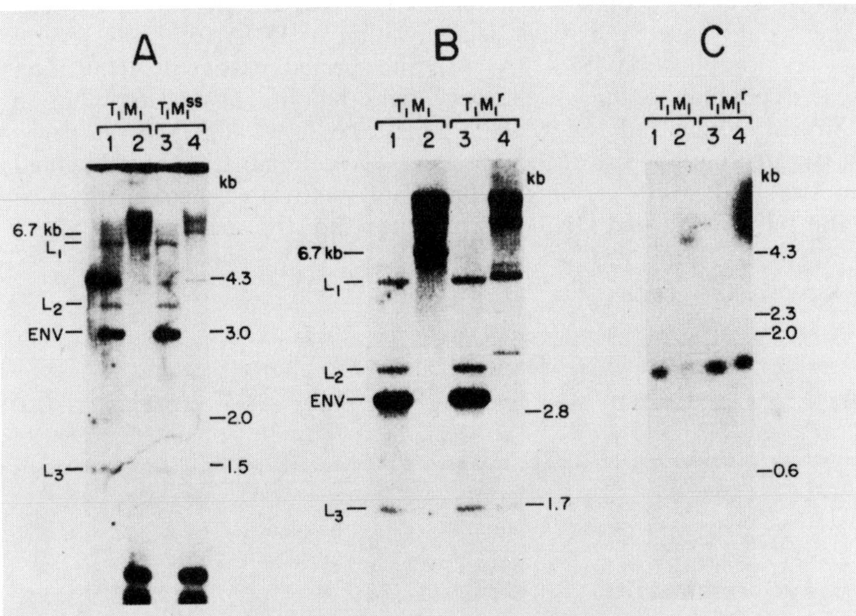

Figure 2. Methylation of MMTV sequences in T_1M_1 cell lines. Cellular DNA from the designated cell lines was cleaved to completion with various restriction enzymes, and 10-μg samples were analyzed by Southern blot hybridization to LTR and envelope DNA probes. (A) DNAs extracted from T_1M_1 (lanes 1 and 2) and $T_1M_1^{ss}$ (lanes 3 and 4) were cut with MspI (lanes 1 and 3) and HpaII (lanes 2 and 4) and hybridized to the LTR and envelope probes. Completion of HpaII digestion is shown by the low molecular weight bands generated by digestion of traces of PMB9 added to the DNA of lanes 2 and 4. Molecular weights (kb) of the Pst I fragments of the LTR and envelope (ENV) and ot pBR322 are shown. (B) DNAs extracted from T_1M_1 (lanes 1 and 2) and $T_1M_1^r$ (lanes 3 and 4) were digested with MspI (lanes 1 and 3) or HpaII (lanes 2 and 4) and hybridized to the LTR and envelope probes. HpaII activity was checked by inclusion of ϕX174 DNA in lanes 2 and 4, generating 2.8- and 1.7-kb bands that were visualized by staining. (C) DNAs extracted from T_1M_1 (lanes 1 and 2) and $T_1M_1^r$ (lanes 3 and 4) were digested with HhaI/MspI (lanes 1 and 3) and HhaI/HpaII (lanes 2 and 4) and hybridized to the LTR probe. L_1, L_2, and L_3, 5′-LTR fragments; ENV, the 3-kb fragment containing the 3′ LTR and the envelope gene (Mermod et al., 1983). Reprinted with permission from Proc. Natl. Acad. Sci. USA *80* (1983) 110–114.

Whether there is a causal link between the demethylation of mCpGs and the constitutive expression of MMTV RNA remains to be demonstrated. In the above system we have two clear cut situations: the short-term induction of MMTV-LTR expression in the absence of DNA demethylation and in the long-term glucocorticoid treated cells, the constitutive expression of MMTV-LTR in presence of demethylated DNA. If the role of methylation of specific CpGs is to assure a high binding affinity of a putative repressor, it is conceivable that the short-term treatment of the cells T_1M_1 with dexamethasone results only in the lowering of the repressor binding activity. This should be sufficient to relieve the repression of MMTV-LTR expression in the absence of DNA demethylation. In the second case, as a consequence of the lowering of the repressor binding activity over a long period of time, one would have the passive demethylation of DNA following several cycles of replication (Figs. 3A, 4B). Site-specific, passive demethylation could possibly occur by the very tight binding of a determination protein (Razin and Riggs, 1980; Riggs and Jones, 1983) at the newly replicated binding site of the repressor. The determination protein could possibly bind with very high affinity non-methylated, hemi-methylated and fully methylated DNA. It would have no strict sequence specificity

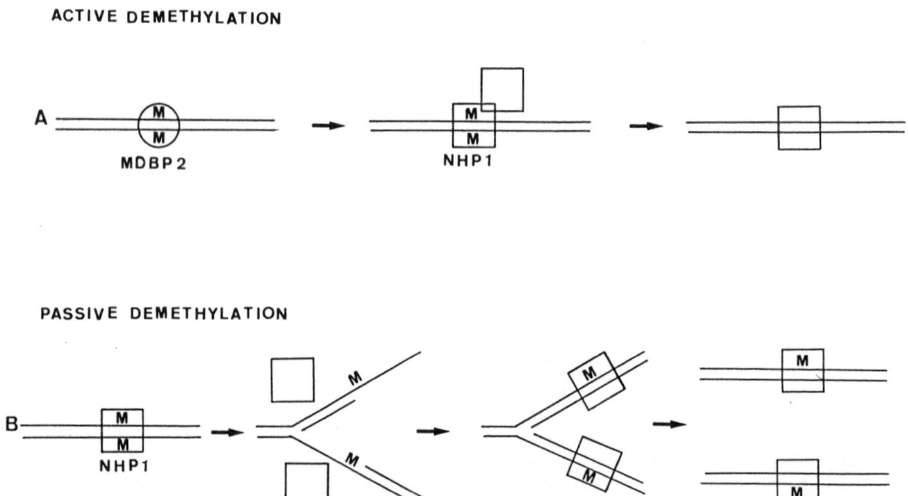

Figure 3. Model of the effect of long-term steroid treatment (or deprivation) of cells on DNA methylation. Long-term treatment of cells with steroid triggers the selective and gradual demethylation of DNA (A). Demethylation occurs through the binding of a high affinity determination protein (squares) that prevents the access of the methylase (triangle) to the newly replicated DNA. (B) shows the effect of long-term deprivation of steroids on the selective methylation of DNA. Methylation occurs through the methylase(s) (traingles) in the progressive absence of the determination proteins (squares). In this model the determination protein is under the control of steroids. In the presence and absence of steroids their binding activity is high and low, respectively.

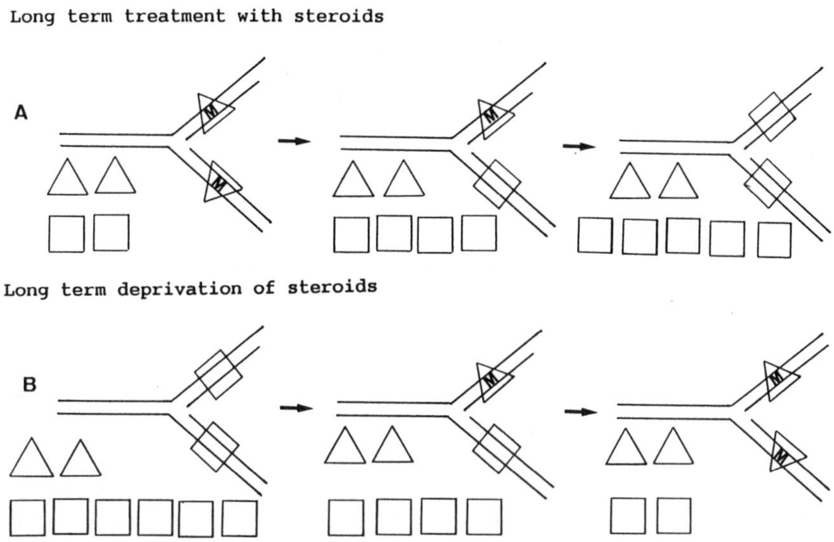

Figure 4. Model of the active and passive demethylation of DNA. (A) Active demethylation occurs through an enzymatic reaction in the absence of DNA replication. Methylcytosine is thought to be removed enzymatically and replaced by cytosine (Razin et al., 1984, 1986). Our working hypothesis is that NHP-1 may be part of the enzymatic complex responsible for the base excision of mC (Hughes et al., 1989). MDBP-2 is a repressor binding with high affinity to methylated DNA. In the absence of the repressor the demethylation complex could bind to the DNA and initiate demethylation. (B) shows the passive demethylation by the tight binding of a determination protein to the sequence containing the CpG. The tight binding of this protein would inhibit the maintenance methylase. The subunit 80 kDa of NHP-1 could possibly serve such a function (see Table 1).

but would require one or several CpGs for high binding affinity. This protein would then inhibit the activity of the maintenance methylase by binding very tightly to the DNA. A change in the concentrations or binding activities of the repressor and the determination protein(s) could be sufficient to trigger either an inhibition of MMTV-LTR expression or an activation of the gene with a concomitant demethylation of mCpGs. We propose that the binding activity of the determination protein is under the control of steroid hormones. It is high in the presence of the hormone and low in its absence (Fig. 3A). The determination protein would have no direct effect on the transcription of the genes.

3 Long-term deprivation of steroid hormones in mouse mammary tumor cells favors the selective methylation of MMTV-LTR DNA

When deprived of steroid hormones over a long period of time a clone of hormone-responsive mouse mammary tumor cells gives rise to a population of hormone-unresponsive cells altered in proliferative response, saturation-density response, cell morphology and level of expression of MMTV-LTR (Darbre and King, 1984). For example, 9 weeks after deprivation of steroids by cultivating the cells in charcoal-Dextran-treated serum, the mammary tumor cells lose their proliferative response to androgens. A somewhat more complex picture was shown for the changes in the saturation density in monolayer cultures. The loss of the recovery of saturation density in monolayer culture after long-term deprivation of steroids was progressive, culminating 30 weeks after the withdrawal of steroids from the cultures.

The time course of the loss of expression of MMTV-LTR during long-term withdrawal of steroids (6–18 weeks) revealed that the MMTV-LTR expression could be restored after several days of treatment with testosterone. However, after 34 weeks of steroid deprivation no MMTV-LTR expression could be recovered even after 8 weeks of additing testosterone to the culture medium (Darbre et al., 1983; Darbre and King, 1984). From the reproducibility of these experiments with cloned cells, the authors concluded that selection of mutated cells is unlikely to be involved since it is highly improbable that identical mutations occur within such a short length of time (Darbre and King, 1984). Changes in the expression of MMTV-LTR were not accompanied by gross genetic alterations. The possible involvement of epigenetic changes such as DNA methylation have been investigated by means of the isochizomeric restriction enzymes HpaII and MspI. As shown in Figure 5, 13 weeks after steroid deprivation, a small increase in the methylation of HpaII-sensitive sites was observed for the MMTV-LTR DNA and total methylation of the HpaII sites was observed after 42 weeks withdrawal of steroid hormones. Addition of androgen at that time did not result in any demethylation of the HpaII sites even after a prolonged (20 weeks) treatment. The degree of methylation of MMTV-LTR DNA was correlated with the loss of expression of MMTV genes (Darbre and King, 1984). It is, therefore, tempting to conclude that a *de novo* methylation of CpGs is responsible for the irreversible step towards androgen insensitivity. Nevertheless, it is unlikely that DNA methylation is the only parameter responsible for the silencing of the MMTV gene and the phenotypic alteration observed during hormone deprivation.

The mechanism for the methylation of MMTV DNA in the absence of steroid hormones is unlikely to be solely the result of a progressive change in the level of the methylase. The two systems described above have several common features, e.g., both the selective demethylation of MMTV DNA in thymoma cells and the methylation of the same gene in mouse mammary tumor cells require several rounds of DNA replication. Steroid hormones are implicated in both cases, and their level determines whether a specific gene will be methylated or demethylated, suggesting that the same mechanism of selective methylation/demethylation may be operational in both cases. Moreover, at least one element of each system is steroid dependent. Figure 3B shows schematically how a

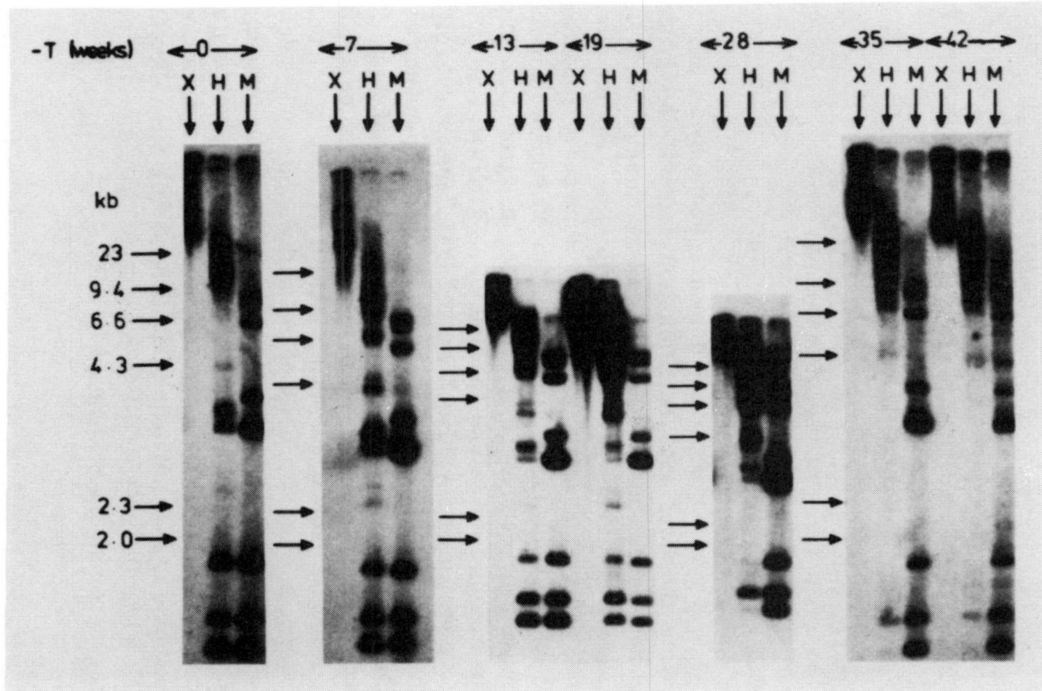

Figure 5. Restriction endonuclease analysis of MMTV-LTR-related sequences in the DNA of cloned +A S115 cells following increasing periods of androgen withdrawal (−T). DNA was either undigested (X) or digested with HpaII (H) or MspI (M) and blots were probed with ^{32}P-labeled MMTV-LTR. Molecular weight standards were provided by a Hind III digest of bacteriophage λ DNA, and their positions are indicated by horizontal arrows (Darbre and King, 1984). Reproduced from J. Cell Biol. *99* (1984) 1410–1415 by copyright permission of the Rockefeller University Press.

Table 1. Properties of NHP-1 (a determination protein?). The following references deal with the Ku protein, an autoantigen present in the blood of patients suffering from Scleroderma-polymyositis (Reeves and Sthoeger, 1989; Mimori et al., 1990; Mimori and Hardin, 1986)

1	Molecular weight of native protein:	170kDa
2	Subunits:	85 and 75 kDa
3	pI 5.0	
4	Only the 80 kDa subunit binds to DNA.	
5	Ubiquitous, is not organ or species specific, can be found in large quantities in many tumor cells.	
6	It is not detected in yeast and *Drosophila.*	
7	Microsequencing of peptides derived from HeLa cells NHP-1 reveals an identity with the Ku protein[1].	
8	High binding affinity for methylated and non-methylated DNA Kd 10^{-11} M.	
9	Not strictly sequence specific, binds to sequences containing a pair of CpG.	
10	Binds only to double-stranded DNA.	
11	The native 170 kDa NHP-1 causes nicks on either side of the mCpG.	
12	Binding activity of NHP-1 is regulated by steroid hormones.	
13	NHP-1 has no effect on the *in vitro* transcription of the avian vitellogenin gene.	

deprivation of the steroid hormone may influence the methylation of a gene by lowering the level of a determination protein, and hence increasing the *de novo* methylation of DNA. Conversely, high concentrations of the steroid hormone increase the level of the determination protein which then blocks the *de novo* methylation of the newly replicated DNA (Fig. 3). In short, our model predicts that the change in the level of the determination protein is under the control of steroids. In this context it is interesting to note that the ubiquitous NHP-1 (a protein related to the Ku protein family) (Hughes et al., 1989; Hughes and Jost, 1989) may have such a function since its binding activity is regulated by steroid hormones (Jost et al., 1991). It is not sequence specific, and it binds sequences containing mCpG or CpG with very high affinity (Kd 10^{-11} M) (see Table 1). How could such a control mechanism of gene methylation be specific? The specificity of the methylation would solely reside in the changes in the stoichiometry of the diverse protein factors binding to the sequences containing mCpGs or CpGs. As we shall see for the vitellogenin gene, a change in the ratio of NHP-1 and the repressor MDBP-2, which bind to the same sequence independently with different affinities (NHP-1 Kd 10^{-11} M, MDBP-2, Kd 10^{-9} M), could determine whether the sequence remains methylated or not.

4 Short-term estradiol treatment of immature chickens triggers the demethylation of mCpGs and the expression of the vitellogenin gene

4.1 The first observation on vitellogenin gene demethylation

In vertebrates, vitellogenin genes, which encode egg yolk proteins, are only expressed in hepatocytes of females and are under the control of estradiol (Tata et al., 1987; Wahli, 1988). A single injection of a pharmacological dose of estradiol to immature chickens, roosters or quail resulted in the activation of the vitellogenin gene (Jost et al., 1990; Copeland and Verninder Gibbins, 1989; Evans et al., 1987) and the demethylation of several mCpG present in the coding and non-coding region of the gene (Jost et al., 1990; Copeland and Verninder Gibbins, 1989; Meijlink et al., 1983; Philipsen et al., 1985). Similar differences in the state of methylation of the vitellogenin gene were observed when adult roosters were compared to egg-laying hens, indicating that the estradiol effect was not an experimental artefact. A 30-Kb region containing 17 HpaII and 18 HhaI sites was investigated by Philipsen et al. Of these 35 sites, 9 were found to be unmethylated in laying hen liver and methylated in immature chicken liver. The mCpGs present in the transcribed region became demethylated only after prolonged transcription of the gene (Philipsen et al., 1985) whereas one HpaII site in the upstream region of the gene became demethylated much sooner. The correlation between commitment and degree of methylation may be obscured by the heterogeneity of liver tissue where about 70% of the cells are the potentially vitellogenin-synthesizing paraenchymal cells (Williams et al., 1979). The slow demethylation of mCpGs in the structural gene suggested that these sites were not actively demethylated but became demethylated upon DNA replication. In sharp contrast, the HpaII site situated in the estrogen response element at nucleotide position −611 became demethylated much sooner, and the demethylation occurred in the absence of DNA synthesis (Wilks et al., 1984). When DNA synthesis was inhibited *in vivo* by cytosine arabinoside or hydroxyurea, demethylation of the mCpG (nucleotide position −611) upon estradiol treatment was still observed. These results suggested that an active demethylation may take place. However, we have no direct evidence of an active enzymatic removal of methyl-cytosine and its replacement by cytosine. If this were the case, we would find *in vivo*, specific nicks on either sides of the mCpGs during the first days post estradiol treatment. No experiment to demonstrate this possibility has been carried out so far. Moreover, a close look at the results of Wilks et al. (1984) shows clearly that cytosine arabinoside did not inhibit liver DNA synthesis to 100%. The residual DNA synthesis occurring in the presence of cytosine arabinoside could still represent a limited and selective replication of DNA. Clearly the question still remains open and should be reinvestigated with more accurate tools. For other systems there is also strong evidence that an active mechanism of demethylation may play an important role; for example, during embryonic development and cell differentiation there is a genome-wide demethylation of mCpGs followed by a remethylation of CpGs (Razin et al., and references therein 1986; 1981; Razin et al., 1984; Young and Tilghman 1984; Bestor et al., 1984).

Although the experiments with restriction enzymes showed a general trend, it is difficult to draw any firm conclusion. In most cases the enzymes detected only a relatively small proportion of all possible mCpGs. Thus, it could not be ruled out that undetected CpGs may have a crucial role for the binding of transcription factors. The restriction enzymes only detect demethylation of both DNA strands; a hemimethylated DNA would not be detected. Church and Gilbert (1984) developed a procedure of genomic sequencing that permits to study strand specific methylation of any CpG or CpXpG present in a gene. This procedure was further developed in our laboratory (Saluz, H. P. and Jost, J. P., 1990) and a simplified version using Taq polymerase was also later reported (Saluz and Jost, 1989). A short overview on the different techniques available for the detection of mCpG are described in Chapter 2 of this book.

4.2 Demethylation of the mCpG in the estrogen-response element and neighboring sequence is estrogen dependent but not organ specific

The first evidence for the demethylation of a CpG located in the estrogen-response element (Fig. 6) was obtained by HpaII/MspI digestion of DNA (Wilks et al., 1982). Demethylation at that site was shown to be estrogen dependent but not organ specific since it was also demethylated in oviduct cells where the vitellogenin gene is not expressed. The kinetics of demethylation in the liver of immature chickens upon injection of a single pharmacological dose of estradiol were studied by genomic sequencing. Figure 7 shows that the first trace of demethylation appeared within a few hours of estradiol treatment (Saluz et al., 1986). The demethylation of the sites a, c, d, (see Fig. 6) started first on the upper strand and was followed 24 h later by the demethylation of the lower DNA strand. Site b was unmethylated in the upper strand, before estradiol treatment. A conservative estimate indicates that the demethylation of the upper DNA strand of the sites a, c,

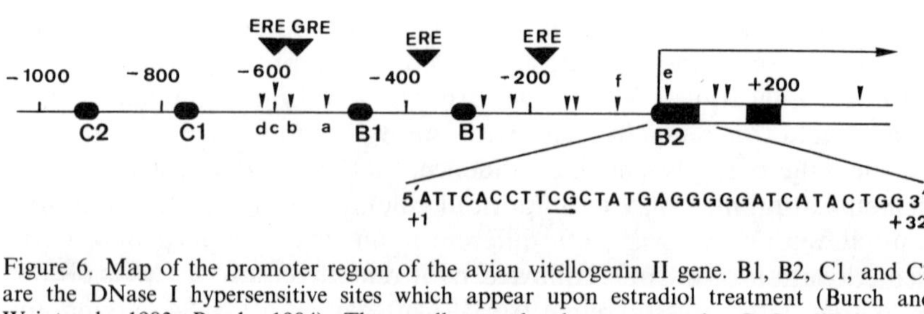

Figure 6. Map of the promoter region of the avian vitellogenin II gene. B1, B2, C1, and C2 are the DNase I hypersensitive sites which appear upon estradiol treatment (Burch and Weintraub, 1983; Burch, 1984). The small arrowheads represent the CpGs. ERE is the estrogen response element and GRE is the glucocorticoid response element (Jost et al., 1991).

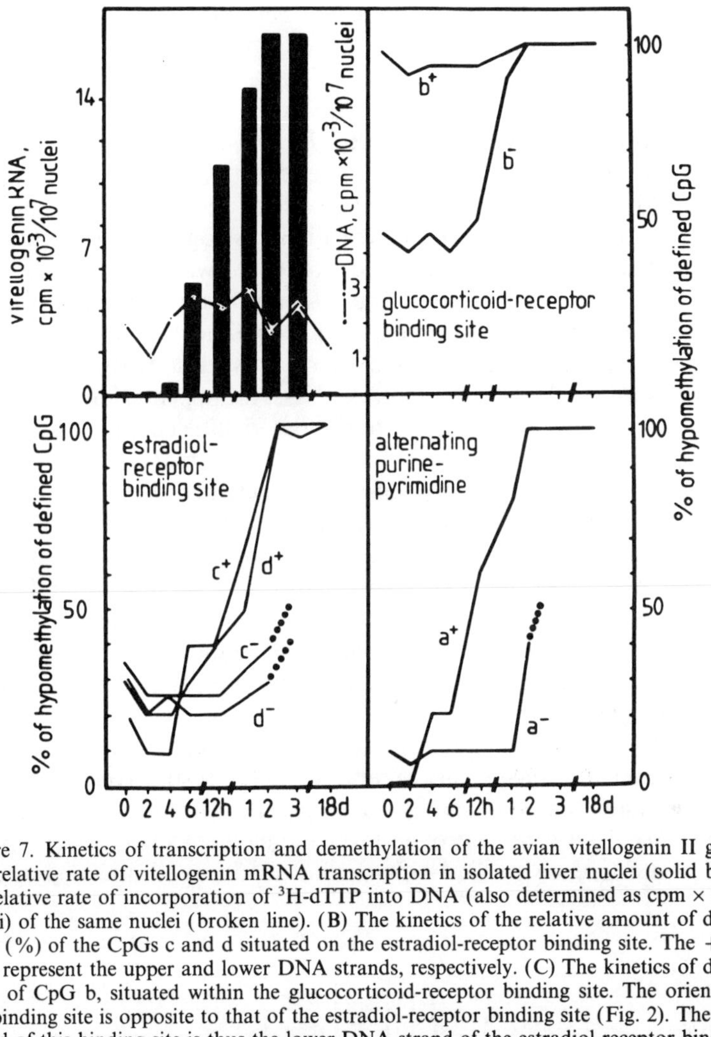

Figure 7. Kinetics of transcription and demethylation of the avian vitellogenin II gene. (A) The relative rate of vitellogenin mRNA transcription in isolated liver nuclei (solid bars) and the relative rate of incorporation of ^{3}H-dTTP into DNA (also determined as cpm $\times 10^{-3}/10^7$ nuclei) of the same nuclei (broken line). (B) The kinetics of the relative amount of demethylation (%) of the CpGs c and d situated on the estradiol-receptor binding site. The + and − signs represent the upper and lower DNA strands, respectively. (C) The kinetics of demethylation of CpG b, situated within the glucocorticoid-receptor binding site. The orientation of this binding site is opposite to that of the estradiol-receptor binding site (Fig. 2). The "upper" strand of this binding site is thus the lower DNA strand of the estradiol-receptor binding site; it therefore carries the − sign. (D) The kinetics of demethylation of CpG a, present on the short stretch of alternating purine-pyrimidine bases (Saluz et al., 1986). Reprinted with permission from Proc. Natl. Acad. Sci. USA *83* (1986) 7167–7171.

d does not occur until the onset of vitellogenin gene transcription. Due to the heterogeneity of the liver tissue (60–70% hepatocytes), the maximal demethylation of the mCpG never reaches 100% but stays around 60–70% (comparison of the signal strength with the neighbouring cytosines) 1–2 days after estradiol treatment. These values correlate well with the number of cells synthesizing vitellogenin 1–2 days after estradiol treatment (Saluz et al., 1986).

4.3 In vivo and in vitro protein-DNA interaction at the estrogen-response element

The nature of protein-DNA interaction *in vivo* was investigated by *in vivo* footprinting with dimethylsulfate. *In vivo* footprinting of hepatocytes from adult hens and roosters treated with 0.5–0.0005% dimethylsulfate (DMS) revealed, at critical concentrations, a protection of distinct guanosine residues within the distal estrogen-response element and the adjacent downstream sequence (McEwan et al., 1991). In addition to the estrogen receptor, another protein different from the estrogen receptor protected the center of the palindrome of the estrogen response element regardless of the state of activation of the vitellogenin gene (McEwan et al., 1991). *In vitro* DNA binding interference studies with partially depurinated and depyrimidated labeled oligonucleotides and proteolytic clipping assays with hen and rooster liver nuclear extracts indicated that the estrogen-response element binding protein is the same in hen and rooster (McEwan et al., 1991). Methylation at the 2 CpGs sites c and d had no effect on the binding of this protein to DNA. This nuclear protein called the non-histone protein 1 (NHP1) has been purified to homogeneity (Hughes et al., 1989) microsequenced and cloned (unpublished results). NHP1 has been found in all animal species (except in yeast and *Drosophila*) and in all organs studied so far. NHP1 is closely related to the Ku protein family. The properties of NHP1 are listed in Table 1. NHP1 from HeLa cells or other sources consists of two peptides with masses of 75 and 85 kDa, respectively. Only the 85 kDa subunit binds to DNA. The precise contact points of the purified NHP-1 with the estrogen-response element (ERE) were established by interference studies with partial depurinated and depyrimidi-

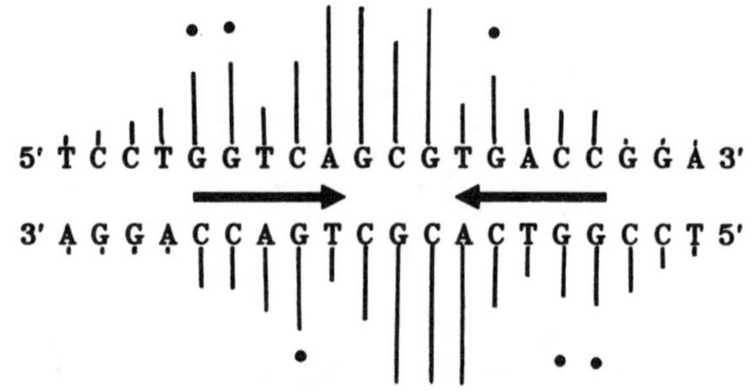

Figure 8. Summary of contact points of NHP-1 with ERE. The lengths of the vertical lines indicate the importance of the corresponding base for the binding of NHP-1. The horizontal arrows emphasize the dyad symmetry of the ERE (Hughes et al., 1989). The filled circles indicate which bases are important for the binding of the ER homodimer (Kumar and Chambon, 1988; Klein-Hitpass et al., 1989). Reprinted with permission from Biochemistry *28*, (1989) 9137–9142. Copyright 1992 American Chemical Society.

nated synthetic ERE oligonucleotide (Fig. 8). The highest binding affinity was observed for the bases AGCG in the center of the palindrome. A suppression of the three bases GCG in the center of the palindrome of the ERE drastically decreased of NHP-1 binding to DNA. NHP-1 binds with very high affinity (Kd 10^{-11} M) to many different sequences containing CpGs. The role of NHP-1 is not yet defined, but it may be involved in the passive (85 kDa subunit) and/or active (85 kDa + 75 kDa subunit) demethylation of DNA (see Figs. 3 and 4). The binding level of NHP-1 is regulated by estrogen (Jost et al., 1991; McEwan et al., 1991) and possible by other steroid hormones as well (Jost, unpublished results). NHP-1 binding activity is very high in the presence of estrogen and lower in its absence.

4.4 Evidence of the existence of a repressor binding to a mCpG in the proximal promoter region of the avian vitellogenin gene

The nature of the *cis*-acting control element of the avian vitellogenin gene was tested by transcription-competition experiments in a homologous system (Vaccaro et al., 1990). Transcription competition

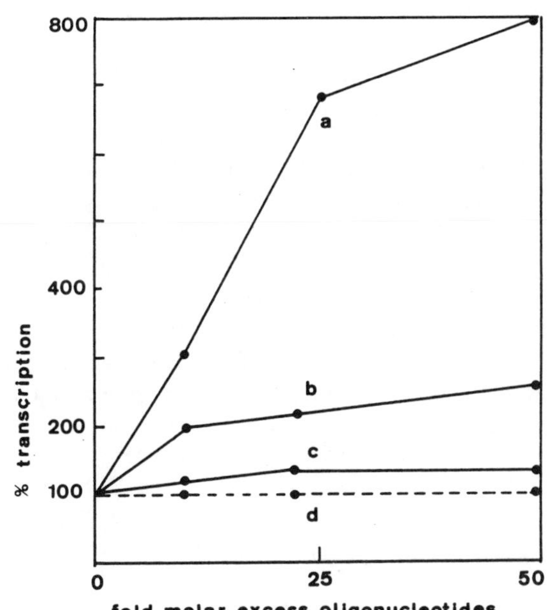

Figure 9. *In vitro* transcription competition curves. Increasing amounts of oligonucleotides a) 5′ ATTCACCTTmCGCTATGAGGGGGATCATACTG 3′ or b) 5′ TCACCTTmCGCTATGAG 3′, c) is the same oligonucleotide but not methylated. The oligonucleotides were used to compete for the binding of the repressor MDBP-2 present in the transcription system. The dotted line is the incubation mixture with no addition of competing DNA.

experiments with an oligonucleotide covering part of the expression-specific DNAse I hypersensitive site B_2 (Fig. 9) containing a CpG at the nucleotide position +10 increased the transcription of the gene, suggesting a repressor binds to this region. The enhancement of transcription was even stronger when the same oligonucleotide was methylated at the corresponding +10 cytosine. When a methylated oligonucleotide covering the entire footprint of the repressor was used, a 8-fold increase in the transcription was observed (Fig. 9). These results suggest that nuclear extracts contain a protein that binds preferentially methylated DNA, and that binding of this protein to this region results in the repression of the gene (Vaccaro et al., 1990; Pawlak et al., 1991).

4.5 Properties of the methylated DNA binding repressor (MDBP-2)

A protein binding preferentially to methylated DNA (MDBP-2) was identified in fractionated liver and other organ nuclear extracts. This protein bound with a Kd of 10^{-9} M to the methylated DNA sequence 5′TTCACCTTmCGCTATGAGGGGGATCATACTGG3′ of the avian vitellogenin II promoter and to the same unmethylated DNA with at least 10–100-fold lower affinity. The binding is not sequence specific and requires only 1 pair of methylated CpGs on a 30-bp-long oligonucleotide for high affinity binding (Jost et al., 1992). Gel mobility shift assays and competition experiments indicated that MDBP-2 did not bind to shorter oligonucleotides of the footprint such as TATTCACCTTmCGTAT or GAGGGGATCTACTGGCA (Pawlak et al., 1991). UV light crosslinking experiments to BrUTP-substituted oligonucleotides and SDS-PAGE indicated that MDBP-2 had an apparent molecular weight of about 40 kDa whereas south-west blots indicated a MW of 21 kDa (Jost, unpublished results). *In vitro* transcription experiments showed that addition of a purified nuclear fraction containing MDBP-2 inhibited the transcription of the vitellogenin gene in a dose-dependent manner (Fig. 10). MDBP-2 is a member of the histone H1 family (Jost et al., 1992).

4.6 Demethylation of the mCpG at the repressor MDBP-2 binding site is estrogen dependent and expression specific

The state of methylation of the CpG in the binding site of the MDBP-2 was studied by genomic sequencing (Saluz et al., 1988). The results in Fig. 11 show that the CpG at nucleotide position −52 (Fig. 6, f+) is unmethylated regardless of the expression of the gene. In contrast, the CpG in the MDBP-2 binding site was only demethylated in an expres-

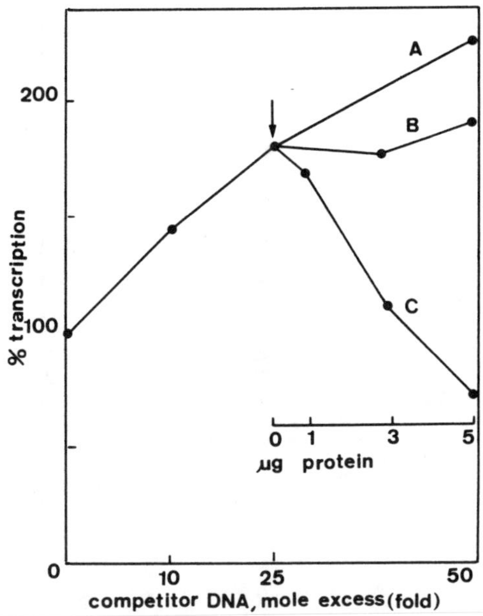

Figure 10. Effect of the purified MDBP-2 on the *in vitro* transcription of the avian vitellogenin II gene. *In vitro* transcription competition in a homologous system was carried out as previously described (Vaccaro et al., 1990). Curve A shows the effect of increasing concentrations of methylated oligonucleotide binding site on the transcription of the vitellogenin gene. In the presence of a 25-fold mole excess of the methylated oligonucleotide (arrow), increasing concentrations of serum albumin (curve B) or purified MDBP-2 (purified 200 ×) from the cell free extract (curve C) were added to the transcription system. The transcription products were analyzed by primer extension with reverse transcriptase and separated on an 11% denaturing polyacrylamide gel. Upon autoradiography the bands corresponding to the full-size transcription product were cut out and counted for radioactivity. The 100% represents the transcription product in the complete system in the absence of any competing oligonucleotide. The X axis labeled: "competitor DNA, mole excess (fold)", is only relevant for curve A, whereas the X axis labeled: "µg protein", is relevant to the curves B and C that start at the arrow (Pawlak et al., 1991). Reprinted with permission from Nucl. Acids Res. *19* (1991) 1029–1034.

sion-specific manner when the gene is expressed in the liver of hens but not in the oviduct of the same animal or in the liver of roosters. The kinetics of demethylation of the mCpG in the binding site of MDBP-2 was investigated in immature roosters treated with a single pharmacological dose of estradiol. Figure 12 shows that the first traces of demethylation were detectable within 4 h after estradiol treatment (Jost et al., 1991) and that demethylation took place on both DNA strands simultaneously. The start of the demethylation preceded the onset of vitellogenin gene transcription but at a slower rate than transcription of the gene, suggesting that demethylation may be a consequence of a change in the chromatin structure.

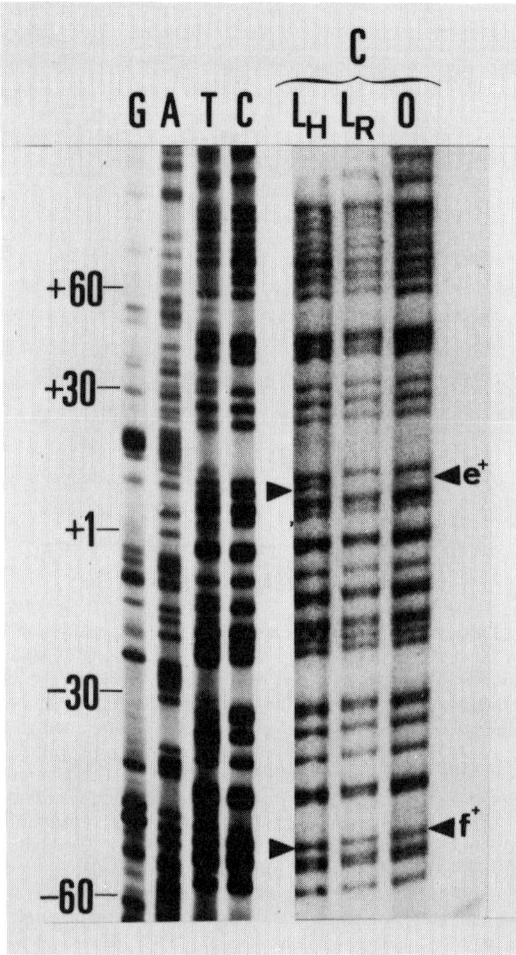

Figure 11. The demethylation of the mCpG in the promoter region of the avian vitellogenin II gene is expression specific. Genomic sequencing was carried out as outlined by Saluz and Jost (1987). C-specific reactions were carried out with total genomic DNA from hen liver (L_H), rooster liver (L_R) and hen oviduct (O). Arrows e^+ and f^+ indicate the position of the CpGs on the upper DNA strand (see also Fig. 7).

4.7 In the liver the repressor binding activity of MDBP-2 is down regulated by estradiol while the binding activity of NHP-1 increases

In vivo genomic footprinting of the DNAse I hypersensitive site B_2 with dimethylsulfate in silent and expressed gene revealed different patterns of DNA protection (Saluz et al., 1988) suggesting that a significant change in protein-DNA interaction occurred in this region upon transition from the silent to the active state of the gene. Gel mobility shift

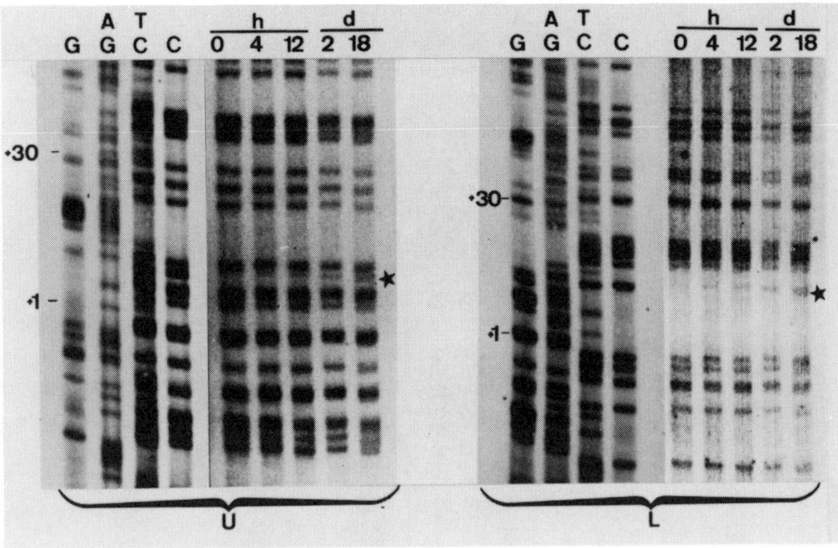

Figure 12. *In vivo* early kinetics of demethylation of the mCpG in the binding site of the repressor MDBP-2 following estradiol injection to immature roosters. The state of methylation of the CpG was determined by genomic sequencing. U and L represent the genomic sequence of the upper and lower strand of DNA. All lanes of the genomic DNA represent the cytosine-specific reaction. The control lanes G, A, T, and C are shown for orientation. The CpG at nucleotide position +10 is marked by the star (Jost et al., 1991). Reprinted with permission from Nucl. Acids Res. *19* (1991) 5771–5775.

assays carried out with fractionated liver nuclear extracts showed 2 different proteins (MDBP-2 and NHP-1) binding with high affinity to this region. However, assays also showed that the proteins do not occupy the same sequence at the same time (Jost et al., 1991). For example, prebinding of MDBP-2 to DNA inhibited the subsequent binding of NHP-1 and vice versa. Incubation of the two protein fractions at the same time with the labeled oligonucleotide did not result in the formation of larger complexes (Fig. 13). These results suggested that MDBP-2 and NHP-1 have distinct functions. This hypothesis was strengthened by the observation that the two proteins are regulated differently. MDBP-2 was down regulated by estradiol whereas the binding activity of NHP-1 was increased by estradiol (Fig. 14). However, recent immunological studies have shown that the level of MDBP-2 antigen remained unchanged (Jost et al., unpublished results). Under normal physiological conditions MDBP-2 binding activity was very high in mature roosters where the vitellogenin gene was methylated and not expressed, while in the liver of egg-laying hens, where the gene was expressed and unmethylated, there was a very low binding activity of MDBP-2 (Fig. 14). In other organs where the gene was not expressed there was also a high binding activity of MDBP-2. Conversely NHP-1

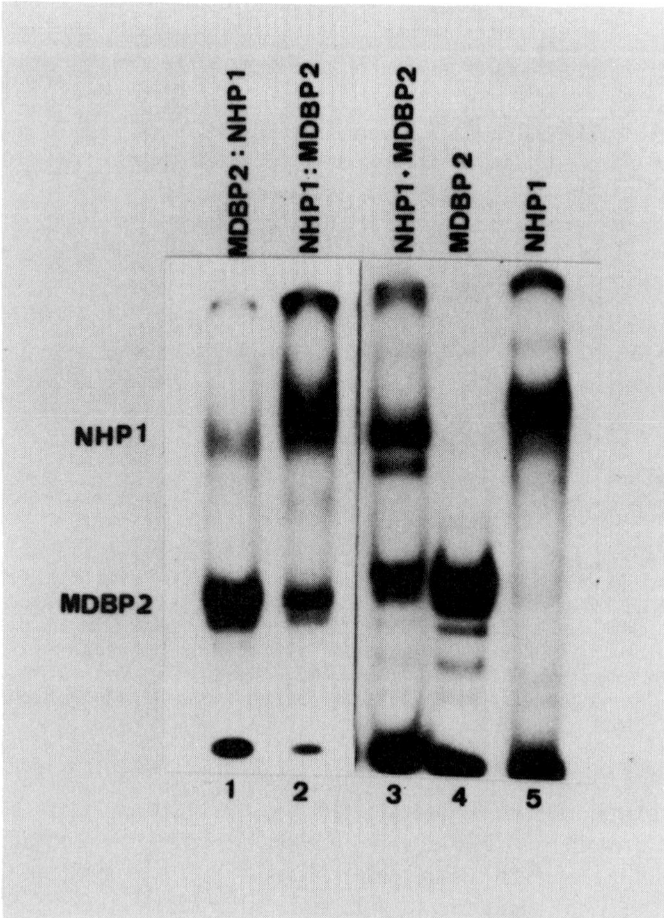

Figure 13. Gel mobility shift assay of the endlabeled olignucleotide 5′ TTCACCTTmCGC-TATGAGGGGGATCATACTGG 3′ with MDBP-2 (lane 4), NHP-1 (lane 5) or a combination of both (lanes 1–3). For lane 1, the oligonucleotide was preincubated for 15 min with MDBP-2 (4 μg), the incubation was continued for another 15 min. For lane 2, the reverse experiment was done, NHP-1 was added first to the reaction mixture followed by MDBP-2. For lane 3, NHP-1 and MDBP-2 were incubated at the same time for 30 min at 25°C. All reaction mixtures were analyzed on a 5% polyacrylamide gel (Jost et al., 1991). Reprinted with permission from Nucl. Acids Res. *19* (1991) 5771–5775.

binding activity was high in the liver of egg-laying hens and lower in the liver of roosters (Fig. 14B). A single injection of estradiol to immature roosters resulted in a very rapid decrease in the binding activity of MDPB-2 and an increase in the binding activity of NHP-1. The sharp decrease in the binding activity of MDBP-2 preceded the onset of vitellogenin gene transcription (Jost et al., 1991) (Fig. 15). From these kinetic experiments it was concluded that the down regulation of the

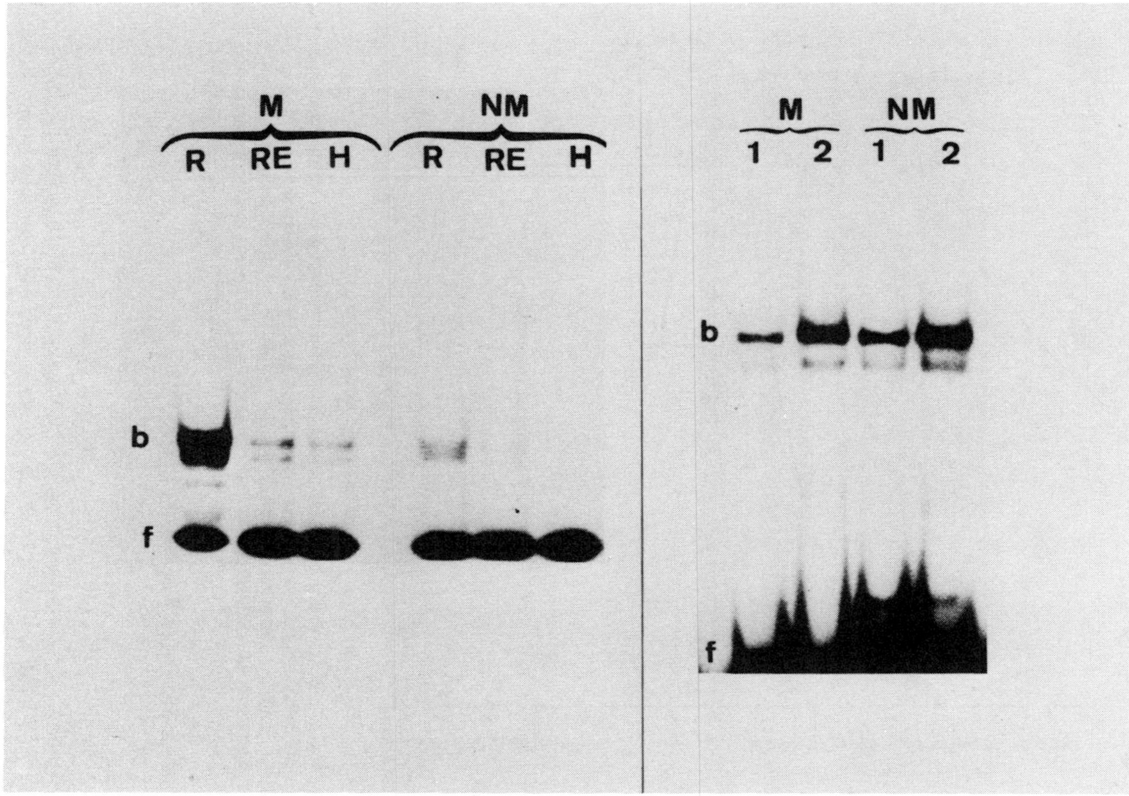

Figure 14. MDBP-2 (left panel) binding activity (gel shift assay) is very high in the liver of adult roosters (R) and low in the liver of egg-laying hens (H). M and NM are the methylated and non-methylated binding sites of MDBP-2. Injection of estrogen to an adult rooster (RE) drastically reduces the binding activity of MDBP-2. NHP-1 (right panel) binding activity is much higher in egg-laying hens (lane 2) than in adult roostes (lane 1). NHP-1 binds equally well to the methylated (M) and non-methylated (NM) binding site of MDBP-2 (Jost et al., 1991). Reprinted with permission from Nucl. Acids Res. *19* (1991) 5771–5775.

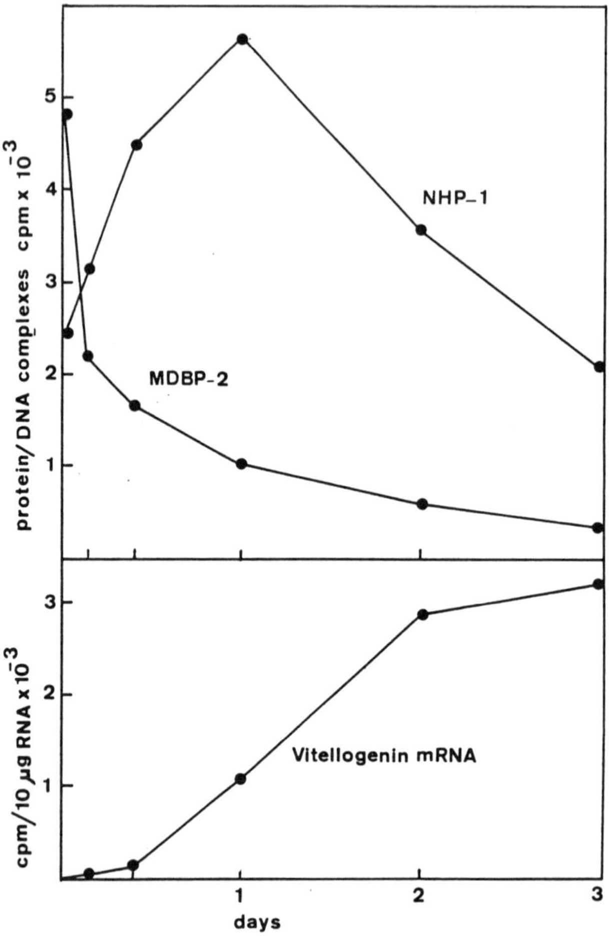

Figure 15. *In vivo* kinetics of the down regulation of MDBP-2 binding activity by estradiol. At the time indicated, nuclei were isolated, fractionated by FPLC on Heparin-Sepharose and tested by gel mobility shift assay. The fractions eluted from Heparin-Sepharose with 0.5 M KCl contained the MDBP-2 binding activity whereas NHP-1 was present in the 0.3 M KCl fraction. Total RNA was isolated from the same livers and tested by dot blot hybridization for the vitellogenin mRNA sequence (Jost et al., 1991). Reprinted with permission from Nucl. Acids Res. *19* (1991) 5771–5775.

repressor by estradiol may have a greater effect on the derepression of the gene than the DNA demethylation. Demethylation could be a consequence of the transcription of the gene, and the function of mCpG in this case might be solely to assure a high binding affinity of the repressor to the DNA binding site. These results could explain why during the secondary stimulation of the vitellogenin gene (where the gene is already demethylated as a consequence of primary stimulation) there is no lag period preceding the onset of gene transcription and the rate and amplitude of transcription is much greater than during the primary stimulation (memory effect) (Jost et al., 1990). The direct implication of the above results is that in the absence of DNA demethylation, the lowering of the methylated DNA binding repressor should be sufficient to activate the gene. This seems to be the case with other systems. Using a different approach Boyes and Bird (1991) and Levine et al. (1991) showed that DNA methylation inhibited the transcription of specific genes indirectly via a mCpG-binding protein. In this case lowering the concentration of MeCP1 by cotransfection or incubation of DNA bearing the methylated CpGs of the binding site was sufficient for *in vivo* as well as *in vitro* gene activation in the absence of DNA demethylation.

4.8 NHP-1 may be involved in the passive and/or active demethylation of mCpGs

As we have previously seen, the binding of NHP-1 to DNA is not strictly sequence specific, i.e. it binds with high affinity to different double-stranded sequences containing a CpG. Since it binds with similar affinities to the same methylated or unmethylated DNA sequence it is tempting to speculate that NHP-1 may bind to the newly replicated DNA during passive demethylation, thus preventing the maintenance methylation of DNA (Figs. 3 and 4). The 85 kDa subunit of NHP-1 which has a high binding affinity for DNA would then play the role of a determination protein. Table 1 shows many of the properties of the NHP-1, some of them to be expected in a determination protein. Since the protein can easily be purified to homogeneity it would be of interest to test whether incubation of hemi-methylated DNA with pure NHP-1 could inhibit *in vitro* maintenance methylation. On the other hand the two subunits 85 and 75 kDa could bind to the methylated DNA and under special conditions initiate repair excision of mCpGs (active demethylation) (Fig. 4). Experiments carried out with highly purified NHP-1 (purified 40,000 fold) bound to the methylated ERE have shown that the protein complex could initiate nicks around the mCpGs in a strand-specific fashion whereas an incubation of the same NHP-1 preparation with a mutated ERE which bound NHP-1 very poorly remained intact (Hughes et al., 1989). Attempts to detect specific methylcytosine glycosylase activity in purified nuclear extracts have been negative so far, and further work is needed to establish firmly the mechanism active demethylation. In this context it is interesting to note that Friend erythroleukemia cells (MEL) when treated with various chemicals causing differentiation, underwent transient genome wide demethylation (Razin et al., 1986). With labeling experiments, Razin et al. (1986) discovered that hypomethylation occurred by an enzymatic reaction leading to cytosine replacement of methylcytosine. The rapid and transient demethylation in MEL cells seems to be necessary, but not sufficient, to trigger the differentiation because inhibition of protein synthesis by cycloheximide prevented differentiation but did not inhibit transient demethylation (Razin et al., 1988). One can infer from these results that the enzymatic system is present in the cells prior to initiation of differentiation. It is also interesting to note that while the DNA is being hypomethylated there is in MEL cells a delay in the methylation of the DNA synthesized in the early period of induction. According to Adams et al. (1990) this could lead to the 30–50% transient hypomethylation. Further experiments carried out with other systems suggested also that the demethylation of mCpG may be an active process (Sullivan and Grainger, 1987; Frank et al., 1990; Yisraeli et al., 1988).

As an alternative mechanism for active demethylation it could be that the unprotected mCpGs in chromatin are first deaminated, creating a G/T mismatch that is then repaired by the G/T mismatch repair mechanism (Wiebauer and Jiricny, 1989; 1990) (see also chapter Wiebauer et al., pp. 510–522 of this book). As described above for the vitellogenin gene, there appears to be two types of kinetics of demethylation. In one set of circumstances both DNA strands are demethylated simultaneously, such as for the MDBP-2 binding site (Jost et al., 1991). Alternatively, in the case of the CpG within the ERE (Saluz et al., 1986) there is a demethylation of one strand first, followed 24 h later by the demethylation of the complementary strand. Similar kinetics have been observed for the demethylation of the actin gene, transfected into myoblasts. There is first a hemi-demethylation of specific mCpGs followed 48 h later by the full demethylation of the same sites (Paroush et al., 1990). The biological significance of such kinetics is at present unknown.

4.9 Are all methylation sites biologically important?

Methylation of CpGs could affect the activity of genes by changing DNA configuration (Rich, 1984) or by influencing directly the interaction of proteins with DNA (see chapter Zacharias, pp. 27–38 of this book). Cytosine methylation can either enhance or inhibit the binding of proteins to DNA (Ehrlich et al., 1990 and references therein). For the avian vitellogenin II gene we have identified two clear-cut cases. The methylation of 2 CpGs situated in the estrogen response element (ERE) in the upstream region of the gene does not interfere with the binding of the estrogen receptor (Jost et al., 1984) or NHP-1 (Hughes et al., 1989; Hughes and Jost, 1989). However, recent results have shown that a 30-bp-long methylated oligonucleotide covering the ERE binds also with high affinity MDBP-2 (H1) (Jost et al., 1992). Methylation of the

CpG at the nucleotide position +10 of the same gene is also required for the high affinity binding of the repressor MDBP-2 to DNA (Pawlak et al., 1991). Several other sequences containing CpGs in the structural part of the gene do not bind any specific factor, except MDBP-2 (H1), provided that the methylated oligonucleotide is at least 30-bp long (Jost, unpublished results). These mCpGs may have an important function for the formation of a higher order structure of the silent chromatin (heterochromatin). Methylated CpG could, for example, increase the binding affinities of subclasses of histones ($H_{1'}$) responsible for the compaction of silent chromatin.

4.10 The spreading of mCpG demethylation is progressive for a given gene

Following estradiol treatment, demethylation of mCpGs begins in the promoter region of the gene and then spreads into the structural part of the gene. The mCpGs situated in the estrogen response element and the MDBP-2 binding site act as a starting site for the progressive demethylation. It is interesting to note that for the reverse effect, for example for the methylation of the late E_2A promoter of adenovirus type 2 integrated into host DNA (Toth et al., 1988; 1990), methylation starts from several seminal sites and spreads through the gene in both directions. What determines these sites to demethylate/methylate? As suggested above, changes in the stoichiometry of different factors, including the determination protein(s) could be instrumental in initiating demethylation/methylation of the CpGs. The spreading of demethylation/methylation could be a consequence of changes in the chromatin structure. In this case, changes in the epigenetic blueprint would not play any role in the tuning of gene activity but may only be important for the long-term switching off and on of specific genes or group of genes.

5 Concluding remarks

The results summarized here show that steroid hormones influence the epigenetic blue print of methylation of certain genes. For the vitellogenin gene, and probably for other systems where methylation plays a role, the first signal for demethylation triggered by steroid hormone is upon binding of the steroid-receptor complex to DNA, a specific change in the binding activity of a protein or group of proteins that bind to a site containing a methylated cytosine. Below a critical concentration of the protein(s) there is competition for the binding site by a determination protein. This protein could either initiate the active demethylation or stay on the DNA to inhibit the maintenance methylation that occurs upon replication. In our system NHP-1 may serve this function since it lacks the sequence specificity, is ubiquitous, and has a very high affinity for methylated and non-methylated DNA. In this case, the specificity of DNA demethylation is directed by the changes in the binding activity of the specific transacting factors (positive or negative). Such a change would also be responsible for the modification of the chromatin structure (as tested by DNase I hypersensitivity for example). In conclusion, demethylation and methylation are a consequence of a change in the chromatin structure brought about by steroid hormones, and upon methylation or demethylation of DNA there is, for a given DNA sequence, a new stable epigenetic blue print that influences the binding affinity and specificity of transacting factors and other nuclear proteins that determine whether the gene is "open" for transcription or not.

Adams, R. L. P., Hanley, A., and Rinaldi, A. (1990) DNA demethylation in erythroleukemia cells. FEBS Lett. *269*, 29–31.

Bestor, T. H., Hellewell, S. B., and Ingram, V. M. (1984) Differentiation of two mouse cell lines is associated with hypomethylation of their genomes. Mol. Cell. biol. *4*, 1800–1806.

Boyes, J., and Bird, A. (1991) DNA methylation inhibits transcription indirectly via a methyl CpG binding protein. Cell *64*, 1123–1134.

Burch, J. B. E., and Weintraub, H. (1983) Temporal order of chromatin structural changes associated with activation of the major chicken vitellogenin gene. Cell *33*, 65–76.

Burch, J. B. E. (1984) Identification and sequence analysis of the 5′ end of the major chicken vitellogenin gene. Nucl. Acids Res. *12*, 1117–1135.

Church, G. M., and Gilbert, W. (1984) Genomic sequencing. Proc. Natl. Acad. Sci. USA *81*, 1991–1995.

Cohen, J. C. (1980) Methylation of milk-borne and genetically transmitted mouse mammary tumor virus proviral DNA. Cell *19*, 653–662.

Copeland, K. F. T., and Verrinder Gibbins, A. M. (1989) Demethylation of the gene expressing a yolk protein precursor in quails. Poult. Sci. *68*, 1678–1687.

Darbre, P., Dickson, C., Peters, G., Page, M., Curtis, S., and King, R. J. B. (1983) Androgen regulation of cell proliferation and expression of viral sequences in mouse mammary tumor cells. Nature *303*, 431–433.

Dabre, P., and King, R. J. B. (1984) Progression to steroid autonomy in S 115 mouse mammary tumor cells: Role of DNA methylation. J. Cell Biol. *99*, 1410–1415.

Evans, M. I., O'Malley, P. J., Krust, A., and Burch, J. B. E. (1987) Developmental regulation of the estrogen receptor and the estrogen responsiveness of five yolk protein genes in the avian liver. Proc. Natl. Acad. Sci. USA *84*, 8493–8497.

Ehrlich, M., Zhang, X. Y., Asiedu, C. K., Khan, R., and Supakar, C. (1990) Methylated DNA-binding protein from mammalian cells, in: Nucleic Acid Methylation, vol. 128, pp. 351–365, UCLA Symposia on Molecular Biology New Series. Eds G. A. Clawson, D. B. Willis, A. Weissbach, and P. A. Jones. Wiley-Liss, New York.

Frank, D., Lichenstein, M., Paroush, Z., Bergman, Y., Shani, M., Razin, A., and Cedar, H. (1990) Demthylation of genes in animal cells. Phil. Trans. R. Soc., London B *326*, 241–252.

Holliday, R. (1990) Mechanisms for the control of gene activity during development. Biol. Rev. *65*, 431–471.

Hughes, M. J., Liang, H., Jiricny, J., and Jost, J. P. (1989) Purification and characterization of a protein from HeLa cells that binds with high affinity to the estrogen response element GGTCAGCGTGACC. Biochemistry *28*, 9137–9142.

Hughes, M. J., and Jost, J. P. (1989) The ubiquitous protein NHP-1 binds with high affinity to different sequences of the chicken vitellogenin II gene. Nucl. Acids Res. *17*, 8511–8520.

Jost, J. P., Saluz, H. P., McEwan. I., Feavers, I. M., Hughes, M., Reiber, S., Liang, H., and Vaccaro, M. (1990) Tissue specific expression of avian vitellogenin gene is correlated with DNA hypomethylation and in vivo specific protein-DNA interactions. Phil. Trans. R. Soc., London B *326*, 231–240.

Jost, J. P., Saluz, H. P., and Pawlak, A. (1991) Estradiol down regulates the binding activity of an avian vitellogenin gene repressor (MDBP-2) and triggers a gradual demethylation of the mCpG of its binding site. Nucl. Acids Res. *19*, 5771–5775.

Jost, J. P., Seldran, M., and Geiser, M. (1984) Preferential binding of the estrogen receptor complex to a region containing the estrogen dependent hypomethylation site preceding the chicken vitellogenin II gene. Proc. Natl. Acad. Sci. USA *81*, 429–433.

Jost, J. P., and Hofsteenage, J. (1992) The repressor MDBP-2 is a member of the Histone H1 family that binds preferentially to methylated DNA in vitro and in vivo. Proc. Natl. Acad. Sci. USA (in press).

Klein-Hitpass, L., Tsai, S. Y., Greene, G. L., Clark, J. H., Tsai, M. J., and O'Malley, B. W. (1989) Specific binding of estrogen receptor to the estrogen response element. Mol. Cell. Biol. *9*, 43–49.

Knuth, M. W., Gunderson, S. I., Thompson, N. E., Strasheim, L. A., and Burgess, R. R. (1990) Purification and characterisation of proximal sequence element-binding protein 1. a transcription activiting protein related to Ku and TREF that binds the proximal sequence element of the human U_1 promoter. J. Biol. Chem. *265*, 17911–17920.

Kumar, V., and Chambon, P. (1988) The estrogen receptor binds to its responsive element as a ligand induced homodimer. Cell *55*, 145–156.

Levine, A., Cantoni, G. L., and Razin, A. (1991) Inhibition of promoter activity by methylation. Possible involvement of protein mediators. Proc. Natl. Acad. Sci. USA *88*, 6515–6518.

McEwan, I. J., Saluz, H. P., and Jost, J. P. (1991) In vivo and in vitro protein-DNA interactions at the distal estrogen response element of the chicken vitellogenin gene: evidence for the same protein binding to this sequence in hen and rooster liver, J. Steroid Biochem. Molec. Biol. *38*, 275–283.

Meijlink, F. C. P. W., Philipsen, J. N. J., Gruber, M., and AB, G. (1983) Methylation of the chicken vitellogenin gene: influence of estradiol administration. Nucl. Acids Res. *11*, 1361–1373.

Mermod, J. J., Bourgeois, S., Defer, N., and Grepin, M. (1983) Demethylation and expression of murine mammary tumor provirus in mouse thymoma cell lines. Proc. Natl. Acad. Sci. USA *80*, 110–114.

Mimori, T., and Hardin, J. A. (1986) Mechanism of interaction between Ku protein and DNA. J. Biol. Chem. *261*, 10375–10379.

Mimori, T., Obosone, Y., Hama, N., Suwa, A., Akizuki, M., Homma, M., Griffith, A. J., and Hardin, J. A. (1990) Isolation and characterization of cDNA encoding the 80 K-Da subunit protein of the human autoantigen Ku (p70/p80) recognized by autoantibodies from patients with scleroderma-polymyositis overlap syndrome. Proc. Natl. Acad. Sci. USA *87*, 1777–1781.

Paroush, Z., Keshet, I., Yisraeli, J., and Cedar, H. (1990) Dynamics of demethylation and activation of the alpha actin gene in myoblasts. Cell *63*, 1229–1237.

Pawlak, A., Bryans, M., and Jost, J. P. (1991) An avian 40 KDa nucleo-protein binds preferentially to a promoter sequence containing one single pair of methylated CpG. Nucl. Acids Res. *19*, 1029–1034.

Philipsen, J. N. J., Gruber, M., and AB, G. (1985) Expression-linked demethylation of 5-methylcytosine in the chicken vitellogenin gene region. Biochim. Biophys. Acta *826*, 186–194.

Prabhakar, B. S., Allaway, G., Srinivasappa, J., and Notkins, A. L. (1990) Cell surface expression of the 70 KD component of Ku, a DNA binding nuclear autoantigen. J. Clin. Invest. *86*, 1301–1305.

Razin, A., Levine, A., Kafri, T., Agostini, S., Gomi, T., and Cantoni, G. L. (1988) Relationship between transient DNA hypomethylation and erythroid differentiation of mouse erythroleukemia cells. Proc. Natl. Acad. Sci. USA *85*, 9003–9006.

Razin, A., and Riggs, A. D. (1980) DNA methylation and gene function. Science *210*, 604–610.

Razin, A., Szyf, M., Kafri, T., Roll, M., Giloh, H., Scarpa, S., Carotti, D., and Cantoni, G. L. (1986) Replacement of 5-methylcytosine by cytosine. A possible mechanism for transient DNA demethylation during differentiation. Proc. Natl. Acad. Sci. USA *83*, 2827–2831.

Razin, A., Webb, C., Szyf, M., Yisraeli, J., Rosenthal, A., Navey-Many, T., Sciaky-Gallili, N., and Cedar, H. (1984) Variations in DNA methylation in vivo and in vitro. Proc. Natl. Acad. Sci. USA *81*, 2275–2279.

Reeves, W. H., and Sthoeger, Z. M. (1989) Molecular cloning of cDNA encoding the p70 (Ku) Lupus autoantigen. J. Biol. Chem. *264*, 5047–5052.

Riggs, A. D., and Jones, P. (1983) 5-methylcytosine, gene regulation and cancer. Adv. Cancer Res. *40*, 1–40.

Ringold, G. M., Cardiff, R. D., Varmus, H. E., and Yamamoto, K. R. (1977) Infection of cultured rat hepatoma cells by mouse mammary tumor virus. Cell *10*, 11–18.

Rich, A. (1984) Left-handed Z-DNA and methylation of d(CpG) sequences, in: DNA Methylation Biochemistry and Biological Significance, pp. 278–292. Eds A. Razin, H. Cedar and A. Riggs. Springer Verlag, New York, Berlin, Heidelberg and Tokyo.

Saluz, H. P., Jiricny, J., and Jost, J. P. (1986) Genomic sequencing reveals a positive correlation between the kinetics of strand-specific DNA demethylation of the overlapping estradiol/glucocorticoid receptor binding sites and the rate of avian vitellogenin mRNA synthesis. Proc. Natl. Acad. Sci. USA *83*, 7167–7171.

Saluz, H. P., Feavers, I. M., Jiricny, J., and Jost, J. P. (1988) Genomic sequencing and *in vivo* footprinting of an expression-specific DNase I hypersensitive site of avian vitellogenin II promoter reveal a demethylation of a mCpG and a change in specific interactions of protein with DNA. Proc. Natl. Acad. Sci. USA *85*, 6697–6700.

Saluz, H. P., and Jost, J. P. (1989) A simple high-resolution procedure to study DNA methylation and in vivo DNA-protein interactions on a single gene copy level in higher eukaryotes. Proc. Natl. Acad. Sci. USA *86*, 2602–2606.

Saluz, H. P., and Jost, J. P. (1990) A laboratory guide for *in vivo* studies of DNA methylation and protein-DNA interactions. Birkhauser Verlag, Basel.

Stallcup, M. R., Ring, J. C., Ucker, D. S., and Yamamoto, K. R. (1979) in: Hormones and Cell Cultures, vol. 6, pp. 919–936. Cold Spring Harbor, Conferences on Cell Proliferation. Eds G. H. Sato and R. Ross. Cold Spring Harbor Laboratory, New York.

Stuiver, M. H., Coenjaerts, F. E. J., and van der Vliet, P. (1990) The autoantigen Ku is indistinguishable from NF IV a protein forming multimeric protein-DNA complexes. J. Exp. Med. *172*, 1049–1054.

Sullivan, C. H., and Grainger, R. M. (1987) Delta-crystallin gene becomes hypomethylated in post-mitotic lens cells during chicken development. Proc. Natl. Acad. Sci. USA *84*, 329–333.

Tata, J. R., and Smith, D. F. (1979) Vitellogenesis: a versatile model for hormonal regulation of gene expression. Recent Prog. Hormone Res. *35*, 47–95.

Toth, M., Lichtenberg, U., and Doerfler, W. (1990) Genomic sequencing reveals a 5-methylcytosine-free domain in active promoters and the spreading of preimposed methylation patterns. Proc. Natl. Acad. Sci. USA *86*, 3728–3732.

Toth, M., Muller, U., and Doerfler, W. (1990) Establishment of de novo DNA methylation patterns. J. Mol. Biol. *214*, 673–683.

Vaccaro, M., Pawlak, A., and Jost, J. P. (1990) Positive and negative regulatory elements of chicken vitellogenin II gene characterized by in vitro transcription competition assays in a homologous system. Proc. Natl. Acad. Sci. USA *87*, 3047–3051.

Wahli, W. (1988) Evolution and expression of vitellogenin genes. Trends Gen. *4*, 227–232.

Wiebauer, K., and Jiricny, J. (1990) Mismatch-specific thymine DNA glycosylase and DNA polymerase beta mediate the correction of G.T. mispairs in nuclear extracts from human cells. Proc. Natl. Acad. Sci. USA *87*, 5842–5845.

Wiebauer, K., and Jiricny, J. (1989) In vitro correction of G.T. mispairs to G.C. pairs in nuclear extracts from human cells. Nature *339*, 234–236.

Wilks, A., Seldran, M., and Jost, J. P. (1984) An estrogen-dependent demethylation at the 5′end of the avian vitellogenin gene is independent of DNA synthesis. Nucl. Acids Res. *12*, 1163–1177.

Williams, D. I., Tseng, M. T., and Rottman, W. (1979) Albumin synthesis and secretion by isolated morphologically characterized avian hepatic parenchymal cells. Life Sci. *23*, 195–206.

Yisraeli, J., Adelstein, R., Melloul, D. Nadel, U., Yaffe, D., and Cedar, M. (1988) Muscle-specific activation of a methylated chimeric actin gene. Cell *46*, 409–416.

Young, P. R., and Tilghman, S. M. (1984) Induction of alpha feto-protein synthesis in differentiating F9 teratocarcinoma cells is accompanied by a genome side loss of DNA methylation. Mol. Cell Biol. *4*, 898–907.

DNA Methylation: Molecular Biology and Biological Significance
ed. by J. P. Jost & H. P. Saluz

Epigenetic inheritance based on DNA methylation

Robin Holliday

CSIRO Division of Biomolecular Engineering, Laboratory for Molecular Biology, PO Box 184, North Ryde, NSW 2113, Australia

1 Introduction: Basic concepts

Normal Mendelian genetic inheritance is based on changes in the sequence of DNA, and these are accurately transmitted through mitosis and meiosis. Changes in sequences are produced by single base pair mutations, as well as deletions, insertions or inversions of stretches of DNA. Genetic systems are based on clonal inheritance within cell lineages, and the phenotype of cells or organisms is ultimately dependent on the information in DNA. In complex organisms another type of inheritance is superimposed on the classical genetic system. The clonal expansion of the single cell zygote leads to diversity of cell types. During the developmental process cells become committed or determined, that is, assigned to a particular role, position or fate in the final organism. As tissues and organs are formed, cells become differentiated or specialised for appropriate cell functions, and once differentiation has occurred it is very stable. Thus, post-mitotic neurons or muscle cells do not alter their phenotype, and the same is also true for differentiated cells capable of cell division, since fibroblasts divide to produce fibroblasts, lymphocytes produce lymphocytes, and so on. Whatever the controls are which maintain these specialised phenotypes, they are clearly very stably inherited. Stem line cells are somewhat similar to determined cells during development. They are not outwardly differentiated, but when they divide they give rise to cells which will later differentiate into one or several cell types, and also to more stem line cells. Stem line cells therefore also have a heritable phenotype, but in a sense they are unstable, because a single mitotic division can give rise to cells with very different cell fates: one remains a stem line cell and the other will become differentiated.

All these somatic events are governed by a heritable process in which the rules are very different from the classical genetic system. Taken together, it is convenient to refer to the somatic system as *epigenetic*. The term *dual inheritance* has also been introduced to cover both genetic and epigenetic heritability (Maynard Smith, 1990). As well as the constrasts between stable and unstable inheritance in the epigenetic system, there are also other striking differences from Mendelian inheritance (see Holliday, 1990a and b). For example, not all events occur in cell lineages, as it is well known that during development groups of cells may together alter their phenotype, or cell fate. Also, external events mediated by morphogenetic signals will induce phenotypic changes in individual cells or groups of cells. Such external influences are conspicuous by their absence in Mendelian genetics. Finally, epigenetic events can be reversed at meiosis. Although germ line cells are to be distinguished from somatic cells, they are also an integral part of the developmental process, and their ultimate fate is to differentiate in the gonads, undergo meiosis and form male and female gametes. It is known that epigenetic events in the cells are removed or erased before the gametes are produced, whereas genetic mutations are simply transmitted from generation to generation.

2 Biochemistry of epigenetic inheritance

In almost all cases, the molecular basis for epigenetic changes during development or for epigenetic controls in the adult are not understood. It would be generally agreed that protein-DNA interactions can provide the necessary specificity, and a large number of sequence specific transcription factors are now being described. It is widely believed that the phenotype of a specialised cell is based on the presence of many of these specific transcription factors, which by repression or activation mechanisms, allow a set of specialised genes to function, whilst at the same time silencing those genes which are active in other types of specialised cells. If this is correct, then the necessary controls must in many cases be stably transmitted through cell division. Moreover, in some types of epigenetic control the genes in homologous chromosomes have very different activities, although the presence of a particular set of transcription factors should interact with homologous chromosomes in the same way. What is required is the cooperative binding of specific proteins at particular DNA sequences, with strict maintenance of the complex through cell division (Weintraub, 1985). This is sometimes referred to as "protein inheritance". Although this general concept seems to be widely accepted, so far very little biochemical evidence that it exists has been published.

An alternative epigenetic system is based on the chemical modification of DNA. In higher organisms the post-synthetic modification of cytosine to 5-methyl cytosine (5-mC) is very widespread. Although the site of modification, the CpG doublet, is not by itself specific, there are now good reasons to believe that there is DNA sequence specificity in the pattern of methylation (Kochanek et al., 1990; Behn-Krappa et al.,

1991), but it may well be that only a small subset of 5-mC residues has important epigenetic or regulatory roles. In this review I will concentrate on the evidence that the pattern of methylation is heritable, and that it therefore provides a potential basis for the epigenetic processes mentioned in the Introduction. In bacteria and protozoa methylation of adenine also occurs, but it is usually assumed that such methylation does not exist in the DNA of higher organisms. It should be borne in mind that specific methylation controls of gene activity might be relatively rare, so that even the very small amount of a modified base such as 6-methyl adenine might have profound effects on regulation, but be difficult to detect by the normal methods of biochemical analysis. Even more extreme suggestions were previously made that specific deamination of particular bases would convert one base pair into another at specialised sites. [Cytosine can be deaminated to uracil and 5-mC to thymidine to produce a G-C to A-T transition; adenine can be deaminated to inosine which pairs with cytosine to produce an A-T to G-C transition (see Scarano, 1971; Holliday and Pugh, 1975).] Such changes may occur in the process of RNA editing (see Benne, 1990), and it is possible that they could provide the basis for heritable epigenetic controls, or conceivably, a high rate of somatic mutation in the generation of antibody diversity. All these possibilities should be borne in mind, but the experimental evidence so far relates only to the conversion of cytosine to 5-mC in DNA.

3 Inheritance of DNA methylation

The properties of DNA methylases are described elsewhere in this volume (Adams, 1992; Bestor, 1992). *In vitro*, the purified enzyme is most active on hemi-methylated DNA substrates, but it does also act on non-methylated DNA. This reaction, when it occurs *in vivo*, is referred to as *de novo* methylation. It is known in many biological contexts that *de novo* methylation is a rare event, so presumably the known enzyme has different properties *in vivo* than *in vitro*, perhaps because it interacts with other proteins. The simple concept of maintenance of methylation depends on the existence of an enzyme system or enzyme complex which specifically recognizes hemi-methylated DNA after replication and methylates the new strand, and the same enzyme does not act at all on non-methylated DNA (Holliday and Pugh, 1975; Riggs, 1975). In the ensuing discussion it should be understood that the evidence for faithful maintenance of methylation comes from examination of DNA or genes in various biological contexts; it is not based on an understanding of the actual biochemical mechanism which occurs in cells.

3.1 X chromosome inactivation

In female mammals one X chromosome is randomly inactivated early in development. Once this switch in activity has occurred it is very stably maintained in somatic cells. X chromosome inactivation provides one of the best examples of an epigenetic switch superimposed on the DNA, which is subsequently stably inherited. Reactivation of the inactive X may occur during the aging of mice, but probably much less frequently in cells from aged human donors (see below).

The CpG islands at the 5′ ends of active genes are normally unmethylated, but for X linked genes coding for glucose 6-phosphate dehydrogenase (G6PD), hypoxanthine guanine phosphoribosyl transferase (HPRT) and phosphate glycerate kinase (PGK), the islands are methylated on the inactive X chromosome (see Riggs, 1992). Recently, direct evidence for the inheritance of methylation of the CpG island of PGK has been obtained, using a method of genomic sequencing which distinguishes between cytosine and 5-mC in genomic DNA (Pfeifer et al. 1990). In a cell hybrid treated with 5-azacytidine, which is known to inhibit DNA methylation, a proportion of CpG doublets in the island of the inactive X chromosome became unmethylated. In three clones examined, the pattern of methylated and unmethylated CpGs was random. When the DNA was re-examined after thirty additional cell divisions, the pattern remained almost identical. Calculations indicated that the fidelity of transmission of a 5-mCpG or a CpG doublet was about 99.9%. This provides one of the most direct demonstrations of heritability of the pattern of DNA methylation. The methylation of islands may not be the only difference between active and inactive X chromosomes. There is evidence that DNA outside islands may be less methylated on the inactive X than the active X (Wolf et al., 1984; Lindsay et al., 1985). Presumably these differences are also maintained, although their functional significance is not understood.

3.2 Transfection experiments

In studies on the effect of DNA methylation on gene expression many experiments have been done in which methylated or unmethylated DNA is transfected into mammalian cells. In cases where a selectable marker is used to isolate transfected clones, the DNA is integrated into the chromosome. During the growth of the single selected cell to a colony and subsequently a population, many sequential replications will have occurred. Experiments show that when the DNA is isolated from such populations, it remains in its original methylated or unmethylated state (see, for example, Wigler et al., 1981; Busslinger et al., 1983). The examination of the DNA usually depends on the use of methylsensitive restriction enzymes, which would not detect a small proportion of molecules with an altered methylation pattern.

In other experiments hemi-methylated DNA was microinjected into cells and shown to become methylated on both strands prior to replication (Sandberg et al., 1991). This demonstrates quite directly the existence of a maintenance function in mammalian cells.

3.3 Mammalian cell lines

The availability of selective procedures makes it possible to detect rare variants amongst a background of many non-growing cells. Thus, if genes are silenced by *de novo* methylation, this frequency can be measured with accuracy, and similarly the frequency of gene reactivation can be measured. There is much evidence that thymidine kinase deficient cells (TK^-) can arise by *de novo* methylation, and they can be reactivated to TK^+ by the demethylating agent 5-azacytidine, which shows they retain an intact TK^+ gene (Tasseron-de Jong, 1989a and b, and see Holliday, 1987). These epigenetic variants, or epimutants, can now be studied by the same procedures as are used in classical somatic cell genetics (see Holliday, 1991), and the frequency with which clear changes in phenotype occur can be measured or estimated, as shown in Table 1. In experiments with CHO cells in which fluctuation tests were carried out, the frequency of TK^- variants arising from TK^+ was 6.0×10^{-5} (Holliday and Ho, 1990), but the spontaneous reactivation of TK^- to TK^+ is considerably lower than this. CHO cells are proline auxotrophs and they are also sensitive to cadmium (Cd^S), because they lack metallothionein activity. Both these phenotypes can be reactivated to the "wild type", that is to pro^+ and cadmium resistance (Cd^R) at very high frequency with azacytidine. However, in both cases the spontaneous rate of reactivation is very low (Table 1). This must mean that the methylated cytosines which prevent gene expression can be maintained with great fidelity. In other experiments genes were silenced by a procedure which depended on electroporation in the presence of 5 methyl dCTP (Nyce, 1991; Holliday and Ho, 1991). In these experiments evidence was obtained that 5 methyl dCTP was incorporated directly into DNA, and as a result cells could be selected in which the genes coding for one or other of three housekeeping enzymes (APRT, HPRT or TK) were not expressed. In all cases these induced epimutants stably maintained their phenotype, but were reverted at high frequency after 5-azacytidine treatment. The revertants are also stable, only very occasionally producing cells without one of these enzyme activities.

The availability of epimutants and of selection procedures makes it possible to look for mutants which have a greatly elevated level of spontaneous reactivation. The selection protocol used screened for

Table 1. Frequencies of spontaneous epigenetic changes in phenotype (epimutants) in CHO cell populations

	Phenotypic change	Selective medium*	Populations	Colonies/viable cells	Frequency
A	$TK^+ \rightarrow TK^-$	BrdU	20	2 fluctuation tests*	6.0×10^{-5}
B	$TK^- \rightarrow TK^+$	HAT	7	$17/5.14 \times 10^6$	3.3×10^{-6}
C	$Cd^S \rightarrow Cd^R$	Cd	7	$0/1.20 \times 10^6$	$< 10^{-6}$
D	$pro^- \rightarrow pro^+$	proline free	6	$0/8.68 \times 10^5$	$< 1.2 \times 10^{-6}$
E	$APRT^- \rightarrow APRT^+$	AAT	1	$0/5.36 \times 10^{-5}$	$< 1.9 \times 10^{-6}$

*For methods used, see Holliday and Ho (1990, 1991). A, believed to be due to *de novo* methylation, since all TK^- isolates tested are revertible by 5-azacytidine; B-E, these phenotypic changes would occur if methylation was lost; in each case 5-azacytidine induces the change at a frequency in the range $10^{-3} - 10^{-1}$.

temperature-sensitive conditional mutants, since it is likely that the loss of methylation would be lethal. After initial EMS mutagenesis an isolate was obtained in which spontaneous reactivation of both TK^- to TK^+ and Cd^S to Cd^R was increased 10^3–10^4 fold at 40°, but not at 34° (Gounari et al., 1987). Using probes for these genes, spontaneous phenotypic changes were shown to be associated with loss of methylation. However, when total methylation was examined, no decline at 40° was seen and the cells grew indefinitely at this temperature. This may mean that the methylation controlling gene expression, at least for the two genes examined, is maintained in a different way from the maintenance of most methylation in the genome. This result pinpoints our ignorance of the biochemical mechanisms by which DNA methylation patterns are maintained *in vivo*.

It is apparent that some genes in cell lines become silenced through *de novo* methylation and this is the likely reason for their well-known "functional hemizygosity" (Siminovitch, 1976). [A better genetic term for the situation where one of the two autosomal genes is active and the other inactive is *allelic exclusion* (Holliday and Ho, 1991).] There are also many examples of viruses integrated into chromosomal DNA which have become methylated and inactive (Doerfler, 1983, 1992; Holliday, 1987; Bednarik, 1992). These silent viral genomes are stably transmitted from cell to cell, but can be reactivated by demethylating agents.

3.4 Controlling elements in maize

There is now much detailed molecular information about transposable elements in maize. The elements can move in the genome through the activity of a transposase encoded by a structural gene within the element. Elements which lose mobility are known in many cases to have a mutation in the transposase gene or have part of the DNA deleted. However, it has now also been shown that the activity of elements can be dramatically changed by DNA methylation. Methylation has been shown to inactivate or alter the transposase activity of the *Ac* element, the *Spm* element and the *Mu* element (reviewed by Federoff et al., 1989 and Dennis and Brettell, 1990). These phenotypes are heritable, showing that once imposed the methylation is transmitted. The importance of these studies is that the phenotypes can be scored in both individual plants and also in their offspring. Thus, in this case, transmission of the methylated state through meiosis can be demonstrated. Although the authors usually claim only a correlation between methylation and a particular phenotype, they are, at least in some cases, studying the inheritance in exactly the same way as they would standard mutations, which are due to changes in DNA sequence. One does not say that a mutation correlates with a phenotype, but rather that it causes the phenotype. Provided the appropriate genetic analysis is done, changes in methylation can also be said to cause an alteration in phenotype, even though the underlying molecular mechanisms may not be understood.

3.5 Genomic imprinting

In standard Mendelian inheritance the genetic contribution from each patient is the same (leaving aside sex chromosomes). Imprinting is a phenomenon, first described in insects, where the chromosomes from different parents have different properties (reviewed by Monk and Surani, 1990; Surani, 1992). Information is therefore in some way added to the normal genome by one parent and not the other, and this information is transmissible. Genomic imprinting is now known to occur in some autosomes of mice, and also in the X chromosomes of mammals. In marsupials, for instance, the paternal X chromosome is always inactivated in female offspring, and in eutherian mammals it is preferentially inactivated in the extra-embryonic tissues. The existence of autosomal imprinting in mice has been demonstrated by genetic methods, and later proved by showing that zygotes with two paternal or two maternal pronuclei were not viable. This shows that the genomes of different parents in some way complement each other.

When studies with transgenic mice were initiated, it became possible with the use of probes to examine the methylation status of the transgene. It soon became apparent that the methylation was often different in the genes from different parents. Since the original DNA was not methylated, the methylation must at some stage have been imposed on the transgene in one sex and not in the other, very likely at meiosis or during gametogenesis. In almost all cases the paternally derived transgene is under-methylated and the maternally derived transgene is hyper-methylated. In the next generation the methylation pattern can be maintained in somatic tissue, but it is erased in the gametes of male animals.

Methylation provides a possible mechanism for adding information to the genome, so it could be responsible for normally occurring imprinting. In no case, however, has this been demonstrated. Nonetheless experiments with transgenic animals demonstrate that *de novo* methylation can be subsequently transmitted to the next generation, as well as in somatic tissues.

3.6 Fragile X syndrome

A common cause of mental retardation is associated with a fragile site in the long arm of the X chromosome. The inheritance of the condition is very unusual, since it is known that an initial genetic change, known as premutation, can be present in males with normal phenotype and transmitted to his daughters who are also normal. However, the male offspring of the daughters are commonly mentally retarded and also have the fragile site in the X chromosome. The pedigrees demonstrate beyond doubt that the premutation is converted at very high frequency to the fully mutant phenotype at some stage during or prior to oogenesis. Recent molecular studies demonstrate that a gene FMR (fragile mental retardation) with a CpG island is at or very close to the fragile site. Moreover, sites in the CpG island are methylated and the gene is not expressed in affected males (Bell et al., 1991; Heitz et al., 1991; Oberlé et al., 1991; Pieretti et al., 1991; Verkerk et al., 1991), whereas the CpG island is not methylated and the gene is expressed in transmitting males and their daughters. It is known that CpG islands are methylated in the inactive X chromosome, so it seems very likely that the basic hypothesis proposed by Laird (1987) is correct, namely, that the inheritance of the fragile X syndrome is due to the failure to reactivate a small part of the inactive X during or prior to gametogenesis. Males which inherit the X chromosome which was previously active in the female germ cell lineage would be unaffected. Daughters which receive an X chromosome with a fragile site from their mother and a normal X chromosome from their father have a variable phenotype, as would be expected for the X chromosome mosaicism in their somatic cells.

These observations indicate that methylation in an altered gene can be transmitted through the female germ line and subsequently in somatic cells. But in addition, the molecular studies also show that the region is extremely unstable at the DNA sequence level. Part of the CpG island is a CGG repeat within the coding region of the gene (coding for polyarginine) and the premutation has been shown to be due to an increase in the repeat number (Fu et al., 1991; Kremer et al., 1991; Verkerk et al., 1991). The repeat number becomes even higher in affected males, and this presumably provides a cytogenetic basis for the fragile site. The relationship between the instability of this CGG repeat, DNA methylation and the failure to reactivate the region remains obscure.

3.7 Gene silencing in duplicated DNA sequences

DNA transformation in the fungi *Neurospora* and *Ascobolus* often results in non-homologous integration of DNA, and therefore copies of a given sequence at two locations in the genome. The surprising discovery was made that such duplications are unstable prior to meiosis (reviewed by Selker, 1990, 1992). In *Neurospora* the DNA can be methylated and also mutated, and all the mutations are due to C-G to T-A transitions. It is likely that methylated cytosines are deaminated at high frequency to thymidine. If so, then it is clear that methylation occurs in CpA, CpT, CpC as well as CpG doublets, but not at equal frequency. Evidence of persisting methylation can be seen when the DNA from progeny is analysed using methyl sensitive restriction enzymes. In *Ascobolus* similar results have been obtained, but there is so far no evidence for transition mutations. Instead, the methylation is maintained in the vegetative progeny and silent genes can reactivate spontaneously or at a higher frequency after treatment with 5-azacytidine (Goyon and Faugeron, 1989). It is likely that premeiotic methylation in both *Neurospora* and *Ascobolus* depend on the pairing of homologous sequences in haploid nuclei prior to their fusion before meiosis. The evidence indicates that neither or both copies of a duplicated sequence are methylated, and when 3 copies are present, 0, 2 or 3, but never 1, are affected (Fincham et al., 1989; Faugeron et al., 1990). (More than one round of pairing between homologous regions could yield 3 methylated sequences.)

In the cases examined duplications are not subject to methylation and/or mutation in vegetative cells. However, a DNA sequence, known as ζ-η, has been shown to be a "portable signal" for methylation. This sequence carries some information which in some way brings about its own methylation. This has led Selker et al. (1987) to question the simple concept of maintenance methylation, since they argue that the persistence of methylation in this region could simply be due to reiterated *de novo* methylation.

In transgenic plants, evidence has been obtained that the existence of duplicated sequences can also lead to "co-suppression" or inactivation, and that this involves methylation (see Matzke and Matzke, 1991). The extent of suppression and methylation can be variable and mosaic in the affected plants. Co-suppression of duplicated sequences in plants, the premeiotic inactivation of duplicated DNA in ascomycete fungi and the silencing of ectopic transgenes in mice probably have commone features, but the basic mechanisms are obscure. What is clear is that DNA methylation can be involved in all cases and that this methylation is heritable.

4 Epigenetic defects in cancer and ageing

The emergence of tumours is the result of several sequential events, with concomitant selection of cells with altered phenotypes. Much attention has been paid to the significance of mutations and chromosome abnormalities in tumour progression, but it is quite likely that epigenetic events based on methylation changes are also important. There is much evidence that the pattern of methylation in tumour cell lines is abnormal (reviewed by Jones, 1992). A recent striking example is the demonstration of the strict conservation of the pattern of methylation in normal cells, and the corresponding variable methylation in tumour lines derived from them (Kochanek et al., 1990, 1991; Achten et al., 1991; Behn-Krappa et al., 1991). Thus, it is clear that during oncogenesis the normal controls of DNA methylation break down. This is also shown by the *de novo* methylation of CpG islands in cell lines. In normal tissues such islands remain unmethylated, but in several immortalised cell lines (which are partly or fully transformed) the islands are often methylated (Antequera et al., 1990). This type of methylation can silence genes, and therefore an obvious possibility in these lines is the silencing of tumour suppressor genes by the methylation of islands, or other sequences in DNA. Also, it is likely that DNA damaging agents can alter methylation patterns without causing mutation (Holliday, 1979; Lieberman et al., 1983) so carcinogens may act by inducing epigenetic defects which are important in tumour progression (see also Jones, 1992; Holliday and Jeggo, 1985). The demethylating agent 5-azacytidine is known to be a potent carcinogen (Carr et al., 1984).

Ageing, like cancer, is also likely to be a multiple hit process, since deleterious changes can occur in DNA, in long lived proteins, or in other cellular components. Again, epigenetic defects may be of significance in the loss of cellular homeostasis, which is an important part of the ageing process. It may well be that the controls which keep cells in the normal differentiated state can be disturbed and as a result cells become partly dedifferentiated or acquire other abnormal phenotypes. The only direct evidence for this so far comes from the experiments on the reactivation of the inactive X during ageing. X inactivation is generally thought to be under very stable epigenetic control, so it is surprising that a quite high frequency of X chromosome reactivation is seen in ageing mice (Wareham et al., 1987; Brown and Rastan, 1988). In human cells, on the other hand, age-related reactivation of a gene of the X chromosome has not been seen (Migeon et al., 1988). It has been suggested that epigenetic controls in short-lived animals, such as mice, might be much less tight than in long-lived animals such as man (Holliday, 1989). It is certainly true that cell transformation *in vitro* and tumour formation *in vivo* occurs much more frequently in mouse than man. The argument can therefore be made that epigenetic defects also arise much more frequently in mouse than man, and these events contribute to the ageing process. Although little direct information is available about methylation and ageing, total methylation declines during ageing of cells *in vitro* (Wilson and Jones, 1983; Fairweather et al., 1987) and probably also *in vivo* (Wilson et al., 1987). If the same is true for the subset of 5-mC residues which are important in the control of gene activity, then the loss of methylation might activate specialised genes in cells where such genes are normally silent. In dividing cells this ectopic expression of genes would be inherited and could contribute to the emergence of senescent cell phenotypes during ageing. In non-dividing cells heritability is not important, but it is also possible that DNA damage could lead to loss of methylation and therefore to the breakdown of normal cellular controls.

The importance of epigenetic defects follows from the fact that the normal pattern of methylation is strictly conserved and inherited. Any deviations from this pattern, which might occur spontaneously at low frequency, and at high frequency after DNA damage, are likely to have significant phenotypic effects. Cancer and ageing both represent examples of the breakdown of normal controls of cell division and metabolism, and it would therefore be surprising if epigenetic defects were not important components of these processes. It is also possible that there are special mechanisms to recognize and eradicate such defects at meiosis, before the germ cells are formed (see Holliday, 1987).

5 Inheritance of acquired characteristics?

The central dogma of molecular biology asserts that information can be transmitted from nucleic acid to nucleic acid and from nucleic acid to protein, but not from protein to nucleic acid (Crick, 1958). In terms of the sequence of bases in DNA and of amino acids in proteins this seems infallible, but it is not when the properties of DNA can be changed by the enzymic modification of one or more bases. If gene expression can be changed by enzyme action, as in the case of *de novo* methylation of a CpG island, then it is conceivable that outside influences could alter the protein metabolism of a cell in such a way as to change the epigenetic information or the *epigenotype*. More specifically, it is possible to envisage a hormone, growth factor or morphogen interacting with a specific receptor, which then binds to its specific sequence of DNA and alters its properties. A well-known example is the action of estradiol and stimulating vitellogenin synthesis in chicken liver cells (Jost, 1992). In this sequence of events the DNA of the gene becomes demethylated and the gene activated. The same type of response could also lead to a change in DNA modification which is subsequently inherited (Holliday, 1990a, b).

Jablonka and Lamb (1989) have speculated along these lines, suggesting that external influences could alter the heritable modification of DNA and that new genetic properties could then be transmitted to the next generation. They discuss several unexplained examples of environmental influences on inheritance. One of the best known is the effect of plant fertilizers on the induction of heritable variation in flax. The seeds from plants grown under different conditions can produce "genotrophs", which are offspring with different physical characteristics from the parent, and these characteristics are transmitted to subsequent generations. It is known in these cases that the amount of repetitive DNA can be altered (reviewed by Cullis, 1983, 1987), but it is also possible that changes in DNA methylation are involved. The disappearance of a *Taq*1 restriction site in 5S repeated sequences, which is assumed to be due to deletion of DNA (Cullis and Cleary, 1986), might instead be due to 6-adenine methylation within the *Taq*1 sequence, which would be resistant to digestion. Several other possible examples of environmentally induced DNA modification or genetic variation are discussed by Jablonka and Lamb (1989). Their argument is strengthened by the evidence that methylation differences in human DNA can be inherited from one generation to the next (Silva and White, 1988).

It is very likely that during development the properties of specific somatic cells are changed by external influences emanating from other cells, and that such altered phenotypes are inherited through somatic cell division. This situation is a normal part of the developmental process. What is novel is the possibility that external influences mediated by somatic cells might influence the germ line and alter the phenotypes of offspring. At present there is not enough information to assess the significance of possible examples of what can be loosely referred to as "Lamarckian inheritance". The pathway of transmission of information and the mechanism of inheritance can be understood in general terms, but in no case is there any detailed information. Even if such mechanisms exist, Maynard Smith (1990) on the basis of quite detailed models, has concluded that inheritance of this type is unlikely to be of evolutionary significance.

6 Conclusions

The alternative modified or unmodified states of cytosine in DNA are inherited in a variety of biological contexts. The accuracy of the transmission of a given pattern of 5-mC is very high, and this can account for the specificity of the pattern which is highly conserved between human individuals (Kochanek et al., 1990; Behn-Krappa et al., 1991). The mechanism for the accurate transmission, or maintenance, is not understood. The DNA methylase must be able to specifically distinguish hemi-methylated and non-methylated sites, that is, 5-mCpG and CpG doublets, and maintain methylation only in the former. Known methylases do not have this ability *in vitro*, perhaps suggesting that there may be a maintenance enzyme complex *in vivo*.

Our lack of understanding of the biochemical mechanism of maintenance of 5-mC does not weaken important biological conclusions. It should be remembered that the whole discipline of genetics was developed in the absence of knowledge of the chemistry of genes, or their means of replication. Similarly we can deduce that an epigenetic system of inheritance exists that is independent of DNA sequence *per se*, although ultimately derived from the DNA genome. It is possible to study genetic and epigenetic inheritance in cultured mammalian cells, and this provides information about changes in methylation, as well as the heritability of methylation (Holliday, 1991). In mammalian cell lines it is evident that genes can be silenced by *de novo* methylation at a fairly low rate, provided of course that there is no selection against the changed phenotype. We do not know how many cytosine residues need to be methylated to silence a gene, but one study indicates that one or two may be sufficient (Tasseron-de Jong, 1989a and b). However, it is likely that in these cell lines many CpG doublets within CG islands will eventually become methylated (Antequera et al., 1990). What then is the likelihood of such methylation being lost spontaneously with reactivation of gene expression? The results in Table 1 indicate that this occurs very rarely, and it is easy to understand why. If individual 5-mCpG doublets change to CpG at a given low frequency, it is likely that the reverse *de novo* methylation occurs at a higher frequency. [The results of Pfeifer et al. (1990) suggest about 5% per cell division.] Thus, the probability of losing several 5-mCpG doublets and activating the gene would be very low. Much more information is needed about the exact molecular relationships between methylated cytosine residues and gene silencing and expression. This information will come from the new methods of "5-base" genomic sequencing (see Saluz, 1992) which distinguish cytosine from 5-mC in a given sequence of DNA.

Methylation patterns are inherited in mouse diploid cells *in vivo*, and presumably also in such cells *in vitro*. However, a given pattern may be tissue specific, and almost nothing is known about the way the changes in pattern are controlled. One must assume that there is regulated *de novo* methylation, so that particular CpGs in particular sequences become methylated. Also, there must be specific demethylation, either by an active enzymic process, or alternatively, by a specific blocking process where a protein bound to DNA would prevent methylation after replication (see Holliday, 1990b). We are also almost totally ignorant of the controls of methylation which operate in gametogenesis or in early development, and also the rules governing the transmission of methylation from one generation to the next.

In existing literature there is frequent reference to the "correlation" between methylation and gene inactivity. Yet we now know in many instances that the change in methylation is concomitant with the change in gene expression. This is exactly analogous to a change in DNA sequence producing a mutant phenotype: the mutation causes the phenotype. In the same way the methylation signal is directly related to the phenotype. The conclusion is not affected by the fact that there may be unknown transcription factors which recognise the presence or absence of methylation. The real importance of DNA methylation is that it provides the basis for a heritable epigenetic system. This makes it possible to add or subtract information to DNA which may be essential components of development and differentiation. Mistakes or defects in this heritable information may be important in tumour progression and also in ageing. The experimental problem is to identify the important genes in which the epigenetic information resides. When that is done, it will become possible to examine normal, or abnormal, changes in DNA methylation in important biological contexts.

Achten, S., Behn-Krappa, A., Jücker, M., Sprengel, J., Hölker, I., Schmitz, B., Tesch, H., Diehl, V., and Doerfler, W. (1991) Patterns of DNA methylation in selected human genes in different Hodgkin's lymphoma and leukemia cell lines and in normal human lymphocytes. Cancer Res. *51*, 3702–3709.

Adams, R. L. P. (1992) – this volume, pp. 120–144.

Antequera, F., Boyes, J., and Bird, A. (1990) High levels of *de novo* methylation and altered chromatin structure at CpG islands in cell lines. Cell *62*, 503–514.

Bednarik, D. P. (1992) – this volume, pp. 300–329.

Behn-Krappa, A., Hölker, I., Sandaradusa de Silva, U., and Doerfler, W. (1991) Patterns of DNA methylation are indistinguishible in different individuals over a wide range of human DNA sequences. Genomics *11*, 1–7.

Bell, M. V., Hirst, M. C., Nakahori, Y., MacKinnon, R. N., Roche, A., Flint, T. J., Jacobs, P. A., Tommerup, N., Tranebjaerg, L., Froster-Iskenius, U., Kerr, B., Turner, G., Lindenbaum, R. H., Winter, R., Pembrey, M., Thibodeau, S., and Davies, K. E. (1991) Physical mapping across the fragile X: hypermethylation and clinical expression of the fragile X syndrome. Cell *64*, 861–866.

Benne, R. (1990) RNA editing in trypanosomes: is there a message? Trends Genet. *6*, 177–181.

Bestor, T. H. (1992) – this volume, pp. 109–119.

Brown, S., and Rastan, S. (1988) Age related reactivation of an X linked gene close to the inactivation centre in the mouse. Genet. Res. *52*, 151–154.

Busslinger, M., Hurst, H., and Flavell, R. A. (1983) DNA methylation and the regulation of globin gene expression. Cell *34*, 197–206.

Carr, B. I., Garrett Reilly, J., Smith, S. S., Winberg, C., and Riggs, A. D. (1984) The tumorigenicity of 5-azacytidine in the male Fischer rat. Carcinogenesis *5*, 1583–1590.
Crick, F. H. C. (1958) On protein synthesis. Symp. Soc. Exp. Biol. XIII, pp. 138–163. Cambridge University Press.
Cullis, C. A. (1983) Environmentally induced DNA changes in plants. CRC Crit. Rev. Plant Sci. *1*, 117–129.
Cullis, C. A. (1987) The generation of somatic and heritable variation in response to stress. Am. Nat. *130* suppl., S62–S73.
Cullis, C. A., and Cleary, W. (1986) Rapidly varying DNA sequences in flax. Can. J. Genet. Cytol. *28*, 252–259.
Dennis, E. S., and Brettel, R. I. S. (1990) DNA methylation of maize transposable elements is correlated with activity. Phil. Trans. Roy. Soc. B. *326*, 217–230.
Doerfler, W. (1983) DNA methylation and gene activity. Annu. Rev. Biochem. *52*, 93–124.
Doerfler, W. (1992) – this volume, pp. 262–299.
Fairweather, S., Fox, M., and Margison, P. (1987) The *in vitro* lifespan of MRC-5 cells is shortened by 5-azacytidine induced demethylation. Exp. Cell Res. *168*, 153–159.
Faugeron, G., Rhounim, L, and Rossignol, J. L. (1990) How does the cell count the number of ectopic copies of a gene in the pre-meiotic inactivation process in *Ascobolus immersus*? Genetics *124*, 585–591.
Federoff, N., Masson, P., and Banks, J. A. (1989) Mutations, epimutations, and the developmental programming of the maize suppressor-mutator transposable element. BioEssays *10*, 139–144.
Fincham, J. R. S., Connerton, I. F., Notarianni, E., and Harrington, K. (1989) Pre-meiotic disruption of duplicated and triplicated copies of the *Neurospora crass am* (glutamate dehydrogenase) gene. Curr. Genet. *15*, 327–334.
Fu, Y.-H., Kuhl, D. P. A., Pizzuti, A., Pieretti, M., Sutcliffe, J. S., Richards, S., Verkerk, A. J. M. H., Holden, J. J. A., Fenwick, Jr., R. G., Warren, S. T., Oostra, B. A., Nelson, D. L., and Caskey, C. T. (1991) Variation of the CGG repeat at the fragile X site results in genetic instability: resolution of the Sherman paradox. Cell *67*, 1047–1058.
Gounari, F., Banks, G. R., Khazaie, K., Jeggo, P. A., and Holliday, R. (1987) Gene reactivation: a tool for the isolation of mammalian DNA methylation mutants. Genes & Dev. *1*, 899–912.
Goyon, C., and Faugeron, G. (1989) Targeted transformation of *Ascobolus immersus* and *de novo* methylation of the resulting duplicated DNA sequences. Mol. Cell. Biol. *9*, 2818–2827.
Heitz, D., Rosseau, F., Devys, D., Saccone, S., Abderrahim, H., Le Paslier, D., Cohen, D., Vincent, A., Toniolo, D., Della Valle, G., Johnson, S., Schlessinger, D., Oberlé, I., and Mandel, J. L. (1991) Isolation of sequences that span the fragile X and identification of a fragile X-related CpG island. Science *251*, 1236–1239.
Holliday, R. (1979) A new theory of carcinogenesis. Br. J. Cancer *40*, 513–522.
Holliday, R. (1987) The inheritance of epigenetic defects. Science *238*, 163–170.
Holliday, R. (1989) X chromosome reactivation and ageing. Nature *337*, 311.
Holliday, R. (1990a) Paradoxes between genetics and development. J. Cell Sci. *97*, 395–398.
Holliday, R. (1990b) Mechanisms for the control of gene activity during development. Biol. Rev. *65*, 431–471.
Holliday, R. (1991) Mutations and epimutations in mammalian cells. Mutat. Res. *250*, 345–363.
Holliday, R., and Ho, T. (1990) Evidence for allelic exclusion in Chinese hamster ovary cells. New Biologist *2*, 719–726.
Holliday, R., and Ho, T. (1991) Gene silencing in mammalian cells by uptake of 5 methyl deoxycytidine 5′ triphosphate. Somatic Cell Mol. Genet. *17*, 537–542.
Holliday, R., and Jeggo, P. A. (1985) Mechanisms of changing gene expression and their possible relationship to carcinogenesis. Cancer Surv. *4*, 557–581.
Holliday, R., and Pugh, J. E. (1975) DNA modification mechanisms and gene activity during development. Science *187*, 226–232.
Jablonka, E., and Lamb, M. J. (1989). The inheritance of acquired epigenetic variations. J. Theor. Biol. *139*, 69–83.
Jones, A. P. (1992) – this volume, pp. 487–509.
Jost, J.-P. (1992) – this volume, pp. 425–451.

Kochanek, S., Toth, M., Dehmel, A., Renz, D., and Doerfler, W. (1990) Inter-individual concordance of methylation profiles in human genes for tumor necrosis factors α and β. Proc. Natl. Acad. Sci. USA *87*, 8830–8834.

Kochanek, S., Radbruch, A., Tesch, H., Renz, D., and Doerfler, W. (1991) DNA methylation profiles in the human genes for tumor necrosis factors α and β in the subpopulations of leukocytes and in leukemia. Proc. Natl. Acad. Sci. USA *88*, 5759–5763.

Kremer, E. J., Pritchard, M., Lynch, M., Yu, S., Holman, K., Baker, E., Warren, S. T., Schlessinger, D., Sutherland, G. R., and Richards, R. I. (1991) Mapping of DNA instability at the fragile X to a trinucleotide repeat sequence p(CCG)*n*. Science *252*, 1711–1714.

Laird, C. D. (1987) Proposed mechanism of inheritance and expression of the human fragile X syndrome of mental retardation. Genetics *117*, 587–599.

Lieberman, M. W., Beach, L. R., and Palmiter, R. D. (1983) Ultraviolet radiation-induced metallothionein-1 gene activation is associated with extensive DNA demethylation. Cell *35*, 207–214.

Lindsay, S., Monk, M., Holliday, R., Huschtscha, L. I., Davies, K. E., Riggs, A. D., and Flavell, R. A. (1985) Differences in methylation on the active and inactive X chromosomes. Ann. Hum. Genet. *49*, 115–127.

Maynard Smith, J. (1990) Models of a dual inheritance system. J. Theor. Biol. *143*, 51–53.

Matzke, M. A., and Matzke, A. J. M. (1991) Differential inactivation and methylation of a transgene in plants by two suppressor loci containing homologous sequences. Plant Mol. Biol. *16*, 821–830.

Migeon, B. R., Axelman, J., and Beggs, A. H. (1988) Effect of ageing on reactivation of the human X-linked HPRT locus. Nature *335*, 93–96.

Monk, M., and Surani, A., eds (1990) Genomic imprinting. Development Suppl. 1990. Company of Biologists, Cambridge.

Nyce, J. (1991) Gene silencing in mammalian cells by direct incorporation of electroporated 5-methyl-2′-deoxycytidine 5′ triphosphate. Somatic Cell Mol. Genet. *17*, 543–550.

Oberlé, I., Rousseau, F., Heitz, D., Kretz, C., Devys, D., Hanauer, A., Boué, J., Bertheas, M. F., and Mandel, J. L. (1991) Instability of a 550-base pair DNA segment and abnormal methylation in fragile X syndrome. Science *252*, 1097–1102.

Pfeifer, G. P., Steigerwald, S. D., Hansen, R. S., Gartler, S. M., and Riggs, A. D. (1990) Polymerase chain reaction-aided genomic sequencing of an X chromosome-linked CpG island: methylation patterns suggest clonal inheritance, CpG site autonomy, and an explanation of active state stability. Proc. Natl. Acad. Sci. USA *87*, 8252–8256.

Peiretti, M., Zhang, F., Fu, Y-H., Warren, S. T., Oostra, B. A., Caskey, C. T., and Nelson, D. L. (1991) Absence of expression of the FMR-1 gene in fragile X syndrome. Cell *66*, 817-822.

Riggs, A. D. (1975) X inactivation, differentiation and DNA methylation. Cytogenet. Cell Genet. *14*, 9–25.

Riggs, A. D. (1992) – this volume, pp. 358–384.

Saluz, H. P. (1992) – this volume, pp. 11–26.

Sandberg, G., Guhl, E., Graessmann, M., and Graessmann, A. (1991) After microinjection hemimethylated DNA is converted into symmetrically methylated DNA before DNA replication. FEBS Lett. *283*, 247–250.

Scarano, E. (1971) The control of gene function in cell differentiation and in embroygenesis. Adv. Cytopharmacol. *1*, 13–24.

Selker, E. U. (1990) Pre-meiotic instability of repeated sequences in *Neurospora crassa*. Annu. Rev. Genet. *24*, 579–613.

Selker, E. U. (1992) – this volume, pp. 212–217.

Selker, E. U., Jensen, B. C., and Richardson, G. A. (1987). A portable signal causing faithful DNA methylation *de novo* in *Neurospora crassa*. Science *238*, 48–53.

Silva, A. J., and White, R. (1988) Inheritance of allelic blueprints for methylation patterns. Cell *54*, 145–152.

Siminovitch, L. (1976) On the nature of heritable variation in cultured somatic cells. Cell *7*, 1–11.

Surani, M. A. (1992) – this volume, pp. 469–486.

Tasseron-de Jong, J. G., der Dulk, H., van de Putte, P., and Giphart-Gassler, M. (1989a) *De novo* methylation as a major event in the inactivation of transfected herpes virus thymidine kinase genes in human cells. Biochim. Biophys. Acta *1007*, 215–223.

Tasseron-de Jong, J. G., Aker, J., der Dulk, H., van de Putte, P., and Giphart-Gassler, M. (1989b) Cytosine methylation in the *Eco*RI site of active and inactive herpes virus thymidine kinase promoters. Biochim. Biophys. Acta *1008*, 62–70.

Verkerk, A. J. M. H., Pieretti, M., Sutcliffe, J. S., Fu, Y-H., Kuhl, D. P. A., Pizzuti, A., Reiner, O., Richards, S., Victoria, M. F., Zhang, F., Eussen, B. E., van Ommen, G-J.B., Blonden, L. A. J., Riggins, G. J., Chastain, J. L., Kunst, C. B., Galjaard, H., Caskey, C. T., Nelson, D. L., Oostra, B. A., and Warren, S. T. (1991) Identification of a gene (FMR-1) containing a CGG repeat coincident with a breakpoint cluster region exhibiting length variation in fragile X syndrome. Cell *65*, 905–914.

Wareham, K. A., Lyon, M. F., Glenister, P. H., and Williams, E. D. (1987) Age related reactivation of an X linked gene. Nature *327*, 725–727.

Weintraub, H. (1985) Assembly and propagation of repressed and derepressed states. Cell *42*, 705–711.

Wigler, M., Levy, D., and Perucho, M. (1981) The somatic replication of DNA methylation. Cell *24*, 33–40.

Wilson, V. L., and Jones, P. A. (1983) DNA methylation decreases in ageing but not in immortal cells. Science *220*, 1055–1057.

Wilson, V. L., Smith, R. A., Ma, S., and Cutler, R. G. (1987) Genomic 5-methyl deoxycytidine decreases with age. J. Biol. Chem. *262*, 9948–9951.

Wolf, S. F., Jolly, D. J., Lunnen, K. D., Friedman, T., and Migeon, B. R. (1984) Methylation of the hypoxanthine phosphoribosyl transferase locus on the human X chromosome: implications for X chromosome inactivation. Proc. Natl. Acad. Sci. USA *81*, 2806–2810.

DNA Methylation: Molecular Biology and Biological Significance
ed. by J. P. Jost & H. P. Saluz

DNA methylation and genomic imprinting in mammals

Hiroyuki Sasaki, Nicholas D. Allen and M. Azim Surani

Department of Molecular Embryology, AFRC Institute of Animal Physiology and Genetics Research, Babraham, Cambridge CB2 4AT, England

1 Introduction

In mammalian sexual reproduction, each parent contributes a haploid set of chromosomes to the offspring. It is usually believed that these two haploid sets are on the whole equivalent in their function: reciprocal crosses show identical phenotypic effects at many genetic loci. Only two clear exceptions to this rule have been recognized: sex-linked inheritance and cytoplasmic inheritance (the latter is actually not linked to the nucleus). Therefore, it seems reasonable to assume that autosomal genes behave in the same manner no matter which parent they are contributed by. However, recent studies have revealed that parental origin does affect gene expression at certain autosomal loci. In other words, these autosomal regions somehow remember their parental origin and regulate gene activity according to that memory. These genes constitute the third category of exceptions to the principle of reciprocity.

The chromosome memory regarding parental origin is conferred by an epigenetic process called "genomic imprinting" or "parental imprinting" which probably occurs during gametogenesis. Examples of such germline-specific chromosome memories or imprints can be seen not only in mammals but also in many other organisms including insects and plants. In fact, the term imprinting was first used by Crouse (1960) to define the mechanism that leads to the selective elimination of paternal X chromosomes and autosomes in the fly, *Sciara*. However, it is not known whether imprinting phenomena observed in different species have evolved for the same reason and/or with the same molecular basis. (Nevertheless, an interesting speculation has been made on the reason for imprinting by Moore and Haig [1991].)

At present little is known about the molecular basis of mammalian genomic imprinting. However, as will be discussed in this article, DNA methylation may be one mechanism which plays a role for this epigenetic phenomenon. Since the first endogenously imprinted mouse genes were identified only recently, most of the data reviewed here comes from studies on related biological phenomena, principally transgene imprinting. Nevertheless, during these five years or so, the issue of epigenetic mechanisms of imprinting has attracted many researchers' attention and considerable progress has been made using this model system. Although most of our discussion will remain speculative with regard to imprinting of endogenous genes, some very recent observations made on endogenous sequences do indeed point to a role of DNA methylation for genomic imprinting.

2 Autosomal imprinting and X chromosome inactivation in mammals

The concept of genomic imprinting in mammals has evolved largely from observations made in mouse embryology and genetics. First, parthenogenetic/gynogenetic and androgenetic embryos, which carry only either maternal or paternal chromosomes respectively, cannot develop normally and die before birth (reviewed by Surani, 1986; Surani et al., 1986, 1990b; Solter, 1988). For example, nuclear transfer experiments in one-cell embryos suggest that the paternal genome is relatively more important for the formation of the extraembryonic tissues, while the maternal genome is more important for the development of the embryo proper. Based on further studies with chimeras it appears that imprinting plays a crucial role in maintaining a balance between proliferation and differentiation of various embryonic cell lineages (Surani et al., 1990b). Second, genetic complementation experiments using intercrosses between mice with either Robertsonian or reciprocal translocations have identified several autosomal regions whose activities are affected by their parental origin (reviewed by Searle and Beechey, 1978, 1985; Cattanach, 1986; Searle et al., 1989; Beechey et al., 1990; Cattanach and Beechey, 1990). These uniparental disomies often die in the course of development, and sometimes the phenotypes of maternal and paternal disomy of the same chromosomal region are complementary. Third, some genetic traits such as hairpin-tail (Thp) (Johnson, 1974, 1975) show different phenotypes depending on whether the mutation is inherited from the father or the mother. The implication of these observations is that certain mouse autosomal regions or genes are somehow imprinted to show differential expression patterns that are specific to their parental origin.

In addition to the above evidence in mice, a number of genetic disorders and chromosome abnormalities in man show unusual inheritance or aberrant manifestations which can be explained reasonably only in the context of imprinting (reviewed by Reik, 1989; Clarke, 1990; Hall, 1990). For example, the juvenile form of Huntington's chorea appears to be inherited preferentially from the male parent. Non-random retention of parental chromosomes in recessive tumor syndromes, such as Wilms' tumor and rhabdomyosarcoma, seems to point to the involvement of imprinting in carcinogenesis (reviewed by Ferguson-Smith et al., 1990; Sapienza, 1991).

What is the molecular basis of imprinting? The primary imprint must be established during gametogenesis, and then the germline-specific imprints must be stably maintained and propagated through fertilization and subsequent rounds of DNA replication and cell division during development. Moreover, reprogramming should be possible upon each passage through the germline in successive generations. This last process may be associated with the restoration of general nuclear totipotency. Therefore heritability, reversibility and the ability to affect gene expression are the requirements for imprinting mechanisms (Surani, 1986; Monk, 1988; Sapienza, 1989). DNA methylation is heritable according to the maintenance methylase concept (Holliday and Pugh, 1975; Riggs, 1975; Wigler, 1981) and has been implicated in the regulation of gene expression (Razin and Riggs, 1980; Cedar, 1988). Furthermore, a difference in the DNA methylation level has been reported in the sperm and egg genome (Monk et al., 1987). For these reasons, DNA methylation has been proposed as a possible molecular mechanism of imprinting.

A precedent for differential methylation of homologous chromosomes can be seen in X chromosome inactivation, which is the compensation mechanism for the sex differences in dosage of X-linked genes. While the embryo proper of eutherian mammals shows random inactivation of the maternal or paternal X chromosomes (reviewed by Lyon, 1972), a preferential inactivation of the paternal X chromosome occurs in all cells of female marsupials (Cooper et al., 1971; Richardson et al., 1971; Sharman, 1971; reviewed by VandeBerg et al., 1987) and in the extraembryonic tissues of developing female mice (Takagi and Sasaki, 1975; West et al., 1977). Therefore, X chromosome inactivation can be affected by imprinting. Cytologically the inactived state is associated with late replication and heterochromatinization (reviewed by Lyon, 1972; Gartler and Riggs, 1983; Grant and Chapman, 1988). At the molecular level, a redistribution of DNA methylation is observed and this is almost certainly involved in maintaining the inactive state in eutherian mammals (see Riggs, this volume; Monk, 1986). Thus the CpG islands, which are normally free of methylation (Bird, 1986), of X-linked genes are methylated when they are on the inactive X, whereas the remaining part of these genes are often less methylated (for example, Wolf et al., 1984; Yen et al., 1984; Toniolo et al., 1984). This inactive X-specific island methylation is seen not only in the embryo proper but also in the extraembryonic lineages (Lock et al., 1987; Singer-Sam et al., 1990), where inactivation is non-random, although a proportion of extraembryonic cells do not show such hypermethylation

(Lock et al., 1987). By contrast, this mechanism is obviously not adopted by marsupials (Kaslow and Migeon, 1987). Since the inactive state is poorly maintained in marsupials, CpG island methylation is viewed as a stabilizing mechanism (Kaslow and Migeon, 1987; Migeon et al., 1989). In fact, it has been shown that methylation of CpG islands on the inactive X occurs after chromosome inactivation (Lock et al., 1987).

What can we learn about autosomal imprinting from these X inactivation studies? It is evident that differential methylation is not the primary imprint for the preferential inactivation of the paternal X chromosome in marsupials and this is probably also true for the extraembryonic tissues of mice. This notion is supported by the observation on an X-linked transgene, that a post-fertilization event and not germline methylation determined the paternal hypermethylation of the transgene in the extraembryonic tissues (Collick et al., 1988). Nevertheless, it is important that differential methylation dependent on parental origin does occur in the extraembryonic tissues, at both endogenous and transgene loci in mice. This implies that DNA methylation may aid maintenance of parental imprints at least on some occasions. Furthermore, the possibility of germline methylation being the mechanism for autosomal imprinting still exists since autosomal imprinting seems to differ from non-random X-inactivation in many respects. For example, X inactivation is basically a chromosome-wide phenomenon whereas autosomal imprinting appears to occur on a domain basis or on a gene by gene basis. In the next section, we will examine the behavior of imprinted transgenes located on mouse autosomes.

3 Imprinted transgene loci

Normally, it is not easy to distinguish the level of methylation of different parental genomes, or specific alleles, coexisting in the same cell. Therefore, transgenes integrated at various sites in the genome have been used to study differential methylation after paternal and maternal inheritance. This has led to the exciting discovery of transgene imprinting (Reik et al., 1987, 1990; Sapienza et al., 1987; Swain et al., 1987; Hadchouel et al., 1987; Sasaki et al., 1991). The essential features of these transgenes are as follows. (Details of these experiments have also been reviewed and discussed in Reik et al., 1990, 1992; Surani et al., 1988, 1990a).

The most striking observation is the parental-origin-dependent differential methylation which can be switched from one form to the other upon passage through the germline of opposite sex. (Although, the case reported by Hadchouel et al. (1987) is an exception. See later.) A typical situation is illustrated in Figure 1. In this case (transgene locus

MPA434), the vector plasmid sequences and the promoter sequences derived from the mouse MT-I gene contained within the transgene construct were totally methylated when derived from the mother but unmethylated when derived from the father (Sasaki et al., 1991). In each case examined, the differential methylation pattern was essentially the same in different somatic tissues except for the testis. This indicates that the parental-origin-dependent differential methylation was indeed somatically heritable and that the pattern was perhaps switched during gametogenesis (see later). Furthermore, tissue-specific expression of the RSV-myc and HBsAg transgenes was inversely correlated with their methylation levels (Swain et al., 1987; Hadchouel et al., 1987). All of these observation strongly support DNA methylation as a possible mechanism of imprinting.

In general, only some of the transgenic lines produced with the same transgene construct show signs of imprinting. Accordingly, chromosomal position-effects should be counted as one causative factor for the phenomenon. However, the frequency of detecting imprinted transgenes among established lines is relatively high (10–20%, Reik et al., 1990) especially when the map produced by the genetic complementation studies is considered for a comparison. In addition, only one out of the three imprinted transgenes examined for their chromosomal locations mapped to a region that has been shown to contain imprinted genes

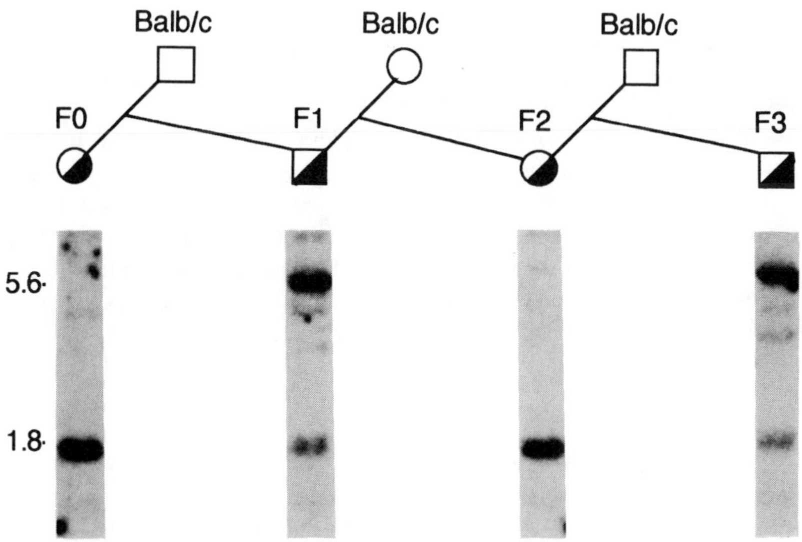

Figure 1. The alternating pattern of high and low methylation of the MPA434 transgene crossed to BALB/c males and females (Sasaki et al., 1991). Tail skin DNA was cut with *Eco*RI plus *Hpa*II and subjected to Southern analysis. An undermethylated transgene gives the 1.8-kb band and a methylated one the 5.6-kb band.

(MPA434, chromosome 11 A5, Sasaki et al., 1991) (for the others, see Hadchouel et al., 1987; Reik et al., 1990). Although a number of possible explanations exist (for example, failure of detecting subtle phenotypes in genetic complementation studies; non-random integration of transgenes), one contributing factor may reside within the transgene sequences. For example, in the case of the RSV-myc and Troponin I transgenes, imprinting appears to be related to some sequence specificity within the constructs since 2/2 and 4/5 of the lines examined showed imprinting, respectively (Chaillet et al., 1991; Sapienza et al., 1987). If imprinting of some of these constructs are indeed position-independent, it may be possible to define the minimal sequence requirements for transgene imprinting. On the other hand, in cases where position-effects do seem to contribute to imprinting, it is interesting to ask whether the endogenous mouse sequences at the insertion sites show imprinting or not (see later).

One major discrepancy between transgene imprinting and endogenous imprinting is the fact that it is always the maternally inherited transgene which is more methylated, with one exception reported by Sapienza et al. (1987). This overwhelming bias is inconsistent with the genetic map produced by the complementation studies. Although several possible reasons exist (for example, injection of male pronucleus for transgenesis; action of sex-linked dosage-sensitive modifier genes, Sapienza, 1990), none of them seems quite satisfactory. Again, some sequence information contained within the transgene constructs might be involved. For example, as pointed out previously (Sasaki et al., 1991), most of the imprinted transgenes carry some prokaryotic sequences (e.g., plasmid vectors, CAT gene, lacZ gene) and these could somehow relate to the observed preponderance of maternal methylation.

Another very important fact was realized when systematic breeding studies were carried out on these imprinted loci: transgene imprinting is controlled by genetic background. As a matter of fact, a number of transgenes which exhibit imprinting do so only when they are maintained on mixed genetic backgrounds. The genes responsible for the background effects are called strain-specific or genotype-specific modifiers and the actions of these modifier genes are summarized in the next section.

4 Modifiers of transgene imprinting

A typical example of a strain-specific background effect on transgene imprinting can be seen in the CAT17 locus. As reported initially, this locus is imprinted when present on a mixed genetic background with the maternal allele being hypermethylated and the paternal allele being undermethylated (Reik et al., 1987). However, if this transgene is crossed to C57BL/6 (B6), it becomes methylated regardless of the mode of transmission, and the imprinting effect is eventually lost after repeated crosses to this strain (Reik et al., 1990). Another case, the pHRD 432 locus, is of particular interest because the strain-specific modifier has been genetically identified and chromosomally mapped (Engler et al., 1991). When on a B6 strain background, this transgene is always highly methylated due to the effect by the B6 allele of a single modifier, *Ssm-1*, located on chromosome 4. However, if this transgene is crossed to DBA/2 (DB) and thus the B6 allele of *Ssm-1* is segregated away, it becomes less methylated, but more rapidly upon paternal transmission than upon maternal transmission. After a continued breeding into DB strain, the parental effect on reprogramming disappears on this non-methylating background. Therefore, in these cases, the parental effect can be seen only in a segregating population but not in a homozygous population. Background effects similar to these have been reported for Troponin I 379, OX1-5, and HBsAg E36 (Sapienza et al., 1989; Reik et al., 1990; Hadchouel et al., 1987).

The modifier genes described above obviously exert their effects after fertilization. Similar modifiers may operate on endogenously imprinted loci since the developmental capacities of both androgenetic embryos and androgenetic aggregation chimeras have been reported to be influenced by their strain background (Latham and Solter, 1991; Mann and Stewart, 1991). However, whether these background effects are caused by changes in imprinted genes' own activities is at present unknown. It is rather reasonable to assume that at least some of the endogenously imprinted loci identified by the embryological and genetic studies are differentially modified in the male and female germline and that this occurs on any homozygous background. A recent report by Fundele et al. (1991) showed that the developmental potential of parthenogenetic embryos are virtually independent of their strain background. Thus far, only two transgenes have been shown to be imprinted on a homozygous strain background. The RSV-myc transgene retains the imprinting pattern after breeding for more than ten generations into the inbred strain FVB/N (Chaillet et al., 1991). The MPA434 locus exhibits a germline-specific imprinting on an essentially homozygous BALB/c (BC) background and after a few crosses to some other strains as well (Sasaki et al., 1991 and unpublished). Interestingly, the parental methylation patterns of these two transgenes have been shown to be established in the germline of parents (see next section).

A related, but different type of background phenomenon has been observed in the TKZ751 locus (Allen et al., 1990). When this transgene is bred on a DB strain background, it is unmethylated and expressed at a high level (Fig. 2, left). By contrast, if this locus is kept on a BC background, it is fully methylated and not expressed. Intriguingly, however, the methylation of a DB transgene occurs only when it is

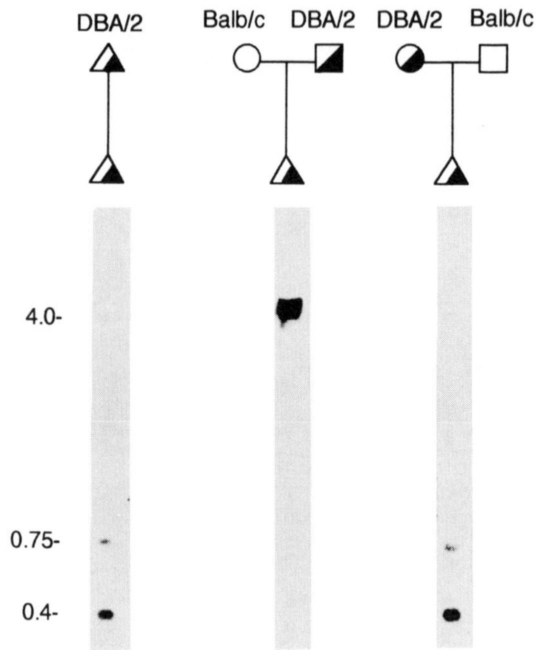

Figure 2. The effect of strain-specific modifiers on the TKZ751 locus (Allen et al., 1990). The transgene is undermethylated when bred on a DBA/2 background (the 0.75-kb and 0.4-kb bands) (left). If the transgene is crossed to a BALB/c male, it is still undermethylated but if crossed to a BALB/c female, it becomes methylated (the 4.0-kb band) (middle and right). This would appear as an imprinting of the transgene, but what is important is the parental origin of the BALB/c modifier allele and not the transgene itself (see text).

crossed to BC females (Fig. 2, middle and right). This situation would appear as a germline-specific imprinting of the TKZ751 locus if only the behavior of the transgene is considered, but this is clearly not the case. Rather, the phenotype of the transgene is dependent on the parental origin of the BC modifier allele. A maternally inherited BC modifier allele will lead to the high methylation phenotype of the transgene, while a paternally derived allele will not. Needless to say, this type of parental origin effect is seen only on a mixed genetic background. Similar observations have been made on the CMZ12 transgene locus (Surani et al., 1990b) and also on the endogenous gene SPARC, located on chromosome 11 (Reik et al., 1990).

In some instances, the epigenetic changes conferred by modifier genes are heritable and cumulative over successive generations. This is best illustrated in the case of the TKZ751 locus (Allen et al., 1990). When an unmethylated DB allele of this transgene is crossed to BC females, it becomes more methylated generation by generation. After three consecutive crosses to BC, the transgene reaches a full methylation level, and this methylation status cannot be changed any more even after repeated crosses to the non-methylating DB background. An extreme of this phenomenon is seen at the HBsAg E36 locus, where an irreversible methylation occurs upon a single cross with a B6 female (Hadchouel et al., 1987). The implication of the heritability observed in these cases is that the modifiers cause methylation changes prior to the germline allocation in early development and that the established germline cannot erase or reprogramme these changes.

To summarize these observations, we are led to believe that there are different categories of transgene imprinting all of which show different phenotypes (or methylation patterns) upon reciprocal crosses but depending on the background conditions (also discussed by Reik et al., 1990; Reik, 1992). The first category is retained on an essentially homozygous background (e.g., RSV-myc, MPA434), and thus may serve as a paradigm for the strictly germline-dependent endogenous imprinting identified by the embryological and genetic complementation studies. The second category is observed in outbred populations, where different modifier alleles segregate, but is lost when brought into a homozygous population (e.g., CAT17). This situation closely resembles the original observation of the "dominance modification" phenomenon in *Drosophila*, where dominant mutations collected in the wild population become much less obvious as the mutant alleles are crossed onto the laboratory fly stocks (Fisher, 1928). The last category is a "secondary", so to speak, imprinting, which is brought about by imprinted modifiers, and here the transgene's own parental origin is completely irrelevant (e.g., TKZ751). This category is only seen in reciprocal crosses between inbred strains and may serve as a model for non-reciprocal effects (on sterility, for example) in inter-specific hybridizations. In the case of TKZ751, this effect involves post-fertilization expression of the modifier gene (Allen et al., 1990; Surani et al., 1990b).

At present the relationships between these three categories are somewhat obscure. For instance, the modifier action in the third category only becomes clear when inbred strains are systematically used and therefore some part of the second category might be proved to belong to the third category. Sapienza (1990) has argued that even the first category can be explained by the action of modifier genes. In his model, the presence of X-linked modifier genes that are dosage sensitive in their activity is hypothesized, and this would result in a double dose of their activity in the female germline as compared with the male germline. Thus modifier genes can be "imprinting genes". However, this model cannot be a complete explanation for imprinting as discussed by Reik (1992).

5 Methylation changes in the germline

As mentioned earlier, there are two transgene loci which are imprinted in a strictly germline-dependent fashion even on a homozygous background. Are methylation differences between the parental alleles already present in the gametes? Do they persist through early development?

The MPA434 locus is a single copy transgene which is more methylated after maternal transmission (Fig. 1). It has previously been shown by the analysis of the prepubertal testis, adult testis, and mature sperm that the transgene is undermethylated in male germ cells at least from the leptotene spermatocyte stage onwards (Sasaki et al., 1991). This study has now been extended to both male and female fetal germ cells and oocytes by using a sensitive PCR method (T. Ueda, T. Higashinakagawa, and H.S., submitted). Figure 3 (top) summarizes the results. In both male and female primordial germ cells, the transgene promoter region is unmethylated regardless of its parental origin. This suggests that the methylation imprint is completely erased in the very early stage of germ cell development. The unmethylated state then

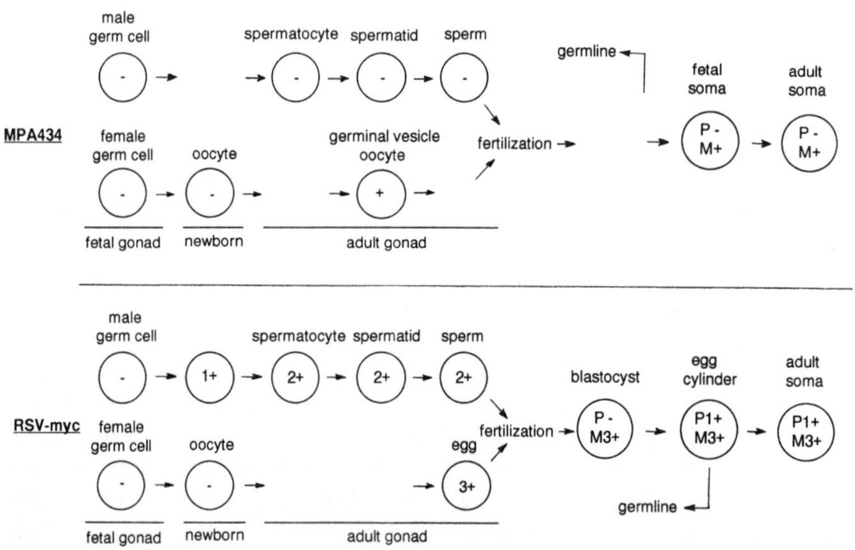

Figure 3. Summary of the methylation changes of the imprinted transgenes during gametogenesis and embryogenesis. Top, the MPA434 locus (T. Ueda et al., submitted); bottom, the RSV-myc locus (Chaillet et al., 1991). Fetal germ cells were collected at day 12.5 and 14.5 of gestation (MPA434) or at day 13.5 and 17.5 of gestation (RSV-myc). Fetal soma (top) was analyzed at embryonic day 14.5. The methylation states are represented by either a minus or a plus sign, and the number of plus signs corresponds to the level of methylation. P = paternal allele; M = maternal allele.

persists through all stages of spermatogenesis. By contrast, in the female gonad, the transgene becomes highly methylated between the newborn oocyte stage and the germinal vesicle-oocyte stage. This observation indicates that a *de novo* methylation event occurs in the growth phase of oocytes arrested at the diplotene stage of meiotic prophase I. Thus, in this case, the parental methylation pattern observed in adult somatic tissues is already established in the gamete of each sex.

A more extended analysis has been carried out with the other transgene locus, RSV-myc, by Chaillet et al. (1991). As shown in Figure 3 (bottom), the observations on germ cells are essentially similar to those of MPA434 except that a gradual methylation occurs during the male germ cell development. This change is first detected at day 17.5 of gestation when meiosis has not yet started. The final methylation level observed in the sperm is somewhat higher than that seen for the paternal allele in somatic tissues. By contrast, the transgenes in ovulated eggs have already acquired the same methylation level as the maternal copy in adult somatic tissues. Importantly, Chaillet et al. have pursued the methylation changes in early embryos following fertilization. In these experiments, it was shown that the maternal pattern established in the egg persists through embryogenesis, while the methylation pattern found in the sperm undergoes further modifications. Thus in the blastocyst stage, the methylation pattern established in the sperm is first lost and then, in the egg cylinder stage, the final somatic methylation level, which is lower than that of the sperm, is acquired.

Based on these studies, we can conclude that the parental methylation patterns are first erased in primordial germ cells and then the corresponding patterns are re-established, completely or partly, in developing gametes. Thus reprogramming does occur in the germline. Interestingly, some of the events observed in these transgenes appear to be closely related to the changes which occur in endogenous mouse sequences. For example, some repetitive sequences are undermethylated in newborn oocytes (Sanford et al., 1984, 1987) but become methylated during oocyte growth and/or maturation (Howlett and Reik, 1991). This change is probably due to the extraordinarily high level of methylase found in germinal vesicle oocytes and ovulated eggs (Howlett and Reik, 1991; Monk et al., 1991). Marked loss of DNA methylation in the blastocyst genome (Monk et al., 1987; Howlett and Reik, 1991) is also consistent with the behavior of the paternal RSV-myc allele. A steep decline in amount and activity of methylase occurs just prior to this stage (Howlett and Reik, 1991; Monk et al., 1991). However, the maintenance of the highly methylated state of the maternal allele across this stage is a unique feature of the two transgenes.

6 Can transgenes detect endogenous imprinted loci?

One original prediction in the transgenic experiments on imprinting was that transgenes may serve as molecular probes for imprinted chromosomal domains. Apparent involvement of chromosomal position-effects in most cases of transgene imprinting appears to support this notion. Do the flanking mouse sequences display parental-origin-dependent methylation patterns?

Two interesting observations have been reported. The OX1-5 locus exhibits parental-origin-dependent methylation changes on a mixed genetic background (Reik et al., 1990). The changes are detected by the methylation-sensitive enzyme *Hap*II but the transgene itself does not contain any *Hap*II sites. Therefore, the differential methylation occurs outside the transgene integrant (Reik et al., 1990). The second locus, *Adp* (Acrodysplasia), provides evidence that both the transgene and the endogenous mouse gene at the insertion site are imprinted (DeLoia and Solter, 1990). This locus is characterized by a dominant insertional mutation which causes deformity of both fore- and hindlimbs. However, only those mice that received the transgene from males are deformed. Transgene expression evidenced by papilloma formation (which is caused by the BPV sequences within the construct) is only detected in these deformed mice. Therefore, it appears that the same imprinting mechanism is controlling both the endogenous gene and the transgene. In either of these cases, the methylation status of the pre-insertion locus in wild-type mice has not been reported.

Recently, two independent transgene insertion loci were cloned and the methylation status of the flanking mouse sequences were characterized. In such an analysis of the MPA434 locus (Sasaki et al., 1991), a total of four endogenous sites adjacent to the transgene were assayed for differential methylation but they were always methylated irrespective of the parental origin. Interestingly, the differential methylation within the transgene was restricted to the MT-I-derived promoter and the plasmid vector sequences. This clearly demonstrates that even within the same transgene some sequence specificity for imprinting exists. Three endogenous sites of the pre-insertion locus were also examined in wild-type mice, but again, no parental-origin-dependent changes were detected. Instead, a different type of allelic methylation variation, that is strain-specific and heritable through the germline (in a Mendelian fashion), was found at two sites. At present it is not known whether there is any connection between the heritable changes and imprinting. To summarize, no evidence for endogenous imprinting was obtained in this case. The main conclusion drawn from the study is that transgenes do not necessarily reflect the methylation behavior of the immediate surrounding regions.

In contrast to the above study, a very interesting picture emerged when the HBsAg E36 was analyzed (C. Pourcel, personal communication). Using a flanking probe, it was possible to distinguish between the maternal and paternal alleles in transgenic heterozygotes based on the restriction patterns. By using this system it was found that the maternal allele is always hypermethylated whether it contains the transgene or not. Furthermore, DNA regions conserved in different species were identified on both sides of the transgene and using these fragments several species of testis-specific transcripts (most probably from post-meiotic germ cells) were detected (X. Wu, M. Hadchouel, H. Farza, L. Amar, and C. Pourcel, personal communication). Therefore, in this case, transgene imprinting is likely to be driven by the endogenous sequences (although chromosome 13, on which the transgene resides, is not imprinted according to the genetic studies). It is possible that activity of the testis-specific gene influences methylation of the adjacent transgene in the germline. Whether differential methylation is seen or not in somatic tissues of wild-type mice is under investigation.

At present, we cannot decide which one of the above two cases is more likely to happen on other occasions. However, from the discussion thus far, the notion has emerged that the relative importance of the transgene sequences versus endogenous sequences for imprinting is considerably different from transgene to transgene. For example, while the RSV-myc construct shows virtually position-independent imprinting, the HBsAg E36 appears to be directed by an endogenous imprinting signal. In most cases of transgene imprinting, it is conceivable that both transgene sequences and endogenous sequences play a role and that imprinting is a result of complex interactions between these sequences.

7 Imprinted endogenous loci

When three independent endogenous mouse genes were shown to be imprinted in 1991, the research on genomic imprinting entered a new phase. It has now become possible to approach the molecular mechanism of endogenous imprinting directly. Firstly, the insulin-like growth factor type-2 receptor (*Igf2r*) gene was found to be expressed only from the maternal chromosome using deletion mutant mice (Barlow et al., 1991). This gene is located in the T-complex region of chromosome 17 and closely linked or identical to the maternal effect locus, *Tme*. Secondly, the insulin-like growth factor II (*Igf2*) gene, which resides on the distal region of chromosome 7 and encodes a ligand for the above receptor, was shown to be imprinted by gene knockout experiments (DeChiara et al., 1991). Only the paternal copy of this gene is active; the opposite of the receptor. This finding was confirmed later by using mouse embryos with a maternal duplication and paternal deficiency of this chromosomal region (Ferguson-Smith et al., 1991). The last gene, the H19 gene, is closely linked to *Igf2* and expressed only from the maternal copy (Bartolemei et al., 1991). This was demonstrated by an RNase protection assay which can distinguish alleles in different mouse subspecies. Thus *Igf2* and *H19*, which are closely linked on the distal region of chromosome 7, are reciprocally imprinted. Significantly, all of these loci are mapped on the chromosomal regions which have been shown to be imprinted by genetic complementation studies.

The search for epigenetic changes associated with the imprinting of these genes has started only recently. Nevertheless, some interesting observations have already been made. For example, we are systematically looking for parental differences in DNA methylation within the *Igf2* locus using embryos maternally disomic for the distal region of chromosome 7. We have found that, at least in this case, extensive CpG island methylation such as those found in X-chromosome inactivation is not involved in repressing the maternal allele (Sasaki et al., 1992). Interestingly, however, it has been pointed out that there is one region which shows methylation differences that are dependent on the parental origin in the vicinity of *Igf2* and that these differences are established in the germline (Chaillet et al., 1991). Our preliminary studies suggest that there are at least three discrete regions which show parental-origin-dependent methylation differences within the imprinted region of chromosome 7 identified by the complementation studies (A. C. Ferguson-Smith, W. Reik, H.S., and M.A.S., unpublished).

In addition to the above findings in mice, a differential methylation that is potentially associated with imprinting has been found in the human genome. Prader-Willi syndrome and Angelman syndrome are thought to be the result of maternal or paternal disomy of chromosome 15q 11-13, respectively (Nicholls et al., 1989; Malcolm et al., 1991). Strikingly, one of the probes isolated from this region showed methylation differences when normal individuals and patients with the above diseases were compared (R. D. Nicholls, E. M. Rinchik, and D. J. Driscoll, personal communication).

Altogether, there is growing evidence that mammalian genomes contain regions that are differentially methylated depending on their parental origin. Although any systematic screening of the entire genome for differential methylation has not been carried out, the fact that these methylation changes are physically linked to imprinted genes seems to suggest a role in endogenous imprinting. The establishment of one of these changes in the germline also seems to support this notion. In the next step, the most important question to be addressed is whether there is any functional relationship between the germline-specific methylation changes and the differential expression patterns.

8 Concluding remarks

Molecular studies on genomic imprinting, which have been carried out mostly with mice, have revealed a number of interesting aspects of DNA methylation. DNA methylation of certain genes are influenced by the action of different alleles of modifier genes. The modifier effect is classified into several categories, each of which may serve as a paradigm for genomic imprinting or one of the related phenomena seen in reciprocal crosses. In the germline, erasure and re-establishment of differential DNA methylation patterns can occur. The essential patterns established in the gametes can then persist through fertilization and subsequent embryonic development with or without further modifications. The parental differences in DNA methylation imprinted in this way are inherited somatically and can be seen in fetal and adult tissues.

At present we cannot conclude whether DNA methylation is involved in endogenous imprinting or not. However, as described above, we have gained a considerable amount of new knowledge on DNA methylation through the studies on imprinted transgenes. The recent identification of endogenously imprinted mouse genes will certainly facilitate the research on genomic imprinting, and the information we have obtained thus far will be useful in formulating guides for future works.

Finally, in this article, we largely ignored the possible role of chromatin structure because of the space. In fact, researchers have pointed out a number of parallels between the modifier effects on transgenes and the position-effect variegation in *Drosophila* and, based on this, the possible involvement of heterochromatinization in imprinting has been suggested. Besides this possibility, there is the reasonable assumption that at least some DNA-binding proteins or protein complexes play roles in the imprinting process. It is very likely that full understanding of the molecular basis of genomic imprinting requires studies on nuclear proteins, higher order structure of chromatin, as well as DNA methylation.

Acknowledgments. We thank Drs. Christine Pourcel and Robert D. Nicholls for communication of results prior to publication. H.S. is supported by the Uehara Memorial Foundation, the Japan Society for Promotion of Sciences, and the Wellcome Trust. N.D.A. is supported by an AFRC TAP grant (IAPGR, CRS1.2).

Allen, N. D., Norris, M. L., and Surani, M. A. (1990) Epigenetic control of transgene expression and imprinting by genotype-specific modifiers. Cell *61*, 853–861.

Barlow, D. P., Stoeger, R., Hermann, B. G., Saito, K., and Schweifer, N. (1991) The mouse insulin-like growth factor type-2 receptor is imprinted and closely linked to the *Tme* locus. Nature *349*, 84–87.

Bartolomei, M. S., Zemel, S., and Tilghman, S. M. (1991) Parental imprinting of the mouse H19 gene. Nature *351*, 153–155.

Beechey, C. V., Cattanach, B. M., and Searle, A. G. (1990) Genetic imprinting map. Mouse Genome *87*, 64–65.

Bird, A. P. (1986) CpG-rich island and the function of DNA methylation. Nature *321*, 209–213.

Cattanach, B. M. (1986) Parental origin effects in mice. J. Embryol. Exp. Morph. (Suppl.) *97*, 137–150.

Cattanach, B. M., and Beechey, C. V. (1990) Autosomal and X-chromosome imprinting. Development (Suppl.), 63–72.

Cedar H. (1988) DNA methylation and gene activity. Cell *53*, 3–4.

Chaillet, J. R., Vogt, T. F., Beier, D. R., and Leder, P. (1991) Parental-specific methylation of an imprinted transgene is established during gametogenesis and progressively changes during embryogenesis. Cell *66*, 77–83.

Clarke, A. (1990) Genetic imprinting in clinical genetics. Development (Suppl.), 131–136.

Collick, A., Reik, W., Barton, S. C., and Surani, M. A. H. (1988) CpG methylation of an X-linked transgene is determined by somatic events postfertilization and not germline imprinting. Development *104*, 253–244.

Cooper, D. W., VandeBerg, J. L., Sharman, G. B., and Poole, W. F. (1971) Phosphoglycerate kinase polymorphism in kangaroo provides further evidence for paternal X inactivation. Nature New Biol. *230*, 155–157.

Crouse, H. V. (1960) The controlling elements in sex chromosome behaviour in *Sciara*. Genetics *45*, 1429–1443.

DeChiara, T. M., Robertson, E. J., and Efstratiadis, A. (1991) Parental imprinting of the mouse insulin-like growth factor II gene. Cell *64*, 849–859.

DeLoira, J. A., and Solter, D. (1990) A transgene insertional mutation at an imprinted locus in the mouse genome. Development (Suppl.), 73–79.

Engler, P., Haasch, D., Pinkert, C. A., Doglio, L., Glymour, M., Brinster, R., and Storb, U. (1991) A strain-specific modifier on mouse chromosome 4 controls the methylation of independent transgenic loci. Cell *65*, 939–947.

Ferguson-Smith, A. C., Cattanach, B. M., Barton, S. C., Beechey, C. V., and Surani, M. A. (1991) Embryological and molecular investigations of parental imprinting on mouse chromosome 7. Nature *351*, 667–670.

Ferguson-Smith, A. C., Reik, W., and Surani, M. A. (1990) Genomic imprinting and cancer. Cancer Surveys *9*, 487–503.

Fisher, R. A. (1928) Two further notes on the origin of dominance. Am. Nat. *62*, 571–574.

Fundele, R., Howlett, S. K., Kothary, R., Norris, M. L., Mills, W. E., and Surani, M. A. (1991) Developmental potential of parthenogenetic cells: role of genotype-specific modifiers. Development *113*, 941–946.

Gartler, S. M., and Riggs, A. D. (1983) Mammalian X-chromosome inactivation. Annu. Rev. Genet. *17*, 155–190.

Grant, S. G., and Chapman, V. M. (1988) Mechanisms of X-chromosome regulation. Annu. Rev. Genet. *22*, 199–233.

Hadchouel, M., Farza, H., Simon, D., Tiollais, P., and Pourcel, C. (1987) Maternal inhibition of hepatitis B surface antigen gene expression in transgenic mice correlates with *de novo* methylation. Nature *329*, 454–456.

Hall, J. G. (1990) Genomic imprinting: review and relevance to human diseases. Am. J. Hum. Genet. *46*, 103–123.

Holiday, R., and Pugh, J. E. (1975) DNA modification mechanisms and gene activity during development. Science *187*, 226–232.

Howlett, S. K., and Raik, W. (1991) Methylation levels of maternal and paternal genomes during preimplantation development. Development *113*, 119–127.

Johnson, D. R. (1974) Hairpin-tail: a case of post-reductional gene action in the mouse egg. Genetics *76*, 795–805.

Johnson, D. R. (1975) Further observations on the hairpin tail (Thp) mutation in the mouse. Genet. Res. *24*, 207–213.

Kaslow, D. C., and Migeon, B. R. (1987) DNA methylation stabilizes X chromosome inactivation in eutherians but not in marsupials: evidence for multistep maintenance of mammalian X dosage compensation. Proc. Natl. Acad. Sci. USA *84*, 6210–6214.

Latham, K. E., and Solter, D. (1991) Effect of egg composition on the developmental capacity of androgenetic mouse embryos. Development *113*, 561–568.

Lock, L. F., Takagi, N., and Martin, G. R. (1987) Methylation of the *Hprt* gene on the inactive X occurs after chromosome inactivation. Cell *48*, 39–46.

Lyon, M. F. (1972) X-chromosome inactivation and developmental patterns in mammals. Biol. Rev. *47*, 1–35.
Malcolm, S., Clayton-Smith, J., Nicholas, M., Robb, S., Webb, T., Armour, J. A. L., Jeffreys, A. J., and Pembrey, M. E. (1991) Uniparental disomy in Angelman's syndrome. Lancet *337*, 694–697.
Mann, J. R., and Stewart, C. L. (1991) Development to term of mouse androgenetic aggregation chimeras. Development *113*, 1325–1333.
McGowan, R., Campbell, R., Peterson, A., and Sapienza, C. (1989) Cellular mosaicism in the methylation and expression of hemizygous loci in the mouse. Genes Dev. *3*, 1669–1676.
Migeon, B. R., deBuer, S. J., and Axelman, J. (1989) Frequent depression of G6PD and HPRT on the marsupial inactive X chromosome associated with cell proliferation in vitro. Exp. Cell Res. *182*, 597–609.
Monk, M. (1986) Methylation and the X chromosome. BioEssays *4*, 204–208.
Monk, M. (1988) Genomic imprinting. Genes Dev. *2*, 921–925.
Monk, M., Adams, R. L. P., and Rinaldi, A. (1991) Decrease in DNA methylase activity during preimplantation development in the mouse. Development *112*, 189–192.
Monk, M., Boubelik, M., and Lehnert, S. (1987) Temporal and regional changes in DNA methylation in the embryonic, extraembryonic and germ cell lineages during mouse embryo development. Development *99*, 371–382.
Moore, T., and Haig, D. (1991) Genomic imprinting in mammalian development: a parental tug-of-war. Trends Genet. *7*, 45–49.
Nicholls, R. D., Knoll, J. H. M., Butler, M. G., Karam, S., and Lalande, M. (1989) Genomic imprinting suggested by maternal heterodisomy in non-deletion Prader Willi syndrome. Nature *342*, 281–285.
Razin, A., and Riggs, A. D. (1980) DNA methylation and gene function. Science *210*, 604–610.
Reik, W. (1989) Genomic imprinting and genetic disorders in man. Trends Genet. *5*, 331–336.
Reik, W. (1992) Genomic imprinting, in: Transgenic Animals in Biology and Medicine, pp. 99–126. Eds F. Grosveld and G. Kollias. Academic Press.
Reik, W., Collick, A., Norris, M. L., Barton, S. C., and Surani, M. A. (1987) Genomic imprinting determines methylation of parental alleles in transgenic mice. Nature *328*, 248–251.
Reik, W., Howlett, S. K., and Surani, M. A. (1990) Imprinting by DNA methylation: from transgenes to endogeneous gene sequences. Development (Suppl.), 99–106.
Richardson, B. J., Czuppon, A. B., and Sharman, G. B. (1971) Inheritance of glucose-6-phosphate dehydrogenase variation in kangaroos. Nature New Biol. *230*, 154–155.
Riggs, A. D. (1975) X chromosome inactivation, differentiation and DNA methylation. Cytogenet. Cell Genet. *14*, 9–25.
Sanford, J. P., Clark, H. J., Chapman, V. M., and Rossant, J. (1987) Differences in DNA methylation during oogenesis and spermatogenesis and their persistence during early embryogenesis in the mouse. Genes Dev. *1*, 1039–2834.
Sanford, J., Forrester, L., and Chapman, V. (1984) Methylation patterns of repetitive DNA sequences in germ cells of *Mus musculus*. Nucl. Acids Res. *12*, 2823–2834.
Sapienza, C. (1989) Genomic imprinting and dominance modification. Ann. NY Acad. Sci. *564*, 24–38.
Sapienza, C. (1990) Sex-linked dosage-sensitive modifiers as imprinting genes. Development (Suppl.), 107–113.
Sapienza, C. (1991) Genomic imprinting and carcinogenesis. Biochim. Biophys. Acta *1072*, 51–61.
Sapienza, C., Paquette, J., Tran, T. H., and Peterson, A. (1989) Epigenetic and genetic factors affect transgene methylation imprinting. Development *107*. 165–168.
Sapienza, C., Peterson, A. C., Rossant, J., and Balling, R. (1987) Degree of methylation of transgenes is dependent on gamete of origin. Nature *328*, 251–254.
Sasaki, H., Hamada, T., Ueda, T., Seki, R., Higashinakagawa, T., and Sakaki, Y. (1991) Inherited type of allelic methylation variations in a mouse chromosome region where an integrated transgene shows methylation imprinting. Development *111*, 573–581.
Sasaki, H., Jones, P. A., Chaillet, J. R., Ferguson-Smith, A. C., Barton, S. C., Reik, W., and Surani, M. A. (1992) Parental imprinting: potentially active chromatin of the repressed maternal allele of the mouse insulin-like growth factor II (*Igf2*) gene. Genes Dev., in press.

Searle, A. G., and Beechey, C. V. (1978) Complementation studies with mouse translocations. Cytogenet. Cell Genet. *20*, 282–303.

Searle, A. G., and Beechey, C. V. (1985) Noncomplementation phenomena and their bearing on nondisjunctional effects, in: Aneuploidy, Aetiology and Mechanisms, pp. 363–376. Eds V. L. Dellarco, P. E. Voytek and A. Hollaender. Plenum, New York.

Searle, A. G., Peters, J., Lyon, M. F., Hall, J. G., Evans, E. P., Edwards, J. H., and Buckle, V. J. (1989) Chromosome map of man and mouse. IV. Ann. Hum. Genet. *53*, 89–140.

Sharman, G. B. (1971) Late replication in the paternally derived X chromosome of female kangaroo. Nature *230*, 231–232.

Singer-Sam, J., Grant, M., LeBon, J. M., Okuyama, K., Chapman, V., Monk, M., and Riggs, A. D. (1990) Use of a *Hpa*II-polymerase chain reaction assay to study DNA methylation in the *Pgk-1* CpG island of mouse embryos at the time of X-chromosome inactivation. Mol. Cell. Biol. *10*, 4987–4989.

Solter, D. (1988) Differential imprinting and expression of maternal and paternal genomes. Annu. Rev. Genet. *22*, 127–146.

Surani, M. A. H. (1986) Evidences and consequences of differences between maternal and paternal genomes during embryogenesis in the mouse, in: Experimental Approaches to Mammalian Embryonic Development, pp. 401–436. Eds J. Rossant and R. A. Pederson. Cambridge University Press, Cambridge.

Surani, M. A., Allen, N. D., Barton, S. C., Fundele, R., Howlett, S. K., Norris, M. L. Reik, W. (1990a) Developmental consequences of imprinting of parental chromosomes by DNA methylation. Phil. Trans. R. Soc. Lond. B *326*, 313–327.

Surani, M. A., Kothary, R., Allen, N. D., Singh, P. B., Fundele, R., Ferguson-Smith, A. C., and Barton, S. C. (1990b) Genomic imprinting and development in the mouse. Development (Suppl.), 89–98.

Surani, M. A., Reik, W., and Allen, N. D. (1988) Transgenes as molecular probes for genomic imprinting. Trends Genet. *4*, 59–62.

Surani, M. A. H., Reik, W., Norris, M. L., and Barton, S. C. (1986) Influence of germline modifications of homologous chromosomes on mouse development. J. Embryol. Exp. Morphol. (Suppl.) *97*, 123–136.

Swain, J. L. Stewart, T. A., and Leder, P. (1987) Parental legacy determines methylation and expression of an autosomal transgene: a molecular mechanism for parental imprinting. Cell *50*, 719–727.

Takagi, N., and Sasaki, M. (1975) Preferential inactivation of the paternally derived X chromosome in the extraembryonic membranes in the mouse. Nature *256*, 640–642.

Toniolo, D., D'Urso, M., Martini, G., Persico, M., Tufano, V., Battistuzzi, G., and Luzatoo, L. (1984) Specific methylation pattern at the 3′ end of the human housekeeping gene for glucose-6-phosphate dehydrogenase. EMBO J. *3*, 1987–1995.

VandeBerg, J. L., Robinson, E. S., Samollow, P. B., and Johnston, P. G. (1987) X-linked gene expression and X-chromosome inactivation: marsupials, mouse and man compared. Isozymes: Curr. Topics Biol. Med. Res. *16*, 225–253.

West, J. D., Frels, W. I., Chapman, V. M., and Papaioannou, V. E. (1977) Preferential expression of the maternally derived X chromosome in the mouse yolk sac. Cell *12*, 873–882.

Wigler, M. H. (1981) The inheritance of methylation patterns in vertebrates. Cell *24*, 285–286.

Wolf, S. F., Dintzis, S., Toniolo, D., Persico, G., Lunnen, K. D., Axelman, J., and Migeon, B. R. (1984) Complete concordance between glucose-6-phosphate dehydrogenase activity and hypomethylation of 3′ CpG clusters: implications for X chromosome dosage compensation. Nucl. Acids Res. *12*, 9333–9348.

Yen, P. H., Patel, P., Chinault, A. C., Mohandas, T., and Shapilo, L. J. (1984) Differential methylation of hypoxanthine phosphoribosyltransferase genes on active and inactive human X chromosomes. Proc. Natl. Acad. Sci. USA *81*, 1759–1763.

DNA Methylation: Molecular Biology and Biological Significance
ed. by J. P. Jost & H. P. Saluz

DNA Methylation and cancer

Charles H. Spruck III, William M. Rideout III
and Peter A. Jones

Urologic Cancer Research Laboratory, Kenneth Norris Jr. Comprehensive Cancer Center, University of Southern California, Los Angeles, CA 90033, USA

1 Introduction

The methylation of cytosine at the 5′ position in the CpG dinucleotide palindrome is the only known epigenetic modification in vertebrate DNA. Although 5-methylcytosine (5-mC) comprises only 1–3% of the bases in the genome, it may play an important role in carcinogenesis through its ability to influence gene expression and increase mutational frequencies in genes involved in carcinogenesis.

The tissue-specific level of 5-mC and pattern of DNA methylation are clearly altered during tumorigenesis. Alterations in tumors and derived cell lines include the *de novo* methylation of CpG islands which are often associated with regions of transcriptional regulation. These abnormalities may be a mechanism for the selective repression of genes in tumors since CpG island methylation has been causally correlated with transcriptional inactivity in experimental systems. In this way, methylation alterations may influence the expression of genes resulting in the promotion of cellular growth, for example by activating oncogenes or repressing tumor suppressor and genes inducing differentiation.

Recent findings show that 5-mC also contributes to the carcinogenic process by playing a major role in the generation of mutations in tumor suppressor genes both in the germline and in somatic cells. Its role as an endogenous mutagen in vertebrate DNA, by its inherent ability to spontaneously deaminate to thymine, is reflected in the frequency of mutations at CpG dinucleotides. In the germline, the CpG sequence represents hotspots for mutations in a variety of genes causing genetic disorders of familial cancer syndromes in humans. Furthermore, the demonstrated hypermutability of 5-mC in generating inactivation mutations of several tumor suppressor genes in somatic tissues, suggests a direct role in carcinogenesis.

2 CpG islands

The modification of cytosine at position 5 occurs exclusively at the CpG dinucleotide in the vertebrate genome. The CpG sequence is underrepresented by a factor of 4 to 5 in vertebrates (Josse et al., 1961; Swartz et al., 1962; Russell et al., 1976) possibly due to the inherent potential of 5-mC to deaminate spontaneously to thymine (Coulondre et al., 1978; Lindahl, 1979). The dinucleotide is not distributed randomly throughout the genome, but instead shows a tendency to aggregate in clusters known as CpG islands (Tykocinski and Max, 1984; Bird et al., 1985). Cooper et al. (1983) showed that approximately 1% of the mouse genome is digested by the restriction endonuclease Hpall into tiny fragments (HTFs) averaging 120 base pairs in length. Later, Bird et al. (1985) demonstrated that these HTFs corresponded to regions of the genome with highly concentrated CpG dinucleotides associated with gene sequences. They estimated that approximately 30,000 CpG islands exist in the haploid genome of the mouse.

CpG islands have been shown to be associated with the 5′ regions of a variety of housekeeping genes and the 5′ and 3′ regions of many tissue-specific genes, and may represent regulatory regions (Bird et al., 1984, 1985; Gardiner-Garden and Frommer, 1987). While many studies have shown a generalized inverse correlation between methylation and gene expression, most CpG islands on autosomal genes remain unmethylated in the germline and in adult tissues, independent of expression status (Bird et al., 1987; McKeon et al., 1982; Lloyd et al., 1987; Spanopolou et al., 1988; Jones et al., 1990a). The lack of cytosine methylation in CpG islands in the germline is believed to be responsible for the absence of suppression of the CpG sequence in these regions. This may reflect the evolutionary consequence of the mutagenicity of 5-mC and the necessity to limit regulatory sequence variations. The majority of the known examples of methylation of CpG islands in normal tissues are those islands associated with genes located on inactive X chromosomes where there is good evidence that methylation plays a role in stabilizing transcriptional incompetency.

3 CpG island methylation and transcriptional repression

The idea that methylation might influence gene control was first postulated by Scarano (1971) and later by Holliday and Pugh (1975) and Riggs (1975). Gene transfer studies have demonstrated directly that the regional methylation of CpG rich promoter sequences can repress transcription (Busslinger et al., 1983; Keshet et al., 1985). Rachal et al. (1989) used a transient CAT expression system to determine the role of methylation in the transcriptional regulation of a 551 base pair CpG island promoter region of the Ha-*ras*-1 oncogene. Generalized CpG methylation by the human placental DNA methyltransferase decreased CAT expression driven by the promoter by greater than 95%. Borello et al. (1987) have further shown that *in vitro* methylation of the promoter can diminish the transforming activity of an activated human Ha-*ras*-1 oncogene upon transfection into mouse NIH3T3 fibroblasts. In addition, Toth et al. (1989) have demonstrated that the regional methylation of the adenovirus E2a promoter between positions +24 and −160 can inactivate transcription. There is thus direct evidence that methylation of CpG rich promoter regions can down-regulate transcription.

Limited evidence has suggested that some CpG islands on autosomes may be methylated in adult somatic tissues. Although not believed to represent a general phenomenon in somatic cells, methylation has been demonstrated in CpG islands located on chromosome region 11p13 (Bonetta et al., 1990), and the sub-telomeric region of human chromosomes (de Lange et al., 1990), as well as the CpG-rich promoter of the interphotoreceptor retinol-binding protein (IRBP) gene (Albini et al., 1990). Methylation within the IRBP island is associated with transcriptional inactivity in lymphocytes (Albini et al., 1990), suggesting a possible role of methylation in autosomal gene regulation. It remains to be demonstrated whether IRBP CpG island methylation is directly responsible for the non-expressing phenotype in lymphocytes, or if it represents a general mechanism of transcriptional repression in other tissues.

Extensive methylation of CpG islands on X-linked genes has been unequivocally demonstrated for the HPRT, G6PD, and PGK genes on the inactive X chromosome in female mammals (Wolf and Migeon, 1985; Toniolo et al., 1988; Pfeifer et al., 1990). The state of methylation in CpG regions is associated with the transcriptional inactivity of the expression of these genes. Venolia et al. (1982) have shown that DNA from the inactive X chromosome remains transcriptionally repressed when transfected into recipient cells showing that transcriptional inactivity is attributable to the inherent properties of the DNA itself and not a transacting diffusible factor. Furthermore, Mohandas et al. (1981) and Lock et al. (1986) have shown that X-linked genes located on the inactive X chromosome can be transcriptionally reactivated by 5-azacytidine (5-Aza-CR) induced demethylation of the 5′ region of the gene, providing further evidence for a causal relationship between these processes.

The CpG island methylation observed for genes harbored on the inactive X chromosome may be responsible for the stability and maintenance of transcriptional inactivity in tissues, since transcriptional repression occurs early during embryogenesis, prior to the methylation events (Lock et al., 1987). DNA methylation may therefore be a mechanism by which transcriptional inactivity can be stabilized and suppressed in similar promoter regions regulated by ubiquitous transcription factors.

These experiments demonstrate, in a coherent way, that methylation of CpG islands leads not only to transcriptional inactivity, but that this mechanism is apparently used for long-term silencing of X-linked genes. As yet, there is no convincing evidence for methylation of CpG islands playing a role in the control of autosomal genes since they have not generally been found to be methylated in normal tissues. Nevertheless, it has been found that extensive methylation of these regions occurs in cell lines and tumors and this, coupled with the demonstrated ability of 5-mC to silence island-containing genes, raises the possibility that this abnormal methylation may play a role in immortality and malignancy.

4 CpG island methylation in carcinogenesis

In contrast to the situation *in vivo*, CpG islands are frequently modified *in vitro*. For example, the metallothionein (Compere and Palmiter, 1981), thymidine kinase (Wise and Harris, 1988), and α-globin genes (Antequera et al., 1989) are heavily modified in cultured cells. Jones et al. (1990) demonstrated that the CpG island of the mouse MyoD1 muscle determination gene acquired *de novo* methylation in all immortalized cell lines examined, although unmethylated in adult tissues and low passage fibroblasts. Interestingly, the island became progressively more methylated when cells were oncogenically transformed with chemicals such as 3-methylcholanthrene or by transfection with an activated *ras* gene (Jones et al., 1990b; Marziasz, 1991). These experiments provide an example of progressive *de novo* methylation of a CpG island during immortalization, and transformation. While the function of this methylation remains to be determined, it seems remarkable that it was observed in 7 independently derived cell lines and their tumorigenic derivatives.

Antequera et al. (1990) have recently attempted to delineate the possible significance of *de novo* CpG island methylation in cultured cell lines. They studied the methylation status of the CpG islands of a variety of housekeeping and tissue specific genes in mouse and human cell lines. They discovered high levels of new methylation in these lines associated with the CpG islands of many tissue-specific genes including human α-globin, retinol-binding protein, and mouse Thy-1 and major histocompatibility complex genes. CpG island methylation was not observed in essential housekeeping genes such as the human triosephosphate isomerase, and mouse HFT9 genes. The results suggest that the long-term culturing of the cell lines resulted in the selective methylation of CpG island regions of many non-essential genes, leaving unmethylated those islands associated with essential genes for cellular survival. In this way, methylation may stabilize the long-term transcriptional repression of the non-essential genes possibly ensuring against the activation of terminal differentiation pathways. In support of this hypothesis, *de novo* CpG island methylation of the thymidine kinase and Thy-1 genes has been observed in resistant variants following selection against expression of these genes in culture (Harris, 1982; Gounari et al., 1987; Sneller and Gunter, 1987). The expression of both genes can be restored following treatment with 5-Aza-CR, suggesting that methylation has suppressed gene expression in response to their detriment to cell survival.

Antequera and colleagues (Antequera et al., 1990) termed these methylation changes MAGI-methylation associated gene inactivation, and estimated that approximately 50% of the 30,000 CpG island associated genes may be repressed by this mechanism. The alterations in overall levels and patterns of DNA methylation in tumors may further represent the selective repression or possibly activation of gene transcription analogous to MAGI in cell lines.

Hypermethylation of the 5′ CpG rich regulatory region of the calcitonin gene, located on chromosome 11p, has been observed in a number of cell lines and tumor types (Baylin et al., 1986; Baylin et al., 1987; Silverman et al., 1989; Nelkin et al., 1991), or in response to viral infection by Epstein-Barr, human T-cell leukemia (HTLV), or SV40 viruses in culture (de Bustros et al., 1988). This *de novo* hypermethylation occurs in human colonic neoplasms (Silverman et al., 1989) where generalized DNA hypomethylation has also been demonstrated (Goelz et al., 1985; Feinberg et al., 1988). In addition, Nelkin et al. (1991) have shown the methylation event to be closely correlated with the stages of disease progression in chronic myelogenous leukemia. Hypermethylation of the 5′ region of the calcitonin gene was found in 6% of those patients with chronic disease, 63% in the accelerated phase, and 92% in the advanced blast crisis.

De Bustros et al. (1988) demonstrated that the hypermethylation occurs over a large region of chromosome 11p. These studies are particularly interesting since this arm of chromosome 11 has been proposed to harbor several putative tumor suppressor genes (Weisman et al., 1987; Saxon et al., 1986) including the Wilms tumor suppressor gene (Call et al., 1990; Gessler et al., 1990). Alterations in cellular methylation during carcinogenesis may include the *de novo* methylation of CpG island regions associated with tumor suppressor genes. Transcriptional repression by these events, along with chromosomal deletions and base alterations, could represent a mechanism of tumor suppressor gene inactivation consistent with Knudson's (1983) two-hit hypothesis. In support of this, Greger et al. (1989) have reported methylation in the CpG island of the RB gene in a retinoblastoma tumor. The commonality and significance of these events in human carcinogenesis remains to be demonstrated.

The evident alterations in levels and patterns of DNA methylation in tumors may be the result of a deregulation of the enzymes involved in the methylation process. Kautiainen and Jones (1986) observed a 4–3000-fold higher level of extractable DNA methyltransferase activity in cultured tumorigenic cells compared to non-tumorigenic cells. El-Deiry et al. (1991) studying DNA methyltransferase in colon cancer demonstrated a 60-fold increase in DNA methyltransferase expression in benign colon polyps, and a greater than 200-fold increase in colon carcinomas compared to normal human colon mucosa. In addition, it was shown that the normal mucosa from patients with benign polyps or colon cancers possessed a 15-fold increase in DNA methyltransferase expression. It is not currently understood if these increases in DNA methyltransferase expression are a general phenomenon in human tumors, or how these increases can account for the generalized hypomethylation observed during carcinogenesis.

5 Methylation changes in non-CpG islands in tumor cells

A large number of publications have described alterations in overall levels of 5-mC and the patterns of DNA methylation in tumor-derived cell lines and primary tumors (see Jones and Buckley, 1990). In addition to other DNA alterations such as mutations, rearrangements, and deletions, changes in cellular methylation may represent an important mechanism of carcinogenesis by contributing to the abnormal regulation of gene expression and genomic instability (Pugh and Holliday, 1978; Holliday, 1979; Ehrlich and Wang, 1981; Riggs and Jones, 1983; Holliday and Jeggo, 1985). However, in contrast to the situation with the CpG island containing genes discussed earlier, many of the changes reported have not been mapped to known regulatory regions so that their exact significance remains obscure.

Decreases in overall levels of 5-mC in a diverse variety of cell lines and uncultured primary tumors have been observed frequently. Feinberg et al. (1988) using HPLC showed an 8% and 10% average reduction in 5-mC content respectively in colon adenomas and adenocarcinomas. The general hypomethylation observed prior to progression in the malignant state implies that alterations in DNA methylation is an early event in colon tumorigenesis. DNA hypomethylation has also been observed in human melanomas (Liteplo and Kerbel, 1987), tumors of the prostate (Bedford and Van Helden, 1987), colon and lung (Feinberg and Volgelstein, 1983; Goelz et al., 1985) among others. The interpretation of the significance of methylation changes in cell lines is often complicated by the fact that appropriate normal control cells are not available for comparison. In addition, alterations in DNA methylation levels and patterns may be a function of cell culturing (Shmookler-Reis and Goldstein, 1982; Wilson and Jones, 1983a).

Hypomethylation may also become more pronounced in the later stages of tumor progression. Bedford and Van Helden (1987) observed decreases in 5-mC content in metastatic prostate carcinomas but not in non-metastatic variants. In addition, Gama-Sosa et al. (1983) have observed hypomethylation associated with metastatic tumor states compared to benign neoplasms. However, demethylation does not appear to represent a universal phenomenon of all tumors and tumor derived cell lines since Pfeifer et al. (1988) failed to detect changes in methylation levels in acute human leukemia.

Alterations in the maintenance of normal cellular methylation may be instrumental in the progression of tumors. Treatment of cultured cells with the base analog 5-Aza-CR, a potent inhibitor of DNA methylase activity, strikingly reduces the level of DNA methylation (Jones and Taylor, 1981; Creusot et al., 1982; Taylor and Jones, 1982). 5-Aza-CR has been shown to transform cells in culture (Benedict et al., 1977) and alter the metastatic properties of tumor derived cell lines (Frost and Kerbel, 1983). Furthermore, Carr et al. (1984) demonstrated that 5-Aza-CR can induce the formation of multiple primary tumors in rats including leukemias, reticuloendotheliosis, and tumors of the testes, skin, and bronchus. It is not understood whether these affects are the direct result of 5-Aza-CR-induced demethylation altering the expression of genes involved in carcinogenesis.

Whether any of these gross epigenetic alterations in tumors influence the expression of genes, altering or stabilizing a transcriptional pattern that is conducive to carcinogenesis, remains to be demonstrated. For example, although hypomethylation of a large number of oncogenes has been observed, few studies have correlated these changes in tumors with alterations in the levels of expression of oncogenes or tumor suppressor genes. The c-Ha-*ras* gene has been shown to be extensively methylated in the germline and somatic tissues where it is actively expressed (Ghazi et al., 1990). Barbieri et al. (1987) demonstrated that the human leukemic cell line K562 actively expresses c-Ha-*ras* with methylation in the exons, introns, and 3′ untranslated portions of the gene, while unmethylated in the 5′ CpG island region. Drug-induced hypomethylation by 5-Aza-CR, in tumors, was shown not to influence c-Ha-*ras* gene expression (Barbieri et al., 1987), indicating that the demethylation in the body of the gene observed in tumors may be inconsequential to regulation. Therefore it is unlikely that the hypomethylation of non-CpG sequences in the coding regions of genes is involved in controlling expression.

6 5-Methylcytosine as an endogenous mutagen

The initiation and progression of tumors is driven to a large extent by the accumulation of DNA damage caused by exogenous and endogenous mutagenic mechanisms. Exogenous processes are well characterized and include physical carcinogens such as ionizing radiation and chemical carcinogens. Chemical carcinogens; in addition to causing base alterations in DNA; possess the ability to induce epigenetic changes in DNA, such as decreases in total 5-mC content and alterations in DNA methylation (Wilson and Jones, 1983b).

Endogenous mutagenic processes that may have a role in carcinogenesis are those which are inherent within cells. These include depurination and depyrimidation of DNA, damage incurred by oxygen free radicals, errors in DNA replication and repair processes, and deaminations of 5-mC, among many others. The mutagenicity of 5-mC in biological systems was first demonstrated by the observation that sites of cytosine methylation corresponded to mutational hotspots in the LacI repressor gene of *E. coli* (Coulondre et al., 1978). Duncan and Miller (1980) have attributed the hypermutability of the base to its propensity to deaminate spontaneously to thymine. Although the rate of 5-mC deamination in DNA has not been determined *in vivo*, methylation at position 5 has been shown to elevate the rate of deamination of the base *in vitro* by an estimated factor of 4–5 in single-stranded DNA at 37°C (Ehrlich et al., 1986).

Frederico et al. (1990) used a sensitive genetic assay and determined the deamination rate of cytosine in double stranded DNA to be 7×10^{-13} per second at physiological conditions of pH 7.4 and 37°C which is considerably slower than its rate in single-stranded DNA. Assuming that the rate of deamination of 5-mC is 4–5-fold higher than cytosine in double-stranded DNA, and that the deamination rates measured *in vitro* are comparable *in vivo*, approximately 3.15×10^{-12} deaminations per second would occur at 5-mC residues. Since 5-mC comprises approximately 1% of the 6×10^9 bases of the human haploid genome, about sixteen 5-mC→T deamination events would occur per day per haploid genome. This probably underestimates the true frequency of these events since deamination rates in single stranded DNA, which are often associated with DNA regions active in transcription, replication or repair, are approximately 150-fold higher compared to double-stranded DNA (Frederico et al., 1990).

The increased frequency of mutation driven by methylation-mediated deaminations of cytosine at these sites may be responsible for the under-representation of the CpG sequence in the vertebrate genome. CpG dinucleotides are present at a frequency of 20% of that expected in the total genome, and 37% of that expected in coding regions (Nussinov, 1981; Beutler et al., 1989). Sved and Bird (1990) estimate that an enhanced mutation frequency at CpG dinucleotides, of 12 times the normal frequency of transitions, resulting in CpG to TpG or CpA base changes, can explain the depletion of CpG during the course of vertebrate evolution. They estimated at this frequency the CpG sequence could be reduced to an assumed state of equilibrium between the rate of loss of CpG and the rate of creation of new CpG sites by other mutational events today in a period of 25 million years. Evolutionary comparison studies have supported these results by demonstrating that CpG sequences display a high degree of associated change during evolution (Savatier et al., 1985; Cooper et al., 1987).

7 T:G mismatch repair

The hypermutability of 5-mC has probably resulted in the evolution of the efficient mechanisms for the repair of T:G base mismatches in both prokaryotes and eukaryotes. A sequence-specific DNA repair pathway in *E. coli* repairs T:G mismatches by the biased excision of thymine from the heteroduplex at DNA cytosine methylase (DCM) recognition sites (Lieb, 1987). Very short patch repair (VSP) appears to be involved in these processes since reintroduction of the Vsp gene into $Vsp^{-}DCM^{+}$ bacteria reduces the frequency of mutations by a factor of 4 at DCM recognition sites (Lieb, 1991). T:G mismatches at other sites are repaired via different enzymatic mechanisms, since the repair shows no strong intrinsic directionality of mismatch base excision unless the repair occurs at a replication fork (Kramer et al., 1984; Radman and Wagner, 1986; Shenoy et al., 1987).

Mismatches of T:G in mammalian cells, the result of 5-mC deaminations at CpG dinucleotides, are repaired by a thymine specific excision, non-sequence dependent pathway (Brown and Jiricny, 1988). SV40 viral DNA harboring T:G mismatches transfected into monkey cells are repaired at a greater than 95% frequency, with 92% of the repairs indicative of thymine excision (Brown and Jiricny, 1988). Wiebauer and Jiricny (1989) have demonstrated that T:G to C:G repair *in vitro* by HeLa cell nuclear extracts occurs by a mechanism of thymine-specific base excision followed by polymerase gap filling analogous to prokaryotic systems, possibly mediated by a 200-kD protein that selectively binds T:G mispairs in DNA (Jiricny et al., 1988).

Despite the efficiency of these repair systems, transitions resulting from deaminations of 5-mC occur frequently in the mammalian genome and the CpG dinucleotide is a hotspot for somatic and germline mutations.

8 CpG in germline mutations

The enhanced mutability of the CpG sequence is evident in the analysis of human germline mutations. Germline mutations resulting in restriction fragment length polymorphisms (RFLPs) have been demonstrated to be associated at a high frequency to enzymes whose recognition sequences contain the CpG dinucleotide (Barker et al., 1984; Cooper and Schmidtke, 1984).

CpG dinucleotides have been implicated as hotspots in germline mutations causing human genetic disorders (Cooper and Youssoufian, 1988; Cooper and Krawczak, 1989, 1990). In a recent review of 139 published point mutations causing human disease, Cooper and Krawczak (1990) calculated that 37.4% of these mutations were localized to CpG dinucleotides. Forty-four mutations or 31.7% of the total were CpG to TpG or CpA which are changes indicative of methylation-mediated cytosine deaminations. From these data, 5-mC was estimated to be 46 times more mutagenic than any other nucleotide. The prominence of 5-mC as an endogenous mutagen is reflected in the fact that although it comprises approximately 1% of the nucleotides in the genome, it contributes to almost 1/3 of all disease caused by germline mutations.

Although many genes are expressed in somatic tissues with methylated cytosine in their coding regions, very little is known of the methylation status in the germline of CpG sequences which have undergone characteristic methylation-mediated mutations. This distinction should be made in order to determine if cytosine methylation within CpG sites is reponsible for its observed hypermutability in human disease. Rideout et al. (1990) have recently examined the methylation status of a CpG dinucleotide at codon 408 of the low density lipoprotein (LDL) receptor gene. This CpG is known to have acquired a germline mutation (CpG → CpA) in one of the founder-defective genes resulting in familial hypercholesterolemia. It was shown through ligation-mediated PCR (LM-PCR) genomic sequencing (Mueller and Wold, 1989) that this CpG is methylated in genomic DNA from white blood cells, and sperm. These results are consistent with the germline mutational type being the result of 5-mC deamination.

9 Germline mutations in tumor suppressor genes

The activation of oncogenes or inactivation of tumor suppressor genes are critical steps in carcinogenesis. These events result from multiple genetic alterations during tumor progression including chromosomal deletions, rearrangements, amplifications, and point mutations. Tumor suppressor genes represent a class of cell cycle regulatory genes that are believed to be involved in the negative regulation of cellular proliferation. Several putative tumor suppressor genes have been isolated including p53 (Lane and Benchimol, 1990), retinoblastoma (Friend et al., 1986; Lee et al., 1987; Fung et al., 1987), Wilms tumor (Gessler et al., 1990; Call et al., 1990), DCC (Fearon et al., 1990), NF-1 (Xu et al., 1990), and DP2.5/APC (Groden et al., 1991; Joslyn et al., 1991; Kinzler et al., 1991).

Table 1. Germline mutations in tumor suppressor genes

Tumor suppressor gene	Hereditary cancer	Transitions at CpG*	Reference
APC/DP2.5	Familial adenomatous	4/8	Groden et al., 1991; Nishisho et al., 1991
p53	Li-Fraumeni syndrome	2/6	Srivastava et al., 1990; Malkin et al., 1990
p53	Ependymoma	0/1	Metzger et al., 1991
RB	Retinoblastoma	3/8	Yandell et al., 1989

*CpG to TpG or CpA indicative of 5-mC deaminations.

Although little is known of the direct function of tumor suppressor genes in cellular growth control, they have been shown to be frequent targets of mutation in tumors. Knudson (1983) postulated that tumorigenesis may proceed through the inactivation of genes involved in the negative regulation of celular growth. Extensive study of the p53 tumor suppressor gene has supported this hypothesis by demonstrating that in a variety of human tumors, both alleles of the gene are often inactivated by chromosomal deletion and mutation (Nigro et al., 1989).

Certain familial predispositions to cancer are believed to be the result of heritable mutations in tumor suppressor genes. Segregating germline mutations have been observed in the retinoblastoma (Yandell et al., 1989), p53 (Srivastava et al., 1990; Malkin et al., 1990; Metzger et al., 1991), and DP2.5/APC genes (Groden et al., 1991; Nishisho et al., 1991) in familial cancer syndromes (Table 1). In the analysis of 23 reported germline mutations in tumor suppressor genes, 9 or 39% are CpG to TpG or CpA mutations at CpG dinucleotides, consistent with 5-mC mediated deaminations. Although the reported data of these mutations is limited, the proportion of mutations at CpG dinucleotides is remarkably similar (31.7%) to that observed in germline mutations causing other human genetic diseases (Cooper and Krawczak, 1990).

10 5-mC in somatic mutations of p53

The role of 5-mC in human carcinogenesis is particularly evident in the analysis of inactivating point mutations of the p53 tumor suppressor gene in somatic cells. Abnormalities associated with the p53 gene represent the most common known genetic alteration in human neoplastic processes (Volgelstein, 1990). RFLP studies have demonstrated that the loss of heterozygosity of chromosome region 17p, which harbors the p53 gene, is common in tumors of the colon, lung, brain, bladder, and breast among others and the remaining p53 allele is frequently inactivated by point mutation (Nigro et al., 1989). In addition, loss of 17p has been shown to be a late stage event in colon (Baker et al., 1990), and bladder carcinogenesis (Olumi et al., 1990; Sidransky et al., 1991) possibly contributing to the progression of these tumors to more malignant states.

Ninety eight percent of the somatic point mutations of p53 in tumors occurs within 4 evolutionarily conserved domains of the gene located between exons 5 through 8 (Hollstein et al., 1991b). Base substitutions within this approximately 550 base pair region, yielding missense variations in the protein, results in the loss of growth suppression function of p53. Raycroft et al. (1991) suggest that this may be attributed to the diminished capacity of the mutant p53 to activate transcription *in vivo*. They demonstrated that various mutant p53/GAL 4 protein fusions are 30 to 40 fold less effective at activating transcription of a chloramphenicol acetyltransferase reporter construct, containing the GAL4 binding site, in HeLa cells than wild type p53/GAL4 fusions.

The nature of the reported inactivating mutations of p53 varies dramatically among tumors and cell lines of different tissue origins (Tables 2 and 3). Mutations in p53 appear more frequently in established cell lines than in tumors, possibly reflecting a selective growth advantage and clonal selection in culture of tumor cells harboring the mutation. Although the total spectrum of mutations differs in tumors and cell lines, comparisons are complicated by variations in the representation of mutations from different tissue origins. The most evident variation in the spectrum of mutations is observed between tumors of the colon and lung. In colon cancers, approximately 46% (24/52) of the mutations in p53 are transitions at CpG sequences consistent with methylation-mediated deamination of 5-mC, whereas in cancers of the lung these changes account for only 7% of the total mutations.

Since cytosine methylation occurs in all animal tissues, deaminations of 5-mC at CpG dinucleotides to TpG or CpA represents an endogenous mutagenic process common to all somatic cells. The prevalence of 5-mC induced mutations in colon cancers and other cancers that do not demonstrate a strong association of incidence to a particular carcinogenic agent may therefore represent a baseline level of the endogenous mutagenic process.

The tissue-specific differences in the spectrum of mutations in the p53 gene likely reflects the etiology of the exposure of these tissues to various carcinogenic agents. For example, benzo(a)pyrene, a constituent of cigarette smoke is known to form adducts with guanines leading to $G \rightarrow T$ transversions in DNA (Mazur and Glickmann, 1988). Since 70% of the inactivating mutations of p53 are transversions in lung cancers, compared to 22% in cancers of the colon, the correlation between tobacco use and lung cancer could explain the prevalence of transversions in the p53 gene in lung tumors.

Table 2. Point mutations in p53 in human tumors

Cancer type	Number samples	Number point mutations	Number transversions (%)	Number transitions at non-CpG sites (%)	Number transitions at CpG (%)	Reference
Brain	50	9	2 (22%)	2 (22%)	5 (56%)	Nigro et al., 1989; Mashiyama et al., 1991
Bladder	18	9	5 (56%)	2 (22%)	2 (22%)	Sidransky et al., 1991
Breast	77	13	10 (77%)	2 (15%)	1 (8%)	Nigro et al., 1989; Prosser et al., 1990; Davidoff et al., 1991; Kovach et al., 1991; Osborne et al., 1991
Colon	77	52	11 (22%)	17 (32%)	24 (46%)	Baker et al., 1989; Nigro et al., 1989; Baker et al., 1990; Ishioka et al., 1991
Endometrial	24	3	0 (0%)	1 (33%)	2 (67%)	Okamoto et al., 1991
Esophageal	89	30	15 (50%)	7 (23%)	8 (27%)	Hollstein et al., 1990; Bennet et al., 1991; Casson et al., 1991; Hollstein et al., 1991a
Gastric	20	5	4 (80%)	1 (20%)	0 (0%)	Kim et al., 1991; Tamura et al., 1991; Yamada et al., 1991
Hepatocellular	69	20	15 (75%)	3 (15%)	2 (10%)	Bressac et al., 1991; Hsu et al., 1991; Murakami et al., 1991
Lung	91	61	43 (70%)	14 (23%)	4 (7%)	Nigro et al., 1989; Chiba et al., 1990; Hensel et al., 1991; Iggo et al., 1990; Takahashi et al., 1991;
Leukemia & lymphoma	88	26	9 (35%)	7 (27%)	10 (38%)	Gaidano et al., 1991; Slingerland et al., 1991
Ovarian	37	18	9 (50%)	8 (44%)	1 (6%)	Marks et al., 1991; Mazars et al., 1991
Prostate	2	1	1 (100%)	0 (0%)	0 (0%)	Issacs et al., 1991
Sarcoma	13	7	2 (29%)	1 (14%)	4 (57%)	Menon et al., 1990; Mulligan et al., 1990; Stratton et al., 1990
Totals	655	254	126 (50%)	65 (25%)	63 (25%)	

Table 3. Point mutations in p53 in cell lines

Cancer type	Number samples	Number point mutations	Number transversions (%)	Number transitions at non-CpG sites (%)	Number transitions at CPG (%)	Reference
Brain	3	1	1 (100%)	0 (0%)	0 (0%)	Saylors et al., 1991
Breast	8	8	2 (25%)	4 (50%)	2 (25%)	Nigro et al., 1989; Bartek et al., 1990; Kovach et al., 1991
Colon	21	11	0 (0%)	2 (18%)	9 (82%)	Nigro et al., 1989; Baker et al., 1990; Rodrigues et al., 1990
Esophageal	4	2	1 (50%)	1 (50%)	0 (0%)	Hollstein et al., 1990
Gastric	10	6	1 (17%)	2 (33%)	3 (50%)	Kim et al., 1991; Yamada et al., 1991
Lung	24	18	9 (50%)	7 (33%)	3 (17%)	Nigro et al., 1989; Sameshima et al., 1990; Cote et al., 1991; Lehman et al., 1991; Takahashi et al., 1991
Leukemia & lymphoma	55	43	7 (16%)	15 (35%)	21 (49%)	Cheng et al., 1990; Rodrigues et al., 1990; Farrell et al., 1991; Gaidano et al., 1991; Slingerland et al., 1991
Prostate	5	3	2 (67%)	1 (33%)	0 (0%)	Issacs et al., 1991
Sarcoma	10	5	2 (40%)	1 (20%)	2 (40%)	Romano et al., 1989; Stratton et al., 1990
Thyroid	1	1	0 (0%)	0 (0%)	1 (100%)	Wright et al., 1991
Uterine	6	6	3 (50%)	0 (0%)	3 (50%)	Hensel et al., 1991; Scheffner et al., 1991; Yaginuma et al., 1991
Various	5	5	3 (60%)	1 (20%)	1 (20%)	Gusterson et al., 1991; Murakami et al., 1991
Totals	152	110	31 (28%)	34 (31%)	45 (41%)	

The p53 gene is a preferred subject for studies of molecular epidemiology (Jones et al., 1991), since a large region of the gene is sensitive to inactivation by a wide variety of mutational types. The p53 gene has been used as an epidemiological tool in the study of hepatocellular carcinomas (HCC) in regional areas of Southern Africa (Bressac et al., 1991) and Qidong, China (Hsu et al., 1991) where aflatoxins are risk factors, In these studies, 11/13 mutations in the p53 gene were G→T transversions specifically at the third base position of codon 249. Furthermore, p53 mutations in HCCs from geographic regions where aflatoxin is not a risk factor are not localized to codon 249 (Ozturk et al., 1991). The mutational hotspot is thought to be the result of the direct action of aflatoxin B_1 since the carcinogen has been shown to specifically induce G→T transversions preferentially in G+C rich regions of DNA (Foster et al., 1983; Muench et al., 1983). Puisieux et al. (1991) have shown that *in vitro*, aflatoxin B_1 selectively forms adducts with the third base position of codon 249 in p53, and benzo(a)pyrene specifically targets bases in the gene that have undergone characteristic transversions in lung tumors, associating these mutational hotspots directly to the effects of environmental carcinogens.

The predominance of transitions at CpG sequences in the p53 gene likely represents the true endogenous mutability of 5-mC since no other etiological agent is known to effect the rate of mutation specifically at these sites. A recent study by Wink et al. (1991) has shown that the deamination rates of cytosine and 5-mC can be greatly enhanced by exposure to the bioregulatory agent, and the cigarette smoke constituent nitric oxide. However, since this agent favors deaminations at cytosines and 5-mCs relatively equally, nitric oxide alone cannot account for the hotspot for mutation in vertebrate DNA specifically at CpG methylation sites. Rideout et al. (1990) using LM-PCR demonstrated that the CpG dinucleotides at codon positions 175, 273 and 282 in the p53 gene, which correspond to hotspots for CpG to TpG or CpA mutations, are methylated in white blood cells, sperm, and urothelial cell DNA, suggesting that methylation may be responsible for the elevated mutation frequency at these sites. An interesting result in the analysis of the mutational spectrum of tumors is the low frequency (6%) of CpG transitions in p53 in ovarian cancers. Driscoll and Migeon (1990) have shown that the erythropoietin autosomal gene as well as a variety of X-linked genes are unmethylated in human fetal ovaries. Therefore, if this methylation state is conserved throughout adult development the low frequency of CpG transitions in these tumors may be the result of the absence of cytosine methylation.

The hypermutability of 5-mC is clearly evident by considering that the 22 CpG dinucleotides in the 550 base pair conserved region of exons 5 through 8 of the p53 gene, though they represent only 8% of the base sequences, account for approximately 25% of the inactivation mutations in all tumors, as high as 46% in some tumors such as colon, and 41% in derived cell lines. The frequency of CpG to TpG or CpA alterations in the somatic mutations of p53 (25%) is comparable to that observed in germline mutations causing genetic disorders (31.7%) and familial cancer (39%) indicating a common mutagenic mechanism in these processes, and suggests that 5-mC probably acts as an endogenous mutagen in both somatic and germline tissues.

Albini, A., Tofferetti, J., Zhu, Z., Chader, G. J., and Noonan, D. M. (1990) Hypomethylation of the interphotoreceptor retinoid-binding protein (IRBP) promoter and first exon is linked to expression of the gene. Nucl. Acids Res. *18*, 5181–5187.

Antequera, F., Macleod, D., and Bird, A. P. (1989) Specific protection of methylated CpGs in mammalian nuclei. Cell *58*, 509–517.

Antequera, F., Boyes, J., and Bird, A. (1990) High levels of *de novo* methylation and altered chromatin structure at CpG islands in cell lines. Cell *82*, 503–514.

Baker, S. J., Fearon, E. R., Nigro, J. M., Hamilton, S. R., Preisinger, A. C., Jessup, J. M., Van Tuinen, P., Ledbetter, D. H., Barker, D. F., Nakamura, Y., White, R., and Volgelstein, B. (1989) Chromosome 17 deletions and p53 gene mutations in colorectal carcinomas. Science *244*, 217–221.

Baker, S. J., Preisinger, A. C., Jessup, J. M., Paraskeva, C., Markowitz, S., Willson, K. V., Hamilton, S., and Volgelstein, B. (1990) p53 Gene mutations occur in combination with 17p allelic deletions as late events in colorectal tumorigenesis. Cancer Res. *50*, 7717–7722.

Barbieri, R., Piva, R., Buzzoni, D., Volinia, S., and Gambari, R. (1987) Clustering of undermethylated CCGG and GCGC sequences in the 5′ region of the Ha-*ras*-1 oncogene in human leukemic K-562 cells. Biochem. Biophys. Res. Comm. *145*, 96–104.

Barker, D., Schafer, M., and White, R. (1984) Restriction sites containing CpG show a higher frequency of polymorphism in human DNA. Cell *36*, 131–138.

Bartek, J., Iggo, R., Gannon, J., and Lane, D. P. (1990) Genetic and immunochemical analysis of mutant p53 in human breast cancer cell lines. Oncogene *5*, 893–899.

Baylin, S. B., Hoppener, J. W. M., de Bustros, A., Steenbergh, P. H., Lips, C. J. M., and Nelkin, B. D. (1986) DNA methylation patterns of the calcitonin gene in human lung cancers and lymphomas. Cancer Res. *46*, 2917–2922.

Baylin, S. B., Fearon, E. R., Volgelstein, B., de Bustros, A., Sharkis, S. J., Burke, P. J., Staal, S. P., and Nelkin, B. D. (1987) Hypermethylation of the 5′ region of the calcitonin gene is a property of human lymphoid and acute myeloid malignancies. Blood *70*, 412–417.

Bedford, M. T., and van Helden, P. D. (1987) Hypomethylation of DNA in pathological conditions of the human prostate. Cancer Res. *47*, 5274–5276.

Benedict, W. F., Banerjee, A., Gardner, A., and Jones, P. A. (1977) Induction of morphological transformation in mouse C3H/10T1/2 clone 8 cells and chromosomal damage in hamster $A(T_1)C1$-3 cells by cancer chemotherapeutic agents. Cancer Res. *37*, 2202–2208.

Bennet, W. P., Hollstein, M. C., He, A., Zhu, S. M., Resan, H., Trump, B. F., Metcalf, R. A., Welsh, J. A., Midgley, C., Lane, D. P., and Harris, C. C. (1991) Archival analysis of p53 genetic and protein alterations in Chinese esophageal cancer. Oncogene *6*, 1779–1784.

Beutler, E., Gelbart, T., Han, J., Koziol, J. A., and Beutler, B. (1989) Evolution of the genome and the genetic code: selection at the dinucleotide level by methylation and polyribonucleotide cleavage. Proc. Natl. Acad. Sci. USA *86*, 192–196.

Bird, A. P. (1984) DNA methylation – How important in gene control? Nature *307*, 503–504.

Bird, A., Taggart, M., Frommer, M., Miller, O. J., and Macleod, D. (1985) A fraction of the mouse genome that is derived from islands of non-methylated CpG-rich DNA. Cell *40*, 91–99.

Bird, A. P., Taggart, M. H., Nicholls, R. D., and Higgs, D. R. (1987) Non-methylated CpG-rich islands at the human α globin locus: implications for evolution of the α-globin pseudogene. EMBO J. *6*, 999–1004.

Bonetta, L., Kuehn, S. E., Huang, A., Law, D. J., Kalikin, L. M., Koi, M., Reeve, A. E., Brownstein, B. H., Yeger, H., Williams, B. R. G., and Feinberg, A. P. (1990) Wilms tumor locus on 11p13 defined by multiple CpG island-associated transcripts. Science *250*, 994–997.

Borrello, M. G., Pierotti, M. A., Bongarzone, I., Donghi, R., Mondellini, P., and Porta, G. D. (1987) DNA methylation affecting the transforming activity of the human Ha-*ras* oncogene. Cancer Res. *47*, 75–79.

Bressac, B., Kew, M., Wands, T., and Ozturk, M. (1991) Selective G to T mutations of p53 gene in hepatocellular carcinoma from southern Africa. Nature *350*, 429–431.

Brown, T. C., and Jiricny, J. (1988) Different base/base mispairs are corrected with different efficiencies and specificities in monkey kidney cells. Cell *54*, 705–711.

Busslinger, M., Hurst, J., and Flavell, R. A. (1983) DNA methylation and the regulation of globin gene expression. Cell *34*, 197–206.

Call, K. M., Glaser, T., Ito, C. Y., Buckler, A. J., Pelletier, J., Haber, D. A., Rose, E. A., Kral, A., Yeger, H., Lewis, W. H., Jones, C., and Housman, D. E. (1990) Isolation and characterization of a zinc finger polypeptide gene at the human chromosome 11 Wilms' tumor locus. Cell *60*, 509–520.

Carr, B. I., Reilly, J. G., Smith, S. S., Winberg, C., and Riggs, A. (1984) the tumorigenicity of 5-azacytidine in the male fischer rat. Carcinogenesis *5*, 1583–1590.

Casson, A. G., Mukhopadhyay, T., Cleary, K. R., Ro, J. Y., Levin, B., and Roth, J. A. (1991) p53 Gene mutations in Barrett's epithelium and esophageal cancer. Cancer Res. *51*, 4495–4499.

Cheng, J., and Haas, M. (1990) Frequent mutations in the p53 tumor suppressor gene in human leukemia T-cell lines. Mol. Cell. Biol. *10*, 5502–5509.

Chiba, F., Takahashi, T., Nav, M. M., D'Amico, D., Curiel, D. T., Mitsudomi, T., Buchhagen, D. L., Carbone, D., Piantadosi, S., Koga, H., Reissman, P. T., Slamon, D. J., Holmes, E. G., and Minna, J. D. (1990) Mutations in the p53 gene are frequent in primary, resected non-small cell lung cancer. Oncogene *5*, 1603–1610.

Compere, S. J., and Palmiter, R. D. (1981) DNA methylation controls the inducibility of the mouse metallothionein-1 gene in lymphoid cells. Cell *25*, 233–240.

Cooper, D. N., Taggart, M. H., and Bird, A. P. (1983) Unmethylated domains in vertebrate DNA. Nucl. Acids Res. *11*, 647–658.

Cooper, D. N., and Schmidtke, J. (1984) DNA restriction fragment length polymorphisms and heterozygosity in the human genome. Hum. Genet. *66*, 1–16.

Cooper, D. N., Gerber-Huber, S., Nardelli, D., Schubiger, J. L., and Wahli, W. (1987) The distribution of the dinucleotide CpG and cytosine methylation in the vitellogenin gene family. J. Mol. Evol. *25*, 107–115.

Cooper, D. N., and Youssoufian, H. (1988) The CpG dinucleotide and human genetic disease. Hum. Genet. *78*, 151–155.

Cooper, D. N., and Krawczak, M. (1989) Cytosine methylation and fate of CpG dinucleotides in vertebrate genomes. Hum. Genet. *83*, 181–189.

Cooper, D. N., and Krawczak, M. (1990) The mutational spectrum of single base-pair substitutions causing human genetic disease: patterns and predictions. Hum. Genet. *85*, 55–74.

Cote, R. J., Jhanwar, S. C., Novick, S., and Pellicer, A. (1991) Genetic alterations of the p53 gene are a feature of malignant mesotheliomas. Cancer Res. *51*, 5410–5416.

Coulondre, C., Miller, J. H., Farabaugh, P. J., and Gilbert, W. (1978) Molecular basis of base substitution hotspots in *Escherichia coli*. Nature *274*, 775–780.

Creusot, F., Acs, G., and Christman, J. K. (1982) Inhibition of DNA methyltransferase and induction of Friend erythroleukemia cell differentiation by 5-azacytidine and 5-aza-2′-deoxycytidine, J. Biol. Chem. *257*, 2041–2048.

Davidoff, A. M., Kerns, B.-J., Iglehart, D., and Marks, J. R. (1991) Maintenance of p53 alterations throughout breast cancer progression. Cancer Res. *51*, 2605–2610.

De Bustros, A., Nelkin, G. D., Silverman, A., Ehrlich, G., Poiesz, B., and Baylin, S. B. (1988) The short arm of chromosome 11 is a "hotspot" for hypermethylation in human neoplasm. Proc. Natl. Acad. Sci. USA *85*, 5693–5697.

De Lange, T., Shive, L., Myers, R. M., Cox, D. R., Naylor, S. L., Killery, A. M., and Varmus, H. E. (1990) Structure and variability of human chromosome ends. Mol. Cell. Biol. *10*, 518–527.

Driscoll, D. J., and Migeon, B. R. (1990) Sex difference in methylation of single-copy genes in human meiotic germ cells: implications for X chromosome inactivation, parental imprinting, and origin of CpG mutations. Somat. Cell. Mol. Genet. *16*, 267–282.

Duncan, B. K., and Miller, J. H. (1980) Mutagenic deamination of cytosine residues in DNA. Nature *287*, 560–561.

Ehrlich, M., and Wang, R. V.-H. (1981) 5-methylcytosine in eukaryotic DNA. Science *212*, 1350–1357.

Ehrlich, M., Norris, K. F., Wang, R. Y.-H., Kuo, K. C., and Gehrke, C. W. (1986) DNA cytosine methylation and heat-induced deamination. Biosci. Rep. *6*, 387–393.

El-Deiry, W. S., Nelkin, B. D., Celano, P., Yen, R.-W. C., Falco, J. P., Hamilton, S. R., and Baylin, S. B. (1991) High expression of the DNA methyltransferase gene characterizes human neoplastic and progression stages of colon cancer. Proc. Natl. Acad. Sci. USA *88*, 3470–3474.

Farrell, P. J., Allan, G. J., Shanahan, F., Vousden, K. H., and Crook, T. (1991) p53 is frequently mutated in Burkitt's lymphoma cell lines. EMBO J. *10*, 2879–2887.

Fearon, E. R., Cho, K. R., Nigro, J. M., Kern, S. E., Simons, J. W., Ruppert, J. M., Hamilton, S. R., Presinger, A. C., Thomas, G., Kinzler, K. W., and Voglelstein, B. (1990) Identification of a chromosome 18q gene that is altered in colorectal cancers. Science *247*, 49–56.

Feinberg, A. P., and Volgelstein, B. (1983) Hypomethylation distinguishes genes of some human cancers from their normal counterparts. Nature *301*, 89–92.

Feinberg, A. P., Gehrke, C. W., Kuo, K. C., and Ehrlich, M. (1988) Reduced genomic 5-methylcytosine content in human colonic neoplasia. Cancer Res. *48*, 1159–1161.

Foster, P. L., Eisenstadt, E., and Miller, J. N. (1983) Base substitution mutations induced by metabolically activated aflatoxin B_1. Proc. Natl. Acad. Sci. USA *80*, 2695–2698.

Frederico, L. A., Kunkel, T. A., and Shaw, B. R. (1990) A sensitive genetic assay for the detection of cytosine deamination: Determination of rate constants and the activation energy. Biochemistry *29*, 2532–2537.

Friend, S. H., Bernards, R., Rogeli, S., Weinberg, R. A., Rappaport, J. M., Albert, D. M., and Dryja, T. P. (1986) A human DNA segment with properties of the gene that predisposes to retinoblastoma and osteosarcoma. Nature *323*, 16–22.

Frost, P., and Kerbel, R. S. (1983) On a possible epigenetic mechanism(s) of tumor cell heterogeneity: The role of DNA methylation. Cancer Metastasis Rev. *2*, 375–378.

Fung, Y. K., Murphree, A. L., T'ang, A., Qian, J., Hinrichs, S. H., and Benedict. W. F. (1987) Structural evidence for the authenticity of the human retinoblastoma gene. Science *236*, 1657–1661.

Gaidano, G., Ballerini, P., Gong, J. Z., Inghirami, G., Neri, A., Newcomb, E. W., Magrath, I. T., Knowles, D. M., and Dalla-Favera, R. (1991) p53 mutations in human lymphoid malignancies: association with Burkitt lymphoma and chronic lymphocytic leukemia. Proc. Natl. Acad. Sci. USA *88*, 5513–5417.

Gama-Sosa, M. A., Slagel, V. A., Trewyn, R. W., Oxenhandler, R., Kuo, K., Gehrke, W., and Ehrlich, M. (1983) The 5-methylcytosine content of DNA from human tumors. Nucl. Acids Res. *11*, 6883–6894.

Gardiner-Garden, M., and Frommer, M. (1987) CpG islands in vertebrate genomes. J. Mol. Biol. *196*, 261–282.

Gessler, M., Poustka, A., Cavence, W., Neve, R. L., Orkin, S. H., and Bruns, G. A. (1990) Homozygous deletion in Wilms tumors of a zinc-finger gene identified by chromosome jumping. Nature 343, 774–778.

Ghazi, H., Magewu, A. N., Gonzales, F., and Jones, P. A. (1990) Changes in the allelic methylation patterns c-H-*ras*-1, insulin and retinoblastoma genes in human development. Development Supplement, 115–123.

Goelz, S. E., Volgelstein, B., Hamilton, S. R., and Feinberg, A. P. (1985) Hypomethylation of DNA from benign and malignant human colon neoplasms. Science *228*, 187–190.

Gounari, F., Banks, G., Khazaie, K., Jeggo, P., and Holliday, R. (1987) Gene reactivation: A tool for the isolation of mammalian DNA methylation mutants. Genes Dev. *1*, 899–912.

Greger, V., Passarge, E., Hopping, W., Messmer, E., and Horsthemke, B. (1989) Epigenetic changes may contribute to the formation and spontaneous regression of retinoblastoma. Hum. Genet. *83*, 155–158.

Groden, J., Thliveris, A., Samowitz, W., Carlson, M., Gelbert, L., Albertsen, H., Joslyn, G., Stevens, J., Spirio, L., Robertson, M., Sargeant, L., Krapcho, K., Wolff, E., Burt, R., Hughes, J. P., Warrington, J., McPhearson, J., Wasmuth, J., Le Paslier, D., Abderrahim, H., Cohen, D., Leppart, M., and White, R. (1991) Identification and characterization of the familial adenomatous polyposis coli gene. Cell *66*, 589–600.

Gusterson, B. A., Anbazhagan, R., Warren, W., Midgely, C., Lane, D. P., O'Hare, M., Stamps. A., Carter, R., and Jayatilake, H. (1991) Expression of p53 in premalignant and malignant squamous epithelium. Oncogene *6*, 1785–1789.

Harris, M. (1982) Induction of thymidine kinase in enzyme-deficient Chinese hamster cells. Cell *29*, 483–492.

Hensel, C. H., Xiang, R. H., Sakaguchi, A. Y., and Naylor, S. L. (1991) Use of the single strand conformation polymorphism technique and PCR to detect p53 gene mutations in small cell lung cancer. Oncogene *6*, 1067–1071.

Holliday, R. (1979) A new theory of carcinogenesis. Br. J. Can. *40*, 513–522.

Holliday, R., and Jeggo, P. A. (1985) Mechanisms for changing gene expression and their possible relationship to carcinogenesis. Cancer Surveys *4*, 554–581.

Holliday, R., and Pugh, J. E. (1975) DNA modification mechanisms and gene activity during development. Science *187*, 226–232.

Hollstein, M. C., Metcalf, R. A., Welsh, J. A., Montesano, R., and Harris, C. C. (1990) Frequent mutation of the p53 gene in human esophageal cancer. Proc. Natl. Acad. Sci. USA *87*, 9958–9961.

Hollstein, M. C., Peri, L., Mandard, A. M., Welsh, J. A., Montesano, R., Metcalf, R. A., Bak, M., and Harris, C. C. (1991a) Genetic analysis of human esophageal tumors from two high incidence geographic areas: frequent p53 base substitutions and absence of ras mutations. Cancer Res. *51*, 4102–4106.

Hollstein, M., Sidransky, D., Volgelstein, B., and Harris, C. C. (1991b) p53 mutations in human cancers. Science *253*, 49–53.

Hsu, I. C., Metcalf, R. A., Sun, T., Welsh, J. A., Wang, N. J., and Harris, C. C. (1991) Mutational hotspot in the p53 gene in human hepatocellular carcinomas. Nature *350*, 427–428.

Iggo, R., Gatter, K., Bartek, J., Lane, D., and Harris, A. L. (1990) Increased expression of mutant forms of p53 oncogene in primary lung cancer. Lancet *335*, 675–679.

Ishioka, C., Sato, T., Gamoh, M., Suzuki, T., Shibata, H., Kanamaru, R., Wakul, A., and Yamazaki, T. (1991) Mutations of the p53 gene, including an intronic point mutation, in colorectal tumors. Biochem. Biophys. Res. Comm. *177*, 901–906.

Issacs, W. B., Carter, B. S., and Ewing, C. M. (1991) Wild-type p53 suppresses growth of human prostate cancer cells containing mutant p53 alleles. Cancer Res. *51*, 4716–4720.

Jiricny, J., Hughes, M., Corman, N., and Rudkin, B. B. (1988) A human 200-kDa protein binds selectivity to DNA fragments containing G-T mismatches. Proc. Natl. Acad. Sci. USA *85*, 8860–8864.

Jones, P. A., and Taylor, S. M. (1981) Hemimethylated duplex DNAs prepared from 5-azacytidine treated cells. Nucl. Acids Res. *9*, 2933–2947.

Jones, P. A., and Buckley, J. D. (1990a) The role of DNA methylation in cancer. Adv. Cancer Res. *54*, 1–23.

Jones, P. A., Wolkowitz, M. J., Rideout III, W. M., Gonzales, F. A., Marziasz, C. M., Coetzee, G. A., and Tapscott, S. J. (1990b) *De novo* methylation of the MyoD1 CpG island during the establishment of immortal cell lines. Proc. Natl. Acad. Sci. USA *87*, 6117–6121.

Jones, P. A., Buckley, J. D., Henderson B. E., Ross, R. K., and Pike, M. C. (1991) From gene to carcinogen: a rapidly evolving field in molecular epidemiology. Cancer Res. *51*, 3617–3620.

Joslyn, G., Carlson, M., Thliveris, A., Albertsen, H., Gelbert, L., Samowitz, W., Groden, J. Stevens, J., Spirio, L., Robertson, M., Sargeant, L., Krapcho, K., Wolff, E., Burt, R., Hughes, J. P., Warrington, J., McPhearson, J., Wasmuth, J., Le Paslier, D., Abderrahim, H., Cohen, D., Leppert, M., and White, R. (1991) Identification of deletion mutations and three new genes at the familial polyposis locus. Cell *66*, 601–603.

Josse, J., Kaiser, A. A., and Kornberg, A. (1961) Enzymatic synthesis of deoxyribonucleic acid VII. Frequencies of nearest neighbor base-sequences in deoxyribonucleic acid. J. Biol. Chem. *236*, 864–875.

Kautainien, T. L., and Jones, P. A. (1986) DNA methyltransferase levels in tumorigenic and non-tumorigenic cells in culture. J. Biol. Chem. *261*, 1594–1598.

Keshet, I., Yisraelli, J., and Cedar, H. (1985) Effect of regional DNA methylation on gene expression. Proc. Natl. Acad. Sci. USA *82*, 2560–2564.

Kim, J.-H., Takahashi, T., Chiba, I., Park, J.-G., Birrer, M. J., Roh, J. K., Lee, H. D., Kim, J.-P., Minna, J. D., and Gazdar, A. F. (1991) Occurrence of p53 gene anbormalities in gastric carcinoma tumors and cell lines. J. Natl. Cancer Inst. *83*, 938–943.

Kinzler, K. W., Nilbert, M. C., Li-Kuo, S., Volgelstein, B., Bryan, T. M., Levy, D. B., Smith, K. J. Preisinger, A. C., Hedge, P., McKechnie, D., Finniear, R., Markham, A., Groffen, J., Boguski, M. S., Altschul, S. F., Horii, A., Ando, H., Miyoshi, Y., Miki, Y., Nishisho, I., and Nakamura, Y. (1991) Identification of FAP locus genes from chromosome 5q21. Science *235*, 661–665.

Knudson, A. G. (1983) Model hereditary cancers of man. Prog. Nucl. Acid Res. Mol. Biol. *29*, 17–25.

Kovach, J. S., McGovern, R. M., Cassady, J. D., Swanson, S. K., Wold, L. E., Volgelstein, B., and Sommer, S. S. (1991) Direct sequencing from touch preparation of human carcinomas: analysis of p53 mutations in breast carcinomas. J. Natl. Cancer Inst. *83*, 1004–1009.

Kramer, B., Kramer, W., and Fritz, H. J. (1984) Different base/base mismatches are corrected with different efficiencies by the methyl-directed DNA mismatch-repair system of *E. coli*. Cell *38*, 879–887.

Lane, D. P., and Benchimol, S. (1990) p53: oncogene or anti-oncogene? Genes Dev. *4*, 1–8.

Lee, W. H. Bookstein, R., Hong, F., Young, L. J., Shen, J. Y., and Lee, E. Y. (1987) Human retinoblastoma susceptibility gene: cloning, identification, and sequence. Science *235*, 1394–1399.

Lehman, T. A., Bennet, W. P., Metcalf, R. A., Welsh, J. A., Ecker, J., Modali, R. V., Ullrich, S. Romano, J. W., Appella, E., Testa, J. R., Gerwin, B. I., and Harris, C. C. (1991) p53 mutations, *ras* mutations, and p53-heat shock 70 protein complexes in human lung carcinoma cel lines. Cancer Res. *51*, 4090–4096.

Lieb, M. (1987) Bacterial genes mutL, mutS, and dcm participate in repair of mismatches at 5-methylcytosine sites. J. Bacteriol. *169*, 5241–5246.

Lieb, M. (1991) Spontaneous mutation at a 5-methylcytosine hotspot is prevented by very short patch (VSP) mismatch repair. Genetics *128*, 23–27.

Lindahl, T. (1979) DNA glycosylases, endonucleases for apurinic/apyrimidinic sites, and base excision-repair. Progr. Nucl. Acids Res. Mol. Biol. *22*, 135–192.

Liteplo, R. G., and Kerbel, R. S. (1987) Reduced levels of DNA 5-methylcytosine in metastatic variants of the human melanoma cell line MeWo. Cancer Res. *47*, 2265–2267.

Lock, L. F., Melton, D. W., Caskey, C. T., and Martin, G. R. (1986) Methylation of the mouse hprt gene differs on the active and inactive X chromosomes. Mol. Cell. Biol. *6*, 914–924.

Lock, L. F., Takagi, N., and Martin, G. R. (1987) Methylation of the HPRT gene on the inactive X occurs after chromosome inactivation. Cell *48*, 39–46.

Lloyd, J., Brownson, C., Tweedie, S., Charlton, J., and Edwards, Y. (1987) Human muscle carbonic anhydrase: Gene structure and DNA methylation patterns in fetal and adult tissues. Genes Dev. *1*, 594–602.

Malkin, D., Li, F. P., Strong, L. C., Fraumeni, J. F. J., Nelson, C. E., Kim, D. H., Kassel, J., Gryka, M. A., Bischoff, F. Z., Tainsky, M. A., and Friend, S. H. (1990) Germline p53 mutations in familial syndrome of breast cancer, sarcomas, and other neoplasms. Science *250*, 1233–1238.

Marks, R., Davidoff, A. M., Kerns, B. J., Humphrey, P. A., Pence, J. C., Dodge, R. K., Clarke-Pearson, D. L., Iglehart, D., Bast, Jr., R. C., and Berchuck, A. (1991) Overexpression and mutation of p53 in epithelial ovarian cancer. Cancer Res. *51*, 2979–2984.

Marziasz, C. M. (1991) Ph.D. Thesis, University of Southern California.

Mashiyama, S., Murakami, Y., Yoshimoto, T., Sekiya, T., and Hayashi, K. (1991) Detection of p53 gene mutations in human brain tumors by single-strand conformation polymorphism analysis of polymerase chain reaction products. Oncogene *6*, 1313–1318.

Mazars, R., Pujol, P., Maudelonde, T., Jeanteur, P., and Theillet, C. (1991) p53 mutations in ovarain cancer a late event? Oncogene *6*, 1685–1690.

Mazur, M., and Glickman, B. (1988) Sequence specificity of mutations induced by benzo(a)pyrene-7,8-diol-9,10-epoxide at endogenous APRT gene in CHO cells. Somat. Cell Mol. Genet. *14*, 393–400.

McKeon, C., Ohkubo, H., Pastan, I., and de Crombrugghe, B. (1982) Unusual methylation pattern of the α2(1) collagen gene. Cell *29*, 203–210.

Menon, A. G., Anderson, K. M., Riccardi, V. M., Chung, R. Y., Whaley, J. M., Yandell, D. W., Farmer, G. E., Freiman, R. N., Lee, J. K., Li, F. P., Barker, D. F., Ledbetter, D. H., Kleider, A., Martuza, R. L., Gusella, J. F., and Seizinger, B. R. (1990) Chromosome

17p deletions and p53 gene mutations associated with the formation of malignant neurofibrosarcomas in von Recklinghausen neurofibromatosis. Proc. Natl. Acad. Sci. USA *87*, 5435–5439.

Metzger, A. K., Sheffield, V. C., Duyk, G., Daneshvar, L., Edwards, M. S. B., and Cogen, P. H. (1991) Identification of a germ-line mutation in the p53 gene in a patient with an intracranial ependymoma. Proc. Natl. Acad. Sci. USA *88*, 7825–7829.

Mohandas, T., Sparkes, R. S., and Shapiro, L. J. (1981) Reactivation of an inactive human X-chromosome: Evidence for X-inactivation by DNA methylation. Science *211*, 393–396.

Mueller, P. R., and Wold, B. (1989) *In vivo* footprinting of a muscle specific enhancer by ligation medicated PCR. Science *246*, 780–786.

Muench, K. F., Misra, R. P., and Humayun, M. Z. (1983) Sequence specificity in aflatoxin B_1-DNA interactions. Proc. Natl. Acad. Sci. USA *80*, 6–10.

Mulligan, L. M., Matlashewski, G. J., Scrable, H. J., and Cavenee, W. K. (1990) Mechanisms of p53 loss in human sarcomas. Proc. Natl. Acad. Sci. USA *87*, 5863–5867.

Murakami, Y., Hayashi, K., Kirohashi, S., and Sekiya, T. (1991) Aberrations of the tumor suppressor p53 and retinoblastoma genes in human hepatocellular carcinomas. Cancer Res. *51*, 5520–5525.

Nelkin, B. D., Przepiorka, D., Burke, P. J., Thomas, E. D., and Baylin, S. B. (1991) Anbormal methylation of the calcitonin gene marks progression of chronic myelogenous leukemia. Blood *77*, 2431–2434.

Nigro, J. M., Baker, S. J., Preisinger, A. C., Jessup, J. M., Hostetter, R., Cleary, K., Bigner, S. H., Davidson, N., Baylin, S., Devilee, P., Glover, T., Collins, F. S., Weston, A., Modali, R., Harris, C. C., and Volgelstein, B. (1989) Mutations in the p53 gene occur in diverse human tumour types. Nature *342*, 705–708.

Nishisho, I., Nakamura, Y., Miyoshi, Y., Miki, Y., Ando, H., Horii, A., Koyama, K., Utsunomiya, J., Baba, S., Hedge, P., Markham, A., Krush, A. J., Petersen, G., Hamilton, S. R., Nilbert, M. C., Levy, D. B., Bryan, T. M., Preisinger, A. C., Smith, K. J., Su, L., Kinzler, K. W., and Volgelstein, B. (1991) Mutation of chromosome 5q21 genes in FAP and colorectal cancer patients. Science *235*, 665–669.

Nussinov, R. (1981) Nearest neighbour nucleotide patterns; structural and biological implications. J. Biol. Chem. *256*, 8458–8462.

Okamota, A., Sameshima, Y., Yamada, Y., Teshima, S.-I., Terashima, Y., Terada, M., and Yokota, J. (1991) Allelic loss on chromosome 17p and p53 mutations in human endometrial carcinoma of the uterus. Cancer Res. *51*, 5632–5636.

Olumi, A. F., Tsai, Y. C., Nichols, P. W., Skinner, D. G., Cain, D. R., Bender, L. I., and Jones, P. A. (1990) Allelic loss of chromosome 17p distinguishes high grade from low grade transitional cell carcinomas of the bladder. Cancer Res. *50*, 7081–7083.

Osborne, R. J., Merlo, G. R., Mitsudomi, T., Venesio, T., Liscia, D. S., Cappa, A. P. M., Chiba, I., Takahashi, T., Nau, M. M., Callahan, R., and Minna, J. D. (1991) Mutation in the p53 gene in primary human breast cancer. Cancer Res. *51*, 6194–6198.

Ozturk, M., and collaborators (1991) p53 mutation in hepatocellular carcinoma after aflatoxin exposure. Lancet *338*, 1356–1359.

Pfeifer, G. P., Steigerwald, S., Boehm, T. L. J., and Drahovsky, D. (1988) DNA methylation levels in acute human leukemia. Cancer Lett. *39*, 185–192.

Pfeifer, G. P., Tanguay, R. L., Steigerwald, S. D., and Riggs, A. D. (1990) *In vivo* footprint and methylation analysis by PCR-aided genomic sequencing: Comparison of active and inactive X chromosomal DNA at the CpG island and promoter of human PGK-1. Genes Dev. *4*, 1277–1287.

Prosser, J., Thompson, A. M., Cranston, G., and Evans, H. J. (1990) Evidence that p53 behaves as a tumor suppressor gene in sporadic breast tumours. Oncogene *5*, 1573–1579.

Pugh, J. E., and Holliday, R. (1978) Do chemical carcinogens act by altering epigenetic controls through DNA repair rather than by mutations? Heredity *40*, 329 (Abstract).

Puisieux, A., Lim, S., Groopman, J., and Ozturk, M. (1991) Selective targeting of p53 gene mutational hotspots in human cancers by etiologically defined carcinogens. Cancer Res. *51*, 6185–6189.

Rachal, M. J., Yoo, H., Becker, F. F., and Lapeyre, J.-N. (1989) *In vitro* DNA cytosine methylation of cis-regulatory elements modulates c-Ha-*ras* promoter activity *in vivo*. Nucl. Acids Res. *17*, 5135.

Radman, M., and Wagner, R. (1986) Mismatch repair in *Escherichia coli*. Annu. Rev. Gen. *20*, 523–538.

Raycroft, L., Schmidt, J. R., Yoas, K., Hao, M., and Lozano, G. (1991) Analysis of p53 mutants for transcriptional activity. Mol. Cell. Biol. *11*, 6067–6074.

Rideout, III, W. M., Coetzee, G. A., Olumi, A. F., and Jones, P. A. (1990) 5-Methylcytosine as an endogenous mutagen in the human LDL receptor and p53 genes. Science *249*, 1288–1290.

Riggs, A. D. (1975) X inactivation, differentiation, and DNA methylation. Cytogenet. Cell Genet. *14*, 9–25.

Riggs, A. D., and Jones, P. A. (1983) 5-Methylcytosine, gene regulation and cancer. Advances in Cancer Res. *40*, 1–30.

Rodrigues, N. R., Rowan, A., Smith, M. E. F., Kerr, I. B., Bodmer, W. F., Gannon, J. V., and Lane, D. P. (1990) p53 mutations in colorectal cancer. Proc. Natl. Acad. Sci. USA *87*, 7555–7559.

Romano, J. W., Ehrhart, J. C., Duthr, A., Kim, C. M., Appella, E., and May, P. (1989) Identification and characterization of a p53 gene mutation in a human osteosarcoma cell line. Oncogene *4*, 1483–1488.

Russell, G. J., Walker, P. M. B., Elton, R. A., and Subak-Sharpe, J. H. (1976) Doublet frequency analysis of fractionated vertebrate nuclear DNA. J. Mol. Biol. *108*, 1–23.

Sameshima, Y., Akiyama, T., Mon, N., Mizoguchi, H., Toyoshima, K., Sugimura, T., Terada, M., and Yokota, J., (1990) Point mutation of the p53 gene resulting in splicing inhalation in small cell lung carcinoma. Biochem. Biophys. Res. Comm. *173*, 697–703.

Savatier, P., Trabuchet, G., Fauré, C., Chebloure, Y., Gouy, M., Verdier, G., and Nigon, V. M. (1985) Evolution of the primate beta-globin gene region. High rate of variation in CpG dinucleotides and in short repeated sequences between man and chimpanzee. J. Mol. Biol. *182*, 21–29.

Saxon, P. J., Srivatsan, E. S., and Stanbridge, E. J. (1986) Introduction of human chromosome 11 via microcell transfer controls tumorigenic expression HeLa cells. EMBO J. *5*, 3461–3466.

Saylors, R. L., III, Sidransky, P., Friedman, H. S., Bigner, D. D., Volgelstein, B., and Brodeur, G. M. (1991) Infrequent p53 gene mutations in medulloblastomas. Cancer Res. *51*, 4721–4723.

Scarano, E. (1971) The control of gene function in cell differentiation and in embryogenesis. Adv. Cytopharmacol. *1*, 13.

Scheffner, M., Münger, K., Byrne, J. C., and Howley, P. M. (1991) The state of the p53 and retinoblastoma genes in human cervical carcinoma cell lines. Proc. Natl. Acad. Sci. USA *88*, 5523–5527.

Shenoy, S., Ehrlich, K. C., and Ehrlich, M. (1987) Repair of thymine-guanine mismatched base-pairs in bacteriophage M13mp18 DNA heteroduplexes. J. Mol. Biol. *197*, 617–626.

Shmookler-Reis, R. J., and Goldstein, S. (1982) Interclonal variation in methylation patterns for expressed and non-expressed genes. Nucl. Acids Res. *10*, 4293–4304.

Sidransky, D., Von Eschenbach, A., Tsai, Y. C., Jones, P., Summerhayes, I., Marshall, F., Paul, M., Green, P., Hamilton, S. R., Frost, P., and Volgelstein, B. (1991) Identification of p53 gene mutations in bladder cancers and urine samples. Science *252*, 706–709.

Silverman, A. L., Park, J.-G., Hamilton, S. R., Gazdar, A. F., Luk, G. D., and Baylin, S. B. (1989) Abnormal methylation of the calcitonin gene in human colonic neoplasms. Cancer Res. *49*, 3468–3473.

Slingerland, J. M., Minden, M. D., and Benchimol. S. (1991) Mutation of the p53 gene in human acute myelogenous leukemia. Blood *77*, 1500–1507.

Sneller, M. C., and Gunter, K. C. (1987) DNA methylation alters chromatin structure and regulates Thy-1 expression in EL-4T cells. Immunology *138*, 3505–3512.

Spanopoulou, E., Giguere, V., and Grosveld, F. (1988) Transcriptional unit of the murine Thy-1 gene: Different distribution of transcription initiation sites in brain. Mol. Cell. Biol. *8*, 3847–3856.

Srivastava, S., Zou, Z. Q., Pirollo, K., Blatner, W., and Chang, E. H. (1990) Germ-line transmission of a mutated p53 gene in a cancer-prone family with Li-Fraumeni syndrome. Nature *348*, 20–27.

Stratton, M. R., Moss, S., Warren, W., Patterson, H., Clark, J., Fisher, C., Fletcher, C. D. M., Ball, A., Thomas, M., Gusterson, B. A., and Cooper, C. S. (1990) Mutation of the p53 gene in human soft tissue sarcomas: association with abnormalities of the RB1 gene. Oncogene *5*, 1297–1301.

Sved, J., and Bird, A. (1990) The expected equilibrium of the CpG dinucleotide in vertebrate genomes under a mutation model. Proc. Natl. Acad. Sci. USA *87*, 4692–4696.

Swartz, M. N., Trautner, T. A., and Kornberg, A. (1962) Enzymatic synthesis of deoxyribonucleic acid. J. Biol. Chem. *237*, 1961–1967.

Takahashi, T., Takahashi T., Suzuki, H., Hida, T., Sekido, Y., Ariyoshi, Y., and Ueda, R. (1991) The p53 gene is very frequently mutated in small-cell lung cancer with a distinct nucleotide substitution pattern. Oncogene *6*, 1775–1778.

Tamura, G., Kihana, T., Nomura, K., Terada, M., Sugimura, T., and Hiroshasi, S. (1991) Detection of frequent p53 gene mutations in primary gastric cancer by cell sorting and polymerase chain reaction single-strand conformation polymorphism analysis. Cancer Res. *51*, 3056–3058.

Taylor, S. M., and Jones, P. A. (1982) Changes in phenotypic expression in embryonic and adult cells treated with 5-azacytidine. J. Cell. Physiol. *111*, 187–194.

Toniolo, D., Martini, G., Migeon, B. R., and Dono, R. (1988) Expression of the GGPD locus on the human X chromosome is associated with demethylation of three CpG islands within 100 Kb of DNA. EMBO J. *1*, 401–406.

Toth, M., Lichtenberg, V., and Doerfler, W. (1989) Genomic sequencing reveals a 5-methylcytosine-free domain in active promoters and the spreading of preimposed methylation patterns. Proc. Natl. Acad. Sci. USA *86*, 3728–3732.

Tykocinski, M., and Max, E. (1984) CG clusters in MHC genes and $5\frac{1}{2}$ demethylated genes. Nucl. Acids Res. *12*, 4385–4396.

Venolia, L., Gartler, S. M., Wassman, E. R., Yen, P., Mohandas, T., and Shapiro, L. J. (1982) Transformation with DNA from 5-azacytidine-reactivated X chromosomes. Proc. Natl. Acad. Sci. USA *79*, 2352–2354.

Volgelstein, B. (1990) Cancer. A deadly inheritance. Nature *348*, 681–682.

Weissman, B. E., Saxon, P. J., Pasquale, S. R., Jones, G. R., Geiser, A. G., and Stanbridge, E. J. (1987) Introduction of a normal human chromosome 11 into a Wilms tumor cell line controls its tumorigenic expression. Science *236*, 175–180.

Wiebauer, K., and Jiricny, J. (1989) *In vitro* correction of G·T mispairs to G·C pairs in nuclear extracts from human cells. Nature *339*, 234–236.

Wilson, V. L., and Jones, P. A. (1983a) DNA methylation decreases in aging but not in immortal cells. Science *220*, 1055–1057.

Wilson, V. L., and Jones, P. A. (1983b) Inhibition of DNA methylation by chemical carcinogens *in vitro*. Cell *32*, 239–246.

Wink, D. A., Kasprzak, K. S., Maragos, C. M., Elespuru, R. K., Misra, M., Dunams, T. M., Cebula, T. A., Koch, W. H., Andrews, A. W., Allen, J. S., and Keefer, L. K. (1991) DNA deaminating ability and genotoxicity of nitric oxide and its progenitors. Science *254*, 1001–1003.

Wise, T. L., and Harris M. (1988) Deletion and hypermethylation of thymidine kinase gene in V79 Chinese hamster cells resistant to bromodeoxyuridine. Somat. Cell. Mol. Genet. *14*, 567–581.

Wolf, S. F., and Migeon, B. R. (1985) Clusters of CpG dinucleotides implicated by nuclease hypersensitivity as control elements of housekeeping genes. Nature *314*, 467–469.

Wright, P. A., Lemoine, N. R., Govetzki, P. E., Wyllie, F. S., Bond, J., Hughes, C., Röher, H.-D., Williams, E. D., and Winford-Thomas, D. (1991) Mutation of the p53 gene in a differentiated human thyroid carcinoma cell line, but not in primary thyroid tumours. Oncogene *6*, 1693–1697.

Xu, G. F., O'Connell, P., Viskochil, D., Cawthon, R., Robertson, M., Culver, M., Dunn, D., Stevens, J., Gesteland, R, White, R., and Weiss, R. (1990) The neurofibromatosus type I gene encodes a protein related to GAP. Cell *62*, 599–608.

Yaginuma, Y., and Westphal, H. (1991) Analysis of the p53 gene in human uterine carcinoma cell lines. Cancer Res. *51*, 6506–6509.

Yamada, Y., Yoshida, T., Hayashi, K., Sekiya, T., Yokota, J., Hirohashi, S., Nakatami, K., Nakano, H., Sugimura, T., and Terada, M. (1991) p53 gene mutations in gastric cancer metastases and in gastric cancer cell lines derived from metastases. Cancer Res. *51*, 5800–5805.

Yandell, D. W., Campbell, T. A., Dayton, S. H., Petersen, R., Walton, D., Little, J. B., McConkie, R. A., Buckley, E. G., and Dryja, T. P. (1989) Oncogenic point mutations in the human retinoblastoma gene: their application at genetic counseling. N. Engl. J. Med. *321*, 1689–1695.

DNA Methylation: Molecular Biology and Biological Significance
ed. by J. P. Jost & H. P. Saluz

The repair of 5-methylcytosine deamination damage

Karin Wiebauer, Petra Neddermann, Melya Hughes and Josef Jiricny
IRBM, Via Pontina Km 30,600, 00040 Pomezia, Italy

1 Introduction

The fact that sites of cytosine methylation in DNA are mutagenic hotspots, and that the genome of higher eukaryotes is thus depleted of the modified sequence CpG is common knowledge – certainly in the field of DNA methylation. The reason why 5-methylcytosine is an endogenous mutagen would also appear to be clear: 5-methylcytosine deaminates to thymine, giving thus rise to a C → T transition mutation. Before accepting this hypothesis point blank, however, we ought to remember that cytosine and adenine are also hydrolytically deaminated to uracil and hypoxanthine, respectively, and would thus also be expected to give rise to transition mutations. Yet, neither C nor A are considered to be mutagenic hotspots, most likely because their deamination is known to be efficiently repaired. Does this mean that the hydrolytic deamination of 5-methylcytosine is not corrected? Let us consider both, the hydrolysis and the repair aspects in greater detail.

DNA is a complicated chemical entity with many diverse functional groups. As such, it is a target for attack by numerous reagents, both endogenous and exogenous, which can modify it in several ways. Even if we were to limit the discussion of these various modification reactions solely to the mutagens relevant to the readers of this book, namely, S-adenosylmethionine (SAM) and water, we would still come out with an impressive and frightening list. Thus, for example, SAM, required in the formation of 6-methyladenine and 5-methylcytosine, also modifies purines to yield 7-methylguanine and 3-methyladenine. These side-reactions lead to the loss of the respective purines by hydrolysis to generate apurinic (AP) sites in the DNA, which have a known miscoding potential. Water can bring about the hydrolysis of all exo-amino groups of the DNA bases. The deamination of cytosine to uracil is very well known. However, another mutagenic hydrolytic event, the conversion of adenine to hypoxanthine also proceeds at an appreciable rate (Lindahl, 1982).

The above-mentioned modifications are either lethal (e.g. 3-methyladenine), mutagenic (e.g. uracil, hypoxanthine) or premutagenic (7-methylguanine). It is obvious that if the cell is to survive and if the genetic information encoded in the DNA is to be preserved for successive generations, the cell must ensure that it can deal with this chemical assault. Indeed, a large number of repair pathways, dedicated to the removal of damage to DNA by both exogenous and endogenous mutagens, have been described (Lindahl, 1982; Sancar and Sancar, 1988).

Base damage in general, and deamination in particular, is corrected by a base-excision repair pathway, the initial step of which is carried out by DNA glycosylases. These enzymes are capable of substrate recognition in single-stranded (ss), as well as in double-stranded (ds) DNA, and act by cleaving the N–C bond between the base and the pentose ring of the sugar-phosphate backbone. The common trait of all these enzymes is the fact that they recognize modified or fragmented DNA bases, i.e. structures that do not belong in the DNA. [The two notable exceptions to this rule are the recently-characterized MutY protein of *E. coli* – a G/A mismatch specific glycosylase (Au et al., 1989) and a human G/T-specific glycosylase (Wiebauer and Jiricny, 1989) discussed in more detail below.]

One base modification reaction is in a class of its own: the spontaneous hydrolytic deamination of 5-methylcytosine. This reaction generates thymine in a way similar to the formation of uracil from cytosine (Scheme 1). Clearly, both these deamination events are mutagenic, leading to C→T transitions. The rates of hydrolysis would also appear to be of the same order of magnitude, with 5-methylcytosine deaminating, according to some recent data, slightly faster than cytosine (Ehrlich et al., 1986). Rough and, most probably, conservative estimates would put the number of deamination events at around 10–15 per haploid genome/day. This is a significant number and one would expect that, as in the case of the other base deaminations, a DNA repair pathway would have evolved that could counteract the loss of 5-methylcytosine residues from the DNA.

In double-stranded DNA, the deamination of adenine in an A/T base pair results in the formation of an H/T mispair (where H = hypoxanthine). Similarly, a U/G mispair will form at sites of cytosine deamination. Despite the fact that these structures bring about a significant distortion of the double helix, neither uracil DNA glycosylase, nor hypoxanthine DNA glycosylase would appear to utilize these distortions for damage recognition; both enzymes can remove U or H respectively from ss as well as from ds DNA (Lindahl, 1982).

In the case of the hydrolytic deamination of 5-methylcytosine, a G/T mispair arises in the DNA. In contrast to the above-mentioned H/T and U/T mispairs, however, any processing of the G/T mismatch would be

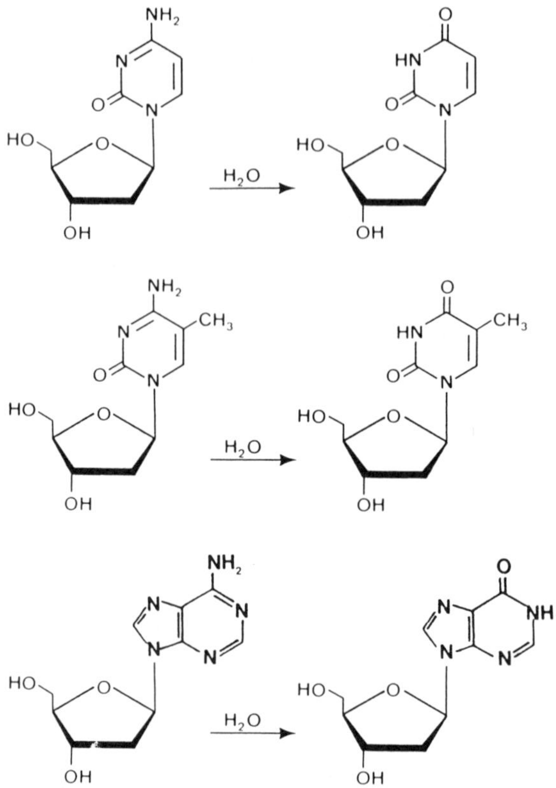

Scheme 1. Hydrolytic deamination of cytosine, 5-methylcytosine and adenine.

dependent on the recognition in ds DNA, because in ss DNA a thymine arising from 5-methylcytosine deamination cannot be distinguished from any other thymine. This mode of recognition imposes an important constraint on the repair system, however, as the G/T mispair is also the most frequent polymerase error arising in DNA during replication. The presence of a G/T mispair in the DNA thus presents the repair system with a dilemma: a G/T arising from the deamination of 5-methylcytosine must in all cases be corrected to a G/C. In contrast, the other types of a G/T mispair, structurally indistinguishable from the former, must be corrected either to a G/C or to an A/T, depending on whether the polymerase misincorporated a G opposite a T or a T opposite a G, i.e. the repair directionality must always favor the parent, template strand.

This article discusses the recent progress in the study of the mechanisms of G/T mismatch repair, and will attempt to provide an answer to the question of why sites of cytosine methylation are mutagenic hotspots, despite the existence of an efficient repair pathway.

2 G/T mismatch repair in *E. coli*

Our knowledge of the mechanisms of mismatch correction is based mainly on the data originating from numerous studies of this process in bacterial systems. The existence of *E. coli* mutator phenotypes has allowed us to estimate, at least approximately, the number of genes (or genetic loci) involved in mismatch repair in this organism. Early experiments of Wagner and Meselson (1976) established that mismatch correction *in vivo* involves repair tracts up to several kilobases long. Subsequent studies had shown that this repair process was dependent on the presence of hemimethylated d(GATC) sites that arise during DNA replication, and that all or nearly all types of mismatches are addressed by this pathway both *in vivo* and *in vitro* (see Claverys and Lacks, 1986; Radman and Wagner, 1986; Modrich, 1987; Meselson, 1988; Modrich, 1989; Jiricny, 1991 for reviews). Consequently, it was deduced that this so-called Long Patch Mismatch Repair or Methyl-Directed Mismatch Repair pathway, was dedicated to the correction of biosynthetic errors. The system uses the d(GATC) sites for strand discrimination, directing mismatch correction to the newly-synthesized strand, which remains transiently unmethylated due to the lag of the Dam methylase behind the polymerase. It is outside the scope of this article to discuss the mechanism of this repair process, but the reader can judge its complexity from the fact that to date the products of the *mut*S, *mut*L, *mut*H, *mut*U (*uvr*D), *ssb*I and *dam* genes have been demonstrated to be involved, in addition to the DNA polymerase I and DNA ligase (Lahue et al., 1989). The finding that, contrary to expectation, in some phage lambda crosses two closely-spaced markers were not co-repaired, led Lieb (1983) to propose that some mispairs were addressed by an alternative repair pathway. She named it VSP repair (for Very Short Patch) and it soon transpired that this mode of mismatch correction was limited solely to the repair of the G/T mispair in a sequence CCWGG (where W is A or T), which was known to be the recognition site of the bacterial deoxycytidine methylase encoded by the *dcm* gene (Lieb, 1985). The VSP repair thus appeared to be dedicated to the correction of G/T mispairs arising through the spontaneous hydrolytic deamination of 5-methylcytosine at this site. More detailed genetic analysis of this pathway showed that it does not require any of the other components of the long-patch repair system (although functional MutS and MutL proteins augment the efficiency of VSP repair) and that it is thus indeed an independent metabolic pathway, dedicated to the neutralization of the mutagenic threat of 5-methylcytosine deamination (Lieb, 1985; Jones et al., 1987). The discovery that *dcm*-mutants were also *vsp*-mutants gave rise to the hypothesis that the methylase gene is involved in the VSP repair pathway (Lieb, 1987). It was not until fine genetic analysis of the *dcm* locus by Bhagwat and colleagues

(Bhagwat, 1988; Sohail et al., 1990) revealed the existence of a second ORF within the *dcm* gene (frameshifted by +1 with respect to the latter) as well as demonstrating quite unambiguously (by using fine deletion mapping of the locus) that solely the second gene, termed *vsr*, is necessary for VSP repair.

In a recent report by Fritz and coworkers (Hennecke et al., 1991), the mechanism of the repair process was elucidated. The *vsr* protein was shown to be a mismatch-specific endonuclease, which initiates the repair process by nicking ds DNA within the sequence C$\underline{T}$WGN or N$\underline{T}$WGG (where N = any nucleotide), 5′ from the underlined, mismatched thymidine residue. The presence of a cut at this site is thought to initiate a pol I-catalyzed nick-translation reaction, which would remove the mismatched TMP together with several neighboring nucleotides by means of its 5′ → 3′ exonuclease activity to generate the short repair stretch typical of this enzyme. DNA ligase can safely be assumed to seal the nick on the repaired DNA strand, following the dissociation of the polymerase.

3 G/T mismatch repair in higher eukaryotes

Research into the mechanisms of mismatch correction in higher eukaryotes has been severely hampered by the lack of mutant cell lines, deficient in this mode of repair. To date, we have no idea how many genes are involved in this process. Until recently, all available evidence came in the form of data obtained from transfection experiments involving mismatch-containing heteroduplex molecules. In 1985, Hare and Taylor reported that the analysis of progeny viral plaques, following the transfection of an SV40 heteroduplex carrying a single mispair (A/C and G/T in a 1:1 ratio), revealed the existence of a mismatch repair pathway in CV-1 (African green monkey kidney) cells. In addition, secondary signals such as methylation of both, adenine and cytosine, and strand-specific nicks, appeared to influence the directionality of repair in their system. As these latter signals should not be required in the repair of G/T mismatches arising from the deamination of 5-methylcytosine (these must in all instances be corrected to G/C), we assumed that the data of Hare and Taylor related to a general mismatch repair pathway, most probably analogous to the long-patch repair of *E. coli*. This prediction would appear to have been substantiated by the *in vitro* experiments of Brooks et al. (1989) and Holmes et al. (1990), who demonstrated that in extracts from *Xenopus laevis* oocytes, respectively HeLa and *Drosophila melanogaster* cells, several mismatch types were addressed. Moreover, nicks were shown to have been used as strand discrimination signals in this process (Holmes et al., 1990).

We decided to use a similar system to search for the existence of a G/T-specific repair process in CV-1 cells. We ligated a synthetic oligonucleotide duplex, containing a single mismatch in a defined orientation, into the intron of the large T antigen gene of SV40 and transfected these well-characterized heteroduplexes, this time lacking potential strand-differentiating signals, into CV-1 cells. Analysis of progeny viral DNA revealed that the heteroduplexes containing the G/T mispair were addressed extremely efficiently, and that the repair bias went almost exclusively in favor of G/C (Brown and Jiricny, 1987, see also Table 1). As none of the other mismatch-containing heteroduplexes were corrected with a similar efficiency or directionality (Brown and Jiricny, 1988), we took these results as evidence for the existence of a G/T-specific mismatch repair process in CV-1 cells, which most likely evolved to counteract the loss of 5-methylcytosine residues through deamination. The G/T mismatch was corrected with similar efficiency and directionality in cell lines of human origin: normal human fibroblasts, Xeroderma pigmentosum type A and Bloom's syndrome (Brown et al., 1989).

In later experiments, using a synthetic 90-mer oligonucleotide duplex containing a single G/T or A/C mismatch (Fig. 1) and extracts from Hela cells, we were able to show that only the former substrate was processed under conditions where the long-patch repair is inactive (Wiebauer and Jiricny, 1989). In contrast to the *E. coli*. Vsr-mediated repair (see above), the mammalian extracts catalyze the removal of the mispaired thymine by a G/T mismatch-specific DNA glycosylase, to generate an apyrimidinic site opposite the guanine. The mammalian cells have thus chosen a path of G/T repair different from bacteria: the use of a mismatch-specific glycosylase for excision of the mispaired thymine ensures an absolute strand-specificity, and dispenses with the need for secondary signals that would otherwise be required to dictate the directionality of the repair event. We are at present not certain how

Table 1. Efficiency and directionality of mismatch correction in SV40 heteroduplexes transfected into CV-1 cells (Brown and Jiricny, 1987, 1988)

Mispair	Unrepaired (%)	Repaired to (%) G/C	Repaired to (%) A/T
G/T	4	92	4
A/C	22	41	37
A/G	61	27	12
C/T	28	60	12
G/G	8	92	—
A/A	36	66	—
T/T	61	—	39
C/C	34	—	64

5' ACGTTGTAAAACGACGGCCAGTGAATTCCCGGGGATCCGTC**R**ACCTGCAGCCAAGCTTGGCGTAATCATGGTCATAGCTGTTTCCTGTGT 3'
3' TGCAACATTTTGCTGCCGGTCACTTAAGGGCCCCTAGGCAG**Y**TGGACGTCGGTTCGAACCGCATTAGTACCAGTATCGACAAAGGACACA 5'

Figure 1. 90-mer oligonucleotide duplex used in the *in vitro* mismatch repair experiments (Wiebauer and Jiricny, 1989, 1990). **R** = G or A, **Y** = C or T; the sites of incision of the G/C duplex by *Sal*I (▼), *Acc*I (▽) and *Hin*cII (∨) are shown.

the repair proceeds. As the first incision of the DNA strand takes place at the 3′-side of the AP-site, it seems likely that it is mediated by an AP-lyase activity associated with the thymine glycosylase. Following the first incision, the baseless phosphodeoxyriboside may be removed by an AP-endonuclease to generate the observed single nucleotide gap

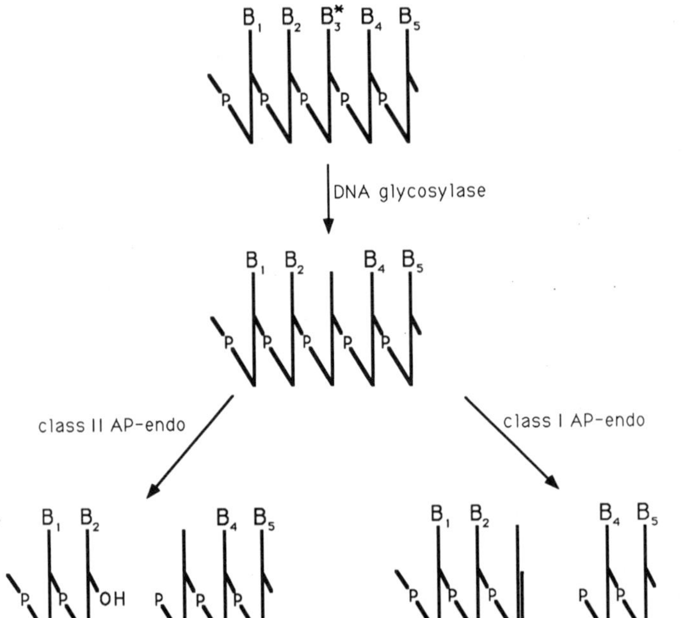

Scheme 2. Alternative base excision repair pathways. The asterisk (*) denotes a modified or a mispaired base. The experimental evidence available to date shows that the endonucleolytic incision at free AP-sites proceeds via the "left" pathway (AP-endo class II, DRPase). DNA hydrolysis 3′-from an AP-site (via the "right" pathway) proceeds most likely by a concerted action of a single enzyme, which possesses both, DNA glycosylase and DNA lyase activities.

(Scheme 2, right-hand pathway). This is, in turn, filled-in by DNA polymerase β and the remaining nick is sealed by a DNA ligase (Wiebauer and Jiricny, 1990).

4 Why is 5-methylcytosine a mutagenic hotspot?

The commonly accepted hypothesis of DNA modification is that sites of cytosine methylation are mutagenic hotspots. Indeed, experimental evidence, both direct and inferred, would appear to substantiate this. However, the mismatch correction data demonstrated quite conclusively that both the above organisms have evolved specialized repair pathways to overcome the potential mutagenic effect of the hydrolytic deamination of 5-methylcytosine. Can these apparently opposing sets of data be reconciled given our present knowledge? In order to provide an answer to this question, we must examine the correction efficiency of the G/T mispair in a broader context, namely that of the mutagenicity of the spontaneous base deamination process in general.

As already mentioned, all bases carrying exocyclic amino groups can (and do) spontaneously deaminate. When unrepaired, adenine deamination gives rise to A/T → G/C transitions and both cytosine and 5-methylcytosine deaminations lead to G/C → A/T transitions. Examination of spontaneous mutation spectra of many target genes shows that, indeed, these types of mutations are the ones most frequently found (Schaaper et al., 1986). In addition, in the DNA of higher eukaryotes, the methylated sequence, CpG, is significantly underrepresented, whereas the dimers TpG and CpA, which would be expected to arise after replication through a G/T mismatch, are found more frequently than anticipated (Sved and Bird, 1990). This evidence, albeit circumstantial, does strongly implicate unrepaired deamination of 5-methylcytosine as the culprit responsible for this phenomenon.

Coulondre and Miller (1978) showed that in the mutation spectrum of the *lac*I gene the most commonly encountered changes were three C → T transitions, which all mapped to *dcm*-modified sites. They proposed that the deamination of 5-methylcytosine must be responsible for the observed hypermutability of these sites in *E. coli*. In a follow-up of these initial experiments, Duncan and Miller (1980) were able to show that these hotspots disappeared in an *ung*-background, due to the rise in the mutability of the other cytosines, as their deamination product, uracil, could no longer be removed from the DNA in this uracil DNA glycosylase mutant. They proposed that as the product of 5-methylcytosine deamination, thymine, is not a substrate for this enzyme, it remains in the DNA and becomes fixed as a mutation during the following round of replication. We now know that, contrary to the prediction of Miller and colleagues, 5-methylcytosine deamination damage is actively repaired, both in *E. coli* and in higher eukaryotes. What underlying mechanism could then be behind the high mutagenicity of 5-methylcytosine?

In an *in vivo* transfection experiment, Zell and Fritz (1987) elegantly showed that a G/T mispair in a *dcm* site was corrected with nearly a 100% efficiency by the VSP repair system. However, their experiments also demonstrated that under certain circumstances, i.e. when the G/T mispair was in a hemi-methylated DNA heteroduplex, the VSP system had to compete for repair with the d(GATC)-dependent long-patch mismatch repair pathway. The net effect was an apparent reduction in efficiency of the G/T → G/C correction in cases where the d(GATC)-directed repair dictated the opposite strand bias (i.e. G/T → A/T). Although it could be argued that long-patch mismatch correction functions only on newly-replicated DNA, whereas the spontaneous deamination of 5-methylcytosine occurs in DNA where both strands are equally modified, we cannot exclude the possibility that a few percent of the *dcm*-associated G/T mispair could be corrected the "wrong way" by the *mut*HLS (long-patch) system, thus giving rise to the apparently high mutagenicity at this site. In an analogous situation, the *mut*HLS repair pathway can also address G/U mispairs. However, this effect becomes easily apparent only in an *ung*$^-$-background, as the U is normally excised with very high efficiency by the uracil DNA glycosylase (Shenoy et al., 1987). Thus, apparently, the frequency of C → T transitions, found at unmethylated and methylated cytosines in the mutational spectra of the *lac*I gene of *E. coli* (Schaaper et al., 1986; Duncan and Miller, 1980, resp.), could be considered to be dependent on three distinct events: (i) the rate of deamination of the two bases, (ii) the rate of repair by the two dedicated short-patch repair processes (G/U and G/T) and (iii) by the competition between these latter repair processes and the long-patch mismatch correction system. The direct experiments addressing this problem have, however, not been carried out as yet.

Our *in vivo* data from SV40 heteroduplex transfection experiments would testify to the existence of a similar competition also in mammalian cells. In the case of the G/T mismatch, 96% of the transfected SV40 heteroduplexes were corrected, with the remaining 4% undergoing replication in their unrepaired state. Of the 96%, 92% were corrected to G/C (Table 1). As the heteroduplex lacked specific strand-discrimination signals, the repair to A/T (4%) must have been carried out by the long-patch mismatch repair system that most probably utilized unspecific nicks, arising in the closed-circular DNA during the transfection process, to initiate the repair. The random nature of these nicks would imply that if 4% of the G/T heteroduplex were corrected to A/T, another 4% had to be corrected to G/C by the same repair pathway. To redress the balance therefore, we can reinterpret the results of the SV40 transfection experiments as follows: 4% unrepaired, 88% G/T → G/C specific, 8% long-patch repair. Thus, as in the case of *E. coli*, the mammalian cell would appear to have two competing mismatch repair pathways and the observed hyper-mutagenicity of 5-methylcytosine could therefore be attributed to the fact that, in some instances, the G/T mispair arising as the result of a deamination is corrected to A/T by the "wrong" pathway. And how about G/U in mammalian cells? Brown and Brown-Luedi (1989), using SV40 heteroduplexes identical to those in the G/T studies, examined the efficiency of repair of the G/U mispair *in vivo*. They found 100% repair to G/C, suggesting that, at least in this experimental system, the uracil DNA glycosylase-mediated repair is much more efficient than the G/T-specific process. Indeed, the latter pathway would appear to be slower also *in vitro*. In HeLa extracts, correction of a 90-mer G/U substrate to G/C could be accomplished within a few minutes, during which time only a small proportion (<10%) of the G/T duplex was processed (Fig. 2).

In summary therefore, the observed hypermutagenicity at sites of cytosine methylation would appear to be brought about by the high rate of deamination of this modified base, coupled with the fact that the repair of the G/T mispair arising through this spontaneous hydrolytic process, allows a few percent to go uncorrected or to be corrected by the "wrong" mismatch repair system. Does the cell really need a repair pathway that is inefficient? In this case the answer is certainly "Yes". In the absence of the specific G/T repair, each deamination event would give rise to a point mutation in 50% of the progeny if uncorrected. Obviously, better an inefficient repair than no repair at all.

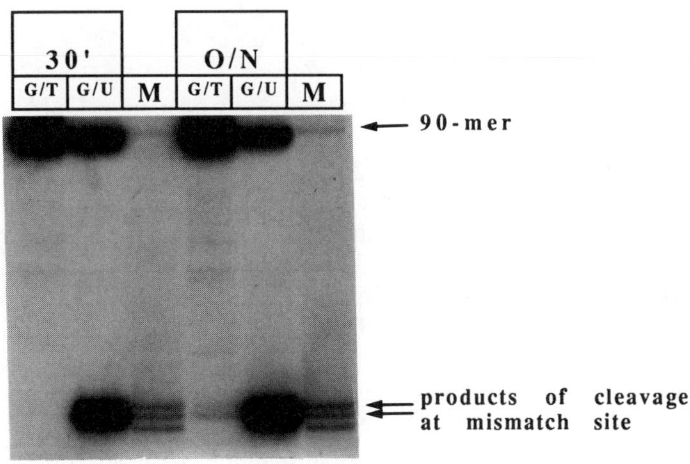

Figure 2. Differences in the efficiency of incision of the T and U strands in 90-mer G/T and G/U duplexes, respectively. The substrates were labelled at the 5′-end of the T or U-containing strand. M, marker (G/C duplex digested with *Sal*I, *Acc*I and *Hin*cII – for symbols see Figure 1).

5 5-Methylcytosine deamination and cancer

The possible repercussions of an inefficient correction of deamination-associated G/T mispairs are discussed in great detail in the chapter by Jones and colleagues. The number of C → T transitions that result in the activation of an oncogene or, alternatively, in the inactivation of a tumor suppressor gene are many and, no doubt, more remain yet to be discovered. However, the loss of a 5-methylcytosine residue through deamination may go beyond a simple point mutation.

Let us consider the following scenario. In a terminally-differentiated cell, a gene whose product is not needed for the maintenance of a particular phenotype has been transcriptionally silenced by DNA methylation. In many instances, such an inactivation involves the modification of most or all CpGs in a long stretch of DNA, not only in the body of the gene, but also several kilobases of the upstream and downstream flanking sequences. In this case, the loss of a single CpG would most likely have no effect whatsoever. However, some genes would appear to have been inactivated by the methylation of only one or a few CpGs (Hergersberg, 1991 for review). In such an instance, the loss of a critical CpG may lead to an erroneous reactivation of this gene and thus to the expression of a protein that is not tolerated in this differentiated phenotype. For example, were this protein to be a growth hormone, its expression may result in an uncontrolled cell division and thus, conceivably, may lead to cancer. (See also chapter on DNA methylation and cancer, pp. 487–509 of this book.)

6 Conclusions

It would appear that, despite the existence of an active DNA repair pathway that corrects G/T mismatches arising through the spontaneous hydrolytic deamination of 5-methylcytosine back to G/C, the sites of cytosine methylation remain significantly mutagenic. As discussed briefly above, this mutagenic process could have serious repercussions for the mammalian cell. Unfortunately, in the absence of a cell line mutant deficient in this pathway, the real biological importance of the G/T-specific DNA repair cannot be assessed. For this reason, we have set out to purify the G/T mismatch-specific thymine DNA glycosylase. The activity has an apparent molecular mass of 30–50 kDa (Neddermann and Jiricny, unpublished), and is distinct from the G/T mismatch binding protein characterized by us previously (Jiricny et al., 1988; Hughes and Jiricny, 1992). We have to await the results of the purification and the sequencing of this protein, in order to be able to attempt to inactivate the gene encoding this activity by gene targetting. Only then will it be possible to see how serious a threat to the organism the deamination of 5-methylcytosine really poses.

Au, K. G., Clark, S., Miller, J. H., and Modrich, P. (1989) *Escherichia coli mut*Y gene encodes an adenine glycosylase active on G.A mispairs. Proc. Natl. Acad. Sci. USA *86*, 8877–8881.

Bhagwat, S. A., Sohail, A., and Lieb, M. (1988) A new gene involved in mismatch correction in *Escherichia coli*. Gene *74*, 155–156.

Brooks, P., Dohet, C., Almouzni, G., Méchali, M., and Radman, M. (1989) Mismatch repair involving localized DNA synthesis in extracts of *Xenopus* eggs. Proc. Natl. Acad. Sci. USA *86*, 4425–4429.

Brown, T. C., and Jiricny, J. (1987) A specific mismatch repair event protects mammalian cells from loss of 5-methylcytosine. Cell *50*, 945–950.

Brown, T. C., and Jiricny, J. (1988) Different base/base mispairs are corrected with different efficiencies and specificities in monkey kidney cells. Cell *54*, 705–711.

Brown, T. C., and Brown-Luedi, M. L. (1989) G/U lesions are efficiently corrected to G/C in SV40 DNA. Mut. Res. *227*, 233–236.

Brown, T. C., Zbinden, I., Cerutti, P. A., and Jiricny, J. (1989) Modified SV40 for analysis of mismatch repair in simian and human cells. Mut. Res. *220*, 115–123.

Claverys, J.-P., and Lacks, S. A. (1986) Heteroduplex deoxyribonucleic acid base mismatch repair in bacteria. Microbiol. Rev. *50*, 133–165.

Coulondre, C., and Miller, J. H. (1977) Genetic studies of the lac repressor. IV. Mutagenic specificity in the lacI gene of *Escherichia coli*. J. Mol. Biol. *117*, 577–606.

Duncan, B. K., and Miller, J. H. (1980) Mutagenic deamination of cytosine residues in DNA. Nature *287*, 560–563.

Ehrlich, M., Norris, K. F., Wang, R. Y.-H., Kuo, K. C., and Gehrke, C. W. (1986) DNA cytosine methylation and heat-induced deamination. Biosci. Rep. *6*, 387–393.

Hare, J. T., and Taylor, J. H. (1985) One role for DNA methylation in vertebrate cells is strand discrimination in mismatch repair. Proc. Natl. Acad. Sci. USA *82*, 7350–7354.

Hennecke, F., Kolmar, H., Brundl, K., and Fritz, H.-J. (1991) The usr gene product of *E. coli* K-12 is a strand- and sequence-specific DNA mismatch endonuclease. Nature *353*, 776–778.

Hergersberg, M. (1991) Biological aspects of cytosine methylation in eukaryotic cells. Experientia *47*, 1171–1185.

Holmes, J. Jr., Clark, S., and Modrich, P. (1990) Strand-specific mismatch correction in nuclear extracts of human and *Drosophila melanogaster* cell line. Proc. Natl. Acad. Sci. USA *87*, 5837–5841.

Hughes, M. J., and Jiricny, J. (1992) The purification of a human mismatch binding protein and identification of its associated ATPase and helicase activities. J. Biol. Chem., in press.

Jiricny, J., Hughes, M., Corman, N., and Rudkin, B. B. (1988) A human 200-kDa protein binds selectively to DNA fragments containing GT mismatches. Proc. Natl. Acad. Sci. USA *85*, 8860–8864.

Jiricny, J. (1991) Mismatch repair in eukaryotic systems, in: Nucleic Acids and Molecular Biology, Vol. 5, pp. 72–83. Eds. F. Eckstein and D. M. J. Lilley. Springer-Verlag, Berlin, Heidelberg.

Jones, M., Wagner, R., and Radman, M. (1987) Mismatch repair of deaminated 5-methylcytosine. J. Mol. Biol. *194*, 155–159.

Lahue, R. S., Au, K. G., and Modrich, P. (1989) DNA mismatch correction in a defined system. Science *245*, 160–164.

Lewis, J., and Bird, A. (1991) DNA methylation and chromatin structure. FEBS Letts *285*, 155–159.

Lieb, M. (1983) Specific mismatch correction in bacteriophage lambda crosses by very short patch repair. Mol. Gen. Genet. *191*, 118–125.

Lieb, M. (1985) Recombination in the lambda repressor gene: evidence that very short patch (VSP) mismatch correction restores a specific sequence. Mol. Gen. Genet. *199*, 465–470.

Lieb, M. (1987) Bacterial genes *mut*L, *mut*S, and *dcm* participate in repair of mismatches at 5-methylcytosine sites. J. Bacteriol. *169*, 5241–5246.

Lindahl, T. (1982) DNA repair enzymes. Annu. Rev. Biochem. *5*, 61–87.

Marinus, M. G. (1984) Methylation of prokaryotic DNA, in: DNA Methylation, Biochemistry and Biological Significance, pp. 81–109. Eds A. Razin, H. Cedar and A. D. Riggs. Springer-Verlag, New York.
Meselson, M. (1988) Methyl-directed repair of DNA mismatches, in: The Recombination of Genetic Material, pp. 91–113. Ed. K. B. Low. Academic Press Inc., San Diego.
Modrich, P. (1987) DNA mismatch correction. Annu. Rev. Biochem. *56*, 435–466.
Modrich, P. (1989) Methyl-directed DNA mismatch correction. J. Biol. Chem. *264*, 6597–6600.
Radman, M., and Wagner, R. (1986) Mismatch repair in *Escherichia coli*. Ann. Rev. Genet. *20*, 523–538.
Sancar, A., and Sancar, G. B. (1988) DNA repair enzymes. Annu. Rev. Biochem. *57*, 29–67.
Shenoy, S., Ehrlich, K. C., and Ehrlich, M. (1987) Repair of thymine·guanine and uracil·guanine mismatched base-pairs in bacteriophage M13mp18 DNA heteroduplexes. J. Mol. Biol. *197*, 617–626.
Schaaper, R. M., Danforth, B. N., and Glickman, B. W. (1986) Mechanisms of spontaneous mutagenesis: an analysis of the spectrum of spontaneous mutation in the *E. coli lacI* gene. J. Mol. Biol. *189*, 273–284.
Sohail, A., Lieb, M., Dar, M., and Bhagwat, A. S. (1990) A gene required for very short patch repair in *Escherichia coli* is adjacent to the DNA cytosine methylase gene. J. Bact. *172*, 4214–4221.
Sved, J., and Bird, A. (1990) The expected equilibrium of the CpG dinucleotide in vertebrate genomes under a mutation model. Proc. Natl. Acad. Sci. USA *87*, 4692–4696.
Wagner, R., and Meselson, M. (1976) Repair tracts in mismatched DNA heteroduplexes. Proc. Natl. Acad. Sci. USA *73*, 4135–4139.
Wiebauer, K., and Jiricny, J. (1989) *In vitro* correction of G/T mispairs to G/C pairs in nuclear extracts from human cells. Nature *339*, 234–236.
Wiebauer, K., and Jiricny, J. (1990) Mismatch-specific thymine DNA glycosylase and DNA polymerase *β* mediate the correction of G/T mispairs in nuclear extracts from human cells. Proc. Natl. Acad. Sci. USA *87*, 5842–5845.
Zell, R., and Fritz, H.-J. (1987) DNA mismatch-repair in *Escherichia coli* counteracting the hydrolytic deamination of 5-methyl-cytosine residues. EMBO J. *6*, 1809–1815.

DNA Methylation: Molecular Biology and Biological Significance
ed. by J. P. Jost & H. P. Saluz

Gene methylation patterns and expression

Agnes Yeivin and Aharon Razin

Department of Cellular Biochemistry, Hebrew University, Hadassah Medical School, Jerusalem, Israel

1 Introduction

Several lines of evidence strongly indicate that methylation of cytosine residues within CpG dinucleotides are implicated in the regulation of gene expression in eukaryotes. Transfection experiments have shown that methylation could inhibit gene transcription while inhibition of DNA methylase by the cytosine analog 5-azacytidine could induce gene reactivation. Furthermore, analysis of the methylation pattern of genes in different tissues has revealed in most cases the existence of a direct correlation between hypomethylation and gene activity (Razin and Cedar, 1991).

In a review published in 1984, Yisraeli and Szyf presented the methylation patterns of 30 eukaryotic genes, on the basis of data published up to the end of 1983. Five paradigmatic groups which reflect the level of correlation between methylation patterns and gene expression were defined: the first group includes genes which are fully unmethylated when expressed and fully methylated in tissues that do not express the gene. The second group includes genes in which some tissue-specific hypomethylation occurs exclusively in active tissues, while other hypomethylation events are site-specific and occur in non-expressing tissues as well. The third group represents primarily housekeeping genes which are invariably undermethylated in all tissues including sperm. The fourth group contains genes that remain fully methylated in all tissues including the tissue where the gene is expressed. All genes in which hypomethylation is observed in both active and non-active tissues, with no obvious correlation with the state of activity of the gene, were compiled in the fifth group. About two thirds of the genes for which at least a partial methylation pattern was established exhibited a certain degree of correlation between hypomethylation and gene activity.

Methylation patterns of additional 45 genes that were studied and reported in the past 7 years are brought up and discussed here (see Table 2).

Genes, such as *Xenopus laevis* tRNA (Talwar et al., 1984), rat α 1(I) collagen (Waye et al., 1989), rat gamma-glutamyl transpeptidase (Baik et al., 1991) and human MIC2 (Goodfellow et al., 1982) will not be discussed here since the respective reports were lacking sufficient information.

2 Methods of methylation analysis and mode of presentation

Most methylation patterns presented here were constructed using data obtained by restriction enzyme analysis. This analysis consists of digestion of genomic DNA with methylation-sensitive enzymes electrophoresis of the produced restriction fragments, blotting and hybridization to specific probes. The major limitation of this method resides in the limited number of sites that can be analyzed by the available restriction enzymes. This analysis reveals only 10–15% of all CpG sites. Some genes have been analyzed by genomic sequencing by which the methylation state of all CpG sites can be determined. However the original sequencing method (Church and Gilbert, 1984) is very laborious and even the improved methods (Saluz and Jost, 1986) including the ligation-mediated genomic sequencing method (Pfeifer et al., 1989) have been used so far by only a limited number of laboratories. Although genomic sequencing provides a comprehensive picture by-and-large, the restriction enzyme analysis proved to be fairly representative in those cases where a reasonable number of sites are present in the region of interest. Another limitation of the available methods stems from the fact that relatively large quantities of DNA are required for the analysis (in the microgram range). Therefore, these methods were not sensitive enough when nanogram amounts of DNA are available, like in the case of mammalian oocytes or early embryonic tissues. A more sensitive method based on the polymerase chain reaction (PCR) has been recently developed to solve the problem of small amounts of DNA (Singer-Sam et al., 1990).

This assay is based on the use of methylation-sensitive restriction enzymes to digest minute amounts of DNA prior to amplification of the DNA by the polymerase chain reaction. The method can be used to determine the methylation state of individual CpG sites in DNA extracted from a few cells.

A very promising method that should improve the methylation analysis by genomic sequencing has been reported recently (Frommer et al., 1992). This method is based on PCR using the dideoxy sequencing procedure. This method has the advantage over the previous genomic sequencing methods of displaying 5 methylcytosine residues as positive bands rather than blank space. (For details see chapter written by Saluz and Jost in this book.)

The methylation patterns presented here are displayed as follows: Each gene is presented by its structural map, with shaded boxes representing exons. The level of methylation of each individual site analyzed is represented by the length of a vertical line descending from the restriction site. A full length line indicates 100% hypomethylation; the absence of a line indicates that the site is fully methylated; vertical lines of intermediate length represent partial methylation (25%, 50% or 75% respectively). The various methylation sites are symbolized as follows: HpaII ●; HhaI □; AvaI ▲; HaeII ◆; BstUI △; SmaI ■; XhoI ■; NciI ●; AhaII ▼; CpG site analyzed by genomic sequencing +; CpG island ▭. When a number of adjacent sites could not be distinguished they were designated by ○. Sites that were not analyzed are denoted by ×.

Table 1. CpG islands in tissue-specific and housekeeping genes. Xa and Xi are the X chromosome in its active and inactive form respectively

Species	Gene	Figure	Island location	State of methylation
Tissue-specific genes				
Human	Muscle carbonic anhydrase	11	5′ end	unmethylated
	ApoAI	1	3′ end	unmethylated
	ApoAIV	1	3′ end	methylated in non-expressing tissues
	Tc receptor β chain	14	Jβ2 region	unmethylated
	Myeloperoxydase	12	Exons 3–6	methylated in non-myeloid cells unmethylated in promyelocytes and in terminally differentiated myeloid cells
Mouse	MyoD1	27	5′ end	unmethylated in tissues (methylated in cell lines)
	Thy 1	32	5′ end	unmethylated
	Protamine 1	31	5′ end	unmethylated
	Tc receptor β chain	33	J2 region	unmethylated, partially methylated in macrophages (non expressing)
	Int1	25	entire sequence	unmethylated
	Hox2.1	23	entire sequence	unmethylated
	ApoAI	19	3′ end	unmethylated
Chicken	α2(I) collagen	Yisraeli and Szyf (1984)	5′ end	unmethylated
Housekeeping genes				
Human	HPRT	7	5′ end	unmethylated on Xa;
	PGK	13	5′ end	methylated on Xi
	G6PD	5	5′ and 3′ (?) ends	
Mouse	HPRT	24	5′ end	unmethylated on Xa, methylated on Xi
	PGK 1	29	5′ end	3′-HpaII site methylated on Xi
	DHFR	Yisraeli and Szyf (1984)	5′ end	unmethylated
	HMGCR	22	5′ end	unmethylated
Chinese hamster	Asparagine synthetase	45	5′ end	methylated in non-expressing cell lines
	APRT	Yisraeli and Szyf (1984)	5′ end	unmethylated
Chicken	Thymidine kinase	43	5′ end	unmethylated

Table 2. List of the genes for which methylation patterns exist

Species	Gene	Figure/Reference
Human	Apolipoprotein AI-CIII-AIV (ApoaI-AII-AIV)	1
	Apolipoprotein B (5′ end) (ApoB)	2
	Chorionic somatomammotropin	Yisraeli and Szyf, 1984
	Estrogen receptor (5′ end)	3
	Factor IX	4
	Globin gene cluster	Yisraeli and Szyf, 1984
	Glucose 6-phosphate dehydrogenase (G6PD)	5
	Growth hormone	Yisraeli and Szyf, 1984
	HLA-Drα	6
	Hypoxanthine phosphoribosyl transferase (5′ end) (HPRT)	7
	Immunoglobulin light Cκ genes	8
	Immunoglobulin Cμ genes	9
	Interphotoreceptor retinoid binding protein (5′ end) (IRBP)	10
	Muscle carbonic anhydrase	11
	Myeloperoxydase	12
	Phosphoglycerate kinase (5′ end) (Pgk)	13
	T cell receptor β chain DJC-1 and DJC-2 regions	14
	Thyroglobulin	15
	Tumor necrosis factor α (TNFα)	16
	Tumor necrosis factor β (TNF β)	17
Mouse	Albumin (5′ half)	18
	Apolipoprotein AI (ApoAI)	19
	Dihydrofolate reductose (DHFR)	Yisraeli and Szyf, 1984
	β Fcγ-receptor (Fcγ RII)	20
	α fetoprotein	21
	Hydroxymethyl glutaryl CoA reductase (HMGCR) (5′ end)	22
	Hox-2.1	23
	Hypoxanthine phosphoribosyl transferase (HPRT)	24
	*Immunoglobulin heavy chain, Cγ, δ, μ	Yisraeli and Szyf, 1984
	*Immunoglobulin κ heavy chain region	Yisraeli and Szyf, 1984
	Int 1	25
	c-Myc	26

Table 2. (continued)

Species	Gene	Figure/Reference
	MyoD1	27
	Ornithine carbamoyl transferase (5′ end)	28
	Phosphoglycerate kinase 1 (Pgk1) (5′ end)	29
	Phosphoglycerate kinase 2 (Pgk2)	30
	Protamine 1 (5′ end)	31
	Thy-1.1 (5′ end)	32
	T cell receptor β chain	33
Rat	*Albumin	Yisraeli and Szyf, 1984
	α actin	Yisraeli and Szyf, 1984
	α casein	Yisraeli and Szyf, 1984
	α2(I) collagen (5′ end)	34
	*α fetoprotein	Yisraeli and Szyf, 1984
	Growth hormone	35
	Insulin I	Yisraeli and Szyf, 1984
	Insulin II	Yisraeli and Szyf, 1984
	Myosin light chain 2 (MLC2)	37
	*Phosphoenolpyruvate carboxykinase (PEPCK)	38
	Prolactin	36
	S14	39
	Seminal vesicle secretory protein IV (SVSIV)	40
Chicken	α-2(I) collagen	Yisraeli and Szyf, 1984
	*δ1 crystallin	41
	δ-crystallin II	Yisraeli and Szyf, 1984
	α globin cluster	Yisraeli and Szyf, 1984
	β globin cluster	Yisraeli and Szyf, 1984
	β-like globin cluster	Yisraeli and Szyf, 1984
	Lysozyme (5′ end)	42
	Ovalbumin	Yisraeli and Szyf, 1984
	Thymidine kinase (Tk)	43
	*Vitellogenin	44
Chinese hamster	Adenine phosphoribosyl transferase (APRT)	Yisraeli and Szyf, 1984
	Asparagine synthetase	45

Rabbit	β-like globin cluster	Yisraeli and Szyf, 1984
Xenopus	Albumin	Yisraeli and Szyf, 1984
	Globin	Yisraeli and Szyf, 1984
	rRNA	Yisraeli and Szyf, 1984
	Vitellogenin A-1	Yisraeli and Szyf, 1984
	Vitellogenin A-2	Yisraeli and Szyf, 1934

*Genes previously presented (Yisraeli and Szyf, 1984) and further analyzed in later studies. The methylation of CpG sequences of mouse μ chain genes has been assayed in non-expressing tissues (liver, embryo) and in various B lymphoma and IgM secreting cell lines. Undermethylation of specific sites was found to correlate with gene expression (Blackman et al., 1985). The methylation pattern of the murine κ immunoglobulin locus has been analyzed by Nelson et al. (1984) in pre B cells and by Kelley et al. (1988) in mature B cells. Rat albumin and α fetoprotein genes have been analyzed by Tratner et al. (1987) in hepatoma cell lines that show various patterns of expression.

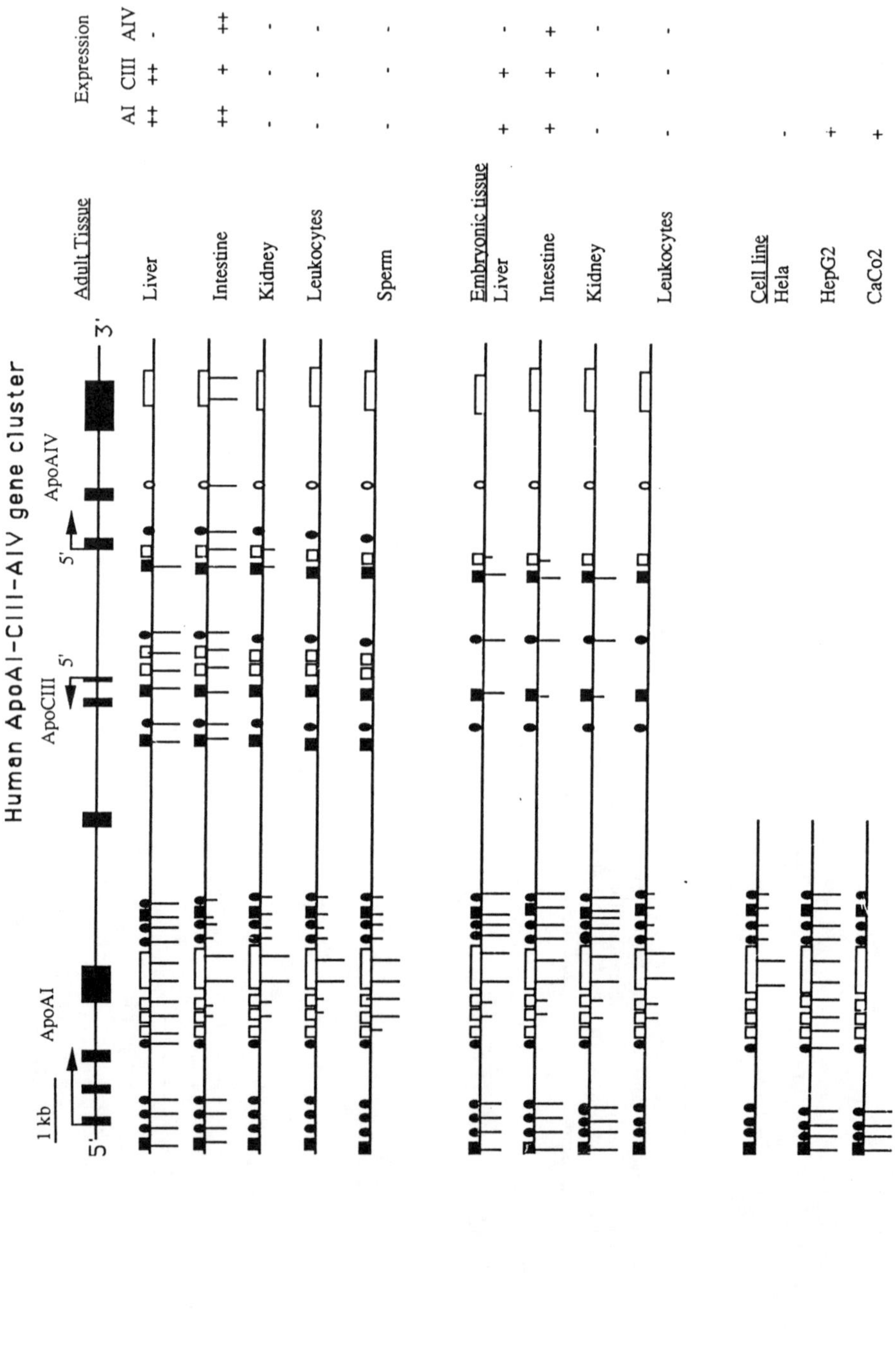
Human ApoAI-CIII-AIV gene cluster
1 kb
5'
3'
ApoAI
ApoCIII
ApoAIV
Adult Tissue
Liver
Intestine
Kidney
Leukocytes
Sperm
Embryonic tissue
Liver
Intestine
Kidney
Leukocytes
Cell line
Hela
HepG2
CaCo2
Expression
AI CIII AIV

Figure 1. *Human Apolipoprotein AI-CIII-AIV gene cluster*. Analysis of the methylation patterns was achieved by double digestion and hybridization to specific probes (Shemer et al., 1990, 1991). Designation of restriction sites is as detailed in the "Methods of methylation analysis and mode of presentation" section. Expression data were taken from Shemer et al. (1990), Breslow (1988) and Karathanasis et al. (1986) for ApoAI, CIII and AIV, respectively. HepG2 – Hepatoma cell line; CaCo2 – Colon Carcinoma cell line.

Human Apolipoprotein B gene – (5′ end)

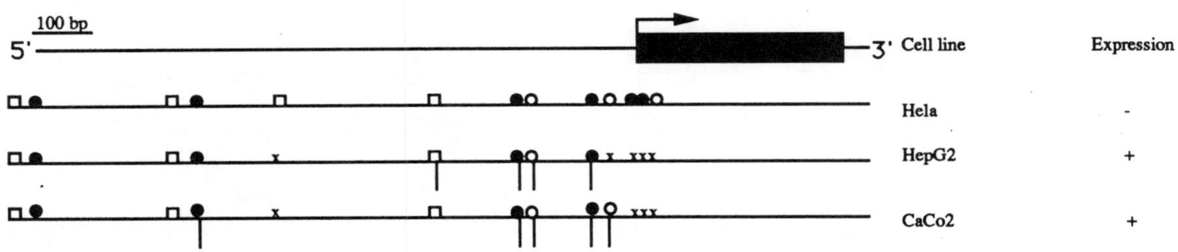

Figure 2. *Human Apolipoprotein B gene-5′ end.* Analysis of promoter and 5′ flanking region was performed by double digestion and hybridization to specific probes (Levy-Wilson and Fortier, 1989). The tissue-specific expression of the gene was demonstrated at the RNA and protein level by Demmer et al. (1986). HepG2 – Hepatoma cell line; CaCo2 – Colon carcinoma cell line.

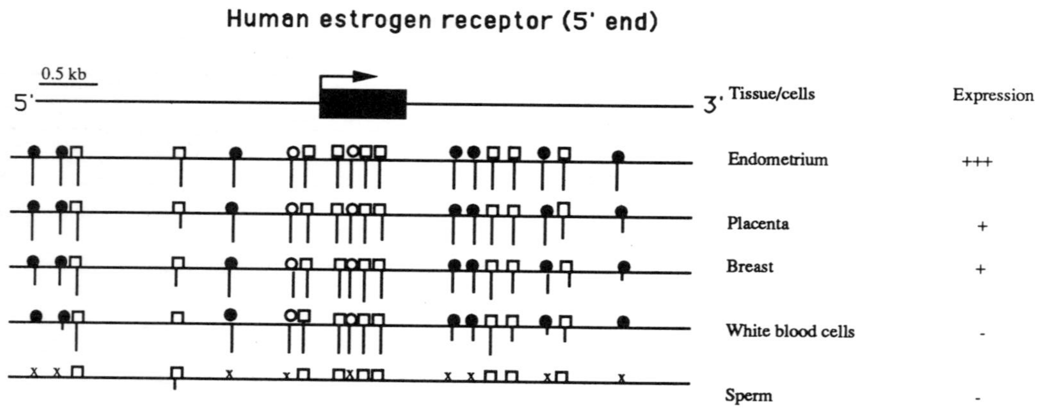

Figure 3. *Human estrogen receptor 5′ end.* The analysis has been performed by double digestion, Southern blotting and hybridization to regional probes. The sites located in the first exon which remain unmethylated in all tissues except sperm may be part of a CpG island (Piva et al., 1989). Relative expression data based on Piva et al., 1989.

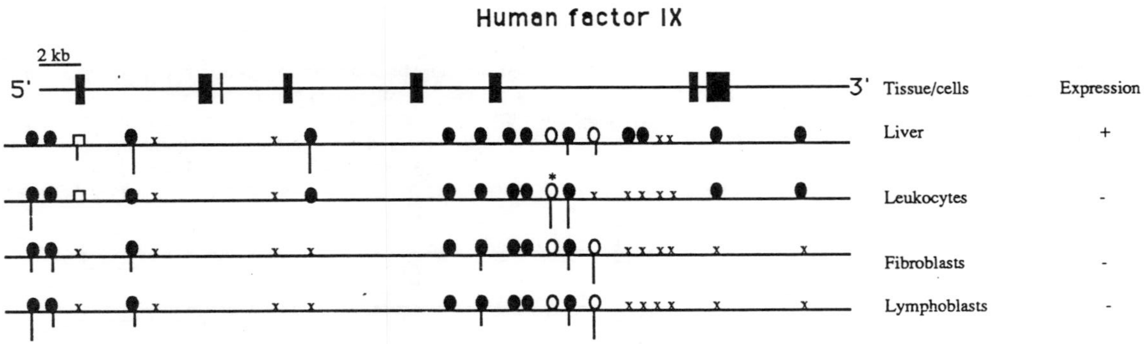

Figure 4. *Human factor IX gene*. A tissue-specific X-linked gene which is inactive on the inactive X chromosome. The methylation pattern was analyzed in males and females and no difference was found in the methylation pattern of the gene on the active and inactive chromosomes (except for HpaII sites (2 sites) (*) which are methylated on inactive X and unmethylated on active X). The methylation level was estimated by visual examination of the raw data (Ruta-Cullen et al., 1986). The gene was shown to be expressed exclusively in the liver (Wion et al., 1985). No obvious correlation with methylation was observed.

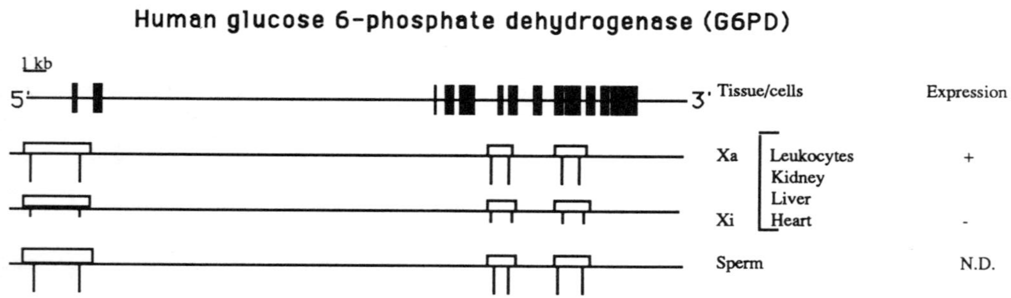

Figure 5. *Human glucose 6-phosphate dehydrogenase (G6PD).* The 5′ end CpG island includes 18 HhaI sites and 20 HpaII sites (Toniolo et al., 1988). The 3′ CpG islands include about 14 HhaI and 19 HpaII sites (Wolf et al., 1984; Toniolo et al., 1984). Beside these regions other HpaII sites were mapped but could not be analyzed. The methylation state of the 5′ CpG rich island on the inactive X (Xi) is heterogenous (variable methylation from chromosome to chromosome). The analysis was performed by double digestion and hybridization to specific probes (Toniolo et al., 1988). The same pattern of methylation was assessed for the 3′ CpG clusters (unmethylated on the active X (Xa) and extensively methylated but heterogenously on Xi) (Wolf et al., 1984; Toniolo et al., 1984). In 1988, Toniolo et al. reported that the two 3′ islands were not associated with the gene but located at the 5′ ends of two unrelated X-linked genes (P_3 and GdX). Battistuzzi et al. (1985) investigated the level of gene activity and the methylation pattern of the 3′ end of the gene in eight different fetal and adult tissues. They found a subset of HpaII sites located downstream of the 3′ CpG island identified in previous studies, which exhibit positive correlation between degree of methylation and level of gene expression.

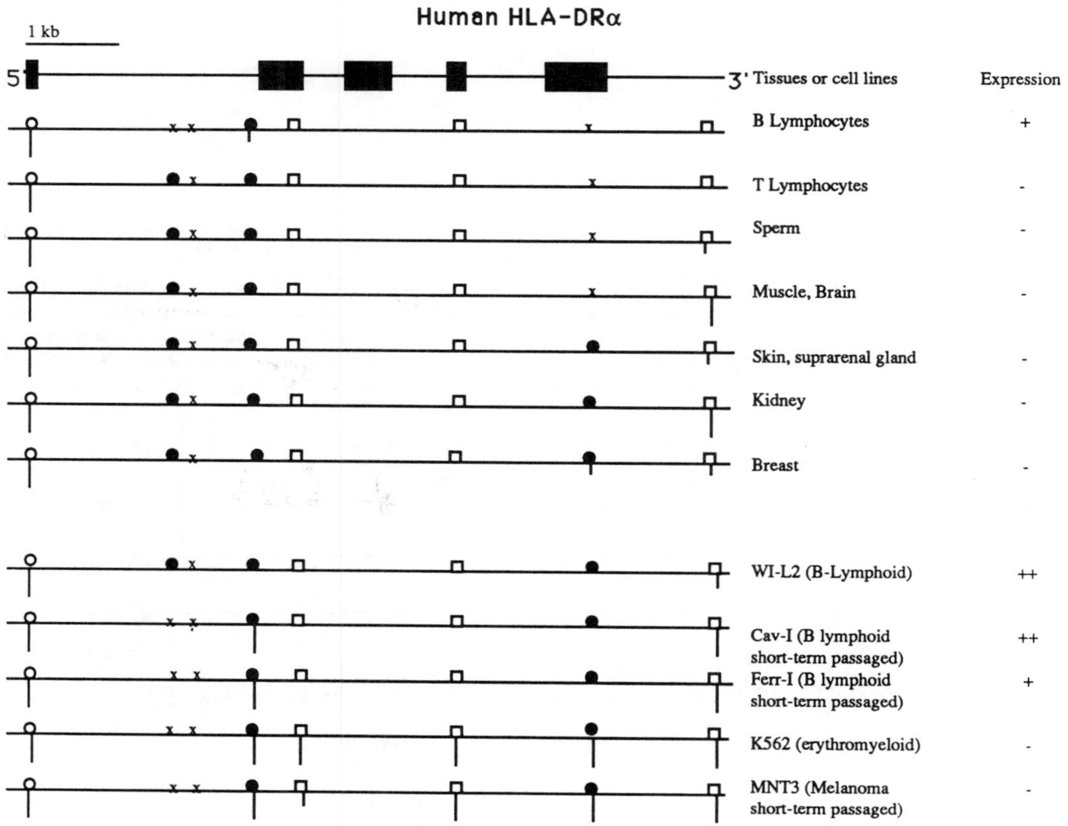

Figure 6. *Human HLA-DRα*. Analysis of methylation was performed by double digestion and hybridization to specific probes. The level of expression was determined on the basis of mRNA assays (Northern blot), and protein levels (indirect immunofluorescence and ELISA) (Gambari et al., 1987 in human cell lines; Barbieri et al., 1990, in human tissues). The gene can be transcribed even though extensively methylated; unmethylation of CpG sites near the promoter may be necessary but not sufficient for expression of HLA Drα.

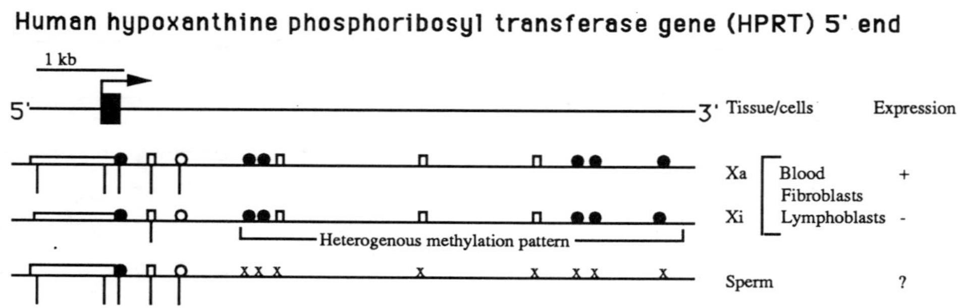

Figure 7. *Human hypoxanthine phosphoribosyltransferase gene (HPRT) 5′ end.* Analysis of methylation was performed by double digestion and hybridization to specific probes. The methylation pattern of the inactive allele is heterogenous. The analyzed sites, 3′ to CpG island, are frequently unmethylated while the island is methylated on the inactive X chromosome. There are at least 12 HpaII and 15 HhaI sites in the 1 kb preceding the first intron. The different pattern of methylation of the inactive allele is at least in part, due to changes in *de novo* DNA methylation since clonal populations of lymphoblasts and fibroblasts derived from single cells presented a similar heterogeneity (Wolf et al., 1984a; Yen et al., 1984).

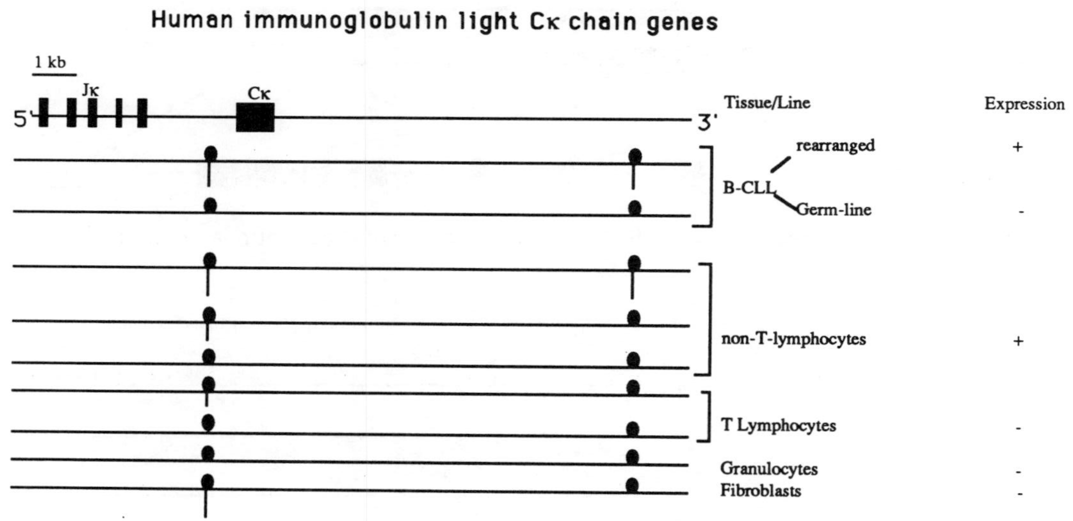

Figure 8. *Human immunoglobulin light Cκ chain genes*. Analysis of DNA from four different Cκ chain-secreting B-Chronic Lymphocytic Leukemia Cells (CLL); two DNA samples from non T-lymphocytes, T lymphocytes and granulocytes and four samples of fibrolast DNA were analyzed by HpaII digestion and Southern blotting (Bianchi et al., 1988). In non T-lymphocytes, methylation of part of the Cκ genes can be explained by the fact that only 60% of B cells express light κ chain genes while 40% produce light λ chain genes.

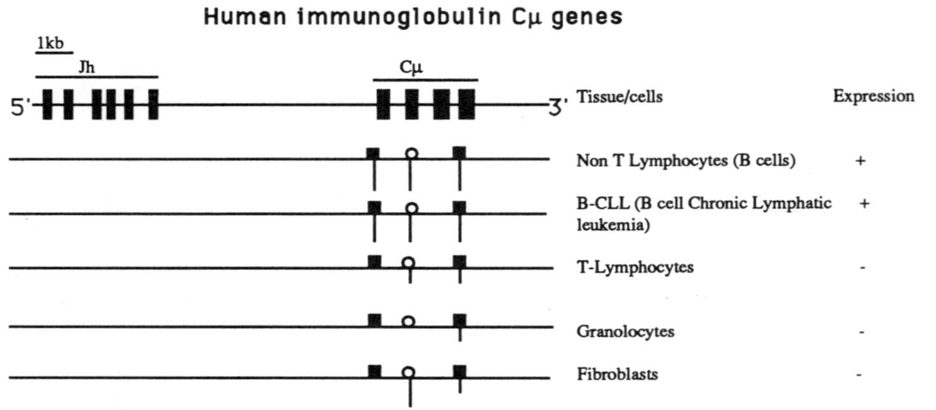

Figure 9. *Human immunoglobulin Cμ genes.* In B-CLL and non T lymphocytes the allele analyzed is the rearranged one (productive allele) (Bianchi et al., 1987).

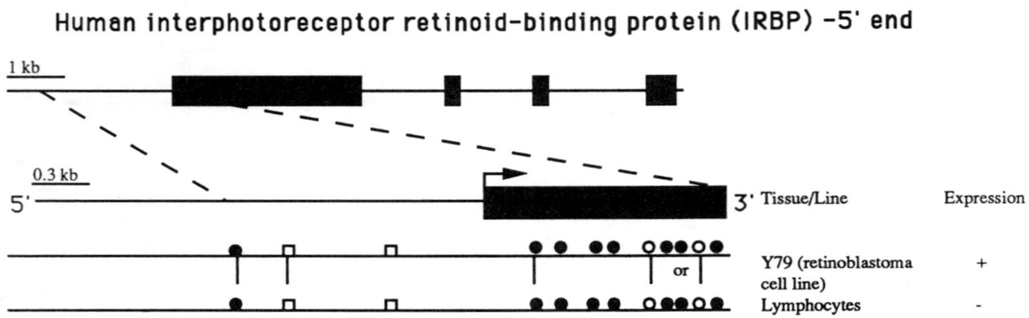

Figure 10. *Human interphotoreceptor retinoid-binding protein (IRBP)-5′ end.* Promoter region was analyzed by double digestion of DNA and hybridization to specific probes (Albini et al., 1990). IRBP expression data are from Van Veen et al. (1986).

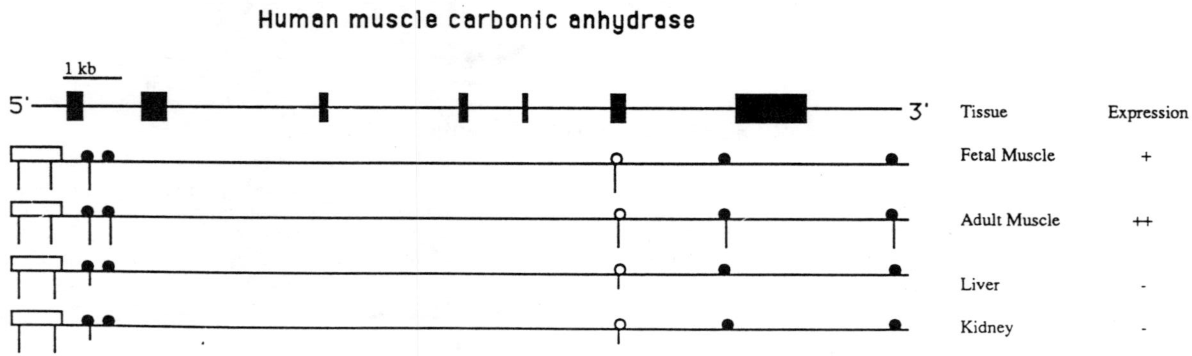

Figure 11. *Human muscle carbonic anhydrase*. Analysis of methylation was performed by double digestion and hybridization to specific probes (Lloyd et al., 1987). Expression of the CAIII gene in fetal and adult tissues has been investigated by Jeffery et al. (1980).

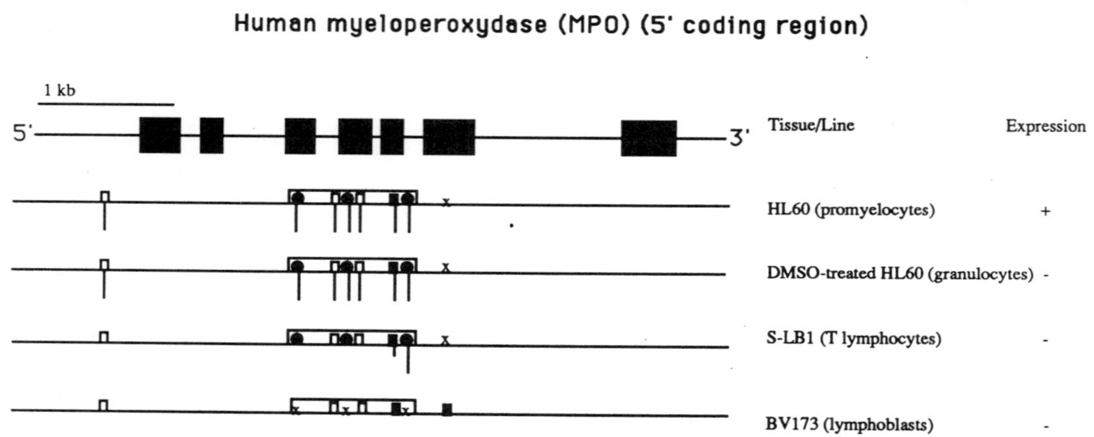

Figure 12. *Human myeloperoxydase (MPO) (5′ coding region).* The methylation pattern was analyzed by single or double digestion of DNA and Southern blot analysis. Levels of expression was determined by Northern blot analysis. Additional cell lines representing the spectrum of hematopoietic cell differentiation stages were analyzed for the methylation pattern of the 5′ region of MPO. Myeloid cells arrested at stages earlier than promyelocytes showed an intermediate methylation status (Lubbert et al., 1991).

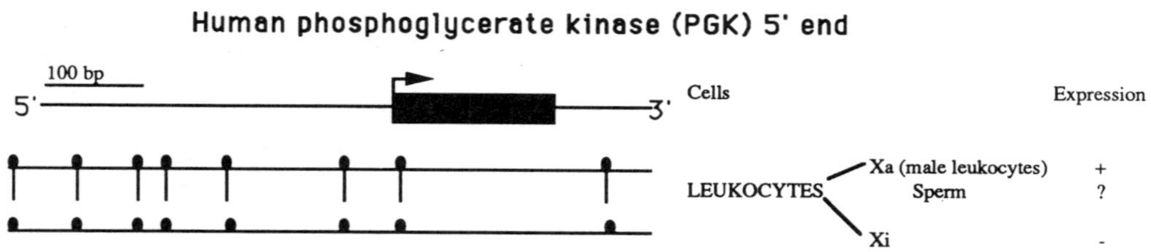

Figure 13. *Human phosphoglycerate kinase (PGK) 5′ end.* Analysis of CpG island was by double digestion and use of a 6% acrylamide gel in order to detect small fragments. An analysis of methylation status of HpaII sites in the body of the gene has also been performed (Keith et al., 1986). The analysis, whether performed by PCR-aided genomic sequencing (Pfeifer et al., 1990) or by digestion with methylation sensitive restriction enzymes (Hansen et al., 1988; Hansen et al., 1990) has shown a direct correlation between PGK1 activity and demethylation of 5′ CpG sites. Xa – active X chromosome; Xi – inactive X chromosome.

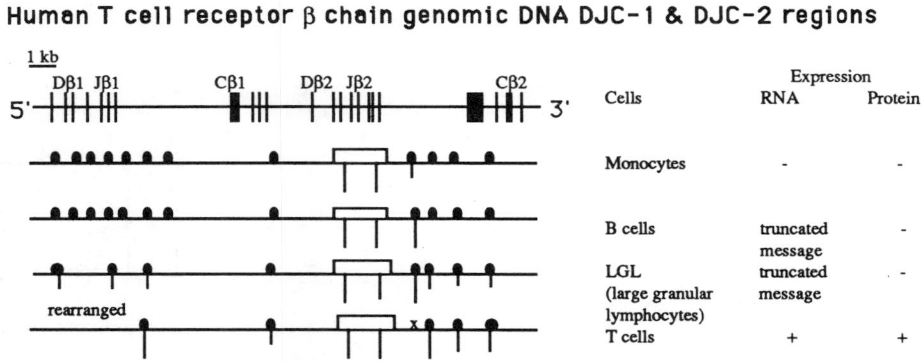

Figure 14. *Human T cell receptor β chain genomic DNA DJC-1 & DJC-2 regions.* Analysis of methylation was by double digestion and hybridization to specific probes of DNA extracted from leukocytes isolated from 8 healthy donors (Sakamoto and Young, 1988; Sakamoto et al., 1989a). On the basis of the methylation pattern, the authors hypothesize that T cells and LGL share an initial stem cell and differentiation pathway up to a certain branch point from which T cells-TcRβ undergoes further demethylation while LGL-TcRβ does not. Expression data were obtained from Young et al. (1986).

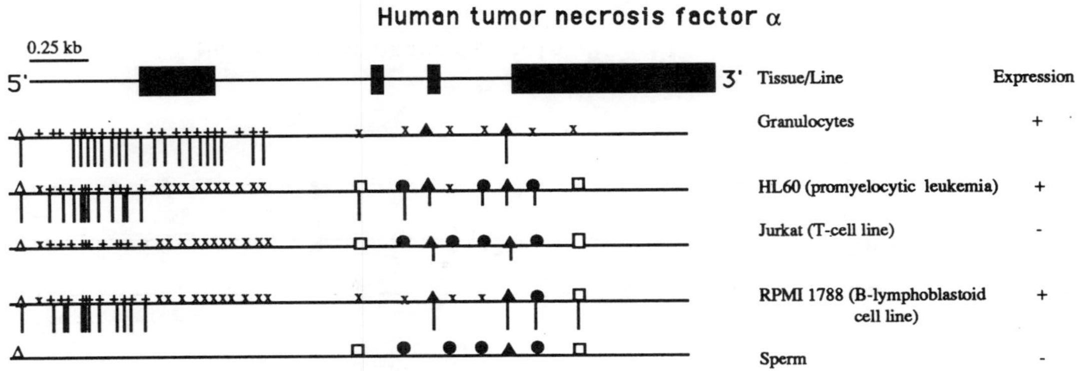

Figure 15. *Human thyroglobulin.* Analysis of methylation was by digestion with methylation-sensitive enzymes and hybridization to specific probes (Libert et al., 1986). When placenta DNA was analyzed, the 4 HpaII sites were found to be unmethylated although this tissue is not known to express the gene. It is known (Rossant et al., 1986) that numerous genes are hypomethylated in extra-embryonic tissues, irrespective of their expression.

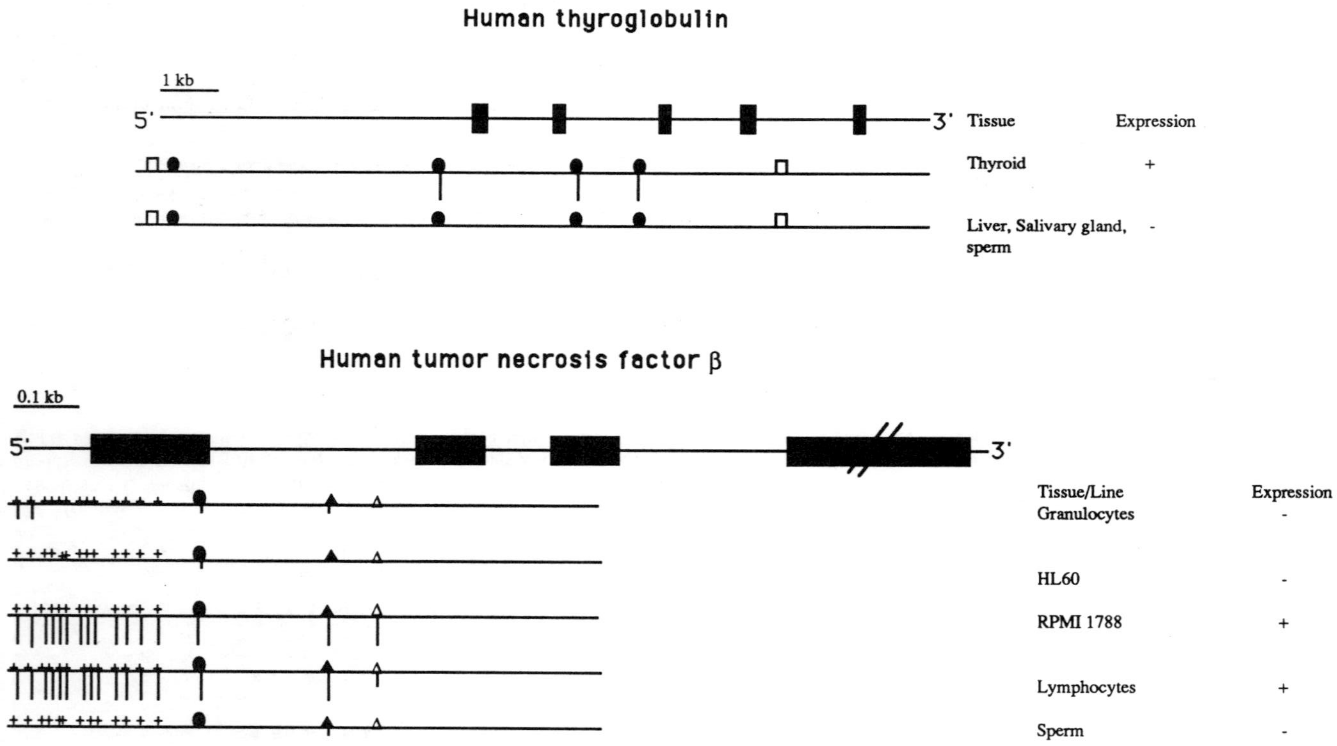

Figures 16–17. *Human tumor necrosis factors α and β*. Methylation analysis at the promoter region was by genomic sequencing; analysis at the body of the gene was by methyl-sensitive restriction enzymes. The RNA levels of TNFα and TNFβ were measured by Northern blot analysis (Kochanek et al., 1990).

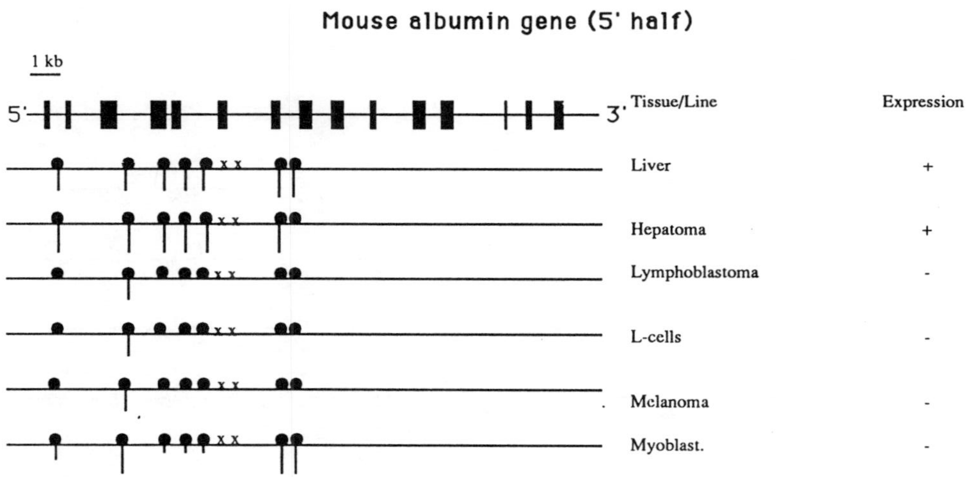

Figure 18. *Mouse albumin gene (5′ end).* Only HpaII sites on the 5′ half of the gene was analyzed in this study. The possible role of methylation in the mouse albumin expression was further investigated, using the phenomenon of gene activation by cell hybridization. The sites at the extreme 5′ end of the genes were undermethylated in all the isolated hybrid clones (mouse lymphoblastoma X rat hepatoma), regardless of albumin production. In contrast, in hybrids of mouse fibroblasts (L-cells) with hepatoma cells, the sites were demethylated only in albumin producing clones. The fact that the gene is relatively undermethylated in myoblasts, although it is not active in these cells is in accord with the rule that hypomethylation is necessary but not sufficient for gene expression. Albumin production was assayed by immunocytochemical staining (Sellem et al., 1985).

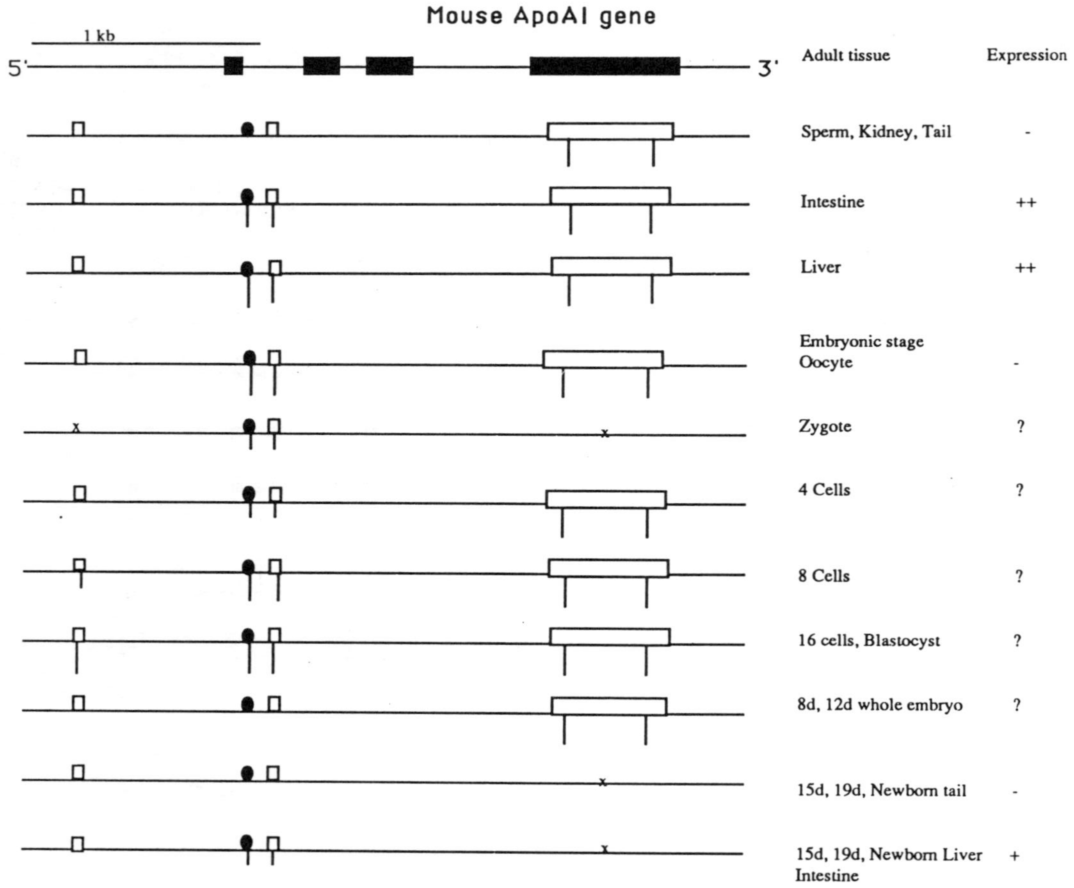
Mouse ApoAI gene
1 kb
5'
3'
Adult tissue
Expression
Sperm, Kidney, Tail
-
Intestine
++
Liver
++
Embryonic stage
Oocyte
-
Zygote
?
4 Cells
?
8 Cells
?
16 cells, Blastocyst
?
8d, 12d whole embryo
?
15d, 19d, Newborn tail
-
15d, 19d, Newborn Liver
Intestine
+

Figure 19. *Mouse ApoAI gene*. Methylation status was analyzed by double digestion of DNA, and Southern blotting. Analysis of the methylation pattern of the gene in the germline and early stages of embryogenesis was performed by a PCR assay. Expression data was obtained by Northern analysis (Shemer et al., 1991a). □ designates a CpG island (see discussion).

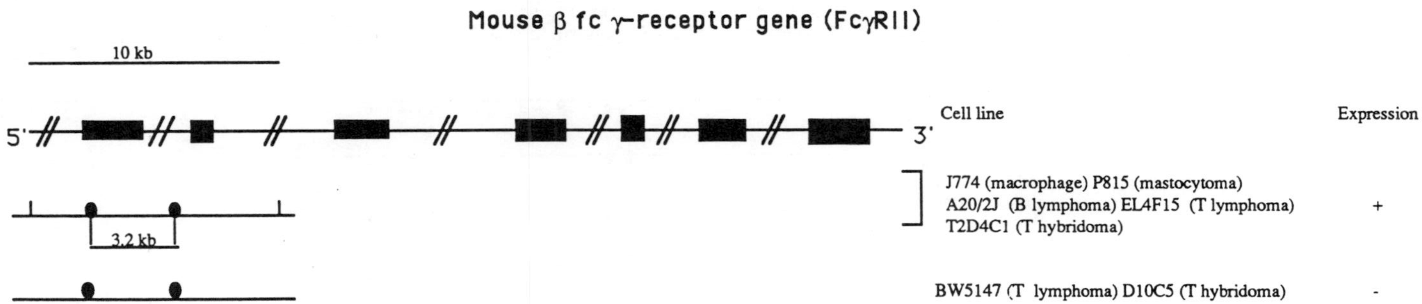

Figure 20. *Mouse β Fc γ-receptor gene (FCγRII)*. Only two 5′-HpaII sites have shown cell specific methylation. Other sites in the 5′ end were found to be methylated in both expressing and non-expressing lines. The two differentially methylated sites have not been precisely mapped but are known to be 3.2 kb apart within a 10-kb EcoRI fragment spanning the first 2 exons. The relative level of expression was determined by measuring mRNA level by Northern blotting, and the percentage of cells forming rosettes with SRBC coated with rabbit IgG antibodies (Bonnerot et al., 1988).

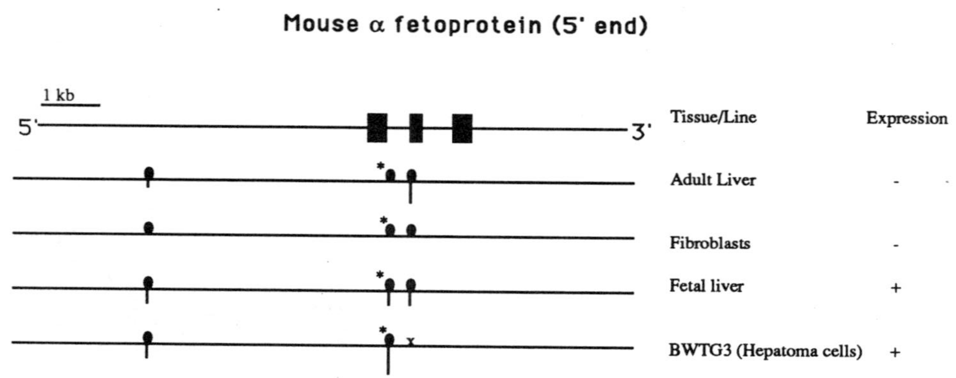

Figure 21. *Mouse α fetoprotein (5′ end).* The analysis was performed by single or double digestion of DNA followed by Southern blotting and hybridization to a regional probe. The authors also showed in transfection assays that *in vitro* methylation at the intronic HpaII site (*) down regulates the expression of a reporter gene (Opdecamp et al., 1992).

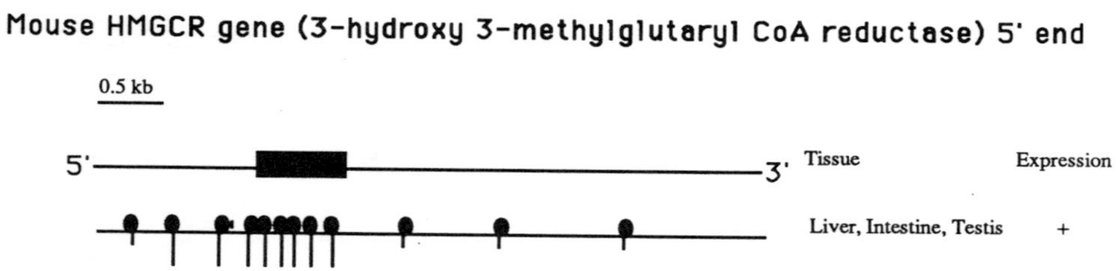

Figure 22. *Mouse HMGCR gene (3-hydroxy 3-methylglutaryl CoA reductase) 5′ end.* The HpaII sites in the promoter region are located within a CpG island. DNA was cleaved with HpaII or MspI and analyzed by Southern blotting. Quantitative scanning densitometry of autoradiograms was used to determine the HpaII/MspI ratio. The study was performed with transgenic mice bearing the same region placed upstream of the bacterial CAT gene. It was found that the methylation-free status of the CpG cluster was progressively lost with increasing transgene copy number; this phenomenon could suggest the role of saturable DNA-binding factors in the protection of 5′ CpG island from methylation (Mehtali et al., 1990).

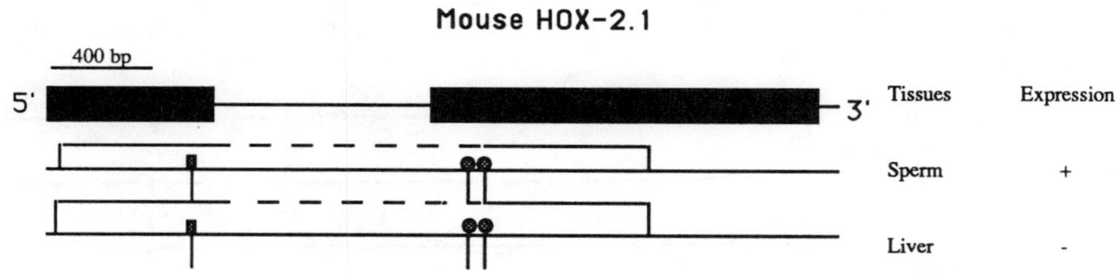

Figure 23. *Mouse Hox-2.1 gene*. The 5′ end of the transcribed region contains numerous HpaII and HhaI sites. The intron region has not been characterized (Ariel et al., 1991). Expression data was taken from Krumlauf et al. (1987). ▭ designates a CpG island.

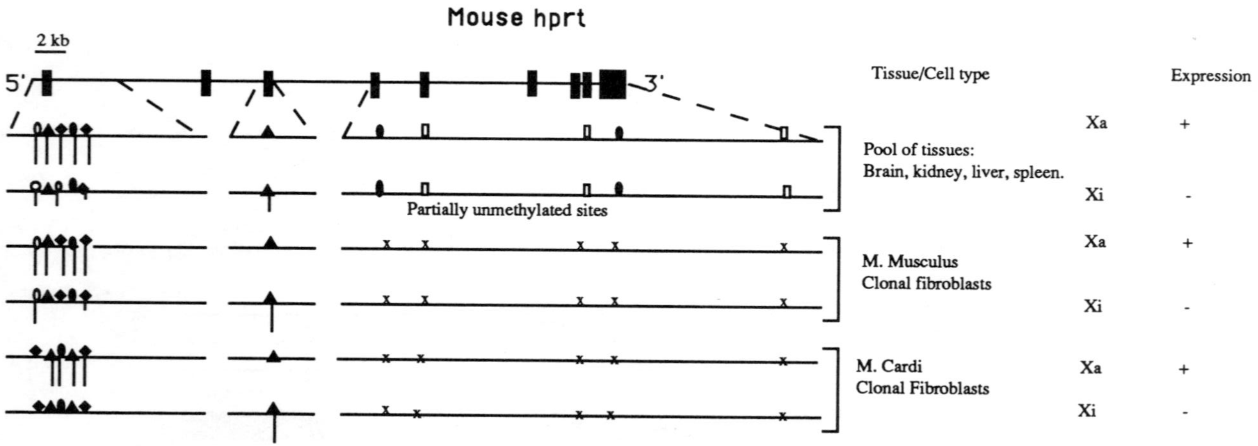

Figure 24. *Mouse hprt gene*. The 5′ end sites analyzed in this study are part of a CpG rich island. The methylation pattern of hprt on the inactive and the active X chromosomes was determined by analysis of DNA from clonal fibroblastic cell lines established from a female embryo derived from a mating of the *Mus musculus* and *Mus cardi* mouse species. The correlation between hypomethylation of the 5′ end sites and gene activity was further demonstrated by the finding that these sites are extensively demethylated in hprt genes reactivated either spontaneously or by 5 aza C treatment (Lock et al., 1986). In contrast no change in the methylation status of the 3′ end sites could be detected upon reactivation of the gene. Xa – active X chromosome; Xi – inactive X chromosome.

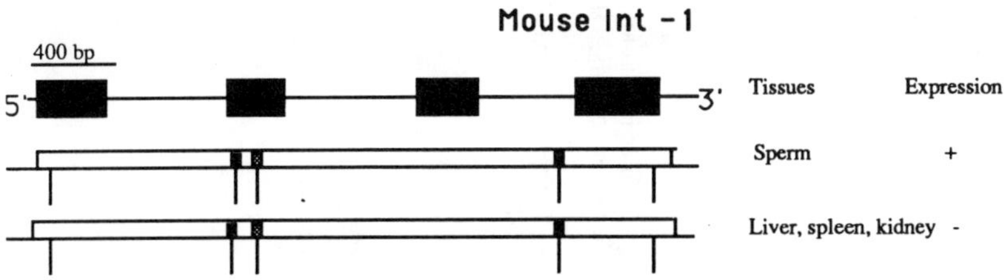

Figure 25. *Mouse Int-1 gene*. In addition to the SmaI and XhoI sites which are mapped, more than 60 HpaII and HhaI sites are distributed within the CpG island that spans the entire gene (Ariel et al., 1991). The expression of Int-1 was investigated by Shackleford et al. (1987). ▭ represents a CpG island.

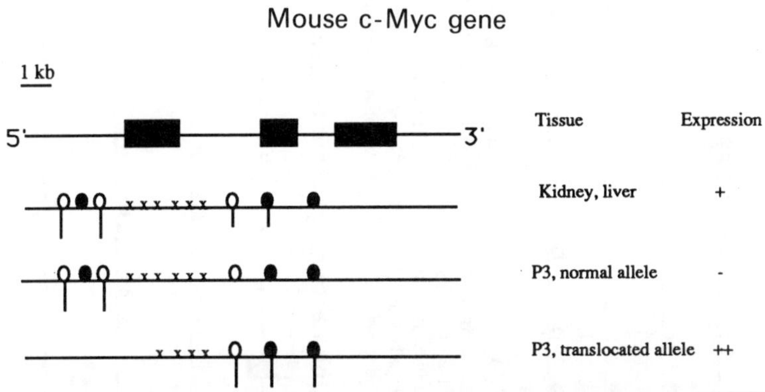

Figure 26. *Mouse c-Myc gene*. The analysis of the methylation pattern was performed by Southern blots of double or triple restriction enzyme digestions. Expression data are taken from Stanton et al. (1983). The normal and translocated alleles of P3 cells were separated by centrifugation in a sucrose gradient, and analyzed separately (Dunnick et al., 1985).

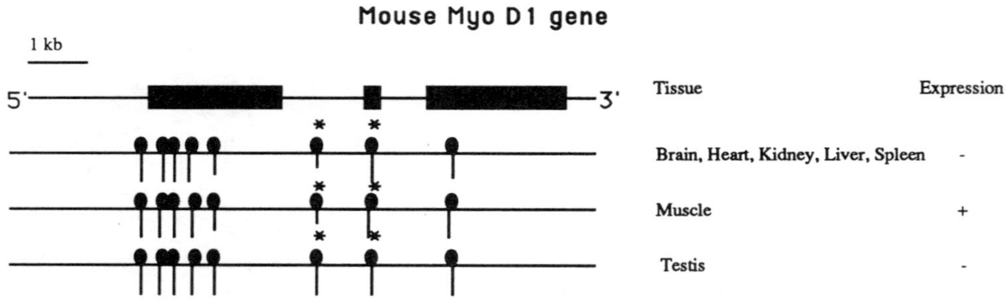

Figure 27. *Mouse Myo D1 gene*. The six HpaII sites at the 5′ end are part of a CpG island. Some of these sites underwent methylation in immortal and transformed cell lines (Jones et al., 1990). Restriction enzyme analysis and ligation-mediated PCR were used to map methylation in the mouse MyoD1 gene. Davis et al. (1987) have isolated MyoD-cDNA from mouse fibroblasts (10T1/2) which were converted into myoblast upon treatment with 5 azaC. They have shown by transfection assays that this gene is sufficient to convert 10T1/2 into stable myoblasts. Jones et al. (1990) showed that the CpG island of MyoD1 gene is methylated in 10T1/2 at least at two HpaII sites (designated *) and unmethylated in somatic tissues. The methylation pattern of the gene suggests that demethylation precedes activation of the gene which is required for myogenesis.

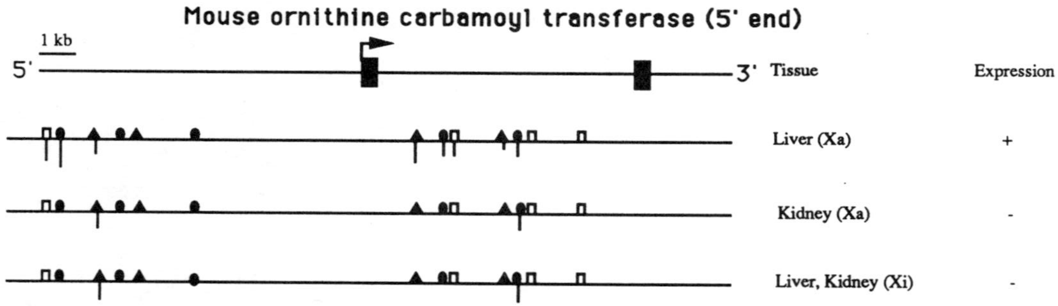

Figure 28. *Mouse ornithine carbomoyl transferase gene (5′ end).* Analysis by double digestion and hybridization with specific probes (Mullins et al., 1987). Expression of this X-linked tissue-specific gene was investigated by Veres et al. (1986). Xa – active X chromosome; Xi – inactive X chromosome.

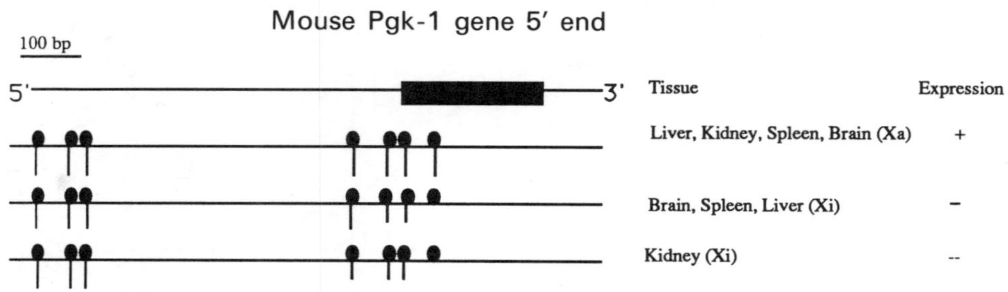

Figure 29. *Mouse Pgk-1 gene 5′ end.* The 7 HpaII-sites analyzed are part of a CpG island. The methylation pattern was determined by Southern blot analysis. The methylation status of the most 3′ HpaII site was further assessed by an HpaII-PCR assay. Expression was assayed by Northern blotting and Pgk-1 specific activity measurement (Singer-Sam et al., 1990). Xa – active X chromosome; Xi – inactive X chromosome.

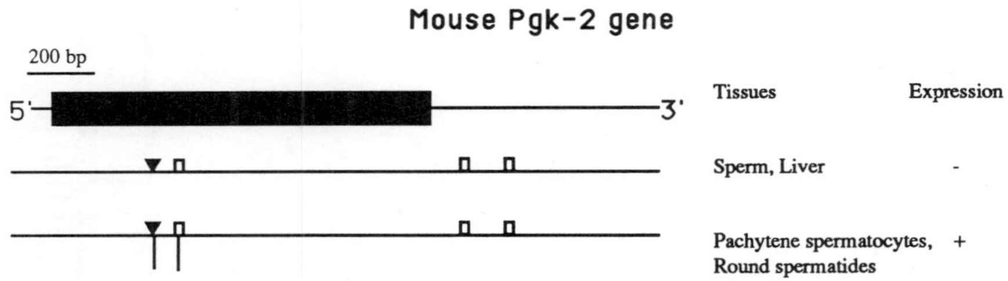

Figure 30. *Mouse Pgk2 gene.* An autosomal gene homologous to the X-linked Pgk1. The methylation status of the sites presented was analyzed by Southern blotting (Ariel et al., 1991). The gene becomes active in meiotic cells (4N) (McCarrey et al., 1987). Two additional postmeiotically expressed mouse genes encoding nuclear chromatin proteins – transition protein 1 (TP1) and protamine 2 (Prm2) – have been analysed by Trasler et al. (1990). TP1 undergoes gradual demethylation during spermatogenesis, while Prm2 becomes progressively more methylated.

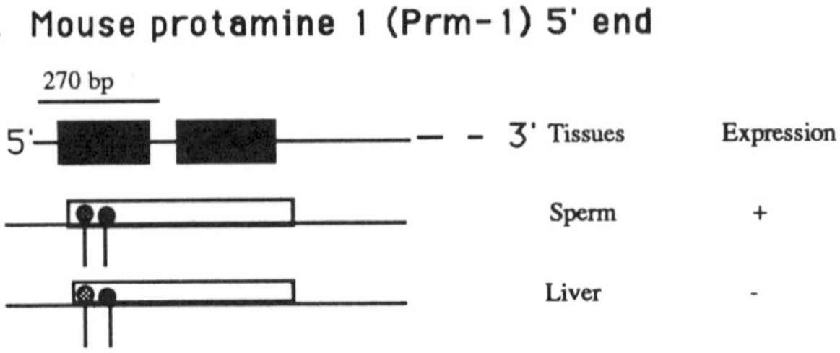

Figure 31. *Mouse protamine 1 (Prm-1)-5′ end.* The only sites analyzed within the island were the NciI and HpaII sites (Ariel et al., 1991). Expression of the gene has been studied by Hecht et al. (1986). ▭ designates a CpG island.

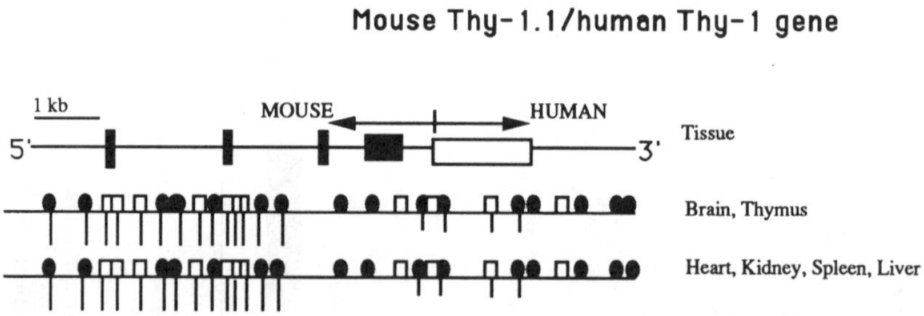

Figure 32. *Mouse Thy-1.1/Human Thy-1 hybrid gene.* The 5′-HpaII and HhaI sites spanning the first 2 exons are part of a CpG island. The hybrid gene was inserted unmethylated into the male pronucleus of fertilized eggs. Four transgenic mice tested positive for the gene were analyzed. The methylation pattern was determined by double digestion of DNA from tissues of adult transgenic mice and 14.5 embryos and Southern blotting. The expression of the Thy-1 transgene was analyzed by S1 nuclease mapping (Kolsto et al., 1986). Szyf et al. (1990) have identified a 214-bp sequence upstream to the transcription start of the Thy-1 promoter capable of protecting exogenous DNA from *de novo* methylation, when transfected into mouse embryonic stem cells (ES). This is one of a group of tissue-specific genes that contain a CpG island at their 5′ end which is unmethylated regardless of expression; other genes in this group are α globin (Bird et al., 1987), human carbonic anhydrase (see Fig. 11) and chicken α2 (I) collagen (Yisraeli and Szyf, 1984).

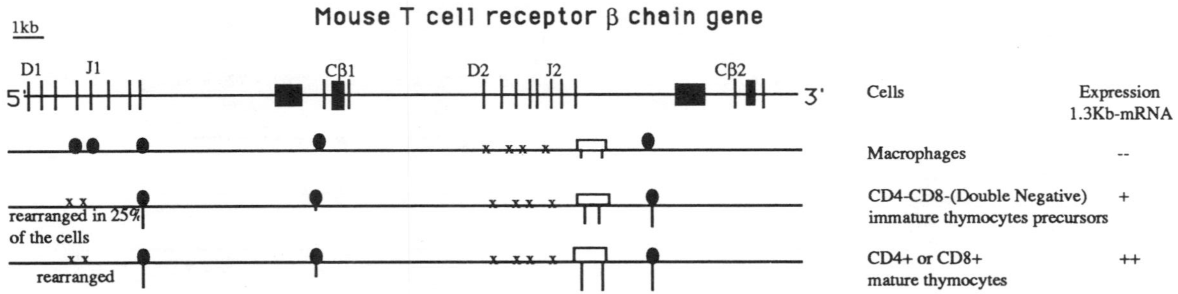

Figure 33. *Mouse T cell receptor β chain gene*. Methylation was analyzed by HpaII digestion and Southern blotting. The expression was determined by Northern blot analysis. The $CD4^-$ $CD8^-$ cells express both the functional 1.3-kb mRNA and a truncated 1-kb message in a 1:4 ratio. The mature thymocytes (CD4+ and CD8+) express predominantly the 1.3-kb mRNA (Sakamoto et al., 1989). □ represents a CpG island.

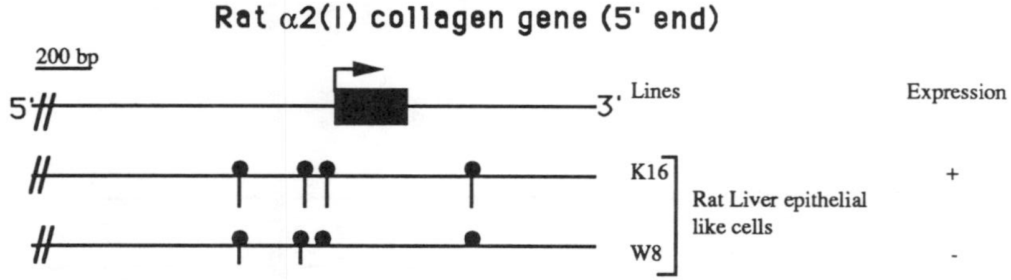

Figure 34. *Rat α2(I) collagen gene (5′ end).* The W8 cell line is derived from K16 and was transformed chemically by 2-N-(acetoxyacetyl) aminofluorene (AAF) treatment. The expression was examined at the level of transcription and translation. The methylation status was determined by double digestion and Southern blot analysis, using a specific probe spanning the first exon and intron. The methylation of the 3′ region was analyzed and did not show any correlation with transcription (Smith and Marsilio, 1988). Previous studies on the chicken collagen gene (chick pro α 2(I)) have shown the presence of a 5′ end CpG island which remains hypomethylated, regardless of the expression level in all the tested tissues (Yisraeli and Szyf, 1984). Fernandez et al. (1985) have shown that the 3′ region of the chick proα1(II) collagen gene is more methylated in non-expressing tissues (fibroblasts or erythrocytes) compared to chondrocytes (expressing cells) but the state of methylation was not altered when chondrocytes were exposed to dedifferentiating agents such as 5BudR or retinoic acid.

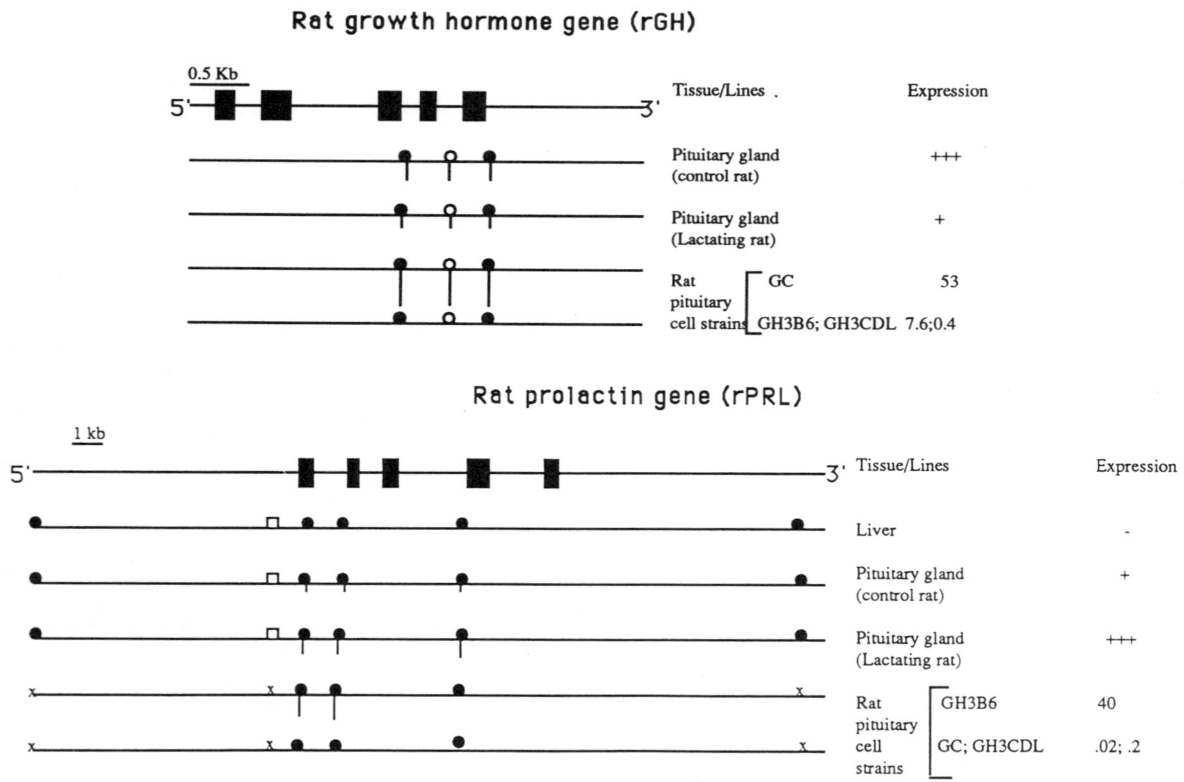

Figures 35–36. *Rat growth hormone gene (rGH) and rat prolactin (rPRL).* The methylation patterns of the gene in liver and pituitary gland DNA were analyzed by Kumar and Biswas (1988) and in rat-pituitary cell strains (GC, GH_3B_6, GH_3CDL) by Laverriere et al. (1986). The relative level of expression of rPRL and rGH in the tissues was determined by Northern blot analysis. In the cell strains rPRL and rGH were measured by RIA and expressed as micrograms produced per mg cell proteins in 48 h. The methylation status of the different sites was determined by double digestion of the DNA, and Southern blotting. The PRL gene is expressed in the lactotrophs and the GH in the somatotrophs of the pituitary gland. The expression is therefore tissue specified and is also modulated by ovarian hormones during pregnancy and lactation.

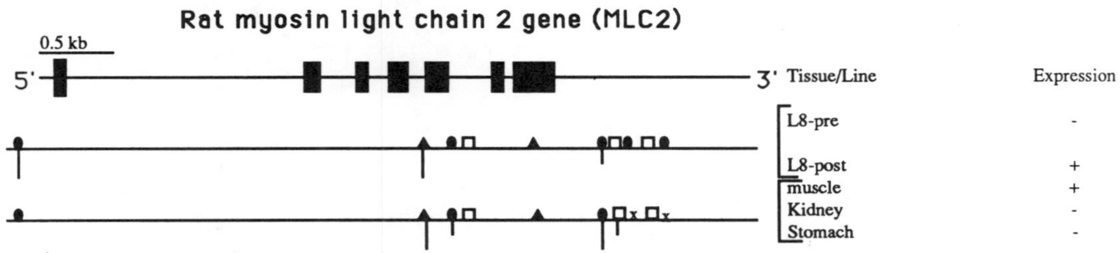

Figure 37. *Rat myosin light chain 2 gene (MLC2).* L8 is a rat myogenic cell line. Before induction (L8-pre), the culture consists of proliferating mononucleated cells and is transformed to multinucleated fibers after induction (Shani et al., 1984). Methylation analysis was by Southern blotting. Expression data is taken from Shani et al. (1981).

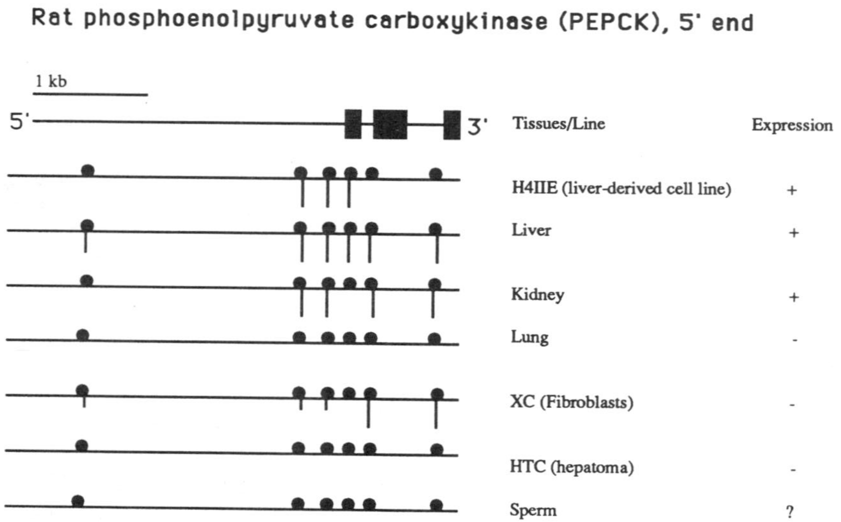

Figure 38. *Rat phosphoenolpyruvate carboxykinase (PEPCK), 5′ end.* The methylation state was analyzed by Southern blot analysis after digestion of the DNA with HpaII. Beside DNA methylation, the chromatin structure and nucleosomal arrangement were analyzed. Five hypersensitive sites (HS) were identified, four of which appear only in expressing cells (Ip et al., 1989). Expression data are from MacDonald et al. (1978).

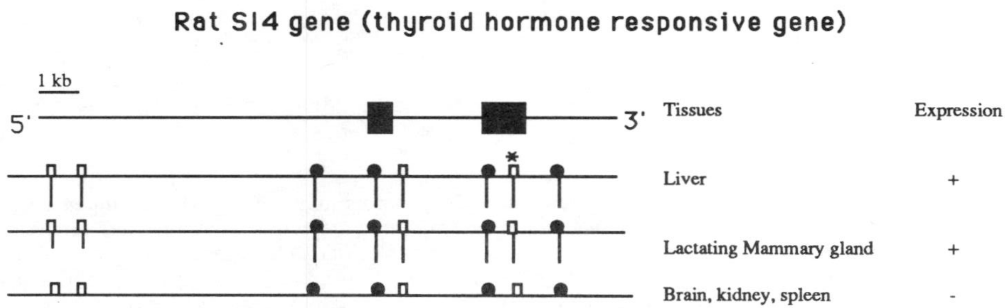

Figure 39. *Rat S14* gene *(thyroid hormone responsive gene).* The methylation status of the HhaI site at the extreme 3′ end of the gene (designated by *) is influenced by the presence of the hormone in the liver. This site seems to undergo demethylation in the transition from hypothyroidism to euthyroidism (Jump et al., 1987). mRNA expression was investigated by Jump et al. (1984) by Northern blot analysis.

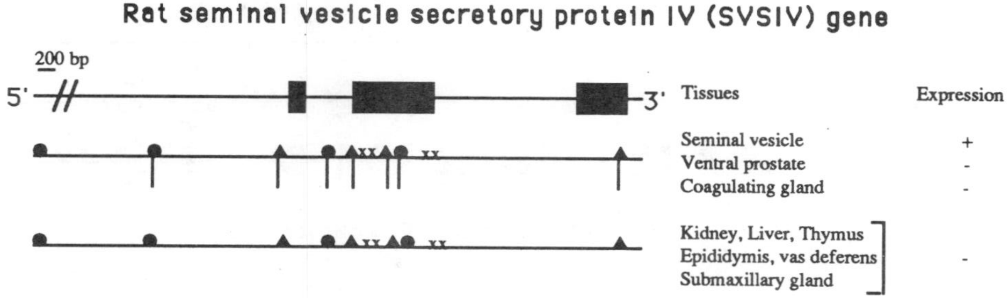

Figure 40. *Rat seminal vesicle secretory protein IV gene (SVSIV).* Methylation was analyzed by Southern blotting. The expression of SVSIV in ventral prostate was assayed by radioimmunoassay, Northern blot and run-on assay (Kandala et al., 1985). No protein, message or active transcription could be detected in ventral prostate. Seminal vesicle, ventral prostate and coagulating gland are located next to one another near the base of the urethra but they have distinct embryological origins. Since all three organs are responsive to androgen, the authors suggest a similar demethylation phenomenon as the one occurring at the 5′ HpaII site in the chicken vitellogenin gene (Wilks et al., 1982) upon exposure to estrogen (Kandala et al., 1985).

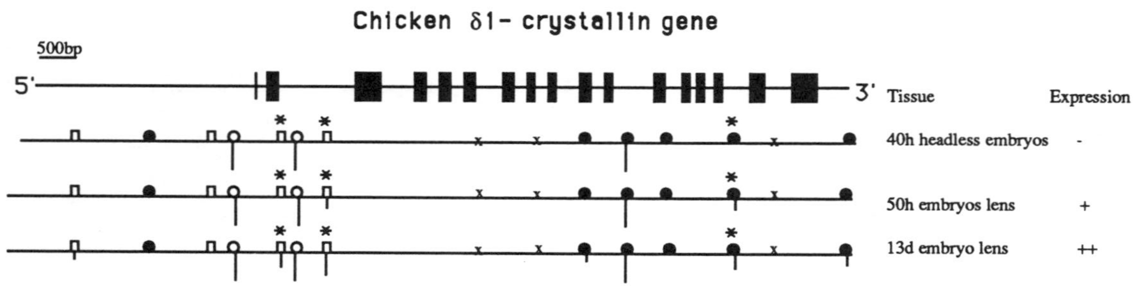

Figure 41. *Chicken δ1-crystallin gene*. The methylation status of the gene has been previously described in different tissues (Yisraeli and Szyf, 1984) and the gene was assigned to group II on the basis of work of Grainger et al. (1983) which showed the occurrence of a specific hypomethylation at 50 h of development when δ-crystallin transcripts begin to accumulate at high levels in the lens. In a more recent study (Sullivan et al., 1989), the methylation pattern was determined in chicken embryo lens cells at different stages of development. The HpaII site (*) that undergoes hypomethylation beginning at 50 h (Grainger et al., 1983) as well as two additional specifically demethylating HhaI sites (*) were identified in this study. The percentage of methylation at the sites where methylation is partial was determined by visual examination of the raw data. Sullivan and Grainger (1986) proved that hypomethylation of the gene in chicken lens occurs in postmitotic cells, implying that demethylation is independent of DNA replication. Shinohara and Piatigorsky (1976) quantitated the amount of δ crystallin mRNA. They found about 0.5 pg of mRNA per embryo head at 48 h of development.

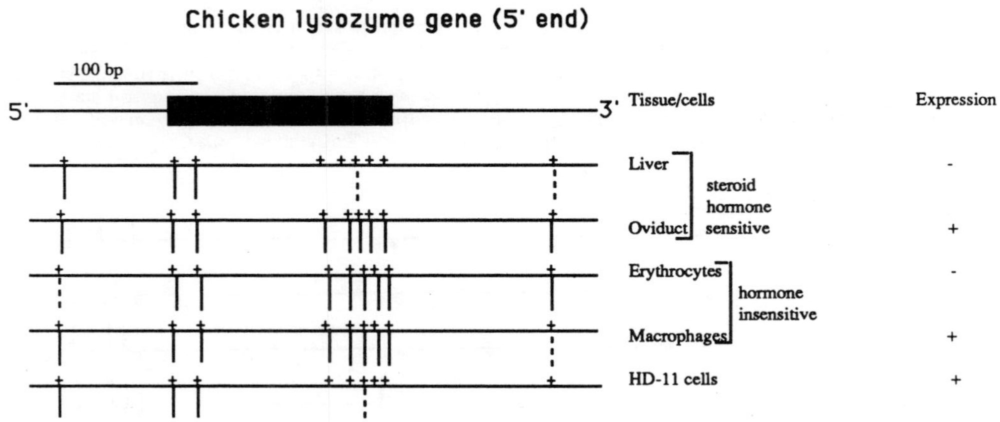

Figure 42. *Chicken lysozyme gene (5′ end).* Methylation analysis has been performed by genomic sequencing of both DNA strands (Wolfl et al., 1991). The region studied (−450−+250) includes 2 hormone responsive elements, one DNase I-hypersensitive site, several A+T-rich sequence motifs, and putative alternative start sites. Transcriptional activity in different tissues and cells was analyzed by Palmiter (1972) and Steiner et al. (1987). + designates CpG sites analyzed by genomic sequencing. Vertical dotted line represents hemimethylated site. HD cells are derived from chicken macrophages.

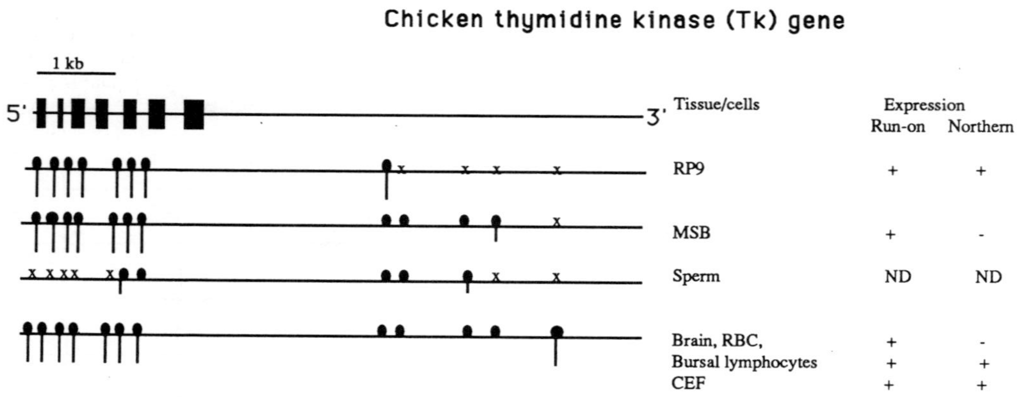

Figure 43. *Chicken thymidine kinase (Tk) gene*. The analysis was performed by double digestion of DNA and hybridization to specific probes. RP9 are logarithmically growing hematopoietic cells derived from transformed B cells. MSB cells consist of T cells transformed by Marek disease virus (MDV). CEF cells are chicken embryonic fibroblasts. The steady state level of mRNA was monitored by Northern blot analysis and was in correlation with the mitotic activity of the cells (except for MSB cells wherein some kind of mutation probably impairs proper processing of cellular transcripts). The rate of transcription was determined by a nuclear run on assay. The results of this study imply that the principal control of chicken Tk gene expression is post-transcriptional in nature. The methylation pattern was obtained by HpaII digestion and Southern blotting and was found to correlate with the rate of transcription as determined by nuclear run-on assay except for the sites located 3′ to the poly A site (Groudine and Casimir, 1984).

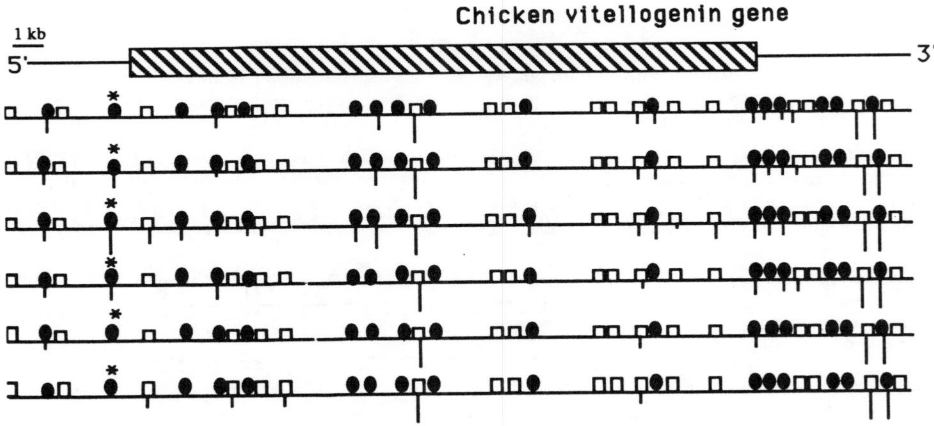

Figure 44. *Chicken vitellogenin gene*. The methylation pattern was determined by double digestion of DNA and Southern blot analysis using regional probes. Autoradiograms were scanned by densitometry to determine the level of methylation (Philipsen et al., 1985). In previous studies, reviewed by Yisraeli and Szyf (1984), only 6 HpaII sites in this gene have been analyzed; all of them were found methylated except one in the 5′ flanking region (*) which undergoes an estrogen-dependent demethylation. Three additional CpG sites adjacent to this HpaII site were analyzed using genomic sequencing of DNA isolated from liver of hormone-induced roosters (Saluz et al., 1986). These sites become hypomethylated following estradiol treatment in liver and are similarly hypomethylated in liver and oviduct of laying hens, indicating a correlation between demethylation and the presence of an activated estradiol-receptor complex. In a more recent study Saluz et al. (1988) have identified by genomic sequencing a CpG site within the promoter region (position +10), the methylation state of which correlates with gene expression. This site was found methylated in DNA from oviduct and erythrocytes of chickens. It is worth noting that about 70% of the liver cells are the potentially vitellogenin synthesizing parenchymal cells. Tissue heterogeneity thus, may obscure to a certain degree the correlation between expression and methylation level. The expression data are taken from Jost et al. (1973). ▧ represents the entire coding sequence of the gene.

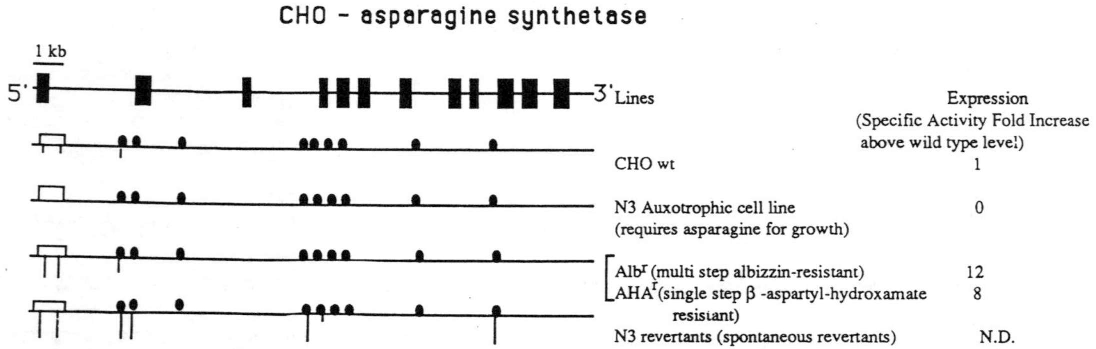

Figure 45. *CHO-asparagine synthetase (AS) gene*. The 5′ CpG rich island contains 71 CpG dinucleotides, 14 of which are HpaII sites. The N3 mutant cell line has been obtained after treatment with the mutagen EMS (ethylmethane sulfomate). Alb^r & AHA^r were also isolated after EMS mutagenesis. All the Alb^r lines have multiple copies of AS (between 3 to 9 copies). The amplification alone cannot account for the elevation in AS activity. The AHA^r lines have only 1 copy of the gene. The expression was determined by Northern blot analysis (Andrulis and Barrett, 1989).

4 Conclusions

Most genes presented in this review exhibit a correlation between hypomethylation of the promoter region and transcription. In several cases the gene is fully unmethylated in expressing cells while the non-active gene is fully methylated; however in many cases some site-specific hypomethylation occurs in both active and non-active tissues. Among the genes described here the methylation patterns of only five tissue-specific genes, human HLA-DRα (Fig. 6), human factor IX (Fig. 4), rat MLC2 (Fig. 37), rat SVSIV (Fig. 40) and chicken lysozyme (Fig. 42) do not correlate with expression. The reason for this exception to the rule is yet unclear.

All housekeeping genes are characterized by a CpG island at the 5′ end of the gene. Nevertheless, several tissue-specific genes have been reported to possess a CpG-rich cluster as well (see Table 1). In general, CpG islands on autosomal genes remain unmethylated in all tissues, regardless of gene activity. The only exceptions reported so far, include the 3′ CpG island of the human ApoAIV gene which is modified in non-expressing tissues (Fig. 1) and the mouse T cell receptor gene whose island is partially methylated in macrophages (non-expressing cells) (Fig. 33).

In contrast to the situation *in vivo*, methylation of CpG-rich sequences occurs more frequently in immortal cell lines (Antequera et al., 1990). This situation is well illustrated by the CHO-asparagine synthetase (Fig. 45), the mouse EndoA (Tamai et al., 1991) and mouse EndoB (Oshima et al., 1988) genes whose 5′-CpG island is methylated in non-expressing cell lines. Similarly, the MyoD1-CpG island which is unmethylated in adult mouse tissues undergoes modification in the mouse fibroblasts cell line 10T1/2 (Jones et al., 1990), and a 3′ CpG island in the human ApoAI gene is methylated in colonic carcinoma cell line CaCo2 where the gene is transcriptionally active (Shemer et al., 1990). On the inactive X chromosome of the female somatic tissues CpG islands of housekeeping genes are modified, whereas their counterparts on the active allele are fully unmethylated. DNA methylation has been therefore implicated in the process of X-inactivation or at least in the maintenance of the inactive state (Lock et al., 1987). The understanding of the significance of methylation patterns of CpG islands of some tissue specific genes of the mouse. Thy 1 (Fig. 32), Prm 1 (Fig. 31), Int 1 (Fig. 25), and Hox 2.1 (Fig. 23) must await further study.

Acknowledgments. Data produced in this laboratory is a representation of projects supported by the Israel U.S. Binational Foundation, National Institutes of Health, U.S.A., Israeli Ministry of Health. We are grateful to Ms. Sara Ivry for her contribution in preparing this manuscript.

Andrulis, I. L., and Barrett, M. T. (1989) DNA methylation patterns associated with asparagine synthetase expression in asparagine-overproducing and auxotrophic cells. Mol. Cell. Biol. *9*, 2922–2927.

Albini, A., Toffenetti, J., Zhu, Z., Chader, G. J., and Noonan, D. M. (1990) Hypomethylation of the interphotoreceptor retinoid-binding protein (IRBP) promoter and first exon is linked to expression of the gene. Nucl. Acids Res. *18*, 5181–5187.

Antequera, F., Boyes, J., and Bird, A. (1990) High levels of de novo methylations and altered chromatin structures at CpG islands in cell lines. Cell. *62*, 503–514.

Ariel, M., McCarrey, J., and Cedar, H. (1991) Methylation patterns of testis-specific genes. Proc. Natl. Acad. Sci. USA *88*, 2317–2321.

Baik, J. H., Griffiths, S., Giuli, G., Manson, M., Siegrist, S., and Guellaen, G. (1991) DNA methylation patterns of the rat gamma-glutamyl transpeptidase gene in embryonic, adult and neoplastic liver. Carcinogenesis *12*, 1035–1039.

Barbieri, R., Nastruzzi, C., Volinia, S., Villa, M., Piva, R., Giacomini, P., Natali, P. G., and Gambari, R. (1990) Methylation pattern of the HLA-DRα gene in human tissues. J. Immunogen. *17*, 51–66.

Battistuzzi, G., D'urso, M., Toniolo, D., Persico, G. M., and Luzzatto, L. (1985) Tissue-specific levels of human glucose-6-phosphate dehydrogenase correlate with methylation of specific sites at the 3′ end of the gene. Proc. Natl. Acad. Sci. USA *82*, 1465–1469.

Bianchi, N. O., Peltomaki, P., Bianchi, M. S. Knuutila, S., and de la Chapelle, A. (1987) The pattern of methylation in rearranged and germ-line human immunoglobulin constant μ genes. Biochem. Biophys. Acta *909*, 245–250.

Bianchi, N. O., Peltomaki, P., Bianchi, M. S., Knuutila, S., and de la Chapelle, A. (1988) Demethylation of two specific DNA sequences in expressed human immunoglobulin light kappa constant genes. Somat. Cell. Mol. Genet. *14*, 13–20.

Bird, A. P., Taggart, M. H., Nicholls, R. D., and Higgs, D. R. (1987) Non methylated CpG-rich islands at the human α-globin locus: implications for evolution of the α-globin pseudogene. EMBO J. *6*, 999–1004.

Blackman, M. A., and Koshland, M. E. (1985) Specific 5′ and 3′ regions of the μ-chain gene are undermethylated at distinct stages of B-cell differentiation. Proc. Natl. Acad. Sci. USA *82*, 3809–3813.

Bonnerot, C., Daeron, M., Varin, N., Amigorena, S., Hogarth, P. M., Even, J., and Fridman, W. H. (1988) Methylation in the 5′ region of the murine βFcγ receptor II. J. Immunol. *141*, 1026–1033.

Breslow, J. L. (1988) Apolipoprotein genetic variation and human disease. Physiol. Rev. *68*, 85–132.

Church, G. M., and Gilbert, W. (1984) Genomic sequencing. Proc. Natl. Acad. Sci. USA *81*, 1991–1995.

Davis, R. L., Weintraub, H., and Lassar, A. B. (1987) Expression of a single transfected cDNA converts fibroblasts to myoblasts. Cell *51*, 987–1000.

Demmer, L. A., Levin, M. S., Elovson, J., Reuben, M. A., Lusis, A. J., and Gordon, J. I. (1986) Tissue-specific expression and developmental regulation of the rat apolipoprotein B gene. Proc. Natl. Acad. Sci. USA *83*, 8102–8106.

Dunnick, W., Baumgartner, J., Fradkin, L., Schultz, C., and Szurek, P. (1985) Methylation of plasmacytoma c-myc genes. Gene *39*, 287–292.

Fernandez, M. P., Young, M. F., and Sobel, M. E. (1985) Methylation of type II and type I collagen genes in differentiated and dedifferentiated chondrocytes. J. Biol. Chem. *260*, 2374–2378.

Frommer, M., McDonald, L. E., Millar, D. S., Collis, C. M., Watt, F., Grigg, G. W., Molloy, P. L., and Paul, C. L. (1992) A genomic sequencing protocol that yields a positive display of 5-methylcytosine residues in individual DNA strands. Proc. Natl. Acad. Sci. USA *89*, 1827–1831.

Gambari, R., Barbieri, R., Piva, R., Tecce, R., Fisher, P. B., Giacomini, P., and Natali, P. G. (1987) Regulation of the expression of class II genes of the human major histocompatibility complex in tumor cells. Ann. N.Y. Acad. Sci. USA *511*, 292–296.

Goodfellow, P. J., Mondello, C., Darling, S. M., Pym, B., Little, P., and Goodfellow, P. M. (1988) Absence of methylation of a CpG-rich region at the 5′ end of the MIC2 gene on the active X, the inactive X, and the Y chromosome. Proc. Natl. Acad. Sci. USA *85*, 5605–5609.

Grainger, R. M., Hazard-Leonards, R. M., Samaha, F., Hougan, L. M., Lesk, M. R., and Thomsen, G. H. (1983) Is hypomethylation linked to activation of δ-crystallin genes during lens development? Nature *306*, 88–91.
Groudine, M., and Casmir, C. (1984) Post-transcriptional regulation of the chicken thymidine kinase gene. Nucl. Acids Res. *12*, 1427–1446.
Hansen, R. S., Ellis, N. A., and Gartler, S. M. (1988) Demethylation of specific sites in the 5′ region of the inactive X-linked human phosphoglycerate kinase gene correlates with the appearance of nuclease sensitivity and gene expression. Mol. Cell. Biol. *8*, 4692–4699.
Hansen, R. S., and Gartler, S. M. (1990) 5-Azacytidine-induced reactivation of the human X chromosome linked PGK1 gene is associated with a large region of cytosine demethylation in the 5′ CpG island. Proc. Natl. Acad. Sci. USA *87*, 4174–4178.
Hecht, N. B., Bower, P. A., Waters, S. H., Yelick, P. C., and Distel, R. J. (1986) Evidence for haploid expression of mouse testicular genes. Exp. Cell. Res. *164*, 183–190.
Ip, Y. T., Granner, D. K., and Chalkley, R. (1989) Hormonal regulation of phosphoenolpyruvate carboxykinase gene expression is mediated through modulation of an already disrupted chromatin structure. Mol. Cell. Biol. *9*, 1289–1297.
Jeffery, S., Edwards, Y. H., and Carter, N. (1980) Distribution of CAIII in fetal and adult human tissue. Biochem. Genet. *18*, 843–849.
Jones, P. A., Wolkowicz, M. J., Rideout III, W. M., Gonzales, F. A., Marziasz, C. M., Coetzee, G. A., and Tapscott, S. J. (1990) De novo methylation of the MyoD1 CpG island during the establishment of immortal cell lines. Proc. Natl. Acad. Sci. USA *87*, 6117–6121.
Jost, J. P., Keller, R., and Dierks-Ventling, C. (1973) Deoxyribonucleic acid and ribonucleic acid synthesis during phosvitin induction by 17β-estradiol in immature chicks. J. Biol. Chem. *248*, 5262–5266.
Jump, D. B., Narayan, P., Towle, H. C., and Oppenheimer, J. H. (1984) Rapid effects of triiodothyronine on hepatic gene expression. J. Biol. Chem. *259*, 2789–2797.
Jump, D. B., Wong, N. C. W., and Oppenheimer, J. H. (1987) Chromatin structure and methylation state of a thyroid hormone responsive gene in rat liver. J. Biol. Chem. *262*, 778–784.
Kandala, J. C., Kistler, W. S., and Kistler, M. K. (1985) Methylation of the rat seminal vesicle secretory protein IV gene. J. Biol. Chem. *260*, 15959–15964.
Karathanasis, S. K., Yunis, I. L., and Zannis, V. I. (1986) Structure, evolution, and tissue-specific synthesis of human apolipoprotein AIV. Biochemistry *25*, 3962–3970.
Keith, D. H., Singer-Sam, J., and Riggs, A. D. (1986) Active X chromosome DNA is unmethylated at eight CCGG sites clustered in a guanine-plus-cytosine-rich island at the 5′ end of the gene for phosphoglycerate kinase. Mol. Cell. Biol. *6*, 4122–4125.
Kelley, D. E., Pollok, B. A., Atchison, M. L., and Perry, R. P. (1988) The coupling between enhancer activity and hypomethylation of κ immunoglobulin genes is developmentally regulated. Mol. Cell. Biol. *8*, 930–937.
Kochanek, S., Toth, M., Dehmel, A., Renz, D., and Doerfler, W. (1990) Interindividual concordance of methylation profiles in human genes for tumor necrosis factors α and β. Proc. Natl. Acad. Sci. USA *87*, 8830–8834.
Kolsto, A. B., Kollias, G., Giguere, V., Isobe, K. I., Prydz, H., and Grosveld, F. (1986) The maintenance of methylation-free islands in transgenic mice. Nucl. Acids Res. *14*, 9667–9678.
Krumlauf, R., Holland, P. W. H., McVey, J. H., and Hogan, B. L. M. (1987) Developmental and spatial patterns of expression of the mouse homeobox gene, Hox 2.1. Development *99*, 603–617.
Kumar, V., and Biswas, D. K. (1988) Dynamic state of site-specific DNA methylation concurrent to altered prolactin and growth hormone gene expression in the pituitary gland of pregnant and lactating rats. J. Biol. Chem. *263*, 12645–12652.
Laverriere, J. N., Muller, M., Buisson, N., Tougard, C., Tixier-Vidal, A., Martial, J. A., and Gourdji, D. (1986) Differential implication of deoxyribonucleic acid methylation in rat prolactin and rat growth hormone gene expressions: a comparison between rat pituitary cell strains. Endocrinology *118*, 198–205.
Ledent, C., Parmentier, M., and Vassart, G. (1990) Tissue-specific expression and methylation of a thyroglobulin-chloramphenicol acetyltransferase fusion gene in transgenic mice. Proc. Natl. Acad. Sci. USA *87*, 6176–6180.

Levy-Wilson, B., and Fortier, C. (1989) Tissue specific undermethylation of DNA sequences at the 5′ end of the human apolipoprotein B gene. J. Biol. Chem. *264*, 9891–9896.
Libert, F., Vassart, G., and Christophe, D. (1986) Methylation and expression of the human thyroglobulin gene. Biochem. Biophys. Res. Commun. *134*, 1109–1113.
Lloyd, J., Brownson, C., Tweedie, S., Charlton, J., and Edwards, Y. H. (1987) Human muscle carbonic anhydrase: gene structure and DNA methylation patterns in fetal and adult tissues. Genes Develop. *1*, 594–602.
Lock, L. F., Melton, D. W., Caskey, C. T., and Martin, G. R. (1986) Methylation of the mouse hprt gene differs on the active and inactive X chromosomes. Mol. Cell. Biol. *6*, 914–924.
Lock, L. F., Takagi, N., and Martin, G. R. (1987) Methylation of the hprt gene on the inactive X occurs after chromosome inactivation. Cell *48*, 39–46.
Lubbert, M., Miller, C. W., and Koeffler, H. P. (1991) Changes of DNA methylation and chromatin structure in the human myeloperoxydase gene during myeloid differentiation. Blood *78*, 345–356.
MacDonald, M. J., Bentle, L. A., and Lardy, H. A. (1978) P-enolpyruvate carboxykinase ferroactivator. Distribution and the influence of diabetes and starvation. J. Biol. Chem. *253*, 116–124.
McCarrey, J. R., and Thomas, K. (1987) Human testis-specific PGK gene lacks introns and possesses characteristics of a processed gene. Nature *326*, 501–505.
Mehtali, M., Lemeur, M., and Lathe, R. (1990) The methylation-free status of a housekeeping transgene is lost at high copy number. Gene *9*, 179–184.
Mullins, L. J., Veres, G., Caskey, C. T., and Chapman, V. (1987) Differential methylation of the ornithine carbamoyl transferase gene on active and inactive mouse X chromosomes. Mol. Cell. Biol. *7*, 3916–3922.
Nelson, K. J., Mather, E. L., and Perry, R. P. (1984) Lipopolysaccharide-induced transcription of the kappa immunoglobulin locus occurs on both alleles and is independent of methylation status. Nucl. Acids Res. *12*, 1911–1923.
Opdecamp, K., Riviere, M., Molne, M., Szpirer, J., and Szpirer, C. (1992) Methylation of an α fetoprotein gene intragenic site modulates gene activity. Nucl. Acids Res. *20*, 171–178.
Oshima, R. G., Trevor, K., Shevinsky, L. H., Ryder, O. A., and Cecena, G. (1988) Identification of the gene coding for the EndoB murine cytokeratin and its methylated stable inactive state in mouse nonepithelial cells. Genes Develop. *2*, 505–516.
Palmiter, R. D. (1972) Regulation of protein synthesis in chick oviduct. Independent regulation of ovalbumin, conalbumin, ovomucoid, and lysozyme induction. J. Biol. Chem. *247*, 6450–6461.
Pfeifer, G. P., Steigerwald, S. D., Mueller, P. R., Wold, B., and Riggs, A. D. (1989) Genomic sequencing and methylation analysis by ligation mediated PCR. Science *246*, 810–813.
Pfeifer, G. P., Steigerwald, S. D., Hansen, R. S., Gartler, S. M., and Riggs, A. D. (1990) Polymerase chain reaction-aided genomic sequencing of an X chromosome-linked CpG island: Methylation patterns suggest clonal inheritance, CpG site autonomy, and an explanation of activity state stability. Proc. Natl. Acad. Sci. USA *87*, 8252–8256.
Philipsen, J. N. J., Gruber, M., and Ab, G. (1985) Expression-linked demethylation of 5-methylcytosines in the chicken vitellogenin gene region. Biochim. Biophys. Acta *826*, 186–194.
Piva, R., Kumar, L. V., Hanau, S., Maestri, I., Rimondi, A. P., Pansini, S. F., Mollica, G., Chambon, P., and del Senno, L. (1989) The methylation pattern in the 5′ end of the human estrogen receptor gene is tissue specific and related to the degree of gene expression. Biochem. Int. *19*, 267–275.
Razin, A., and Cedar, H. (1991) DNA methylation and gene expression. Microbiol. Rev. *55*, 451–458.
Rossant, J., Sanford, J. P., Chapman, V. M., and Andrews, G. K. (1986) Undermethylation of structural gene sequences in extraembryonic lineages of the mouse. Dev. Biol. *117*, 567–573.
Ruta Cullen, C., Hubberman, P., Kaslow, D. C., and Migeon, B. R. (1986) Comparison of factor IX methylation on human active and inactive X chromosomes: Implications for X Inactivation and transcription of tissue-specific genes. EMBO J. *5*, 2223–2229.
Sakamoto, S., and Young, H. W. (1988) Modification of T cell receptor β chain gene: the J2 region but not the J1 region of the T cell receptor β chain is hypomethylated in human B cells and monocytes. Nucl. Acids Res. *16*, 2149–2163.

Sakamoto, S., Mathieson, B. J., Komschlies, K. L., Bhat, N. K., and Young, H. A. (1989) The methylation state of the T cell antigen receptor β chain gene in subpopulations of mouse thymocytes. Eur. J. Immunol. *19*, 873–879.

Sakamoto, S., Ortaldo, J. R., and Young, H. A. (1989a) Analysis of the methylation state of the T cell receptor β chain gene in T cells and large granular lymphocytes. J. Biol. Chem. *264*, 251–258.

Saluz, H. P., and Jost, J. P. (1986) Optimized genomic sequencing as a tool for the study of cytosine methylation in the regulatory region of the chicken vitellogenin II gene. Gene *42*, 151–157.

Saluz, H. P., Jiricny, J., and Jost, J. P. (1986) Genomic sequencing reveals a positive correlation between the kinetics of strand-specific DNA demethylation of the overlapping estradiol/glucocorticoid-receptor binding sites and the rate of avian vitellogenin mRNA synthesis. Proc. Natl. Acad. Sci. USA *83*, 7167–7171.

Saluz, H. P., Feavers, I. M., Jiricny, J., and Jost, J. P. (1988) Genomic sequencing and *in vivo* footprinting of an expression-specific DNase I-hypersensitive site of avian vitellogenin II promoter reveal a demethylation of a mCpG and a change in specific interactions of proteins with DNA. Proc. Natl. Acad. Sci. USA *85*, 6697–6700.

Saluz, H. P., and Jost, J. P. (1989) A simple high-resolution procedure to study DNA methylation and in vivo DNA-protein interactions on a single-copy gene level in higher eukaryotes. Proc. Natl. Acad. Sci. USA *86*, 2602–2606.

Sellem, C. H., Weiss, M. C., and Cassio, D. (1985) Activation of a silent gene is accompanied by its demethylation. J. Mol. Biol. *181*, 363–371.

Shackleford, G. M., and Varmus, H. E. (1987) Expression of the proto-oncogene int-1 is restricted to postmeiotic male germ cells and the neural tube of mid-gestational embryos. Cell *50*, 89–95.

Shani, M., Zevin-Sonkin, D., Saxel, O., Carmon, Y., Karcoff, D., Nudel, U., and Yaffe, D. (1981) The correlation between the synthesis of skeletal muscle actin myosin heavy chain and myosin light chain and the accumulation of corresponding mRNA sequences during myogenesis. Dev. Biol. *86*, 483–492.

Shani, M., Admon, A., and Yaffe, D. (1984) The methylation of 2 muscle-specific genes: restriction enzyme analysis did not detect a correlation with expression. Nucl. Acids Res. *12*, 7225–7234.

Shemer, R., Walsh, A. M., Eisenberg, S., Breslow, J. L., and Razin, A. (1990) Tissue specific methylation patterns and expression of the human apolipoprotein AI gene. J. Biol. Chem. *265*, 1010–1015.

Shemer, R., Eisenberg, S., Breslow, J. L., and Razin, A. (1991) Methylation patterns of the human ApoAI-CIII-AIV gene cluster in adult and embryonic tissue suggest dynamic changes in methylation during development. J. Biol. Chem. *266*, 23676–23681.

Shemer, R., Kafri, T., O'Connell, A., Eisenberg, S., Breslow, J. L., and Razin, A. (1991a) Methylation changes in the apoAI gene during embryonic development of the mouse. Proc. Natl. Acad. Sci. USA *88*, 10300–10304.

Shinohara, T., and Piatigorsky, J. (1976) Quantitation of δ-crystallin messenger RNA during lens induction in chick embryos. Proc. Natl. Acad. Sci. USA *73*, 2808–2812.

Singer-Sam, J., Yank, T. P., Mori, N., Tanguay, R. L., Lebon, J. M., Flores, J. C., and Riggs, A. D. (1990) DNA methylation in the 5′ region of the mouse PGK-1 gene and a quantitative PCR assay for methylation. Eds G. Clawson, D. Willis, A. Weissbach and P. Jones. Nucleic Acid Methylation *128*, 285–298.

Smith, B. D., and Marsilio, E. (1988) Methylation of the α2(I) collagen gene in chemically transformed rat liver epithelial cells. Biochem. J. *253*, 269–273.

Stanton, L. W., Watt, R., and Marai, K. B. (1988) Translocation breakage and truncated transcripts of c-Myc oncogene in murine plasmacytomas. Nature *303*, 401–406.

Steiner, C., Muller, M., Baniahmad, A., and Renkawitz, R. (1987) Lysozyme gene activity in chicken macrophages is controlled by positive and negative regulatory elements. Nucl. Acids Res. *15*, 4163–4178.

Sullivan, C. H., and Grainger, R. M. (1986) δ-Crystallin genes become hypomethylated in postmitotic lens during chicken development. Proc. Natl. Acad. Sci. USA *83*, 329–333.

Sullivan, C. H., Norman, J. T., Borras, T., and Grainger, R. M. (1989) Developmental regulation of hypomethylation of δ-crystallin genes in chicken embryo lens cells. Mol. Cell. Biol. *9*, 3132–3135.

Szyf, M., Tanigawa, G., and McCarthy Jr., P. L. (1990) A DNA signal from the Thy-1 gene defines de novo methylation patterns in embryonic stem cells. Mol. Cell. Biol. *10*, 4396–4400.

Talwar, S., Pocklington, M. J., and Maclean, N. (1984) The methylation pattern of tRNA genes in Xenopus laevis. Nucl. Acids Res. *12*, 2509–2517.

Tamai, Y., Takemoto, Y., Matsumoto, M., Morita, T., Matsushiro, A., and Nozaki, M. (1991) Sequence of the EndoA gene encoding mouse cytokeratin and its methylation state in the CpG-rich region. Gene *104*, 169–176.

Toniolo, D., D'urso, M., Martini, G., Persico, M., Tufano, V., Battistuzzi, G., and Luzzatto, L. (1984) Specific methylation pattern at the 3′ end of the human housekeeping gene for glucose 6-phosphate dehydrogenase. EMBO J. *3*, 1987–1995.

Toniolo, D., Martini, G., Migeon, B. R., and Dono, R. (1988) Expression of the G6PD locus on the human X chromosone is associated with demethylation of three CpG islands within 100 kb of DNA. EMBO J. *1988*, 401–406.

Transler, J. M., Hake, L. E., Johnson, P. A., Alcivar, A. A., Millette, C. F., and Hecht, N. B. (1990) DNA methylation and demethylation events during meiotic prophase in the mouse testis. Mol. Cell. Biol. *10*, 1828–1823.

Tratner, I., Nahon, J. L., Sala-Trepat, J. M., and Venetianer, A. (1987) Albumin and α-fetoprotein gene transcription in rat hepatoma cell lines is correlated with specific DNA hypomethylation and altered chromatin structure in the 5′ region. Mol. Cell. Biol. *7*, 1856–1864.

van Veen, T., Katial, A., Shinohara, T., Barrett, D. J., Wiggert, B., Chader, G. J., and Nickerson, J. M. (1986) Retinal photoreceptor neurons and pinealocytes accumulate mRNA for interphotoreceptor retinoid-binding protein (IRBP). FEBS Lett. *208*, 133–137.

Veres, G., Craigen, W. J., and Caskey, C. T. (1986) The 5′ flanking region of the ornithine transcarbamylase gene contains DNA sequences regulating tissue-specific expression. J. Biol. Chem. *261*, 7588–7591.

Waye, M. M. Y., Robinson, R., Orfanides, A. G., and Aubin, J. E. (1989) Loss of alpha type I collagen gene expression in rat clonal bone cell lines is accompanied by DNA methylation. Biochem. Biophys. Res. Commun. *162*, 1146–1452.

Wilks, A. F., Cozens, P. J., Mattaj, I. W., and Jost, J. P. (1982) Estrogen induces a demethylation of the 5′ end region of the chicken vitellogenin gene. Proc. Natl. Acad. Sci. USA *79*, 4252–4255.

Wion, K. L., Kelly, D., Summerfield, J. A., Tuddenham, E. G. D., and Lawn, R. M. (1985) Distribution of factor VIII mRNA and antigen in human liver and other tissues. Nature *317*, 726–729.

Wolf, S. F., Dintzis, S., Toniolo, D., Persico, G., Lunnen, K. D., Axelman, J., and Migeon, B. R. (1984) Complete concordance between glucose 6-phosphate dehydrogenase activity and hypomethylation of 3′ CpG clusters: implications for X chromosome dosage compensation. Nucl. Acids Res. *12*, 9333–9348.

Wolf, S. F., Jolly, D. J., Lunnen, K. D., Friedman, T., and Migeon, B. R. (1984a) Methylation of the hypoxanthine phosphoribosyltransferase locus on the human X chromosome: implications for X-chromosome inactivation. Proc. Natl. Acad. Sci. USA *81*, 2806–2810.

Wolfl, S., Schrader, M., and Wittig, B. (1991) Lack of correlation between DNA methylation and transcriptional inactivation: the chicken lysozyme gene. Proc. Natl. Acad. Sci. USA *88*, 271–275.

Yen, P. H., Patel, P., Chinault, A. C., Mohandas, T., and Shapiro, L. J. (1984) Differential methylation of hypoxanthine phosphoribosyltransferase genes on active and inactive human X chromosomes. Proc. Natl. Acad. Sci. USA *81*, 1759–1763.

Yisraeli, J., and Szyf, M. (1984) Gene methylation patterns and expression, in: DNA Methylation: Biochemistry and Biological Significance, pp. 353–378. Eds A. Razin, H. Cedar and A. D. Riggs. Springer-Verlag, New York.

Young, H. A., Ortaldo, J. R., Heberman, R. B., and Reynolds, C. W. (1986) Analysis of T cell receptors in highly purified rat and human large granular lymphocytes (LGL): lack of functional 1.3 kb β-chain m-RNA. J. Immunol. *136*, 2701–2704.

Subject Index